# 24th Association for the Advancement of Animal Breeding and Genetics Conference 2021



Online
2 – 4 November 2021

ISBN: 978-1-7138-3940-8

**Printed from e-media with permission by:**

Curran Associates, Inc.
57 Morehouse Lane
Red Hook, NY 12571

**Some format issues inherent in the e-media version may also appear in this print version.**



Printed with permission by Curran Associates, Inc. (2022)

For permission requests, please contact AAABG – Association for the Advancement of Animal Breeding and Genetics at the address below.

AAABG – Association for the Advancement of Animal Breeding and Genetics
University of New England
Armidale NSW 2351
Australia

Phone: 61 2 677 320 55
Fax: 61 2 677 332 66

AGBU@une.edu.au

**Additional copies of this publication are available from:**

Curran Associates, Inc.
57 Morehouse Lane
Red Hook, NY 12571 USA
Phone: 845-758-0400
Fax: 845-758-2633
Email: curran@proceedings.com
Web: www.proceedings.com

# Association for the Advancement of Animal Breeding and Genetics

Proceedings of the Twenty-fourth Conference

On-line with local hubs in Australia (Brisbane Qld, Armidale NSW, Melbourne Vic, Adelaide, SA and Perth WA) and New Zealand (Dunedin)

2 – 4 November 2021

ISSN Number: 1328-3227

Produced by:
Association for the Advancement of Animal Breeding and Genetics
C/- AGBU
University of New England
Armidale NSW 2351

http://www.aaabg.org

## PRESIDENT'S MESSAGE

On behalf of the organising committee, I extend a very warm welcome to the 24th Conference of the Association for the Advancement of Animal Breeding and Genetics. The format of this conference is necessarily different from earlier conferences, in response to the impact of the COVID-19 pandemic and associated restrictions. Delegates are participating from hubs in a number of locations, but also on-line.

The theme of the 24th Conference is 'Widening the range of technologies used in animal breeding and genetics' with an emphasis on the many new tools that are now available to improve animal breeding programs. These tools range from novel methods for developing additional phenotypes to innovative molecular approaches to increase the accuracy of genetic selection.

Despite the changed conditions that many delegates have found themselves working under during the last 18 months, I am delighted to note that this has not slowed down the contribution of papers to AAABG. Overall, we have well over 140 papers for presentation at the 24th conference. These include:

- 28 full-article papers published in a special edition of the journal *Animal Production Science*
- Over 115 short communication papers published in the AAABG conference proceedings that will appear on the AAABG Home page.

All delegates registered for the 24th AAABG conference will have full digital access to all papers, with the full text of special edition articles available via a password. The short communication papers, will also be available at a later date from the AAABG Home Page. In addition, all recordings of presentations of talks will be available, once initially presented, to all registered delegates, for a period of up to 3 months after the conference.

The central day of this year's conference, is a Breeder day. The organising committee decided to put it in the middle of the 3-day conference, to highlight the importance of addressing one of the main objectives of AAABG, which is '*to promote communication among all those interested in the application of genetics to animal production, particularly breeders and their organisations, consultants, extension workers, educators and geneticists.*' All organising committees grapple with how to effectively meet this objective. Our attempt, taking place during the COVID-19 pandemic, although not as large as originally planned for at the Adelaide Convention Centre, should provide some learnings for future conferences to consider.

The organising committee is grateful for the ongoing support of sponsors, especially for staying with us following a change to the conference format. We are also indebted to the Adelaide Convention Centre and the Adelaide Wine Centre, who returned all funds paid following cancellation of our venue bookings. Our thanks also to organisers of hubs, who have taken on responsibility for making arrangements at a local level so that delegates have the option to meet together close to their home location to participate in the conference.

Finally, I thank members of the conference organising committee and Dr Sue Hatcher, our AAABG Editor, for all their hard work in making the 24th AAABG Conference possible. I am honoured to have had the privilege of being President of AAABG and leading the organising committee for the 24th Conference.

Forbes Brien
*President*

## ASSOCIATION FOR THE ADVANCEMENT OF ANIMAL BREEDING AND GENETICS

**2021**

### TWENTY FOURTH CONFERENCE COMMITTEE



### CITATION OF PAPERS

Papers in this publication should be cited as appearing in the Proceedings of the Association for the Advancement of Animal Breeding and Genetics.

(Abbreviation: Proc. Assoc. Advmt. Anim. Breed. Genet.)

Example citation:
Bowley F.E., Amer P.R. and Meier S. (2013) New approaches to genetic analysis of fertility traits in New Zealand dairy cattle. *Proc. Assoc. Advmt. Anim. Breed. Genet*. 20: 37-40.

## REVIEWERS

All papers, invited and contributed, were subject to peer review, by two referees. We acknowledge and thank those listed below for reviewing the AAABG papers contained in these proceedings.

Pamela Alexandre
Panoraia Alexandri
Hassan Aliloo
Michael Aldridge
Peter Amer
Paul Arthur
Michelle Axford
Ireti Balogun
Robert Banks
Amy Bell
Tracie Bird-Gardiner
Timothy Bilton
Hugh Blair
Cynthia Bottema
Forbes Brien
Daniel Brown
Kim Bunter
Annika Bunz
Andrew Byrne
Nadine Chapman
Yizhou Chen
Evans Cheruiyot
Bronwyn Clarke
Schalk Cloete
Lucy Coleman
Natalie Connors
James Copley
Neil Cullen
Hans Daetwyler
Lino De La Cruz Colin
Sara De Lasa Heras Saldana
Ken Dodds
Sonja Dominik
Esther Donkersloot
Christian Duff
Katie Eager
Katherine Egerton-Warburton
Bailey Engle
Madeline Facy
Mohammad Ferdosi
Marina Fortes
Mehrnush Forutan
John Gibson
Johan Greeff
Boyd Gudex
Philip Gurman
Mekonnen Haile-Mariam
Rhiannon Handcock
Sue Hatcher
Mark Henryon
John Henshall
Susanne Hermesch
Melanie Hess
Stefan Hiendleder
Brad Hine
Patricia Johnson
Majid Khansefid
Mehar Khatkar
Brian Kinghorn
Harry Lamb
Stephen Lee
Sigrid Lehnert
Li Li
Yutao Li
Paul MacLean
Iona MacLeod
Mette Madsen
Emily Mantilla Valdivies
Coralia Manzanilla-Pech
Russell McCulloch
Rudi McEwin
Aaron McMillan
Lorna McNaughton
Karin Meyer
Catriona Millen
Lee-ann Monks
Kirsty Moore
Suzanne Mortimer
Anupama Mukherjee
Carel Muller
Cornelius Nel
Joanne Newton
Bang Nguyen
Tuan Nguyen
Brendon O'Rourke
Peter Parnell
Jen Peart
Gertje Petersen
Laurie Piper
Judith Pitchford
Wayne Pitchford
Raul Ponzoni
Torsten Pook
Laercio Porto-Neto
Jennie Pryce
Ian Purvis
Cheryl Quinton
Imtiaz Randhawa
Kieran Ransom
Toni Reverter
Caeli Richardson
Elizabeth Ross
Amali Samaraweera
Bruno Santos
Kaitlyn Sarlo Davila
Boris Sepulveda-Molina
Ric Sherlock
Lucia Sider
Jen Smith
Katarzyna Stachowicz
Luke Stephen
Melissa Stephen
Bolormaa Sunduimijid
Imke Tammen
Andrew Tan
Jose Antonio Torres Vazquez
Irene Van Den Berg
Julius van der Werf
Alison Van Eenennaam
Laura Vargovic
Peter Wahinya
Sam Walkom
Jessica Wallace
Bradley Walmsley
Dominic Waters
Franziska Weik
Ruidong Xiang
Luna Zhang
Jinghang Zhou

## SPONSORS OF THE 24TH AAABG CONFERENCE 2021


The financial assistance of the following organisations is gratefully acknowledged.

### *Gold Sponsor*

WEATHERBYS
SCIENTIFIC AUSTRALIA

### *Silver Sponsors*

awi Australian Wool Innovation Limited

illumina®

### *Bronze Sponsors*

Angus Australia

zoetis

mla MEAT & LIVESTOCK AUSTRALIA

SA CATTLE INDUSTRY FUND

intertek Total Quality. Assured.

SA SHEEP INDUSTRY FUND

NEOGEN®

### *Supporting Partners*

THE UNIVERSITY of ADELAIDE

DAVIES LIVESTOCK RESEARCH CENTRE

## THE ASSOCIATION FOR THE ADVANCEMENT OF ANIMALBREEDING AND GENETICS[1] INCORPORATED[2]

**OBJECTIVES**

(i) to promote scientific research on the genetics of animals;
(ii) to foster the application of genetics in animal production;
(iii) to promote communication among all those interested in the application of genetics to animal production, particularly breeders and their organisations, consultants, extension workers, educators and geneticists.

**To meet these objectives, the Association will:**

(i) hold regular conferences to provide a forum for:
   (a) presentation of papers and in-depth discussions of general and industry-specific topics concerning the application of genetics in commercial animal production;
   (b) scientific discussions and presentation of papers on completed research and on proposed research projects;
(ii) publish the proceedings of each Regular Conference and circulate them to all financial members;
(iii) use any such other means as may from time to time be deemed appropriate.

**MEMBERSHIP**

Any person interested in the application of genetics to animal production may apply for membership of the Association and, at the discretion of the Committee, be admitted to membership as an Ordinary Member.

Any organisations interested in the application of genetics to animal production may apply for membership and, at the discretion of the Committee, be admitted to membership as a Corporate member. Each such Corporate Member shall have the privilege of being represented at any meeting of the Association by one delegate appointed by the Corporate Member.

**Benefits to Individual Members.**

- While it is not possible to produce specific recommendations or "recipes" for breeding plans that are applicable for all herd/flock sizes and management systems, principles for the development of breeding plans can be specified. Discussion of these principles, consideration of particular case studies, and demonstration of breeding programs that are in use will all be of benefit to breeders.
- Geneticists will benefit from the continuing contact with other research workers in refreshing and updating their knowledge.
- The opportunity for contact and discussions between breeders and geneticists in individual members' programs, and for geneticists in allowing for detailed discussion and appreciation of the practical management factors that often restrict application of optimum breeding programs.

**Benefits to Member Organisations.**

- Many of the benefits to individual breeders will also apply to breeding organisations. In addition, there are benefits to be gained through coordination and integration of their efforts.

---

[1] AAABG was formerly known as the Australian Association for Animal Breeding and Genetics. Following the 1995 OGM the name was changed when it became an organisation with a joint Australian and New Zealand membership.

[2] The Association for the Advancement of Animal Breeding and Genetics is incorporated in South Australia.

Recognition of this should follow from understanding of common problems, and would lead to increased effectiveness of action and initiatives.

- Corporate members can use the Association as a forum to float ideas aimed at improving and/or increasing service to their members.

**General Benefits**.

- Membership of the Association may be expected to provide a variety of benefits and, through the members, indirect benefits to all the animal industries.
- All members should benefit through increased recognition of problems, both at the level of research and of application, and increased understanding of current approaches to their solution.
- Well-documented communication of gains to be realised through effective breeding programs will stimulate breeders and breeding organisations, allowing increased effectiveness of application and, consequently, increased efficiency of operation.
- Increased recognition of practical problems and specific areas of major concern to individual industries should lead to increased relevance of applied research.
- All breeders will benefit indirectly because of improved services offered by the organisations which service them.
- The existence of the Association will increase appreciably the amount and use of factual information in public relations in the animal industries.
- Association members will comprise a pool of expertise – at both the applied and research levels – and, as such, individual members and the Association itself must have an impact on administrators at all levels of the animal industries and on Government organisations, leading to wiser decisions on all aspects of livestock improvement, and increased efficiency of animal production.

**CONFERENCES**

One of the main activities of the Association is the Conference. These Conferences will be structured to provide a forum for discussion of research problems and for breeders to discuss their problems with each other, with extension specialists and with geneticists.

## ASSOCIATION FOR THE ADVANCEMENT OF ANIMAL BREEDING AND GENETICS

### FELLOWS OF THE ASSOCIATION

"Persons who have rendered eminent service to animal breeding in Australia and/or New Zealand or elsewhere in the world, may be elected to Fellowship of the Association…"

*Elected February 1990*
R.B.M Dun
F.H.W. Morley
A.L. Rae
H.N Turner

*Elected September 1992*
K. Hammond

*Elected July 1995*
C.H.S. Dolling
J.R. Hawker
J. Litchfield

*Elected February 1997*
J.S.F. Barker
R.E. Freer

*Elected June 1999*
J. Gough
J.W. James

*Elected July 2001*
J.N. Clarke
A.R. Gilmour
L.R. Piper

*Elected September 2005*
B.M. Bindon
M.E. Goddard
H.-U. Graser
F.W. Nicholson

*Elected September 2007*
K.D. Atkins
R.G. Banks
G.H. Davis

*Elected September 2009*
N.M. Fogarty
A.R. Fyfe
J.C. McEwan
R. Mortimer
R.W. Ponzoni

*Elected September 2011*
B.P. Kinghorn
A. McDonald

*Elected October 2013*
H.M. Burrow
P.F. Fennessy
G. Nicol
P. Parnell

*Elected October 2015*
P.F. Arthur
D. Johnson
K. Meyer
B. Tier
R. Woolaston

*Elected October 2019*
S.A. Barwick
H.T. Blair
S.W.P. Cloete
I.W. Purvis

*Elected November 2021*
F.D. Brien
D. Garrick
J. Greeff
B. Hayes
J.E. Pryce
J.H.J van der Werf

## HONORARY MEMBERS OF THE ASSOCIATION

"Members who have rendered eminent service to the Association may be elected to Honorary Membership…"

*Elected September 2009*
W.A. Pattie
J.R.W. Walkley

## HELEN NEWTON TURNER MEDAL TRUST

The Helen Newton Turner Medal Trust was established in 1993 following an anonymous donation to the Animal Genetics and Breeding Unit to perpetuate the memory of Helen Newton Turner and to encourage and inspire those engaged in animal genetics.

**Helen Newton Turner AO**



### THE HELEN NEWTON TURNER MEDAL

The Medal is named after Dr Helen Newton Turner whose career with CSIRO was dedicated to research into the genetic improvement of sheep for wool production.

The Helen Newton Turner Medal was first awarded in 1994 to Associate Professor John James and a list of all recipients to date is given below. The Helen Newton Turner Medallist is chosen by Trustees from the ranks of those persons who have made an outstanding contribution to genetic improvement of Australian livestock.

The recipient of the Medal is invited to deliver an Oration on a topical subject of their choice.

**Medallists**

| | | | | | |
|---|---|---|---|---|---|
| 1994 | J.W. James | 2001 | G.A. Carnaby | 2011 | R. Banks |
| 1995 | L.R. Piper | 2003 | F.W. Nicholas | 2013 | M. Goddard |
| 1997 | J. Litchfield | 2005 | K. Hammond | 2015 | A.R. Gilmour |
| 1998 | J.S.F. Barker | 2007 | L. Corrigan | 20147 | A. Collins |
| 1999 | C.W. Sandilands | 2009 | R. Hawker | 2019 | K.D. Atkins |

The Oration of the 2019 Medal recipient, Dr. Kevin Atkins, is reproduced in the AAABG Special Issue of Animal Production Science that accompanies these proceedings.

## THE HELEN NEWTON TURNER BRIGHT FUTURES AWARD

In 2021, the Trust established a new award, the Helen Newton Turner Bright Futures Award to recognise the achievements of an up-and-coming individual who is showing evidence of establishing a reputation for excellence in the field of animal genetics within Australia.

## TABLE OF CONTENTS

**Breeders Papers**

*Proc. Assoc. Advmt. Anim. Breed. Genet: 24: 1-5*

# BREEDING AUSTRALIAN CATTLE FOR PRODUCTION IN THE YEAR 2050

**B. J. Hayes[1], K.E. Kemper[2] and L. Prada E Silva[1]**

[1] Queensland Alliance for Agriculture and Food Innovation, University of Queensland, St Lucia, QLD, 4072 Australia

[2] Institute for Molecular Biosciences, University of Queensland, St Lucia, QLD 4072, Australia.

## SUMMARY

Australia's beef and dairy industries will need to adapt very rapidly to projected climate change by 2050 if they are to remain productive. This review first highlights lessons learned in decades past in balancing productivity and adaptation, illustrated in a series of Vercoe and Frisch papers, that should not be overlooked as the industries push forward. New strategies made possible with genomics, including appropriately balanced selection for heat tolerance in dairy cattle to maximise productivity in projected future climates, are then described. Finally, a couple of novel strategies using genomic information, including sentinel herds, precision adaption traits, and chromosome segment stacking, are proposed.

## INTRODUCTION

According to the International Panel on Climate Change Report (2021), the rate of change in temperature is accelerating and a 2°C increase from pre-industrial times is expected by 2050 for most regions of Australia. This will significantly impact beef and dairy production systems in Australia. For example, by 2050 the currently temperate dairying regions in southern Australia are expected to have average daily temperatures, humidity and rainfall more similar to Tenterfield in New South Wales (https://www.climatechangeinaustralia.gov.au/en/projections-tools/climate-analogues/). In northern Australia, summer temperatures are predicted to regularly exceed 50°C. Although the production systems are suited to current climatic conditions, this paper asks 'How can the Australian cattle industry adapt to a rapidly changing climate?' We address this question from the perspective of an animal breeder – with the knowledge that 28 years is a relatively short timeframe for genetic change (7 generations at most). The objective of this paper is to review strategies proposed to achieve this, and to propose some new strategies made possible with the availability of genomics.

This review is inspired by the framework described by Frisch and Vercoe (1984) for balancing adaptation and productivity potential. Frisch and Vercoe (1984), building on decades of elegant experiments with *Bos taurus* and *Bos indicus* cattle breeds and their crosses at Belmont Research Station in Rockhampton QLD, (Vercoe and Frisch 1969, etc), concluded:

*"Each breed was best suited to one particular environment. The general principle that arises from this is that resistance to environmental stresses will affect the capacity of breeds to express growth potential and will result in a change of ranking in different environments that may arise in different years, different seasons or at different locations."*

*"The task of the animal breeder to combine the adaptation traits of the Brahman with the growth potential of Bos taurus breeds."*

While Frish and Vercoe (1984) were largely concerned with growth, the other major component of productivity, fertility was being addressed at the same time in the same place by Turner (e.g. Turner *et al.* 1983).

## CROSSBREEDING AND COMPOSITES TO OPTIMISE ADAPTATION AND PRODUCTIVITY

The insights of Vercoe, Frisch and Turner, and others, supported a major innovation in Australian cattle breeding, the creation and widespread introduction of composite breeds, both from Australia and overseas (e.g. Africander), in an attempt to balance adaptation and productivity. Composite cattle now dominate pastoral company holdings and are very widespread in Northern Australia.

Creation of composite breeds was also attempted for dairy cattle, with the Australian Friesian Sahiwal and the Australian Milking Zebu (Tierney *et al.* 1986, Hayman 1974). However, these composites could not compete on milk production with Holsteins from southern Australia, particularly following deregulation.

## WITHIN BREED SELECTION FOR ADAPTATION AND PRODUCTIVITY

Another pathway to improved productivity in harsh environments is within breed selection. Frisch and Vercoe (1986) recognised this possibility *"In the case of the Brahmans there is no genetic alternative to selection unless other breeds that have higher resistance to environmental stresses, and ultimately, higher productivity, can be identified and imported. This avenue should be vigorously explored".* In other words, in particularly harsh environments, even composite cattle may not be sufficiently adapted, and selection within the most adapted breed becomes the only alternative. There are herds in Australia which exemplify this. For example, the Collins Belah Valley herd which has had sustained selection for high fertility, on both number of calves over the lifetime of cows and since it has been available, BREEDPLAN days to calving (the number of days to calving following bull in date). The genetic trend for days to calving in this herd is pronounced (Figure 1) and many of the cow herd produce a calf every year. It is interesting to note that these gains have been achieved by selecting for a productivity trait – days to calving, without selection for specific adaptation traits, in harsh environments. However, adaptation is of course indirectly strongly selected for, otherwise the cows would not survive to produce a calf.

It should be pointed out that while major gains for fertility have been achieved in the Collins herd, in other Brahman and indeed Northern Herds, gains for fertility have been modest at best (Figure 1). This is in part due to the difficulty of recording fertility in extensive conditions. Genomic prediction offers new opportunities for selecting for fertility traits in these situations. This requires large reference populations to be established where animals are both recorded for fertility and genotyped for genome wide DNA markers, ideally in the harsh commercial environment which most cattle experience or will experience in the future. In Australia, both the Repronomics project (Johnston *et al.* 2019) and the Northern Genomics project (Hayes *et al.* 2019) have been set up with this aim.

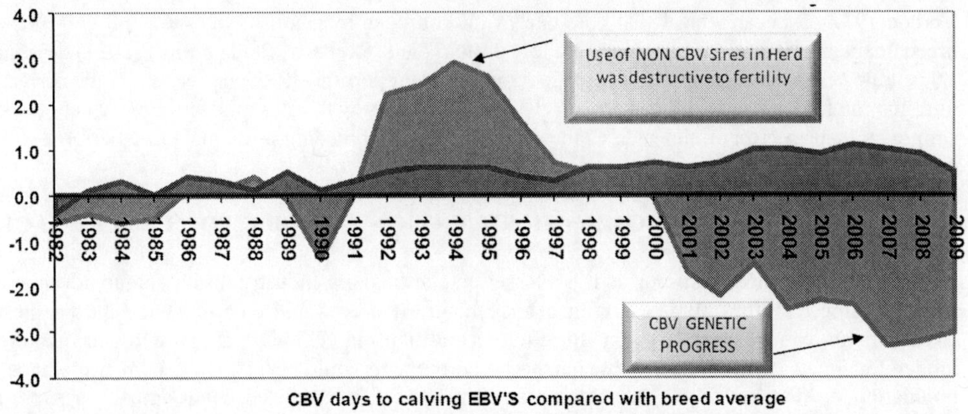


**Figure 1. Genetic trend for days to calving for the Brahman breed and the Collins Brahman herd (source ABRI)**

**BREEDING DIRECTLY FOR FUTURE ADAPTATION – CAN WE BREED FOR 2050?**

It could be argued, as Vercoe and Frisch (1990) and Frisch and Vercoe (1984) effectively did, that selection on productivity in a challenging environment automatically selects for the appropriate level of adaptation – if the animals were not adapted, they would not perform (as demonstrated by the Collins herd). While it is challenging to breed cattle for productivity in harsh environments, it is even more challenging to breed for productivity when the environment is in flux or the projected future environment is not within the currently (feasible) production areas? The challenge then becomes, what is the best approach for a breeder to maximise adaptation and profitability in future environments?

An interesting case study here is heat tolerance. Nguyen *et al.* (2016) described a web-based tool which allowed dairy farmers to appropriately weight heat tolerance in their selection decisions given projected climate on their farm in 2050 (using CSIRO and BoM projections (2015)). The heat tolerance trait they used was for each cow the regression of test day milk production on temperature humidity index on that test day (Nguyen *et al.* 2016). The GEBV were validated in a chamber experiment by Garner *et al.* (2017), in which cows high and low for heat tolerance GEBV were measured for milk production before and after a simulated heat wave event. To answer the question how much emphasis in a selection index should be placed on heat tolerance, in order to maximise future profitability, Nguyen *et al.* (2016) first estimated genetic correlations between key dairy traits. Milk production and heat tolerance were negatively correlated (-0.85) while fertility (six week in calf rate) and heat tolerance were positively correlated (0.39). Nguyen *et al.* (2016) then derived selection index weights (per dairy farm) for traits in the Balanced Performance Index (the Australian dairy selection index, Byrne *et al.* 2016), and heat tolerance, given projected future climates for individual dairy farms, as predicted by CSIRO and BoM (2015). Using this approach, Nguyen *et al.* (2016) could show predicted future profitability was higher as a result of including heat tolerance in the selection index.

Another approach to selecting for productivity in increasingly harsh conditions might be to dissect adaptation more precisely, and select for these precision traits. For example, differences in nitrogen metabolism are associated with animal adaptation to environmental stress. Heat stress increases water intake and nitrogen losses, with *Bos indicus* cattle being less vulnerable to increased nitrogen losses because of their ability to restrict urine output during high temperatures (Vercoe 1976). Together with the lower fasting metabolic rate of *Bos indicus* cattle (Frisch and

Vercoe 1977), it became clear that both energy and nitrogen metabolism were integral part of the breed adaptation to hot environments and low protein diets. Recently, Prada e Silva (2021) devised an isotope test for tail hair that could determine the proportion of dietary versus body derived nitrogen, and demonstrated this was correlated with both growth rates and fertility. Selection for simple and cheap proxy traits such as these may accelerate breeding of animals that perform well on low quality diets and are adapted to very harsh conditions.

## NOVEL USES OF GENOMIC INFORMATION TO BREED CATTLE FOR PRODUCTION IN 2050.

The Australian beef industry, and to a lesser extent the dairy industry, has a unique advantage in responding to future climates. With a huge range of breeds and environments, the harshest environments can be used a proxy for future conditions in regions that currently have more moderate climates. The cattle in these harshest environments could be a "2050 genomic reference" population, to develop GEBV for performance in the future predicted environment, as part of Australia's effort to ensure the ongoing productivity of the cattle industry. This could occur even though the studs where the selection is made are in much more moderate environments, as is common with the stud sector.

Finally, genomic information could also be used to deliver a twist on the Vercoe and Frisch vision of *combining the adaptation traits of the Brahman with the growth potential of Bos taurus breeds*. The genomic information can be used to produce breeding values for growth, fertility traits, nitrogen use efficiency, and adaptation traits for the *haplotypes* (unique chunks of the genome) in populations, rather than the individuals in those populations. For example regions of the genome associated with adaptation were recently identified by Kim *et al.* (2020), by identifying genome regions that had been selected for 1000s of years in *Bos indicus* x *Bos taurus* composite cattle in harsh environments in Africa. This haplotype information, combined with a breeding strategy to rapidly "stack" the most desirable haplotypes (Kemper *et al.* 2015) could be used to breed the "ultimate" composite cattle anticipating future climates.

## CONCLUSION

Future climates will force many changes on the beef and dairy industries in Australia. This review has highlighted lessons learned in decades past, i.e. the Frisch and Vercoe opus of 1984, that should not be overlooked as the industries push forward. The use genomic tools offer new opportunities to realise their vision for optimising adaptation and productivity.

## ACKNOWLEDGEMENTS

The authors enjoyed extensive conversations with Mr Alf Collins Snr and Dr Matthew Kelly during the writing of this manuscript. We thank them for their contributions.

## REFERENCES

Chadeau-Hyam M., Hoggart C., O'Reilly P., Whittaker J., De Iorio M. and Balding D. (2008) *BMC Bioinformatics* **9**: 364.

Byrne T.J., Santos B.F.S., Am P.R., Martin-Collado D., Pryce J.E., Axford M. (2016) *J. Dairy Sci.* **99**: 8146.

CSIRO and Bureau of Meteorology (2015) *Technical Report, CSIRO and Bureau of Meteorology, Australia*

Frisch J.E. and Vercoe J.E. (1977) *Animal Science* **25**: 343.

Frisch J.E. and Vercoe J.E. (1984) *J. Agric. Sci. Camb.* **103**: 137.

Garner J.B., Douglas M.L., Williams S.R., Wales W.J., Marett L.C., Nguyen T.T., Reich C.M., Hayes B.J. (2016) *Sci Rep.* **29**: 34114.

Hayes B.J., Fordyce G. and Landmark S. (2019) *Proc. Assoc. Advmt. Anim. Breed. Genet.* **23**: 282.

Hayman R.H.. (1974) *Wld. Anim. Rev.* **11**: 31.

International Panel on Climate Change 2021. 6th Report. https://www.ipcc.ch/2021/08/09/ar6-wg1-20210809-pr/

Kemper K.E., Bowman P.J., Pryce J.E., Hayes B.J. and Goddard M.E. (2012) *J Dairy Sci.* **95**: 4646.

Kim K., Kwon T., Dessie T., *et al.* (2020) *Nat Genet.* **52:**1099.

Koufariotis L., Hayes B.J., Kelly M. *et al.* (2018) S*ci Rep* **8**: 17761.

Moore K.L., Girard C.J., Grant T.P., Johnston D.J. (2019) *Proc. Assoc. Advmt. Anim. Breed. Genet.* **23**: 366.

Nguyen Thuy T.T., Hayes B.J., Pryce J.E. (2016) *Anim. Prod. Sci.* **57**: 1488.

Prada e Silva L. (2021) MLA Report. P.PSH.1016.

Tierney M.L., Reason G.K; Schmidt T.V. and Alexander G.I. (1986) *Proc. Aust. Soc. Anim. Prod.* **16:**

Vercoe J. E. (1976). In "Progress in Biometeorology'. pp. 428-35.

Vercoe J.E. and Frisch J.E. (1992) *AJAS* **5**: 401.

*Proc. Assoc. Advmt. Anim. Breed. Genet: 24: 6-9*

# ADVANCING LIVESTOCK WELL-BEING – THE ROLE OF GENETIC IMPROVEMENT

**S. Dominik**

CSIRO Agriculture and Food, Armidale, NSW, 2350 Australia

## SUMMARY

Livestock well-being can be defined as a wicked problem. It is difficult to approach and ever evolving due to its multifaceted characteristics and multiple stakeholders of influence. Well-being consists of two areas: health and welfare. This review outline some of the areas that contribute to health and welfare research and explores the role of genetic improvement in advancing livestock well-being as an overarching concept. It is concluded that due to the complex qualities of the problem, an interdisciplinary approach is required to create lasting change.

## INTRODUCTION

A variety of definitions exist for animal welfare, health, and well-being (Lerner 2008). In a research context, health and welfare have evolved as separate streams of science. Welfare research explores the effects of the environment and husbandry procedures on an animal's' physical and mental well-being. Animal health research is concerned about functional mechanisms that contribute to disease resilience and the effect of disease on animals' physical well-being. Here animal well-being is defined as the term that encompasses animal health and welfare.

The ethical treatment of livestock is increasingly the focus of animal welfare groups, advocacy ranging from requesting alternatives to animal husbandry practices to elimination. Such scrutiny could pose a threat to the social licence of livestock farming (Martin and Shepheard 2011), potentially threatening livestock producer's livelihood. The red meat industry is taking a pro-active approach with "world class animal health, welfare, biosecurity and production practices" being one of the six priorities in Red Meat 2030, which sets the direction for the red meat industries for the next decade (Red Meat Advisory Council 2019). However, the improvement of livestock well-being is a "wicked" problem (Rittel and Webber 1973). A wicked problem has ten inherent characteristics: 1) it is difficult to define; 2) it is hard to measure success; 3) it can only be improved rather than solved; 4) approaches have to be made up; 5) multiple explanations exists, all stemming from individual opinions; 6) it is interconnected and a symptom of another problem; 7) mitigation strategies do not have a ultimate test of success; 8) it has little scope for learning through trial and error; 9) every wicked problem is unique and; 10) planners are liable for their consequences. Approaching a wicked problem requires an understanding of the complexity, interconnectedness and of the multiplicity of stake holders involved. Selective breeding, adapting the animal, is an integral part of the approach, next to management interventions, adapting the environment, to move towards improved livestock well-being. This review is exploring the role of genetic improvement in advancing mitigation strategies to the wicked problem of livestock well-being.

## ASSESSMENT OF ANIMAL WELL-BEING

The traditional definition of animal well-being, based on the framework of The Five Freedoms (Table 1), was adopted into RSPCA Australia policy in 1993. It outlines key aspects of animal well-being, including physical and mental requirements of animals. Whilst it is relatively easy to assess the vigour and health of an animal, it is exceedingly difficult to assess the mental state which is a subjective perception by the animal and the assessor.

Grandin and Johnson (2009) argue that the concept of freedom is difficult, and it is necessary to understand the underlying emotions. They give the example that it might be assumed that chickens that are kept in predator safe barns are free of fear. Chickens have evolved to be only free of fear when they can hide to lay their eggs and it is irrelevant if that is indoors or outdoors. To understand emotions in animals research has been conducted into "affective state" (Boissy and Lee 2014). Affective state has two dimensions 1) the extent to which the state is negative or positive 2) the level of arousal which can be high or low (Mendl *et al.* 2010). Methodology to assess affective state provides a useful model for experimental validation of the mental state of animals (Graunke *et al.* 2013; Monk *et al.* 2018). However, currently methods do not exist that can provide such an assessment at a scale necessary for inclusion in a breeding program.

**Table 1. The Five Freedoms**

| Principle | Implementation |
|---|---|
| Freedom from hunger and thirst | by ready access to fresh water and a diet to maintain full health and vigour |
| Freedom from discomfort | by providing an appropriate environment including shelter and a comfortable resting area |
| Freedom from pain, injury and disease | by prevention through rapid diagnosis and treatment |
| Freedom to express normal behaviour | by providing sufficient space, proper facilities and company of the animal's own kind |
| Freedom from fear and distress | by ensuring conditions and treatment which avoid mental suffering |

Although it is obvious that several commercial husbandry procedures (e.g. castration, tail docking), are associated with pain, the level and duration of pain that an animal experiences following such procedures is difficult to assess objectively. Pain models based on physiological and behavioural responses have been developed (Landa 2012). Objective measures (e.g. cortisol levels) are difficult to obtain, because they are influenced by multiple factors and are expensive, while observation of behaviour is also species specific, with some species not expressing pain very overtly (Landa 2012). However, models exist and have underpinned the successful development of advanced pain relief options for sheep (Smith *et al.* 2017; Colditz *et al.* 2019).

Remote animal sensing technology provide opportunities not just for precision farm management, but also for the collection of animal behaviour data at high frequency at the level of the individual animal (Handcock *et al.* 2009), which can be used for novel trait development of welfare traits and possibly assist in disentangling social interaction effects (Pérez-Enciso and Steibel 2021). Biosensors and wearable technology can be used to collect data for animal health related traits, such as stress, heat load and disease occurrence which can inform management and could also be used as phenotypes for genetic improvement (Neethirajan 2017)

## ADAPTING THE ANIMAL TO THE ENVIRONMENT

Whilst management strategies to improve livestock well-being, such as pain relief and appropriate husbandry systems, adapt the environment to the animal, genetic improvement provides the parallel mechanism to adapt the animal to the environment. It has been demonstrated that selection for production traits with little consideration to well-being traits can lead to unfavourable correlated responses in trait complexes related to animal well-being, such as reproduction, metabolism and health traits (Rauw *et al.* 1998). The challenge for genetic improvement strategies of livestock well-being is the integration of often novel and difficult to measure traits into existing breeding programs. Fundamental research is required on trait measurements, the establishment of genetic and phenotypic relationships with other traits and determining an economic value for welfare traits, because it is difficult to attach a monetary value.

Genetic and genomic strategies have been developed to improve livestock well-being and will continue to have significant impact, as highlighted by the following examples in sheep and cattle. Australian Wool Innovation Limited collaborated with the CSIRO and the Western Australia Department of Agriculture and Food to explore the genetic background of breech strike resistance to provide tools to industry to cease the practice of mulesing for breech flystrike control (Smith *et al.* 2009; Greeff *et al.* 2014). The research projects identified dag, breech wrinkle and breech cover as suitable indirect selection criteria for breech flystrike and since 2009 estimated breeding values for these traits have been reported by Sheep Genetics, the Australian sheep performance recording system (http://www.sheepgenetics.org.au/Home). Direct selection on breech flystrike is feasible in the future through genomic selection (Dominik *et al.* 2021).

Single genes for five recessive conditions in Angus cattle have been identified and genetic tests provide information on the carrier status of bulls for informed purchasing decisions (https://www.angusaustralia.com.au/registrations/dna/genetic-conditions/). Rather than promoting the eradication of the recessive alleles, Angus Australia developed a policy for the use of carrier bulls, which has seen a drop in allele frequency from 7% to 2% whilst minimising the effect on the genetic gain for production traits (Teseling and Parnell 2013).

Angus Australia has also been fostering the improvement of general disease resistance, termed immune competence, in the Angus breed (Angus Australia 2019). Immune competence is moderately heritable and yields accurate genomic breeding values that can be used as a long-term strategy to improve livestock well-being. The approach will see fewer animals affected by disease which reduces the reliance on antibiotics in the Angus industry (Hine *et al.* 2019).

Genetic improvement in cattle well-being related traits, such as temperament, calving ease and structural soundness, have been advanced through the inclusion of these traits in BREEDPLAN the Australian beef cattle genetic evaluation system. Selection for temperament was introduced to into BREEDPLAN Version 4.2 in 2002 in form of a docility breeding value. The phenotype and genetic background of the trait can be objectively assessed using flight speed and crush score (Fordyce *et al.* 1982). In sheep, a clear linkages exist between the temperament and mothering behaviour in sheep that can be exploited for genetic improvement (Brown *et al.* 2016).

Next to selection on breeding values, genomic selection, marker assisted selection, also simple mass selection is often still applied to welfare related traits. For traits that affect longevity, breeding values might not be available. In sheep this could include traits such as leg conformation, shoulder confirmation, fleece rot and flystrike amongst others. Genetic gains can still be achieved if these traits are moderately heritable, but the practice compromises genetic gain towards the overall breeding objective because these cannot be balanced as part of the selection index. At the other end of the spectrum of selection strategies, precise gene editing (PGE) holds great promise for the improvement of livestock well-being, but its application is still debated. Great impact on animal well-being has been achieved in the cattle industry with genetic dehorning which alleviates the need for surgical procedures. The genetic test has been refined over the last 12 years to increase its effectiveness for prediction of the poll status (Randhawa *et al.* 2020), but PGE would be an even more effective strategy to reduce the frequency of the poll allele in a population (Mueller *et al.* 2019). Other applications of PGE have been demonstrated e.g. resistance to porcine reproductive and respiratory syndrome in pigs (Chen *et al.* 2019) and foetal sexing of layer chicks to avoid euthanasia of male chicks after birth (Doran *et al.* 2017).

## CONCLUSIONS

Traditionally selection has furthered improvement in a number of trait complexes that are related to livestock well-being. Building on existing genetic and genomic tools a multidisciplinary effort is required to take advantage of behavioural and biological data from sensors to work towards a solution of the ever-evolving challenge of improving livestock well-being. Gene editing may provide novel opportunities to improve livestock well-being, but it also increases the level of complexity of the wicked problem.

## REFERENCES


Angus Australia (2019) https://www.angusaustralia.com.au/content/uploads/2019/05/ImmuneDEX-Report-May-2019NEW.pdf.

Boissy A., Lee C. (2014) Revue Scientifique et Technique de l'OIE **33**: 103.

Brown D.J., Fogarty N.M., Iker C.L., Ferguson D.M., Blache D., and Gaunt G.M. (2016) *Anim. Prod. Sci.* **56**, 767.

Chen J., Wang H., Bai J., Liu W., Liu X., Yu D., Feng T., Sun Z., Zhang L., Ma L., Hu Y., Zou Y., Tan T., Zhong J., Hu M., Bai X., Pan D., Xing Y., Zhao Y., Tian K., Hu X., and Li N. (2019) *Int. J. Biol. Sci.* **15**: 481.

Colditz I., Paull D., Lloyd J., Johnston L., and Small A. (2019) *Aust. Vet. J.* **97**.

Dominik S., Reverter A., Porto-Neto L.R., Greeff J., and Smith J. (2021) *Livest. Prod. Sci.* **24**.

Doran T., Challagulla A., Cooper C., Tizard M. and, Jenkins K. (2017) *Nat. Inst. Biosci. J.* **1**.

Fordyce G., Goddard M.E. and Seifert G.W. (1982) *Proc. Aust. Soc. Anim. Prod.* **14**: 329.

Grandin T. and Johnson C. (2009) 'Animals make us human.' Houghton Miffley Hardcourt, Boston, US.

Graunke K.L., Nürnberg G., Repsilber D., Puppe B. and Langbein J. (2013) *PLoS ONE* **8**: e74579.

Greeff J.C., Karlsson L.J.E., and Schlink A.C. (2014) *Anim. Prod. Sci.* **54**: 125.

Handcock R., Swain D., Bishop-Hurley G., Patison K., Wark T., Valencia P., Corke P. and O'Neill C. (2009) *Sensors* **9**: 3586.

Hine B.C., Bell A.M., Niemeyer D.D.O., Duff C.J., Butcher N.M., Dominik S., Ingham A.B. and Colditz I.G. (2019) *J. Anim. Sci.* **97**: 4053.

Landa L. (2012) *Vet. Med. Czech* **57**: 185.

Lerner H. (2008) PhD Thesis, Linkoepings University.

Martin P. and Shepheard M. (2011) In 'Defining the social license of farming: issues, challenges and new directions in agriculture' editors J. William, P. Martin pp. 3-12, CSIRO Publishing: Collingwood, Victoria.

Mendl M., Burman O.H.P. and Paul E.S. (2010) Proceedings of the Royal Society B: *Biological Sciences* **277**.

Monk J.E., Doyle R.E., Colditz I.G., Belson S., Cronin G.M. and Lee C. (2018) *PLoS ONE* **13**.

Mueller M.L., Cole J..B, Sonstegard T.S. and Van Eenennaam A.L. (2019) *J. Dairy Sci.* **102**: 4215.

Neethirajan S. (2017) *Sens. Biosensing Res.* **12**: 15.

Pérez-Enciso M. and Steibel J.P. (2021) *Genet. Sel. Evol.* **53**, 22.

Randhawa I.A.S., Burns B.M., McGowan M.R., Porto-Neto L.R., Hayes B.J., Ferretti R., Schutt K.M. and Lyons R.E. (2020) *G3- Genes Genom. Genet.* **10**: 539.

Rauw W.M., Kanis E., Noordhuizen-Stassen E.N. and Grommers F.J. (1998) *Livest. Prod. Sci.* **10**, 539.

Red Meat Advisory Council (2019) http://rmac.com.au/misp2030-staging/wp-content/uploads/2019/10/RedMeat2030.pdf.

Rittel H.W.J. and Webber M.M. (1973) *Policy Sci.* **4**, 155.

Smith J., Brewer H.G. and Dyall T. (2009) *Proc. Assoc. Advmt. Anim. Breed. Genet.* **18**: 334.

Smith R., Graham M., Colditz I., and Small A. (2017) *Int. Sheep Vet. Congr.*

Teseling C.F. and Parnell P.F. (2013) *Proc. Assoc. Advmt. Anim. Breed. Genet.* **20**:. 558.

*Proc. Assoc. Advmt. Anim. Breed. Genet: 24: 10-13*

# MEASUREMENT ENABLED PRECISION INTERVENTIONS: A FUTURE OPPORTUNITY FOR LIVESTOCK FARMING

**M.R. Hutchinson [1,2]**

[1] Adelaide Medical School, University of Adelaide
[2] Australian Research Council Centre of Excellence for Nanoscale BioPhotonics

## SUMMARY

When the opening ceremony of the 2032 Brisbane Olympic games is televised, it will be the first time an Olympics will be viewed by humans on Mars and the Earth's moon. Technological advances that will enable us to become a multiplanetary species, and to sustain and feed a long-term moon base are happening at such an extraordinary rate and enabling truly amazing feats of humanity. Whilst these may, at first glance, seem far-fetched ambitions, the scientific advances that are enabling these endeavours have immediate relevance to farming practices that will bring new efficiencies, productivity, new products, and ultimately new value to the producers who adopt them, and societies who purchase them. An example of these innovations will be in the explosion of measurement enabled precision interventions, based on the generation of decision enabling data. Critically, these advances will occur in human medicine before they are tested in livestock. However, the fundamental biological discoveries are still relevant and therefore innovations in human medicine can be rapidly translated to livestock. This measurement enabled understanding is extremely relevant to livestock production, where knowledge of the "mind and body" state of a dynamic and living organism is crucial. Here, a short summary of the types of technologies and biological breakthroughs they afford will be provided, with specific focus on humanities greatest health burden: PAIN.

## INTRODUCTION

Diseases of the central and peripheral nervous system account for more than three quarters of the years humans live with disabilities (Institute for Health Metrics and Evaluation 2020). These diseases do not always result in death. However, disorders of the brain and spinal cord account for the greatest societal burden of disability because patients live a long time with these disorders in chronic pain or with mental health problems. Why is it that medicine and the scientific research that advances the quality use of medicines have failed to address these problems of the brain and spinal cord? The simplest and yet most important answer is that these disorders are very complex and involve multiple physiological systems. These disorders arise from developmental adaptations and involve a complex presentation of gene and environment interactions. As such, we currently lack knowledge of the "mind and body" state of the dynamic and living organism in humans. Sound familiar? We face the same shortcomings in available technology to quantify optimal strategies for livestock care and production.

Unfortunately, the acceleration of the use of precision medicine in fields like cancer which are allowing for personalisation of treatments based on measurements (mechanistic biomarkers) have not yet developed for complications of the brain and spinal cord. Some advances have been made, such as the use of advanced imaging platforms such as fMRI, PET and measurements from EEG. However, despite these isolated advances, most diseases of the central nervous system are still approached with subjective diagnosis and empirical treatment selection. Moreover, even for human medicine, these complex imaging approaches will not economically scale to diagnose and manage three quarters of the world's human population who are afflicted by disorders of the central nervous system. Let alone translate to farm based, livestock production tools. Therefore, new methods and solutions are needed that create measurement enabled precision interventions that have translational, and scalability engineered into them from the beginning.

Let's take PAIN as an example where future innovations are needed. In livestock production, acute pain is experienced due to management procedures, such as castration and tail docking, injuries from fighting or housing conditions, diseases such as mastitis or other infections, and at time of birth. These acute injuries and acute painful experiences can transition into the persistence of pain, which has a profound impact on the wellbeing and resilience of the animal that cause increased costs and reduced productivity. Pain in animals is an experience that we are unable to reliably diagnose or quantify. Even when animals in pain are identified we are still left ineffective in verifying the success of interventional treatments. These limitations arise from our inability to objectively measure pain. This means that opportunities to add beneficial chronic pain resilience genetic traits to breeding values have yet to intentionally begin. We still use old medicines based on empirical data for the management of acute pain associated with injury, illness, and husbandry practices. We need new measurement data of the "mind and body" state of the dynamic and living organism to advance the management of pain in livestock.

Within the Australian Research Council Centre of Excellence for Nanoscale BioPhotonics we have established a decadal research program to meet this challenge. To create scalable, and hence economical fast solutions, we have harnessed light-based imaging and sensing tools to capture information from biological processes. The advantage of light based measurements is that it is capable of adapting over scales ranging from events occurring at the single molecule (nano), through to secondary and tertiary biological structures (micro) to subcellular and cellular anatomy (macro) (Hutchinson 2020). Moreover, light can be safe and non-destructive and is therefore uniquely positioned to be able to provide the desired measurement and imaging outcomes. The breakthroughs that are allowing the new use of light in these measurement technologies is afforded through optimisation of the technologies that enhance the light matter (in this case biology) interactions, creating customised light, sculpting how the light enters or interacts with matter, making structures and surfaces that allow sensing of events to occur when and where we want them and of course imaging beyond where light normally can go was needed (Hutchinson 2020). Given the brain and spinal cord are classically viewed as "dark", the measurements and imaging here can be conducted on very low background noise. Finally, the cost to scale these technologies are rapidly decreasing and therefore they are becoming increasingly accessible to a broader range of scientists right through to handheld technologies in the field with producers.

This measurement enabled innovation is not new to the red meat industry, with the Meat and Livestock Australia implementing the Meat Standards Australia many years ago. As of mid-2021 the Meat and Livestock Australia have called for proposals on how to develop and implement the Lifetime Animal Wellbeing Index. It is critical to start efforts to improve outcomes in complex traits like pain and wellbeing from a basis of excellence in measurement. Without the development and implementation of measurement tools, significant efforts and resources can be expended that ultimately achieve suboptimal outcomes (i.e. a waste of time and money). The recent human medicine equivalent of this has been the greater than 5000% increase in opioid use in the management of pain states that have proven to be non-responsive to opioids. In fact, the Centres of Disease Control have defined opioids as contraindicated for all chronic pain (non-cancer pain) conditions owing to the substantial worsening of the individual's quality of life and concerning fatality rates.

Unlike human medicine, many experiences of livestock husbandry care are highly standardised. This makes for establishment of a quantified best practice highly achievable. An example of a potential routine trigger for the classical presentation of a persistence of pain in animals is amputation. Whilst on the decline in livestock, surgical removal of body parts is still widespread. This practice itself causes pain, resulting from the resection (cutting) of peripheral nerves and the possible formation of traumatic neuromas and causes significant ongoing sensitisation at the level of the brain and spinal cord to mechanical stimuli. Light touch transitions to a painful response. Imagine grass blades running across sunburnt skin. The parallel amputation and the associated changes in brain and spinal cord function in humans is considered to be significantly painful (Hutchinson and Terry 2019). We cannot ever know what an animal feels or thinks and therefore we avoid anthropomorphising these states. Instead, we can use a reductionist scientific approach to examine at the molecular and cellular level events and anatomical structures of the sensory system in animals. We can then use comparative histology and classical biology to infer possible functional consequences. Using these approaches, it is possible to see the hallmarks of chronic pain in animals. This can be seen in cellular adaptations in both the injury site and within the brain and spinal cord. Importantly, these same changes are associated with the phenomenon of residual stump pain and phantom limb pain in humans. This is a prevalent experience as painful symptomatic neuromas following amputations are observed in up to a quarter of amputees. These types of measurements are now afforded through high volume and rapid light based sensing equipment, or alternatively cheap and disposable field based assays that can use light and cameras on mobile phone devices (Orth *et al.* 2018).

It is important to also realise that the field of medical neuroscience is rapidly evolving. The international effort that the opioid pandemic has triggered has resulted in thousands of new studies that have identified hundreds of previously untested targets that could provide chronic pain solutions for humans. And of course, because of the highly conserved nature of these systems, many have the potential to be applied to livestock. One major area of growth has been the realisation that a solely neuronal or electrical view of brain and spinal cord function is wrong. We now view the brain and spinal cord as capable of immune functions, literally speaking the molecular language of the immune cells that circulate around our body. This has triggered a revolution in the pain field, as pain which was once thought of as solely a neuronal wiring problem has given way to an integrated neuro and immune hypothesis of exaggerated pain (Hutchinson and Terry 2019).

Glial cells (immune-like cells of the brain and spinal cord), and peripheral immune cells circulating through the brain and spinal cord are now understood to be integral to creating and maintaining the neuroexcitatory states that underpin persistent pain (Grace *et al.* 2021). This has immense implications. Firstly, all the nerve block agents we use to "stop pain" may work to stop the "electrical signalling" of injury, but may do nothing to stop the immune signalling of pain which is able to bypass all nerve blocks and communicate directly to the brain and spinal cord to establish the foundation of chronic pain. Interestingly, the greater prevalence of exaggerated pain in females also appears to have its origins in this neuroimmune involvement, through estrogenic priming of immune functions. We know that male and female immunology differs with females more likely to have autoimmune disease. We also know that women experience up to 12 times the rate of chronic pain (Grace *et al.* 2021). Hence, the persistent pain problem, and the neuroimmune contributors are likely to be even more relevant in livestock owing to the predominance of female animals in production. Therefore, it is critical to understand this immune to brain and brain to immune communication between the peripheral immune, spinal immune and brain immune systems which create and maintain chronic pain states in livestock (Hutchinson and Terry 2019). Moreover, while neuronal processes are critical for the conduction of heightened pain, there is an anatomically distributed immune signal that triggers conduction of the exaggerated pain response (Marsh *et al.* 2021). This breakthrough provides us now with the first opportunity to diagnose pain through a blood sample in livestock. To date, the translational benefits of these discoveries in the fundamental neuroscience of pain have passed directly to the human clinical setting, without changes in animal husbandry practices. This is a critical missed opportunity.

## CONCLUSION

We are at the cusp of a measurement science watershed moment, where quantification of "mind and body" state of the dynamic and living organism, such as pain and some emotional states will be possible in humans. Very similar technologies can be used in livestock to make these crucial measurements. However, we cannot wait decades for these innovations to spontaneously occur. If humanity reaches Mars before these opportunities have been translated to livestock production, we will have failed. We need to cultivate specific opportunities, and the relationships that develop from them, to allow for the tough questions to be asked and breakthrough ideas to be tested. If we can accelerate these translational opportunities in the future, then streams of research in neuroscience, immunology, pharmacology and biophotonics will emerge to equip the Australian livestock industry with world-first platform technologies that will be able to, for example objectively quantify pain in livestock. These new technologies will then be used to enhance livestock production practices. For example, it will be possible to rapidly identify new drug targets for their ability to block the persistence of pain, underpin productivity gains and an iteratively improve production and business practices. These capabilities will then all contribute to a greater understanding of how breed selection and defined genetic traits contribute to minimising chronic pain in livestock. Given this futuristic technology is on our doorstep, imagine what the plenary of the 2032 Association for the Advancement of Animal Breeding and Genetics meeting might be… Will we be talking about breeding selection technologies and traits that have be deployed in establishing the first interplanetary transfer of livestock? Only time and your imagination will tell….

## ACKNOWLEDGEMENTS

MRH is funded by an ARC Future Fellowship (FT180100565), and research program is funded by the Australian Research Council Centre of Excellence for Nanoscale BioPhotonics (CE140100003), Meat and Livestock Australia, Sheep Industry Fund and the Davies Livestock Research Centre, University of Adelaide.

## REFERENCES

Grace P.M., Tawfik V.L., Svensson C.I., Burton M.D., Loggia M.L. and Hutchinson, M.R. (2021)\. *J. Neurosci.* **41**: 855.

Hutchinson M.R. (2020) *Brain Behav. Immun.* https://doi.org/10.1016/j.bbi.2020.04.029

Hutchinson M.R. and Terry R. (2019) *Animal* **13**: 3000.

Institute for Health Metrics and Evaluation (2020) GBD compare data visualization [WWW Document]. URL http://vizhub.healthdata.org/gbd-compare (accessed 1.7.21).

Marsh L., Hutchinson M.R., McLaughlan C., Musolino S.T., Hebart M.L., Terry R., Verma P.J., Hiendleder S. and Whittaker A.L. (2021) *Animals* (Basel) **11**: 2054.

Orth A., Wilson E.R., Thompson J.G. and Gibson B.C. (2018) *Sci. Rep.* **8**: 3298.

*Proc. Assoc. Advmt. Anim. Breed. Genet: 24: 14*

# THE FUTURE OF GENOTYPING

**J.C. McEwan[1], K.G. Dodds[1], T.P. Bilton[1], R. Brauning[1], S.J. Rowe[1], M.K. Hess[1], K.M. McRae[1], A.F. McCulloch[1], R.A. Anderson[1]and S.M. Clarke[1]**

[1]AgResearch, Invermay Agriculture Centre, Mosgiel, New Zealand

## SUMMARY

The future of genotyping is sequencing. A wide variety of technologies are available, but cost-effective low coverage methods need further development. The downstream bioinformatics and statistics also have to be able handle the incomplete and noisy data produced. DNA sequencing-based technologies separate into 3 classes: whole genome or skim sequencing, random sampling of the genome typically 0.05X coverage or less, and amplicon or equivalent technology for defined segments and variants. There will be a place for all three, possibly combined into a single sequencing assay. However, the most cost effective for genomic selection currently are random sampling methods: best known as genotyping by sequencing (GBS). Currently, most flavors of GBS are based on restriction enzyme reduced representational sequencing or RE-RRS. In the longer-term methods based on random selective primers and PCR may prevail. Increasingly, opportunities will be taken to contemporaneously explore DNA methylation, structural variation and the host microbiome. Nanopore and long read technologies will also be used, in part, to reduce infrastructure costs and reduce turn-a-round time. There is still a niche for array-based technologies, at least for the next decade, but if they are to persist beyond that date the ability to manufacture small runs of chips, cost effectively, coupled with further cost reductions will be required. As sequencing costs decline, research emphasis will shift to better DNA sampling and DNA extraction techniques.

*Proc. Assoc. Advmt. Anim. Breed. Genet: 24: 15-18*

# THE CONTRIBUTION ANIMAL BREEDING CAN MAKE TO INDUSTRY CARBON NEUTRALITY GOALS

**S.J. Rowe[1], S.M. Hickey[2], P.L. Johnson[1], T.P. Bilton[1], A. Jonker[3], W. Bain[1], B. Veenvliet[1], G. Pile[1], B. Bryson[4], K. Knowler[4], N.C. Amyes[2], S.A. Newman[1] and J.C. McEwan[1]**

[1]AgResearch, Invermay Agricultural Centre, Mosgiel, New Zealand
[2]AgResearch, Ruakura Agricultural Centre, Hamilton, New Zealand
[3]AgResearch, Grasslands Research Centre, Palmerston North, New Zealand
[4]AgResearch, Woodlands Research Farm, Invercargill, New Zealand

## SUMMARY

In New Zealand, ~84% of methane and ~30% of total greenhouse gas emissions are derived from enteric emissions from grazing ruminants. Here, we show a simple estimation of the impact of implementing a national breeding scheme in sheep to reduce methane emissions, based on real on-farm figures. We demonstrate how a modest change in breeding for environmental traits can lead to substantial changes in ruminant emissions. Currently there is uncertainty around how methane emissions at the farm gate will be valued. Our estimates show that breeding although slow, is a viable option to make real progress towards carbon neutrality with a very high rate of return on investment and a very modest cost per tonne $CO_2$ equivalents saved regardless of accounting method.

## INTRODUCTION

Methane is a greenhouse gas associated with climate change and approximately 84% of methane emissions in New Zealand are produced from grazing livestock (MfE 2020). Reducing methane emissions from livestock is therefore of environmental and economic importance. Ruminant animals primarily produce methane as a by-product of the complex microbial fermentation process in their rumen that breaks down feed to volatile fatty acids, which are absorbed through the gut wall and are a major source of energy for the animal (Matthews *et al.* 2019). Although the mechanism by which the host controls this fermentation process is not well understood, heritable individual variation has been shown to exist and methane mitigation has been shown to be possible through breeding (Pinare-Paino *et al.* 2013, Jonker *et al.* 2018, Lassen *et al.* 2016). The impact of changes in methane emissions is generally measured by converting to carbon dioxide global warming equivalents ($CO_2e$). The Global Warming Potential (GWP) of a greenhouse gas is its ability to trap extra heat in the atmosphere over time expressed as $CO_2e$. This is most often calculated over 100 years and is commonly referred to as $GWP_{100}$. While $GWP_{100}$ is the accepted metric for describing the warming impact of greenhouse gases, it uses a single scaling factor that doesn't account for methane being a short-lived gas in the atmosphere (Lynch *et al.* 2020). Recently, the use of warming potential or $GWP_{w.e.}$, based on warming equivalents, has been proposed. This accounts for the behaviour of change of methane emission in the atmosphere, and on its contribution to global warming over time.

Over the past decade, two small sheep flocks (n = 100 ewes) have been selected for divergent methane yield, with low-methane sheep currently emitting 10-12% less methane than the high-methane animals (Rowe *et al.* 2019). Furthermore, methane emissions have been measured in a research flock run under commercial conditions and methane breeding values included in a maternal selection index at a hypothetical cost of NZD$100 per $CO_2e$ using $GWP_{100}$. The development of portable accumulation chambers (PAC) to phenotype sheep and the demonstrated success, together with funding from the Pastoral Greenhouse Gas Research consortium has been used to successfully roll out methane measures on commercial farms. Research breeding values for methane emissions have been implemented within the national breeding scheme - Sheep Improvement Limited (Newman, 2009 Beef and Lamb New Zealand 2020). The realised genetic progress made in methane breeding values has been used to estimate the impact of measuring a proportion of the New Zealand commercial flock.

## MATERIALS AND METHODS

**Animals.** Data to estimate genetic gain was available from multiple sources. Genetic parameters were based on 15,000 methane records from sheep measured through portable accumulation chambers from flocks across New Zealand. Flock records for methane emissions recorded from 2009 and performance traits recorded from 1995 to 2020 were also available for a 750 ewe, composite maternal flock (research flock 2638) where selection on an index that included methane was implemented in 2018.

Over the last two years over 5,000 sheep have been measured across ram breeders' farms for methane emissions (grams $CH_4$ per day) using PAC. The number of sheep measured per farm varied from 84 to 268. This demonstrates that given current infrastructure, that ~5,000 measures are physically possible across the national flock annually. Based on this, we assumed that ~2.5% of all lambs born into the stud tier (n=~200,000) could be measured annually.

For predicting the impact of breeding for low methane, we assumed a 20-year time horizon with a lowering of emissions through a combination of phenotyping and genomic prediction. We assumed no reduction of emissions for the first 5 years in the commercial flock due to genetic lag effects and a starting adoption rate of 0.3, which is the current penetration of genotyping in NZ sheep flocks, increasing adoption by 0.1 per year until 100% adoption in the breeding tier in year 8. We assumed 200,000 lambs born each year into the stud tier with 2.5% of these (n=5,000) measured for methane and the remainder genotyped. Commercial production based on 2020 figures assumed 16.85 million breeding ewes producing 8390 kt $CO_2$ equivalent. Methane was valued at NZD$50, $100 and $200 per tonne $CO_{2e}$.

## RESULTS AND DISCUSSION

Figure 1 shows that in a research flock (2638) farmed under commercial conditions, implementation of methane into the index using a $GWP_{100}$ and NZD$100/tonne CO2e led to a reduction of methane emissions by 1-2% per year whilst still maintaining genetic gain in the maternal index (without accounting for any gains made in methane).

**Figure 1 Genetic trends for New Zealand Maternal Worth (NZMW) index (cents) (left) and breeding values for methane emissions (PACCH$_4$gBV measured in g $CH_4$ per day) (right) in research flock 2638 and average recorded maternal flock**

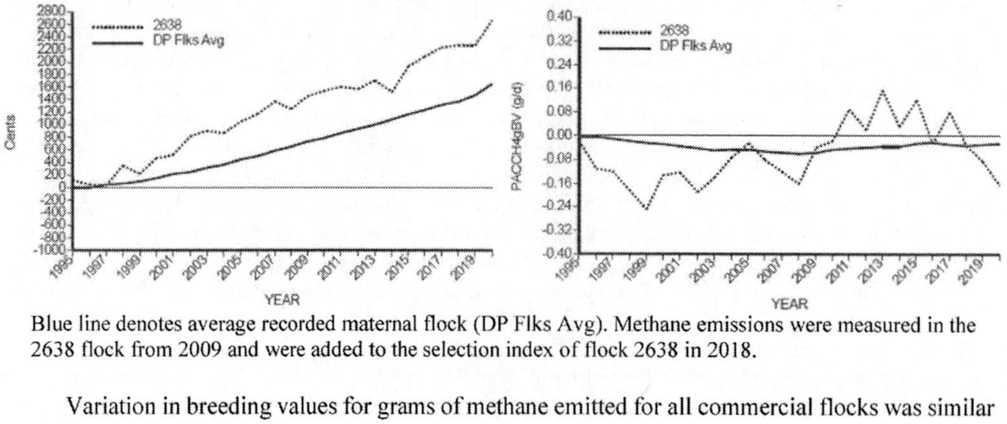


Blue line denotes average recorded maternal flock (DP Flks Avg). Methane emissions were measured in the 2638 flock from 2009 and were added to the selection index of flock 2638 in 2018.

Variation in breeding values for grams of methane emitted for all commercial flocks was similar across farms, commonly ranging from ~-1.5g to +1.5g grams per day. This suggests that there is sufficient variation segregating in the general population for genetic improvement.

The commercial and physical impact of a national breeding scheme to lower methane emissions was estimated as 0.58%/year using genomic selection. After 20 years, annual methane production in 2040 was predicted to have reduced by 7.5% per annum saving a total of 4490 kt of $CO_2$e over the 20-year period (Figure 2) with a cumulative saving of $CO_2$e assuming $GWP_{100}$.

**Figure 2 Predicted annual and cumulative savings in methane emissions measured in carbon dioxide equivalents (kt) based on a 0.58% genetic gain per year**

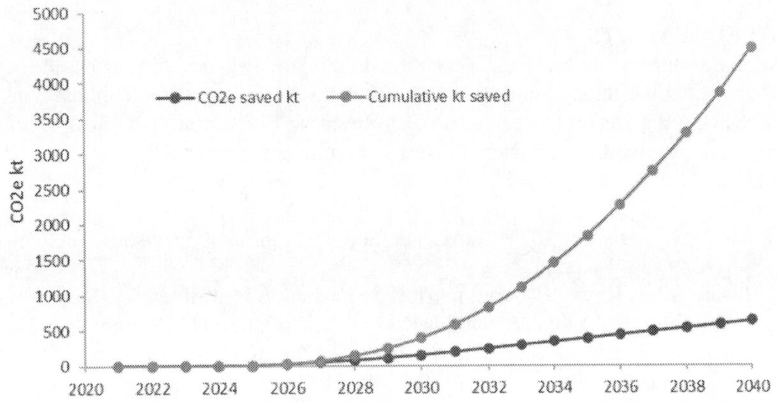


Assuming 5,000 sheep measured annually, methane measures at a cost of NZD$30 per measure and genotyping at a cost of $25 per animal, the total cost of implementing the scheme over the 20-year period including additional genotyping of stud animals by breeders was estimated at NZD$31 million. This includes capital expenditure of NZD$200,000 on PAC measuring equipment. The marginal cost of reducing a tonne of $CO_2$e was estimated at $1.72. This includes losses made in years 1 to 5 due to genetic lag and low adoption rates in the early years. Assuming a tax rate of 28%, and cost of borrowing at 10%, the internal rate of return on a total investment of ranges from 42 to 80% for $GWP_{100}$ to 64-111% for $GWP_{w.e.}$ depending on the costs per tonne of $CO_2$e (Table 1).

**Table 1. Internal rate of return for 0.58% genetic gain per year expressed at a national level**

| | $tonne CO2e | ΔG/year% | NPV(10%) $M[1] | IRR[2] |
|---|---|---|---|---|
| $GWP_{100}$ | 50 | 0.58 | 41.6 | 42% |
| | 100 | 0.58 | 94.1 | 60% |
| | 200 | 0.58 | 199.3 | 80% |
| GWPw.e. | 25 | 0.58 | 111.3 | 64% |
| | 50 | 0.58 | 233.6 | 85% |
| | 100 | 0.58 | 478.3 | 111% |

[1]NPV=Net Present Value, [2]IRR =Internal rate of return

## CONCLUSIONS

We have demonstrated a 1-2% reduction per annum in our commercial research flock since methane breeding values were included in the index, whilst maintaining genetic gain for all other traits. We have also estimated that if we achieve less than one half of this reduction in the breeding tier in the national flock, given likely adoption rates, and including genetic lags in the deployment of improved livestock, we can use breeding to make a substantial contribution to methane mitigation at a very low cost to the industry. These benefits have been made achievable by the development of low-cost high throughput phenotyping for methane combined with the widespread adoption of genomic selection. Sheep account for 1/3 of enteric emissions in New Zealand with cattle contributing most of the remainder. There is currently a barrier to achieving reduction of methane emissions in cattle due to a lack of high throughput measurement technology. If this barrier were overcome, benefits from breeding would be potentially much greater than for the sheep industry and more straightforward to achieve, given widespread use of artificial insemination and the much lower effective population size.

**ACKNOWLEDGEMENTS**

This work was funded by the New Zealand Ministry for Primary Industries, the AgResearch Strategic Scientific Investment Fund, the New Zealand Agricultural Greenhouse Gas Research Centre (NZAGRC), the Pastoral Greenhouse Gas Research Consortium (PGgRc), Beef and Lamb New Zealand and various sheep breeders in New Zealand.

**REFERENCES**

Beef and Lamb New Zealand. (2020) https://beeflambnz.com/news-views/low-methane-emitting-sheep-reality-nz

Jonker A., Hickey S.M., Rowe S.J., Janssen P.H., Shackell G.H., Elmes S., Bain W.E., Wing J., Greer G.J., Bryson B., MacLean S., Dodds K.G., Pinares-Patiño C.S., Young E.A., Knowler K., Pickering N.K. and McEwan J.C. (2018) *J. Anim. Sci.* **96**: 3031.

Lassen J. and LØvendahl P. (2016) *J. Dairy Sci.* **99:** 1959.

Lynch J., Cain M., Pierrehumbert R. and Allen M. (2020) *Environ Res Lett.* **15**: 4.

Matthews C., Crispie F., Lewis E., Reid M., O'Toole P.W. and Cotter P.D. (2019). *Gut Microbes* **10**: 115.

MfE (2020). New Zealand`s Greenhouse Gas Inventory 1990-2018. In Ref. MFE 1496. Ministry for the Environment, Wellington, New Zealand.

Newman S-A.N., McEwan J.C. and Young M.J. (2009). *Proc. Assoc. Advmt. Anim. Breed. Genet.* **18**: 624.

Pinares-Patiño C.S., Hickey S.M., Young E.A., Dodds K.G., MacLean S., Molano G., Sandoval E., Kjestrup H., Harland R., Hunt C., Pickering N.K. and McEwan J.C. (2013). *Animal.* Jun;7 Suppl **2**: 316.

Rowe S.J., Hickey S.M., Jonker A., Hess M.K., Janssen P., Johnson T., Bryson B., Knowler K., Pinares-Patiño C., Bain W., Elmes S., Young E., Wing J., Waller E., Pickering N. and McEwan J.C. (2019). *Proc. Assoc. Advmt. Anim. Breed. Genet.* **23**: 306.

*Proc. Assoc. Advmt. Anim. Breed. Genet: 24: 19-21*

# WHAT CAN RESEARCH ON THE GENETICS OF HUMAN WELL-BEING TELL US ABOUT IMPROVING LIVESTOCK WELL-BEING?

**P.M. Visscher[1] and N.R. Wray[1]**

[1] Institute for Molecular Bioscience, The University of Queensland, St Lucia 4072

## SUMMARY

Genetics of common disease and subjective well-being in human populations has been given a major boost in the last 15 years through genome-wide association studies (GWAS). Tens of thousands of loci have been identified that are robustly associated with one or more of these traits, with strong evidence for pleiotropy. Limitations for cross-species comparisons of traits, genes and pathways are the arbitrariness of defining phenotypes and the current lack of resolution in gene and variant mapping. Developments in both livestock and human genetic studies imply that better comparisons will be feasible in the near future. Methods underlying genomic prediction are converging between livestock and humans.

## INTRODUCTION

The title of this short paper was suggested by the Organisers of the AAABG 2021 conference. The initial response to the question is 'not a lot', because humans can be asked directly about their well-being where inference in livestock has to be made otherwise, for example by measuring proxy phenotypes. Also, 'Improving' well-being in humans is by changing the environment whereas in livestock it is by genetic selection. Nevertheless, there may be room for cross-fertilisation of the two disciplines on the topic of well-being, and we have tried to draw a few conclusions by reviewing what experimental designs and data have worked best in humans.

## MATERIALS AND METHODS

Well-being can be defined in many ways. An operational measure in humans might be absence of disease or a long and healthy life. One definition that geneticists and economists are both interested in is called 'subjective well-being', which is asking individuals to rate themselves on well-being on a quantitative score. "Disease count" has been used as a quantitative trait in human genetic studies and is a sum of the presence/absence of a number of common diseases and disorders, measured in biobanks or obtained from electronic health records.

## RESULTS AND DISCUSSION

Livestock and humans share biology and a number of studies have tested whether genes that explain variation for a particular trait in one species also explain variation is a similar trait in another. For the trait stature or size, this is clearly the case (Pryce *et al.* 2011; Kemper *et al.* 2012; Bouwman *et al.* 2018; Raymond *et al.* 2020). For other traits where the corresponding traits are less easy to define aligning "well-being" traits across species may not be straightforward.

To our knowledge, the only study that has directly tried to integrate genetic analyses of livestock and humans for a well-being trait was a recent study using the trait of temperament, measured as flight time after release from a weighing box in beef cattle (Costilla *et al.* 2020). In fact, the study was initiated with the misplaced hypothesis that studying traits in livestock (where temperament can be measured objectively) could inform genetic analyses of behavioural traits in humans. It came as a surprise to us that human studies (using self-report questionnaire data) are very much more powered than cattle data sets to detect trait-marker associations, because of the superior experimental sample sizes. In that study we tested for enrichment of genetic associations for flight time in orthologous gene sets associated with the human behavioural trait of neuroticism (quantitative measure of an anxiety-related emotional state), and three disorders of the brain: schizophrenia, an adult onset disorder with lifetime risk ~1%; autism spectrum disorder, a childhood onset disorder lifetime risk ~1% of the population; developmental delay disorders, which are less common with more severe early childhood behavioural traits and *de novo* mutations. We found evidence supporting shared genetic signals between flight time and autism spectrum disorders, perhaps because the age when the cattle were measured (10 – 14 months) equates best on the onset of these disorders in humans. In this study, we also integrated the cattle temperament associated genes, with gene expression measured in human tissues from the GTEx consortium. In summary, the study was less illuminating than we had hoped, but large sample sizes for human behavioural traits continue to accumulate, and the study provides a template for the analyses that can be conducted.

GWAS have been conducted on disease account, subjective well-being and 'life satisfaction' on 100,000s of thousands of samples, all reporting multiple genome-wide significant loci (Okbay *et al.* 2016a; Zhu *et al.* 2018). Subjective well-being is genetically strongly negatively correlated with neuroticism and depression (OkbaY *et al.* 2016a) and utilising that information improves polygenic prediction accuracy (Turley *et al.* 2018). From functional enrichment analyses, all subjective well-being and traits correlated with it all point to the brain. More generally, there is clear evidence in humans that genetic variation that is associated with behaviour is correlated with risk of many diseases and disorders, likely in a causal manner. For example, a polygenic predictor (= EBV in animal breeding, see Wray *et al.* (2019)) for 'educational attainment' (= the number of years of schooling), which is an imperfect proxy for intelligence, is negatively correlated with risk of dementia and neuroticism (Okbay *et al.* 2016b). Although more research is needed, it appears that there are both behavioural and physiological pathways to many diseases.

One known limitation of GWAS is that neither the causative mutations (polymorphisms) nor the target gene are identified, and that makes it hard to make meaningful species comparisons. However, sample size of exome and whole genome sequence studies are increasing, with the advantage that associated rare mutations that have predicted pathogenic effects (e.g., non-synonymous coding mutations) are likely to be causal and the target gene is known. This will allow a better comparison of genes and pathways related to traits across species. Similarly, the identification of additional dominant and recessive mutations in cattle for a number of syndromes (Reynolds *et al.* 2021) may lead to improved identification or prediction of pathogenic mutations in humans.

In addition to comparisons of genetic variation for well-being and disease traits in livestock and humans, there is increasing interest in humans in the application of polygenic (genomic) prediction. Although many researchers in human genetics don't realise that genomic prediction has its origin in animal breeding, as pointed out by Wray *et al.* (2019), there is increasing convergence in (Bayesian) methods to maximise accuracy. Even though the primary purpose of genomic prediction is in identifying people in the population who are at high risk of developing disease, so that preventative or therapeutic interventions can be better targeted, there is also a growing interest in using genomic predictors in the context of IVF and embryo selection. Within-family genomic selection in humans! Not surprisingly, the theory presented on the expected gains and (in)accuracy of prediction in the context of embryo selection using polygenic scores (Karavani *et al.* 2019; Turley *et al.* 2021) could come straight out of an animal breeding textbook. Finally, for some traits (such as human height) the discovery (training) datasets are becoming so large that using statistically significant (GWS) loci only in the prediction is approaching BLUP and other Bayesian approaches that use all genetic variants.

## CONCLUSIONS

Well-being and disease studies in humans are characterised by ever-larger genome studies, many of which are now reaching millions of individuals. GWAS is slowly moving from using SNP arrays and imputation to the use of whole-genome sequence data, thereby facilitating a better identification of causal variants and target genes and, ultimately, better prediction accuracy. This, combined with more discovery of specific causative mutations and target genes in livestock species, will allow better comparisons of genes and traits across species.

## REFERENCES

Bouwman A.C., Daetwyler H.D., Chamberlain A.J., Ponce C.H., Sargolzaei M. *et al.* (2018) *Nat Genet.* **50:** 362.

Costilla R., Kemper K.E., Byrne E.M., Porto-Neto L.R., Carvalheiro R. *et al.* (2020) *Genet Sel Evol* **52:** 51.

Karavani E., Zuk O., Zeevi D., Barzilai N., Stefanis N.C. *et al.* (2019) *Cell* **179:** 1424 e1428.

Kemper K.E., Visscher P.M. and Goddard M.E. (2012) *Genome Biol.* **13:** 244.

Okbay A., Baselmans B.M., De Neve J.E., Turley P., Nivard M.G. *et al.* (2016a) *Nat Genet.* **48:** 624.

Okbay, A., Beauchamp J.P., Fontana M.A., Lee J.J., Pers T.H. *et al.* (2016b) *Nature* **533:** 539.

Pryce J.E., Hayes B.J., Bolormaa S. and Goddard M.E. (2011) *Genetics* **187:** 981.

Raymond B., Yengo L., Costilla R., Schrooten C., Bouwman A.C. *et al.* (2020) *PLoS Genet.* **16:** e1008780.

Reynolds E.G.M., Neeley C., Lopdell T.J., Keehan M., Dittmer K. *et al.* (2021) *Nat Genet.* **53:** 949.

Turley P., Meyer M.N., Wang N., Cesarini D., Hammonds E. *et al.* (2021) *N. Engl. J Med.* **385:** 78.

Turley P., Walters R.K., Maghzian O., Okbay A., Lee J.J. *et al.* (2018) Nat. Genet. **50:** 229.

Wray N.R., Kemper K.E., Hayes B.J., Goddard M.E. and Visscher P.M. (2019) *Genetics* **211:** 1131.

Zhu Z., Zheng Z., Zhang F., Wu Y., Trzaskowski M. *et al.* (2018) *Nat. Commun.* **9:** 224.

*Proc. Assoc. Advmt. Anim. Breed. Genet: 24: 22-25*

# ASSESSMENT OF GENOMIC PREDICTIONS FOR FEEDLOT AND CARCASE TRAITS IN AUSTRALIAN ANGUS STEERS

**P.A. Alexandre[1], Y. Li[1], B.C. Hine[2], C.J. Duff[3], A.B. Ingham[1], L.R. Porto-Neto[1] and A. Reverter[1]**

[1]CSIRO, Agriculture and Food, Queensland Bioscience Precinct, St Lucia, QLD, Australia.
[2]CSIRO, Agriculture and Food, F.D. McMaster Laboratory, Armidale, NSW, Australia.
[3]Angus Australia, Armidale, NSW, Australia.

## SUMMARY

Improving feedlot performance, carcase weight and quality is a primary goal of the beef industry globally. Here we used data from 3,408 Australian Angus steers from seven birth cohorts (2011 to 2017) with genotypes for 45,152 SNPs. We report genetic parameter estimates and accuracies of genomic estimated breeding values (GEBV) for feedlot and carcase traits, namely feedlot average daily gain (ADG), carcase weight (CWT) and carcase Meat Standard Australia marbling score (MBL). Prediction accuracies were estimated based on traditional method as well as method LR. The average prediction accuracies across cohorts assessed with the traditional method were 0.28 (ADG), 0.49 (CWT) and 0.50 (MBL), while method LR accuracies were 0.47 (ADG), 0.64 (CWT) and 0.59 (MBL). We found a strong correlation (0.74, P-value<0.001) between traditional accuracies and method LR accuracies. Heritability estimates were moderate to large (0.29 for ADG, 0.53 for CWT and 0.41 for MBL). The metrics of GEBV quality and heritabilities reported here suggest good potential for accurate genomic selection of Australian Angus for feedlot performance and carcase characteristics.

## INTRODUCTION

Genomic selection represents a revolution in animal breeding. It enables the identification of superior animals through the estimation of genomic breeding values (GEBVs) for relevant quantitative traits (Goddard *et al.* 2010; Hayes *et al.* 2013). But the accuracy of GEBVs depends on several aspects including the size of the reference population and heritability of the trait (Goddard and Hayes 2009).

In this sense, Legarra and Reverter (2018) have proposed the method LR, which provides estimates of accuracy and biases by comparing genomic predictions based on partial and whole data. This method has been successfully applied to data from several different species (Aliakbari *et al.* 2020; Chu *et al.* 2019; Macedo *et al.* 2020; Silva *et al.* 2019).

Here we used method LR and a traditional method to evaluate the accuracy of genomic estimates in Australian Angus cattle. Angus is the dominant breed in the Australian cattle herd with an estimated 5.6 million females influenced by Angus genetics, accounting for 48% of the national female heard (Angus Australia 2019). Considering its importance, we aimed at determining the potential for accurate genomic selection of Australian Angus for feedlot performance and carcase characteristics.

## MATERIALS AND METHODS

The dataset used for this study was collected as part of the Australian Angus Sire Benchmarking Program (ASBP). It includes phenotypes, genotypes, and fixed effect information of 3,408 Australian Angus steers from seven year of birth cohorts (YOB, 2011 to 2017) and imputed genotyped for 45,152 autosomal SNPs. The steers represent 12 breeding properties and 294 sires with an average of 11.5 progeny per sire, ranging from 1 to 27. The number of animals and sires (in brackets) in YOB cohorts 2011 to 2017 are respectively 361 (35), 514 (48), 579 (44), 274 (25), 569 (49), 575 (63) and 536 (56).

Three phenotypes were analysed, including feedlot average daily gain (ADG, 1.59 ± 0.33kg/d), carcase weight (CWT, 432.99 ± 65.60kg) and Meat Standard Australia marbling score (MLB, 494.66 ± 122.54). Variance components, heritabilities and genetic correlations were estimated using Qxpak5 (Pérez-Enciso and Misztal 2011). The linear mixed model used to analyse all traits contained the fixed effects of contemporary group (CG), including property of origin, year and month of birth, and date of measurement, age of dam (AOD) at birth of calf in years and the linear covariate of age at measurement. Contemporary groups were different for each phenotype due to the different measurement dates. The random additive polygenic and residual effects were fitted with assumed distributions $N(\mathbf{0}, \mathbf{G} \otimes \mathbf{V_G})$ and $N(\mathbf{0}, \mathbf{I} \otimes \mathbf{V_R})$, respectively, where **G** represents the genomic relationship matrix (GRM) generated using the first method of VanRaden (2008), $\mathbf{V_G}$ is the genetic covariance matrix, **I** is an identity matrix, $\mathbf{V_R}$ is the residual covariance matrix and $\otimes$ represents the Kronecker product. The analyses were undertaken in two stages. First, one multivariate (3-variate) analysis was performed with all traits. The resulting GEBV ($\hat{\mathbf{u}}_\mathbf{w}$) from this multivariate analysis, based on the whole dataset, was used as the calibration in the computation of accuracy and bias. Next, a series of single-trait analyses were undertaken where the values from animals from a given YOB cohort were treated as missing. The resulting GEBV ($\hat{\mathbf{u}}_\mathbf{p}$) from these univariate analyses based on partial data were used as the validation.

To ascertain the quality of the resulting GEBV in the validation population we used: 1) Traditional accuracy, calculated as the Pearson correlation between a GEBV and its associated phenotype adjusted for fixed effects for individuals in the validation population, divided by the square root of heritability (Bolormaa *et al.* 2013); 2) Method LR Bias, calculated as the difference between the average GEBV of individuals in the validation population minus that using the calibration data; 3) Method LR Dispersion, measured for individuals in the validation population from the slope of the regression of $\hat{\mathbf{u}}_\mathbf{w}$ on $\hat{\mathbf{u}}_\mathbf{p}$; and 4) Method LR accuracy, computed for individuals in the validation population according to Legarra and Reverter (2018) as follows:

$$\mathrm{ACC_{LR}} = \sqrt{\frac{cov(\hat{\mathbf{u}}_w, \hat{\mathbf{u}}_p)}{(1 + \bar{F} - 2\bar{f})\sigma^2_{g,\infty}}}$$

Where $\bar{F}$ is the average inbreeding coefficient, $2\bar{f}$ is the average relationship between individuals, and $\sigma^2_{g,\infty}$ is the genetic variance at equilibrium in a population under selection which, assuming the individuals in the validation population are not under selection, can be estimated by the additive genetic variance estimated from the partial dataset.

## RESULTS AND DISCUSSION

Heritability estimates were 0.30 for ADG, 0.53 for CWT and 0.41 for MBL which are well within reported values in literature. For instance, Somavilla *et al.* (2017) using Bayesian GBLUP to evaluate feedlot ADG in Nellore cattle found a heritability of 0.31. For the carcase traits, Su *et al.* (2017) working with Hereford and admixed Simmental reported heritabilities of 0.48 and 0.43 for marbling score and 0.51 and 0.34 for CWT, respectively.

Genetic correlations were high and positive between ADG and CWT (0.64) and close to zero between those 2 traits and MBL (0.05 and 0.04, respectively). These results corroborate literature that have reported low correlations between live/carcass weight and traits such as fat deposition and marbling (Nkrumah *et al.* 2007).

The metrics of GEBV quality are presented in Table 1. Traditional accuracies were 0.28 (ADG), 0.49 (CWT) and 0.50 (MBL), while method LR accuracies were 0.47 (ADG), 0.64 (CWT) and 0.59

(MBL). This is in accordance with the literature that reports greater accuracy for carcase traits than for live animal body composition traits (Boerner *et al.* 2014) and increased accuracy for traits with a higher heritability (Fernandes Júnior *et al.* 2016). We found a strong correlation (0.74, P<0.001) between traditional accuracy and Method LR accuracy (Figure 1). Values of bias for all the traits were fairly close to zero, showing an absence of bias. In the absence of bias, the expected value of dispersion is 1, which was observed for all traits.

**Table 1. Traditional accuracy ($ACC_T$) and method LR accuracy ($ACC_{LR}$), bias ($Bias_{LR}$) and dispersion ($Disp_{LR}$) of GEBV for feedlot average daily gain (ADG), carcase weight (CWT) and marbling score (MBL), based on a 7-way cross-validation schema**

| | ADG | | | | CWT | | | | MBL | | | |
|---|---|---|---|---|---|---|---|---|---|---|---|---|
| | Mean | SD | Min | Max | Mean | SD | Min | Max | Mean | SD | Min | Max |
| $ACC_T$ | 0.28 | 0.11 | 0.08 | 0.42 | 0.49 | 0.07 | 0.40 | 0.58 | 0.50 | 0.06 | 0.43 | 0.60 |
| $ACC_{LR}$ | 0.47 | 0.04 | 0.42 | 0.53 | 0.64 | 0.05 | 0.57 | 0.67 | 0.59 | 0.05 | 0.53 | 0.67 |
| $Bias_{LR}$ | 0.00 | 0.01 | -0.01 | 0.01 | 0.27 | 0.61 | -0.54 | 1.20 | -0.08 | 1.71 | -2.14 | 2.13 |
| $Disp_{LR}$ | 0.97 | 0.15 | 0.74 | 1.17 | 0.99 | 0.09 | 0.83 | 1.10 | 0.98 | 0.09 | 0.88 | 1.13 |

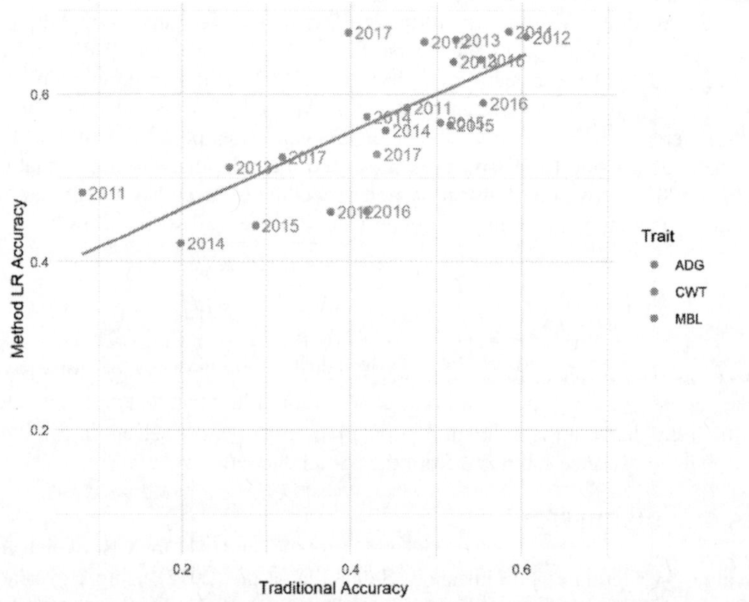


**Figure 1. Relationship between traditional accuracy and Method LR accuracy for feedlot average daily gain (ADG), carcase weight (CWT) and carcase marbling score (MBL) according to the 7-way cross-validation scheme based on year of birth cohorts**

The relationship between heritability and GEBV accuracy is also reflected in the phenotypic differences between validation animals in the highest and lowest GEBV quartile (Table 2). Based on SD units, ADG shows a Q1-Q4 difference of 0.35, CTW shows 0.93 and MBL 0.89. This demonstrates that the higher the GEBV accuracy, the higher the genetic gain expected when selecting elite bulls to sire the next generation.

**Table 2. Difference between highest and lowest quartile for adjusted phenotypes (feedlot average daily gain - ADG, carcase weight - CWT and marbling score - MBL) based on GEBV ranking**

| Cohort | ADG | CWT | MBL |
|---|---|---|---|
| 2011 | 0.00 | 33.57 | 103.47 |
| 2012 | 0.14 | 33.25 | 116.36 |
| 2013 | 0.08 | 34.44 | 99.20 |
| 2014 | 0.10 | 25.90 | 78.60 |
| 2015 | 0.08 | 28.36 | 85.45 |
| 2016 | 0.13 | 31.51 | 86.56 |
| 2017 | 0.09 | 20.53 | 60.19 |
| Average | 0.09 | 29.65 | 89.98 |
| Average/SD* | 0.35 | 0.94 | 0.89 |

*Standard deviation of adjusted phenotypes

## CONCLUSIONS

The metrics of GEBV quality based on method LR, including accuracy, bias, and dispersion, as well as the heritabilities reported here, suggest good potential for accurate genomic selection of Australian Angus for the analysed traits. Further analyses are being undertaken to include other relevant feedlot and carcass traits.

## REFERENCES

Aliakbari, A., Delpuech, E., Labrune, Y., Riquet, J., Gilbert, H. (2020) *Genet. Sel. Evol.* **52**: 1–15.

Angus Australia (2019). Available at: https://www.angusaustralia.com.au/research/australian-beef-breeding-insights-survey/

Boerner, V., Johnston, D. J., and Tier, B. (2014) *Genet. Sel. Evol.* **46**: 1–11.

Bolormaa, S., Pryce, J. E., Kemper, K., Savin, K., Hayes, B. J., Barendse, W., et al. (2013) *J. Anim. Sci.* **91**: 3088–3104.

Chu, T. T., Bastiaansen, J. W. M., Berg, P., Romé, H., Marois, D., Henshall, J., et al. (2019) *Genet. Sel. Evol.* **51**: 50.

Fernandes Júnior, G. A., Rosa, G. J. M., Valente, B. D., Carvalheiro, R., Baldi, F., Garcia, D. A., et al. (2016) *Genet. Sel. Evol.* **48**: 1–8.

Goddard, M. E., Hayes, B. J., and Meuwissen, T. H. E. (2010) *Genet. Res.* **92**: 413–421.

Hayes, B., Bowman, P., Chamberlain, A., and Goddard, M. (2009) *J. Dairy Sci.* **92**: 433–443.

Hayes, B. J., Lewin, H. a, and Goddard, M. E. (2013) *Trends Genet.* **29**: 206–14.

Legarra, A., and Reverter, A. (2018) *Genet. Sel. Evol.* **50:** 53.

Macedo, F. L. L., Reverter, A., and Legarra, A. (2020) *J. Dairy Sci.* **103**: 529–544.

Nkrumah, J. D., Keisler, D. H., Crews, D. H., Basarab, J. A., Wang, Z., Li, C., et al. (2007) *J. Anim. Sci.* **85**: 2147–55.

Pérez-Enciso, M., and Misztal, I. (2011) *BMC Bioinformatics* **12:** 202.

Silva, R. M. O., Evenhuis, J. P., Vallejo, R. L., Gao, G., Martin, K. E., Leeds, T. D., et al. (2019) *Genet. Sel. Evol.* **51:** 42.

Somavilla, A. L., Regitano, L. C. A., Rosa, G. J. M., Mokry, F. B., Mudadu, M. A., Tizioto, P. C., et al. (2017) *Genes|Genomes|Genetics* **7**: 1855–1859.

Su, H., Golden, B., Hyde, L., Sanders, S., and Garrick, D. (2017). *J. Anim. Sci.* **95**: 4718–4727.

VanRaden, P. M. (2008) *J. Dairy Sci.* **91**: 4414–23.

*Contributed paper*

# GENETIC ASSOCIATIONS BETWEEN ULTRASOUND AND CARCASE MUSCLE DIMENSION MEASURES IN SHEEP

**P. Alexandri[1, 2], S.F. Walkom[1, 2] and D.J. Brown[1, 2]**

[1] Animal Genetics Breeding Unit*, University of New England, Armidale, NSW, 2350 Australia
[2]Advanced Livestock Measurement Technologies project, Meat & Livestock Australia, 2060 Australia

## SUMMARY

This study investigated the genetic relationship between eye muscle width and depth recorded via ultrasound on live animals and on carcases in two populations of Australian and New Zealand sheep. Genetic correlations between ultrasound and carcase muscle dimensions were estimated within populations. Carcase eye muscle dimensions have sufficient genetic variation to be included in sheep breeding programs. Genetic correlations between carcase eye muscle depth (CEMD) and width (CEMW), and between CEMW and ultrasound eye muscle depth (PEMD) in Australian sheep were lower than expected. On the other hand, high genetic correlations were observed between ultrasound depth and width recorded in different ages on New Zealand Merinos. These differences indicate further research about CEMW is required and the implications of current selection practises has on carcase eye muscle dimensions.

## INTRODUCTION

Lean meat yield is an important driver of profit for producers, processors and retailers of sheep meat. Ultrasound scanned eye muscle depth is moderately heritable and strongly correlated genetically with eye muscle depth in the carcase. Consequently, the majority of genetic gain in the depth of the eye muscle and in turn lean meat yield has been achieved by seed stock breeders selecting upon the ultrasound trait in the live animal (Brown and Swan 2016). The strong genetic correlations between ultrasound scanned eye muscle depth and width, previously observed in several studies (Safari *et al.* 2005), has meant that Sheep Genetics (Brown *et al.* 2007) has provided breeding values only for muscle depth. This is in part also due to the greater difficulty in measuring eye muscle width via ultrasound.

There are several studies that have reported on the genetic relationship between ultrasound muscle dimensions (Brito *et al.* 2017) and ultrasound and carcase measurements (Safari *et al.* 2005; Greeff *et al.* 2008; Mortimer *et al.* 2010), but often with low records. In the following study the genetic relationship between ultrasound and carcase eye muscle measurements was investigated in two different data sets: > 25,000 Australian Merino and Merino-cross sheep where eye muscle dimensions were measured both with ultrasound post weaning and on the carcase; and >30,000 New Zealand Merinos with ultrasound measurements at different ages. The objective of this study was to update the understanding of the relationship between these measurements and determine the impact selection decisions may have on the dimensions of the eye muscle in the carcase.

## MATERIALS AND METHODS

**Australian Dataset.** Data from Australian Merino and Merino-cross sheep were collected between 2007 and 2019 from 35 commercial flocks, 8 Information Nucleus Flocks and the MLA Resource Flock (van der Werf *et al.* 2010). Ultrasound muscle scanners accredited through Sheep

* A joint venture of NSW Department of Primary Industries and the University of New England

Genetics (MLA) scanned eye muscle depth (PEMD) at the C site over the 12th rib, 45 mm from the midline at post weaning age (mean age 213±45 days). Carcase traits were measured using the procedures described in Mortimer *et al.* (2017b). The carcases were cut between the 12th and 13th ribs and eye muscle (*M. longissimus thoracis et lumborum*, LL) depth (CEMD) and eye muscle width (CEMW) were measured with vernier callipers. Mean animal age for carcase traits was 263 (±76) days.

**New Zealand Dataset.** Data from New Zealand Merinos were collected between 2009 and 2019. Animals were ultrasound scanned at the C site over the 12th rib and measured for eye muscle depth and width at post weaning (7 – 10 months, PEMD, PEMW), yearling (10 – 13 months, YEMD, YEMW) and hogget age (13 – 18 months, HEMD, HEMW). For both data sets live weight was recorded at the time of scanning and was used to adjust the ultrasound measurements for weight. Summaries for each trait are presented on Table 1.

**Table 1. Number of records, mean (standard deviation), coefficient of variation (CV) and number of sires and dams. CEMD: carcase eye muscle depth, CEMW carcase eye muscle width, PEMD and PEMW: post weaning ultrasound eye muscle depth and width, YEMD and YEMW: yearling ultrasound eye muscle depth and width, and HEMD and HEMW: hogget ultrasound eye muscle depth and width**

| Dataset | Trait | Records | Mean (SD) | CV | Sires | Dams |
|---|---|---|---|---|---|---|
| Australian | PEMD | 25,628 | 25.4 (4.8) | 18.8 | 1,651 | 12,799 |
| | CEMD | 26,284 | 31.0 (4.7) | 15.3 | 1,874 | 12,747 |
| | CEMW | 26,282 | 60.6 (5.5) | 9.0 | 1,874 | 12,747 |
| New Zealand | PEMD | 3,251 | 26.1 (2.8) | 10.7 | 169 | 3,251 |
| | YEMD | 6,591 | 27.9 (3.6) | 12.8 | 339 | 4,038 |
| | HEMD | 21,616 | 27.8 (3.8) | 13.5 | 752 | 11,118 |
| | PEMW | 5,616 | 68.8 (6.0) | 8.8 | 144 | 2,760 |
| | YEMW | 6,596 | 71.6 (6.2) | 8.7 | 342 | 4,040 |
| | HEMW | 21,087 | 71.1 (6.9) | 9.7 | 733 | 10,629 |

**Statistical analysis.** Within each dataset, variance components and genetic parameters for each trait were estimated using a linear mixed model and REML methods with ASReml software (Gilmour *et al.* 2015). Fixed effects included type of birth, contemporary group, sex (male or female) and the age of dam. The quadratic function of live weight (post weaning, yearling, hogget) and hot carcase weight were included to adjust the ultrasound and the carcase traits respectively. All models included the random effects of animal, genetic group (Swan *et al.* 2016) and sire × flock interaction. Maternal effects were not fitted since preliminary analysis showed they were non-significant. For Australian data set age of the animal was included as a fixed effect. For both datasets the animal effect represented the additive genetic variance. Contemporary group was defined by breed, flock, management group, sex, date of measurement and – for carcass data – kill group. Phenotypic variance was calculated as the sum of the additive genetic, sire × site and the residual variance. For each dataset, phenotypic and genetic covariance for all traits and correlations between traits were estimated using bivariate analysis in ASReml.

## RESULTS AND DISCUSSION

Variance components and heritability estimates for ultrasound and carcase traits for each of the data sets are shown in Table 2. For the Australian dataset, heritability estimates were moderate for carcase traits ranging from 0.19 (±0.02) for CEMD to 0.27 (±0.02) for CEMW; higher heritability (0.32±0.02) was observed for PEMD. Similar heritabilities for CEMD and CEMW have been observed in previous studies (Greeff *et al.* 2008; Huisman *et al.* 2016; Mortimer *et al.* 2017b). Heritability for PEMD for both data sets was higher than previously reported (Safari *et al.* 2005; Greeff *et al.* 2008; Mortimer *et al.* 2017a). Higher heritabilities were observed for the New Zealand Merino ultrasound traits: ranging from 0.23 (±0.03, YEMW) to 0.45 (±0.04, PEMD) (Table 2). Increased heritabilities have been observed in the past when live weight was used to adjust measurements (Mortimer *et al.* 2014).

**Table 2. Estimates of phenotypic ($\widehat{\sigma}_p$), additive ($\widehat{\sigma}_a$) and residual ($\widehat{\sigma}_\varepsilon$) variance and heritability ($h^2$) for ultrasound and carcase eye muscle traits. Standard error in parentheses**

| Dataset | Trait | $h^2$ | $\widehat{\sigma}_p$ | $\widehat{\sigma}_a$ | $\widehat{\sigma}_\varepsilon$ | $\widehat{\sigma}_{sire\ x\ site}$ |
|---|---|---|---|---|---|---|
| | **PEMD** | 0.32 (0.02) | 4.95 (0.46) | 1.59 (0.1) | 3.28 (0.08) | 0.08 (0.02) |
| **Australian** | **CEMD** | 0.19 (0.02) | 10.12 (0.09) | 1.92 (0.18) | 8.06 (0.16) | 0.14 (0.05) |
| | **CEMW** | 0.27 (0.02) | 14.81 (0.14) | 3.93 (0.3) | 10.55 (0.25) | 0.33 (0.08) |
| | **PEMD** | 0.45 (0.04) | 3.15 (0.07) | 1.35 (0.14) | 1.76 (0.10) | 0.03 (0.02) |
| | **YEMD** | 0.34 (0.04) | 3.42 (0.07) | 1.13 (0.18) | 2.16 (0.13) | 0.13 (0.04) |
| **New Zealand** | **HEMD** | 0.31 (0.02) | 3.78 (0.04) | 1.16 (0.10) | 2.49 (0.07) | 0.13 (0.02) |
| | **PEMW** | 0.29 (0.03) | 10.01 (0.22) | 2.86 (0.40) | 7.09 (0.32) | 0.06 (0.04) |
| | **YEMW** | 0.23 (0.03) | 9.48 (0.19) | 2.20 (0.42) | 7.01 (0.33) | 0.27 (0.11) |
| | **HEMW** | 0.27 (0.02) | 10.56 (0.12) | 2.82 (0.26) | 7.46 (0.20) | 0.27 (0.06) |

Estimates of genetic and phenotypic correlations between carcase traits and post weaning ultrasound eye muscle depth for the Australian dataset are shown in Table 3. The genetic correlation between PEMD and CEMD was strong (0.77±0.04), but for the same animals CEMD was only moderately correlated with CEMW (0.38±0.05). Moreover, the correlation between CEMW and PEMD was low (0.17±0.04).

In contrast, for the New Zealand dataset, the correlations between ultrasound traits exhibited high genetic correlations between muscle depth and width at the same age (0.92±0.03 to 0.99±0.02) as well as between traits recorded at different ages (0.78±0.15 to 0.90±0.07, Table 4).

**Table 3. Estimates of genetic (below diagonal) and phenotypic (above diagonal) correlations and their standard errors (parentheses) between carcase traits and ultrasound post weaning eye muscle depth for Australian dataset (see Table 1 for abbreviations)**

| | PEMD | CEMD | CEMW |
|---|---|---|---|
| **PEMD** | | 0.23 (0.01) | 0.06 (0.01) |
| **CEMD** | 0.77 (0.04) | | 0.09 (0.01) |
| **CEMW** | 0.17 (0.04) | 0.38 (0.05) | |

High correlations between PEMD and CEMD have previously been reported by Greeff *et al.* (2008) (0.77) and Mortimer *et al.* (2010) (0.82). Moderate positive genetic correlations between CEMD and CEMW found in this study were similar to Safari *et al.* (2005) (0.23) and Greeff *et al.* (2008) (0.41). Based on these results, carcase eye muscle depth appears to be a genetically different trait to carcase eye muscle width. These low correlations in carcase measures contradict the New Zealand ultrasound results for corresponding traits as well as previous studies using ultrasound eye muscle dimensions at post weaning age, where correlations between eye muscle depth and width ranged between 0.78 in Australia (Safari *et al.* 2005) and 0.82 in New Zealand (Brito *et al.* 2017). Lower genetic correlations between ultrasound and carcase measurements could be a result of ultrasound limitations to accurately predict muscle dimensions. Hopkins *et al.* (2007) showed that ultrasound muscle depth measurements are subject to more errors in heavier sheep. Moreover, it would be beneficial for future investigations to include accurate animal age records since limitations might also include potential failure to properly account for age variation.

**Table 4. Estimates of genetic and phenotypic correlations between ultrasound eye muscle depth and width for different ages (post weaning, yearling and hogget) for New Zealand Merino (standard error in parentheses)**

| | Genetic | | | Phenotypic | | |
|---|---|---|---|---|---|---|
| | **PEMD** | **YEMD** | **HEMD** | **PEMD** | **YEMD** | **HEMD** |
| **PEMW** | 0.92 (0.03) | 0.84 (0.16) | 0.88 (0.09) | 0.61 (0.01) | 0.15 (0.94) | 0.64 (0.23) |
| **YEMW** | 0.78 (0.15) | 0.99 (0.02) | 0.87 (0.07) | 0.57 (0.46) | 0.68 (0.01) | 0.49 (0.03) |
| **HEMW** | 0.90 (0.07) | 0.80 (0.07) | 0.95 (0.01) | 0.60 (0.21) | 0.48 (0.03) | 0.70 (0.01) |

## CONCLUSIONS

The high genetic correlation between ultrasound PEMD and CEMD means that ultrasound should continue to be used as a selection trait to improve CEMD. However, whilst ultrasound measures of EMD and EMW are strongly correlated with each other, their correlations with carcase measurements are weaker. In particular, further research is required to determine if current selection practices are changing the dimensions of the eye muscle within the carcase and increase the need for a CEMW breeding value.


## ACKNOWLEDGEMENTS

This work was conducted as part of the Advanced Measurement Technologies for globally competitive Australian meat project, funded by the Australian Government Department of Agriculture and Water Resources as part of its Rural R&D for Profit programme. The authors thank the teams behind the Sheep CRC for Sheep Industry Innovation information nucleus and MLA resource flock.


## REFERENCES


Brito L.F., McEwan J.C., Miller S., Bain W., *et al.* (2017) *Small Rumin. Res.* **154:** 81.

Brown D.J., Huisman A.E., Swan A.A., U. G. H., Woolaston R. R. B., A. J., Atkins K. D. and Banks R.G. (2007) *Proc. Assoc. Advmt. Anim. Breed. Genet.* **17:** 187.

Brown D.J. and Swan A.A. (2016) *Anim. Prod. Sci.* **56:** 690.

Gilmour A.R., Gogel B.J., Cullis B.R., Welham S.J. and Thompson R. (2015) ASReml User Guide Release 4.1. VSN International Ltd, Hemel Hempstead, UK

Greeff J.C., Safari E., Fogarty N.M., Hopkins D. ., *et al.* (2008) *J Anim Breed Genet* **125:** 205.

Hopkins D.L., Stanley D.F. and Ponnampalam E.N. (2007) *Aust. J. Exp. Agric.* **47:** 1304.

Huisman A.E., Brown D.J. and Fogarty N.M. (2016) *Anim. Prod. Sci.* **56:** 95.

Mortimer S.I., Hatcher S., Fogarty N.M., van der Werf J.H.J., *et al.* (2017a) *J. Anim. Sci.* **95:** 1879.

Mortimer S.I., Hatcher S., Fogarty N.M., van der Werf J.H.J., *et al.* (2017b) *J. Anim. Sci.* **95:** 2385.

Mortimer S.I., Swan A.A., Brown D.J., and van der Werf J.H.J. (2014) *WCGALP Proceedings*:345.

Mortimer S.I., van der Werf J.H.J., Jacob R.H., Pethick D.W., *et al.* (2010) *Anim. Prod. Sci.* **50:** 1135.

Safari E., Fogarty N.M. and Gilmour A.R. (2005) *Livest. Prod. Sci.* ***92:*** 271

Swan A.A., Brown D.J. and van der Werf J.H.J. (2016) *Anim. Prod. Sci.* **56:** 87.

van der Werf J.H.J., Kinghorn B.P. and Banks R.G. (2010) *Anim. Prod. Sci.***50***:* 998.

*Contributed paper*

# MERITS OF DEVELOPING A GENETIC EVALUATION FOR THE AUSTRALIAN DAIRY SHEEP AND GOAT INDUSTRIES

**P. Alexandri, R.G. Banks, D.J. Brown and S.F. Walkom**

Animal Genetics Breeding Unit*, University of New England, Armidale, NSW, 2350 Australia

## SUMMARY

The Australian dairy sheep and goat industries have been constrained by the size of the national flock and the geographical spread of flocks across the country. The absence of a national genetic evaluation system to underpin meaningful genetic improvement has contributed to the production performance of Australian dairy sheep and goat milk production being lower compared to the more developed dairy sheep and goat industries of Europe. Implementing a national genetic evaluation scheme will aid the development and future progress of the Australian dairy sheep and goat industries through identification and selection of the genetically superior animals. This study investigated the advantages of a genetic evaluation program for traits of value in Australian dairy sheep and goats, and outlined potential gains from implementing a breeding program.

## INTRODUCTION

The potential to develop breeding programs within the Australian dairy goat and milk sheep industries is constrained by the small size of the national flock, geographical spread across the country in a variety of different environments and management practises, and the ability to measure phenotypes and record pedigree (Lindsay and Skerritt 2003). Despite these challenges, both the number of Australian milking goats and production volume increased more than 62% from 2012 to 2018 (Zalcman and Cowled 2018). However, average production, expressed in number of kilos per lactation, is still lower than what is achieved in Europe. The dairy sheep industry in Australia is estimated to be growing around 10% annually (Cameron 2014) but it relies mostly on low yielding non-dairy crossbred ewes with low lactation persistency compared to animals used in other countries with more advanced dairy sheep industries (Lindsay and Skerritt 2003). Most Australian dairy ewes are East Friesian crossbred (Morrisey *et al.* 2007), and while their average flock milk yield is around 40% lower compared to Europe, there is lots of variation within and between flocks, with the performance of a number of ewes being on par with European animals. Variation in milk yield within the flock is a necessary prerequisite for genetic gain, but standardised recording protocols are needed for effective genetic evaluation which in turn will enable genetic selection. This study investigated the advantages and potential genetic gains of a genetic evaluation system for traits of value in Australian dairy sheep and goats.

## MATERIALS AND METHODS

**Review of genetic parameters.** Milk yield represents the majority of the total income in the dairy sheep and goat industries (Carta *et al.* 2009). For both dairy sheep and dairy goats almost all of the milk produced is used for cheese production, and thus milk content traits (fat and protein) are also important as they affect cheese yield and flavour. Therefore, increasing milk yield and improving milk quality (mostly though increasing fat and protein content) are the most important components of the breeding objective for most breeds (Ramon *et al.* 2010).

For dairy goats, milk yield heritability ranged between 0.18 and 0.34 (Analla *et al.* 1996;

---

* A joint venture of NSW Department of Primary Industries and the University of New England

Scholtens *et al.* 2019), but can be higher when estimated per lactation (e.g. 0.30 for first lactation does; Arnal *et al.* 2020). A similar pattern was observed for fat and protein yield (Analla *et al.* 1996; Bélichon *et al.* 1999; Scholtens *et al.* 2019; Arnal *et al.* 2020). Genetic correlations between milk, fat and protein yield were high, with estimates reported between 0.77 and 0.95 (Bélichon *et al.* 1999; Manfredi and Ådnøy 2012). However, genetic correlations between milk yield and milk content (daily fat and protein percentage) were negative (-0.48 to -0.12) (Bélichon *et al.* 1999; Maroteau *et al.* 2014).

For most dairy sheep breeds, milk yield standardised for lactation length was originally the only selection criterion, while milk content traits were added later. Yield traits (milk, fat and protein) were reported to have moderate heritability (between 0.16 and 0.32) with high genetic correlations between the yield traits (between 0.77 and 0.93) (Barillet and Boichard 1987; Carta *et al.* 2009). Similar to dairy goats, genetic correlations between milk yield and milk content for dairy sheep was negative but varied for different sheep breeds (-0.43 to -0.64) (Carta *et al.* 2009).

**Prediction of genetic gains.** Predictions of potential genetic gain based on a defined breeding objective for within-flock selection was undertaken using MTINDEX (van der Werf 2019). It was assumed that milking does/ewes entered their first lactation at 12 months and they were maintained in the herd/flock for four lactations (generation interval of 2.5 years). All milking females were recorded for lactation yield with test day records available for fat and protein percentage. Two breeding objectives were tested; (1) single trait selection based solely on total milk yield and (2) multiple trait selection placing equal economic weighting on milk yield, total fat and total protein yield. Potential gains were modelled for small, medium and large herds/flocks with flock size modelled by varying the assumptions associated with progeny and half-sib records. The number of paternal half-sibs and progeny was assumed to be 10, 30 or 100 respectively for small, medium and large herd/flocks. The modelling utilised variance components and genetic correlations for yield traits compiled from literature: a) for dairy goats, heritability of 0.25 for milk, fat and protein yield and genetic correlations between 0.77 and 0.89 (Bélichon *et al.* 1999); b) for dairy sheep, heritability of 0.28 for milk, fat and protein yield and genetic correlations between 0.82 and 0.92 (Barillet and Boichard 1987). Heritabilities and genetic correlations between traits were similar for the two species, however there were big differences in phenotypic variance; in dairy sheep, fat and protein yield phenotypic variance was a lot lower (Barillet and Boichard 1987).

## RESULTS AND DISCUSSION

**Genetic Gains in Dairy Goats.** Potential gains in milk yield were predicted to be between 1.08 and 1.32 kg of milk per animal per year (Figure 1A), depending on farm size. Considering a national flock of ~ 30,000 milking goats (Zalcman and Cowled 2018), this equals to an additional 32k to 40k kg of milk per annum, which at Australian goat milk prices ($1.20 - $1.50/L, *J Cameron pers. comm.*) is worth between $38.4k and $60k per year to the industry. With cheese being a valuable product of the industry, selection targeting fat and protein content could increase fat and protein yields by 38.4k – 40.8k and 33.6k – 36k kg per year, respectively.

**Genetic Gains in Dairy Sheep.** Potential gain from selection in dairy sheep was 1.04 to 1.28 kg of milk per animal per year (Figure 1B), depending on farm size. Australian national flock size for dairy sheep is smaller than that for dairy goats, estimated at ~5,000 milking ewes. Potential gain for dairy sheep is likely to be between 5.2k and 6.4k kg of milk per annum, and this is worth between $10.4k and $12.8k per year to the industry (sheep milk valued at $2/L, *J Cameron pers. comm.*). Selecting for fat and protein content could increase fat and protein yields by 1.3k and 1.26k kg per year, respectively. Differences between predicted fat and protein yield gains for dairy sheep and goats reflect differences in the observed phenotypic variance for these traits in the two species (Barillet and Boichard 1987; Bélichon *et al.* 1999).

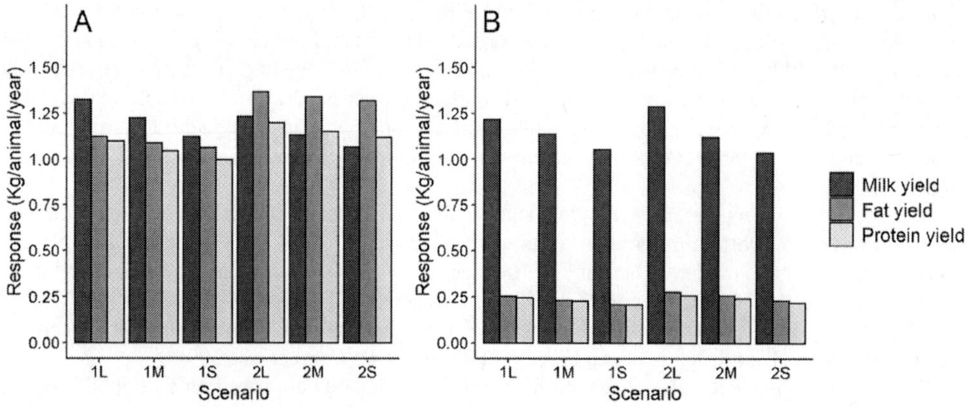


**Figure 1. Estimated genetic gains for dairy goats (A) and sheep (B) under scenarios for a small, medium and large flock (S, M, L) with the objective to maximise only milk yield (1) or to maximise milk, fat and protein yield (2)**

**Implementation in Industry.** For most livestock industries, prices received do not increase in real terms, but both fixed and variable costs do, resulting in what is usually known as the "cost-price squeeze". The effect of this is that profit from an enterprise declines in real terms over time. The only way to offset this decline is to make continual productivity improvement(s). Genetic improvement is attractive because its effects are permanent and cumulative, and provided that improvement is sufficiently rapid, the cost-price squeeze can be offset.

For highly productive dairy sheep and goat breeds, genetic improvement has been achieved through extensive recording and a pyramidal breeding structure (Barillet *et al.* 2001; Carta *et al.* 2009). As a result, some European breeds have achieved remarkable genetic progress under genetic evaluation. For example, genetic gain for French Lacaune between 1980 and 1994 was estimated at around 6 kg/year (Ugarte and Gabina 2004). Average production for improved European dairy sheep and goat breeds was very different to average production of Australian animals. For example, average milk production for dairy goats in France was 963 kg/goat per annum (IDELE 2020a), while Australian dairy goats have been reported to produce 519 kg/goat (Zamuner *et al.* 2020). Similarly, Australian dairy sheep produce 168 kg/ewe (*J Cameron pers. coms*), while European breeds like Lacaune produce 330 kg/ewe per annum (IDELE 2019). Importing and including high performance animals would be beneficial for the local dairy sheep and goat industry, but is currently prohibited under Australian quarantine regulations (Lindsay and Skerritt 2003). Consequently, the Australian industries will need to utilise the genetic variation that exists within the current population to drive genetic gains.

The principal implication from this study was that genetic improvement in dairy goats and sheep is quite feasible, and will generate significant benefits to producers. Achieving effective genetic improvement will require some changes in the current practice, principally in ensuring that records of milking performance are systematically collected, along with pedigree and fixed effects, and that the resulting data is appropriately analysed to produce EBVs, all of which will come at a cost to the operation (Banks and Walkom 2016). Previous investigations (Lindsay and Skerritt 2003) noted that despite reported support and enthusiasm from producers there has been very little progression to develop a genetic evaluation at the national level. This may be in part due to a lack of public support and R&D funding in these industries compared to other industries (Banks and Walkom 2016).

Overall, without national support the costs are too great for the smaller operations to bother recording and the larger dairies are potentially too time poor to lead a national genetic evaluation program.

## CONCLUSIONS

This study provided an example of the genetic gains that could be achieved by the Australian dairy sheep and goat industries through appropriate recording schemes and genetic evaluation. The rate of genetic gain could be further improved with the use of genomic selection to increase the accuracy of breeding values and reduce generation interval. This will require groups like the Australian Dairy Goat Society to develop a vision and strategy for long-term genetic improvement, and seek to assist the R&D and extension funding to support it. It is quite likely that some individuals will move to implement modern genetic improvement systems of their own accord, but the "trickle-down" flow of superior genetics from such a fragmented approach will be slow, and in the absence of industry-wide genetic improvement, goat and sheep dairying will remain a relatively small-scale industry in Australia.

## REFERENCES


Analla M., Jiménez-Gamero I., Muñoz-Serrano A., Serradilla J.M. and Falagán A. (1996) *J. Dairy Sci.* **79**: 1895.

Arnal M., Larroque H., Leclerc H., Ducrocq V. and Robert-Granie C. (2020) *J. Dairy Sci.* **103:** 4517.

Banks R.G and Walkom S.F. (2016) Rural Industries Research and Development Corporation No. 15/109.

Barillet F. and Boichard D. (1987) *Genet. Sel. Evol.* **19**: 459.

Barillet F., Marie C., Jacquin M., Lagriffoul G. and Astruc J.M. (2001) *Livestock Prod. Sci.* **71**: 17.

Bélichon S., Manfredi E. and Piacère A. (1999) *Genet. Sel. Evol.* **31:** 529.

Cameron A.W.N. (2014) Rural Industries Research and Development Corporation No. 14/083.

Carta A., Casu S. and Salaris S. (2009) *J. Dairy Sci.* **92**: 5814.

IDELE (2020a) 'Caprins 2020: productions lait et viande.' Available at http://idele.fr/domaines-techniques/economie/economie-des-filieres/publication/idelesolr/recommends/chiffres-cles-caprins-2020.html. [Accessed 02 March 2021].

IDELE (2019) 'Milk recording results of sheep - France 2019.' Available at http://idele.fr/domaines-techniques/economie/economie-des-filieres/dossiers-economie-de-lelevage/dossiers-2018/publication/idelesolr/recommends/milk-recording-results-of-cattle-goats-and-sheep-france-2019.html. [Accessed 02 March 2021].

Lindsay D. and Skerritt J. (2003) Rural Industries Research and Development Corporation No. 02/150.

Manfredi E. and Ådnøy T. (2012) *INRA Prod. Anim.* **25**: 233

Maroteau C., Palhière I., Larroque H., Clément V., Ferrand M., *et al.* (2014) *J. Dairy Sci.* **97**: 3142.

Morissey A.D, Cameron A.W.N., Caddy D.J. and Tilbrook A.J. (2007) *J. Dairy Sci.* **90**: 5056.

Ramon M., Legarra A., Ugarte E., Garde J.J. and Perez-Guzman M.D. (2010) *J. Dairy Sci.* **93**: 3303.

Scholtens M.R., Lopez-Villalobos N., Garrick D., Blair H., Lehnert K. and Snell R. (2019) *Anim. Sci. J.* **91**: e13310.

Ugarte E. and Gabina D. (2004) *Arch. Tierz., Dummerstorf* **47**: 10.

van der Werf J. (2019) Available at http://www-personal.une.edu.au/~jvanderw/software.htm. [Accessed 02 February 2021].

Zalcman E. and Cowled B. (2018) *Aust. Vet. J.* **96:** 341.

Zamuner F., DiGiacomo K., Cameron A.W.N. and Leury B.J. (2020) *J. Dairy Sci.* **103**: 954.

*Contributed paper*

# GENETIC PROGRESS FOR ENVIRONMENTAL OUTCOMES - HOW DO WE GET IT?

**P.R. Amer**

AbacusBio Limited, Dunedin 9016, New Zealand

## SUMMARY

Genetic improvement of traits in animals that affect environmental outcomes provides a rich opportunity for animal breeding researchers. Despite a growing imperative for action driven by a diverse range of stakeholders, real world application remains challenging. This paper highlights the difficulties that need to be overcome in terms of cost and practicality of novel selection criteria, change drivers including regulatory policy and farmer attitudes, and the implications of alternative breeding objective definitions. Clear and detailed policy direction along with cost effective monitoring of impacts will be pivotal to effective deployment of genetic mitigations.

## INTRODUCTION

People of the world with a comfortable level of income and security are becoming increasingly concerned about the ability of the world to sustain the resource demand pressures and adverse external impacts of livestock farming. Ruminant livestock are a particular target because of their role in methane production, but all livestock farming systems are under pressure. The focus of this paper is primarily on greenhouse gas (GHG) emissions and nutrient leaching. A substantial body of research is underway to develop mitigation strategies, and some of this research has identified potential opportunities to apply directional genetic selection for traits that contribute to mitigation. However, there remains a persistent drag on application. The objective of this paper is to describe the factors that are limiting genetic selection for environmental traits, and to attempt to map out a path to overcome these limits. First, though we identify the selection criteria available and the critical consideration of perspective, to make sure the desired outcomes of genetic improvement for environmental traits are achieved.

## SELECTION CRITERIA

A detailed consideration of traits that can be selected for and which might influence environmental outcomes is beyond the scope of this study. However, Wall *et al.* (2010) identified three types of traits. The most obvious category of traits directly targets biological functions of the animal that lead to improved outcomes (Type 1). Good examples of these traits are those that quantify methane emissions after accounting for food eaten. Productivity traits that dilute maintenance (Type 2), and survival traits, and traits that reduce the need for replacement animals and the emissions associated with them (Type 3) are additional options for selection criteria. An emerging consideration relates to the potential for farm systems changes expected to result from new environmental policies imposed by regulatory authorities. Here, focus of genetic changes in some existing traits could facilitate a more profitable transition to more environmentally friendly farming systems through a reorientation of the breeding goal. This new focus would target functional traits such as fertility.

## PERSPECTIVE

People of the world with income and personal security levels at or below the borderline for what is deemed necessary for basic human rights tend not to rate adverse environmental impacts as a priority. Despite the quality of life of these people often being adversely impacted by a severe deterioration in the environment they live, the acute priorities of basic nutrition, shelter, health, and security dominate their existence. This is a substantial proportion of the world's population. Thus, it is important to accept that the drive to improve environmental traits in livestock systems supporting disadvantaged people will be for general production efficiency, potentially including adaptability to climate change. It is likely therefore that, genetic gain for environmental traits will only obtain meaningful focus in developed livestock industries where efficient breeding programs are already in existence. Part of the driver will be a desire by well off consumers to make purchases of livestock products with a low environmental impact per unit of product.

Global demand for livestock products has been growing at a very fast rate, largely driven by the emerging middle classes of developing countries. If we accept the rights of these emerging middle classes to make dietary choices, such as consumption of livestock products for personal gain and health that are still well below the levels of consumption of many developed countries, then there is an important perspective of food demand that shapes the direction of change. Of particular concern when taking a global perspective, is the potential for reductions in livestock product output from developed countries due to a focus on the environment, that would likely lead to a shift of livestock production into less efficient farming regions where environmental damage per unit of product is even greater.

Strong drivers for environmental improvements exist at both national and regional levels. The focus here is typically on reducing the total amount of environmental pollution on a regional basis. For attributes of production that affect nutrient leaching, the focus is on a reduction in the total level of pollution, with an elevated focus in more sensitive catchments. For greenhouse gas traits, there is a strong drive to achieve commitments to reductions in the total national inventory of emissions.

## STRATEGIC OPTIONS

The following opportunities exist for breeding for environmental outcomes:

1. **Status quo** - Continue genetic gain on current trait change trajectories, quantify the environmental impacts, and argue that the reduced emissions intensity (emissions per unit of product) is sufficient.
2. **Artificial evolution** - Develop novel selection criteria that make it cheaper and easier for breeding programs to make genetic change in traits that improve both farm profit and which also improve environmental outcomes (Amer, 2012).
3. **Index manipulation** - Modify trait change trajectories through changes in breeding objectives and index weightings (Cottle *et al.* 2011) for existing traits to achieve:
   - reductions in emissions intensity
   - reductions in gross emissions per animal.
4. **Novel traits** – Develop new selection criteria targeting animal physiology changes that will lead to improved environmental outcomes (Pickering *et al.* 2015; Beatson *et al.* 2019).
5. **Facilitate system change** – Modify the breeding goal to target genetic changes that fit to future farming systems that have favourable environmental attributes.

While options 1 & 2 above are legitimate alternatives many livestock industries are under pressure to make more substantial and pro-active changes. The most obvious candidate traits will be new measurement techniques and technologies that improve feed efficiency. However, pastoral farmers' responses to improved feed efficiency are most likely to be through increased stocking rates to make sure all available pasture is used efficiently, and so pollution mitigation at a gross national output level (both leaching and GHG emission) from gains in feed efficiency could be minimal. The value of index manipulation to decrease selection emphasis on traits that increase gross emissions must also be treated with caution. The traits most unfavourably associated with gross emissions per animal are typically milk yield and animal growth rate. Genetic progress in these traits has been a long-standing driver of genetic gains in livestock efficiency, and so an emissions intensity philosophy which tends to favour rather than penalise these traits may in some cases lead to better long-term outcomes (Amer *et al.* 2018). The contra argument (i.e. for penalising traits with high emissions) is that a shift in selection emphasis away from genetic gain in milk yield and growth rate traits could fit more closely with legislated reductions in farming system intensity. For example, seasonal calving pasture-based farming systems with limited supplementation. This option is under serious consideration for the dairy industries for New Zealand and Ireland. For nutrient leaching mitigation, a shift towards less intensive farming systems in sensitive catchments is an obvious solution. With less intensive systems, genetic traits that enhance the cost reductions required to offset the reductions in product revenue will potentially increase in relative value to maintain economic viability.

**POLICY**

Regulatory policy provides an important but complex backdrop to livestock production. In particular, over the next decade, new policy mechanisms are likely to emerge that will have profound effects on advanced livestock industries. History can provide many examples of how agricultural policies have had the opposite effects to what was intended. For example, agricultural policies targeting support for smaller family farms sometimes disproportionally benefit the larger corporate farms who are better able to navigate the bureaucracy and exploit new technological opportunities. Of particular interest to animal breeders, is the question of how these policies might change preference drivers for trait improvements. At this point in time, key details of these policies are not sufficiently clear to be helpful in informing future breeding directions.

The primary target of policies to reduce environmental emissions from livestock industries should be to shift farm practice and land use from the most polluting to less polluting alternatives. For countries with relatively low greenhouse gas emissions intensity, there is potential leakage of emissions to less efficient competing industries when policies targeting emissions result in reduced domestic industry output. Because of the pressure on countries to do their bit in reducing their national GHG emissions it seems inevitable that countries with major livestock product exports will be forced to impose policies which either cap or reduce livestock product output. This will force international prices for livestock products to rise, which will be beneficial if it reduces livestock consumption in wealthy countries, but could also limit the supply of cheap, safe, and nutritious food to the growing middle classes in countries with emerging economies.

**INCENTIVES**

In a growing number of livestock industries, national concerns about environmental issues are threatening the social licence to farm. This creates a dilemma, as do date, there have been only quite limited or more commonly no direct financial incentives placed on farmers or farm practices which reduce pollution. Instead policies tend to force reductions in overall pollution levels, often by limiting production. Farmers in New Zealand are feeling substantial pressure to make changes that reduce environmental consequences of what they do. In a recent survey of stakeholders involved in dairy cattle breeding in New Zealand, a significant proportion (approximately 50%) stated that they would be prepared to give up 10% of genetic progress in profitability traits to achieve meaningful gains in each of nutrient leaching and GHG emissions traits. In the absence of an antagonistic correlation with the profit breeding goal, this 10% sacrifice leads to approximately 40% of the maximum possible genetic progress in an environmental trait.

Crude accounting methods used in regulation pose a significant risk of creating adverse outcomes, particularly when considering incentives for genetic improvement mitigations. For example, if farms' emissions are quantified based on per animal constants, a hidden incentive is created to increase output per animal. Processor level deployment of carbon equivalent costs, for example per tonne of meat or milk solids produced have limited effectiveness, as they will fail to incentivise the shifts in farm practices required for mitigation.

Requirements of New Zealand farmers to have individual environmental budgets for GHG emissions, and for nutrient leaching are rapidly becoming mandatory. National data bases that quantify commercial herd or flock genetic merit for novel environmental traits would be a potential tool to directly incentivise uptake of more environmentally friendly genotypes. Monitoring of breeding purchases (e.g., semen, bulls and rams) by commercial farmers with potential use of DNA-based auditing of a sample of commercial animals would be the most cost-effective way to achieve this. Extensions of national data infrastructures to incorporate aggregate information (e.g. herd level daily milk production from processors, fertilizer applications from contractors) would complement per animal performance records to facilitate accurate monitoring of the farm emissions profile (Zhang *et al.* 2021). These should provide more granular and incentivising policies than what could be achieved by taxing output (processor level obligation) or by counting animals in a way that assumes all animals have the same environmental impact. A strong science backing will be required to support the case for reducing a farm's environmental budget based on observed genetic change for mitigation traits. New international standards for carbon accounting could then more accurately reflect changes in emissions per animal. This in turn would create incentives at both national and international level.

## CONCLUSIONS

Genetic improvement undoubtedly has a significant role to play in addressing the substantial environmental challenges facing livestock farming systems. Research on environmental mitigation traits is likely to grow, and both national and commercial breeding goals for livestock industries with advanced breeding infrastructure will increasingly shift towards recognition of their associated environmental implications. However, there is a risk that clumsy implementations will lead to unintended consequences. The consequences of reduction in the genetic gains of production efficiency that have historically driven a huge reduction in the environmental footprint of livestock products, in response to strong demand for action from the growing middle classes to reduce gross per animal emissions outputs, need to be carefully considered. Multi-disciplinary teams that go beyond the science and technology to consider breeding strategies, policy mechanisms and farmer adoption and behaviour will be critical to achieving genetic progress in environmental outcomes. National infrastructures supporting performance recording and genetic improvement are a logical platform to build more sophisticated monitoring systems. New developments should include genomic-based auditing systems, be well supported by science, and provide knowledge flows and training with incentives for both genetic and management mitigations at farm level. These mitigations could then be recognised in regional and national inventories. There is strong evidence, at least in NZ, that many farmers would already be prepared to modify their selection decisions to improve environmental traits if the tools were available to do so.

## REFERENCES

Amer P. (2012) *Animal* **6**: 551

Amer P, Hely F, Quinton C, and Cromie A. (2018). *Animal* **12**: 5

Beatson P, Meier S, Cullen N, and Edding H. (2019). *Animal* **13**: 2164

Cottle D, Nolan J, and Wiedermann S. (2011). *Anim. Prod. Sci.* **51**, 491

Pickering N, Oddy V, Basarab J, Cammack K, Hayes B, Hegarty R, Lassen J, McEwan J, Miller S, Pinares-Patiño C, and de Haas Y (2015). *Animal* **9**: 1431.

Wall E, Simm G and Moran D (2010). *Animal* **4**: 366.

Zhang X, Amer P, Stachowicz K, Quinton C and Crowley J. (2021). *Animal* **15**: 100325:

*Contributed paper*

# TRAIT PRIORITISATION METHODS USED IN ANIMALS ALSO WORK IN PLANTS

**I.A. Balogun[1,2], T.J. Byrne[3], B.F.S. Santos[2], P.R. Amer[2], P.F. Fennessy[2], D. Martin-Collado[4], B. Abolore[5], B. Okoye[6], B. Teeken[5], O.D. Olaosebikan[5], T. Madu[6], D. Owoade[5], A. Ogunade[5], M. Ejechi[6], N. Onyemauwa[6], S. Onwuka[6], B. Ukeje[6], P. Kulakow[5], C. Egesi[5,6,7], J. Onyeka[6], I. Rabbi[5], J. Jannink[7,8] and H.A. Tufan[7]**

[1] Department of Human Nutrition, University of Otago, New Zealand
[2] AbacusBio Limited, Dunedin 9016, New Zealand
[3]AbacusBio International Limited, EH25 9RG Edinburgh, United Kingdom
[4] AgriFood Institute of Aragon – IA2 (CITA-University of Zaragoza), 50013 Zaragoza, Spain
[5]International Institute of Tropical Agriculture (IITA), Ibadan, Nigeria
[6]National Root Crops Research Institute (NRCRI), Umudike, Nigeria
[7] Cornell University, Ithaca, New York, USA
[8]United States Department of Agriculture (USDA), Ithaca, NY, United States of America

## SUMMARY

We applied methods used successfully in prioritising trait improvements in animal breeding to cassava, to demonstrate that they are also relevant to plant breeding. Preference survey methodology based on 1000minds® was adapted and utilised to assess cassava trait improvement preferences of smallholder cassava farmers and other actors in the cassava value chain. We then establish how preference surveys can be employed to quantify and translate preferences into terms (trait units and scale) that align with estimated breeding values in plant breeding. Trait economic values were calculated according to the preference for each trait, relative to the preference for a monetary value included in the survey. Typologies of preferences were identified according to cassava traits preferences, and the resultant economic values differed between the typologies. This presents the potential for plant breeders to consider economic gains and cluster groups based on traits preferences in the development of breeding objectives.

## INTRODUCTION

Trait selection and breeding goal establishment are important when developing breeding objectives in crop and animal breeding. Animal breeding has made more advances than plant breeding in the use of economic values for index selection. This difficulty in crop breeders using economic theory to develop economic weights (Sölkner *et al.* 2008) is attributed to the absence of formal frameworks for derivation of economic weights. Participatory breeding, which involves including farmers and other value chain actors in the development of breeding objectives, has been employed in breeding programs for several crops; however, a challenge in participatory breeding has been an inability to transfer farmers' and other actors' descriptions of, and expressed preferences for, traits into quantitative terms that would allow them to be combined with estimated breeding values in a formal selection index. This challenge increases the risk of the breeding program releasing varieties that do not meet the requirements of the farmers and markets.

1000minds® (https://www.1000minds.com/) is a preference survey tool that employs an adaptive conjoint analysis methodology to minimise user burden. A detailed description of the algorithm of 1000minds can be found in Hansen and Ombler (2009). The 1000minds® method has been applied in the breeding of pasture plants (Smith and Fennessy 2011), sheep (Byrne *et al.* 2012), and dairy cattle (Martin-Collado *et al.* 2015) to assess farmers' preferences for trait improvements. Analysis of the outputs from 1000minds surveys enables the derivation of economic values and provides insights into trait preference heterogeneity across farmers and other supply chain actors.

This paper describes the application of methods and tools used in animal breeding to crops and shows how survey approaches can be employed to assign economic values to traits for genetic improvement when developing breeding objectives.

## MATERIALS AND METHODS

We applied the 1000minds® survey tool (https://www.1000minds.com/) to prioritise trait improvements for cassava in Nigeria. The 1000minds software asks a series of choice questions, where respondents are repeatedly required to select their preference between two trait improvement alternatives. The survey was conducted in four geopolitical zones in Nigeria: the north-central, south-east, south-south and south-west zones. The traits included in the survey were selected in consultation with experts and through literature research. Prior to the survey, focus group discussions (FGDs) were carried out with farmers and other cassava value chain actors, and in addition to discussing the traits to include in the surveys, they were used to establish benchmarks, units, and economic equivalents for cassava traits. Table 1 presents the parameters used to calculate equivalent levels for the traits in the 1000minds survey. Economic equivalents were calculated as the economic effect on increment per unit change in each of the traits independently.

The survey included 11 cassava traits and was administered to 792 smallholder cassava farmers and other actors in the cassava value chain. A demographic questionnaire was administered alongside the 1000minds® survey to explore the sociodemographic factors. The 1000minds output contains rankings of traits and preference percentage. These preference percentages were employed in the calculation of economic values.

**Derivation of Economic values.** An economic value is defined as the marginal impact of a one-unit change in a genetic trait. Trait economic values were calculated according to the preference (%) for each trait relative to the preference (%) for the trait expressed in monetary terms in the survey, 'price per 100kg bag' (Byrne *et al.* 2012).

**Table 1. Parameters used to calculate economic equivalence of levels for 1000minds preference survey traits**

| Inputs | Value |
|---|---|
| Average price per 100kg bag[1] | 2,500 |
| Average fresh roots yield (Number of100kg bags) per acre | 40 |
| Total crop value/acre | 100,000 |
| Average crop duration (days) | 270 |
| Average ground storage(days) | 365 |
| Price difference per change in root size | 1,000 |
| Price difference per change in root colour | 1,000 |
| Gari price per kg of gari | 200 |
| Gari price per 100 kg of gari | 15,000 |
| Average gari value (number of bags per acre) | 13 |
| Total gari value/acre[1] | 200,000 |
| Price difference across change in taste per 100kg[2] | 1,000 |
| Price difference across change in texture per 100kg[2] | 1,000 |
| Price difference across change in colour per 100kg[2] | 1,000 |
| Price difference across change in swelling per 100kg[2] | 1,000 |

[1]Prices of cassava in Ibadan, Nigeria at the time of the survey

[2] Assumes NGN 1,000 (Nigerian currency) between lowest and highest score (i.e., NGN 250/ score change) for a 5-point scale.

Thus, the economic value per trait unit was calculated to reflect a unit change in the trait according to this equation:

$$\text{pEV}_{qi} = \left[\frac{P_{iq}}{\alpha_i}\right] \times \beta_q, \quad (1)$$

where for trait $i$ and individual respondent $q$, $P$ is the preference (%) for each trait, $\alpha$ is the number of units represented in the trait level (to convert to the desired final trait unit), and $\beta$ is the monetary value per preference (%) for individual $q$. Values for $\beta$ were calculated as:

$$\beta_q = \left[\frac{amv}{Pmv_q}\right], \quad (2)$$

where $amv$ is the number of units represented in the level for the monetary trait and, for individual respondent $q$, $Pmv$ is the preference (%) for the monetary trait.

## RESULTS AND DISCUSSION

**The Economic Values.** Average economic values (NGN per trait unit) are presented in Table 2. The economic values presented are based on the cost of cassava at the time of the survey. Price of cassava varies greatly in Nigeria and so we used the current price of a bag of cassava at the time of the survey to derive economic values. Economic values are presented per trait unit, as defined in the survey. The calculation of economic values is based on equation (1) and equation (2). Given the preference for the trait 'maturity time' = 7.53% ($P_{1q}$), and the preference (%) for the (monetary) trait 'price per 100kg bag' ($amv$) = 8.00% ($Pmv_q$), by way of example, applying equation (1) to the (non-monetary) trait, we deduce that 28 days (4 weeks) of maturity ($\alpha_1$) (Table 2) is worth 7.5% and thus 1 day of maturity is worth 0.27% (7.5%/ 28 days). Similarly, for the (monetary) trait using equation (2), NGN 250 is worth 8.00% and thus 1% is worth NGN 31.22 (NGN 250/ 8.00%). Given 1-day maturity is worth 0.3% and 1% of monetary trait is worth NGN 31.22, then the economic value for maturity time can be calculated as NGN 8.40 per day (i.e., 0.27% × NGN 31.22).

**Table 2. Trait economic values for all respondents**

| Traits | Surveyed Unit | Per unit | Mean Trait ranks+ | Average preference % | Economic value (NGN/ survey trait unit) |
|---|---|---|---|---|---|
| Yield | Per 4 bags | Per 1 bag | 4.8 | 10.6 | 82 |
| Ground storage | Per 5 weeks | Per 1 day | 5.4 | 9.7 | 9 |
| Gari colour | Per 1 score | Per 1 score | 6.0 | 9.0 | 281 |
| Dry matter content | Per 5% | Per 1% | 6.2 | 8.8 | 55 |
| Gari taste | Per 1 score | Per 1 score | 6.5 | 8.3 | 259 |
| Root size | Per 25% | Per 1% | 6.7 | 8.1 | 10 |
| Gari swelling | Per 1 score | Per 1 score | 6.8 | 8.0 | 248 |
| Gari texture | Per 1 score | Per 1 score | 7.0 | 7.7 | 239 |
| Root colour | Per 1 score | Per 1 score | 7.1 | 7.6 | 236 |
| Maturity | Per 4 weeks | Per 1 day | 7.1 | 7.5 | 8 |
| Disease resistance | Per 10 % | Per 1% | 7.7 | 6.8 | 21 |

Order of trait ranks are from highest to lowest. +Smaller numbers indicate higher ranks.

**Economic values differed by cluster groups**. While Table 2 shows the population level trait preferences, heterogeneity exists in preferences for improvements in cassava traits (e.g., Martin-Collado *et al.* 2015), further analysis of this heterogeneity showed that three typology cluster groups could be established based on preferences for different combinations of cassava traits. The cluster analysis highlights traits that are important for different groups. An example is in the preference for disease resistance. While disease resistance ranked as the least preferred for improvement by the overall population (Table 2), a group of farmers exist that ranked disease resistance very high compared to other traits (data not shown)

**Breeding objective challenges in plant breeding.** In this paper we show how animal breeding trait prioritisation tools can be applied in a plant breeding setting. However, it is important to highlight some of the challenges plant breeders may face in adapting animal breeding tools: (1) The units reported in this study may not reflect the units of the trait breeding values as they are evaluated in the breeding program. This is because units presented to survey participants were developed and presented in ways the respondents can relate to. Plant breeders often use scales (e.g., 0-9 scores for diseases score) that are abstract when considered in terms of the economic impact on farm and/ or are very different to what farmers use (e.g., farmers probably use % crop lost, or % of diseased plants). This makes it difficult to calculate economic values, because a 0-9 score, for example, bears no resemblance to a unit that has an economic impact attached. This is less common in animals. (2) Another difference between plant and animal breeding is in the interactions of genetic traits with environmental variables (G x E). These G x E interactions are more influential in plants than animals; thus, plant breeders need to accommodate critical G x E interactions when developing breeding objectives. (3) The cluster groups (typologies) of preferences identified can be applied in targeting different market segments for breeding, however, complex factors such as breeding costs/benefits, variety replacement targets, and investment priorities need to be considered and integrated into the tools for these tools to be adoption by plant breeders.

## CONCLUSION

This study has shown that traits prioritisation methods that have been successful in animal breeding are also relevant and useful for plant breeding. Many of the challenges and nuances associated with index development are common between plants and animals, although for plants, there are some additional challenges created by the strong influence of G x E interactions, potentially exaggerating differences in trait preferences across different typology cluster groups.

## ACKNOWLEDGEMENTS

This research was undertaken as part of the NextGen Cassava Breeding Project funded by the Bill and Melinda Gates Foundation and the UK Foreign, Commonwealth and Development Office as investment INV-007637. The work was also supported by a Callaghan innovation PhD Fellowship. We would like to appreciate the cassava breeding units at IITA and NRCRI Nigeria, as well as farmers and other cassava value chain actors that participated in the study.

## REFERENCES

Byrne T.J., Amer P.R., Fennessy P.F., Hansen P. and Wickham B.W. (2012) *Animal.* **6**: 778.

Hansen P. and Ombler F. (2008) *J Multi-Crit Decis Anal* **15:** 87.

Martin-Collado D., Byrne T.J., Amer P.R., Santos B.F.S., Axford M. and Pryce J.E. (2015) *J. Dairy Sci.* **98:** 4148.

Nielsen H.M., Amer P.R. (2007) *Animal* **1:**1254.

Smith K.F. and Fennessy P.F. (2011). *Crop and Pasture Science*. **62:** 355.

Sölkner J., Grausgruber H., Okeyo A.M., Ruckenbauer P. and Wurzinger M. (2008). *Euphytica.* **161:** 273.

*Contributed paper*

# IMPACT OF BREEDING FOR DIVERGENT METHANE YIELD ON MILK COMPOSITION IN BREEDING EWES

**T.P. Bilton[1], S.M. Hickey[2], P.H. Janssen[3], A. Jonker[3], M.K. Hess[1], B. Bryson[4], W. Bain[1], E. Waller[1], K.M. McRae[1], S. Muetzel[3], M. Agnew[3], J.C. McEwan[1] and S.J. Rowe[1]**

[1]AgResearch, Invermay Agricultural Centre, Mosgiel, New Zealand
[2]AgResearch, Ruakura Agricultural Centre, Hamilton, New Zealand
[3]AgResearch, Grasslands Research Centre, Palmerston North, New Zealand
[4]AgResearch, Woodlands Research Farm, Invercargill, New Zealand

## SUMMARY

Previous research into breeding sheep based on methane yield has shown that low emitting animals appear to have neutral or superior economic and environmental value compared to high emitting animals. However, the impact of breeding for methane yield on milk composition has not been studied in depth. We investigated differences in detailed fatty acid (FA) profiles and rumen volatile fatty acids (VFA) associated with methane selection line across two lactation years in lactating ewes from a sheep flock selected for divergent methane yield. Changes in FA profiles due to selection line were observed, with increased polyunsaturated fatty acids levels in the milk and VFAs associated with less hydrogen formation in rumen samples from lower methane emitting animals. There was evidence that these differences were partly driven by changes in the rumen microbial profile. These results have important implications in screening for, and processing milk from, low methane emitting animals in industry.

## INTRODUCTION

Methane is a greenhouse gas associated with climate change and approximately 84% of methane emissions in New Zealand are produced from grazing livestock (MFE 2020). Reducing methane emissions from livestock is therefore of environmental and economic importance and is achievable by breeding for animals that emit less methane. Ruminant animals primarily produce methane as a by-product of the complex microbial fermentation process in their rumen that breaks down feed to VFAs, which are absorbed through the gut wall and are a major source of energy for the animal (Matthews *et al.* 2019). The mammary gland also uses these VFA in the de novo synthesis of milk fatty acids (FA) (Negussie *et al.* 2017). Changes in a herd's methane emission levels via breeding is therefore likely to be associated with changes in FA composition.

Over the past decade, a sheep flock has been selected for divergent methane yield, with low-methane sheep emitting 10-12% less methane than the high-methane animals (Rowe *et al.* 2019). Using lactating ewes from this flock, we analysed milk FA profiles and rumen fluid VFA to investigate changes associated with methane selection line, and whether these changes relate to changes in the rumen microbial profile.

## MATERIALS AND METHODS

**Animals.** This study selected 60 out of 100 ewes from the high methane line (HML) and 60 out of 100 ewes from the low methane line (LML) in a divergent methane yield sheep flock (Rowe *et al.* 2019) that were lambing from September each spring. This selection was made in two lactation years (2018 and 2019), with 25 HML and 23 LML ewes retained in the flock and selected in the study in both years leaving 192 unique ewes. The average difference in methane breeding values for these ewes between the two lines was 1.98 g $CH_4$ per kg dry matter intake (DMI) for 2018 and 2.21 g $CH_4$ per kg DMI for 2019 (average methane value was $16 \pm 1.45$ g $CH_4$ per kg DMI).

**Traits.** Milk and rumen samples were collected after lambing in October 2018 and 2019 at two time-points two weeks apart. These time-points were approximately 2 and 4 weeks post lambing if the ewe lambed late (last week of September or October) and approximately 4 and 6 weeks post-lambing if the ewe lambed early (early September). An 8-mL sample of milk was processed and methyl esters of the fatty acids measured using gas chromatography as described by Agnew *et al.* (2019). Rumen fluid were collected via oral stomach tubing and was divided into three 2-mL samples that were processed using the method described by Jonker *et al.* (2019) to obtain volatile fatty acid (VFA) profiles, and into a 30 mL sample for DNA extraction and sequencing to generate a rumen microbial profile as described by Hess *et al.* (2020).

**Statistical Analysis.** Univariate linear mixed models (LMMs) were fitted for each trait using ASREML v4.1 (Gilmour *et al.* 2015). Model equations were:

$$\log_{10}(y) = \mu + cdat*bg + age + nll + lwt + line + p_e \quad (1)$$

$$\log_{10}(y) = \mu + cdat*bg + age + nll + lwt + M \quad (2)$$

where y is the trait of interest, cdat is the collection date of the sample, bg indicates if the ewe lambed late or early, age is the ewe's age (years) at sampling (2, 3, 4+), nll is the number of live lambs (1, 2, 3+), lwt is the ewe's liveweight (kg) at sampling, line is the methane line (low or high), $p_e$ is the permanent environment random effect, and M is the reference-based microbial relationship matrix as described by Hess *et al.* (2020). Most ewes were 2 (47%) or 3 (33%) years old and had 1 (31%) or 2 (58%) lambs. The trait values were log transformed to improve variance homogeneity. Model (1) was fitted to investigate the effect of selection line on each trait while Model (2) was fitted to estimate the microbiability (proportion of variance explained by the rumen microbial profile).

## RESULTS AND DISCUSSION

**Milk FAs.** Results from fitting univariate LMMs on the milk FAs are given in Table 1. FA percentages for each individual polyunsaturated fatty acids (PUFA) (e.g., C18:2 n6, CLA) and the total PUFA value were significantly greater in the LML compared to the HML for both years, with differences ranging between 4.3% to 13.5%. The total saturated fatty acids (SFA) value was significantly smaller in the LML for both years, with a difference around -1.1% to -1.3%, although changes in individual SFA (e.g., C12:0, C17:0) were not consistent across years. There was little evidence of changes in the monounsaturated fatty acids (MUFA). The repeatabilities were moderate across the FA, but greater for the PUFA and the total SFA value across both years.

**Rumen VFAs.** Results from fitting univariate LMMs on the rumen fluid VFA are given in Table 2. Percentages of caproic and propionic acid were on average significantly greater in the LML than in the HML in both lactation years, while changes in the other VFA were inconsistent across years. The two VFA ratios were consistently smaller in the LML compared to the HML and significant at the 5% threshold, indicating that the percentage of acetic and butyric relative to propionic and valeric was smaller in the LML. This is consistent with stoichiometric principles, as formation of acetic and butyric acids is connected with hydrogen formation (utilised by methanogens to form methane) while propionic and valeric acids are associated with less hydrogen formation (Janssen 2010). The repeatabilities were between 0.23 and 0.39 for all VFA, except for caproic acid which had very low repeatability. Similar results in terms of ruminal VFA composition and repeatabilities were found in growing methane selection line sheep fed pasture as in this trial (Jonker *et al.* 2020).

**Microbiability:** Estimates of microbiability for milk FA and rumen VFA are given in Table 3. The microbiability for all the milk PUFA and all the rumen VFA ranged between 0.21 to 0.54 and was greater than 2 standard errors from zero across both years. This was not the case for the milk SFA and MUFA. These results suggest that differences between the selection lines in rumen VFA and milk PUFA are, at least partially, driven by changes in the rumen microbial profile.

## CONCLUSIONS

This study shows that breeding for methane impacts milk FA and rumen fluid VFA profiles and suggests that changes in these profiles are partially driven by changes in the rumen microbial profile. These results suggest there is potential for milk FA and rumen VFA to be used as a proxy measure for methane, but the results also have implications on milk processing, as changes in FA profiles affects the quality and type of products produced from the milk.

**Table 1. Fatty acid (FA) composition of milk samples from low and high selection lines**

| FA (%) | 2018 | | | 2019 | | |
|---|---|---|---|---|---|---|
| | Mean ± s.e. | % diff‡ | Repeatability | Mean ± s.e. | % diff‡ | Repeatability |
| Total SFA[1] | 43.4 ± 2.36 | **-1.3%**† | 0.53 ± 0.08 | 44.8 ± 2.25 | **-1.1%**† | 0.44 ± 0.09 |
| C12:0 | 4.35 ± 1.07 | 1.5% | 0.36 ± 0.09 | 4.33 ± 1.20 | -2.9% | 0.19 ± 0.11 |
| C14:0 | 7.83 ± 1.21 | -0.5% | 0.34 ± 0.10 | 8.06 ± 1.40 | -2.1% | 0.26 ± 0.10 |
| C15:0 | 0.84 ± 0.10 | 0.7% | 0.41 ± 0.09 | 0.93 ± 0.10 | **2.1%**† | 0.12 ± 0.11 |
| C16:0 | 17.6 ± 1.67 | **-1.6%**† | 0.51 ± 0.08 | 18.1 ± 1.79 | -1.2% | 0.39 ± 0.09 |
| C17:0 | 0.65 ± 0.18 | -1.9% | 0.31 ± 0.10 | 0.66 ± 0.12 | -0.6% | 0.28 ± 0.10 |
| C18:0 | 12.1 ± 2.30 | -2.8% | 0.26 ± 0.10 | 12.6 ± 2.70 | 0.4% | 0.28 ± 0.10 |
| C20:0 | 0.12 ± 0.02 | -2.8% | 0.14 ± 0.10 | 0.12 ± 0.03 | -1.3% | 0.30 ± 0.10 |
| Total MUFA[2] | 18.3 ± 3.75 | -1.6% | 0.28 ± 0.12 | 19.1 ± 3.51 | -1.0% | 0.11 ± 0.11 |
| C14:1 | 0.04 ± 0.04 | -0.7% | 0.20 ± 0.23 | 0.07 ± 0.03 | -4.7% | 0.51 ± 0.08 |
| C16:1 | 0.48 ± 0.11 | -1.0% | 0.26 ± 0.10 | 0.49 ± 0.13 | **-5.3%**† | 0.45 ± 0.09 |
| C17:1 | 0.24 ± 0.08 | -1.4% | 0.32 ± 0.10 | 0.23 ± 0.05 | **-4.4%**† | 0.33 ± 0.09 |
| C18:1 c9 | 17.3 ± 3.65 | -2.2% | 0.21 ± 0.10 | 18.0 ± 3.47 | -0.8% | 0.11 ± 0.11 |
| C18:1 c11 | 6.94 ± 1.63 | 4.5% | 0.70 ± 0.07 | 6.94 ± 1.13 | 1.0% | 0.34 ± 0.09 |
| Total PUFA[3] | 3.85 ± 0.66 | **5.4%*** | 0.57 ± 0.07 | 3.92 ± 0.54 | **7.2%*** | 0.53 ± 0.08 |
| C18:2 n6 | 0.63 ± 0.14 | **4.8%*** | 0.38 ± 0.09 | 0.57 ± 0.13 | **9.4%*** | 0.44 ± 0.09 |
| C18:3 n3 | 0.98 ± 0.24 | **7.1%*** | 0.62 ± 0.06 | 0.97 ± 0.22 | **13.5%*** | 0.48 ± 0.08 |
| CLA | 2.24 ± 0.61 | **4.5%**† | 0.61 ± 0.07 | 2.38 ± 0.42 | **4.3%**† | 0.56 ± 0.07 |

[1]SFA = saturated fatty acids (Total = C12:0 + C14:0 + C15:0 + C16:0 + C17:0 + C18:0 + C20:0)
[2]MUFA = monounsaturated fatty acids (Total = C14:1 + C16:1 + C17:1 + C18:1 c9 + C18:1 c11)
[3]PUFA = polyunsaturated fatty acids (Total = CLA + C18:2 n6 + C18:3 n3)
†Significant at 5% threshold, *Significant at 0.1% threshold, ‡Difference (Low – High)

**Table 2. Volatile fatty acids (VFA) in rumen fluid samples from low and high selection lines**

| VFA (%) | 2018 | | | 2019 | | |
|---|---|---|---|---|---|---|
| | Mean ± s.e. | % diff‡ | Repeatability | Mean ± s.e. | % diff‡ | Repeatability |
| Acetic | 65.9 ± 2.73 | -0.4% | 0.22 ± 0.10 | 66.1 ± 2.77 | **-1.0%*** | 0.23 ± 0.10 |
| Butyric | 10.2 ± 1.41 | 0.2% | 0.25 ± 0.10 | 9.94 ± 1.35 | **2.5%**† | 0.31 ± 0.10 |
| Caproic | 0.30 ± 0.12 | **6.8%**† | 0.00 ± 0.00 | 0.31 ± 0.12 | **12.7%*** | 0.01 ± 0.11 |
| Isobutyric | 1.33 ± 0.37 | -2.0% | 0.36 ± 0.09 | 1.19 ± 0.32 | 0.6% | 0.30 ± 0.10 |
| Isovaleric | 1.49 ± 0.52 | -2.8% | 0.39 ± 0.09 | 1.29 ± 0.43 | 0.2% | 0.29 ± 0.10 |
| Propionic | 19.5 ± 1.28 | **1.3%**† | 0.37 ± 0.09 | 19.9 ± 1.31 | **1.5%**† | 0.35 ± 0.10 |
| Valeric | 1.24 ± 0.28 | -0.1% | 0.11 ± 0.11 | 1.21 ± 0.32 | **4.5%**† | 0.25 ± 0.11 |
| A/P[1] | 3.40 ± 0.33 | **-11.6%**† | 0.31 ± 0.10 | 3.34 ± 0.33 | **-17.2%*** | 0.26 ± 0.10 |
| (A+B)/(P+V)[2] | 3.69 ± 0.32 | **-11.4%**† | 0.35 ± 0.09 | 3.62 ± 0.33 | **-16.6%*** | 0.30 ± 0.10 |

[1]A/P = Acetic/Propionic
[2](A + B)/(P + V) = (Acetic + Butyric)/(Propionic + Valeric)
†Significant at 5% threshold, *Significant at 0.1% threshold, ‡ Difference (Low – High)

**Table 3. Microbiability estimates (% ± s.e) for milk fatty acids and rumen volatile fatty acids**

| FA (%) | 2018 | 2019 | FA (%) | 2018 | 2019 |
|---|---|---|---|---|---|
| Total SFA | 0.18 ± 0.07 | 0.13 ± 0.07 | Total PUFA | 0.33 ± 0.09 | 0.34 ± 0.09 |
| C12:0 | 0.11 ± 0.07 | 0.21 ± 0.09 | C18:2 n6 | 0.21 ± 0.09 | 0.40 ± 0.08 |
| C14:0 | 0.04 ± 0.05 | 0.13 ± 0.08 | C18:3 n3 | 0.28 ± 0.09 | 0.46 ± 0.09 |
| C15:0 | 0.01 ± 0.04 | 0.05 ± 0.06 | CLA | 0.26 ± 0.08 | 0.27 ± 0.09 |
| C16:0 | 0.16 ± 0.07 | 0.00 ± 0.00 | VFA (%) | 2018 | 2019 |
| C17:0 | 0.07 ± 0.07 | 0.27 ± 0.09 | Acetic | 0.38 ± 0.09 | 0.44 ± 0.09 |
| C18:0 | 0.07 ± 0.06 | 0.17 ± 0.09 | Butyric | 0.48 ± 0.08 | 0.36 ± 0.09 |
| C20:0 | 0.07 ± 0.06 | 0.17 ± 0.08 | Caproic | 0.32 ± 0.08 | 0.33 ± 0.09 |
| Total MUFA | 0.08 ± 0.08 | 0.22 ± 0.09 | Isobutyric | 0.54 ± 0.09 | 0.32 ± 0.09 |
| C14:1 | 0.00 ± 0.00 | 0.11 ± 0.08 | Isovaleric | 0.52 ± 0.08 | 0.29 ± 0.09 |
| C16:1 | 0.01 ± 0.05 | 0.00 ± 0.00 | Propionic | 0.28 ± 0.09 | 0.46 ± 0.09 |
| C17:1 | 0.15 ± 0.09 | 0.16 ± 0.09 | Valeric | 0.41 ± 0.08 | 0.42 ± 0.09 |
| C18:1 c9 | 0.16 ± 0.09 | 0.21 ± 0.09 | A/P | 0.31 ± 0.09 | 0.48 ± 0.09 |
| C18:1 c11 | 0.41 ± 0.10 | 0.09 ± 0.07 | (A+B)/(P+V) | 0.30 ± 0.09 | 0.47 ± 0.09 |

Abbreviations for FA, VFA, SFA, MUFA, PUFA, A/P and (A+B)/(P+V) are as in Tables 1 and 2.

## ACKNOWLEDGEMENTS

This work was funded by the New Zealand Agricultural Greenhouse Gas Research Centre (NZAGRC), the Pastoral Greenhouse Gas Research Consortium (PGgRc) and the Ministry of Business, Innovation & Employment (MBIE) via the "Mapping the New Zealand Ruminotype Landscape" programme.

## REFERENCES

Agnew M.P., Craigie C.R., Weralupitiya G., Reis M.M., Johnson P. and Reis M.G. (2019). *Metabolites* **9**: 189.

Gilmour A.R., Gogel B.J., Cullis B.R., Welham S.J. and Thompson R. (2015). ASReml User Guide Release 4.1, VSN International Ltd, Hemel Hempstead, HP1 1ES, UK.

Hess M.K., Rowe S.J., Van Stijn T.C., Henry H.M., Hickey S.M., Brauning R., McCulloch A.F., Hess A.S., Kirk M.R., Kumar S., Pinares-Patiño C., Kittelmann S., Wood G.R., Janssen P.H. and McEwan J.C. (2020). *PLoS One* **15**: e0219882.

Janssen P.H. (2010). *Anim/ Feed Sci. Technol.* **160**: 1.

Jonker A., Hickey S.M., McEwan J.C., Rowe S.J., Janssen P.H., MacLean S., Sandoval E., Lewis S., Kjestrup H., Molano G., Agnew M., Young E.A., Dodds K.G., Knowler K. and Pinares-Patiño C.S. (2019). *J. Anim. Sci.* **97**: 2711.

Jonker A., Hickey S., Boma P., Woyimo Woju C., Sandoval E., MacLean S., García Rendón Calzada, M., Yu W., Lewis S., Janssen, P.H., McEwan J.C., Rowe S. (2020). *Anim. Prod. Sci.* **61**: 300

Matthews C., Crispie F., Lewis E., Reid M., O'Toole P.W. and Cotter P.D. (2019) *Gut Microbes* **10**: 115.

MFE (2020). New Zealand`s Greenhouse Gas Inventory 1990-2018. In Ref. MFE 1496. Ministry for the Environment, Wellington, New Zealand.

Negussie E., de Haas Y., Dehareng F., Dewhurst R.J., Dijkstra J., Gengler N., Morgavi D. P., Soyeurt H., van Gastelen S., Yan T. and Biscarini F. (2017). *J. Dairy Sci.* **100**: 2433.

Rowe S.J., Hickey S.M., Jonker A., Hess M.K, Janssen P., Johnson T., Bryson B., Knowler K., Pinares-Patiño C., Bain W., Elmes S., Young E., Wing J., Waller E., Pickering N. and McEwan J.C. (2019). *Proc. Assoc. Advmt. Anim. Breed. Genet.* **23**: 306.

*Contributed paper*

# PHENOTYPIC TRADE-OFFS BETWEEN LAMBS AND WOOL REFLECT WEAK ANTAGONISTIC GENETIC CORRELATIONS BETWEEN REPRODUCTIVE AND WOOL TRAITS

**K.L. Bunter[1] and A.A. Swan[1]**

[1] Animal Genetics Breeding Unit*, University of New England, Armidale, NSW, 2351 Australia

## SUMMARY

Rearing lambs while producing wool is an annual source of competition for available nutritional resources supplied to breeding ewes. Genetic correlations between wool and reproductive traits were estimated from industry data, comparing models that did or did not account for the effects of reproductive level on wool traits recorded at different wool age stages (yearling, hogget, adult). Small to moderate antagonistic correlations between wool and reproductive traits tended to decrease in magnitude when birth-rear type of the individual (yearling and hogget stages) or reproductive output (adult ewes) were accounted for in models for wool traits. Increased recording of reproductive performance would make it possible to more accurately compare young animals for wool traits as well as genetically improve both trait sets (ewe reproduction and wool traits) for Merinos.

## INTRODUCTION

For breeding ewes, reproductive performance and wool production are annual outputs that compete for access to common, potentially limiting, nutritional resources. In addition, progeny born as singles have, on average, superior wool attributes relative to twins (Hocking-Edwards *et al.* 2011). In the Australian Merino industry, the majority of animals are recorded for wool traits as yearlings or hoggets (i.e. between 12 and 18 months of age), prior to their first joining. Data for adult wool traits is subsequently predominantly from breeding ewes. At both time points, variability introduced by litter size at birth-rearing (progeny) or reproductive status (ewe) could potentially affect estimates of the genetic correlations between wool and reproductive traits. Derivation of component traits for reproductive performance (Bunter *et al.* 2021) enable these associations to be investigated further across industry flocks. This paper examines how birth-rear type of offspring and previous reproductive status of ewes affect wool trait values and estimates of genetic correlations between these trait groups.

## MATERIALS AND METHODS

Data and pedigree recorded from 2000 onwards for greasy fleece weight (GFW) and fibre diameter (FD) were extracted from the Sheep Genetics database for the subset of flocks that had some reproductive data for conception (CON), litter size (LS) and ewe rearing ability (ERA) traits. Wool and reproductive records were merged by animal-year of recording. Wool records were classified by age-stage groupings (Y: yearling; H: hogget and A: adult). Wool and ewe reproductive data were concurrent for the adult wool stage data only. Reproductive performance in the year prior to the adult stage shearing was derived from reproductive data, or described as unknown. The complete pedigree contained about 740k animals.

Contemporary groups for wool traits within stage were defined by flock-year-date of shearing-breeder subgroup (Brown *et al.* 2007), and contemporary groups for reproductive traits were as previously described by Bunter *et al.* (2021). Age at recording for wool traits was accounted for

* A joint venture of NSW Department of Primary Industries and the University of New England

using linear regression (Y, H stages) or by fitting age in years as a class effect (A stage). Animals over 6 years old at the time of recording were grouped in the 7yo age group. Sex was accounted for to accommodate wool records for males. An additional model term for Y, H and A stage wool traits was birth-rear type group (BTRTG: SS, MS, MM, SU, MU or UU, where S=single, M=multiple and U=unknown). Previous reproductive outcome (RSTAT: 4 levels for ewes: 0, 1 or 2+ lambs reared, or unknown) was fitted for ewe A stage wool traits only. Models containing these additional terms were compared to base models without these terms. Least squares means for BTRTG and RSTAT for each wool trait were obtained using the GLM procedure of SAS software (2002-2012). Heritabilities were estimated assuming an animal model for the alternative systematic effect models in univariate analyses. Additional random effects included maternal effects for wool traits (all stages) and a permanent environmental effect to accommodate repeated records for adult reproductive and wool traits. Correlations between wool and reproductive traits were estimated from a series of bivariate analyses using ASREML (Gilmour *et al.* 2014).

## RESULTS AND DISCUSSION

Approximately 82% (70%) of animals had sires (dams) known and 6% of adult shearing records for ewes had previous reproductive status known. Data characteristics and heritabilities (Table 1) were generally consistent with expectation, with the exception of higher heritabilities for reproductive traits relative to the data sample of Bunter *et al.* (2021), which contained more flocks characterised by less historical pedigree and data. Maternal effects were about 2% for FD and ranged between 6-9% for GFW across stages, and were consistent within trait stage across models.

**Table 1. Estimates of additive genetic ($h^2$) and permanent environmental effects ($pe^2$) relative to the phenotypic variance ($\sigma^2_p$) for reproductive (CON, LS, ERA) or wool traits under the Base model, or after accounting for birth-rear type group (+ BTRTG) or previous reproductive status (RSTAT)**

| Trait | N | Mean(SD) | Base model | | | Base + BTRTG | | | Base + RSTAT | | |
|---|---|---|---|---|---|---|---|---|---|---|---|
| | | | $h^2$ | $pe^2$ | $\sigma^2_p$ | $h^2$ | $pe^2$ | $\sigma^2_p$ | $h^2$ | $pe^2$ | $\sigma^2_p$ |
| CON | 48899 | 0.91 (0.28) | 0.09 | 0.09 | 0.075 | - | - | - | - | - | - |
| LS | 89165 | 1.34 (0.49) | 0.09 | 0.02 | 0.206 | - | - | - | - | - | - |
| ERA | 51781 | 0.82 (0.35) | 0.03 | 0.05 | 0.117 | - | - | - | - | - | - |
| YGFW | 370089 | 3.24 (1.12) | 0.28 | na | 0.319 | 0.31 | na | 0.309 | - | - | - |
| YFD | 375031 | 16.6 (1.86) | 0.60 | na | 1.23 | 0.60 | na | 1.22 | - | - | - |
| HGFW | 170254 | 4.50 (1.25) | 0.37 | na | 0.401 | 0.38 | na | 0.396 | - | - | - |
| HFD | 201061 | 17.9 (1.90) | 0.61 | na | 1.50 | 0.62 | na | 1.49 | - | - | - |
| AGFW | 196896 | 5.04 (1.49) | 0.35 | 0.20 | 0.529 | 0.36 | 0.19 | 0.524 | 0.36 | 0.19 | 0.523 |
| AFD | 167028 | 17.9 (2.42) | 0.69 | 0.05 | 1.56 | 0.69 | 0.05 | 1.55 | 0.69 | 0.05 | 1.55 |

-: not fitted; all se <0.01; CON: conception; LS: litter size; ERA: rearing ability; greasy fleece weight and fibre diameter for yearling (YGFW, YFD), hogget (HGFW, HFD) and adult (GFW, FD) stages

**Systematic effects.** Lambs born and reared as singles (SS) had heavier fleeces than MM lambs at Y (GFW: 3.21±0.004 vs 2.93±0.005 kg), H (GFW: 4.54±0.01 vs 4.36±0.01 kg) and A stages (GFW: 5.30±0.03 vs 5.16±0.03 kg). Single born lambs also had lower FD than MM lambs at Y (16.8±0.01 vs 16.9±0.01), H (17.9±0.01 vs 18.1±0.01) and A stages (18.0±0.06 vs 18.2±0.06). Animals born as multiples and reared as single were intermediate. These effects result from permanent changes to lamb development arising from competition for resources during gestation and lactation. Phenotypic selection for increased fleece weight and finer micron would therefore favour SS over MM lambs, in the absence of accounting for BTRTG, particularly when based on yearling wool data. Clark and Thompson (2021) showed that BTRTG affects classing results, due to the effects of BTRTG on both weight and wool traits.

Across parities, ewe fleece weights were reduced by about 150g per additional lamb weaned in the previous year (i.e. GFW LSM: no lambs: 4.95±0.03 kg; one lamb: 4.80±0.03 kg; two lambs: 4.65±0.03 kg) but FD did not significantly differ (p=0.07) with the previous years' reproductive performance. These effects result from the trade-off between ewe wool growth and partitioning of ewe resources (feed intake) towards successful gestation and lactation outcomes (Freer *et al.* 1997). Hocking-Edwards *et al.* (2011) suggested that appropriate nutritional management of twinning ewes during pregnancy can offset detrimental effects for wool traits for both ewes and their offspring.

**Genetic parameters.** The significant effects of birth-rear type on fleece weights (FW) or fibre diameter (FD: ewes and rams) and previous reproductive status (adult ewes) on AGFW suggest that correlations between wool and reproductive traits are antagonistic. However, ewe BTRTG categories are cross-classified with lamb BTRTG categories, enabling separation of genetic from non-genetic effects for wool traits associated with litter size in multiple generation data.

*Model comparisons.* Variance in wool traits explained by BTRTG decreased with age/stage (ie A < H < Y stages) and was collectively proportionally more collectively for FW (2-7%) relative to FD (0-2%) (Table 1). Sheep born as multiples have lighter fleeces and broader micron, on average, due to permanent developmental effects (Hocking-Edwards *et al.* 2011). In contrast, RSTAT explained very little variance for adult wool traits (Table 1) and did not alter estimates of genetic correlations with reproductive traits (Table 2). This could be because RSTAT was unknown for the majority of ewe A stage wool records. For other trait combinations, the models fitted for wool traits affected estimates of genetic correlations between wool with reproductive traits.

*Genetic (ra) correlations between wool and reproductive traits.* Genetic correlations were unfavourable (ra: -0.22) between GFW and CON, but this was less evident for Y and H fleece weights (Table 2). A more effective correction for RSTAT may be important for accurate estimates of the genetic correlation between GFW and CON. Genetic correlations between GFW at all stages and LS were negligible, providing BTRTG was included in wool trait models. Genetic correlations between GFW at any stage and ERA were negligible under all models. Fibre diameter at all stages had low positive genetic correlations with LS and ERA (ra: 0.15 to 0.22); antagonistic correlations of FD with CON were evident for HFD only. Overall, breeding objectives that aim to increase fleece weight (less so) and reduce fibre diameter (more so) have detrimental implications for ewe reproductive performance and lambs reared. However, genetic correlations were relatively low, indicating scope for concurrent improvement in both reproductive performance and wool traits with appropriate data recording and selection criteria. Safari *et al.* (2007) previously reported an antagonistic genetic correlation for NLW with CFW (-0.26±0.05), negligible with FD (0.06±0.04). Results here suggest genetically broader FD sheep are more likely to have higher genetic merit for LS and ERA, which is consistent with lower litter size and lamb survival typically observed for fine- relative to medium- or strong-wool sheep (Hatcher *et al.* 2009).

Positive residual (not shown) and phenotypic correlations were evident between YGFW and LS, suggesting that unidentified non-genetic factors increasing ewe YGFW increased future litter size. Negative correlations between permanent environmental effects for ewe wool and reproductive traits (not shown) indicate that persistent high reproductive performance has negative consequences for FD and GFW within individual ewes. Overall, parameters derived for Y & H stage wool data, recorded prior to any joining event provide consistent results: wool traits are affected by reproductive performance levels but the genes that control these trait groups are largely independent, with the absolute magnitude of genetic correlations typically less than 0.2.

## CONCLUSIONS

Merino breeders would benefit from recording reproduction accurately, as knowledge of birth-rear type is important for accurate comparisons of young animals (e.g. YGFW, YFD) and comparison of adult ewe fleeces are also affected by variation in reproductive output. In each case, animals with or resulting from low reproductive output would be favoured for fleece or weight traits if this information were unknown. However, genes affecting reproductive traits are largely independent of genes affecting greasy fleece weight or fibre diameter (ra: magnitude negligible or generally <±0.2), making it possible to change both trait sets with accurate recording of reproduction. Where correlations were not negligible, they were low and antagonistic, suggesting indirect selection pressure against improved reproduction from selection on uncorrected weight or fleece attributes.

**Table 2. Additive genetic (ra), residual (re) and phenotypic (rp) correlations between wool (Trait 1) and reproductive (Trait 2) traits when BTRTG (Y & H stages) and previous reproductive performance (RSTAT, A stage only) are added to Base models for wool traits**

| Trait 1* | Trait 2** | Base model | | | Base + BTRTG | | | Base + BTRTG + RSTAT | | |
|---|---|---|---|---|---|---|---|---|---|---|
| | | ra | re | rp | ra | re | rp | ra | re | rp |
| YGFW | CON | -0.09 | -0.01 | -0.02 | -0.07 | -0.00 | -0.01 | - | - | - |
| | LS | **-0.33** | 0.16 | 0.06 | -0.03 | 0.10 | 0.07 | - | - | - |
| | ERA | 0.01 | 0.02 | 0.01 | 0.04 | 0.02 | 0.02 | - | - | - |
| HGFW | CON | 0.04 | -0.06 | -0.04 | 0.06 | -0.07 | -0.04 | - | - | - |
| | LS | **-0.10** | 0.03 | 0.00 | 0.04 | -0.00 | 0.01 | - | - | - |
| | ERA | 0.06 | -0.03 | -0.02 | 0.06 | -0.03 | -0.01 | - | - | - |
| AGFW | CON | **-0.21** | 0.10 | -0.05 | **-0.21** | 0.10 | -0.05 | **-0.21** | 0.10 | -0.05 |
| | LS | **-0.10** | 0.04 | 0.00 | -0.03 | 0.04 | 0.01 | -0.02 | 0.04 | 0.01 |
| | ERA | -0.04 | 0.05 | -0.02 | -0.02 | 0.05 | -0.02 | -0.02 | 0.04 | -0.02 |
| YFD | CON | 0.06 | 0.01 | 0.02 | 0.06 | 0.01 | 0.02 | - | - | - |
| | LS | **0.19** | 0.01 | 0.05 | **0.14** | 0.03 | 0.05 | - | - | - |
| | ERA | **0.22** | 0.02 | 0.04 | **0.23** | 0.02 | 0.04 | - | - | - |
| HFD | CON | **0.14** | -0.00 | 0.03 | **0.13** | -0.00 | 0.03 | - | - | - |
| | LS | **0.21** | -0.07 | 0.01 | **0.14** | -0.04 | 0.01 | - | - | - |
| | ERA | **0.22** | 0.03 | 0.05 | **0.20** | 0.04 | 0.05 | - | - | - |
| AFD | CON | 0.02 | 0.03 | 0.01 | 0.02 | 0.03 | 0.01 | 0.02 | 0.03 | 0.01 |
| | LS | **0.18** | 0.02 | 0.03 | **0.15** | 0.02 | 0.03 | **0.15** | 0.03 | 0.06 |
| | ERA | **0.19** | 0.03 | 0.02 | **0.18** | 0.03 | 0.02 | **0.18** | 0.03 | 0.02 |

Estimates ra >2×SE from 0 are in bold; *greasy fleece weight and fibre diameter for yearling (YGFW, YFD), hogget (HGFW, HFD) and adult (AGFW, AFD) stages; **conception (CON), litter size (LS) and rearing ability (ERA) traits for adult ewes

**ACKNOWLEGEMENTS**

The authors would like to acknowledge contributions by breeders to the Sheep Genetics database, along with funding by MLA and AWI towards ongoing research.

**REFERENCES**

Brown D.J., Huisman A.E., Swan A.A., Graser H.U., Woolaston R.R., Ball A.J., Atkins K.D. and Banks R.G. (2007). *Proc. Assoc. Advmnt. Anim. Breed. Genet.* **17:** 187.

Bunter K.L., Swan A.A., Gurman P.M. and Brown D.J. (2021). *Anim. Prod. Sci.* **61**: 333.

Clark and Thompson (2021). *Proc. 33rd Aust. Assoc. Anim. Sci. Conf., Fremantle, WA, Australia.*

Freer M, Moore A.D. and Donnelly J.R. (1997) *Agricultural Systems*. **54**: 17.

Gilmour A. R., Gogel B. J., Cullis B. R., Welham S. J. and Thompson R., 2014. ASReml User Guide Release 4.1 Functional Specification. VSN International Ltd, Hemel Hempstead, HP1 1ES, UK.

Hatcher S., Atkins K.D. and Safari E. (2009) *J. Anim. Sci.* **87**: 2781.

Hocking-Edwards J.E., Copping K.J. and Thompson A.N. (2011). *Anim. Prod. Sci.* **51**: 813

Safari E., Fogarty N.M., Gilmour A.R., Atkins K.D., Mortimer S.I., Swan A.A., Brien F.D., Greeff J.C. and van der Werf, J.H.J (2007). *J. Anim. Breed. Genet.* **124**: 65.

SAS Institute Inc. (2002-2012) Cary, NC, USA.

*Contributed paper*

# USING SELECTED SEQUENCE VARIANTS TO IMPROVE GENOMIC PREDICTION OF HEAT TOLERANCE IN DAIRY CATTLE

**E.K. Cheruiyot[1,2], M. Haile-Mariam[1], B.G. Cocks[1,2], I.M. MacLeod[1] and J.E. Pryce[1,2]**

[1]Agriculture Victoria, Department of Jobs, Precincts and Regions, Bundoora, VIC 3083, Australia

[2]School of Applied Systems Biology, La Trobe University, Bundoora, VIC 3083, Australia

## SUMMARY

Genomic breeding values for heat tolerance were first developed and released to the Australian dairy industry in 2017, to allow farmers to select animals that better tolerate hot and humid conditions. It is desirable to improve the reliability of these genomic predictions to help accelerate the genetic improvement for this trait. Whole-genome sequence data may contain causative mutations, or variants in high linkage disequilibrium with causal mutations for traits. This study investigated the potential improvements in the accuracy of genomic prediction for heat tolerance when adding informative markers to the 50k industry SNP panel used routinely by DataGene for Australian dairy genomic evaluations. We selected informative sequence variants from a genome-wide association study (GWAS) of heat tolerance phenotypes of 20,623 Holstein cows (each cow with ~15 million imputed sequence variants) and augmented the 50k SNP panel with these SNPs for genomic prediction using a Holstein bull reference (N = 3,323) and Holstein cow validation set (N = 8,484). The accuracy of genomic prediction of heat tolerance for reduction in milk, fat, and protein yield under hot and humid conditions increased by 0.1%, 4%, and 6% units, respectively when informative markers were integrated with 50k SNP data. Since genetic gain is linearly related to EBV accuracy, this lift in accuracy is important for driving the genetic improvement of heat tolerance.

## INTRODUCTION

Heat tolerance is the ability of an animal to maintain production and reproductive performance under hot and humid conditions. The desire to breed for heat tolerance is growing worldwide due to the increasing effect of global warming on animal production. Considerable research has been conducted so far in many countries, including Australia, where the first breeding values for heat tolerance were released to the dairy industry in 2017 (Nguyen *et al.* 2017).

Since genetic gain is linearly related to the accuracy of estimated breeding values (EBVs), even a small lift in the accuracy of the heat tolerance EBV is important to the dairy industry. Besides increasing the size of the reference population, one way to boost the accuracy is to increase the density of markers used for genomic predictions. However, increasing the marker set from lower density SNP panels to whole-genome sequence have, in most cases, yielded limited, or no appreciable increase in the accuracies for various traits in cattle (e.g., Van Binsbergen *et al.* 2015). A promising alternative, in which a boost of accuracy has been realized in previous reports (e.g., Moghaddar *et al.* 2019), has been to augment standard industry SNP panels (i.e., 50k or 600K arrays) with a small set of informative or causal mutations for a trait. To fully maximize predictions, this approach requires careful selection of informative markers. Thanks to the 1000 Bull Genomes project (Hayes and Daetwyler 2019), it is now possible to use this sequence database to impute genotyped animals up to whole genome sequence. This may facilitate accurate selection of highly informative variants for use in genomic predictions, especially for complex traits such as heat tolerance.

In this study, we selected informative variants for heat tolerance from a genome-wide association study (GWAS) using milk production records of 20,623 Holstein cows, each having over 15 million imputed sequence variants. We then investigated the accuracy of prediction when sets of these selected variants were added to the standard industry 50k SNP array, by training the prediction in a bull reference set, and validating it in an independent set of Holstein cows.

## MATERIALS AND METHODS

**Phenotypes.** The phenotypes used in this study were obtained from DataGene (DataGene Ltd., Melbourne, Australia; https://datagene.com.au/) and included test-day milk, fat, and protein yields for Holstein cows and bulls, collected from dairy herds between 2003 and 2017 that were matched with climate data (daily temperature and humidity) obtained from weather stations across Australia's dairying regions. The distribution of dairy herds and weather stations, data filtering, and the calculation of environmental covariate (i.e., temperature-humidity index or **THI**) used in this work were described in our earlier studies (Nguyen *et al.* 2016, Cheruiyot *et al.* 2020).

**Calculation of heat tolerance phenotypes for cows and bulls.** The rate of decline (slope) in milk, fat, and protein yield due to heat stress events was estimated using reaction norm models as described by Cheruiyot *et al.* 2020. In these models, data on milk, fat, or protein yield were adjusted for fixed effects, including herd test day, year season of calving, parity, age at calving, jointly for parity and DIM, and jointly for stage of lactation and THI. Random effects fitted in the model included a random regression on a linear orthogonal polynomial of THI, where the intercept represents the level of mean milk yield and the linear component represents the change in milk yield (slope) due to heat stress for each cow (i.e., trait deviations (**TD**)) and a residual term. Slope solutions for each bull's daughters were averaged to obtain slope traits for bulls (i.e., daughter trait deviations (**DTD**)).

**Genotypes and study design.** Two genotype data sets were available: 50k SNP array and ~15 million imputed whole-genome sequence variants. The number of Holstein animals with genotypes and heat tolerance phenotypes were 29,107 ♀/3,323 ♂. We split the Holstein cows into two: 1) QTL discovery set (N = 20,623; comprising older cows born before 2013) for selecting informative markers for heat tolerance, and 2) genomic prediction validation set (N = 1,223; young cows born after 2012). We used Holstein bulls as a training set for genomic prediction. We ensured that none of the cows in the QTL discovery set were daughters of the bulls in the training set to avoid parent-daughter pairs between the two datasets to minimise close genetic relationships.

**QTL discovery analysis and selection of informative SNPs.** We performed single-trait GWAS analysis to test associations between individual SNP and cows' slope traits (milk, fat, and protein) using GCTA software (Yang *et al.* 2011). The models used for analyses are described by Cheruiyot *et al.* (https://www.biorxiv.org/content/10.1101/2021.02.04.429719v1.full).

Following the GWAS, we selected informative variants defined as 'top SNPs' for each slope trait as follows: for SNPs passing the GWAS threshold of $-log_{10}(p\ value)$ = 2; we chose the most significant SNP from within each 100 kb window and sliding 50 kb to the next window along each chromosome. We then removed one SNP of any pair of the selected 'top SNPs' in strong LD ($r^2 > 0.95$).

**Genomic prediction.** We used BayesR (Erbe *et al.* 2012) to estimate prediction accuracies for 50k SNP panel and compared the resulting accuracies with those obtained from adding 'top SNPs' to the 50k SNP set (i.e., 50k + 'top SNPs') using BayesRC method (MacLeod *et al.* 2016). The BayesR model fitted to the training bulls (N = 3,323) for 42,572 variants from 50k SNP panel was: $y = X\beta + Zg + Wv + e$, where **y** = vector of heat tolerance slope phenotypes; **X** = design matrix; **β** = vector of fixed effect solutions; **Z** = design matrix relating phenotypes to GBV; g = vector of GBV ~ $N(0, \mathbf{I}\sigma_g^2)$, where $\sigma_g^2$ is the additive genetic variance for the trait; **W** = design matrix of SNP genotypes; **v** = vector of SNP effects, modelled to have four possible normal distributions corresponding to zero, small, medium and large effects, respectively; **e** = vector of residual errors

$N(0, \mathbf{E}\sigma_e^2)$, where **E** is a diagonal matrix calculated as $diag(1/w_i)$, with $w_i$ being a weighting factor for $i$th sire calculated based on the available number records following Garrick *et al.* (2009).

We then used the BayesRC method to analyse 50k + 'top SNPs' dataset; an extension of the BayesR model that allows pre-allocation of variants to 2 or more classes (MacLeod et al., 2016) and hence a different posterior mixture distribution within each class if the class is enriched for informative SNPs. In our case the SNPs from 50k array (42,572) were allocated to class I and the selected 'top SNPs' to a separate class II, because the latter may be enriched with causal mutations for heat tolerance. For both BayesR and BayesRC models, we performed five MCMC replicate chains, each with 40,000 iterations of which 20,000 were discarded as burn-in for all the traits. We ran the analysis for 2 random validation sets of 600, and 623 Holstein cows.

**Calculating accuracy of genomic prediction.** For each of the three validation cow sets (described above), the accuracy of prediction was calculated as: $Accuracy(Val_i) = \frac{r_{GBV,phen}}{\sqrt{h^2}}$, where $Val_i$ = Holstein cow validation set; $r_{GBV}, phen$ = correlation of GBV and phenotypes (i.e., slope traits); $h^2$ = genomic heritability calculated for each trait using variance component estimates of Holstein cows (N = 29,107) for 50k SNP array (45,504 SNPs) data based on –reml option of GCTA software (Yang *et al.* 2011). The bias of prediction was assessed as the regression coefficient of the phenotypes (pre-corrected for fixed effects) on the GBV for animals in the validation set.

## RESULTS AND DISCUSSION

In this study, we used a large dataset of Holstein cows (N = 20,623) to select informative markers from a GWAS and then tested them for increased genomic prediction of heat tolerance phenotypes.

The genomic heritability estimates (± standard errors) for the heat tolerance milk (**HTMYslope**), fat (**HTFYslope**) and protein (**HTPYslope**) yield slope traits from Holstein cows that used to calculate the accuracy of predictions were 0.23 ± 0.01, 0.21 ± 0.01, and 0.20 ± 0.01, respectively. The number of informative markers for heat tolerance (i.e., 'top SNPs') selected from GWAS ($p < 0.01$) was highest for HTPYslope (9,633) followed by HTFYslope (9,352), and HTMYslope (9,207) traits. Similarly, the total number of markers used in the BayesRC analyses (i.e., 50k + top SNPs) were 51,750, 51,894, 52,168, for HTMYslope, HTFYslope and HTPYslope traits, respectively. We chose a cut-off of $p < 0.01$, which is comparatively relaxed, to capture both markers with small and large effect sizes for heat tolerance.

Figure 1 shows the accuracy and bias of genomic predictions in the Holstein validation cows. For the BayesR model using only 50k SNP data, we found the highest accuracy of prediction for HTFYslope (0.49 ± 0.01), followed by HTMYslope (0.49 ± 0.01) and HTPYslope (0.39 ± 0.01). The bias across all study traits was > 1.0 (Figure 1) indicating 'deflation' or under prediction, meaning less variance among predicted than observed values.

When the selected 'top SNPs' were added to the standard 50k SNP array and analysed using the BayesRC model, we found a consistent increase in the prediction accuracy across all the traits with values of 0.001, 0.04, and 0.06 for HTMYslope, HTFYslope and HTPYslope traits, respectively (Figure 1). This increase in accuracy is notable for HTFYslope and HTFYslope traits and likely to be associated with the pre-selected markers (potentially functionally linked with heat tolerance) and the method used (BayesRC). The bias of prediction for BayesRC was comparable that for BayesR. In this study, we investigated the potential benefits of sequence variants selected from a single breed (Holsteins) on the accuracy of genomic predictions for the same breed (within-breed prediction). The value of sequence variants selected in across-breed population (combined Holsteins and Jersey) on genomic prediction of other breeds (Jersey and crossbred cattle) will be investigated in a further study.

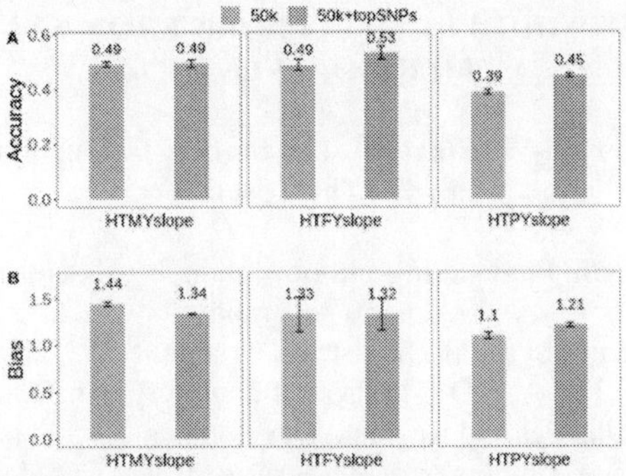


**Figure 1. Accuracy of genomic prediction in Holsteins cows based on BayesR (50k; grey) and BayesRC (50k+topSNPs; blue) models for heat tolerance milk (HTMYslope), fat (HTFYslope), and protein (HTPYslope) yield slope traits. Vertical lines are the standard errors of prediction estimated from 2 random validation sets of 600, and 623 Holstein cows**

## CONCLUSION

Overall, our results show that the accuracy of genomic prediction for reduction in milk, fat, and protein yields under hot and humid conditions can be improved by 0.1%, 4%, and 6% units, respectively when selected informative sequence variants are added to the industry-implemented 50k SNP panel.

## ACKNOWLEDGEMENTS

The study was supported by DairyBio (Melbourne, Australia), funded by Dairy Australia, The Gardiner Foundation and Agriculture Victoria. We are grateful to the 1000 Bull Genomes project for access to the sequence data and thank Dr Bolormaa Sunduimijid (Agriculture Victoria Research) for imputation of sequence genotypes. Thanks to DataGene Ltd and Australian dairy farmers for the phenotype and genotype data.

## REFERENCES

Cheruiyot, E. K., T. T. T. Nguyen, M. Haile-Mariam, B. G. Cocks, M. Abdelsayed, and J. E. Pryce. (2020) *J Dairy Sci* **103**:2460.

Erbe, M., B. Hayes, L. Matukumalli, S. Goswami, P. Bowman, C. Reich, B. Mason, and M. Goddard. (2012) *J Dairy Sci* **95**:4114.

Garrick, D. J., J. F. Taylor, and R. L. Fernando. (2009) *Genet Sel Evol* **41**:55.

Gilmour, A., B. Gogel, B. Cullis, S. Welham, and R. Thompson. (2015) ASReml user guide release 4.1 structural specification. Hemel hempstead: VSN International Ltd.

Hayes, B. J. and H. D. Daetwyler. (2019) *Annu Rev Anim Biosci* **7**:89.

MacLeod, I., P. Bowman, C. Vander Jagt, M. Haile-Mariam, K. Kemper, A. Chamberlain, C. Schrooten, B. Hayes, and M. Goddard. (2016) *BMC genomics* **17**:144.

Moghaddar, N., M. Khansefid, J. H. van der Werf, S. Bolormaa, N. Duijvesteijn, S. A. Clark, A. A. Swan, H. D. Daetwyler, and I. M. MacLeod. (2019) *Genet Sel Evol* **51**:72.

Nguyen, T. T., P. J. Bowman, M. Haile-Mariam, J. E. Pryce, and B. J. Hayes. (2016) *J Dairy Sci* **99**:2849.

Nguyen, T. T., P. J. Bowman, M. Haile-Mariam, G. J. Nieuwhof, B. J. Hayes, and J. E. Pryce. (2017) *J Dairy Sci* **100**:7362.

Yang, J., S. H. Lee, M. E. Goddard, and P. M. Visscher. (2011) *Am J Hum Genet* **88**:76.

Van Binsbergen, R., M. P. Calus, M. C. Bink, F. A. van Eeuwijk, C. Schrooten, and R. F. Veerkamp. (2015) *Genet Sel Evol* **47**:1.

*Contributed paper*

# SPLIT PATERNITY IS HIGH IN TWINS BORN FROM SYNDICATE-MATED MERINO EWES

**B.E. Clarke[1], K.L. Egerton-Warburton[2], J.L. Smith[3], B. Paganoni[4], G.B. Martin[5] and A.N. Thompson[1]**

[1] College of Science, Health, Engineering and Education, Murdoch University, Murdoch, WA, 6150 Australia
[2] NSW Department of Primary Industries, Orange, NSW, 2800 Australia
[3] CSIRO, Agriculture and Food, F.D. McMaster Laboratory, Armidale, NSW, Australia
[4] Department of Primary Industries and Regional Development, Bunbury, WA, 6230 Australia
[5] UWA Institute of Agriculture, University of Western Australia, Crawley, WA 6009, Australia

## SUMMARY

Split paternity rates in multiple born lambs of syndicate-mated Merino flocks have previously not been reported, primarily due to the cost of genotyping. Pedigree data from litters born to genetically diverse syndicate mated ewes in three Merino Lifetime Productivity flocks across Australia were analysed to examine rates of split paternity, or heteropaternal superfecundation. Over three joinings at three sites, 1082 twin or triplet litters were marked, of which 577 were heteropaternal (53.3%). There was no effect of age of dam, year of birth, sire or maternal grandsire on heteropaternity rates. These high rates of heteropaternity confirm the need to genotype all progeny from multiple births in syndicate mated flocks to ensure accurate genetic evaluation.

## INTRODUCTION

Split paternity, or heteropaternal superfecundation, occurs when females are successfully mated by multiple males during the same oestrus cycle. Decreased costs of determining paternity using DNA means that it is now feasible to obtain pedigrees from syndicate mating which may have some practical advantages over single-sire mating. Being able to syndicate mate sheep and determine parentage using DNA requires less paddocks than single sire joining and reduced labour in terms of feeding, labour for lambing rounds (preparation for lambing through identification of dams and mothering up) together with less disturbance during the lambing period as well less risk of failed matings with infertile single-sire mated rams.

Very little data about heteropaternal superfecundation has been published for either sheep or cattle. In cattle, where twinning rates are traditionally very low (1-5% depending on breed), McClure *et al.* (2017) examined rates of heteropaternal superfecundation in Irish herds, citing data from the Irish Cattle Breeding Federation database. They reported that, with an average of 1.7% twin rates, 0.98% of these were heteropaternal. By contrast, in syndicate-mated sheep, surprisingly high levels of heteropaternity, detected by DNA technology, have recently been reported for flocks in Ireland under relatively intensive conditions (Berry *et al.* 2020). The only information on the occurrence or frequency of heteropaternity in multiple-born lambs in naturally mated Merinos showed even higher levels (46-59%) in lambs born in 2012-2015 as part of the Merinolink Genomic Validation Project (Martin 2016).

We predicted that heteropaternity would also be high in extensively run Merino flocks in different parts of Australia with varying genotypes. We tested this in three naturally mated Merino flocks in Western Australia and New South Wales over 2-3 years. We also examined the effect of heteropaternity on total weaning weight of litters to see if there was any difference in the weaning weights of single-sired twin lambs compared to heteropaternal twins.

## MATERIALS AND METHODS

Pedigree data were collected from F2 Merino lambs born between 2018 and 2020 in the Merino Lifetime Productivity (MLP) Project (Ramsay *et al.* 2019). These data are from three MLP sites located at Pingelly in WA and at Trangie (Macquarie) and Armidale (New England) in NSW and consist of lambing records of ewes born to genetically diverse sires representative of the Australian Merino population. Lambs included in this analysis were from paddock-mating of MLP ewes to a team of Merino rams at each site. The ewe to ram ratio was approximately 50 ewes per ram and 8 to 16 rams were used in each flock, depending on flock size. All flocks had a five-week joining period and paddock sizes for joining ranged from 10 to 65 hectares. The ewes were pregnancy scanned for litter size about 80-90 days after the start of joining.

Tissue samples were taken from all F2 lambs alive at marking and tested using an 800K SNP chip. No DNA or data were collected on lambs that did not survive to marking. Parentage was verified against the ewe and ram genotypes, all of which had been previously genotyped using a 50K SNP chip. Lambs that were assigned a birth type as twin or triplet and rear type as single were removed from the analysis as birth type was inferred from pregnancy scanning results and there were no data on litters that were scanned multiple but had less lambs survive to rearing. The pedigree assigned using tissue samples collected at marking and rear type at marking were used to determine if a lamb was a single or multiple and if heteropaternal or not. Data were available for 1,082 litters where sire, dam and lamb's pedigree were available.

**Statistical analysis.** Estimates of heteropaternal effects were assessed by fitting General Linear Mixed Models (GLMM; Genstat VSN International 2017). The approach used a logit transformation and binomial distribution. Using additive models, logits were predicted as a function of syndicate group (a combination of flock, year of birth and joining group) as a fixed effect, and sire and maternal grandsire as random effects.

Estimates of differences between litters for total weaning weight (sum of littermate weaning weights) were restricted to twin-born litters (triplets were excluded). Total weaning weight was analysed using the restricted maximum likelihood method (REML, Genstat, VSN International 2017). Syndicate group and heteropaternal status were fitted as fixed effects, and sire and maternal grandsire were fitted as random terms.

## RESULTS AND DISCUSSION

The average percentage of heteropaternal lambs across the three MLP sites over 2 or 3 joinings was 53.3% or 577 of the 1082 multiple born litters (Table 1). This included three of the four sets of triplets. The frequency of heteropaternity was greater than previous estimates (Berry *et al.* 2020), but similar to that reported by Martin (2016) who studied 349 litters. The estimates of heteropaternity in our data are possibly underestimated as only those litters where at least two lambs survived to marking are included in the analysis. It is reasonable to expect that heteropaternal lambs could have greater variation in weight and other behaviours at birth and hence have lower survival than single sired twins. There were no significant differences between fixed effects of dam year of birth, syndicate group, or sire and maternal grandsire, suggesting no differences between maiden or later joinings in the production of heteropaternal litters.

Over two or three joinings, the Merino flocks in this study showed higher levels of heteropaternity than a similar study of six commercial but smaller flocks of crossbred Irish sheep (Berry *et al.* 2020). This is surprising because our study included a lower ratio of rams to ewes (1:50 v 1:22-39) and was on a much larger and more extensive scale with larger paddocks. The lower ram to ewe mating ratios and larger paddock sizes would be expected to provide less interaction between rams and ewes during the oestrus period (Croker and Lindsay 1972). Nevertheless, the consistently high rates of heteropaternity (42% to 65%) across the three genetically and environmentally diverse flocks suggest these could be typical for rates of heteropaternity in twin lambs in syndicate-mated Merino flocks across Australia. While factors that influence this surprisingly high level of heteropaternity in syndicate mating are largely unknown this reinforces the need to determine parentage of all multiple born lambs included in genetic evaluation using DNA.

**Table 1. Percentages of heteropaternal lambs marked between 2018 and 2020, at each of three MLP sites**

| Site | Dam year of birth | Progeny year of birth | No. heteropaternal litters | No. multiple marked litters | Percent heteropaternal lambs |
|---|---|---|---|---|---|
| Pingelly | 2016 drop | 2018 | 31 | 48 | 64.6% |
| | 2016 drop | 2019 | 60 | 108 | 55.6% |
| | 2017 drop | 2019 | 46 | 86 | 53.5% |
| | 2016 drop | 2020 | 74 | 142 | 52.1% |
| | 2017 drop | 2020 | 66 | 109 | 60.6% |
| Macquarie | 2017 drop | 2019 | 47 | 83 | 56.6% |
| | 2017 drop | 2020 | 62 | 113 | 54.9% |
| | 2018 drop | 2020 | 27 | 63 | 42.9% |
| New England | 2017 drop | 2019 | 18 | 43 | 41.9% |
| | 2017 drop | 2020 | 107 | 219 | 48.9% |
| | 2018 drop | 2020 | 39 | 68 | 57.4% |
| **Total** | | | **577** | **1082** | **53.3%** |

There were no significant differences between single sire and heteropaternal litters in total litter weaning weight ($P = 0.764$) despite Australian Sheep Breeding Values for weaning weight ranging from –1.6 to 9.9 kg (Merino Select analysis run date 21/02/2021) in rams used over the three sites (up to 6.9 kg range within site). There were, however, significant differences between syndicate group ($P < 0.001$) for total weaning weight.

## CONCLUSIONS

Extensively run Merino flocks have not previously been candidates for large scale flock genotyping but, as the technology becomes more cost-effective for ram breeding and commercial sheep flocks, it is becoming feasible to genotype progeny from large syndicate-mated flocks. High rates of heteropaternal litters unequivocally confirm the need to genotype all offspring of syndicate matings to ensure pedigree is correctly assigned. The increased use of genotyping in Merino flocks that syndicate mate will allow for greater access to genetic evaluation with accurate pedigree data that has previously not been available for syndicate mated flocks.

## ACKNOWLEDGEMENTS

This project is funded by the three MLP sites and Australian Wool Innovation. Australian Wool Innovation gratefully acknowledges the funds provided by the Australian government to support research, development and marketing of Australian wool. The Pingelly MLP site gratefully acknowledges the generous financial support provided by Murdoch University and the University of Western Australia and knowledges AMSEA and the Pingelly Breeder Site Committee for their role in the Merino Lifetime Productivity Project. The MLP Macquarie site gratefully acknowledge the generous financial support provided by the site host NSW Department of Primary Industries. The MLP New England site gratefully acknowledges the generous financial support provided by the site host, CSIRO. Staff and Site Committee volunteers at the respective MLP sites are also thanked for their contributions to data collection.

## REFERENCES

Berry, D. P., Bohan, A., O'Brien, A. C., Campion, F. C., McHugh, N. & Wall, E. (2020). *Anim. Genet.* **51**: 579.

Croker, K. P. & Lindsay, D. R. (1972). *Aust. J. Exper. Agric. Anim. Husb.* **12**: 13.

Martin, S. (2016) 'MerinoLink Limited Genomic Validation Project Final Report', Merinolink

McClure, M. C., McClure, J. C. & McCarthy, J. (2017). *Anim. Genet.* **48**(6): 721.

Ramsay, A. M. M., Swan, A. A. & Swain, B. C. (2019). *Proc. Assoc. Advmt. Anim. Breed. Genet.* **23**: 512.

VSN International (2017). Genstat Reference Manual (Release 19), VSN International, Hemel Hempstead, UK.

*Contributed paper*

# VISUAL CLASSING GRADES ARE HERITABLE AND VISUALLY CLASSED MERINO SHEEP BORN AND REARED AS TWINS ARE GRADED LOWER THAN SINGLES

**B.A. Metcalfe[1], J.C. Greeff[2], A.N. Thompson[1] and B.E. Clarke[1]**

[1] School of Science, Health, Engineering and Education, Murdoch University, Murdoch, WA, 6150 Australia
[2]Department of Primary Industries and Regional Development, South Perth, WA, 6151 Australia

## SUMMARY

Selection of Merino sheep for use in breeding programs includes the combination of visual assessment and measurement of production traits. Genetic evaluation of these assessments and traits takes into account non-genetic effects to improve the accuracy of breeding value predictions. The hypothesis tested in this paper was that visual classing assessment of sheep on a traditional three-point and a novel five-point visual scoring system are heritable and both are also affected by non-genetic effects such as birth and rearing type. Using data from the first two years of classing ewes in the Merino Lifetime Productivity project at Pingelly, WA, moderate heritability estimates were observed for both scoring systems (0.24±0.08 for three-point and 0.17±0.07 for five-point). Both traits were moderately repeatable (0.31-0.36±0.03). Birth and rear type impacted visual scoring grades significantly, indicating that they should be accounted for when visually classing Merino sheep.

## INTRODUCTION

Merino sheep breeding routinely combines objectively measured production selection and visual assessment to improve the quality and quantity of wool produced as well as improve structure and conformation. Merino sheep classing by visual selection is based on a number of subjectively assessed traits such as wool quality, quantity and conformation, assessed by professional sheep classers using different scoring systems to class animals into different categories (Brown *et al.* 2002; Robinson *et al.* 2007). When used in conjunction with estimated breeding values for production traits, visual classing can add value to selecting superior animals for replacement, with greater accuracy and efficiency than using a single method alone (Mortimer *et al.* 2010).

Genetic evaluation of Merino production data includes accounting for any fixed effects or non-genetic factors, such as birth type (whether the animal was born as a single or multiple), rear type (whether the animal was raised as a single or multiple), the age of the dam, the date of birth of the animal and whether that animal was born to a maiden or experienced mature ewe (Hadfield and Kruuk 2007; Brown *et al.* 2016). These non-genetic factors influence the phenotype of the animal and can often influence how it is classed visually. For example, twin born and reared lambs are typically smaller and produce broader and less wool than single counterparts (Swan *et al.* 2008, Thompson *et al.* 2011a,b). Accurate estimates of these fixed effects need to be included when estimating breeding values, to ensure accurate estimates of genetic merit.

Research has shown that visually assessed classer grades have a heritability between 0.12 and 0.2 and have favourable genetic and phenotypic correlations with liveweight, wool quality and structural traits (Mortimer *et al.* 2009). These subjectively measured traits are also significantly influenced by birth type, rear type and other non-genetic factors. In addition, Clarke and Thompson (2021) found that classers were influenced by subjective assessments of liveweight, clean fleece weight and fibre diameter when grading of animals. In this study non-genetic factors had a significant effect on classing outcomes such that at the first seven month old professional classing 69% of the culls were twins and only 31% were singles. Conversely 70% of the tops were singles and only 30% were twins. The current study uses an expanded data set from the Merino Lifetime

Productivity Project (Ramsay *et al.* 2019), covering more sires, seasons and repeated measures. It The hypothesis tested that classing grades are affected by non-genetic factors and that both three and five grade classing systems are heritable.

## MATERIALS AND METHODS

The data analysed in this study were collected from a total of 1103 sheep born in 2016 and 2017 as part of the Merino Lifetime Productivity (MLP) Project in Pingelly, Western Australia (Ramsay *et al.* 2019). The sheep were ewes from 29 different sires and dams originating from 3 sources born over 4 years. Dams were evenly distributed to sire groups, taking into consideration ewe age, condition score and weight. The minimum number of ewes joined to each sire was 90. At approximately day 90 of pregnancy, ewes were scanned for litter size using ultrasound and divided according to whether they were single or multiple bearing. Multiple-bearing ewes were managed separately to single-bearing ewes to provide for their increased nutritional requirements based on the recommendations for pregnancy management for Merino ewes (Young *et al.* 2016). Lambing occurred in late June, with marking, tagging and DNA sampling, taken late July. All ewe progeny from each year of birth, were run together from weaning until pregnancy scanning as maidens (22 months).

All progeny were evaluated subjectively using two different visual assessments that were completed by independent sheep classers yearly, prior to shearing (at approximately 8 and 20 months of age) and according to the site breeding objective. This first classing system, called the Australian Merino Sire Evaluation Association (AMSEA) grade, sorts animals into three categories either Tops (1), Flocks (2) or Culls (3) of approximate split 25%, 50%, 25%, based on visual assessment of all traits that are present in the project's breeding objective: in this paper this system will be referred to as GRADE. The second system used a five grade system consisting of Top (1), First (2), Flock (3), Sale (4) or Cull (5), referred to as a professional classer grading (PROF) system with approximate split 2%, 10%, 58%, 20%, 10%. The classer was unaware of the progeny's parentage and birth type at classing.

**Statistical Analysis.** Fixed effects, variance components and genetic parameters were estimated using general linear mixed models and residual maximum likelihood methods with ASReml (Gilmour *et al.* 2009). An animal model was fitted and the animals' year of birth, age of dam at lambing (in years), birth type (litter size, how many lambs were evident at pregnancy scanning), rear type (how many lambs from the little survived to weaning), shearing number (first or second time being shorn) and dam source (where the dam was bred) were fitted as fixed effects. Birth type referred to litter size from pregnancy scanning records (coded as 1, 2 or 3), while rear type was the litter size at weaning (coded as 1 or 2 as no triplets survived). For each trait the fixed effects were tested for significance. Following each analysis, all effects that were not significant were removed from the model, until only significant fixed variables were left, (using a significance level of 5%). Animal was fitted as a random additive genetic effect and as an environmental effect to account for repeated measures on the same animal. The direct heritability was estimated by dividing the additive genetic variance with the total phenotypic variance, whereas the environmental variance component from the repeated measures was added to the direct additive genetic variance, which was then divided by the total variance to estimate the repeatability for each trait s

## RESULTS AND DISCUSSION

Birth type recorded a significant effect on both PROF in both sets of data (yearling and combined) as well as for GRADE when the second shearing measure was included in the analysis (Table 1). Rearing type was highly significant effect for PROF and GRADE and an interaction between these factors of birth and rear type for GRADE. These significance levels reflect findings by Mortimer *et al.* (2009). Dam age however was not significant effect for PROF, while it was significant for GRADE which is consistent with the findings of Mortimer *et al.* (2009). This significance declined when the second year of data was included, most likely due to the effects of many non-genetic factors reducing with the age of an animal (Asadi Fozi *et al.* 2005).

Predicted means for the visually classed traits were also calculated for the combined measures at yearling and first adult shearing (Table 1). Single born and reared lambs had lower predicted means for GRADE and PROF, meaning they are less likely to be culled. These differences were more evident in the yearling data set but had improved accuracy with lower standard errors in the combined data (Table 1). These results emphasise the importance of accounting for birth type, rear type and dam age when selecting animals based on phenotype. Animals should be classed separately based on their birth and rearing type, multiple born and raised animals shouldn't be compared phenotypically to singles as they incur phenotypic disadvantages due to non-genetic effects they are exposed to.

**Table 1. Predicted least square means (with standard errors) for significant fixed effects for visually assessed traits GRADE and PROF and significance of various fixed effects at combined yearling and first shearing ($P < 0.05$)**

| | | GRADE* | PROF** |
|---|---|---|---|
| **Birth Type** | | | |
| | **1** | 1.88 ± 0.09 | 3.13 ± 0.15 |
| | **2** | 2.17 ± 0.09 | 3.50 ± 0.14 |
| | **3** | 2.10 ± 0.22 | 3.14 ± 0.34 |
| **Rear Type** | | | |
| | **1** | 2.01 ± 0.08 | 3.19 ± 0.13 |
| | **2** | 2.07 ± 0.08 | 3.33 ± 0.13 |
| **Shearing** | | | |
| | **1** | 2.12 ± 0.08 | 3.31 ± 0.13 |
| | **2** | 1.89 ± 0.08 | 3.15 ± 0.13 |

*GRADE Visual classing grade on a 3-point scale (Top, Flock, Cull)
**PROF Professional visual classing grade on a 5-point scale (First, Top, Flock, Sale, Cull)

**Table 2. Variances and estimates of Heritability, Repeatability with standard errors for the visual traits measured at yearling age and at first adult shearing of 1100 Merino ewes**

| Age Stage | Variance Component | GRADE* | PROF** |
|---|---|---|---|
| Yearling | Heritability | 0.21 ± 0.10 | 0.18 ± 0.08 |
| Combined yearling and first adult | Heritability | 0.24 ± 0.08 | 0.17 ± 0.07 |
| Combined yearling and first adult | Repeatability | 0.31 ± 0.03 | 0.36 ± 0.03 |

*GRADE Visual classing grade on a 3-point scale (Top, Flock, Cull)
**PROF Professional visual classing grade on a 5-point scale (First, Top, Flock, Sale, Cull)

Heritability estimates for visually assessed traits of GRADE (3-point scale) and PROF (5-point scale) were moderate at 0.21±0.10 and 0.18±0.08 respectively at yearling age and 0.24±0.08 and 0.17±0.07 for the combined years data (Table 2). The heritability for GRADE was similar to that reported by Mortimer *et al.* (2009) confirming that visual grade is a heritable trait and can be used in a selection program. The repeatability across years was estimated at 0.36±0.03 and 0.31±0.03 for PROF and GRADE, respectively. Fulloon *et al.* (2001) found GRADE to have a 0.34 repeatability supporting the finding from our study. This indicates that both GRADE and PROF are heritable and repeatable traits. The estimate of heritability for professional five-point scale (PROF) is a novel finding allowing for accurate selection of PROF which was previously unavailable to producers.

## CONCLUSIONS

Non-genetic factors, in particular birth type and rear type, were found to affect the visually assessed traits significantly. This indicates there is a bias towards single born and raised lambs when visual selection is used. By accounting for non-genetic effects in visual selection, phenotypic gains can be increased. It is therefore recommended that multiple- born and raised lambs shouldn't be visually classed or compared alongside single born and raised counterparts as multiple-born animals incur phenotypic disadvantages.

Both AMSEA classer grade and professional grade, were found to have moderate heritability estimates and favourable repeatability estimates. Professional grade (five point grading system), will provide more discriminatory grading of animals as there are more classes that the traditional three point scale. The novel estimates calculated in this study for heritability and repeatability mean professional grade can now be accurately selected for to provide genetic gains in breeding programs and producers wanting a greater range of classing points.

## ACKNOWLEDGEMENTS

This project is funded by Murdoch University, the University of Western Australia and Australian Wool Innovation. Australian Wool Innovation gratefully acknowledges the funds provided by the Australian government to support research, development and marketing of Australian wool. We acknowledge AMSEA and the Pingelly Breeder Site Committee for their role in the Merino Lifetime Productivity Project. Scholarships for this honours thesis were provided by the Australian Wool Education Trust, the SIBI Trust and the State Government's Science and Agribusiness Connect.

## REFERENCES

Brown, D.J., Ball, A., Mortimer, R. and Oppenheimer, M. (2002). *Int. J. Sheep Wool Sci.* **50**: 378.

Brown J.A., Lewis, G.S. and Taylor J.B. (2016). *Sheep Goat Res J* **31**: 1.

Clarke, B.E. and Thompson A.N. (2021) *Anim. Prod. Aust.* **33:** clii.

Asadi Fozi, M., van der Werf, J.H.J. and Swan, A.A. (2005) *Aust. J. Agric. Res.* ***56***: 789.

Fulloon, O.A., Swan, A.A., Piper, L.R. and van der Werf J. (2001) *Proc. Assoc. Advmt. Anim. Breed. Genet* **14**: 513.

Gilmour AR, Gogel BJ, Cullis BR, Thompson R (2009) *VSN International Ltd: Hemel Hempstead, UK*

Hadfield J.D. and Kruuk L.E.B. (2007) *J. Evol. Biol.* ***20***(5): 1890.

Mortimer, S.I., Robinson, D.L., Atkins, K.D., Brien, F.D., Swan, A.A., Taylor, P.J. and Fogarty, N.M. (2009) *Anim. Prod. Sci.***49**: 32.

Mortimer, S.I., Atkins, K.D., Semple, S.J. and Fogarty, N.M. (2010) *Anim. Prod. Sci.* **50:** 976

Ramsay, A.M.M., Swan, A.A. and Swain, B.C. (2019) *Proc. Assoc. Advmt. Anim. Breed. Genet.* **23**: 512.

Robinson, D.I., Mortimer, S.I., Swan, A.A. and Purvis, I.W. (2007) *Proc. Assoc. Advmt. Anim. Breed. Genet.* **17**: 336.

Swan A.A., Purvis I.W. and Piper L.R. (2008) *Aust. J. Exp. Agricul.* ***48*** (9): 1168.

Thompson A.N., Ferguson M.B., Campbell A., Gordon D.J., Kearney G.A., Oldham C.M. and Paganoni B.L. (2011a) *Anim. Prod. Sci.* **51**:784.

Thompson A.N., Ferguson M.B., Gordon D.J., Kearney G.A., Oldham C.M. and Paganoni B.L. (2011b) *Anim. Prod. Sci.* **51**: 794.

Young J., Thompson A.N., Curnow M., Oldham C.M. and Behrendt R. (2016) *Anim. Prod. Sci.* **56:** 669.

*Contributed paper*

# RESPONSES TO HEAT IN EWES FROM INDIGENOUS AND COMMERCIAL SOUTH AFRICAN SHEEP BREEDS: PRELIMINARY RESULTS

**S.W.P. Cloete[1,2], D.A. van der Merwe[1], S. Steyn[1], A. Muller[1], J. Melgar[3], C.K. Cornwallis[3] and T.S. Brand[1,2]**

[1]Department of Animal Sciences, Stellenbosch University, Matieland, 7602 South Africa; [2]Directorate Animal Sciences, Western Cape Department of Agriculture, Elsenburg, 7607 South Africa; [3]Department of Biology, Lund University, Lund, Sweden

## SUMMARY

Average temperatures are likely to increase, resulting in hotter and dryer conditions in South Africa. The impact of these changes on animal production and welfare is not well-defined. Two trials were conducted to determine the homeothermic response of eight sheep breeds. In 2016, the study included seven breeds, namely Dohne, Dormer, Dorper, Meatmaster (MM), Merino, South African Mutton Merino (SAMM) and White Dorper (WD). The WD was replaced by an unimproved indigenous breed, the Namaqua Afrikaner (NA), in 2017. Ranges of ewes per breed were 10-14 in 2016 and 12-15 in 2017. On days forecast to be hot at noon, these animals were assessed under cool conditions (19-24°C) in the morning and hot conditions at noon (30-33°C) by monitoring individual average eye temperature using thermal imaging (only in 2016) as well as counting flank movements to derive respiration rate (both years). The increased heat load in the afternoon markedly increased both traits. Breed interacted with the time of the day. During 2016, there were suggestions that the hair breeds (Dorper, WD and MM) were able to maintain lower basal respiration rates in the morning compared to the other breeds. Respiration rate in the generally cooler 2017 study increased by more than threefold from the morning to the afternoon in Merino, Dohne, SAMM and Dormer ewes, more than twofold in the Dorper and MM and by only 84% in the NA breed. These results suggest that hair sheep and hardy indigenous breeds may cope better with the anticipated higher heat load in the future. There is still marked scope for further research on ovine adaptation to heat stress conditions in South Africa.

## INTRODUCTION

Sheep form an integral component of most livestock production systems throughout the world, the species being able to adapt to a wide variety of environments. The adaptability and success of sheep is confirmed thereby that they are the world's most diverse mammalian livestock species (Cloete 2012). A list of sheep breeds by region confirmed that the ovine species is indeed globally successful and represented in widely divergent farming landscapes throughout the world.

It is generally accepted that the western parts of Southern Africa will become hotter and drier under the impact of climate change (Meissner *et al.* 2013). Considerable areas of South Africa are already marginal owing to constraints of climate and soil (Cloete and Olivier 2010). Given the ability of sheep to adapt to marginal conditions, the species plays an important role in both the commercial and smallholder animal agricultural sectors. Under the increasing challenge posed by external drivers, such as temperature change, sheep and goats were reported to be more resilient than other livestock species (Rust and Rust 2013). The South African ovine genetic resource encompasses specialist wool and meat breeds, terminal sire breeds, dual-purpose breeds as well as unimproved, indigenous fat-tailed types (Cloete and Olivier 2010). By the number of weaning weight records, the most important South African breeds are the Merino, Dohne, SAMM, Dorper, Dormer and MM (Cloete *et al.* 2014). Although it is only found in conservation flocks at present (Qwabe 2011), the unimproved indigenous NA breed performed well in fitness traits when compared to commercial breeds (Cloete *et al.* 2016).

Against this background, it is important to assess these breeds for their ability to withstand high temperatures. It is also important to quantify potential differences between breeds, as well as between individuals within breeds to understand the mechanisms underlying the ability of sheep to maintain homeothermy under heat stress conditions.

## MATERIALS AND METHODS

Two studies were conducted on the Langgewens research farm of the Western Cape Department of Agriculture in the Swartland district, where it is common for the diurnal maximum temperature to exceed 30°C during the summer. The homeothermic response of seven sheep breeds, namely the Dohne, Dormer, Dorper, MM, Merino, SAMM and WD were assessed, by monitoring individual eye temperature using thermal imaging as well as respiration rate over four sessions. The Merino, Dohne, SAMM and Dormer originated from breeds developed in temperate regions, whereas the Dorper, WD and MM were composite hair breeds with temperate and heat-adapted breeds as parents. Respiration rate was determined by counting flank movements over a 30 second interval and then express it as breaths per minute (bpm). Sheep were monitored for two sessions during the cooler mornings and for two sessions during hotter afternoons over a three-day period from 31 October to 2 November 2016.

Experienced stockmen released ewes in groups of three to four from a crush into an outside yard where they could be approached to approximately 3 to 4 meters. Individual ewes had numbered tags tied around their necks to allow identification from a distance. Average eye temperature was recorded by an operator equipped with a thermal camera while a second operator counted the flank movements of individual sheep. A scribe recorded the respiration rate of individual ewes, while also acting as a time-keeper. When all sheep in a group were processed, the group was moved to a separate holding yard before the next group was assessed. This routine was followed until all ewes were processed. Several temperature forecast services were used to identify days for breath counting and eye temperature recording with a likely spread of temperatures well in the thermo-neutral zone (19-24°C ambient temperature according to the weather station) in the morning, to increase to a range where some individuals/breeds may experience heat stress (>30°C ambient temperature according to the weather station; see Marai *et al.* 2007) in the afternoon.

The second study involved the same breeds with the exception of the WD, which was replaced by the NA. Apart from this change, the same basic procedure was followed during 7 and 8 November 2017. The thermal camera was not available at this stage and the recordings were restricted to respiration rate. The mean (±s.d.) sizes of the breed groups were 12.4±1.3 (range 10-14) during 2016 and 13.0±1.1 (range 12-15) during 2017. All ewes were purchased from reputable breeders within each breed, but possible family relationships were unknown. The ewes were already on the farm for at least 7 months (including the Mediterranean winter) when assessed.

Mixed model methods were used to analyse the data with ASReml4 (Gilmour *et al.* 2015) within years (2016 and 2017). The model fitted was the following:

$$y_{ijkl} = mu + b_i + t_j + b_it_j + ewe_{ijk} + e_{ijkl}$$

with $y_{ijkl}$ = the i$^{th}$ eye temperature or respiration rate observation on the ijk$^{th}$ ewe; mu = the overall mean; $b_i$ = the i$^{th}$ breed (as described within years); $t_j$ = the j$^{th}$ time of day (morning or afternoon); $b_it_j$ = the breed x time of day interaction; $ewe_{ijk}$ = the random effect of the ijk$^{th}$ ewe and $e_{ijkl}$ = the random error term. The between-ewe variance component so derived was used to estimate the repeatability of the trait under consideration. Random ewe effects were then interacted with the time of the day to assess the variance associated with the re-ranking of ewes under hotter conditions in the afternoon.

## RESULTS AND DISCUSSION

Mean (±SD) temperatures derived from weather station data indicated morning temperatures during recording of 23.6±1.6°C during 2016 and 18.9±2.6°C during 2017. Corresponding means for the afternoon recording were respectively 32.6±1.4°C and 30.2±1.2°C. The 2017 recording were thus done under somewhat cooler conditions, especially in the mornings.

Ewe breed, time of day and the interaction between these fixed effects were significant in 2016 (Table 1). The average eye temperature increased from 35.5±0.1°C in the morning to 36.7±0.1°C at noon (P<0.01). However, these responses were not similar for all breeds (Table 1). Eye temperature increased by around 2% for WD, Dormer, Dorper and Merino ewes, but by much more (3.7 to 8.7%) in the case of SAMM, Dohne and MM ewes. Respiration rate similarly increased from 75±2 bpm in the morning to 122±2 bpm at noon (P<0.01). In this case there was evidence of differentiation according to origin, as the breeds from temperate regions (Dormer, SAMM, Merino and Dohne) generally exhibited smaller increases of 43 to 58% from morning to noon, compared to 83 to 100% observed in hair sheep (WD, Dorper and MM). These results stem from the fact that the heat-adapted hair sheep generally had lower basal respiration rates of 54 to 60 bpm in the mornings, compared to 77 to 96 bpm for the breeds originating from temperate regions. The between-ewe variance component went to the boundary of parameters space (zero) for average eye temperature while the repeatability of respiration rate amounted to 0.26±0.06. Interacting ewe with the time of the day resulted in estimates of 0.22±0.07 for the repeatability and 0.17±0.08 for the re-ranking term. The regression of respiration rate on eye temperature yielded a coefficient of 5.0±1.1 breaths per minute for one °C increase in eye temperature (r=24; P<0.01).

**Table 1. Estimated means (±s.e.) for respiration rate and average eye temperature of the respective breeds during cool (morning) and hot (noon) periods during 2016**

| Trait and time | Breed | | | | | | |
|---|---|---|---|---|---|---|---|
| | WD | Dormer | SAMM | Dorper | Merino | Dohne | MM |
| Average eye temperature (°C) | | | | | | | |
| Morning | 35.6±0.32 | 35.9±0.29 | 34.4±0.29 | 36.1±0.31 | 36.2±0.34 | 35.0±0.29 | 35.1±0.29 |
| Noon | 36.3±0.31 | 36.6±0.30 | 37.4±0.31 | 36.9±0.31 | 36.8±0.34 | 36.6±0.29 | 36.4±0.28 |
| Increase | 1.97 | 1.94 | 8.72 | 2.22 | 1.66 | 4.27 | 3.70 |
| Respiration rate (bpm) | | | | | | | |
| Morning | 58±4.9 | 96±4.7 | 77±4.7 | 60±4.8 | 94±5.3 | 89±4.7 | 54±4.5 |
| Noon | 108±4.8 | 137±4.7 | 118±4.7 | 120±4.8 | 133±5.3 | 132±4.7 | 104±4.5 |
| Increase | 83.1 | 42.7 | 53.2 | 100.0 | 58.3 | 48.3 | 92.6 |

The increase from morning to noon is expressed relative to the mean for the morning

During 2017, overall respiration rate increased by approximately three-fold from morning to noon as temperatures increased (from 33±1 bpm in the morning to 95±1 bpm at noon (P<0.01). The interaction of breed with time of day was again highly significant (P<0.01). The respiration rate of ewes was quite similar in the cooler mornings, ranging from 31 bpm (Dorpers and Dohnes) to 38 bpm in Merinos (Table 2). The smaller differences between breeds could be related to the lower morning temperatures during 2017. Responses to the higher heat loads at noon were again highly breed-specific. The respiration rate of the unimproved fat-tailed NA increased by 84% from the morning session to the noon session. The respiration rate of the other hair sheep (Dorper and MM) increased by more than 2-fold, while the respiration rate of the breeds from temperate origin increased by more than 3-fold. The repeatability of respiration rate amounted to 0.18±0.06. When the ewe x time of the day interaction was added, most of the variance repartitioned toward the interaction (re-ranking) term, yielding respective estimates of 0.08±0.07 and 0.28±0.09.

The ability of adapted, indigenous genotypes to better cope with heat stress across species was reviewed by Cloete (2012). It was evident that indigenous sheep breeds were better able to cope with heat stress in Egypt and India. The NA, in particular, was described in the literature as a slender breed with long legs to assist in the dissipation of excess heat (Qwabe 2011; Snyman *et al.* 2013). The ability of this breed to cope with heat conditions as well as its resistance to external parasites (Cloete *et al.* 2016) indicate that it may play an important role under challenging and poorly resourced conditions (Molotsi *et al.* 2020). Although other hair sheep (WD, Dorper and MM) also performed better than the temperate breeds for respiration rate, they were not quite as well adapted as the NA.

**Table 2. Estimated means (±s.e.) for respiration rate of the respective breeds during cool (morning) and hot (noon) periods during 2017**

| Trait and time | Breed | | | | | | |
|---|---|---|---|---|---|---|---|
| | NA | Dormer | SAMM | Dorper | Merino | Dohne | MM |
| Respiration rate (bpm)* | | | | | | | |
| Morning | 32±3.7 | 32±3.9 | 34±3.7 | 31±3.8 | 38±3.7 | 31±3.7 | 32±3.4 |
| Noon | 58±3.9 | 121±3.8 | 112±3.7 | 87±3.8 | 122±3.8 | 94±3.7 | 74±3.4 |
| Increase | 84.4 | 378.1 | 329.4 | 280.6 | 321.1 | 303.2 | 231.3 |

The increase from morning to noon is expressed relative to the mean for mornings

## CONCLUSION

Adapted livestock such as particularly the NA, but also the MM and Dorper, may cope better under challenging climate change scenarios than breeds from temperate regions such as the Merino, Dohne, SAMM and Dormer. An easily recorded indicator trait such as respiration rate could be considered as a tool to improve within-breed heat tolerance by selection under low-input systems. The provision is that future studies should allow a better understanding of the interaction of random ewe effects with the ambient conditions, represented in this study by cooler mornings and hotter afternoons.

## ACKNOWLEDGEMENTS

Financial support of the Western Cape Agricultural Research Trust, Cape Wools SA and Red Meat Research and Development SA is gratefully acknowledged.

## REFERENCES

Cloete S.W.P. and Olivier J.J. (2010) *The International Sheep and Wool Handbook, Nottingham University Press, Manor Farm, Thrumpton, Nottingham NG11 OAX, UK.* 95.

Cloete S.W.P. (2012) Encyclopedia of Sustainability, Science and Technology. Springer Science and Business Media. DOI 10.1007/978-1-4419—0851-3. 1740.

Cloete S.W.P., Olivier J.J., Sandenbergh L. and Snyman M.A. (2014) *S. Afr. J. Anim. Sci.* **44**: 308.

Cloete S.W.P., Cloete J.J.E. and Scholtz A.J. (2016) *Vet. Parasitol.* **230**: 33.

Gilmour A.R., Gogel B.J., Cullis B.R., Welham S.J. and Thompson R. (2002) ASREML-user Guide Release 1.0 VSN International Ltd, Hemel Hempstead, UK.

Marai I.F.M., El-Darawany A.A., Fadiel A. and Abdel-Hafez M.A.M. (2007) *Small Rumin. Res.* **71**: 1.

Meissner H.H., Scholtz M.M. and Engelbrecht F.A. (2013) *S. Afr. J. Anim. Sci.* **43**: 298.

Molotsi A.H., Dube B. and Cloete S.W.P. (2020) *Diversity* **12**: 14.

Qwabe S.O. (2011) M.Sc. (Agric.) thesis, University of Pretoria, Pretoria, South Africa.

Rust J.M. and Rust T. (2013) *S. Afr. J. Anim. Sci.* **43**: 255.

Snyman M.A., van Marle-Kőster E., Qwabe S.O. and Visser C. (2013) *Grootfontein Agric.* 13. http://gadi.agric.za/Agric/Vol13No1_2013/namaqua.php. Accessed on 28 April 2020.

*Contributed paper*

# GENETIC PARAMETERS FOR REPRODUCTION IN INTENSIVELY AND EXTENSIVELY MANAGED DOHNE MERINO FLOCKS IN SOUTH AFRICA

**S.W.P. Cloete[1], C.L. Nel[1], J.J. Olivier[2] and G.J. Delport[3]**

[1] Department of Animal Sciences, Stellenbosch University, Matieland, 7602 South Africa; [2] Hillwood Village, Pinehurst, 7607 South Africa;[3]Dohne Merino Breeders' Association, Stutterheim, 3083 South Africa

## SUMMARY

The study used 61,974 ewe-year records of 26,254 ewes aged 20 months and older for number of lamb born per ewe lambed (NLB), number of lambs weaned per ewe lambed (NLW) and ewe rearing ability (ERA) obtained from intensively managed South African Dohne Merino flocks. Corresponding numbers for extensively managed flocks numbered respectively 14,067 and 5,181. Reproductive output of intensively managed flocks was higher at respectively 1.49 vs. 1.28 NLB and 1.32 vs. 1.19 NLW when compared to extensively managed flocks. In contrast, ERA was slightly lower at 0.894 vs. 0.932 in intensively managed flocks. Estimates of heritability in intensively manage flocks were 0.028 for NLB, 0.016 for NLW and 0.002 for ERA. Corresponding values in extensively managed flocks were respectively 0.066, 0.040 and 0.008. Genetic correlations of NLB with NLW and of NLW with ERA were positive, while genetic correlations of NLB with ERA were not significant. Genetic correlations between performance in intensive and extensive flocks, as derived from pedigree information, amounted to 0.999 for NLB, 0.840 for NLW and 0.595 for ERA. However, large SEs for the latter two correlations made it impossible to make firm recommendations. Further research is indicated.

## INTRODUCTION

The South African Dohne Merino was developed from a cross between the Merino and the then German Merino, presently known as the South African Mutton Merino (Van Wyk *et al.* 2008). The Dohne Merino contributed approximately 28% of the weaning weight records to the National Small Stock database during 2010 to 2011 and has shown steady growth over the decade from 2003 to 2012 (Cloete *et al.* 2014). The breed has also been exported to other sheep producing countries, including Australia (Li *et al.* 2013). The genetics of yearling live weight and wool traits in Dohnes were studied by Van Wyk *et al.* (2008). However, the only account of genetic parameters for reproduction traits in the breed involved reproduction totaled over a number of lambing opportunities (Oliver and Cloete 2011).

The Dohne Merino breed is known to be farmed with under widely different conditions, ranging from extensive to very intensive (Jordaan 2013). The enhanced mating systems of intensive management complicate deriving genetic parameters for reproductive performance, since the observed phenotype could also depend on the environment at mating. Also, an alternative to the composite trait selection for reproduction by selecting for number of lambs weaned per mating in South Africa (see review by Brien *et al.* 2014) has been suggested by Bunter *et al.* (2020).

We thus studied the genetics of repeated reproduction records of South African Dohne Merinos. Knowledge of the managerial practices on individual farms allowed us to allocate specific flocks to either an intensive or extensive group. In the intensive group, routine management included the synchronisation of ewes, including the administration of fertility-enhancing drugs, followed by laparoscopic insemination and lambing under controlled conditions in lambing pens. The extensive group, on the other hand, were mated naturally on rangeland without any human intervention. The ewes lambed under the same conditions, and the daily recording of new births was the only human intervention at lambing.

## MATERIALS AND METHODS

**Traits, recordings and numbers:** Data from registered Dohne Merino breeders were available to aid in the allocation of farms to the respective groups defined in the Introduction. Selection for reproduction in South African sheep at present hinges on analysis involving repeated records for number of lambs born (NLB) and number of lambs weaned (NLW) (Brien *et al.* 2014). Since many flocks do not provide complete mating lists to allow for the identification of barren ewes, these traits are expressed per ewe lambed for individual ewes available at lambing. This means that NLB equates to the definition used by Bunter *et al.* (2020) for litter size. Additional to NLB and NLW, ewe rearing ability (ERA) was constructed from the aforementioned traits, as described by Bunter *et al.* (2020). The latter trait is not considered in South African small stock selection at present. The intensive group was represented by 61,974 repeated records of 26,264 ewes lambing in 386 contemporary groups. Analyses on the extensive group used data of 14,067 repeated records of 5,181 ewes in 100 contemporary groups.

**Statistical analyses:** All data were analysed with ASReml4 (Gilmour *et al.* 2015). Contemporary group and ewe age in months were fitted to account for environmental differences among records. Ewe age was modelled as a fixed linear component as well as a cubic spline to account for random deviations from linearity. Random components included additive ewe effects as well as ewe permanent environmental effects. Initially, single-trait analyses were conducted within management groups, to define operational models for two-trait analyses to obtain genetic correlations among traits. Finally, the intensive and extensive datasets were merged to obtain genetic correlations of the expression of the traits in one environment with performance in the other environment. The pedigree file used in all analyses contained 44,145 individuals, 2,179 sires and 23,546 dams.

## RESULTS AND DISCUSSION

Overall, means for the reproduction traits were higher in the intensively managed flocks than in the extensive flocks (Table 1). The exception was ERA which was somewhat better in the extensive grouping, but at a substantially lower birth rate. Derived coefficients of variation ranged from 24.5% for ERA in extensive flocks to 47.4% for NLW in intensive flocks. These estimates are broadly consistent with comparable figures reported in the extensive review by Safari *et al.* (2005) and by Cloete *et al.* (2017).

**Table 1. Descriptive statistics, phenotypic variances ($\sigma^2_P$) and single-trait heritability ($h^2$) and repeatability (t) estimates for number of lambs born per ewe lambed (NLB), number of lambs weaned per ewe lambed (NLW) and ewe rearing ability (ERA) in intensively and extensively managed Dohne Merino flocks**

| Group and trait | Mean ± SD | Range | $\sigma^2_P$ | $h^2$ ± SE | t ± SE |
|---|---|---|---|---|---|
| Intensive (n = 61974) | | | | | |
| NLB | 1.488 ± 0.569 | 1 – 3 | 0.293 | 0.028 ± 0.004 | 0.061 ± 0.004 |
| NLW | 1.318 ± 0.625 | 0 – 3 | 0.356 | 0.016 ± 0.004 | 0.048 ± 0.004 |
| ERA | 0.894 ± 0.272 | 0 – 1 | 0.068 | 0.002 ± 0.002 | 0.019 ± 0.004 |
| Extensive (n = 14067) | | | | | |
| NLB | 1.282 ± 0.486 | 1 – 3 | 0.201 | 0.066 ± 0.012 | 0.095 ± 0.009 |
| NLW | 1.185 ± 0.508 | 0 – 3 | 0.236 | 0.040 ± 0.010 | 0.064 ± 0.009 |
| ERA | 0.932 ± 0.229 | 0 – 1 | 0.052 | 0.008 ± 0.006 | 0.019 ± 0.009 |

NLB and NLW in intensively managed flocks increased to 48 months of age ($P < 0.05$; Figure 1). NLB subsequently remained similar but NLW declined from 60 to 84 months. Extensively managed flocks had an increased NLB and NLW up to 60 months, stabilised at 72 months and declined to 84 months. ERA increased from 24 to 36 months in extensively managed flocks and stabilised thereafter (Figure 1). In contrast, ewe age groups from 36 to 60 months had the best ERA in intensively managed flocks, with lower ($P < 0.05$) figures for 24-month-old ewes and those aged 72 months and older. Trends for NLB and NLW are generally similar with previous results for Dohne Merinos reported by Fourie and Heydenrych (1983).

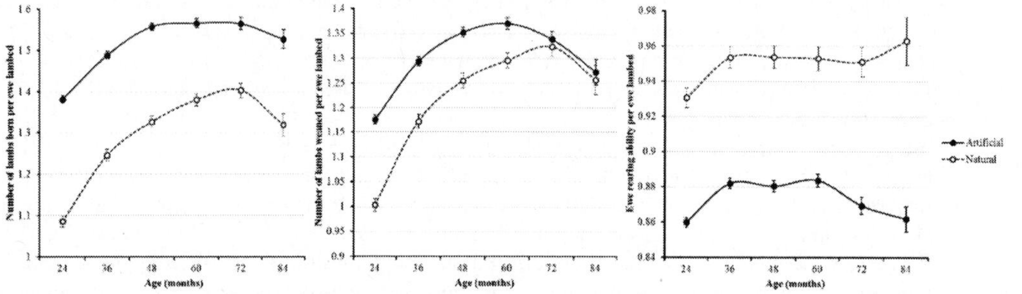


**Figure 1. The effect of ewe age in months on the reproduction traits studied**

Derived single-trait $h^2$ and repeatability estimates for NLB and NLW were significant in both management groups, although estimates in intensively managed flocks were somewhat lower ($P < 0.05$; Table 1). ERA was lowly heritable with a low repeatability estimate of around 0.02 in both groups. Recent results from Australian Merino industry (Bunter *et al.* 2020) and resource flocks (Dominik & Swan 2016; Cloete *et al.* 2017) supported a low $h^2$ for reproduction traits with estimates generally below 0.10 for NLB and NLW, and below 0.05 for ERA.

Phenotypic variance components from two-trait analyses in both intensively and extensively managed flocks in Table 2 were very close to single-trait values in Table 1. Estimates of $h^2$ were, however, slightly higher in intensive flocks than those in Table 1. In contrast, $h^2$ estimates of extensive flocks were only about half those in Table 1 for NLB and NLW ($P < 0.05$). Duplicate $h^2$-estimates from different trait combinations were all within 0.001 from each other. Two-trait repeatability estimates were very close to the corresponding single-trait estimates, suggesting different repartitioning between $h^2$ and the animal permanent environment between management systems. Reasons for this result are not evident. In intensive flocks, genetic correlations were positive between NLB and NLW, negative in direction but not different from zero between NLB and ERA and positive between NLW and ERA (Table 2). The latter traits were uncorrelated in extensive flocks. Phenotypic correlations between NLB and NLW were positive in direction but smaller in magnitude than the corresponding genetic correlations. The negative phenotypic relationship between ERA and NLB (Table 2) reflected higher mortality rates of multiples, but the magnitude was low and in close agreement with previous estimates (Bunter *et al.* 2019; 2020).

Genetic correlations between the same trait expressed in either the intensive or extensive environment amounted to $0.999 \pm 0.314$ for NLB, $0.840 \pm 0.459$ for NLW and $0.575 \pm 0.761$ for ERA. The magnitude of these values suggested that NLB and NLW do not have to be considered as different traits. However, the SEs associated with these traits were high, and only the genetic correlation of unity for NLB reached a level of twice the corresponding SE. Genetic variances of small magnitude together with a low level of genetic connectedness between extensive or intensively managed flocks could contribute to the poor accuracy of these estimates. It is common for top ranking sires only to be used for laparoscopic artificial insemination, thus confounding management groups by sire families. Genomic information could provide an alternative platform to link distant families in future studies, which could improve the accuracy of these estimates.

**Table 2. Two-trait phenotypic variances ($\sigma^2_p$), heritability ($h^2$), repeatability (t) and correlation estimates for number of lambs born per ewe lambed (NLB), number of lambs weaned per ewe lambed (NLW) and ewe rearing ability (ERA)**

| Group and trait | Trait | | |
|---|---|---|---|
| | NLB | NLW | ERA |
| Intensive | | | |
| $\sigma^2_p$ | 0.2932 | 0.3562 | 0.0683 |
| NLB* | **0.033 ± 0.005** | 0.741 ± 0.002 | -0.078 ± 0.004 |
| NLW* | 0.865 ± 0.040 | **0.031 ± 0.005** | 0.560 ± 0.003 |
| ERA* | -0.147 ± 0.140 | 0.364 ± 0.116 | **0.018 ± 0.005** |
| T | 0.061 ± 0.004 | 0.045 ± 0.004 | 0.020 ± 0.004 |
| Extensive | | | |
| $\sigma^2_p$ | 0.2015 | 0.2362 | 0.0515 |
| NLB* | **0.030 ± 0.012** | 0.772 ± 0.004 | -0.092 ± 0.009 |
| NLW* | 0.995 ± 0.110 | **0.017 ± 0.011** | 0.519 ± 0.006 |
| ERA* | -0.196 ± 0.435 | -0.019 ± 0.611 | **0.012 ± 0.011** |
| T | 0.095 ± 0.009 | 0.057 ± 0.008 | 0.020 ± 0.009 |

* Heritability estimates in bold on the diagonal, genetic correlations below and phenotypic correlations above the diagonal

## CONCLUSIONS

This study indicated that reproductive traits in South Africa Dohne Merinos are lowly heritable, estimates for ERA not reaching significance in all instances. Repartitioning variances to $h^2$ and animal PE in two-trait analyses stood to reason in intensive flocks but the reason for the lower two-trait $h^2$ estimates for NLB and NLW in extensive flocks is difficult to explain. Genetic correlations for the same trait in the two environments were high in absolute values. However, interpretation was complicated by large SEs, suggesting the need for further research.

## REFERENCES

Brien F.D., Cloete S.W.P., Fogarty N.M., Greeff J.C., *et al.* (2014) *Anim. Prod. Sci.* **54**: 667.

Bunter K.L., Swan A.A., Gurman P.M., Boerner V., McMillan A.J. and Brown D.J. (2019) *Proc. Assoc. Advmt. Anim. Breed. Gen.* **23**: 560.

Bunter K.L., Swan A.A., Gurman P.M. and Brown D.J. (2020) *Anim. Prod. Sci.* **61**: 333.

Cloete S.W.P., Olivier J.J., Sandenbergh L. and Snyman M.A. (2014) *S. Afr. J. Anim. Sci.* **44**: 308.

Cloete S.W.P., van Wyk J.B. and Olivier J.J. (2017) *Proc. Assoc. Advmt. Anim. Breed. Gen.* **22**: 569.

Dominik S. and Swan A. A. (2016) *Anim. Prod. Sci.* **58**: 207.

Fourie A.J. and Heydenrych H.J. (1983) *S. Afr. J. Anim. Sci.* **3**: 164.

Gilmour A.R., Gogel B.J., Cullis B.R., Welham S.J. and Thompson R. (2016) ASREML - User Guide Release 4.1. VSN International Ltd, Hemel Hempstead, HP11ES, UK.

Jordaan W. (2013) M.Sc. Agric. thesis, Stellenbosch University, Stellenbosch, South Africa.

Li L., Brown D.J. and Gill J.S. (2013) *Proc. Assoc. Advmt. Anim. Breed. Gen.* **20**: 241.

Olivier J.J. and Cloete S.W.P. (2011) *Proc. Assoc. Advmt. Anim. Breed. Gen.* **19**: 167.

Safari E., Fogarty N.M. and Gilmour A.R. (2005) *Livest. Prod. Sci.* **92**: 271.

Van Wyk J.B., Swanepoel J.W., Cloete S.W.P., *et al.* (2008) *S. Afr. J. Anim. Sci.* **38**: 31.

*Contributed paper*

# ADDRESSING SCUR PHENOTYPING CHALLENGES WITH THE SOUTHERN MULTI-BREED PROJECT

**N.K. Connors[1], B.J. Walmsley[1] and K.A. Donoghue[2]**

[1] Animal Genetics and Breeding Unit*, University of New England, NSW 2350, Australia
[2] NSW Department of Primary Industries, Agricultural Research Centre, Trangie, NSW 2823, Australia

## SUMMARY

The genetic basis of polled or horned phenotypes in beef cattle is now well documented, however horned animals will continue to be born in the national herd for some time. Animal welfare will continue to be compromised due to the need to dehorn animals with horn buds. While scurs don't necessarily require removal, the inability to distinguish between horned or scurred animals at the age of dehorning mean they are dehorned nonetheless. Targeted breeding of polled herds in industry is increasing with genetic poll tests available, but without understanding the genetic basis of scurs, horn buds and thus dehorning practices will remain. The difficulty in identifying the genetic basis of scurs remains the lack of a reference population with accurate phenotypes, driven largely by the difficulty in phenotyping horns and scurs at usual dehorning age. This paper describes the challenges and preliminary results of a phenotyping project using the Southern Multibreed project herd, which will form a reference population with poll/horn/scur phenotypes, accompanied by full pedigree recording and genomics data.

## INTRODUCTION

Carcase bruising from horns is estimated to cost the Australian meat industry $30 million per year (CSIRO 2014). These economic losses promote the accepted management practice of horn bud removal (i.e. dehorning) at an early age (less than 6 months) (Medugorac *et al.* 2012). However, dehorning can still lead to economic losses due to wound healing, translating into short term weight loss and increased mortality rates (Prayaga 2007; Bunter *et al.* 2013). Despite being commercially necessary, dehorning procedures are painful. The increasing importance of social license in agriculture and the impact dehorning has on animal welfare may see less acceptance of these practices in the future (Williams and Page 2014). The alternative to dehorning horned cattle is to breed polled cattle.

Horns in cattle form as a free-floating bud, which later fuses to the skull as a fixed bony extension, while scurs appear as small and only loosely attached horns, and polled cattle are naturally hornless (Seichter *et al.* 2012). While polled breeding is now targeted in industry, scurred animals will remain in the population until scur genetics are understood, requiring continued dehorning practices due to the inability to distinguish between horned or scurred cattle at the time of dehorning.

The genetic basis of polled cattle is now largely accepted as an autosomal dominant trait (Mariasegaram *et al.* 2010; Medugorac *et al.* 2012; Seichter *et al.* 2012; Rothammer *et al.* 2014; Utsunomiya *et al.* 2019) with two alleles (Celtic $P_C$, and Friesian $P_F$) forming the basis for genetic testing enabling direct selection for polledness. The genetic basis for horns is the absence of any polled alleles, while the genetic basis of scurs appears to be more complex. The inheritance model initially proposed was that scurs is a sex-influenced trait characterized by two alleles, Sc (scurs)

* AGBU is a joint venture of NSW Department of Primary Industries and University of New England

and sc (no scurs), where one Sc in males results in scurs, while females require two Sc alleles to be scurred (White and Ibsen 1936). Due to growth in the same head position, horns mask the expression of scurs; it is proposed that homozygous polled alleles also inhibit scur growth, unless the animal also possesses homozygous scur alleles (Sc/Sc) (Long and Gregory 1978). More recent studies have offered varied and contradicting conclusions on scur genetics, including as autosomal recessive and not sex-influenced (Capitan *et al.* 2009), mapped to bovine chromosome 19 (Asai *et al.* 2004), autosomal dominant inheritance mapped to chromosome 4 (Capitan *et al.* 2011), and oligogenic and age-dependent penetrance (Gehrke *et al.* 2020). Additionally, studies have shown the diversity in scur phenotypes, with small scabs and scaly patches, to tiny loose buds, to long pendulous loose horns, along with differences in head shape and bumps (Capitan *et al.* 2011; Gehrke *et al.* 2020).

One dimension of the complexity of scur studies lies in the difficulty of obtaining accurate phenotypes, given the variation due to age and breed, and the need to adjust management to enable recording prior to dehorning to avoid misclassification. Furthermore, the number of scurred animals in the population is relatively unclear for similar reasons. Observations from the Australian microsatellite haplotype poll test showed scur phenotype rates of approximately 5% (Connors *et al.* 2018); however phenotypes were industry supplied and subject to bias (Connors *et al.* 2019). Most studies on scurs have been performed in the absence of poll genetic testing, which can clarify an animal's horn genetics in addition to its phenotype. Additionally, obtaining large numbers of phenotypes has been performed within dairy breeds rather than beef breeds, likely due to the differences in management practices. Here, preliminary findings of a beef cattle poll/horn/scur phenotyping study using the Southern Multibreed Project (Walmsley *et al.* 2021) as a reference population are presented. While this study is preliminary, this paper will set the scene for the establishment of a reference dataset of multibreed populations with full pedigree recording, poll genetic testing, SNP genotypes, and phenotypes at marking and when older, for generations.

## MATERIALS AND METHODS

The Southern Multibreed Project has cattle populations at five sites across NSW, including Trangie Agricultural Research Centre (TARC), Grafton Primary Industries Institute, Glen Innes Agricultural Research & Advisory Station, Elizabeth Macarthur Agricultural Institute, and Tocal Agricultural Centre. Each site has varied cattle populations in number and breed, which includes Brahman, Charolais, Shorthorn, Angus, Hereford, and Wagyu, and some Brahman-Hereford and Brahman-Angus F1 crossbreds. Calves at each site were marked at age 8-12 weeks, along with horned/polled/scurred phenotypes, sex, and breed recorded. Phenotypes were classified as the following:

- (i) smooth poll cone = no buds, smooth poll, pointed cone shape of skull;
- (ii) poll broad cone = no buds, smooth poll, broad cone shape of skull;
- (iii) poll = no buds, smooth poll, flatter shape of skull (i.e. no cone);
- (iv) poll frontal bumps = no buds, poll, bumps felt on skull;
- (v) buds = small keratin buds present (diameter measured);
- (vi) horns = small keratin horns present, >3cm length (diameter measured);
- (vii) scurs = scaly, scabby patches, no keratin (diameter measured).

Each animal was photographed and the skull was palpated for head shape and to ensure buds were felt beneath longer coats; horns were already of considerable length such that they are seen through the coat. Buds, horns and scurs were measured in diameter at the skull. Data was recorded manually on-site, entered into Excel afterwards, and cross-checked with project records.

## RESULTS AND DISCUSSION

The number of animals phenotyped in this study was 1309, consisting of 646 male and 663 female calves. The phenotypes recorded are summarised in Table 1, and examples of the phenotypes are pictured in Figure 1. Simply, the number of polled animals was 985, buds/horned was 298, and 26 scurs. The majority of the scurred phenotypes were observed in crossbred calves, which may indicate an effect of the indicus lines on scur development. All but four Wagyu calves had buds or horns, and all Angus calves were polled. Differences in horn growth between sexes is difficult to determine at this early stage, though of note are the number of horned calves compared to budded calves for Wagyu and Hereford, with more males phenotyped as horned (i.e. longer growth) suggesting males grow horns quicker than females, and agreeing with current anecdotes of the horn growth-promoting effects of testosterone. The number of scurs in males and females appears to be similar, showing no agreement with current theory of sex-influenced genetics. It is important to note possible scabby patches phenotyped as scurs may have grown into buds over time, testament of the difficulty in phenotyping. Further observation of these calves will be performed at weaning to determine any late scur/bud growth in poll phenotyped animals. With poll genomic testing now common, testing calves at birth may enable informed disbudding choices, by only disbudding genetically horned calves, which would leave scur phenotypes to grow over time.

**Table 1. Summary of phenotypes for each breed and sex**

| | | Phenotypes | | | | | | | |
|---|---|---|---|---|---|---|---|---|---|
| Breed | Sex | Smooth Poll Cone | Poll Broad Cone | Poll | Poll Frontal Bumps | Buds | Horns | Scurs | Total |
| Angus | F | 201 | | 6 | 1 | | | | 392 |
| | M | 173 | 9 | 2 | | | | | |
| Brahman | F | 6 | | 1 | | 4 | 8 | 2 | 41 |
| | M | 4 | | 3 | | 7 | 3 | 3 | |
| Charolais | F | 26 | 14 | 2 | 2 | 12 | 1 | | 115 |
| | M | 9 | 19 | 10 | 5 | 15 | | | |
| Hereford | F | 109 | 9 | 12 | 1 | 15 | 3 | 1 | 299 |
| | M | 68 | 11 | 22 | 18 | 12 | 14 | 4 | |
| Shorthorn | F | 82 | 4 | 1 | 2 | 2 | | | 182 |
| | M | 62 | 13 | 4 | 4 | 5 | 3 | | |
| Wagyu | F | 2 | | | 2 | 83 | 5 | | 184 |
| | M | | | | | 61 | 31 | | |
| Cross | F | 23 | | 1 | 4 | 3 | 4 | 9 | 96 |
| | M | 21 | | 7 | 10 | 7 | | 7 | |
| Total | | 786 | 79 | 71 | 49 | 226 | 72 | 26 | 1309 |

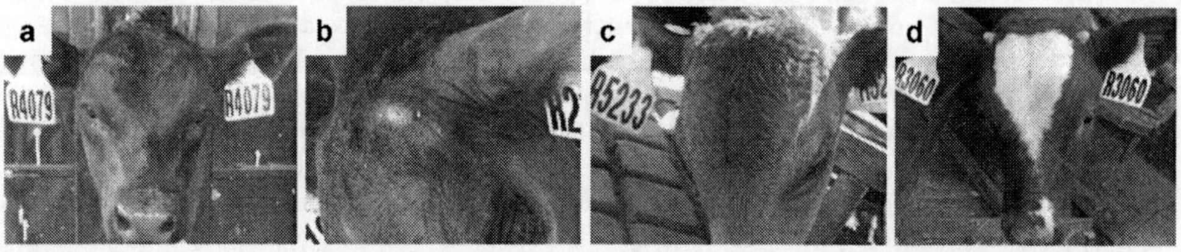


**Figure 1. Phenotype examples. (a) smooth poll cone, (b) scur, (c) buds, and (d) horns**

## CONCLUSIONS

This study describes the preliminary findings of phenotype collection for potential scur research. The population had good distribution of phenotypes across breed and sex. Genotypes on calves and parents, including poll genotype, will be available later in the year where concordance of genotypes and phenotype can be analysed. Use of the Southern Multibreed project as a reference population will provide generations of phenotype, genotype, and pedigree across multiple breeds, for further analysis of the complex scurs trait.

## ACKNOWLEDGEMENTS

The Southern Multibreed project is an initiative co-funded by NSW Department of Primary Industries, University of New England and Meat and Livestock Australia Donor Company. The authors give thanks to the efforts of the research station managers and their staff, our skilled team of technicians, ultrasound scanners, AI, DNA lab, data, project managers and scientists. Individual breeders, participating breed societies and the southern beef industry are thanked for their support.

## REFERENCES

Asai M., Berryere T.G., Schmutz S.M. (2004) *Anim Genet* **35:** 34.

Bunter K.L., Johnston D.J., Wolcott M.L., Fordyce G. (2013) *Animal Production Science* **54:** 25.

Capitan A., Grohs C., Gautier M., Eggen A. (2009) *Bmc Genetics* **10:**

Capitan A., Grohs C., Weiss B., Rossignol M.N., Reversé P., Eggen A. (2011) *PLoS One* **6:** e22242.

Connors N.K., Banks R.G., Johnston D.J. (2019) *Proceedings of the 23rd Conference of the Association for the Advancement of Animal Breeding and Genetics* 330.

Connors N.K., Tier B., Johnston D.J. (2018) Current status of Australia's diagnostic poll haplotype test. In 'Proc. World Congr. Genet. App. Livest. Breed. Volume 11 pp. 344.

CSIRO (2014) Annual Report 2013-2014. New Millenium Print. Available at www.csiro.au/annualreport2014.

Gehrke L.J., Capitan A., Scheper C., König S., Upadhyay M., Heidrich K., Russ I., Seichter D., Tetens J., Medugorac I., Thaller G. (2020) *Genetics Selection Evolution* **52:** 6.

Long C.R., Gregory K.E. (1978) *Journal of Heredity* **69:** 395.

Mariasegaram M., Reverter A., Barris W., Lehnert S.A., Dalrymple B., Prayaga K. (2010) *Bmc Genomics* **11:**

Medugorac I., Seichter D., Graf A., Russ I., Blum H., Göpel K.H., Rothammer S., Förster M., Krebs S. (2012) *PLoS One,* **7:** e39477.

Prayaga K.C. (2007) *Australian Journal of Agricultural Research* **58:** 1.

Rothammer S., Capitan A., Mullaart E., Seichter D., Russ I., Medugorac I. (2014) *Genetics, selection, evolution : GSE* **46:** 44.

Seichter D., Russ I., Rothammer S., Eder J., Forster M., Medugorac I. (2012) *Animal Genetics* **43:** 595.

Utsunomiya Y.T., Torrecilha R.B.P., Milanesi M., Paulan S.C., Utsunomiya A.T.H., Garcia J.F. (2019) *Anim Genet* **50:** 187.

Walmsley B.J., Donoghue K.A., Johnston D.J., Clark S.A., Siddell J.P., Walkom S.F., Granleese T., Arthur P.F. (2021) *Proc. Assoc. Advmt. Anim. Breed. Genet. 24: These Proceedings.*

White W.T., Ibsen H.L. (1936) *J Genet.* **32:** 33.

Williams S., Page S. (2014) B.AWW.0189 Final Report: Literature review of pain and welfare impacts associated with on-farm cattle husbandry procedures. Meat & Livestock Australia Limited, North Sydney, NSW.

*Contributed paper*

# ASSESSING THE RELATIONSHIP BETWEEN BEEF CATTLE FERTILITY TRAITS AND NOVEL ENVIRONMENTAL DESCRIPTORS IN NORTHERN AUSTRALIA

**J.P. Copley[1], B.N. Engle[1], S. Speight[1], G. Fordyce [1], E.M. Ross [1], K.P. Voss-Fels[1] and B.J. Hayes[1]**

[1] Queensland Alliance for Agriculture and Food Innovation (QAAFI), the University of Queensland, St Lucia, QLD, 4069 Australia

## SUMMARY

Fertility is a major challenge for beef producers in the harsh and diverse environmental conditions of Northern Australia. Two of the most significant environmental challenges facing breeding females are low and variable nutrition and exposure to severe heat stress. This paper aimed to define novel environmental descriptors that can be used to account for these effects when modelling fertility traits. Nutrition descriptors were based on rainfall records and average daily liveweight gain (ADWG). Heat stress descriptors were based on daily values for a temperature humidity index (THI). Three fertility phenotypes were collected as part of the Northern Genomics project; puberty (CL600), first pregnancy (PD1) and second pregnancy (PD2). The aims of this study were to examine the relationship between fertility traits and continuous environmental descriptors and confirm the importance of puberty attainment to subsequent reproductive success. Animals that were pubertal at CL600 were found to have increased odds of success in subsequent pregnancy recording. Environmental descriptors based on THI successfully defined an environmental gradient as the cumulative exposure to chronic heat stress. ADWG successfully modelled nutritional availability for PD1 and PD2 but descriptors based upon total rainfall were not successful.

## INTRODUCTION

The Northern Australia beef industry contains 60% of Australia's national herd and is defined by diverse and challenging environmental conditions (McLean *et al.* 2014). These conditions negatively impact female fertility, an important driver of profitability. The environment represents more than location or a contemporary group, it is the sum-total of the temperature, rainfall, pasture availability and other factors to which animals are exposed. The environmental descriptors examined in this study were designed to account for environmental variation as a continuous variable based upon available weather information and animal weight gain performance. Also examined was the role of early attainment of puberty on subsequent pregnancy. Overall herd productivity is enhanced by large numbers of maiden heifers becoming pubertal, conceiving and subsequently calving early in the breeding season, giving maximum opportunity for lifetime reproductive success. The objectives of this study were to assess the relationship between fertility traits and novel, continuous environmental descriptors measured in Northern Australia.

## MATERIALS AND METHODS

**Phenotypes.** The project utilized heifer data (n = 24,768) from 54 commercial herds from across Northern Australia, collected as part of the ongoing Northern Genomics project. The animals represented a diverse range of breeds including: Brahman, Angus, Belmont Red, Charolais, Droughtmaster, Shorthorn, Limousin, Santa Gertrudis, Boran and Wagyu (Hayes *et al.* 2019). Heifer reproductive maturity (CL600) was measured when approximately 50% of the contemporary herd was sexually mature, using a one-time ovarian scan via ultrasound to detect the presence of a *Corpus Luteum* (CL; n = 25,176), following the procedures outlined by Hayes *et al.* (2019). This trait was nominally measured at approximately 600 d of age. Heifers which displayed a CL or were pregnant, were deemed 'pubertal', the rest 'non-pubertal'. Heifer pregnancy status was measured as foetal age in weeks (n = 20,989) at approximately 2.5 yrs of age, following the heifer's first breeding season, and the subsequent pregnancy (n = 10,072) at approximately 3.5 yrs of age (PD1 = first pregnancy and PD2 = second pregnancy). Animals were also recorded for liveweight (Wt), hip height (HH), body condition score (BCS) and a tail hair sample taken.

**Environmental Descriptors.** Weather data for each collaborating property was obtained using the NASAPOWER package in R 3.5.3 (Sparks 2018). NASAPOWER data is a publicly available global climatology database with a 0.5° by 0.5° arc of longitude and latitude (Sparks 2018). Based upon the coordinates provided for each collaborator, daily observations of rainfall, temperature and relative humidity data were downloaded for the 20 years preceding the date of trait recording.

The temperature (T) and relative humidity (RH) was used to calculate a daily temperature humidity index (THI) via using the formula from Wijffels *et al.* (2013).

$$THI = 0.8 * T + ((RH * 0.01) * (T - 14.4)) + 46.4$$

To assess the impact of the severity of the environment in which the heifers were exposed, the number of days where THI was equal to or exceeded different thresholds (65-79) in the 6 months prior to trait recording (CL600) or conception date (PD1) was assessed. A THI value of 79 was used as it is considered to be the threshold of severe heat stress (Moran 2005; McGowan *et al.* 2014). Chronic heat stress was also modelled by calculating the area under the curve (AUC) of daily THI in the 120 d surrounding (60 d prior and 60 d post) trait recording/conception date.

Rainfall descriptors were based on the daily precipitation records. Three separate descriptors were calculated based on key dates in the breeding cycle: conception date (PD1) and trait recording date (CL600). Total rainfall (mm) in the 365 d prior to trait recording, total rainfall (mm) in the 120 d prior to trait recording and the standard deviation of rainfall in the 120 d prior to trait recording compared to the 20 yr average of the same period and location.

ADWG was calculated as the average daily gain from CL600 to PD1 (kg/day). The effect of ADWG was not modelled for CL600 as no Wt data prior to CL600 measurement was available.

**Statistical Analysis.** The *B.indicus* percentage and heterozygosity was calculated using the methods outlined by Hayes *et al.* (2019). The environmental descriptors were fitted as fixed effects in a generalized linear model. The equation of each generalized linear model was:

$$Fertility\ trait \sim HH + BCS + Wt + CG + BI\% + Het + \ Environmental\ Descriptor$$

CL600, PD1and PD2 were modelled as binary traits (0 = 'non-pubertal', 1 = 'pubertal' OR 0 = 'non-pregnant', 1 = 'pregnant') using a logistic regression. Statistical analysis was conducted using the glm.db package in R (Ripley *et al.* 2013). Additional analysis to examine the effect of puberty at CL600 on pregnancy was completed using a least squares mean test via the emmeans package in R (Lenth *et al.* 2020). The relationship between CL600 score and all environmental measures was also modelled using logistic regression (0 = "non-pubertal", 1 = "pubertal") using the glm.db package in R (Ripley *et al.* 2013).

## RESULTS AND DISCUSSION

**Effect of CL600 on pregnancy.** Heifers that were pubertal at CL600 had increased log odds of pregnancy success at both PD1 (0.56) and PD2 (0.75) compared to non-pubertal heifers ($P < 0.05$). This study has reinforced the usefulness of early puberty as a heritable trait capable of being measured earlier in life and which has a positive and significant relationship to later-in-life pregnancy traits (Johnston *et al.* 2014; Corbet *et al.* 2018). Heifers that are pubertal at the commencement of joining conceive early, calve early and readily reconceive.

**Effect of environmental descriptors on CL600.** The effect of the number of days over THI 65 and 70 was significant and negative (P < 0.05. The cumulative daily THI for 120 d surrounding trait recording was significant and negative ($P < 0.05$). Each additional day of severe heat stress (THI >= 79) to which heifers were exposed was significant ($P < 0.05$) to puberty attainment but suggested a positive relationship, a results contrary to expectations. The results suggests that an increased heat load, particularly chronic heat stress in the period prior to ovarian scanning will reduce the proportion of pubertal females. The reason for the lack of a relationship between acute heat stress and diminished puberty outcome is unclear. The secretion of hormones by the structures of the reproductive tract during estrus are subject to interference under high heat stress conditions, an effect that has been accounted for by these descriptors (Wolfenson and Roth 2018). The rainfall total and deviation of total rainfall from the long-term average and 12 mo rainfall in the preceding 12 mo were both not significant to the outcome of CL600. This result suggests that the rainfall descriptors did not conclusively account for the variability of nutrition.

**Effect of environmental descriptors on PD1.** The number of days to which animals were exposed to THI thresholds of 65 and 70 in the 120 d surrounding conception date had a significant impact ($P < 0.05$) and was associated with reduced pregnancy success. Area under the curve of daily THI measurements and the number of days over 75 were both not significant. Days over 79 again had a positive effect ($P < 0.05$). The results suggest that the THI descriptors adequately model chronic heat stress and the detrimental impact it has on early pregnancy and cyclicity (Gilad *et al.* 1993). Total rainfall and deviation from average were not significant to pregnancy outcome. The effect of ADWG was significant ($P < 0.05$), with increased ADWG being associated with improved pregnancy results.

**Effect of environmental descriptors on PD2.** THI descriptors had a universally negative relationship to second pregnancy and were significant ($P < 0.05$) for; AUC, and days over 75. This was consistent with results for PD1 and CL600 in that a stronger relationship was found to chronic heat stress rather than acute or severe heat stress. The lack of a significant relationship between acute heat stress (days over 79) may be due to several factors, including adaptation and confounding seasonal effects as peak THI typically coincides with the wet season, and thus peak nutritional availability, in Northern Australia. The relationship of PD2 outcome to ADWG and rainfall echoed the results from PD1, ADWG was significant ($P < 0.05$) while rainfall was not.

## CONCLUSIONS

This study showed that heifers which were pubertal at CL600 had improved pregnancy outcomes at PD1 and PD2. This underscores the importance of breeding heifers that are early maturing. Heifers that are pubertal at the start of the joining period maximise the available time to conceive early in the joining period which in turn increase the available time to re-conceive.

THI-based descriptors to measure chronic heat stress had significant relationships with puberty attainment and heifer pregnancy. ADWG was found to have a significant relationship, in the expected direction, with first and second pregnancy. However, descriptors based upon rainfall had no significant relationship. The descriptors based upon ADWG and chronic heat stress satisfied the primary objective of the study, to define an environmental gradient based on these descriptors. Further refinement of rainfall descriptors is required.

## ACKNOWLEDGEMENTS

The authors would like to thank the collaborating herds for participating in the Northern Genomics Project and the various DAF staff for their assistance. We also gratefully acknowledge the funding from The University of Queensland and the MLA Donor Company through project P.PSH.P0833.

## REFERENCES

Corbet N., Allen J., Laing A., Fordyce G., McGowan M. and Burns B. (2018) *J. Anim. Prod. Sci.* **58(9):** 1735.

Gilad E., Meidan R., Berman A., Graber Y. and Wolfenson D. (1993) *J. Repro.* **99(2)**: 315.

Hayes B., Fordyce G. and Landmark S. (2019) *Proc. Assoc. Advmt. Anim. Breed. Genet.* **23:** 282.

Johnston D., Barwick S., Fordyce G., Holroyd R., Williams P., Corbet N. and Grant T. (2014) *J. Anim. Prod. Sci.* **54(1)**: 1-15.

Lenth R., Beurkner P., Herve M, Love J, Riebl H and Singmann H. (2020) 'Estimated Marginal Means, aka Least-Squares Means'. R package version 1.3.2.

McGowan M., McCosker K., Fordyce G., Smith D., O'Rourke P., Perkins N., Barnes T., Marquet L., Morton J. and Newsome T. (2014) 'North Australian beef fertility project: Cash Cow. Final Report'. Meat and Livestock Australia, Sydney.

McLean I., Holmes P. and Counsell D. (2014) 'The Northern beef report: 2013 Northern beef situation analysis'. Meat and Livestock Australia, Sydney.

Moran J. (2005) 'Tropical dairy farming: feeding management for small holder dairy farmers in the humid tropics'. Csiro publishing.

Ripley B., Venables B., Bates D. M., Hornik K., Gebhardt A., Firth D. and Ripley M.B. (2013) *J. Cran. R.* **538:** 113.

Sparks, A. H. (2018) 'nasapower: a NASA POWER global meteorology, surface solar energy and climatology data client for R'.

Wijffels G., Gaughan J., Revell D. and Allingham P. (2013) 'Heat Stress Nutrition: A comprehensive review and R&D program'. Meat and Livestock Australia, Sydney.

Wolfenson D. and Roth Z. (2018) *Anim. Front.* **9(1):** 32.

*Contributed paper*

# GENOTYPING DAIRY CATTLE WITH SKIM-WHOLE-GENOME SEQUENCING AND IMPUTATION

**H.D. Daetwyler[1,2], J. Li[3], C.J. Vander Jagt[1], I.M. MacLeod[1], J. Pickrell[3], M. Vasquez[3], J.Hoff[3] and A.J. Chamberlain[1]**

[1] Agriculture Victoria, AgriBio, Centre for AgriBioscience, Bundoora, VIC, 3083 Australia
[2] School of Applied Systems Biology, La Trobe University, Bundoora, VIC, 3083 Australia
[3] Gencove, New York, NY, 10016, USA

## SUMMARY

Cost effective genotyping tools are essential for wide-spread use of genomics in research and industry. While the majority of large-scale industry implementations of genomic selection have relied on single nucleotide polymorphism (SNP) arrays, genotyping using skim-whole-genome sequencing (SWGS) is becoming more accurate and, due to large reductions in sequencing cost, SWGS genotyping is becoming price competitive with SNP arrays. In SWGS genotyping, a sample is sequenced to 0.5 or 1x read depth and imputed to full WGS with a reference population sequenced at higher read depth (e.g. 1000 Bull Genomes Project). Imputation software, such as Beagle, can directly impute SNPs from SWGS to high fold coverage WGS, but they were not designed to do so. Gencove has developed an imputation algorithm especially for this task, loimpute. We compared the genotyping and imputation accuracy of Beagle4.0 and loimpute in a sample of 31 Holstein, 55 Jersey, and 39 Jersey-Holstein crosses. Animals were sequenced to approximately 10-fold coverage and variants and genotypes were identified as part of 1000 Bull Genomes Run8. Each animal's sequence was then randomly down-sampled to 0.5 and 1-fold coverage, aligned to the reference assembly, and imputed either with Beagle4.0 or with loimpute. Imputed genotypes were compared to observed full-sequence genotypes via correlation and proportion correct (concordance). The mean per marker genotype correlation of the 16 million imputed SNP across all breeds was 0.78 (0.5x) and 0.84 (1x) for Beagle and 0.92 (0.5x) and 0.93 (1x) for loimpute. While the Beagle pipeline could be likely further improved, the results demonstrate that a purpose-built imputation method is required to perform accurate SWGS genotyping. The method is attractive as it can provide sequence density genotypes at a cost price point comparable to low or medium-density SNP arrays.

## INTRODUCTION

The large scale implementation of genomic breeding approaches in industry (e.g. genomic selection) requires genotyping tools that are accurate and cheap. The lower the cost of genotyping, the more widespread the adoption of genomic selection. Therefore, the continued development and refinement of genotyping methods is crucial to realising genetic gain from genomics.

Whole-genome sequencing has always underpinned genotyping platform development through the discovery of genetic marker diversity, such as single nucleotide polymorphisms (SNP), from which a subset of markers can be chosen for routine genotyping. Whole-genome sequencing requires the preparation of a library that cuts DNA into segments (i.e. sequence reads) and attaches a barcode to each segment. Once barcoded, samples can be mixed and sequenced together and the data for each sample can be separated afterwards. This multi-plexing approach coupled with vastly increased sequence output of recent technologies are the primary reasons for the large reduction in sequencing costs over time. The amount of sequencing per position of the genome is called read depth (e.g. read depths of 8 to 20x are common in livestock populations).

The most widely used genotyping method in large livestock populations have been SNP chips, which are microarrays that can provide genotypes on a few to many thousands of SNP. SNP chips are generally highly accurate, amenable to high-throughput methods, and deliver near complete data at the loci queried. Low to medium density SNP chips with approximately <10,000 and 50,000 markers are currently available at prices that warrant wide-spread use when compared to impact on farm profitability (e.g. Newton *et al.* 2018). Nevertheless, decreasing genotyping costs further would no doubt increase the use of genomic selection.

Another way to genotype individuals is through genome sequencing directly. The reduced cost of sequencing now makes routine genotyping with whole-genome sequence feasible when sequence depth per sample is kept to 1x read depth or less, so-called skim whole-genome sequencing (SWGS). Due to the low read depth, there are relatively few loci with enough reads to call genotypes accurately and the set of loci called differs for each individual in a population. SWGS could be improved by imputing missing genotypes and improving genotype accuracy of loci with insufficient reads. Several imputation programs are available, such as Beagle, Minimac3, and FImpute, but most have not been developed specifically for imputing SWGS. Gencove have developed an imputation algorithm (1oimpute) for SWGS adapting an methods by Li and Stephens (2003) to routinely impute SWGS genotype data.

Here we present a comparison of SWGS genotyping using 1oimpute and Beagle4.0 imputation in three dairy cattle breed groups, Holstein, Jersey and Holstein-Jersey crossbreds, sequenced at 0.5 and 1x read depth.

## MATERIALS AND METHODS

**Whole-genome sequencing and processing.** Thirty-one Holstein, 55 Jersey, and 39 Holstein-Jersey crossbred bulls were whole-genome sequenced to an average depth of 10x. Raw sequence fastq data were provided to Gencove and each animal's sequences were downsampled to 0.5 and 1x read depth. Full, 0.5 and 1x sequences were quality controlled and aligned with BWA to the ARS-UCD-1.2 reference assembly (Rosen *et al.* 2020) to produce binary alignment (bam) files. Full sequences were included in Run8 of the 1000 Bull Genomes Project (Hayes & Daetwyler 2019) and processed as described in Daetwyler *et al.* (2017).

**Genotype calling and imputation.** Two parallel pipelines were implemented by Gencove and Agriculture Victoria (AgVic) for a total of four scenarios: Gencove 1oimpute 0.5 and 1x read depth and AgVic Beagle at 0.5 and 1x read depth.

Gencove used their imputation software 1oimpute, which implements the Li and Stephens model for a set of reads in each animal's bam file and a known set of phased variants in a reference panel (Li & Stephens 2003). The diploid genotype probabilites are estimated using a Hidden Markov Model (Wasik *et al.* 2019). Gencove used a multi-breed reference panel of 946 animals (including 184 Holstein and 15 Jersey) for each breed (Snelling *et al.* 2020). AgVic performed variant calling on SWGS bam files using GATK3.8 according to the 1000 Bull Genomes Project guidelines. The 1000 Bull Genomes Project Run8 multi-breed taurus dataset with 4109 animals (including 1200 Holstein and 120 Jersey) was used as the AgVic reference for imputation. Random missing genotypes in the reference set were imputed with Beagle4.0 (Browning & Browning 2009) and filtered to only include biallelic SNP whose alleles occur at least 4 times. SWGS genotypes were then imputed with Beagle4.0 utilising genotype probabilities (Browning *et al.* 2018), and imputed animals were removed from reference sets.

**Imputation accuracy evaluation.** The accuracy of imputation was calculated as the Pearson correlation and concordance of imputed SWGS genotypes (coded as 0, 1, 2) from each respective pipeline and raw full sequence genotypes from the 1000 Bull Genomes Run8. Concordance was calculated as the proportion of imputed genotypes matching full sequence genotypes. Further, these statistics were summarised in minor allele frequency (MAF) bins of 0.0-0.03, 0.03-0.06, 0.06-0.1, 0.1-0.2, 0.2-0.3, 0.3-0.4, 0.4-0.5. Comparisons were restricted to the set of SNP imputed by both 1oimpute and Beagle5.1 and passing the GATK quality tranche threshold of 99.9.

**RESULTS AND DISCUSSION**

The SWGS process led to approximately 1.6 million SNP. This is substantially more than a 50,000 marker SNP chip, but the SWGS SNP would be called with lower accuracy. The number of SNP imputed across all bovine autosomes by both loimpute and Beagle was 16,488,621 and the set of overlapping loci between the two pipelines were >95%.

**Table 1. Mean correlation and concordance per SNP of imputed and observed genotypes in Holstein (HOL), Jersey (JER) and Holstein-Jersey crossbreds (HOLJER) from loimpute (G) and Beagle (B) pipelines.**

| Read Depth | 0.5x Read Depth | | | | | | 1x Read Depth | | | | | |
|---|---|---|---|---|---|---|---|---|---|---|---|---|
| Breed | HOL | | JER | | HOLJER | | HOL | | JER | | HOLJER | |
| Method | G | B | G | B | G | B | G | B | G | B | G | B |
| Correlation | 0.95 | 0.79 | 0.90 | 0.78 | 0.90 | 0.76 | 0.95 | 0.84 | 0.91 | 0.84 | 0.92 | 0.83 |
| Concordance | 0.98 | 0.88 | 0.96 | 0.89 | 0.96 | 0.87 | 0.98 | 0.91 | 0.97 | 0.92 | 0.96 | 0.91 |

SD across autosomes ~0.01

The loimpute pipeline achieved substantially higher mean correlations between imputed and observed genotypes across all 16 million SNP tested, with a difference of ~0.15 (Table 1). This trend was also observed when using concordance as the evaluation measure, though the advantage of loimpute over Beagle was slightly less at ~0.1 (Table 1). This is quite a marked improvement that would surely result in improved downstream analyses. Imputation performance was quite similar across the three breeds for both piplelines. Interestingly, loimpute managed to still outperform Beagle even though Beagle had approximately 7 times the number of Holstein and Jersey animals in its reference. We did also test Beagle5.1, but it performed very poorly (correlation reduced by ~0.2) as it does not utilise genotype probabilities. Slightly better imputation was observed when animals were sequenced at 1x versus 0.5x, although the difference was small, and suggests that 0.5x is likely sufficient for the loimpute algorithm. Both imputation methods provide metrics per SNP on their confidence in genotype accuracy, which can be used to filter data further.

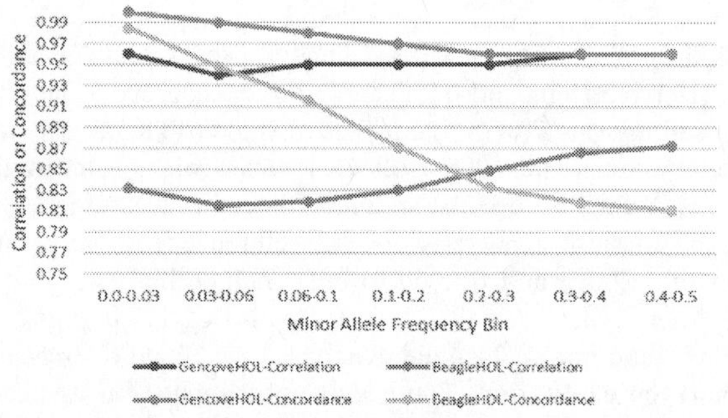


**Figure 1. Mean correlation and concordance in minor allele frequency bins for Gencove loimpute and Beagle imputation for Holstein bulls with 1x sequence read depth.**

It is well known that conventional imputation algorithm performance is substantially reduced for alleles with low frequency in the population (e.g. van Binsbergen *et al.* 2014). This was confirmed for Beagle, where the correlation between imputed and observed genotypes in Holsteins was ~0.83 for loci with MAF < 0.03 (Figure 1). The reverse was observed for Beagle concordance, which was highest for the same low MAF bin. This occurs solely because most of the time, the most likely genotype will be correct and demonstrates the weakness of concordance as a measure of imputation accuracy, especially for low MAF SNP. In contrast, loimpute imputation correlations and concordance were consistently high (~0.95) across all MAF bins. Due to the high level of accuracy achieved by loimpute, both correlations and concordance were higher than Beagle across all MAF, though concordance did reach near 1.00 for low MAF SNP, indicating a small bias in this measure also for loimpute. Correlations and concordance followed similar levels and patterns across MAF for Jersey and crosses (data not shown).

The Beagle pipeline was not built specifically for imputing SWGS data with high proportion of missing genotypes and called genotypes with high uncertainty with different SNP called for each animal. Further improvement may be possible by filtering the SWGS genotypes for loci with read depth >5x. While this would further increase the proportion missing, it would provide more certain SNP genotypes to initiate the Beagle Hidden Markov Model. However, it seems unlikely that Beagle could achieve similar performance to loimpute even with these improvements. Recently, a new SWGS imputation method (GLIMPSE) has been published (Rubinacci *et al.* 2021), which seems competitive in accuracy with loimpute and testing with this method is underway.

The loimpute pipeline produces accurate genotypes at millions of loci and seems to overcome a traditional imputation bottleneck of accurately imputing lower MAF SNP. Industry application with the specific loci currently available on most SNP chips is therefore feasible, and for research applications, it is particularly useful to have access to many accurate genotypes across the MAF spectrum.

## CONCLUSIONS

Substantially higher imputation accuracy was observed with loimpute than with Beagle. While the Beagle pipeline could be likely further improved, the results demonstrate that a purpose-built imputation method is required to perform accurate SWGS genotyping. The loimpute SWGS method is attractive as it can provide sequence density genotypes at a cost price point comparable to low or medium-density SNP chips.

## ACKNOWLEDGEMENTS

The authors thank DairyBio, a joint venture project between Agriculture Victoria, Dairy Australia and The Gardiner Foundation, for funding and the 1000 Bull Genomes Project for use of whole-genome sequence data.

## REFERENCES

Browning B.L. & Browning S.R. (2009) *Am. J. Hum. Genet.* **84**, 210-23.
Browning B.L., Zhou Y. & Browning S.R. (2018) *Am. J. Hum. Genet.* **103**, 338-48.
Daetwyler H.D., Brauning R., …, Kijas J.W. (2017) In: *Proc of AAABG*, Townsville, AUS.
Hayes B.J. & Daetwyler H.D. (2019) *Ann. Rev. Anim. Biosci.* **7**, null.
Li N. & Stephens M. (2003) *Genetics* **165**, 2213 - 33.
Newton J.E., Hayes B.J. & Pryce J.E. (2018) *J. Dairy Sci.* **101**, 6159-73.
Rosen B.D., Bickhart D.M., …, Medrano J.F. (2020) *GigaSci.* **9**, giaa021.
Rubinacci S., Ribeiro D.M., …, Delaneau O. (2021) *Nat Genet* **53**, 120-6.
Snelling W.M., Hoff J.L., …, Pickrell J.K. (2020) *Genes (Basel)* **11**.
van Binsbergen R., Bink M.C., …, Veerkamp R.F. (2014) *Genet. Sel. Evol.* **46**, 1-13.
Wasik K., Berisa T., …, Cox C. (2019) *bioRxiv*, 632141.

*Contributed paper*

# THE IMPACT OF GENOTYPE BY ENVIRONMENT INTERACTION ON BREEDING VALUES FOR 150-DAY WEIGHT IN KATAHDIN SHEEP IN MEXICO

**L. De La Cruz[1, 2], S.F. Walkom[1] and A.A. Swan[1]**

[1] Animal Genetics Breeding Unit*, University of New England, Armidale, NSW, 2350 Australia
[1, 2] National Institute of Forestry, Agriculture and Livestock Research in Mexico. Pachuca of Soto, Hidalgo, Mexico.

## SUMMARY

With the objective of evaluating the impact of genotype by environment interaction (G by E) on breeding values for 150-day weight in Mexican Katahdin sheep, data from a total of 41,323 lambs, the progeny of 1,862 sires, were used to estimate genetic correlations between seven regional environments representing the majority of Mexico. Estimates of heritability within environments ranged from 0.24 ± 0.04 for Pacific Central to 0.42 ± 0.11 for North East. Genetic correlations across environments averaged 0.51 across all pairs, ranging from 0.07 ± 0.61 to 0.86 ± 0.27, indicating the presence of G by E interaction. A validation study predicted progeny performance within each environment with and without sire by flock effects from sire breeding values (EBVs) calculated from single trait BLUP analyses of data in the remaining environments. The regression of offspring performance on sire EBV were predictable across environments, but at lower levels than the expected value, in the absence of G by E, of 0.5. Fitting sire by flock improved the predictability with the regression coefficient increasing from 0.31 to 0.36.

## INTRODUCTION

Katahdin sheep are a composite breed, developed in Maine, USA from crosses between hair and wool breeds (Wildeus 1997). This breed is low maintenance, highly prolific, does not require shearing, and is relatively resistant to internal parasites (Vanimisetti *et al.* 2007). The Katahdin breed plays an important role in the Mexican sheep industry as a maternal breed and makes a major contribution to sheep meat production in the country. Currently Katahdin sheep are dispersed across the diverse environments of Mexico, and have been dominating stud book registrations, with 87,807 animals in the database of the Mexican National Ovinocultores Union (UNO). However, recording of performance is limited, which has affected the development of any systematic breeding programs to improve the breed through selection. The objective of this study was to evaluate performance for 150-day weight, with a particular focus on the importance of G by E across the diverse Mexican environments where the breed is represented.

## MATERIALS AND METHODS

**Data structure.** Records from 41,323 lambs, the progeny of 1,862 sires and 15,340 dams, were used to conduct genetic analyses of the performance for 150-day weight (W150) in Katahdin sheep across seven environments: North Central (NC), North East (NE), Pacific Central (PC), Central (C), Gulf Central (GC), Pacific South, (PS), and South East (SE). A summary of the number of animals, sires and dams represented at each environment is shown in Table 1.

* A joint venture of NSW Department of Primary Industries and the University of New England

**Table 1. Data structure and descriptive statistics for 150-day weight (W150, kg) recorded in Katahdin sheep for all of Mexico and by environment**

| Component | Mexico | NC | NE | PC | C | GC | PS | SE |
|---|---|---|---|---|---|---|---|---|
| Number of records | 41,323 | 4,839 | 3,118 | 10,977 | 12,206 | 2,140 | 5,153 | 2,890 |
| Mean | 40.9 | 42.1 | 38.1 | 44.1 | 41.9 | 40.8 | 31.9 | 42.1 |
| Standard deviation | 8.8 | 9.5 | 7.3 | 7.8 | 8.6 | 7.2 | 6.2 | 7.1 |
| Variation coefficient (%) | 21.5 | 22.6 | 19.2 | 17.7 | 20.5 | 17.7 | 19.5 | 16.9 |
| Number of dams | 15,340 | 1,751 | 1,264 | 3,871 | 4,545 | 787 | 1,977 | 1,272 |
| Number of sires | 1,862 | 350 | 176 | 556 | 734 | 151 | 161 | 215 |

NC: North Central; NE: North East; PC: Pacific Central; C:Central; GC: Gulf Central; PS: Pacific South; SE: South East.

**Statistical analysis.** Fixed effects fitted were sex (males, females), birth type and rearing type (single, twin, and triplet) and age of dam in years (factor: eight levels). Age at measurement was included as a linear regression. Variance components and heritabilities for each environment were estimated using univariate sire model. Random effects included sire genetic (G), maternal permanent environmental (PE) and contemporary group (CG), which was defined by flock of birth, year of birth and season. Genetic correlations across environments were estimated using bivariate analyses between each pair of environments. Bivariate models included the same fixed and random effects, with all analysis performed using ASReml software (Gilmour *et al.* 2018). The general bi-variate form of the variance structures including the residual term (R) was:

$$G=\begin{bmatrix}\sigma^2_{s_i} & \sigma_{s_{ij}}\\ \sigma_{s_{ji}} & \sigma^2_{s_j}\end{bmatrix};$$

$$PE=\begin{bmatrix}\sigma^2_{m_i} & 0\\ 0 & \sigma^2_{m_j}\end{bmatrix};\ R=\begin{bmatrix}\sigma^2_{e_i} & 0\\ 0 & \sigma^2_{e_j}\end{bmatrix};\ CG=\begin{bmatrix}\sigma^2_{cg_i} & 0\\ 0 & \sigma^2_{cg_j}\end{bmatrix}$$

Genetic correlations between environments $i$ and $j$ were derived from the G matrix as: $r_{g_{ij}} = \sigma_{s_{ij}}/(\sigma_{s_i}\sigma_{s_j})$. The CG, PE, and R matrixes have a diagonal structure because contemporary groups, dams, and progeny can only be represented in a single environment. Heritabilities were estimated for environment $i$ as $h_i^2 = 4\sigma^2_{s_i}/\sigma^2_{P_i}$, with $\sigma^2_{P_i}$ the phenotypic variance calculated as the sum of all components, excluding CG.

Validation analyses were used to study the impact of ignoring G by E correlations in the genetic evaluation analyses. For each of the seven environments defined as "targets", we estimated EBVs from "training" data combining the other six environments using a single trait animal model including and excluding the sire by flock interaction, not considering G by E effects (apart from any variation explained by sire by flock effects). Adjusted progeny performance in the target environment was then regressed on sire EBVs from the training analysis, i.e. for sires with progeny in both target and training data. The expected value of this regression in the absence of G by E is 0.5, and regressions of adjusted offspring performance on sire breeding values were calculated from linear models including the fixed effects of sex, birth type, rearing type, age of the dam and age at measurement (W150), along with contemporary groups treated as a random effect.

## RESULTS AND DISCUSSION

The mean weights at W150 for NC, NE, PC, C, GC, PS and SE were 42.1, 38.1, 44.1, 41.8, 40.7, 31.9 and 42.1 kg, respectively (Table 1). The highest level of performance was observed for Pacific Central (PC) and the lowest for Pacific South (PS). The number of the records shows the distribution of this breed across the country, with the largest numbers located in the central area (Central, Pacific Central) and the least in the Gulf Central.

Genetic parameters used to calculate EBVs for the validation study were based on analyses of the whole data set. Parameters estimated from the animal model without sire by flock were 26.4 ± 0.22 for the phenotypic variance, 37.2 ± 1.27 for the CG variance, with corresponding heritability and maternal permanent environment ratios of 0.14 ± 0.01 and 0.06 ± 0.01. For the model with sire by flock fitted phenotypic variance was 27.3 ± 0.25, CG was 34.7 ± 1.23, with heritability, maternal permanent environment and sire by flock ratios of 0.11 ± 0.01, 0.06 ± 0.01 and 0.07 ± 0.01 respectively.

Sire model variance components within environments are presented in Table 2. Estimates of heritability ranged between 0.25 ± 0.04 to 0.42 ± 0.11, averaging 0.32 across all regions. Lower heritability estimates (0.20 ± 0.02) have previously been reported in Katahdin lambs weighed at approximately 120 days of age (Ngere *et al.* 2017). Estimates of the ratio of maternal permanent environment effects ranged from 0.05 ± 0.01 to 0.10 ± 0.01 (averaging 0.08).

**Table 2. Estimates of phenotypic variance ($\sigma^2_p$), heritability ($h^2$), maternal permanent environmental effects ($m^2$) for 150-day weight (W150, kg) within environments in Katahdin sheep**

| Component | NC | NE | PC | C | GC | PS | SE |
|---|---|---|---|---|---|---|---|
| $\sigma^2_p$ | 30.6 ± 0.77 | 23.7 ± 0.86 | 31.4 ± 0.52 | 28.3 ± 0.45 | 20.9 ± 0.78 | 12.7 ± 0.36 | 20.1 ± 0.71 |
| $h^2$ | 0.31 ± 0.07 | 0.42 ± 0.11 | 0.24 ± 0.04 | 0.25 ± 0.04 | 0.28 ± 0.10 | 0.33 ± 0.09 | 0.42 ± 0.10 |
| $m^2$ | 0.10 ± 0.01 | 0.08 ± 0.02 | 0.08 ± 0.01 | 0.09 ± 0.01 | 0.08 ± 0.02 | 0.05 ± 0.01 | 0.08 ± 0.02 |

The highest number of common sires and their progeny were between Pacific Central and Central environments, where more linkage is facilitated by greater sharing of genetic material between farms located relatively closely together (Table 3).

Genetic correlations across environments were positive, ranging between 0.07 ± 0.61 and 0.86 ± 0.27, with an average weighted by the inverse of standard errors of 0.51 (Table 3). The precision of estimates was highly variable, driven by differences in the number of common sires and progeny between regions. However, given the average correlation of 0.51 and considerable variation around the average, there is evidence that G by E interaction will affect selection across regions. We speculate that these interactions could have both biological and industry structural origins. The "biological" refers to true genetic adaptation to diversity in environmental conditions while the "industry structural" refers to the structure of the breed in Mexico into a large number of different breeding associations within regions that have different aims and ways of operating. Development of a single genetic evaluation system across environments would help to address this structural issue.

The validation analysis implemented an evaluation model across environments without assuming G by E interaction i.e., a single trait for all environments. The model was tested with and without sire by flock interactions, which would in theory partly correct G by E interactions. Regression of offspring performance in each environment on sire breeding values calculated in the other environments (Table 4) showed that breeding values were predictable across environments but at a level lower than the expectation of 0.5 in the absence of G by E. Encouragingly, fitting sire by flock improved the weighted average regression (predictability) from 0.31 to 0.36. The reductions in predictability relative to the expectation of 0.5 were slightly higher than proportional to the average genetic correlation between environments: a regression of 0.31 would suggest a genetic correlation of 0.62 (0.31/0.5) and a regression of 0.36 a correspondingly higher correlation of 0.72.

Several environments in Table 4 showed good levels of predictability from data recorded in other environments, including NC, NE, C, and GC, whereas predictability in other environments was lower, including PC, SE, and PS. The two latter regions were smaller and characterised by poor connectivity with other regions. Improved linkage would be beneficial for these regions.

**Table 3. Number of common sires (and progeny) in upper triangle, genetic correlations for 150-day weight (W150, kg) between environments in lower triangle**

| Environment | NC | NE | PC | C | GC | PS | SE |
|---|---|---|---|---|---|---|---|
| NC | | 30 (842) | 96 (4,741) | 85 (3,552) | 28 (723) | 21 (937) | 32 (845) |
| NE | 0.86 ± 0.27 | | 39 (1,777) | 39 (1,633) | 19 (352) | 11 (395) | 17 (245) |
| PC | 0.48 ± 0.19 | 0.21 ± 0.37 | | 147 (7,511) | 44 (1,796) | 31 (1,447) | 45 (2,213) |
| C | 0.84 ± 0.17 | 0.19 ± 0.33 | 0.41 ± 0.18 | | 50 (198) | 29 (1,630) | 46 (2,110) |
| GC | 0.56 ± 0.37 | 0.64 ± 0.40 | 0.50 ± 0.31 | 0.31 ± 0.32 | | 16 (463) | 30 (754) |
| PS | 0.74 ± 0.28 | 0.31 ± 0.58 | 0.77 ± 0.27 | 0.35 ± 0.33 | 0.07 ± 0.61 | | 15 (250) |
| SE | 0.72 ± 0.31 | 0.77 ± 0.33 | 0.12 ± 0.34 | 0.23 ± 0.34 | 0.38 ± 0.47 | 0.45 ± 0.52 | |

**Table 4. Number of sires and progeny used to calculate regression coefficients of sire breeding values from training data on progeny performance for 150-day weight in validation data from each environment with and without sire by flock interaction fitted in the training data**

| Environment | Sires | Progeny | Regression coefficient | |
|---|---|---|---|---|
| | | | No sire flock | Sire flock |
| NC | 124 | 1689 | 0.57 ± 0.09 | 0.64 ± 0.11 |
| NE | 58 | 691 | 0.32 ± 0.10 | 0.35 ± 0.11 |
| PC | 198 | 5189 | 0.17 ± 0.06 | 0.25 ± 0.07 |
| C | 201 | 4335 | 0.38 ± 0.06 | 0.42 ± 0.06 |
| GC | 69 | 599 | 0.53 ± 0.18 | 0.55 ± 0.20 |
| PS | 39 | 738 | 0.15 ± 0.09 | 0.23 ± 0.13 |
| SE | 61 | 920 | 0.06 ± 0.13 | 0.11 ± 0.14 |
| Average | | | 0.31 | 0.36 |

## CONCLUSIONS

While this study showed evidence of G by E interaction across regions of Mexico, there was still evidence of predictability of breeding values across regions, albeit at a lower level than expected in the absence of G by E. Development of a national genetic evaluation system for the Katahdin breed in Mexico will stimulate greater linkage between associations, allowing breeders to benefit from across-flock selection even if there is a biological component of G by E in some circumstances.

## ACKNOWLEDGMENTS

The Mexican National Ovinocultores Union for providing the data set. The first author's PhD studies are supported by the Mexican National Council for Science and Technology (CONACYT).

## REFERENCES

Gilmour A.R., Gogel B.J, Cullis B.R, Thompson R. and Butler D. (2018) ASReml User Guide Release 4.0. VSN International Ltd, Hemel Hempstead, UK

Ngere L., Burke J., Notter D. and Morgan J. (2017) *J. Anim. Sc.* **95:** 3396.

Vanimisetti H.B., Notter D.R. and Kuehn L.A. (2007) *J. Anim. Sc.* **85:** 60.

Wildeus S. (1997) *J. Anim. Sc.* **75:** 630.

*Contributed paper*

# GENOME-WIDE ASSOCIATION ANALYSIS FOR TEMPERAMENT IN AUSTRALIAN SHEEP

**S. de las Heras-Saldana, N. Moghaddar, S.A. Clark and J.H.J. van der Werf**

School of Environmental and Rural Science, University of New England, Armidale, NSW, 2351 Australia

## SUMMARY

In livestock, temperament traits such as flight speed and agitation are important indicators of management, survival, and welfare. Multiple studies have reported a low to moderate heritability for these traits. Identifying the genomic regions associated with temperament could help to find candidate genes and processes involved in defining these traits and this could be helpful in genomic prediction of phenotype or breeding value. This study aimed to identify genomic regions associated with flight speed and agitation. We used imputed whole-genome sequences from animals with records for flight speed (n = 8,737) and agitation (n = 8,586). The heritability for agitation was 0.18 ± 0.03 and 0.14 ± 0.02 for flight speed respectively. Three and five QTL regions were associated with agitation (on Chr3, Chr4, and Chr20) and flight speed (on Chr 1, Chr13, Chr15, and Chr26), respectively. The identification of these genomic regions provides further knowledge on the genetic mechanism involved in temperament traits providing alternative tools to improve sheep breeding programs. Further analysis is needed to find links between agitation and flight speed with production traits.

## INTRODUCTION

Animal temperament has been proposed as a potential indicator of the physical, physiological, and psychological state of the animal in production systems which also defines welfare. Temperament can be assessed through behavioural traits such as agitation and flight speed, with both traits having shown a low to moderate heritability in sheep (Dodd *et al.* 2014). However, a better understanding of the whole genome associated regions and underlying genes involved could help to understand behavioural traits. Previously, a study on single nucleotide polymorphism (SNPs) on only four genes (*SLC6A4*, *TPH2*, *OXTR,* and *HTR2A*) identified SNPs on *TPH2* and *HTR2A* associated with behaviour in sheep (Ding *et al.* 2020), suggesting that a genome-wide association study (GWAS) can provide further information to better understand the involved biological process. This study aimed to identify the genetic regions and candidate genes associated with temperament traits such as flight speed and agitation in sheep.

## MATERIALS AND METHODS

**Animals and phenotypes.** In total, 8,771 genotyped animals were used from the Information Nucleus Flock with records for flight speed (N = 8,737) and agitation (N = 8,586) obtained between 2008 and 2010. Lambs were produced by artificial insemination across eight farms within Australia and there were pure Merino or Merino crosses. A comprehensive description of the breeds is provided in van der Werf *et al.* (2010). A complete description of the recorded traits is provided in Dodd *et al.* (2014). In summary, studied traits were measured at post-weaning age. Lambs were subjected to an isolation test to record agitation by measuring with an agitation meter the number of vibrations caused by movement within the isolation box over a 30 second period. The flight speed corresponded to the speed at which the lamb crosses a specific distance.

**Genotypes.** Low-density genotypes (50k) were imputed to high-density and finally to sequence level to keep ~31 million SNPs after quality control to remove SNPs with minor allele frequency less than 0.01, deviation from Hardy-Weinberg equilibrium ($P < 10^{-10}$), and missing genotypes >

5%. A detailed description of the imputation is provided in Bolormaa *et al.* (2019).

**Statistical analysis.** Each trait was normalized using square root and $\log_{10}$ for agitation and flight speed, respectively. Genetic parameters and genetic correlations for agitation and flight speed were estimated in ASReml v4 (Gilmour *et al.* 2015) using the pedigree in an animal model and bivariate model, respectively. The model fitted fixed effects as age, birth type (BT), month, flock (N = 8), year (N = 3), sex, management group (MG), interactions, and an error term (e). The animal id and breed proportion (GG) were fit as random for flight speed (model 1) and agitation (in addition to dam; model 2). Based on the animal model, phenotypes we adjusted for mentioned fixed and random effects.

$$y = \mu + BT + month + age + flock + year + GG + MG*year*flock + e \quad (1)$$

$$y = \mu + BT + month + age + flock + year + sex + MG + dam + GG\ flock*year + e \quad (2)$$

The adjusted phenotypes and imputed sequences were used to perform a GWAS in GEMMA (Zhou *et al.* 2012) software with the model $\boldsymbol{y} = \boldsymbol{X\beta} + \boldsymbol{Za} + \boldsymbol{e}$, where $\boldsymbol{y}$ is a vector of phenotype, $\boldsymbol{X}$ is the incidence matrix for the fixed effects, $\boldsymbol{\beta}$ is the vector of fixed effects (SNPs), $\boldsymbol{Z}$ is the incidence matrixes to relate random additive genetic effects with the phenotypes, $\boldsymbol{a}$ correspond to the vector of direct additive genetic effects effect with $\boldsymbol{a} \sim N(0, \boldsymbol{G}\sigma_\alpha^2)$ where $\boldsymbol{G}$ is a genomic relationship matrix and $\sigma_\alpha^2$ is the additive genetic variance; and $e$ is a vector of residual effects. A normal distribution was assumed for the additive genetic effects. QTLs were identified based on a false discovery rate < 0.1, which corresponded to a threshold of $-\log_{10}(9\text{x}10^{-08}) \geq 7$, and a 1 Mb window from the significant SNPs.

The percentage of genetic variance captured by the top significant SNPs was calculated as $2p_iq_i\alpha_i^2/\sigma^2 * 100$, where $\sigma^2$ is the additive genetic variance, $p$ and $q$ are the allele frequency for the SNP, and $\alpha_i^2$ s the additive effect of the SNP. An additional threshold of $-\log_{10}(1\text{x}10^{-05}) \geq 5$ was used to identify the candidate genes around significant SNPs in a window of 1 Mb. The candidate genes were used in a pathway and gene ontology (GO) analysis performed with ClueGo v2.5.6 (Bindea *et al.* 2009) plugin. The function of candidate genes was further investigated in the literature.

## RESULTS AND DISCUSSION

Moderate to low heritabilities were observed (Table 1) for agitation (0.18 ± 0.03) and flight speed (0.14 ± 0.02). Similar heritabilities were previously reported in an overlapping population of sheep for agitation ($h^2$ = ~0.20; Lennon *et al.* 2009; Dodd *et al.* 2014), and flight speed ($h^2$ = 0.11; Dodd *et al.* 2014) and cattle (for flight speed $h^2$ = 0.21; Valente *et al.* 2016). A positive genetic correlation ($r_g$) of 0.41 was detected between the studied traits, but this was higher than previously reported by Dodd *et al.* 2014 ($r_g$ = 0.20).

**Table 1. Heritability and genetic variance for temperament traits in sheep**

| Trait | $h^2 \pm$ SE | $V_g$ | $V_P$ |
|---|---|---|---|
| Agitation | 0.18 ± 0.03 | 0.14 ± 0.02 | 0.77 ± 0.02 |
| Flight speed | 0.14 ± 0.02 | 0.11 ± 0.02 | 0.81 ± 0.03 |

$h^2$: heritability; $V_g$: genetic variance; $V_P$: phenotypic variance; SE: standard error

From the GWAS results, there were three QTLs regions (Figure 1A) identified for agitation that account for 9 % of the total genetic variation on the chromosomes Chr3, Chr4, and Chr20 (Table 2). Within these regions, 15 unannotated and 22 annotated genes were identified from which the top 15 genes are *MED27, RAPGEF1, UCK1, POMT1, PRRC2B, PPAPDC3, FAM78A, NUP214, AIF1L, CDK14, FZD1, MTERF1, AKAP9, EDN1*, and *HIVEP1*. For flight speed, five QTLs (Figure 1B) were detected on Chr1 (with two QTLs), Chr13, Chr15, and Chr26 accounting for 16 % of the genetic variance (Table 2). These regions contained 35 annotated genes and 26 unannotated genes, being *CD47, IFT57, HHLA2, MYH15, KIAA1524, DZIP3, GUCA1C, MORC1, ELP4, PAX6, FCRLA, FCRLB, DUSP12, ATF6*, and *CENPA* the top 15 genes in a window of within 1MB from the most significant SNPs.

Most of the candidate genes were previously reported mainly in human studies to schizophrenia (*FCRLA, UHMK1, RGS4, RGS5*, and *DCDC5*; Campbell *et al.* 2008; Stefanis *et al.* 2008), depression (*DUSP12, OLFML2B, ZFP64*, and *DCDC1*; Wray *et al.* 2012), Alzheimer's disease (*ATF6, HSD17B7*, and *DZIP3*; Montibeller *et al.* 2018, Xu *et al.* 2018) and other mental conditions (i.e. stress, bipolar, and anxiety disorders). For agitation candidate genes, fewer studies previously reported their function but similarly, some genes were associated with schizophrenia (*RAPGEF1, FZD1*, and *AKAP9*, Igolkina *et al.* 2018, Lui *et al.* 2020) or Alzheimer (*NEDD9*; Li *et al.* 2008).

**Table 2. Significant SNPs associated with agitation and flight speed**

| Traits | Chr | Mb | AF | beta | p-value | % Vg |
|---|---|---|---|---|---|---|
| Flight speed | 1 | 171 | 0.38 | -0.09 | 5.92E-09 | 3.56 |
| | 1 | 11 | 0.02 | -0.31 | 8.12E-09 | 3.14 |
| | 13 | 79 | 0.14 | -0.11 | 6.03E-08 | 2.71 |
| | 15 | 60 | 0.15 | -0.12 | 1.17E-08 | 3.40 |
| | 26 | 37 | 0.03 | -0.23 | 5.34E-08 | 2.80 |
| Agitation | 3 | 5 | 0.02 | 0.25 | 6.77E-08 | 2.06 |
| | 4 | 8 | 0.20 | -0.10 | 1.45E-08 | 2.32 |
| | 4 | 8 | 0.14 | -0.12 | 1.13E-08 | 2.32 |
| | 20 | 43 | 0.34 | 0.084 | 8.64E-08 | 2.23 |

Chr: chromosome; Mb: megabase pair; AF; allele frequency; $\%V_g$: percentage of genetic variance.

A)

B)

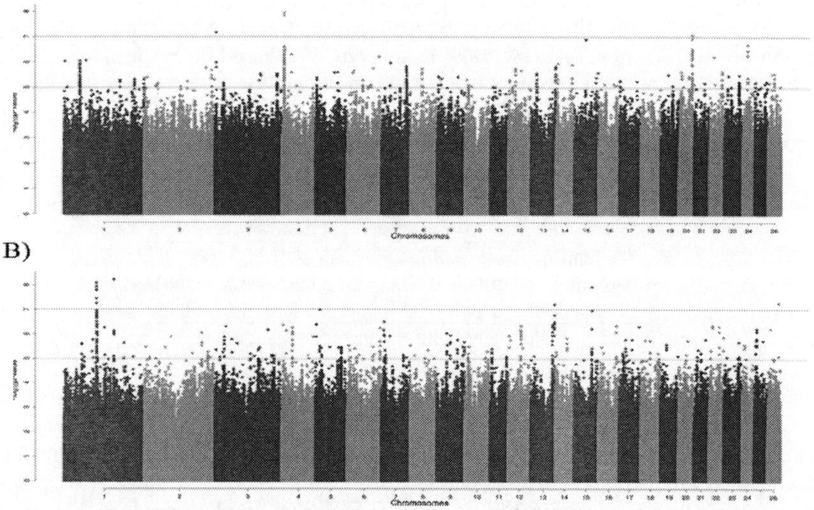

**Figure 1. Manhattan plot of GWAS for (a) agitation and (b) flight speed indicating with lines the FDR < 0.1 threshold (blue) and the suggestive threshold (red)**

The candidate genes located within 1 Mb of significant SNPs ($-\log_{10}(1x10^{-05})\geq 5$) were identified for agitation (582 genes) and flight speed (907 genes) where 53 genes overlapped for both traits. The three most represented gene ontology terms for flight speed were found to be intracellular, regulation of cellular process, and cytoplasmic part; while for agitation the signal transduction, regulation of signaling, and regulation of response to stimulus were the most enriched.

## CONCLUSIONS

Genomic regions associated with flight speed and agitation were identified in this study. The significant SNPs in these regions are close to genes previously associated in multiple studies in humans with schizophrenia disorders, depression, and Alzheimer's disease. Further knowledge on the genetic mechanism of behaviour and other important complex diseases can be provided from GWAS on non-model organisms such as sheep. From a production perspective in sheep, the genetic relationships between temperament and carcass traits are required together with economic values to assess the relevance of including these behavioural traits in selection programs, but a consistent recording of these phenotypes is needed to ensure the advantages in genomic selection.

## ACKNOWLEDGEMENTS

The authors acknowledge MLA for providing access to the datasets used in the present study.

## REFERENCES

Bindea G., Mlecnik B., Hackl H., Charoentong P., Tosolini M., Kirilovsky A., and Galon J. (2009) *Bioinformatics* **25**: 1091.

Bolormaa S., Chamberlain A.J., Khansefid M., Stothard P., Swan A.A., Mason B., and MacLeod I.M. (2019) *Genet. Sel. Evol.* **51**: 1.

Campbell D.B., Lange L.A., Skelly T., Lieberman J., Levitt P., and Sullivan P.F. (2008) *Schizophr. Res.***101**: 67.

Ding L., Maloney S., Wang M., Rodger J., Chen L., and Blache D. (2020) *Brain and Behav*, e12714.

Dodd C.L., Edwards J.H., Hazel S.J., and Pitchford W.S. (2014). *Livest. Sci.* **160**: 12.

Gilmour A.R., Gogel B.J., Cullis B.R., Welham S., and Thompson R. (2015) *Hemel Hempstead: VSN international ltd.*

Igolkina A.A., Armoskus C., Newman J.R., Evgrafov O.V., McIntyre L.M., Nuzhdin S.V., and Samsonova, M.G. (2018) *Front. Mol. Neurosci.* **11**: 192.

Lennon, K. L., Hebart, M. L., Brien, F. D., & Hynd, P. I. (2009) *Proc. Assoc. Advmt. Anim. Breed. Genet.* **18**: 96.

Li Y., Grupe A., Rowland C., Holmans P., Segurado R., Abraham R., and Williams J. (2008) *Hum. Mol. Genet.* **17**: 759.

Liu X., Low S.K., Atkins J.R., Wu J.Q., Reay W.R., Cairns H.M., and Cairns M.J. (2020) *Aust. N. Z. J. Psychiatry* **54**: 902.

Montibeller L., and De Belleroche J. (2018) *Cell Stress Chaperones*: **23**: 897.

Stefanis N.C., Trikalinos T.A., Avramopoulos D., Smyrnis N., Evdokimidis I., Ntzani E.E., and Stefanis C.N. (2008) *Behav. Brain Funct.* **4**: 1.

Wray N.R., Pergadia M.L., Blackwood D.H. R., Penninx B.W.J. H., Gordon S.D., Nyholt D.R., and Sullivan P.F. (2012) *Mol Psychiatry* **17**: 36.

Xu M., Zhang D.F., Luo R., Wu Y., Zhou H., Kong L.L., and Yao Y.G. (2018) *Alzheimer's & Dementia*: **14**: 215.

Zhou X., and Stephens M. (2012) *Nat. Genet.* **44**: 821.

*Contributed paper*

# A GENOMIC COMPARISON OF AUSTRALIAN, NEW ZEALAND AND NORWEGIAN DAIRY GOAT POPULATIONS

**K.G. Dodds[1], J.H. Jakobsen[2], M. Wheeler[3], T. Blichfeldt[2], R. Brauning[1], H.J. Baird[1] and J.C. McEwan[1]**

[1] AgResearch, Invermay Agricultural Centre, Mosgiel, New Zealand
[2] The Norwegian Association of Sheep and Goat Breeders, Box 104, 1431 Ås, Norway
[3] AgResearch, Ruakura Agricultural Centre, Mosgiel, New Zealand

## SUMMARY

Six dairy goat populations that had been genotyped using genotyping-by-sequencing (GBS) were compared. The populations were an industry data set from Norway along with four herds from New Zealand (NZ) and one from Australia. The Norwegian population was found to be the most genetically diverged from the others. One of the NZ populations was also quite distinct while the other NZ populations appear to be genetically similar to each other and closer to the Australian population than the other NZ population. It may be useful to combine these three NZ populations and the Australian population to provide better genomic evaluation.

## INTRODUCTION

AgResearch have been providing genotyping services for several dairy goat enterprises with clients in New Zealand, Australia, and Norway. A genotyping-by-sequencing (GBS) platform has been used to provide a medium-density (~60k) SNP profile. In most cases the genetic background of the populations is poorly recorded. This study characterises and compares these populations based on genotypes from the common GBS platform used. One outcome of a comparison is that it would inform the likely usefulness of combining populations for genomic evaluation.

## MATERIALS AND METHODS

**Animals**. The animals used in this study were from herds in the Norwegian Association of Sheep and Goat Breeders ('Norway', www.nsg.no, Norway), Meredith Dairy ('Aus1', Victoria, Australia), Northland ('NZ1', Northland, New Zealand) and three other New Zealand herds ('NZ2', 'NZ3', 'NZ4'). The Norway population descends from the Norwegian Landrace breed with some recent infusion of French Alpine (Ådnøy 2014). NZ1 is primarily Saanen (Wheeler *et al.* 2018) while Aus1 is a composite of Saanen, Toggenberg and British Alpine (Wheeler *et al.* 2018) with similar likely origins for NZ2, NZ3, NZ4. To approximately balance numbers across the groups, younger animals were removed from some groups (those with birth years from 2018, 2014 and 2017 for Norway, Aus1 and NZ1 respectively).

**GBS genotypes**. The animals were genotyped by genotyping-by-sequencing (GBS) using the methods described by Dodds *et al.* (2015) and Wheeler *et al.* (2018). Prior to this study, sequence reads from a set of 5,395 goats, that were available at the time, from a range of sources (including 3,702 from Aus1, 1,458 from NZ1 and 201 from NZ3 but none from Norway) were used to detect variants. The variants were discovered using UNEAK (Lu *et al.* 2013) on the adapter-trimmed sequences and without using a reference genome. These variants were placed into a catalogue which was used to allow counts of reference and alternate alleles for each variant in any GBS'd sample using TagDigger (Clark and Sacks, 2016).

Only SNPs that mapped on to the autosomal chromosomes of the goat reference assembly (ARS1, https://www.ncbi.nlm.nih.gov/assembly/GCF_001704415.1) were retained. In addition, SNPs with a raw (not adjusted for read depth) Hardy-Weinberg disequilibrium value < 0.05 (Dodds *et al.* 2015) or a depth adjusted Hardy-Weinberg test p-value < $10^{-100}$ (Dodds *et al.* 2018a) were removed. Animals that were genotyped multiple times but had inconsistent genotypes (n=10) or that had a mean read depth < 0.3 (n=191) were removed from the initial set of 8,340 goats.

**Population structure.** A genomic relationship matrix (GRM) was calculated using the method of Dodds *et al.* (2015) which accounts for the read depth in a genotype call. The overall allele frequencies, calculated on the total number of reads for each allele, were used in these calculations. The GRM was then used to perform a principal component analysis. The mean relatedness by group pair was also calculated and plotted using the heatmap function in R (R core team, 2020), which also performs hierarchical clustering. The fixation index ($F_{st}$) between groups was calculated by the depth-adjusted method of Dodds *et al.* (2018b) using KGD software (www.github.com/AgResearch/KGD) with default settings.

## RESULTS AND DISCUSSION

The SNP catalogue contained 60,225 SNPs. After filtering 51,680 SNPs and 8539 animals remained. These SNPs had a 73.9% call rate and mean read depth of 2.71.

The first two principal components are shown in Figure 1. The first component explains 69.3% of the variance and separates out the Norwegian goats from the others. This is consistent with the Norwegian group being genetically isolated for over 1000 years until 2007. The effect of the recent use of semen from French Alpine goats in Norway can be seen by the small cluster (on the left of the main Norway cluster).

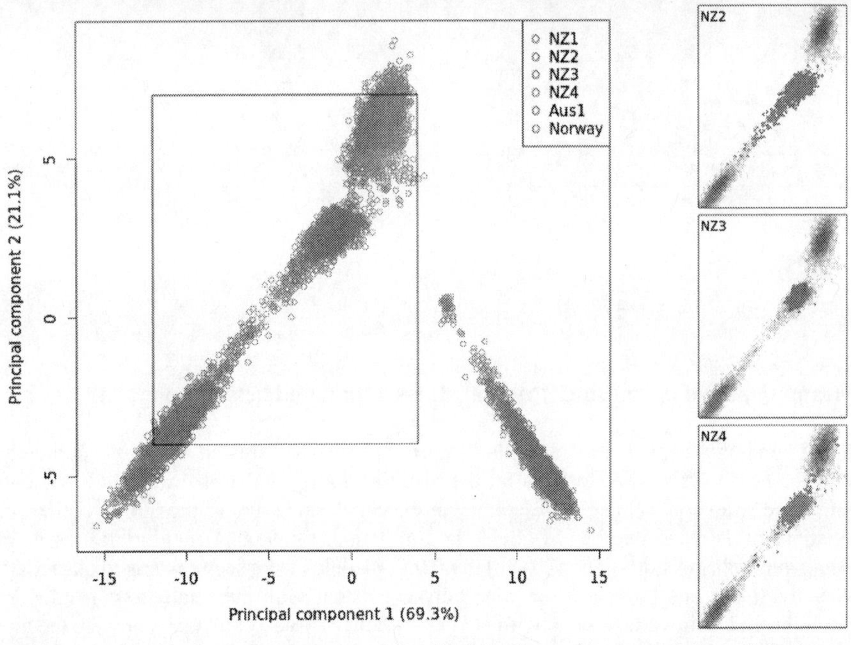


**Figure 1. Principal components plot of the GRM coloured by population. Subplots on the right show three of the populations plotted over the others for the region in the box**

The Australian and New Zealand populations fall into three partially overlapping groups (Figure 1). These groups almost form a linear arrangement. NZ1 forms a group almost on its own at one end. This population is from Northland where climatic & environmental conditions have necessitated a region-specific breeding programme. Aus1 forms a group at the other end with the other three NZ populations in between. Some of NZ4 overlaps with Aus1. Both NZ2 and NZ4 have animals near or just overlapping NZ1, while NZ3 is tightly clustered within the central group of NZ populations. The studies of Brito *et al.* (2017) and Oget *et al.* (2019), which also include some alpine dairy goat breeds but using a 50k SNP chip, suggest other populations that could be investigated using a common genotyping platform.

The clustering based on mean population pair relatedness (Figure 2) also indicates that NZ4 and NZ2 are genetically similar, although it clusters Aus1 with this pair before adding NZ3. The Norway population is added last and appears to have the highest within population similarity. This could be due to the isolation of the Norwegian from other goat populations but may also be partly due to Norway being the most outlying population, a minority (~25%) of the data analysed and that SNPs were ascertained in non-Norwegian goats.

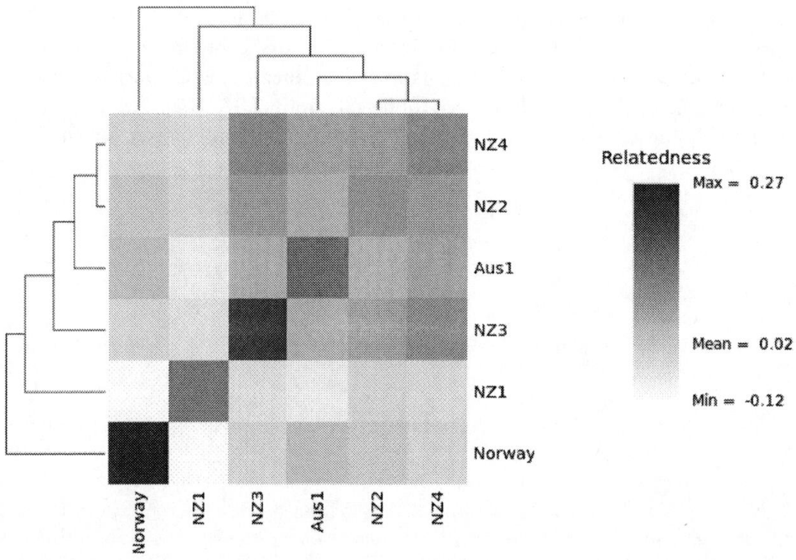


**Figure 2. Heatmap plot of the mean GRM relatedness within and between populations**

As NZ2, NZ3 and NZ4 appeared to be similar (Figure 1), they were treated as a single population ('NZ234') for the $F_{st}$ analysis. The mean $F_{st}$ for the resulting four populations was 0.066 while the pairwise values are shown in Table 1. These have an expected value, under the null hypothesis of no differentiation, of 1/(mean number of alleles in comparison) ≈ 0.0004 (negligible), as it was estimated using methods in Dodds *et al.* (2018b) that 10,000 alleles were seen (averaged over SNPs) in the 8539 individuals. The $F_{st}$ results were broadly consistent with the relatedness results. For example, the highest $F_{st}$ pair were Norway with NZ1 ($F_{st}$=0.061, Table 1) and this pair were the most distantly related (Figure 2) while the lowest $F_{st}$ pair were Aus1 with NZ234 ($F_{st}$=0.018, Table 1) and these groups were the most closely related (Figures 1 and 2).

The SNP minor allele frequencies (MAFs) were also calculated for each of the $F_{st}$ populations and the numbers of SNPs that had no variation (MAF=0) in each group are shown in Table 1. The two most divergent populations (Norway and NZ1) had the highest numbers of MAF=0 SNPs. The high number of non-polymorphic SNPs for Norway could be expected as no animals from Norway were included in the SNP detection process. Even so, there are only 4% of all SNPs used that were not polymorphic in the Norway population.

**Table 1. Mean $F_{st}$ values between pairs of populations**

| Population | Number of animals | $F_{st}$ between population pairs Aus1 | Norway | NZ1 | NZ234 | Number of SNPs with MAF=0 |
|---|---|---|---|---|---|---|
| Aus1 | 2199 | | 0.044 | 0.047 | 0.018 | 358 |
| Norway | 2107 | | | 0.061 | 0.042 | 1967 |
| NZ1 | 2357 | | | | 0.032 | 890 |
| NZ234* | 1876 | | | | | 122 |

*NZ234 has 1249, 199 and 428 animals from NZ2, NZ3 and NZ4, respectively

## CONCLUSIONS

These results indicate that there is some genetic differentiation between the populations of dairy goats investigated. The Norwegian population appears to be the most divergent, although the introduction of French Alpine into that population has reduced the amount of differentiation. Three New Zealand populations (NZ2, NZ3, NZ4) appear to be quite similar and it is likely that a combined genetic or genomic evaluation of those populations would be useful. If such an evaluation should be widened, it is Aus1 (rather than NZ1) that is most likely to be of benefit. There is unlikely to be much predictive power for genomic evaluations between the Norway population and the other populations studied here.

## ACKNOWLEDGEMENTS

Financed by the Norwegian Research Council and the Norwegian Association of Sheep and Goat Breeders through the research project 296561. We are grateful to the breeders and their breeding organisations for access to their data. Some of the methods used in this study were developed by the "Genomics for Production & Security in a Biological Economy C10X1306" project, which was funded by the Ministry of Business, Innovation and Employment (NZ).

## REFERENCES

Ådnøy T. (2014) *Small Rumin. Res.* **122**:4.

Brito L.F., Kijas J.W., Ventura R.V., Sargolzaei M., Porto-Neto L.R., Cánovas A., Feng Z., Jafarikia M. and Schenkel F.S. (2017) *BMC Genomics* **18**:229.

Clark L.V. and Sacks E.J. (2016) *Source Code Biol. Med.* **11**:1.

Dodds K.G., McEwan J.C., Brauning R., Anderson R.A., Van Stijn T.C., Kristjánsson T. and Clarke S.M. (2015) *BMC Genomics* **16**:1047.

Dodds K., McEwan J., Bilton T., Brauning R. and Clarke S. (2018a) *Australasian Applied Stat. Conf.*:38 (abstract).

Dodds K.G., Symonds J.E., Brauning R., McEwan J.C. and Clarke S.M. (2018b) *New Zealand Molecular Mapping Workshop (Mapnet)*(abstract).

Lu F., Lipka A.E., Glaubitz J., Elshire R., Cherney J.H., Casler M.D., Buckler E.S. and Costich D.E. (2013) *PLoS Genet.* **9**:e1003215.

Oget C., Servin B. and Palhière I. (2019) *Anim. Genet.* **50**:54.

R Core Team (2020) https://www.R-project.org/.

Wheeler M., Dodds K.G., Brauning R., Clarke S.M., Van Stijn T.C., Foote B.J., Foote J.F., Cameron A.W.N. and McEwan J.C. (2018) *Proc. 11th World Congress Genet. Appl. Livest. Prod.*:358.

*Contributed paper*

# THE BENEFIT OF A SLICK HAIR COAT FOR HEAT TOLERANCE IN NEW ZEALAND DAIRY CATTLE

**E.G. Donkersloot, G.M. Worth, A.F. Yeates, M.D. Littlejohn, L.R. McNaughton, R.J. Spelman and S.R. Davis**

Livestock Improvement Corporation Ltd., Newstead, Hamilton 3240, New Zealand

## SUMMARY

The slick genetic variant is a single base deletion in the prolactin receptor of cattle which produces a short hair coat and improves heat tolerance. In 2014, a breeding programme begun in New Zealand to introgress this variant from the Senepol beef breed (*Bos taurus*) into a New Zealand dairy genetic background. Heat tolerance was assessed in lactating heifers (12.5% Senepol, 87.5% NZ crossbred) using maXtec rumen boluses for long-term data collection. In mid-lactation, when the daily THI was at ~70 in late afternoon, rumen temperatures were similar between a group of slick heifers (N=9) and an age and size matched control group (N=9). The differential of rumen temperature, associated with the slick variant, became increasingly evident as the THI increased above ~70, with the maximum temperature difference ranging between 0.5-1.0°C at a THI of ~75. Accumulated milk volume in the slick heifers was ~82% of that in a contemporary group of heifers, reflecting the lower genetic merit of the slick group.

## INTRODUCTION

The slick gene is a term used to describe a major, dominant gene segregating in Senepol and other Criollo beef and dual-purpose breeds, in association with a short, slick hair coat (Olson *et al.* 2003). The causal mutation was identified as a deletion in the final exon of PRLR on BTA 20, leading to a truncation of the protein (Littlejohn *et al.* 2014). Similar mutations leading to various degrees of truncation in PRLR have been identified in several other breeds showing the same slick hair coat (Porto-Neto *et al.* 2018). The advantages of the slick hair coat in cattle are improved heat tolerance (Olson *et al*, 2003; Dikmen *et al*, 2014) and tick resistance (Ibelli *et al.* 2014). Data on heat tolerance of slick dairy cattle is limited. With climate change, the potential utility of the slick variant is obvious, particularly in dairy cattle with high feed intake and therefore elevated heat production. In addition, grazing systems in New Zealand (NZ) add to the heat load through solar radiation, mitigated, in part by cooling winds.

A breeding programme to introgress the slick variant into a dairy genetic background was begun in 2014 by crossing NZ dairy cattle with Senepol sires. The subsequent focus has been on maximising the genetic merit for dairy and reducing the proportion of beef genetics in slick offspring while conducting several trials to understand the benefits of the slick coat.

## MATERIALS AND METHODS

The heifers used were born in spring 2018 and entered their first lactation in spring 2020. Slick genotype was confirmed from ear punches by PCR, as described by Littlejohn *et al.* (2014). The trial was performed on a commercial milking herd at LIC's Innovation farm at Rukuhia, Waikato. The herd was milked twice daily, walking from paddock to the milking shed (~600 metres) at 5.30 AM and 2 PM. All milking cows were managed in the same herd, where the diet was predominantly pasture based with a supplement of maize silage, turnips and grass silage. Weather measurements, ambient temperature, relative humidity and rainfall, were collected hourly from the National Climate Database (Ruakura AgResearch/NIWA weather station, Hamilton, New Zealand). All the experiments reported were approved by the Ruakura Animal Ethics Committee.

Nine heterozygous slick heifers (12.5% Senepol 87.5% NZ Holstein Friesian, Jersey crossbred) calved in August 2020 and joined the commercial milking herd. The slick heifers were sired by 3 different slick carrier sires while a matched (age and size) control group of 9 heifers represented 6 different Friesian-Jersey crossbred sires. All 18 heifers received smaXtec rumen temperature boluses in mid lactation (SmaXtec Classic BolusTX-1442A, SmaXtec animal care GmbH Belgiergasse 3, 8020 Graz, Austria). Rumen temperature data was compared to vaginal temperatures for 2 7-day periods using intra-vaginal data loggers (DST centi-T, accuracy: ±0.1°C, resolution: ±0.032°C; Star-Oddi, Gardabaer, Iceland) attached to a shortened, hormone-free controlled internal drug release (CIDR) insert as described by Tresoldi et al. 2020. Data presented were collected in January 2021 when the animals were in mid-lactation (~140 days in milk). Rumen temperature data was corrected for effects of drinking using the smaXtec proprietary algorithm.

## RESULTS AND DISCUSSION

Rumen temperature was ~39°C in the morning before milking but rose from mid-morning to achieve a small differential between the slick and control groups of ~0.2°C when the Temperature-Humidity Index (THI) was 70 (Figures 1 and 3).

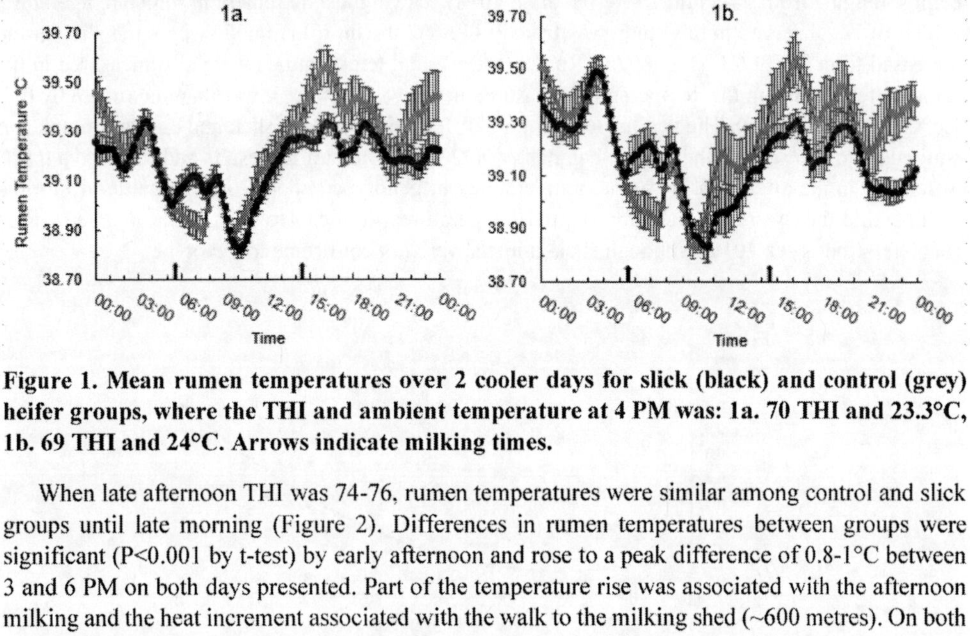


**Figure 1. Mean rumen temperatures over 2 cooler days for slick (black) and control (grey) heifer groups, where the THI and ambient temperature at 4 PM was: 1a. 70 THI and 23.3°C, 1b. 69 THI and 24°C. Arrows indicate milking times.**

When late afternoon THI was 74-76, rumen temperatures were similar among control and slick groups until late morning (Figure 2). Differences in rumen temperatures between groups were significant ($P<0.001$ by t-test) by early afternoon and rose to a peak difference of 0.8-1°C between 3 and 6 PM on both days presented. Part of the temperature rise was associated with the afternoon milking and the heat increment associated with the walk to the milking shed (~600 metres). On both days, the temperatures in the slick animals returned to a baseline more rapidly than the controls. Additionally, the rumen temperature was markedly less variable among the slick animals compared to the controls, particularly during peak rumen temperatures in late afternoon.

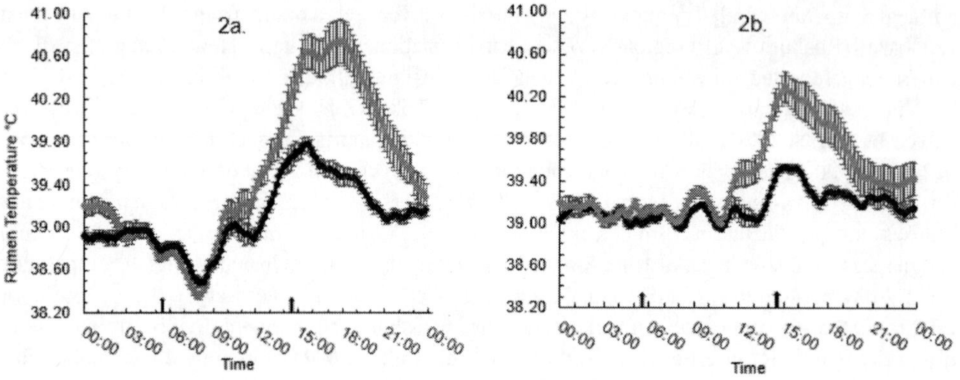


The difference in average peak (3.30-5.30 PM) rumen temperature between the slick and control groups was related to THI, increasing from 0.2°C at a THI of < 71 to 0.5-1.0°C at a THI of 74-76 (Figure 3). Different authors have provided different THI threshold values at which heat stress begins, ranging from 68-74 units (Herbut *et al.* 2018). The highest increment in rumen temperature, at THI of 74, observed in this study was from 39-40.6°C in control animals, where the slick group increased from 39-39.6°C (Figure 2a). Rumen and vaginal temperatures were also measured in the same heifers, showing that temperatures measured in the vagina were lower than the rumen by 0.8-1.0°C (data not shown). The vaginal temperature differential between slick and control groups was equivalent to that seen in the rumen. Dikmen *et al.* 2014 found similar results and reported a 0.5°C difference in the afternoon in vaginal temperatures in genotyped slick, Holstein cattle in Florida. Comparable data were reported for vaginal temperatures in slick Holstein cattle in Puerto Rica (Sánchez-Rodríguez 2019), although slick animals were not confirmed to genotype.

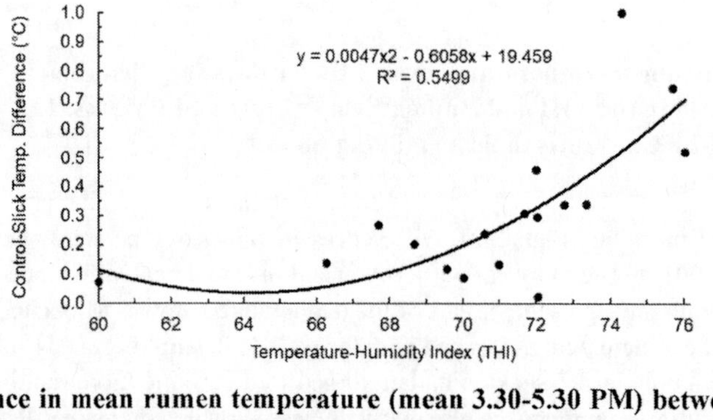


**Figure 3. Difference in mean rumen temperature (mean 3.30-5.30 PM) between control and slick heifers on 16 days in January 2020 as related to THI at 4 PM**

The mechanism by which the slick coat enables maintenance of a lower temperature in the face of higher heat loads remains to be established. Dikmen *et al.* (2014) suggested an association with increased sweating rate and an enhanced ability to dissipate heat at higher environmental temperatures appears to be involved. To measure the full potential of the slick hair coat in NZ, more research is required to develop a customized heat load index. The high heat load through solar radiation in New Zealand, as well as the effect of cooling winds, is currently not considered in the THI.

Milking performance of the slick cows was 18% lower than their non-slick contemporaries, commensurate with their lower overall genetic merit measured as gBW (Table 1).

**Table 1. Comparison of average milk production and genetic merit merit (assessed as genomic breeding worth, gBW) between the slick group (N=9) and a cohort of contemporary milking heifers (N=58) for the 2020/21 season based on monthly herd testing**

| Group | Days in milk | Average accumulated milk yield (l) | Average accumulated fat yield (kg) | Average accumulated protein yield (kg) | gBW* ($) |
|---|---|---|---|---|---|
| Slick (N=9) | 257 | 3350 | 176 | 139 | 118 |
| Contemporaries (N=58) | 272 | 4080 | 215 | 163 | 213 |

## CONCLUSIONS

Even at a relatively low THI (~75) slick heifers at grazing had a substantially lower rumen temperature (0.5-1.0°C) than their non-slick counterparts. The differentials in rumen temperatures between the heifer groups was similar to that observed in vaginal temperatures. The slick genotype has the potential to confer substantial benefit to heat tolerance in dairy cattle, but more research is needed to understand the production and welfare value that the slick genotype brings.

## REFERENCES


Bryant, J.R., López-Villalobos, N., Pryce, J.E., Holmes, C.W. and Johnson, D.L. (2007) *N.Z. J. Agr. Res.* **50**: 327.

Davis, S.R., Spelman, R.J., & Littlejohn, M.D. (2017) *J. Anim. Sci.* **95**: 1788.

Dikmen, S.E., Khan, F.A., Huson, H.J., Sonstegard, T.S., Moss, J.I., Dahl, G.E. and Hansen, P.J. (2014) *J. Dairy Sci* **97**: 5508.

Herbut, P., Angrecka, S. and Walczak, J., (2018) Environmental parameters to assessing of heat stress in dairy cattle—a review. *International journal of biometeorology*, *62*: 2089.

Ibelli, A.M.G., Ribeiro, A.R.B., Giglioti, R., Regitano, L.C.A., Alencar, M.M., Chagas, A.C.S., Paço, A.L., Oliveira, H.N., Duarte, J.M.S., Oliveira, M.C.S., (2012) *Veterinary Parasitology.* **186**: 425– 430.

Littlejohn, M.D., Henty, K.M., Tiplady, K., Johnson, T., Harland, C., Lopdell, T., Sherlock, R.G., Li, W., Lukefahr, S.D., Shanks, B.C. and Garrick, D.J. (2014) *Nat. Comm.* **5**: 1.

Olson, T.A., Lucena, C., Chase Jr, C.C. and Hammond, A.C. (2003) *J. Anim. Sci.* **81**: 80.

Porto-Neto, L.R., Bickhart, D.M., Landaeta-Hernandez, A.J., Utsunomiya, Y.T., Pagan, M., Jimenez, E., Hansen, P.J., Dikmen, S., Schroeder, S.G., Kim, E.S. and Sun, J. (2018) *Frontiers in genetics.* **9**: 57.

Sánchez-Rodríguez, H.L. (2019) J. Agric. *Univ. Puerto Rico.* **103(1)**: 69.

Tresoldi, G., Schütz, K.E., & Tucker, C.B. (2020) *J. Dairy Sci.* **103**: 5414.

*Contributed paper*

# SOUTHERN MULTI-BREED RESOURCE POPULATION: GENERATION OF COHORTS ONE AND TWO

**K.A. Donoghue[1], B.J. Walmsley[2,3], J.P. Siddell[3], T. Granleese[4], L. Penrose[3] and P.F. Arthur[5]**

[1]NSW Department of Primary Industries, Agricultural Research Centre, Trangie, NSW, 2823
[2]Animal Genetics and Breeding Unit*, University of New England, Armidale, NSW, 2351
[3]NSW Department of Primary Industries, Livestock Industries Centre, Armidale, NSW, 2351
[4]NSW Department of Primary Industries, Primary Industries Institute, Grafton, NSW, 2460
[5]NSW Department of Primary Industries, Elizabeth MacArthur Agricultural Institute, Menangle, NSW, 2568

## SUMMARY

This paper describes the generation of the first and second cohort of animals in a large 5-year breeding project being conducted across New South Wales involving five temperate beef breeds and the Brahman breed. Females were joined to 154 sires via artificial insemination and back-up matings in 2019 to generate the first cohort of calves, which were born in 2020. Calves were born and raised in mixed breed groups and will be intensively recorded head-to-head for current BREEDPLAN traits and new economically important traits. Birth data have been collected for Cohort One, with records available for 1,398 calves. Traits recorded at birth included birth weight, calving ease, calf fate, calf bellow and calf vigour. Traits to be collected at weaning include weaning weight, hip height, muscle score, docility (crush-side and flight speed), horn/poll phenotype and worm egg count. Relationships between calf bellow and calf vigour scores at birth and subsequent measures of behaviour and production will be assessed. Steers will be backgrounded at one of two locations before entering Tullimba feedlot and subsequently slaughtered, while females will be retained at each site to be joined naturally. Generation of Cohort Two has begun with females joined to 202 sires via artificial insemination and back-up matings in 2020, with 1,535 calves expected to be born in 2021. Data generated from the project will enhance current within-breed genetic evaluations through collection of data for traits that are currently poorly recorded, and by expanding the suite of traits available for selection.

## INTRODUCTION

Currently, beef producers making selection decisions regarding the genetic merit of potential parents of the next generation of progeny in their herds are only able to utilise estimated breeding values (EBVs) and selection indexes generated from within-breed genetic evaluations (Graser *et al.* 2005). In addition, there is interest to enhance current within-breed genetic evaluations by conducting intensive collection of traits that are currently poorly recorded (such as fertility and eating quality), and by expanding the suite of traits to include behaviour, health and welfare traits. A new project is being conducted over the next 5 years (2020 to 2025) known as the Multi-Breed Genomic Beef Cattle Resource or Southern Multi-Breed (SMB) project. This project will collect phenotypes and genotypes on animals from six breeds that have been managed in mixed breed

---

* AGBU is a joint venture of NSW Department of Primary Industries and University of New England

groups at sites across New South Wales (NSW) for existing BREEDPLAN and new economically important traits, such as early-in-life reproduction. This paper provides a description of the generation of the first two cohorts of animals in the project.

## MATERIALS AND METHODS

**Generation of Cohort One.** The animals used in this project were calves born in the first Cohort of the Southern Multibreed Resource population in five NSW Department of Primary Industries (DPI) research herds (Trangie Agricultural Research Centre, Trangie; Grafton Primary Industries Institute, Grafton; Tocal Agricultural Centre, Tocal; Glen Innes Agricultural Research and Advisory Station, Glen Innes; Elizabeth MacArthur Agricultural Institute (EMAI); Menangle) (Walmsley *et al.* 2021). Calves in Cohort One were from six different breeds (Angus, Brahman, Charolais, Hereford, Shorthorn and Wagyu), and all breeds were managed and recorded in mixed groups at each location. Females at each site were mated via a single synchronised artificial insemination (AI) program in late 2019, with back up bulls introduced five days after insemination. A total of 94 sires were represented in the AI program (Angus=21; Brahman=12; Charolais=10; Hereford=25; Shorthorn=13; Wagyu=13) and 60 sires in the natural joinings (Angus=20; Brahman=3; Charolais=6; Hereford=15; Shorthorn=8; Wagyu=8). Purebred joinings were undertaken at all locations except Grafton, where a small amount of crossbreeding involving Brahman reciprocal matings to Angus and Herefords occurred. Matings were allocated using MateSel based on coancestry to limit inbreeding, with a small amount of emphasis placed on the index (Walkom *et al.* 2021). The joining program was staggered across the five sites, with calving occurring from June (Trangie) to October (EMAI) in 2020. Cohort One calves were fully pedigreed and had a comprehensive suite of traits measured at birth based on BREEDPLAN collection protocols. All Cohort One calves were assessed for horn/poll phenotype at marking time (approximately 10 to 12 weeks of age), with dehorning undertaken for those with the horn phenotype. Calves will be serially assessed for horn/poll phenotype in the project (Conners *et al.* 2021). The number of female and male calves by breed of calf at each location is found in Table 1. There were 1,398 calves born in Cohort One.

**Table 1 Number of female (F) and male (M) calves by breed and site in Cohort One**

| | AA[a] | | BB | | CC | | HH | | SS | | WY | | |
|---|---|---|---|---|---|---|---|---|---|---|---|---|---|
| Site[b] | F | M | F | M | F | M | F | M | F | M | F | M | Total |
| TR | 37 | 38 | - | - | - | - | 37 | 31 | - | - | 29 | 30 | 202 |
| GR | 64 | 58 | 62[c] | 67[d] | - | - | 47 | 54 | - | - | - | - | 352 |
| TO | 48 | 40 | - | - | 6 | 14 | - | - | 44 | 40 | - | - | 192 |
| GI | 38 | 26 | - | - | - | - | 28 | 28 | - | - | 19 | 26 | 165 |
| EMAI | 41 | 49 | - | - | 56 | 57 | 44 | 44 | 52 | 57 | 47 | 40 | 487 |
| Total | 228 | 211 | 62 | 67 | 62 | 71 | 156 | 157 | 96 | 97 | 95 | 96 | 1398 |

[a] AA=Angus; BB=Brahman; CC=Charolais; HH=Hereford; SS=Shorthorn; WY=Wagyu
[b] TR=Trangie; GR=Grafton; TO=Tocal; GI=Glen Innes; EMAI=Menangle
[c] BBxBB (n=22), AAxBB (n=13), BBxAA (n=4), BBxHH (n=12) and HHxBB (n=11) calves
[d] BBxBB (n=19), AAxBB (n=10), BBxAA (n=11), BBxHH (n=22) and HHxBB (n=5) calves

At birth, the calves were tagged, and several traits were recorded following standard BREEDPLAN collection procedures, including birth weight, calving ease and calf fate. In addition, measures of calf bellow and calf vigour were recorded for calves at two locations

(Grafton and EMAI). Calf bellow scores were collected in the following categories: 0 (no bellow); 1 (single bellow less than 1 sec) and 2 (single bellow longer than 1 sec or multiple bellows). Calf vigour scores were collected in the following categories: 1 (extremely weak); 2 (weak); 3 (healthy); 4 (vigorous) and 5 (extremely vigorous). Summary statistics across the project for the traits recorded at birth can be found in Table 2, with variation observed in all traits. All traits collected at calving were recorded by farm staff at the research station.

**Table 2 Summary statistics for birth traits in Cohort One**

| Trait | Number of records | Mean | SD | Minimum | Maximum |
|---|---|---|---|---|---|
| Birth weight (kg) | 1,398 | 38.6 | 7.5 | 18 | 61 |
| Calving ease | 1,398 | 1.1 | 0.5 | 1 | 5 |
| Calf fate | 1,398 | 0.09 | 0.4 | 0 | 2 |
| Calf bellow | 838 | 0.59 | 0.79 | 0 | 2 |
| Calf vigour | 839 | 3.25 | 0.79 | 1 | 5 |

Most calves emitted no bellow (60%) during the recording process, while 21% of calves emitted a single bellow of less than 1 second in duration, and 19% emitted either a single bellow longer than 1 second or multiple bellows (Figure 1). The relationship between calf bellow scores at birth and subsequent measures of behaviour (all animals will have crush-side and flight speed docility measures taken at weaning) will be assessed.

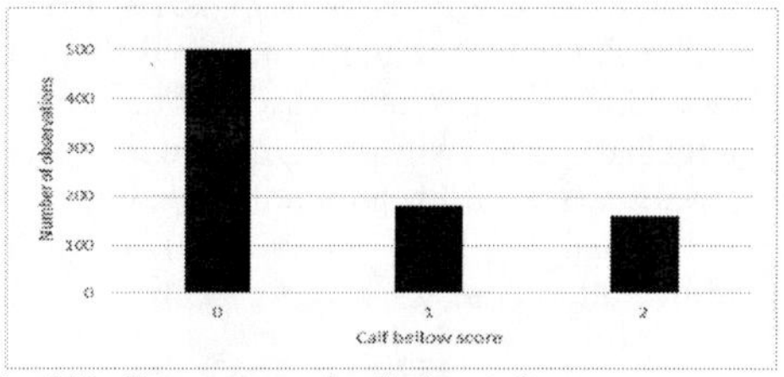


**Figure 1. Distribution of calf bellow scores in Cohort One**

Most calves were healthy or vigorous (87%), with a small number of calves observed to be extremely vigorous (6%) and extremely weak (4%) (Figure 2). The relationship between calf vigour scores at birth and subsequent measures of behaviour and production will be assessed.

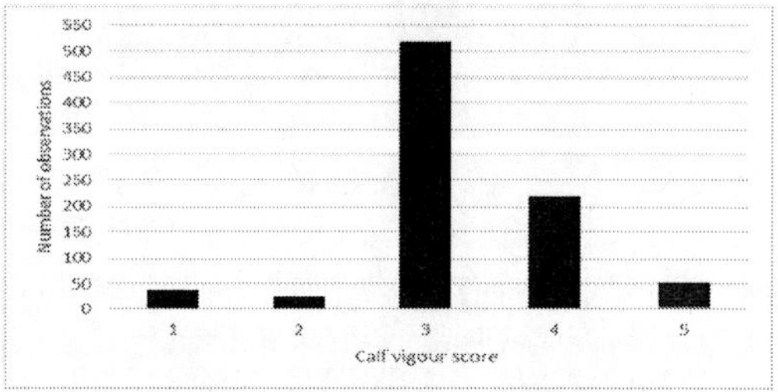


**Figure 2. Distribution of calf vigour scores in Cohort One**

Cohort One animals will be weaned at approximately 8 months of age, with a wide range of traits to be recorded at this time including weaning weight, hip height, muscle score, docility (crush-side and flight speed), horn/poll phenotype (as a repeated record of their phenotype collected at marking) and worm egg count. Steers from Cohort One will be backgrounded at one of two locations before entering Tullimba feedlot and subsequently slaughtered. Cohort One females will be retained at each site to be naturally joined in late 2021. Serial ovarian scans will be undertaken on the females in the lead-up to joining to identify the presence of a corpus luteum as a measure of puberty.

**Generation of Cohort Two.** Females at each site were mated via a single synchronised AI program in late 2020, with back up bulls introduced five days after insemination. A total of 135 sires were represented in the AI program (Angus=36; Brahman=19; Charolais=15; Hereford=29; Shorthorn=16; Wagyu=20) and 67 sires in the natural joinings (Angus=20; Brahman=4; Charolais=8; Hereford=17; Shorthorn=8; Wagyu=10). The joining program was staggered across the five sites, with calving anticipated to occur from June (Trangie) to October (EMAI) in 2021. Females were pregnancy scanned in January/February 2021, and the predicted number of calves in Cohort Two by calf breed and site is contained in Table 3.

**Table 3 Predicted number of calves by breed and site in Cohort Two**

| Site[b] | AA[a] | AAxBB | BB | BBxAA | BBxHH | CC | HH | HHxBB | SS | WY | Total |
|---|---|---|---|---|---|---|---|---|---|---|---|
| TR | 71 | - | - | - | - | - | 62 | - | - | 55 | 188 |
| GR | 90 | 12 | 77 | 46 | 40 | - | 106 | 40 | - | - | 411 |
| TO | 95 | - | - | - | - | 55 | - | - | 92 | - | 242 |
| GI | 77 | - | - | - | - | - | 73 | - | - | 40 | 190 |
| EMAI | 104 | - | - | - | - | 112 | 84 | - | 108 | 96 | 504 |
| Total | 437 | 12 | 77 | 46 | 40 | 167 | 325 | 40 | 200 | 191 | 1535 |

[a] AA=Angus; BB=Brahman; CC=Charolais; HH=Hereford; SS=Shorthorn; WY=Wagyu

[b] TR=Trangie; GR=Grafton; TO=Tocal; GI=Glen Innes; EMAI=Menangle

## CONCLUSIONS

Data collection on the first cohort of animals in the Southern Multibreed resource population has commenced, with birth data recorded the first cohort of calves. Intensive data collection is planned on these animals at weaning, followed by collection of production and carcase data on steers in the feedlot and abattoir, and on heifers prior to joining and subsequent calving. This information, along with genotypes on sires, dams and calves, will be available to enhance current within-breed genetic evaluation as well as provide records for animals from multiple breeds managed in mixed breed groups.


## ACKNOWLEDGEMENTS

The Southern Multi-Breed project is an initiative co-funded by NSW Department of Primary Industries, University of New England and Meat and Livestock Australia Donor Company. The authors give thanks to the efforts of the research station managers and their staff, our skilled team of technicians, ultrasound scanners, AI, DNA lab, data, project managers and scientists. Individual breeders, participating breed societies and the southern beef industry are thanked for their support.


## REFERENCES


Connors, N.K., Walmsley, B.J. and Donoghue, K.A. (2021) *Proc. Assoc. Advmt. Anim. Breed. Genet.* **24**: *These Proceedings.*

Graser, H.-U., Tier, B., Johnston, D.J., and Barwick, S.A. (2005) *Aust. J. Exp. Agric.* **45**:913.

Walkom, S.F., Donoghue, K.A., Arthur, P.F., Clark, S.A., and Walmsley, B.J. (2021) *Proc. Assoc. Advmt. Anim. Breed. Genet.* **24**: *These Proceedings.*

Walmsley, B.J., Donoghue, K.A., Johnston, D.J., Clark, S.A., Siddell, J.P, Granleese, T. and Arthur, P.F. (2021) *Proc. Assoc. Advmt. Anim. Breed. Genet.* **24**: *These Proceedings.*

*Contributed paper*

# THE VALUE OF LIVE-ANIMAL ULTRASOUND SCANNING OF BREEDING CANDIDATES FOR CARCASE TRAITS IN THE AGE OF GENOMICS

**C.J. Duff[1], B.J. Crook[2], A.I. Byrne[1] and M.J. Reynolds[1]**

[1] Angus Australia, 86 Glen Innes Road, Armidale, New South Wales, Australia, and
[2] Agricultural Business Research Institute, University of New England, 2351, Armidale, Australia

## SUMMARY

A common question from Angus seedstock producers is "what is the value of live-animal ultrasound scanning of breeding candidates for carcase traits, particularly young bulls, if they are already genomic tested for genetic evaluation and underpinned by a reference population with carcase data". To help answer this question, 3 ultrasound scan phenotyping scenarios were analysed through the Trans-Tasman Angus Cattle Evaluation (TACE) to produce and compare the subsequent eye muscle area (EMA), intramuscular fat (IMF), rib fat (RIB) and rump fat (RUMP) Estimated Breeding Values (EBVs) and their accuracies. This study shows that ultrasound scanning of genotyped bulls does provide some "value" for breeding programs in terms of increasing accuracy to carcase EBVs across all traits and scenarios. However, the value differs by trait (e.g. more influence on EMA EBV compared to IMF EBV) and by scenario (e.g. more influence from heifer scans, particularly on IMF, RIB and RUMP EBVs, compared to bull scans, because of the differences in genetic parameters for the bull and heifer ultrasound scan traits). Further work is required to understand at a herd and population level the impact of a reduction in ultrasound scan phenotyping, particularly on genotyped bulls, coupled with an increasing number of direct carcase phenotypes in the Angus Australia genomics reference population.

## INTRODUCTION

A common breeding objective for beef producers is to improve carcase traits of animals used in breeding programs. Traditionally, carcase traits have proven expensive and difficult to measure and they cannot be measured on selection candidates. Due to this limitation, breeders use correlated ultrasound scan measurements on the live animal to increase selection accuracy for breeding objective traits related to meat quantity and quality, including eye muscle area (EMA), rib fat (RIB) rump fat (RUMP) and intramuscular fat (IMF). Since becoming available in the mid-1990s, ultrasound scanning for carcase attributes has been widely adopted in beef cattle breeding programs. For example, over 650,000 animals have live-animal ultrasound scan records in the Angus Australia performance database. These phenotypes are included in the Trans-Tasman Angus Cattle Evaluation (TACE) and, as correlated traits, are used to inform the carcase Estimated Breeding Values (EBVs).

A recent alternative method for carcase trait selection is through genomic testing selection candidates and including the genomic profiles in single-step genetic evaluation programs (Johnston *et al.* 2019), such as TACE. The value of the genomic information is directly related to the underlying reference population of phenotypes coupled with genotypes, as described by Goddard *et al.* (2010). With these alternative methods for selection now available, a common question from Angus seedstock producers is "what is the value of live-animal ultrasound scanning of breeding candidates for carcase traits, particularly young bulls, if they are already genomic tested for genetic evaluation and underpinned by a reference population with carcase data". This was modelled for the carcase intra-muscular (IMF) and marbling traits by Duff *et al.* (2019) and concluded that the value of ultrasound scan phenotyping for IMF diminishes as the prediction accuracy of the genomic breeding value (GBV) increases.

This study further explores the answer to this question in the commercial genetic evaluation environment by comparing carcase EBVs and accuracies for defined groups of genotyped Angus breeding cattle under three phenotyping scenarios.

## MATERIALS AND METHODS

In collaboration with the Agriculture Business Research Institute (ABRI), 3 separate research analyses (herein reported as scenarios) of TACE were undertaken to produce a range of Estimated Breeding Values (EBV) and accuracies. TACE is underpinned by the BREEDPLAN software as described by Graser *et al.* (2005), and the single-step component to incorporate genomic information as outlined by Johnston *et al.* (2019). These analyses utilised the phenotype, pedigree and genotype extracts provided by Angus Australia for the mid-August 2020 TACE. The 3 scenarios were:

- Scenario 1: All data available included in the analysis (i.e. standard analysis).
- Scenario 2: As with scenario 1, but with exclusion of bull ultrasound scan phenotypes for eye muscle area (UEMA), rib fat (URIB), rump fat (URUMP) and intramuscular fat (UIMF) recorded from 1st January 2019 onwards.
- Scenario 3: As with scenario 1, but with exclusion of bull, heifer and steer ultrasound scan phenotypes for SEMA, SRIB, SRUMP and SIMF recorded from 1st January 2019 onwards.

The number of ultrasound scan phenotypes, direct carcase phenotypes and genotypes included in each scenario is listed in Table 1, showing scenario 2 and scenario 3 having approximately 20,000 and 40,000 less ultrasound scan records analysed respectively, per trait, compared to scenario 1, while the number of carcase phenotypes and genotypes remained constant. Additionally, approximately 4,000 animals have both a genotype and a direct carcase phenotype, forming an effective segment of the Angus Australia genomics refence population and influencing the EBVs and accuracies of all genotyped animals.

**Table 1. Count of ultrasound scan phenotypes, carcase phenotypes and genotypes included in each scenario based on mid-August 2020 TACE extract**

| | Ultrasound Scan Phenotypes | | | | Direct Carcase Phenotypes | | | | |
|---|---|---|---|---|---|---|---|---|---|
| Scenario | UEMA | UIMF | URIB | URUMP | CEMA | CIMF | CRIB | CRUMP | Genotypes |
| 1 | 643,153 | 594,372 | 642,217 | 642,005 | 7,392 | 13,092 | 5,319 | 14,793 | 95,180 |
| 2 | 622,795 | 573,808 | 621,932 | 622,055 | 7,392 | 13,092 | 5,319 | 14,793 | 95,180 |
| 3 | 603,814 | 554,749 | 603,295 | 602,832 | 7,392 | 13,092 | 5,319 | 14,793 | 95,180 |

The resulting EBVs for EMA, IMF, RIB and RUMP and their accuracies were compared across the 3 scenarios, focussing on young bulls, born in 2018 and 2019, that had genotypes included in all scenarios and additionally had ultrasound scan phenotypes included in scenario 1 (n=9,089).

## RESULTS AND DISCUSSION

For the 3 scenarios, the mean and standard deviation for the carcase EBVs and accuracies, along with EBV correlations are shown in Tables 2 to 5 (one table per trait).

For all carcase EBVs, the mean EBV remained constant across scenarios, with an associated reduction in EBV standard deviation from scenarios 1 to 2 and 2 to 3, being the lowest in all cases for scenario 3. This is also matched with a reduction in EBV accuracy from scenario 1 to 2 and 2 to 3, being again the lowest in scenario 3. The correlations between carcase EBVs were strong and positive in all cases (>0.92) with the weakest correlation observed between scenarios 1 to 3. This is expected as the largest portion of ultrasound scan phenotypes were excluded from scenario 3.

Comparing the carcase traits, the least amount of change was observed for the carcase IMF EBV between scenario 1 and 2, reflected in both for change in EBV accuracy, from 59.5% to 58.6%, and high EBV correlation of 0.989. The carcase trait with the most change in EBV was EMA, between scenario 1 and 2, the change in EBV accuracy was 60.0% to 58.2%, and EBV correlation of 0.957. This is partly explained by differences in the genetic parameters used in TACE for bull ultrasound scan traits, with a bull UIMF heritability of 0.17 being lower than bull UEMA heritability of 0.24. Additionally, bull UIMF has a weaker genetic correlation with CIMF of 0.60, compared to bull UEMA to CEMA correlation of 0.70. In general, this means that bull UIMF phenotypes have less influence on the IMF EBVs and accuracies compared to the bull UEMA influence on the EMA EBVs and accuracies. The results for RIB and RUMP EBVs were closer to those observed for the EMA EBV.

**Table 2. EMA EBV and accuracy means, standard deviations and EBV correlations**

| | EMA EBV ($cm^2$) | | | Accuracy (%) | | |
|---|---|---|---|---|---|---|
| Scenario | 1 | 2 | 3 | 1 | 2 | 3 |
| Mean | +6.2 | +6.2 | +6.1 | 60.0 | 58.2 | 56.7 |
| SD | 2.89 | 2.73 | 2.67 | 3.97 | 4.67 | 4.92 |
| Correlation to Scenario 1 | 1.00 | 0.957 | 0.944 | - | - | - |

**Table 3. IMF EBV and accuracy means, standard deviations and EBV correlations**

| | IMF EBV (%) | | | Accuracy (%) | | |
|---|---|---|---|---|---|---|
| Scenario | 1 | 2 | 3 | 1 | 2 | 3 |
| Mean | +2.3 | +2.3 | +2.3 | 59.5 | 58.6 | 56.9 |
| SD | 1.03 | 1.02 | 1.01 | 4.38 | 4.64 | 4.91 |
| Correlation to Scenario 1 | 1.00 | 0.989 | 0.980 | - | - | - |

**Table 4. RIB EBV and accuracy means, standard deviations and EBV correlations**

| | Rib Fat EBV (mm) | | | Accuracy (%) | | |
|---|---|---|---|---|---|---|
| Scenario | 1 | 2 | 3 | 1 | 2 | 3 |
| Mean | +0.0 | +0.0 | +0.0 | 63.9 | 62.9 | 61.6 |
| SD | 1.52 | 1.45 | 1.43 | 3.75 | 4.05 | 4.34 |
| Correlation to Scenario 1 | 1.00 | 0.964 | 0.947 | - | - | - |

**Table 5. RUMP EBV and accuracy means, standard deviations and EBV correlations**

| | Rump Fat EBV (mm) | | | Accuracy (%) | | |
|---|---|---|---|---|---|---|
| Scenario | 1 | 2 | 3 | 1 | 2 | 3 |
| Mean | -0.4 | -0.4 | -0.4 | 61.5 | 59.8 | 58.6 |
| SD | 1.76 | 1.63 | 1.60 | 3.64 | 4.28 | 4.59 |
| Correlation to Scenario 1 | 1.00 | 0.941 | 0.929 | - | - | - |

While changes in EBV spread, accuracy and correlation between scenarios are informative, for breeding candidate selection, understanding the change in EBVs for traits in the breeding objective can be more useful. To illustrate this, the distribution of change for the EMA EBV and IMF EBV are shown in Figures 1 and 2 respectively. This shows that for the IMF EBV, comparing scenario 1 to 2, 70% of bulls did not change by more than ±0.1 % units (or approximately 1/10$^{th}$ of the IMF EBV SD), while for scenario 1 to 3, this decreases to 51% of bulls. In contrast, for EMA EBV, comparing scenario 1 to 2, 35% of bulls did not change by more than ±0.3 % $cm^2$ units (or approximately 1/10th of the EMA EBV SD), while for scenario 1 to 3, this decreases to 30% bulls. This demonstrates that there is less change and associated re-ranking of breeding candidates for the

IMF EBV, across scenarios, compared to the changes observed for EMA EBV. There is also more re-ranking when comparing scenario 1 to 3, compared to scenarios 1 to 2.

**Figure 1. Change in EMA EBVs comparing scenario 1 to 2 and 1 to 3**

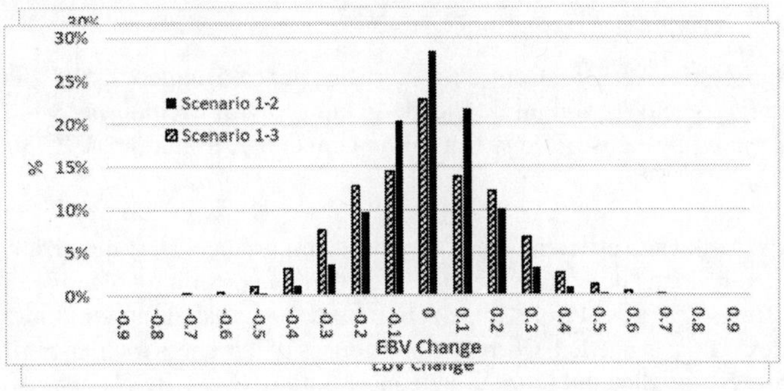


**Figure 2. Change in IMF EBVs (%) comparing scenario 1 to 2 and 1 to 3**

## CONCLUSIONS

This study shows that ultrasound scanning of genotyped bulls does provide some "value" for breeding programs in terms of increasing accuracy to carcase EBVs across all scenarios. However, the value differs by trait (e.g. ultrasound scanning had more influence on EMA EBV compared to IMF EBV) and by scenario (e.g. ultrasound scanning heifers had more influence, particularly on IMF, RIB and RUMP EBVs, compared to bull scans, because of the differences in the genetic parameters for bull and heifer ultrasound scan traits). Before breeding program design advice can be confidently provided, additional research is required, at both a herd and population level, to further understand the cost:benefit relationship and the overall impact of a reduction in ultrasound scan phenotyping, particularity on genotyped bulls, coupled with an increasing number of direct carcase phenotypes in the Angus Australia genomics reference population.

## REFERENCES

Duff CJ, van der Werf JHJ, Parnell P.F and Clark SA (2019) *Proc. Assoc. Advmt. Anim. Breed. Genet.* **23**:496.

Goddard ME, Hayes BJ, & Meuwissen, THE. (2010). *Genetics Research*, **92(5-6)**, 413.

Graser H-U., Tier B., Johnston D.J. and Barwick S.A. (2005) *Aust. J. Exp. Agric.* **45**: 913

Johnston DJ, Ferdosi MH, Connors NK, Boerner V, Cook J, Girard CJ, et al. (2018) WCGALP: 11–16 February 2018; Auckland. 2018.

*Contributed paper*

# REDUCING LAMENESS AND URINARY NITROGEN EXCRETION THROUGH SELECTION ON NEXT GENERATION NATIONAL DAIRY SELECTION INDICES

**J.E. Newton[1], M. Dunstan[2], P.N. Ho[1], M. Khansefid[1], I. van den Berg[1], S.S. Chauhan[2] and J.E. Pryce[1,3]**

[1] Agriculture Victoria, AgriBio, Centre for AgriBioscience, Bundoora, VIC 3083, Australia
[2] Faculty of Veterinary and Agricultural Sciences, University of Melbourne, VIC 3647, Australia
[3] School of Applied Systems Biology, La Trobe University, Bundoora, VIC 3083, Australia

## SUMMARY

Improving cow health and efficiency is of economic importance. New health and efficiency traits need to be considered alongside traits which are already in the breeding objective. Thus, quantifying the correlated responses of novel traits to selection on the Australian national index, the Balanced Performance Index (BPI), is needed. Correlated responses of lameness incidence (LI) and milk mid-infrared spectroscopy predicted blood urea concentration (MIR-BUN), were estimated under selection on the current BPI and on an updated BPI, including these traits in the breeding objective, using the MT Index tool. Not all genetic correlations ($r_g$) for MIR-BUN were available, so missing values were assumed to be zero. Under the current BPI, LI and MIR-BUN increased undesirably, by +0.309%/year and +0.040mmol/L/year, respectively. A sensitivity analysis, varying $r_g$ between LI, MIR-BUN and BPI traits, also found undesirable responses for these traits giving confidence they are moving in undesirable direction. Finally, the economic values required to achieve a desirable response (i.e., reduction) in these traits under selection on the BPI were calculated. We found large economic values of at least –\$350 and –\$91 were required before reductions in LI and MIR-BUN, respectively, were observed. While desired response is achievable, the economic value of LI found in this study exceeds previously reported cost of lameness. A greater emphasis on recording health and efficiency traits, especially in the genomic reference population, will support greater selection response for health and efficiency traits at more moderate economic values. While novel phenotyping approaches like MIR could increase the number of animals with direct and indirect records for traits of interest, like MIR-BUN, further work to understand the underlying biological mechanisms and true economic value of these traits in pasture-based herds is needed.

## INTRODUCTION

Health, welfare and environmental traits are a key focus of breeding value development with a global shift in dairy cattle breeding objectives to incorporate more non-production phenotypes. This can be attributed, in part, to the compilation of datasets with enough health records for genetic parameter estimation of economically important, but lowly heritable traits, e.g., lameness (Khansefid *et al.* 2021). In addition, the development of new phenotypes and indicator traits is being facilitated by emerging technologies like mid infrared spectroscopy (MIR). For example, urinary nitrogen (UN) excretion is of environmental and economic significance to the dairy industry. While UN can be predicted from blood urea nitrogen (BUN) (Kohn *et al.* 2005), BUN is both cost and labour prohibitive to measure in large populations. However, as $r_g$ between BUN and BUN predicted from mid-infrared spectroscopy of milk (MIR-BUN) is >0.95 (Van den Berg *et al.* 2021), BUN and MIR-BUN are genetically analogous to one another so MIR-BUN can be used as an indicator trait for UN excretion.

Currently the BPI, the Australian dairy industry's national selection index, includes production, longevity, fertility, health and conformation traits (Byrne *et al.* 2016, DataGene 2020). The BPI aligns the preferences of Australian dairy farmers across production, functional and type-focused traits along with their economic importance, to drive genetic gains towards the industry's national breeding objectives (NBO) (Byrne *et al.* 2016). As new measures of cow performance and their genetic parameter estimates become available, there is a need to understand how current selection practices are impacting these new traits. Our aim was to estimate the correlated response to selection for lameness and MIR-BUN, under selection on the current BPI and on an updated BPI, including these traits in the breeding objective.

## MATERIALS AND METHODS

The selection index program MTIndex (van der Werf 2008) was used to estimate the correlated response of LI and MIR-BUN under selection on BPI and BPI adjusted to include either trait. The current BPI was constructed in MTIndex and populated with parameters reflective of the current Australian Holstein population sourced from Byrne *et al.* (2016) and DataGene (2020). LI was analysed as a binary trait, where a value of 1 indicated a cow who showed incidence of clinical lameness at any point during lactation, and a value of 0 indicating no incidence. MIR-BUN was measured as a continuous trait, MIR-predicted concentration of urea in blood (mmol/L) during early lactation. Genetic parameters for MIR-BUN and LI, shown in Table 1, were sourced from van den Berg *et al.* (2021) and Khansefid et al. (2021) and supplemented with correlations between EBVs from (Luke *et al.* 2019). Where no estimates were available $r_g$ of 0 were assumed.

**Table 1. Lameness (LI) and MIR-predicted blood urea nitrogen (MIR-BUN), phenotypic standard deviation ($\sigma_p$), accuracy (acc), heritability ($h^2$), and genetic correlations ($r_g$) with standard error in brackets**

| Trait (units) | $\sigma_p$ | acc | $h^2$ | $r_g$ with traits in the BPI[1] | | | | | | | | | |
|---|---|---|---|---|---|---|---|---|---|---|---|---|---|
| | | | | MY | PY | FY | SU | F | SC | OT | MS | UD | PS |
| LI (%) | 1.74 | 0.38 | 0.006 | 0.31 (0.09) | 0.26 (0.09) | 0.15 (0.09) | -0.02 (0.09) | 0.16 (0.08) | - | -0.18 (0.09) | -0.17 (0.10) | -0.05 (0.09) | 0.02 (0.09) |
| MIR-BUN (mmol/L) | 1.79 | 0.37 | 0.22 | -0.16 (0.14) | -0.10 (0.14) | 0.27 (0.14) | 0.38 (0.10[2]) | 0.26 (0.10[2]) | 0.23 (0.10[2]) | - | - | - | - |

[1]Traits: MY: milk yield, PY: protein yield, FY: fat yield, SU: survival, F: fertility, SC: somatic cell count, OT: overall type, MS: mammary system, UD: udder depth, PS: pin set. $r_g$ with feed saved, milking speed and temperament unavailable. [2]$r_g$ from EBV correlations in Luke et al. (2019), standard error of 0.1 assumed

Due to the large standard errors and preliminary nature of $r_g$ available, a sensitivity analysis was performed using $r_g \pm 2$ standard errors with key BPI traits. BPI traits were chosen on the criteria of having high contribution to the index (high economic value) or being of physiological importance to LI or MIR-BUN. The $r_g$ with fertility, survival and protein yield were analysed for both traits; and with overall type for LI; and with fat yield for MIR-BUN. Finally, a desired gains approach was used to estimate the minimum economic value required to achieve selection response in the desired direction for LI or MIR-BUN.

## RESULTS AND DISCUSSION

Under selection on the BPI, correlated responses of a 0.31% increase in LI incidence/year and a 0.04 mmol/L/year increase in MIR-BUN were seen (Table 2). The annual response of the BPI estimated here, $29.98/year, is slightly higher than what is being achieved in the current population (DataGene 2020). This is likely due to factors other than the BPI influencing breeding decisions such as; semen cost and availability, prioritising other selection criteria and the use of overseas indices. The sensitivity analysis of $r_g$ between BPI traits and LI and MIR-BUN showed LI and MIR-

BUN response increased in all scenarios, except when a weaker correlation ($r_g$ – 2SE) with protein yield was assumed (Table 3). Thus, despite the preliminary nature of $r_g$ we are reasonably confident LI and MIR-BUN are increasing under current selection practices. Assuming the breeding goal is to reduce LI and to reduce MIR-BUN to reduce UN, both traits are currently moving in the undesirable direction. To include LI and MIR-BUN in the BPI and reverse this trend large negative economic values would be required, -$350 and -$91, respectively (Table 2). A lower economic value of $187.13-$243.17/cow/calving interval has previously been reported for lameness (Byrne *et al.* 2016). To our knowledge no estimates of the economic value of UN or MIR-BUN exist for Australian conditions.

**Table 2. Annual response to selection on the BPI, BPI + Lameness (LI) and BPI + MIR-predicted BUN and minimum economic values (EV) to achieve desired response in LI and MIR-BUN**

| Breeding objective | Current BPI | | BPI + LI | | BPI + MIR-BUN | |
|---|---|---|---|---|---|---|
| trait | Response | EV ($) | Response | EV ($) | Response | EV ($) |
| BPI ($) | 29.98 | - | 29.38 | - | 28.05 | - |
| Lameness (%) | 0.309 | - | –0.01 | –$350 | - | - |
| MIR-BUN (mmol/L) | 0.04 | - | - | - | –0.0004 | –$91 |

**Table 3: Annual correlated response to selection on the BPI for lameness incidence (LI) and MIR-predicted blood urea concentration (MIR-BUN) for a sensitivity analysis of genetic correlations ($r_g$) ±1 or 2 standard errors[1] (SE) between LI with fertility, survival, protein yield and overall type and MIR-BUN with fertility, survival, protein yield and fat yield**

| Trait | $r_g$ with | Magnitude of $r_g$ | | | |
|---|---|---|---|---|---|
| | | $r_g$ -2SE | $r_g$ -1SE | $r_g$ +1SE | $r_g$ +2SE |
| LI (%) | Fertility | 0.17 | 0.239 | 0.379 | 0.448 |
| | Survival | 0.198 | 0.253 | 0.365 | 0.42 |
| | Protein yield | –0.118 | 0.095 | 0.523 | 0.736 |
| | Overall type | 0.332 | 0.32 | 0.298 | 0.289 |
| MIR-BUN (mmol/L) | Fertility | 0.029 | 0.034 | 0.045 | 0.051 |
| | Survival | 0.032 | 0.036 | 0.044 | 0.048 |
| | Protein yield | –0.003 | 0.018 | 0.061 | 0.083 |
| | Fat yield | 0.024 | 0.032 | 0.048 | 0.056 |

[1]Trait $r_g$ and SE reported in Table 1

This study assumed UN excretion could be improved through selection on MIR-BUN given the relationship between UN and BUN (Kohn *et al.* 2005) and the strong $r_g$ between BUN and MIR-BUN (van den Berg *et al.* 2021). MIR-BUN is an example of novel phenotypes being developed through emerging phenotyping technologies. While the current dataset is small (n = 9158) which contributes to its low accuracy, as MIR-BUN is derived from a milk sample, in future it could be available on all cows with milk records. This and a moderate heritability could offer good opportunities for genomic prediction which could make MIR-BUN a more accessible indicator trait for UN excretion than BUN and a good candidate for including in the BPI. However, selecting for lower MIR-BUN concentrations may not always be desirable. A lower threshold of 1.7mmol/L BUN has been used as a biomarker (indicator) for metabolic health (Luke *et al.* 2019). While an opportunity may exist to improve UN excretion via selection on MIR-BUN without compromising animal health, the appropriate direction of selection pressure on BUN/MIR-BUN remains unclear, especially in pasture-based countries. More knowledge about the range and thresholds of BUN in Australian pasture-based herds and what the biological and economic consequences are for selecting on MIR-BUN is needed. Additionally, if there is a need to select for an optimum range of MIR-BUN, economic values may differ widely by region or herd making it better suited as a standalone EBV, allowing farmers to customise their breeding goals, rather than in the BPI.

As with many health traits, LI response to selection is limited by incomplete recording, low genetic parameter accuracies and low trait heritability. More robust methods for identifying and recording lame cows on farm (especially early detection) could improve management of lameness on-farm, reducing the direct and indirect costs of lameness, as well as improve genetic parameter estimates (Khansefid *et al.* 2021). LI selection response and accuracy could also be improved by combining direct and indirect health traits into a composite health trait (Khansefid *et al.* 2021), as implemented for mastitis resistance in Australian dairy herds (DataGene 2020).

Novel high throughput phenotyping technologies – like MIR – that produce a large amount of data which could be used for multiple purposes are an exciting new opportunity in animal breeding. They offer an opportunity to predict many traits from a single sample, the ability to capture direct and indirect phenotypes for hard or expensive to measure traits (i.e. lameness, BUN) and also develop novel phenotypes. As we continue into this data-rich era it is important to invest in understanding the economic importance and underlying physiological and biological actions of these traits to fully understand the implications for future breeding objectives.

## CONCLUSION

This study shows selection on Australia's national selection index for dairy (known as BPI) is expected to result in more cases of lameness and an increase in urinary nitrogen excretion. While desired response is achievable using large negative economic values on LI and MIR-BUN within the BPI, these values exceed previously reported economic values for LI. True economic value for UN excretion or MIR-BUN is to be investigated. Novel phenotyping approaches like MIR may facilitate the rapid increase in animals with phenotypes for traits of interest. However, further work to understand the true economic and animal health costs associated with these new traits in an Australian dairy context is needed.

## ACKNOWLEDGEMENTS

This study was undertaken as part of the DairyBio program, which is jointly funded by Dairy Australia, Agriculture Victoria and The Gardiner Foundation and was a MAgSc research project at the University of Melbourne

## REFERENCES

Byrne TJ, Santos BFS, Amer PR, Martin-Collado D, Pryce JE, Axford M (2016) *J. Dairy Sci.* **99**: 8146.

DataGene (2020) National Breeding Objective 2020 Options Paper. V2. Melbourne, Australia.

Khansefid M, Haile-Mariam M and Pryce JE (2021) *J. Dairy Sci.* (*in press*)

Kohn RA, Dinnenn MM, Russek-Cohen E (2005) *J. Anim. Sci.* **83**: 879.

Luke TDW, Nguyen TTT, Rochfort S, Wales WJ, Richardson CM, Abdelsayed M, and Pryce JE (2019). *J. Dairy Sci.* **102**(12): 11142.

van der Werf, J. 2008. Multiple Trait Selection Index. https://jvanderw.une.edu.au/software.htm.

van den Berg I, Ho PN, Luke TDW, Haile-Mariam M, Bolormaa S, and Pryce JE (2021). *J. Dairy Sci.* **104**(1): 575.

*Contributed paper*

# GENOMIC PREDICTION OF LIFETIME PRODUCTIVITY IN BRAHMAN COWS

**B.E. Engle[1], A. Collins Snr[2] and B.J. Hayes[1]**

[1] Queensland Alliance for Agriculture and Food Innovation, University of Queensland, St Lucia, QLD, 4072 Australia
[2] Collins Belah Valley Brahman Stud, Marlborough, QLD, 4705 Australia

## SUMMARY

A cow's ability to raise and wean a calf year after year is a primary determinant of profitability in beef production. This lowly heritable trait is impacted by environmental conditions, making it difficult to improve through genetic selection. Given this, there is a heightened interest from industry to develop selection tools that may assist the selection of breeding animals for improved lifetime fertility. The objective of this study was to assess the feasibility and accuracy of a genomic breeding value to capture a cow's potential for lifetime productivity. In this study, yearly mating and calving records were collected from Brahman cows born between 1988-2010. These were used to classify animals based upon whether or not they met a stayability threshold of raising four calves by six years of age (n = 5,516). Relationships between animals was estimated by combining genomic (n = 3,759) and pedigree (n = 11,578) information into an H matrix, and variance components and breeding values were estimated using the blupf90 program suite. Stayability at six years of age was found to be lowly heritable, $h^2 = 0.13$. Despite this, when breeding values were estimated using single step best linear unbiased prediction, a reasonable prediction accuracy was achieved (0.35). This work demonstrates both the potential and limitations of a breeding value for a cow's potential for lifetime productivity, using the intermediate trait, stayability at six years of age.

## INTRODUCTION

A cow's ability to raise and wean a calf year after year is a primary determinant of profitability in beef production. A cow's lifetime productivity, measured in number of calves weaned, is the ultimate female fertility trait representing the cumulative effects of puberty with yearly joining, pregnancy, calving, weaning, and rebreed for as long as a cow remains productive in her given production system. A cow must produce enough offspring to offset the cost of heifer development and yearly maintenance, and generate a profit. Within intensive, temperate production systems common in the United States, this breakeven point is generally considered to be five calves by six years of age (Snelling *et al.* 1995). However, this threshold is likely to be earlier within the more extensive, low-input systems common in Australia's north.

Like many other reproductive traits, this lowly heritable phenotype is greatly affected by variable environmental conditions. For example, (rectal) temperature is unfavourably correlated with both pregnancy rate and days to calving, whereas temperature increases fertility decreases (Burrow 2001). This relationship is exacerbated by the increasingly difficult production environments brought about by climate change and is particularly felt by beef cattle producers throughout Queensland and northern Australia. Consequently, there is a heightened interest from the northern beef industry to develop selection tools that may assist the selection of breeding animals for improved lifetime fertility, especially in tropically adapted beef breeds. However, it is a difficult and expensive trait to measure, as cows must have fully reached maturity and be leaving the herd before their lifetime productivity may be fully characterised.

Assessing stayability, or a cow's probability of surviving to a specific age given the opportunity to first reach that age, is often a viable alternative. Previous work to genomically select for lifetime cow productivity in tropically adapted cattle have focused on intermediate life traits as a proxy for lifelong reproduction, such as weaning rate up to six years of age in Brahman (Johnston *et al.* 2014;

Zhang *et al.* 2014) or ability to produce 4 or more calves by 76 months of age in Nellore cattle (Ramos *et al.* 2020).

Tropically adapted cattle raised in Queensland and northern Australia often have a high *Bos indicus* content, are older at the onset of puberty, and are therefore commonly bred as heifers at approximately two years of age. If a cow were to successfully wean a calf every year within this system, she should raise four calves by six years of age. This is milestone is likely the breakeven point within extensive, low input productions systems, with any calves produced after this point generating net profit. Therefore, the objective of this study was to assess the feasibility and accuracy of a genomic breeding value for a cow's potential for lifetime productivity, or stayability to six years of age.

## MATERIALS AND METHODS

This project used data and samples collected during the course of the commercial management of the herd and before the commencement of the project. Animal ethics approval was not required for these analyses.

For this study, lifetime productivity of a Central Queensland Brahman cow herd was assessed. Born between 1988 and 2010, these cows were part of a stud herd that has been developed with a heavy emphasis on fertility, where failure to produce a calf was the primary culling criterion. As heifers, these cows were first exposed to a natural service bull at approximately 2 years of age and given a 4-5 month joining window. After which, they are expected to maintain a 365-day calving interval, year after year. This is a long joining period, creating significant variation with respect to when cows calve. This may impact a cow's ability to get back in calf within a year, that is those with the younger calves will find it harder to get back in calf. However, this is a typical situation in northern Australia. Stayability to six years of age was measured as a binary threshold trait (n = 5,516), where cows that successfully gave birth to four calves by six years of age were scored as '1' and those that did not reach this milestone were scored as '0', provided that she was given the chance to calve as a heifer with record of bull exposure at two years of age. Only cohorts that included animals that had reached six years of age were considered.

Starting in 2016, all bulls, cows and calves were genotyped. Genotypes on 3,759 animals were generated using the Geneseek TropBeef V2 array, with 50,045 SNP (after quality control, with genotypes with QC score <0.6 set to missing, monomorphic SNP excluded and SNP with all heterozygous calls excluded). All genotypes were imputed to 709,000 SNP from the Bovine HD array (following further QC) using 4,506 cattle genotyped with the Bovine HD array (including a large number of Brahman, Droughtmaster and Santa Gertrudis cattle). Eagle (Loh *et al.* 2016) was used for phasing, and Minimach3 (Das *et al.* 2016) was used for imputation.

In order to incorporate all available stayability phenotypic records, including those that were ungenotyped, a single-step approach was taken. Genomic relationship was estimated by combining all of the available genomic and ten generations of pedigree information (n = 11,578) into an H matrix, using single step procedures in the blupf90 program suite (Legarra *et al.* 2009). Variance components for stayability at six years of age was estimated using restricted maximum likelihood algorithms in the program remlf90, and genomic breeding values were estimated using single-step genomic best linear unbiased predictions in blupf90. Contemporary group was defined by the cow's year and month of birth, the cow's management cohort. Contemporary group was fitted as a random effect in the model as many cohorts were small (range of 9-175 animals per cohort).

The predictive ability of the breeding values were investigated using a forward validation where data from the youngest cohort of cows with stayability phenotypes (born in 2010, n = 246) was dropped from the model and used as the validation population while the remaining, older cows served as the reference set (n = 5,270). Validation accuracy was calculated as the correlation between the estimated breeding value and the actual phenotype, divided by the square root of the heritability.

**RESULTS AND DISCUSSION**

In this particular Brahman herd, approximately 71% of all females successfully calved as heifers (Figure 1). This was slightly lower than the median heifer pregnancy rates reported by McGowan *et al.* (2014) of 80% across northern Australia, but in line with rates observed in the Northern Genomics project of 70% (Copley *et al. submitted for review).* Of those females that successfully calved as heifers, 63% were successfully rebred the subsequent joining season. However, of the cows that successfully calved as heifers, only 41% had three consecutive calves and only 29% gave birth to four consecutive calves. This is higher than the success rate reported by Ramos *et al.* (2020) in Brazilian Nellore cattle, where only 19% of females achieved four calvings by six years of age.

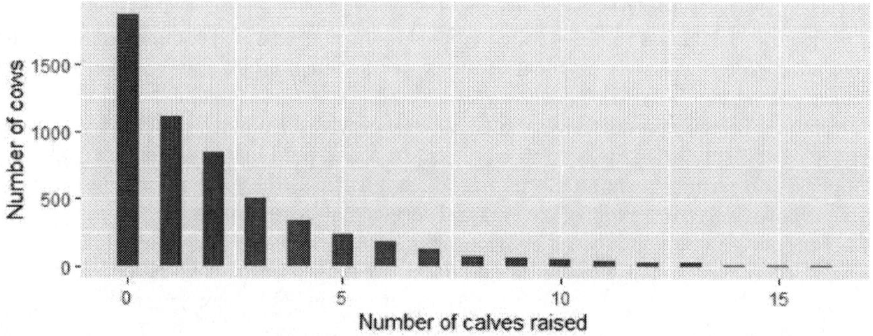


**Figure 1. Distribution of lifetime number of calves produced in this population of Brahman females**

Heritability for stayability to six years of age was low, $h^2 = 0.13$ ($\sigma_a^2 = 0.023$), but expected given the complexity of this trait and the large role environment and management decisions play on longitudinal fertility. This is comparable to heritabilities of similar traits in Brahman, such as average weaning rate up to 6 years of age ($h^2 = 0.11$; Johnston *et al.* 2014). This estimate is higher than the non-genomic heritability for stayability to six years (raised five calves) in American taurine cattle ($h^2 = 0.11$; Snelling *et al.* 1995). However, this result is lower than the heritability for binary stayability at 76 months reported by Ramos *et al.* (2020) in Nellore cattle ($h^2 = 0.14$)

Predictive ability of the estimated breeding values was tested using a forward validation. The forward validation predicted the performance of the youngest cohort of cows with an accuracy of 0.35, adjusting for heritability. The relatively low, but reasonable predictive accuracy of the breeding values highlights the difficulty of creating a single measure for a longitudinal, lifetime trait. Previous efforts to genomically predict lifetime performance in tropically adapted cattle also had reasonable accuracies, with predictions for average weaning rate up to six years of age achieving an accuracy of 0.39 (Zhang *et al.* 2014) and binary stayability at 76 months of age predicting with an accuracy of 0.55 (Ramos *et al.* 2020). As this cohort ages, increasing the number of phenotypic records available, it is expected that the accuracy should improve with the increase in reference size.

Previous efforts to genetically characterise and develop selection tools for lifetime cow productivity have largely focused on intermediate or component traits, such as stayability (Ramos *et al.* 2020) or productivity up to a set age (Zhang *et al.* 2014). Others have focused on maximizing early in life information by including all yearly production records in a random regression analysis (Snelling *et al.* 2018). These approaches tend to result in higher intermediate heritabilities and are more practical than directly selecting for lifetime productivity; phenotypes may be collected earlier in life, making it easier to develop large reference populations for stayability. However, measuring stayability does not provide as much information and will not fully characterise the genetic potential of those cows that may produce above and beyond a set age.

## CONCLUSIONS

This study demonstrated the potential for a genomic breeding value capturing a cow's potential for lifetime productivity, using the intermediate trait, stayability at six years of age. As a lowly heritable trait, the predictive ability of the estimated breeding values was reasonable but low. However, utilising intermediate component traits is a more practical way to genetically select for lifetime productivity than direct selection, increasing the potential for application within commercial production. As the reference population continues to grow, it is expected that the accuracy of these predictions should improve.

## REFERENCES

Burrow H.M. (2001) *Livest. Prod. Sci.* **70**: 213.

Copley J., Engle B.N., Landmark S., Fordyce G., Ross E.M., Wood B., Voss-Fels K.P. and Hayes B.J. (2021) *J. Anim. Sci.* Submitted for review.

Das S, Forer L, Schönherr S, Sidore C, Locke AE et al. (2016) *Nat. Genet.*. **48**: 1284.

Johnston D.J., Barwick S.A., Fordyce G., Holroyd R.G., Williams P.J., Corbet N.J., Grant T. (2014) *Anim. Prod. Sci.* **54**: 1.

Legarra A., Aguilar I., and Misztal I. (2009) *J. Dairy Sci.* **92**: 4656.

Loh P.R., Palamara P.F. and Price A.L. (2016) *Nat Genet.* **48**: 811.

McGowan M., McCosker K., Fordyce G., Smith D., Perkins N., O'Rourke P., Barnes T., Marquart L., Menzies D., Newsome T., Joyner D., Phillips N., Burns B., Morton J., and Jephcott S. (2014) In: *Proceedings of the 28th World Buiatrics Congress, Cairns, Australia*. 185.

Ramos P.V.B., e Silva F.F., da Silva L.O.C., Santiago G.G.,. Menezes G.R.d.O, Soriano Viana J.M., Torres Júnior R.A.A., Gondo A. and Brito L.F.. (2020) *Reprod. Domest. Anim.* **55**: 266.

Snelling W.M., Golden B.L., and Bourdon R.M.. (1995) *J. Anim. Sci.* **73**: 993.

Snelling W.M., Kuehn L.A., Thallman R.M., Bennett G.L., and Golden B.L.. (2018) *J. Anim. Sci.* **97**: 63.

Zhang Y.D., Johnston D.J., Bolormaa S., Hawken R.J., Tier B. (2014) *Anim. Prod. Sci.* **54**: 16.

*Contributed paper*

# THE EFFECTS OF NUMBER OF REFERENCE INDIVIDUALS ON THE ACCURACY OF IMPUTATION FROM LOW AND MEDIUM DENSITIES TO HIGH DENSITY

**M.H. Ferdosi[1], N.K. Connors[1] and M. Khansefid[2]**

[1]Animal Genetics & Breeding Unit, University of New England, Armidale, NSW, 2351 Australia.
[2]AgriBio Centre for AgriBioscience, Agriculture Victoria, Bundoora, VIC 3083, Australia

## SUMMARY

Imputation is a common approach to infer the missing markers for individuals with low marker density (i.e. target population) from a reference population genotyped with higher-density Single-Nucleotide Polymorphism (SNP) panels. Several factors affect the imputation accuracy of untyped, including the number of reference individuals, marker density and population structure. This paper investigates the effects of these factors on the accuracy of imputation by using individuals of a single cattle breed or multiple cattle breeds in the reference population with 600k marker density, as well as assuming the target population was genotyped with low (15k) or medium (30k) marker density. To achieve a within breed imputation accuracy of >90%, we required at least 500 individuals in the reference population when the target population was genotyped with 15k SNP panel. Whereas, if the reference population consisted of a mixture of purebred and multi-breed individuals, the SNP density must be at least 30k in the target population, and there must be more than 900 individuals in the reference population to achieve a similar level of accuracy.

## INTRODUCTION

Genotyping thousands of individuals per month for genomic evaluations has become a common practice in livestock industries in many countries to increase the rate of genatic gain. To reduce the costs of genotyping, industry animals are often genotyped with medium-density panels. Previous studies show that imputing genotypes to high-density and sequence variants can increase genomic prediction accuracy and improve genome-wide association power of Quantitative Trait Loci detection (Moghaddar *et al.* 2019; Khansefid *et al.* 2020). Several factors influence the imputation accuracy of untyped SNPs, such as the number of individuals with high-density markers (i.e. reference population size), the density of markers in the reference and target population, and population structure (Browning and Browning 2011; Ferdosi and Connors 2019; Connors and Ferdosi 2020). The population structure in imputation studies generally refers to the genetic relatedness of individuals within and between reference and target populations. In this study, we investigated the effect of genotyping the target population with higher-density markers and increasing the size of the reference population on the imputation accuracy. Genotypes were imputed from varying medium densities to high density (582k), with reference populations varying in number and breed. The size of the reference population was increased by including more individuals of similar breeds to the target population in the "single-breed reference" or including individuals of multiple breeds in the "multi-breed reference".

## MATERIALS AND METHODS

**Genotypes.** Genotypes were extracted using the BREEDPLAN genomic pipeline (Connors *et al.* 2018; Johnston *et al.* 2018). The individuals and SNPs which had missing rates greater than 5% and the SNPs with minimum allele frequency (MAF) lower than 5% were removed. For multi-breed

---

[1] AGBU is a joint venture of NSW Department of Primary Industries and University of New England

imputation study, the genotypes of 4,458 individuals and 682k SNPs were reduced to 4,363 individuals and 624k SNPs after quality control (QC). The multi-breed populations included Angus (387), Brahman (610), Charolais (730), Hereford (294), Limousin (742), Santa Gertrudis (213), Wagyu (75), Simental (213), Shorthorn (123) and minor breeds (976). For the single-breed imputation study, the relationship between the individuals in the target and reference populations had to be greater than 0.8 (Boerner and Wittenburg 2018). Genotypes of 618 Brahman and 748 Charolais with 682k SNPs were extracted, reducing after QC to 609 Brahmans with 579k SNPs, and 728 Charolais with 582k SNPs.

**Reference and target populations.** A proportion of individuals with high-density genotypes were selected as a reference population, and the genotypes of the remaining individuals were converted to 15k and 30k densities by masking a random set of SNPs. Hence, in the randomly selected individuals for the target populations, some of the known genotypes were converted to missing genotypes and this procedure was repeated 9 times for each scenario.

In the multi-breed imputation study, the reference populations were consisting of 100, 200, 300, 400, 500, 600, 800, 1000 and 2000 individuals. While in the single-breed imputation study, the reference populations only consisted of 100, 200, 300, 400, 500, and 600 individuals.

**Imputation.** FImpute Version 2.2 with default parameters (Sargolzaei *et al.* 2014) was used to impute missing genotypes using single or multiple breeds in the reference without exploiting known pedigree information.

**Imputation accuracy**. Pearson's correlation coefficient between true and imputed genotypes for individuals was calculated to assess the accuracy of imputation in the different scenarios.

## RESULTS AND DISCUSSION

Figure 1 shows the correlation coefficients between true and imputed genotypes in different scenarios. In general, increasing the number of individuals in the reference population and increasing the number of SNPs in the target population from 15k to 30k improved the imputation accuracies for all scenarios. These results were expected and in line with the previous reports (Ferdosi and Connors 2019). Using the same breed in the reference and target populations led to higher imputation accuracy compared to including multiple breeds in the reference. For the purebred individuals with 15k SNPs, there should be more than 500 individuals in the reference population to achieve imputation accuracy higher than 0.9, while with 30k SNPs, 200 individuals in the reference population were sufficient to achieve a similar level of accuracy. For multi-breed, the number of individuals in the reference population and the number of SNPs in the target population needed to be higher compared to single-breed, to achieve a correlation higher than 0.9. Imputation accuracy for a few individuals remained low (shown as outliers in Figure 1) in all scenarios probably because some haplotypes in the target population were undetected or incorrectly detected in the reference population even after including more individuals in the reference. For example, for imputing from 30k SNP to high-density by using 2000 individuals in the multi-breed reference, 54 individuals had imputation accuracy less than 0.85 and 51 of those individuals had a relationship to the relevant breed reference population less than 0.8. This indicates that imputation accuracy tends to be lower in multibreed populations compared to purebreds.

The genotypes from industry animals are used in genomic evaluation and GWAS (i.e. finding QTL) for many traits with diverse range of heritabilities. The accuracy of genomic predictions for the traits with high levels of heritability might be just marginally improved by increasing the marker density through imputation. However, imputation could be still useful to increase the power of QTL detection and especially for improving the accuracy of predictions for the traits with low levels of heritability or when the number of animals in the reference population is limited (Moghaddar *et al.* 2019). Moreover, in terms of practicality, it is much easier to use a same set of SNPs (i.e. imputed to high-density) in genomic evaluation of all traits regardless of their heritability.

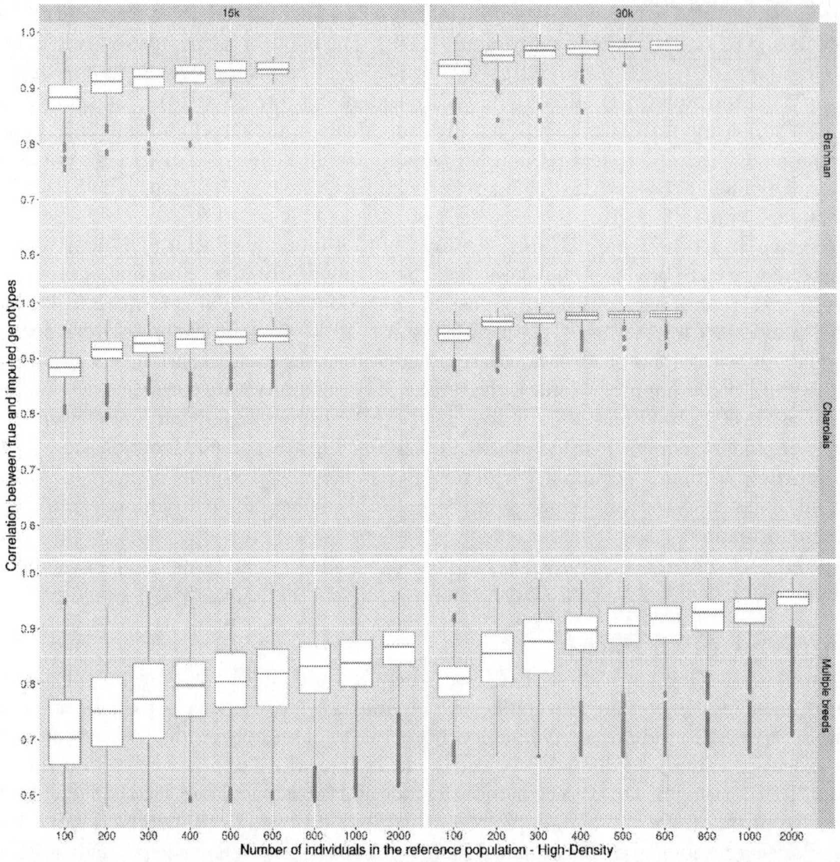


**Figure 1. Correlation between true and imputed genotypes for 3 reference populations and 2 marker densities. The boxplots show the correlation coefficients between true and imputed genotypes in different imputation scenarios. The 15k is low density and 30k is medium density panels**

## CONCLUSIONS

In this study, we explored the effect of the number of SNPs, the number of individuals in the reference and using a single or a multi-breed reference population on the imputation accuracy. The results showed that imputation accuracy was higher when the reference and target populations were of the same breed. In a multi-breed reference population with even a large number of individuals, the imputation accuracy was low, i.e. despite the number of individuals increased in the reference population, the imputation accuracy was lower than purebred scenarios. Increasing the SNP density of the target population to 30k, as well as increasing the number of individuals in the reference population, could improve the imputation accuracy. Algorithms behind the imputation programs are also important and further studies should evaluate how different algorithms affect the imputation accuracies in various scenarios.

## ACKNOWLEDGEMENTS

This study was supported by Meat and Livestock Australia project L.GEN.1704. The authors thank the Australian beef societies and Irish Cattle Breeding Federation for providing the data for this study.

**REFERENCES**

Boerner V., Wittenburg D. (2018) *Front Genet* **9:**

Browning S.R., Browning B.L. (2011) *Nat Rev Genet* **12:** 703.

Connors N.K., Ferdosi M.H. (2020) *6th International Conference on Quantitative Genetics.* 135.

Ferdosi M., Connors N. (2019) *Proc. Association for the Advancement of Animal Breeding and Genetics.* 480.

Khansefid M., Goddard M.E., Haile-Mariam M., Konstantinov K.V., Schrooten C., de Jong G., Jewell E.G., O'Connor E., Pryce J.E., Daetwyler H.D., MacLeod I.M. (2020) *Front Genet* **11:** 598580.

Moghaddar N., Khansefid M., van der Werf J.H.J., Bolormaa S., Duijvesteijn N., Clark S.A., Swan A.A., Daetwyler H.D., MacLeod I.M. (2019) *Genet Sel Evol* **51:** 72.

Sargolzaei M., Chesnais J.P., Schenkel F.S. (2014) *BMC Genomics* **15:**

*Contributed paper*

# EVALUATION OF HAPLOTYPE DIVERSITY OF AUSTRALIAN BEEF POPULATIONS USING MEDIUM-DENSITY SNP GENOTYPES

**M.H. Ferdosi[1], N.K. Connors[1] and M. Khansefid[2]**

[1]Animal Genetics & Breeding Unit, University of New England, Armidale, NSW, 2351 Australia.
[2] AgriBio Centre for AgriBioscience, Agriculture Victoria, Bundoora, VIC 3083, Australia

## SUMMARY

Haplotypes as combinations of multiple markers have more diversity than single markers in the population. In this research we studied the haplotype diversity in four beef breeds (Brahman, Hereford, Santa Gertrudis and Wagyu) in Australia to identify the frequent and rare haplotypes within and between breeds. We found that most of the haplotypes (>90%) with more than one percent frequency within each breed were observed in the other breeds as well. Further, the low within breed haplotype diversity in Wagyu can indicate lower genetic diversity compared to the other breeds.

## INTRODUCTION

Haplotypes could be more informative of genetic diversity than single markers. However, in genomic predictions, defining relationships between individuals using single markers (VanRaden 2008) is more common than by use of haplotype (Hickey *et al.* 2013; Ferdosi *et al.* 2016) for both single and multi-breed genomic evaluations (Khansefid *et al.* 2020). Phasing of the genotypes into haplotypes and partitioning the genome to multiple segments has several benefits. The accuracy of genomic prediction can be increased using haplotypes instead of single markers (Ferdosi *et al.* 2016; Karimi *et al.* 2018). For example, haplotypes have more diversity than single nucleotide polymorphisms (SNPs). Quantitative trait loci (QTL) can be explored better using haplotypes because crossing over between SNPs and QTL can change the linkage disequilibrium (LD) between them across generations. Consequently, the lower relationships between individuals of different breeds could be precisely defined by calculating the proportion of common haplotypes, which is particularly important in multi-breed genomic predictions. In order for haplotypes to be useful in multi-breed genomic prediction, an overlap of haplotypes across breeds is required. Additionally, haplotypes can be used to calculate genomic inbreeding and provide better insight of relationships between individuals of different breeds. This research investigates the overlap of haplotypes across breeds and their use in the calculation of inbreeding and across-breed relationships.

## MATERIALS AND METHODS

**Genomic data.** Genotypes of four beef breeds in Australia were used in this study to assess the haplotype diversities within and across breeds. The individuals with SNP density greater than 30k SNPs were extracted after quality control and before imputation using the BREEDPLAN genomic pipeline (Connors *et al.* 2018; Johnston *et al.* 2018). Included were 12,692 Brahman with 143,829 SNPs, 21,069 Hereford with 51,441 SNPs, 3,563 Santa Gertrudis (SG) with 82,990 SNPs and 59,120 Wagyu with 51,330 SNPs. No SNPs were removed for low minor allele frequencies (MAF) as these SNPs were important for breed distinction.

**Merging the genotypes of four breeds**. Genotypes of these four breeds were combined with custom C++ programming to yield 96,444 individuals with 227,422 unique SNPs.

---

[1] AGBU is a joint venture of NSW Department of Primary Industries and University of New England

**Imputation.** FImpute Version 2.2 with default parameters (Sargolzaei *et al.* 2014) was used to impute missing genotypes using a multi-breed reference but including the pedigree information. The pedigree had 7% and 4% missing sire and dam, respectively. Prior to imputation within each breed SNPs were removed with missing rate greater than 10% resulting in 29,570 SNPs passing this filter and being used in this study.

**Haplotype partitioning.** The phased genotypes were partitioned to the haplotype segments with a length of 10 SNPs without overlap (Ferdosi *et al.* 2016). The total number of unique haplotypes within each breed and number of common haplotypes between breeds were calculated using three scenarios: all haplotypes (ALH), haplotypes with frequencies greater than 1 per cent within breed (High-Frequency Haplotype - HFH) and haplotypes with frequencies less than 1 per cent within breed (Low-Frequency Haplotype - LFH). Further, the percent of individuals covered by each of these scenarios were reported. The haplotype diversities/frequencies in the four studied breeds were plotted and analysed using R (R Development Core Team 2020).

## RESULTS AND DISCUSSION

**Table 1. Number and percent of haplotypes within and between breeds and their population coverage**

| | Common Haplotypes (A) | | | | Mean of percent of population covered by common haplotypes (B) | | | |
|---|---|---|---|---|---|---|---|---|
| | Brahman | Hereford | Santa Gertrudis | Wagyu | Brahman | Hereford | Santa Gertrudis | Wagyu |
| All haplotypes (ALH) | | | | | | | | |
| Brahman | 736,996 | 55% | 48% | 45% | 100% | 78% | 94% | 81% |
| Hereford | 59% | 690,560 | 47% | 49% | 94% | 100% | 91% | 86% |
| Santa Gertrudis | 74% | 67% | 479,930 | 57% | 97% | 90% | 100% | 88% |
| Wagyu | 61% | 62% | 50% | 542,637 | 87% | 92% | 87% | 100% |
| Haplotypes with frequency greater than 1% (HFH) | | | | | | | | |
| Brahman | 54,052 | **92%** | **93%** | **95%** | 63% | 53% | 63% | 58% |
| Hereford | **92%** | 54,094 | **92%** | **96%** | 67% | 69% | 67% | 64% |
| Santa Gertrudis | **94%** | **93%** | 53,535 | **95%** | 53% | 51% | 53% | 52% |
| Wagyu | **91%** | **91%** | **90%** | 56,684 | 82% | 86% | 82% | **93%** |
| Haplotypes with frequency less than 1% (LFH) | | | | | | | | |
| Brahman | 682,944 | 52% | 44% | 41% | 37% | 25% | 32% | 23% |
| Hereford | 56% | 636,466 | 43% | 45% | 26% | 31% | 24% | 22% |
| Santa Gertrudis | 71% | 64% | 426,395 | 52% | 44% | 39% | 47% | 36% |
| Wagyu | 58% | 59% | 45% | 485,953 | 5% | 5% | 5% | 7% |

**(A)** The diagonal elements are the total number of unique haplotypes in each breed. The non-diagonal elements are percentage of common haplotypes between each pair of breeds, where upper (and lower) triangular elements are number of common haplotypes between breed divided by number haplotypes of the breed in that "row" (and column). **(B)** The diagonal elements show the percentage of the genome covered with haplotypes in each breed. The non-diagonal elements are percentage of genome covered with common haplotypes between each pair of breeds, where upper (and lower) triangular elements are the percentage of the genome of the breed in that "row" (and column) covered with common haplotype.

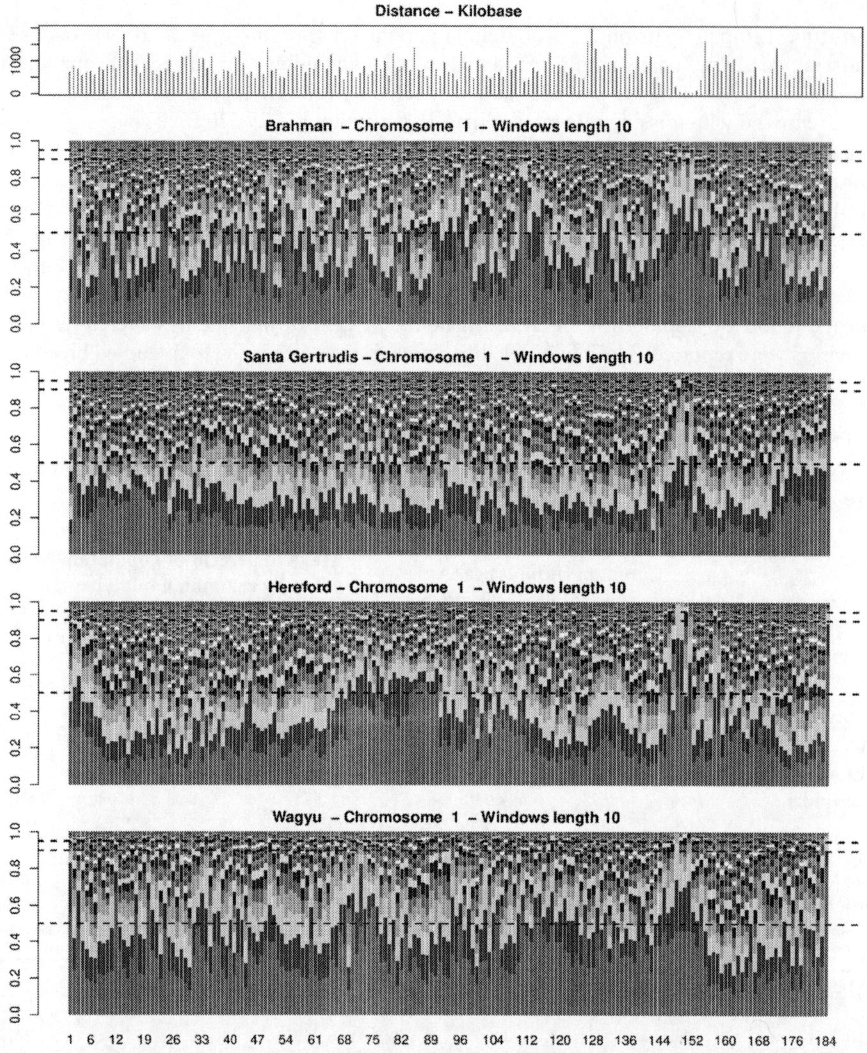


**Figure 1 Haplotype diversity across four breeds in chromosome 1.** The top plot shows the length of haplotypes (base pair) which were constructed by every 10 adjacent SNP. In the rest of bar plots, each bar represents the sorted haplotype frequencies (vertical line). The horizontal dashed lines mark 50%, 90% and 95% haplotype frequencies. The last haplotype was constructed by less than 10 SNPs, and therefore had fewer haplotypes.

The diagonal elements in Table 1–A shows the total number of unique haplotypes in each of the four breeds with Brahman having the highest number. Santa Gertrudis had the smallest number of unique haplotypes but also had the smallest number of genotyped individuals compared to the other breeds, especially Wagyu. Given the imbalance between number of individuals across breeds could affect our results, the haplotype diversity in Wagyu was especially low which could be a reflection of low effective population size due to limited founders originally imported into Australia. Moreover, according to Table 1-B frequent haplotypes (i.e. HFH) in Wagyu cover 93% of the population while such haplotypes cover 53% of the SG population. The number of common HFH between breeds was very high (Table 1-A-HFH) which can imply the potential in using haplotypes to improve multi-breed genomic prediction accuracy. Previous studies have already reported the improvement in Restricted Expectation Maximum Likelihood and accuracy of genomic predictions in cross-validation studies using haplotype-based genomic relationship matrices (Ferdosi *et al.*2016). A similar study using a multi-breed population can shed light on the benefits of using haplotypes instead of single markers in multi-breed genomic prediction as our initial haplotype diversity study indicates noticeable overlap between haplotypes in different breeds. The main supporting reason for the usefulness of using haplotypes in multi-breed genomic predictions is the high possibility that many of the QTL and markers are in different LD or even different phase in different breeds. Hence, haplotypes as a combination of multiple markers, could better track QTL especially from distant ancestors compared to single markers. However, it is also important to assess if the haplotypes have the same effects across different breeds.

Figure 1 shows the haplotype diversity in chromosome 1 across the four beef breeds and demonstrates quite different diversity of haplotypes across chromosome 1, which was seen in other chromosomes as well (not shown). As we expected, the marker distance and the length of haplotype significantly affected the haplotype diversities. For example, close to the end of chromosome 1, the lengths of haplotypes were relatively smaller than the rest of haplotypes which could be a potential reason for the lower haplotype diversity in such regions. Possibly partitioning the genome to haplotypes with relatively equal length or recombination rate instead of using similar number of SNPs in haplotypes, could resolve this issue.

## CONCLUSIONS

In this study we explored the haplotype diversity within and between breeds. We found that low haplotype diversity within a breed could be an indication of lower genetic diversity in the population such as Australian Wagyu. The haplotype diversity showed relatively high relatedness between different breeds which suggests the potential benefits of using haplotype-based relationships in multi-breed genomic predictions. Further study is required to evaluate the benefits of haplotypes on single marker for genomic prediction.

## ACKNOWLEDGEMENTS

MHF, NKC and this study was supported by Meat and Livestock Australia project L.GEN.1704. The authors thank the Hereford, Brahman, Wagyu and Santa Gertrudis societies for providing data for this study.

## REFERENCES

Connors N., Cook J., Girard C., Tier B., Gore K., Johnston D., Ferdosi M. (2017) *Proc.Assoc. Advmt Anim. Breed. Genet.* **22**: 317.

Ferdosi M.H., Henshall J., Tier B. (2016) *Genet Sel Evol* **48:** 75.

Hickey J.M., Kinghorn B.P., Tier B., Clark S.A., van der Werf J.H.J., Gorjanc G. (2013) *J Anim Breed Genet* **130:** 259.

Johnston D., Ferdosi M., Connors N., Boerner V., Cook J., Girard C., Swan A., Tier B. (2017) *Proc. Assoc. Advmt. Anim. Breed.Genet.* **11**: 269.

Karimi Z., Sargolzaei M., Robinson J.A.B., Schenkel F.S. (2018) *Can J Anim Sci* **98:** 750.

Khansefid M., Goddard M.E., Haile-Mariam M., Konstantinov K.V., Schrooten C., de Jong G., Jewell E.G., O'Connor E., Pryce J.E., Daetwyler H.D., MacLeod I.M. (2020) *Front Genet* **11:** 598580.

R Development Core Team (2020) 'A language and environment for statistical computing. R Foundation for Statistical Computing.' (R Foundation for Statistical Computing:

Sargolzaei M., Chesnais J.P., Schenkel F.S. (2014) *Bmc Genomics* **15:**

VanRaden P.M. (2008) *J. Dairy Sci.* **91:** 4414.

*Contributed paper*

# PHENOBANK: A PLATFORM TO FACILITATE COLLABORATION AND GENOMIC SELECTION FOR FEMALE FERTILITY IN BEEF CATTLE

**K.B. Wockner[1], M.S. Tahir[1], J. Kour[1], W.L.A. Tan[1], B.S. Olasege[1], L.R. Porto-Neto[2], B.A. Little[2], C.R. Wilson[5], W. Dollemore[4], A. Reverter[2], B. Hayes[3] and M.R.S. Fortes[1]**

[1]The School of Chemistry and Molecular Biosciences, the University of Queensland, Brisbane, QLD 4072, Australia [2]CSIRO Agriculture and Food, Queensland Bioscience Precinct, 306 Carmody Rd., St. Lucia, Brisbane, QLD, 4067, Australia [3]Queensland Alliance for Agriculture and Food Innovation, Queensland Bioscience Precinct, 306 Carmody Rd., St. Lucia, Brisbane, QLD, 4067, Australia [4]Northern Territory Department of Industry, Tourism and Trade, Katherine Research Station, 449 Stuart Highway, Katherine, NT, 0850, Australia [5]Agricultural Business Research Institute, University of New England, Armidale, NSW, 2351, Australia

## SUMMARY

The PhenoBank project is about creating a platform to facilitate research collaborations and genomic selection in beef cattle. The traits under investigation are female reproduction traits, measured early in life and preferably easy to measure to facilitate adoption. We now have data on more than 9000 cows from many tropical breeds, which have a record for PREG1: a binary trait for the ability to conceive, or not, after the first mating season. These cows were genotyped using SNP chip technology. Using data from 2400 Brahman cows from the PhenoBank we estimated the heritability of PREG1 and conducted a Genome Wide Association Study (GWAS). The estimated heritability was 0.17 (SE 0.03). We identified 59 suggestive SNPs (P-value $< 9.9 \times 10^{-5}$) that mapped to different chromosomes. The SNP with the highest significance (P-value $2.0 \times 10^{-7}$) was on BTA8. SNPs clustered on BTA21 were also significant (P-value $1.1 \times 10^{-6}$). This easy to measure trait can be used for genomic selection. The associated markers need to be confirmed in further studies.

## INTRODUCTION

Female reproduction performance is a major driver of on-farm profitability and currently has immense potential for genetic improvement as many North Australian cattle enterprises have low reproductive rates. Female reproduction traits are often low in heritability and/or expensive to measure, which are hurdles for adoption (Cammack *et al.* 2009). Still, traits such as age at puberty and post-partum anoestrous interval are moderately heritable in tropical beef cattle (Johnston *et al.* 2013). This heritability makes genetic selection for female fertility traits, especially early-in-life traits, a promising strategy for improving beef production in Northern Australian herds. Genomic selection accuracy can be improved by using large phenotype datasets and identifying informative genetic markers. Once established, genomic enhanced Estimated Breeding Values (gEBVs) for easy to measure traits are more likely to be used on extensive cattle enterprises because they will reduce the need for labour intensive data collection (e.g. collecting exact birth dates).

Existing datasets with already-recorded female reproduction traits are the 'fast track' for creating the large reference population needed for robust gEBVs with the potential to transform the rate of genetic gain and the adoption of improved genetics in the beef sector, as it has done in the dairy industry (Meuwissen *et al.* 2013). By linking existing datasets, along with new data, we have created a data warehouse and platform to facilitate genomic selection: the PhenoBank. By using PhenoBank, genetic improvement of fertility traits in *Bos indicus* influenced herds can be further investigated.

## MATERIALS AND METHODS

**PhenoBank Platform.** The PhenoBank database (DB) works with the cloud-based Livestock Information Platform (LIP) developed by CSIRO and Agricultural Business Research Institute

(ABRI). LIP was designed to store and retrieve livestock phenotypes and genotypes. PhenoBank DB has been developed as a pair of customised applications with both Windows and Web user interfaces.

**PhenoBank Data.** The existing phenotype and genotype data contributed to PhenoBank was sourced from the Cooperative Research Centre for Beef Genetics Technologies (Beef CRC), the Northern Territory Department of Industry, Tourism and Trade (NT DITT) breeding herd and the Kamilaroi herd investigated in a CSIRO-led project. Genotypes of the Beef CRC cows were from the Bovine50K v.1 chip (Hawken *et al.* 2012). The NT DITT cows were genotyped in the project with the GGP Bovine50K SNP chip (NEOGEN Inc.), while Kamilaroi cows were genotyped with the GGP TropBeef 35K SNP chip (NEOGEN Inc.). Additionally, we have received phenotypic and genotypic data from 10 industry herds. To contribute to PhenoBank, producers provided the mating outcomes from the first two breeding seasons in a cow's life and its DNA sample for genotyping. New cows for PhenoBank are being genotyped with the GGP TropBeef 35K SNP chip (NEOGEN Inc.).

**Analysis.** We selected, curated and combined data of 2400 Brahman cows for which we defined new phenotypes. These records were sourced from the Beef CRC, NT DITT and Kamilaroi data contributions, and all have an early-in-life, easy-to-measure record for PREG1: a binary trait for the ability to conceive, or not, after the first mating season, outlined in Table 1.

**Table 1. Scoring criteria of early reproductive traits in Brahman heifers**

| No. | Trait | Score | Scoring Criteria |
|---|---|---|---|
| 1 | PREG1 | 1 | Not pregnant as a result of the first mating opportunity (n = 600) |
| | | 2 | Pregnant as a result of the first mating opportunity (n = 1719) |

A reference panel of 546 Brahman animals were genotyped with the BovineHD (770K) SNP chip (NEOGEN) and used to impute genotypes from the medium-density SNP panels. A combination of Eagle v2.4.1 (Loh *et al.* 2016) and Minimac3 (Das *et al.* 2018) were used for imputation. The combined genotypes dataset was passed through final quality control (SNPs with a call rate < 0.9 and multiple allele frequency < 0.05 were discarded) to get over 500,000 SNPs for all cows included in this study.

Each phenotype dataset had particular fixed effects to account for contemporary group effects. Contemporary groups were defined by farm location (animals raised together in the same farm) and by birth year and month, which inform the cow cohort (year) and the birth month class (Aug to Nov = Class A; Dec to April = Class B). For the Beef CRC dataset, farm, cow cohort and birth month class were used as fixed effects. For the NT DITT dataset, cow cohort and birth month class were used as fixed effects. For the Kamilaroi dataset, cow cohort was used as a fixed effect. After adjusting for fixed effects, the three datasets were combined to make a single dataset for pooled analyses. Genome wide association studies (GWAS) were conducted for the combined PREG1 dataset using *SNP & Variation Suite v8.8* (Golden Helix, Inc., Bozeman, MT, www.goldenhelix.com).

## RESULTS AND DISCUSSION

We have created a data platform to facilitate collaborations and genomic selection in beef cattle research. PhenoBank now contains data on more than 9000 cows. PhenoBank presents an opportunity for researchers and producers to engage and collaborate. Datasets can be uploaded, merged, stored securely and shared when desired. The traits under investigation are female reproduction traits, measured early in life and preferably easy to measure to facilitate adoption by beef producers. The accuracy of imputation for CRC, Kamilaroi and NT DITT cows' datasets was

0.95, 0.93 and 0.92, respectively. Linear mixed model analysis of PREG1 resulted in an estimated heritability of 0.17 (SE 0.03). This compares to a heritability of 0.18 for PREG1 after fixed-time artificial insemination in another study on Brahman heifers (Porto-Neto *et al.* 2015).

GWAS for PREG1 identified 59 suggestive SNPs (*P-value* $< 9.9x10^{-5}$) that mapped to different chromosomes. The SNP with the highest significance (*P-value* $2.0x10^{-7}$) was on BTA8. SNPs clustered on BTA21 were also significant (*P-value* $1.1x10^{-6}$) as shown in Figure 1.

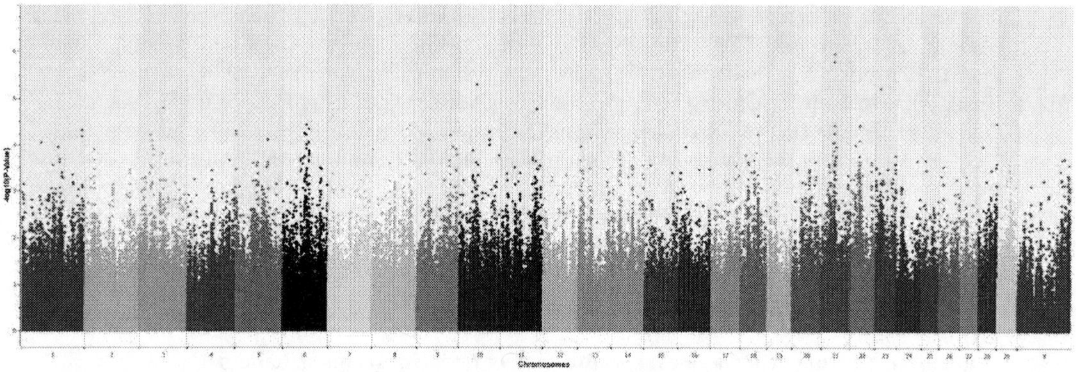

**Figure 1. Manhattan plot of PREG1: genome-wide association results**

## CONCLUSIONS

By re-defining and merging data from previous projects with new data, we are expanding the number of samples available for research use. Using a sample of available PhenoBank data, we were able to show that PREG1 has a heritable component and we identified some potential genomic markers for this trait. Next, we will analyse the rebreeding ability for these cows, as we have the outcomes of the second mating season for most of them.

Our next step for PhenoBank is the imputation of all uploaded genotypes to sequence level data and continued analyses investigating heifer and cow fertility. We will continue to collaborate with North Australian producers and upload their data contributions to PhenoBank to create a world class digital infrastructure for beef cattle genomics. Our goal is to contribute to the sustainability and profitability of Australia's beef industry.

## ACKNOWLEDGEMENTS

The authors gratefully acknowledge the funding received from Meat and Livestock Australia (MLA) for the Female Fertility PhenoBank project (L.GEN.1710) to undertake this work. We also acknowledge all our collaborators for their data contributions to PhenoBank.

## REFERENCES

Cammack K. M., Thomas M. G., and Enns R. M. (2009) *Prof. Anim. Sci.* **25:**517.

Das S., Forer L., Schönherr S., Sidore C., Locke A.E., Kwong A., Vrieze S.I., Chew E.Y., Levy S., McGue M., Schlessinger D., Stambolian D., Loh P-R., Iacono W.G., Swaroop A., Scott L.J., Cucca F., Kronenberg F., Boehnke M., Abecasis G.R., and Fuchsberger C. (2016) *Nat Genet* 2016 **48**: 1284.

Hawken R.J., Zhang Y.D., Fortes M.R.S., Collis E., Barris W.C., Corbet N.J., Williams P.J., Fordyce G., Holroyd R.G., and Walkley J.R.W. (2012) *J Anim Sci* 2012 **90**:1398.

Johnston D.J., Barwick S.A., Fordyce G., Holroyd R.G., Williams P.J., Corbet N.J., and Grant T. (2013) *Anim Prod Sci* **54:** 1.

Loh P-R., Danecek P., Palamara P.F., Fuchsberger C., Reshef Y.A., Finucane H.K., Schoenherr S., Forer L., McCarthy S., Abecasis G.R., Durbin R., and Price A.L. (2016) *Nature* **48**: 1443.

Meuwissen, T., Hayes B., and Goddard, M. (2013). *Annu Rev Anim Biosci* **1:** 221.

NEOGEN Corporation. (2021) GeneSeek® Genomic Profiler™. Accessed July 30, 2021. https://www.neogen.com/categories/genotyping-arrays.

Porto-Neto L.R., Edwards S., Fortes M.R.S., Lehnert S.A., Reverter A., McGowan M. (2015). *J Anim Sci* **93**: 5119.

SNP & Variation Suite ™ (Version 8.8) [Software]. Bozeman, MT: Golden Helix, Inc. Available from http://www.goldenhelix.com.

*Contributed paper*

# GENOME WIDE ANALYSIS OF BOVINE ENHANCERS AND PROMOTERS ACROSS DEVELOPMENTAL STAGES IN LIVER

**M. Forutan[1], C.J Vander Jagt[2], E. Ross[1], A.J. Chamberlain[2], B. Mason[2], L. Nguyen[1], S. Moore[1], J.B. Garner[4], R. Xiang[3] and B.J. Hayes[1]**

[1] Queensland Alliance for Agriculture and Food Innovation, The University of Queensland, St Lucia, QLD 4072, Australia
[2] Agriculture Victoria, Centre for AgriBiosciences, Bundoora, VIC 3083, Australia
[3] Faculty of Veterinary & Agricultural Science, The University of Melbourne, Parkville, VIC 3052, Australia
[4]Agriculture Victoria, Animal Production Sciences, Ellinbank Dairy Centre, Ellinbank, Victoria 3821, Australia

## SUMMARY

Gene transcription is controlled by functional interactions between promoters and enhancers. Cap analysis of gene expression (CAGE) sequencing has allowed for the accurate annotation of most gene promoters (transcription start sites, TSS) and active enhancers. To date, TSSs and enhancer regions in the bovine genome are poorly characterised. To explore bovine developmental-specific patterns of enhancer-TSS usage and model TSS-enhancer interaction, CAGE-seq was applied to 6 bovine liver samples comprised of two different developmental stages (foetal and adult) obtained from 3 cows and their 3 foetuses. We identified approximately 30k and 20k TSSs and enhancer candidates, respectively, across the liver samples. About 231 significant TSS-enhancer interaction candidates were found by looking for closely spaced TSSs and enhancers that have highly correlated expression levels ($r > 0.75$; P-value $< 0.05$). Differential expression between development stages of TSS and enhancer candidates was performed using the Bioconductor package DESeq2 and identified 2050 (6) TSS (enhancer) candidates significantly differentially expressed across developmental stages (P-value $< 0.05$). The resulting catalogue of TSSs and active enhancers enables classification of developmental-specific TSSs-enhancers and modelling their interaction and provides major target regions for investigation of DNA methylation changes with aging. The information will also be useful in refining regions likely to contain causative mutations for complex traits associated with liver gene expression, such as feed efficiency.

## INTRODUCTION

Identifying active regulatory regions in the genome is critical for understanding gene regulation and assessing the impact of genetic variation on phenotype. Although multiple processes are involved in gene expression regulation, the key role of promoters and enhancers has been a central focus of genome annotation for the past decade. Previous studies have confirmed that most genes have an array of close transcription start sites (TSSs) instead of the expected single TSS (FitzGerald *et al.* 2006; Hoskins *et al.* 2011; Djebali *et al.* 2012; Rojas-Duran and Gilbert 2012; Forrest *et al.* 2014), and the transcription of a gene may start from one of several TSSs, a phenomenon known as alternative transcriptional initiation (ATI, Landry *et al.* 2003; de Klerk and Hoen 2015). While promoters specify and enable the positioning of RNA polymerase machinery at TSSs, enhancers modulate the activity of promoters and play a key role in the formation of diverse cell types and respond to changing physiological conditions. Andersson *et al.* (2014) showed that enhancer activity can be detected through the presence of balanced bidirectional capped transcripts using Cap Analysis of Gene Expression (CAGE) (Takahashi *et al.* 2012). Active enhancers produce weak, but consistent, bidirectional transcription of capped enhancer RNA (eRNAs), resulting in a characteristic CAGE tag starting sites (CTSS) pattern of two diverging peaks approximately 400 bp apart. A specific advantage of the CAGE method is that reads mapped to the genome provide accurate location of TSSs and active enhancers and quantify transcription (Kodzius *et al.* 2006; Carninci *et al.* 2007).

To date, enhancer regions in the bovine genome are poorly characterised. To explore bovine tissue-specific patterns of enhancer-TSSs usage, CAGE sequencing was applied to 6 bovine samples comprised of 2 different developmental-stages obtained from 3 cows and their 3 foetuses. To the best of our knowledge, this study is the first bovine TSS-enhancer discovery using CAGE-Seq data.

## MATERIALS AND METHODS

**CAGE library preparation and sequencing.** Two liver samples were collected from one pregnant *Bos indicus* (Brahman cow) and the female cow's foetus (approximately 12 weeks old). Four liver samples were collected from two *Bos taurus* pregnant cows and their female foetuses (approximately 16 weeks old) at the Ellinbank research facility with approval from the DEDJTR Animal Ethics Committee (2014-23). Samples (cows and foetus) were collected from the same anatomical region. The samples were harvested after the cow was slaughtered, immediately snap-frozen in liquid nitrogen, and stored at -80°C until processing (Forutan *et al.* 2021).

**Read processing and alignment.** Sequence read quality was assessed using FastQC (Andrews 2010), including calculation of GC content, and identification of over-represented sequences. The EcoP15I fingerprint was trimmed by cutting the first 9 bases (*CROP:9*) and Illumina adaptor trimmed by cutting the last 14 bases (*HEADCROP:36*) using Trimmomatic (Bolger *et al.* 2014) (version 0.35). Trimmed reads were aligned to *Bos taurus* reference genome (GenBank: ARS-UCD1.2) with Burrows-Wheeler Aligner (BWA. Li and Durbin 2009), version 0.7.13) using the BWA-MEM algorithms. The aligner was run using default parameters, the only exceptions were t=10, and k=10. Also, to alleviate the presence of universal G at the head of the read, which may be present in some of the reads, parameters L (clipping penalty) and B (mismatch penalty) were assigned as 4 and 5, respectively.

**Quality controls and preliminary analyses.** Only primary alignments with a quality of greater than 20 (>99% chance of true) were considered for TSSs and enhancers calling. Further filtering was applied by only selecting CTSS with 3 or more CAGE reads in at least one sample for TSSs calling. Considering that active enhancers produce weak but consistent bidirectional transcription of capped enhancer RNAs (eRNAs), more relaxed filtration was used for enhancer calling (selecting CTSS with 2 or more CAGE reads in at least one sample). The total number of reads before and after quality control and numbers of TSSs and active enhancer candidates across all samples is shown in Table 1.

**TSSs and enhancers calling.** *clusterUnidirectionally* function and the parameter *mergeDist* 20 available in *CAGEfightR* package (Thodberg *et al.* 2019) was used to call TSSs. Ensembl database release 104 for Bos taurus was used for annotation of the signals. Only TSSs overlapping promoter, proximal and 5'UTR regions were used for further analysis. Identification of active enhancer candidates was done using *clusterBidirectionally* function with a balance score > 0.95 in the *CAGEfightR* package. The enhancers not overlapping intergenic and intron regions were removed from the analysis. TSS-enhancer interaction candidates were identified using *findLinks* function from the *InteractionSet* package into an R session (version /4.0.2) by looking for closely spaced TSSs and enhancers that have highly correlated expression within 20 kb distance. Differential TSSs and enhancer usage across developmental stages was performed by using the Bioconductor package DESeq2 (Love *et al.* 2014) and keeping only TSSs expressed in all samples (10,813 TSSs) and enhancers observed to be bidirectional in all samples (21 bidirectional enhancers). The *findStretches* function from *CAGEfightR* package was used to identify groups of closely spaced enhancers, where all enhancers were within a 10 kb distance of another member.

**Data availability.** Bos taurus and Bos indicus raw sequence data are publicly available via European Nucleotide Archive (ENA) under study ID PRJEB43513 and PRJEB44817, respectively.

## RESULTS AND DISCUSSION

Genome-wide association studies (GWAS) have discovered many variants for complex diseases and quantitative traits. However, many implicated variants are classified as non-coding and, they are thought to play a role in gene expression regulation. Functional annotations provide valuable information for prioritizing potential causal variants within complex-trait loci identified through GWAS. Like any specific tissue in the body, the biological features of tissue in foetal and adult stages may be determined mainly at the level of gene expression. So, identification of functional regions such as enhancer and TSSs and differential and quantitative analysis of developmental stage-specific TSS-enhancers expression could be useful to identify informative variants and ultimately improve genomic prediction. In total, 29,940 and 19,264 TSSs and candidate enhancers were detected across all samples, respectively (Table 1). Only 36% of TSSs (10,813) were expressed across all 6 samples. The lower number of enhancers was observed in the adult stage compared to the foetal stage (Table 1). In total, among the 19,264 active enhancer candidates expressed across samples, only a small proportion of enhancer candidates (less than 1%) were expressed across all samples. The enhancers are context-specific and respond to specific physiological, pathological, or environmental conditions which can cause the large variation in number of enhancers observed across samples. About 231 significant TSS-enhancer interaction candidates were found by looking for closely spaced TSSs and enhancers that have highly correlated expression levels ($r > 0.75$; P-value < 0.05). Examination of the differential TSS usage across developmental stages controlling for effect of sub-species revealed 2050 differentially significant TSSs (P-value <0.05). We found 6 developmental enhancers based on the differential enhancer usage analysis (P-value < 0.05), which could be the potential targets of DNA methylation in bovine liver. One of the developmental stage-specific genes in liver is Sulfotransferase isoform 1A1 (*SULT1A1*). *SULT1A1* is the most highly expressed hepatic sulfotransferase and plays the central role in detoxification. Out of five TSSs observed across samples for this gene (Figure 1), two of them were expressed in all samples (TSSs peaks located on positions 26,126,989 bp and 26,127,457 bp) and only the TSS on position 26,126,966 – 26,127,032 bp showed significantly differential expression in foetal stage compared to adult stage (log2FoldChange = -3.291495; adjusted P-value < 0.0006).

**Table 1. Summary of the number of CAGE tags, transcription start site (TSS) and enhancer candidates expressed in bovine liver**

| Stage | Biological samples | Number of CAGE tags | | | Number of TSS | | Number of enhancers | |
|---|---|---|---|---|---|---|---|---|
| | | Total | For TSS calling | For enhancers calling | Total | In promoter, 5'UTR, proximal | Total | In intergenic and intron |
| Adult | *B.taurus* rep1 | 5,850,606 | 1,869,376 | 2,283,311 | 97,639 | | 6,422 | 3,980 |
| | *B.taurus* rep2 | 5,678,632 | 2,167,220 | 2,437,174 | | | | |
| | *B.indicus* rep1 | 5,183,691 | 4,647,556 | 4,796,404 | | | | |
| Foetal | *B.taurus* rep1 | 10,048,108 | 2,607,022 | 3,358,781 | | 140,591 | 22,310 | 17,386 |
| | *B.taurus* rep2 | 9,327,767 | 1,632,971 | 2,476,168 | | | | |
| | *B.indicus* rep1 | 7,448,786 | 5,552,082 | 5,844,322 | | | | |
| Total | | 43,537,590 | 18,476,227 | 21,196,160 | | 162,275 | 24,605 | 19,264 |

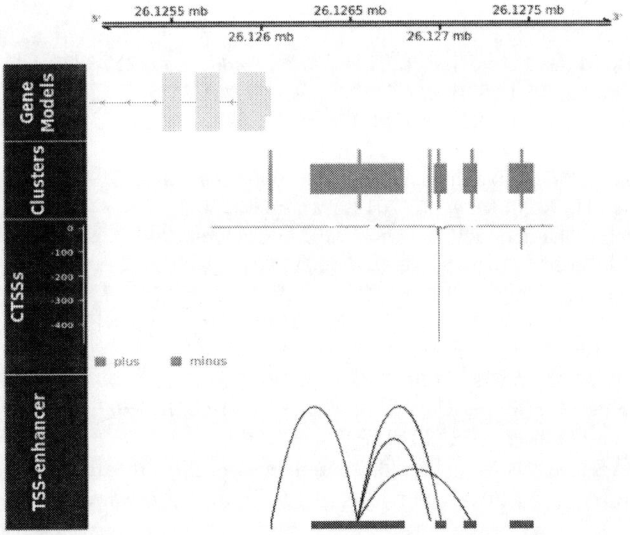


**Figure 1. Plot of position of CAGE tag starting sites (CTSSs), TSSs (orange clusters), and enhancer candidate (pink cluster) of the *SULT1A1* gene in bovine liver. Gene model is plotted based on the Ensembl database (bos_taurus_core_104_12). The links between TSSs and active enhancers is plotted using arches, scaling the height of the arches according to P-values of Kendall correlation**

## CONCLUSIONS

Knowledge of interaction between bovine TSS and enhancer expression would be a useful starting point to predict biological function of specific genes in different developmental stages. In the current study, CAGE-seq was used for the first time to assess TSS-enhancer interactions in bovine liver. Also, we assessed differential TSSs and enhancer usages across developmental-stages in liver tissue for the first time in cattle using CAGE-seq. The results of this study will accelerate future genomic research and will assist in narrowing down candidate genes with differential TSS and enhancer usage across foetal and adult stages in liver. The information will also be useful in refining regions likely to contain causative mutations for complex traits associated with liver gene expression, such as feed efficiency. A limitation with the current study is that only one biological replicate was included for the Bos indicus cow-foetus, so analysis of additional would increase the resolution of the findings.

## ACKNOWLEDGMENTS

We acknowledge financial contributions from Meat and Livestock Australia (P.PSH.0868) for the generation of the *Bos taurus indicus* CAGEseq data. We would also like to acknowledge financial contributions from DairyBio (a joint venture project between Agriculture Victoria and Dairy Australia) and Research Initiative Fund of the Faculty of Veterinary& Agriculture Sciences of The University of Melbourne for the generation of the *Bos taurus taurus* CAGEseq data. We are thankful to Dr. Brian Burns for helping source the *Bos taurus indicus* tissues, and Dr. Bronwyn Venus for collecting *Bos taurus indicus* samples. Thank you to Elise Kho for extracting some the *Bos taurus indicus* RNA.

## REFERENCES

Andrews S. (2010) Available online at: http://www.bioinformatics.babraham.ac.uk/projects/fastqc
Bolger A.M., Lohse M., and Usadel, B. (2014). *Bioinformatics*. **30**: 2114.
Carninci P., Sandelin A., Lenhard B. *et al*. (2006) *Nat. Genet.* **38**:626.
de Klerk E. and Hoen P.A.C't (2015) *Trends Genet.* **31**: 128.
Djebali S., Davis C.A., Merkel A. *et al*. (2012) *Nature*. **489**:101.
FitzGerald P.C., Sturgill D., Shyakhtenko A. *et al*. (2006) *Genome Biol.* **7**: R53.
Forrest A.R., Kawaji H., Rehli M. *et al*. (2014) *Nature*. **507**:462.
Forutan, M., Ross, E., Chamberlain, A.J. *et al*. (2021) *Commun Biol.* **4**: 829.
Hoskins R.A., Landolin J.M., Brown J.B. *et al*. (2011) *Genome Res*. **21**: 182.
Kodzius R., Kojima M., Nishiyori H. *et al*. (2006) *Nat. Methods*. **3**: 211.
Landry J.R., Mager D.L. and Wilhelm B.T. (2003) *Trends Genet.* **19**: 640.
Li H. and Durbin R. (2009) *Bioinformatics*. **25**: 1754.
Love M.I., Huber W., and Anders S. (2014) *Genome Biol.* **15**: 1.
Andersson R., Gebhard C., Miguel-Escalada I. *et al*. (2014) *Nature*. **507**: 455.
Rojas-Duran M.F. and Gilbert W.V. (2012) *RNA.* **18**: 2299.
Takahashi H., Kato S., Murata M. *et al.* (2012) *Methods Mol Biol.* **786**: 181.
Thodberg M., Thieffry A., Vitting-Seerup K. *et al*. (2019) *BMC Bioinform.* **20**: 1.

*Proc. Assoc. Advmt. Anim. Breed. Genet: 23: 131-134*

# IS SEX DETERMINATION IN MERINOS HERITABLE?

**T. Granleese[1], M.R. Mortimer[2] and S.F. Walkom[3]**

[1]NSW Department of Primary Industries, Grafton, NSW, 2460
[2]Centre Plus Merinos, Tullamore, NSW, 2874
[3]Animal Genetics and Breeding Unit*, University of New England, Armidale, NSW, 2351

## SUMMARY

This paper investigates the ability of different linear mixed models to estimate the heritability of sex determination in a sub-set of the Australian Merino population. The dataset used was from Centre Plus Merinos in central-west New South Wales with 25 plus years of full pedigree collection and over 20,000 lambing events where the sex of the progeny were recorded. This study used sex of a lamb as a trait, (i.e. zero phenotype for female and one phenotype for male). We observed a significant, yet normal, amount of phenotypic variation in the sex ratio of progeny for dams, sires, maternal grand sires and maternal grand dams. However, no model was able to estimate significant genetic variation in sex determination and failed to return a heritability above 0.01. Consequently, it can be concluded within this dataset that it would not be possible to select to alter sex determination in Merinos.

## INTRODUCTION

Sex determination in mammals occurs at egg fertilisation with females and males typically having XX and XY chromosomes, respectively. Whether a newly conceived embryo is a male or female is determined by the sperm as dams can only ever pass an X chromosome onto their progeny. Kosswig (1964) thought that sex determination was a polygenic trait in some species of fish. Furthermore, Flanders (1965) purported that winged insects exhibited genetic variation in female behaviour to fertilise or not fertilise eggs which influenced sex ratio. There are no estimates of sex ratio estimation in livestock species. However, in humans, Gellatly (2009) showed a heritability of sex ratio of 0.05 and purported that males tend to produce a sex ratio like that produced by their parents, whereas females do not.

Sex determination is a potentially economically important trait to commercial producers where females are worth significantly more than castrated males. Anecdotally we hear sheep and cattle breeders observe that a cow or ewe only ever has one sex (e.g. "that ewe only ever breeds ram lambs"). This paper investigates whether phenotypic variation exists within a deeply pedigreed and well recorded Merino flock that is highly influential on the breed. If phenotypic variation does exist, we propose to run different types of linear mixed models to investigate whether any genetic variation can be quantified.

## METHODS

**Animals.** Animals from the Centre Plus Merinos flock (601250 flock code), born since 1990, were included in the analysis. All animals without sire and/or dam pedigree were removed as well as any dead at birth (DAB) animals (all DABs were recorded as males). Contemporary grouping was defined as year of birth. No other contemporary grouping was significant enough to fit. In the

* AGBU is a joint venture of NSW Department of Primary Industries and University of New England

case of running a sire, maternal grand sire or dam model, a minimum number of progeny were required to be included in the model (Table 1).

**Measurements.** Phenotype was defined as the sex of each progeny born. Zero for females and 1 for males. Hence an average of 0.5 was expected (Table 1).

**Table 1. Descriptive statistics of each model where direct animal, dam animal, dam dam, service sire, maternal grand sire and maternal grand dam models were run**

| Analysis model type (min. no. progeny) | n | n Sire | n Dams | n MGS* | n MGD^ | Mean | Phen. SD | Min. | Max. |
|---|---|---|---|---|---|---|---|---|---|
| Animal | 23228 | 368 | 6835 | - | - | 0.50 | 0.14 | 0.00 | 1.00 |
| Dam – animal (1) | 23228 | - | 6835 | - | - | 0.50 | 0.35 | 0.00 | 1.00 |
| Dam - dam (7) | 6324 | - | 765 | - | - | 0.50 | 0.18 | 0.00 | 1.00 |
| Service sire (40) | 23120 | 334 | - | - | - | 0.50 | 0.09 | 0.35 | 0.67 |
| Mat. grand sire (50) | 19260 | - | - | 186 | - | 0.50 | 0.06 | 0.38 | 0.65 |
| Mat. grand dam (10) | 7676 | - | - | - | 535 | 0.50 | 0.14 | 0.10 | 1.00 |

*MGS – Maternal grand sires; ^MGD - Maternal grand dams

**Statistical analysis.** Phenotypic variance for each model was assessed prior to any model run to see if the trait was worth investigating (Table 1, Figures 1-4). We also checked to see if average sex ratio sat inside a normal distribution of expectation if sex ratio was random. This is displayed in Figures 1-4 where we can observe distribution sits within a normal bell-curve which suggested enough variation existed to pursue a genetic parameter estimation.

Once phenotypic variance was quantified, we investigated 6 models. These were: 1) animal model where the phenotype of each animal was used; 2) animal model of females where each progeny was a phenotype and multiple progeny were repeated records; 3) dam model similar to a sire model where dam is the random effect estimated; 4) service sire model where the sire of offspring is the estimated random effect; 5) maternal grand sire model similar to sire model; and 6) maternal grand dam model similar to sire model. Contemporary group (defined as year of birth) and conception method (artificial insemination or natural mating) were fitted as fixed effects while age of dam was fitted as a covariate.

Genetic parameters and predicted means were estimated using an animal model in WOMBAT (Meyer 2007). A numerator relationship matrix based on a four-generation pedigree was used.

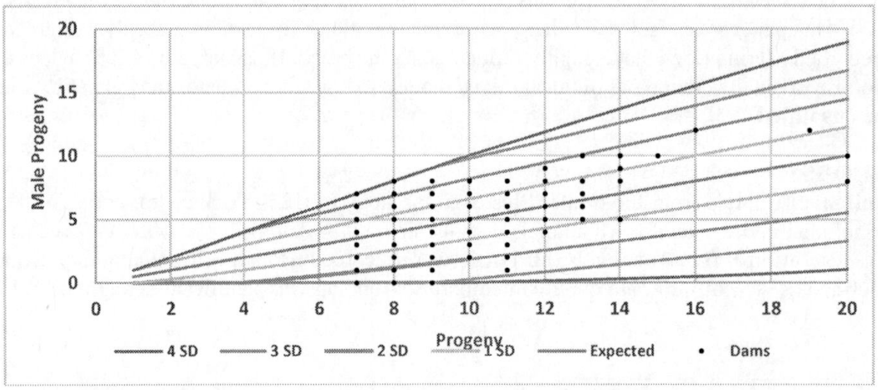


**Figure 1. Number of male progeny vs number of progeny for dams and where each dam sits within an expected normal distribution with a minimum of 7 progeny (n=765)**

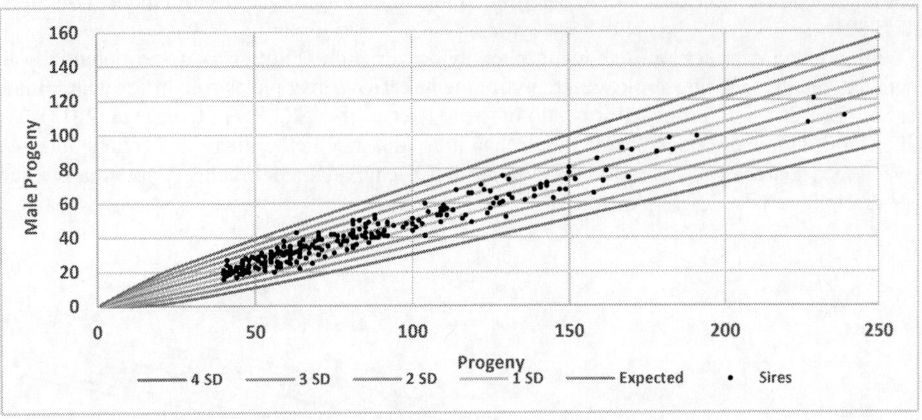


**Figure 2. Number of male progeny vs number of progeny for sires and where each service sire sits within an expected normal distribution with a minimum of 40 progeny (n=343)**

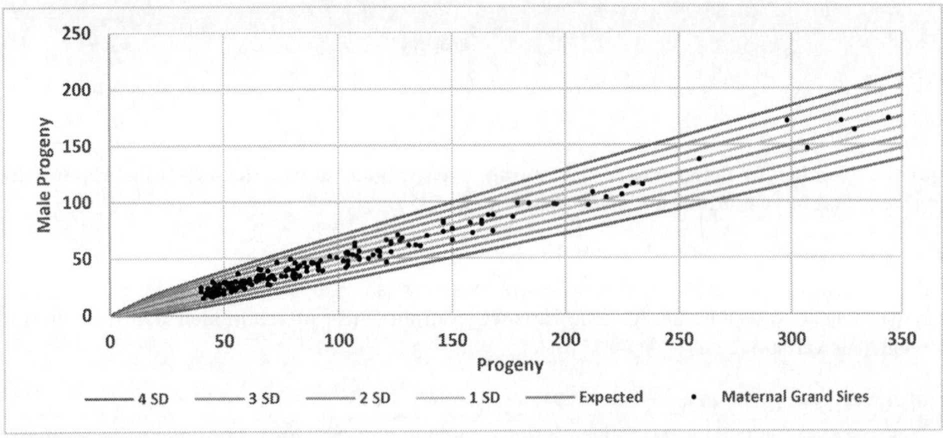


**Figure 3. Number of male progeny vs number of progeny for maternal grand sires with daughters that have a minimum of 40 progeny (n=186)**

## RESULTS AND DISCUSSION

All models converged with negligible genetic variance estimated ($h^2 \leq 0.01$). In model 2, where sex of progeny was used as a phenotype with repeated records, a small but insignificant amount of repeatability (0.02) was estimated. Despite no significant genetic variance being captured by the models, it can be observed in Figures 1-4 that phenotypic variance does exist for dams and sires which suggests that sex ratio is determined by factors outside genetics. If sex determination was random Figures 1-4 demonstrate that the sex ratios sit mostly within the 95% expected rate of a normal distribution with no outliers (i.e. > 4SD above or below the expected).

As there is a reasonable amount of phenotypic variance for all models (Table 1), other genetic sources of variation may be explored. If there were sufficient numbers of genotypes to perform a GWAS for females, a GWAS analysis could be performed. Another avenue of investigation into potential genetic variance of sex ratio determination could be to use a threshold model (Bulmer and Bull 1982).

With sire and dam sex ratio showing phenotypic variation (Figures 1-4) and potentially little genetic interactions playing a role, other environmental effects may play a role in sex determination. Diet has been shown to influence sex ratio in sheep (Green *et al.* 2008, Gulliver *et al.* 2013). These studies looked at whole flock means rather than individuals. Whether there is a genetic interaction between feed sources and sex ratio variation has not been explained, making it potentially a future cross-discipline study.

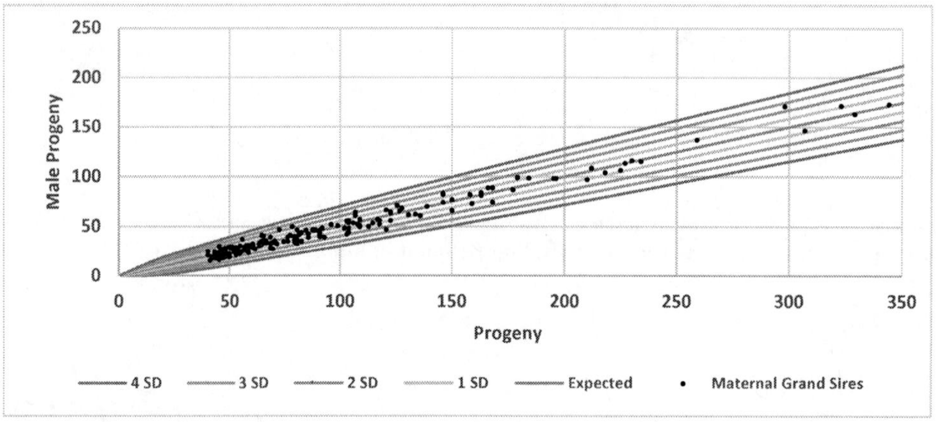


**Figure 4. Number of male progeny vs number of progeny for maternal grand dams with daughters that have a minimum of 10 progeny (n=535)**

## CONCLUSIONS

Phenotypic variation in the Centre Plus Merinos population exists for sex ratio. However, the study was unable to capture any genetic variance from the linear mixed models that were used to assess genetic variation.

## ACKNOWLEDGEMENTS

Thanks are extended to Centre Plus Merinos, Tullamore, NSW for making their data available for analysis.

## REFERENCES

Bulmer M.G., and Bull J.J. (1982) *Evol.,* **36**:13.

Gellatly C., (2009) *Evol. Bio.*, **36**:190.

Green M.P., Spate L. D., Parks T.E. *et al.* (2008) *Repro. Bio. and Endo.*, **6**:21.

Gulliver C.E., Friend M.A., King B.J., Wilkins J.F., and Clayton E.H (2013) *Anim. Prod. Sci.*, **53**:464.

Flanders S.E. (1965) *The American Nat.*, **90**:489.

Kosswig C. (1964) *Experimentia*, **20**:190.

Meyer K. (2007) *J. Zhejiang Uni. SCI. B* **8**: 815.

*Proc. Assoc. Advmt. Anim. Breed. Genet: 24: 135-138*

# MULTIVARIATE ANALYSES USING TWO GENOMIC RELATIONSHIP MATRICES TO WEIGHT PREDICTIVE SNP MARKERS

**P.M. Gurman[1], L. Li[1], A.A. Swan[1], N. Moghaddar[2] and J.H.J. van der Werf[2]**

[1] Animal Genetics Breeding Unit*, University of New England, Armidale, NSW, 2351 Australia
[2] School of Environmental & Rural Science, University of New England, Armidale, NSW, 2351 Australia

## SUMMARY

The Neogen GGP Ovine 50k chip contains approximately 5000 predictive Single-nucleotide polymorphisms (SNPs) that were identified by the Sheep CRC based on their relationship with carcase traits from genome wide association studies. These SNPs have been used in routine MERINOSELECT and LAMBPLAN analyses, equally-weighted with all other SNPs in a single genomic relationship matrix (GRM). This study aimed to examine the impact of fitting all SNPs in one GRM or fitting two GRMs, one with selected predictive SNPs and one with random SNPs, in conjunction with a numerator relationship matrix. Phenotypes on terminal sire breed cross resource flock animals recorded for five carcase and eating quality traits were used for bivariate variance component estimation. Variance components estimates were obtained for models containing only a numerator relationship matrix (NRM), NRM plus a GRM containing only non-selected SNPs, an NRM plus two GRMs containing non-selected and selected SNPs and an NRM plus one GRM containing all SNPs. Log-likelihoods were significantly higher in the models containing two GRMs for all trait pairs. Slightly higher average heritabilities were estimated from the model where the GRM contained all SNPs, except for intramuscular fat and shear force, where the GRM without the predictive SNPs resulted in higher heritabilities. The proportion of genetic variance explained by the genomic relationship matrices ($\lambda$) was estimated to be between 0.59 and 0.86. In terms of the genetic correlations between traits, for many trait-pairs the correlations were similar between the random effects fitted, but for two trait-pairs large differences were observed between the genetic correlations.

## INTRODUCTION

Routine genetic evaluations for Australian terminal sire, maternal and Merino sheep have utilised single-step genomic BLUP (SS-GBLUP) since 2017 (Brown *et al.* 2018). For the genomic relationship matrix used in these analyses, the SNPs used were based on a set that passed quality control from the ISAG 50k sheep panel. In 2019, a new genomic panel for sheep was introduced (GeneSeek Genomic Profiler Ovine 50k, Neogen) which included approximately 5000 additional predictive SNPs that have been significantly associated with specific growth, carcase and eating quality traits in sheep (Moghaddar *et al.* 2019). The union of all SNPs on all genomic panels was chosen (including the predictive SNPs), with imputation of missing SNPs on each panel, followed by imputing all panels to the union set, resulting in 60,410 SNPs used in SS-GBLUP.

The methods commonly used for constructing the genomic relationship matrix (GRM) for GBLUP (VanRaden 2008; Yang *et al.* 2010) assumes that all SNPs have equal weighting. While equal weighting on SNPs is reasonable for random SNPs, it may be appropriate to treat selected SNPs that are associated with specific traits differently. The GRM used in SS-GBLUP is blended with the NRM for these animals based on the parameter $\lambda$, with the currently used value in Australian sheep evaluations set to $\lambda = 0.5$ resulting in the weighted GRM being the mean of the raw GRM

---

* A joint venture of NSW Department of Primary Industries and the University of New England

and NRM (McMillan *et al.* 2017). This paper investigated the impact on covariance matrix estimates of including all SNPs in the same GRM or fitting separate GRMs for regular random SNPs and selected SNPs. The ratio of genetic variance explained by each genetic random effect was investigated, considering trait specific values of $\lambda$. Changes in covariances between genetic effects were also investigated.

## MATERIALS AND METHODS

Data on reference flock animals from both the Sheep CRC Information Nucleus Flock (van der Werf *et al.* 2010) and MLA Resource Flock databases were obtained from the LAMBPLAN terminal sire analysis. Pre-adjusted phenotypic data were used for five traits: post-weaning weight (PWT, kg), carcase eye muscle depth (CEMD, mm), carcase c-site fat (CCFAT, mm), intramuscular fat (IMF, %) and shear force (SF5, Newtons). Phenotypes were only retained for animals with genotypes and where a phenotype was recorded for all six traits, resulting in 9688 animals with data. Phenotypes used were pre-adjusted for birth type, rearing type, age of measurement, age of dam, and hot carcase weight (trait dependant). Contemporary groups were taken from the LAMBPLAN analysis, with PWT belonging to one contemporary grouping (based on breed, flock, management group and sex, $n = 444$) and all carcase traits using different contemporary groupings (based on combinations of breed, flock, management group, sex and kill group, $n = 376$).

The 60410 SNPs available were split into two sets: the random SNPs ($n = 55{,}709$) and the predictive SNPs ($n = 4{,}701$). Three marker sets were then used to construct breed-adjusted genomic relationship matrices (GRMs), using the method described by Gurman *et al.* (2019). These GRMs were labelled $\boldsymbol{G}_r$ for the random SNPs, $\boldsymbol{G}_p$ for the predictive SNPs and $\boldsymbol{G}_{rp}$ for the combined set of SNPs. A corresponding pedigree-based relationship matrix for animals with genotypes was also constructed based on the extended pedigree including all known ancestors. To accommodate variance component estimation using the software package 'mtg2' (Lee *et al.* 2016), animal by animal relationship matrices were constructed for the other random effects to be considered, genetic groups and dam permanent environment. Genetic groups ($n = 89$) were included by constructing a matrix of pedigree-based breed proportions, $\boldsymbol{Q}$, where the rows sum to unity and animals with known parents are the average of their parental group proportions. These proportions were then converted to an animal by animal matrix by $\boldsymbol{QQ}^T$. Similarly, for the dam permanent environment effect, an incidence matrix was constructed relating dams to animals, $\boldsymbol{W}$, which was converted to an animal by animal matrix $\boldsymbol{WW}^T$.

Pairwise bivariate models for all trait combinations were then analysed using various combinations of the genetic random effect matrices described above. The general model fitted was $\boldsymbol{Y} = \boldsymbol{Xb} + \sum_{i=1}^{n} \boldsymbol{Zu}_i + \boldsymbol{e}$ where $\boldsymbol{Y}$ is the data in multivariate form; $\boldsymbol{X}$ is the incidence matrix for the contemporary groups; $\boldsymbol{b}$ is the vector of fixed-effect solutions; $\boldsymbol{Z}$ is the incidence matrix relating animals to breeding value estimates; $\boldsymbol{u}_i$ is the vector of random effect solutions for the $i$th random effect and $\boldsymbol{e}$ represents the residual. The model is also such that $\text{var}(\boldsymbol{Zu}_i) = \boldsymbol{G}_i \otimes \boldsymbol{\Sigma}_i^2$ where $\boldsymbol{G}$ is the random effect matrix for the $i$th effect ($\boldsymbol{G} = \{\boldsymbol{A}, \boldsymbol{G}_r, \boldsymbol{G}_p, \boldsymbol{G}_{rp}, \boldsymbol{QQ}^T, \boldsymbol{WW}^T\}$ and $\boldsymbol{\Sigma}_i^2$ is the estimated covariance matrix for the random effect. For all models presented, genetic group and permanent environment effects of the dam were also included.

## RESULTS AND DISCUSSION

Significantly higher log-likelihood values were found for the models that contained two GRMs. Models that included GRMs had higher heritabilities than the pedigree-only models (Table 1). Further, the highest trait heritabilities were observed in the models that contained $\boldsymbol{G}_{rp}$. The proportion of the total genetic variance explained by the GRMs was between 0.59 and 0.86, with the model containing $\boldsymbol{G}_{rp}$ explaining a slightly higher proportion than the model containing only $\boldsymbol{G}_r$

(Figure 1). The model that contained separate $\boldsymbol{G_r}$ and $\boldsymbol{G_p}$ either explained less variance than the model containing only $\boldsymbol{G_r}$ (see CCFAT and PWT) or less than the model containing $\boldsymbol{G_{rp}}$ (see CEMD, IMF and SF5). These estimates of $\lambda$ are larger than the value of $\lambda$ currently used in MERINOSELECT and LAMPLAN analyses, suggesting that further investigation is required to determine if this finding is consistent for other traits or if $\lambda$ should be trait specific.

**Table 1. Heritabilities calculated from the sum of all genetic effects in each model**

| Random Effect Model | PWT | CEMD | CCFAT | IMF | SF5 |
|---|---|---|---|---|---|
| $\boldsymbol{A}$ | 0.217 | 0.202 | 0.225 | 0.629 | 0.305 |
| $\boldsymbol{A+G_r}$ | 0.283 | 0.225 | 0.252 | 0.636 | 0.313 |
| $\boldsymbol{A+G_{rp}}$ | 0.290 | 0.237 | 0.259 | 0.631 | 0.307 |
| $\boldsymbol{A+G_r+G_p}$ | 0.274 | 0.227 | 0.253 | 0.614 | 0.267 |

Abbreviations: **A**: NRM, $\boldsymbol{G_r}$ GRM calculated from random SNPs, $\boldsymbol{G_p}$ GRM calculated from the predictive SNPs, $\boldsymbol{G_{rp}}$ GRM calculated from all SNPs

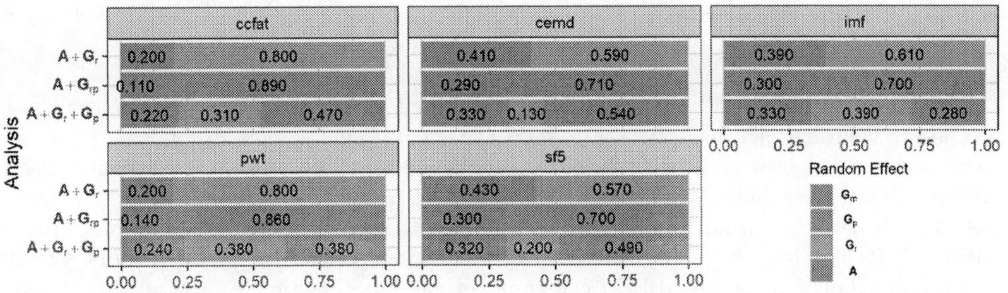


**Figure 1. Proportions of the total genetic variance explained by each random effect. Abbreviations listed in Table 1**

The genetic correlations between traits were not uniform across alternative models for genetic effects (Figure 2). While for most traits the correlations were fairly consistent, some trait pairs show much larger differences in the genetic correlations between models and random effects included, which the most evident of these being those correlations being CCFAT-PWT and CEMD-PWT. For both of these trait pairs, the estimated correlation was slightly negative between CCFAT-PWT and close to zero for CEMD-PWT from the model with only the NRM. When GRMs were added, these NRM correlations were estimated as strongly positive and the GRM correlations strongly negative. It should be noted that these differences largely cancel out when considering the overall genetic correlation. In some cases (CEMD-PWT, CF5-PWT, CEMD-SF5), the correlation estimated for the effects of $\boldsymbol{G_r}$ and $\boldsymbol{G_p}$ were different, suggesting here that the selected and random SNPs are capturing different genetic effects on these traits. Further investigation is required to determine why these differences in correlations occur.

A cross-validation study using the variance components from this study was also conducted to investigate the benefits on predictive ability of using two GRMs or a single GRM with all SNPs together in a large scale BLUP analysis (Li *et al.* 2021).

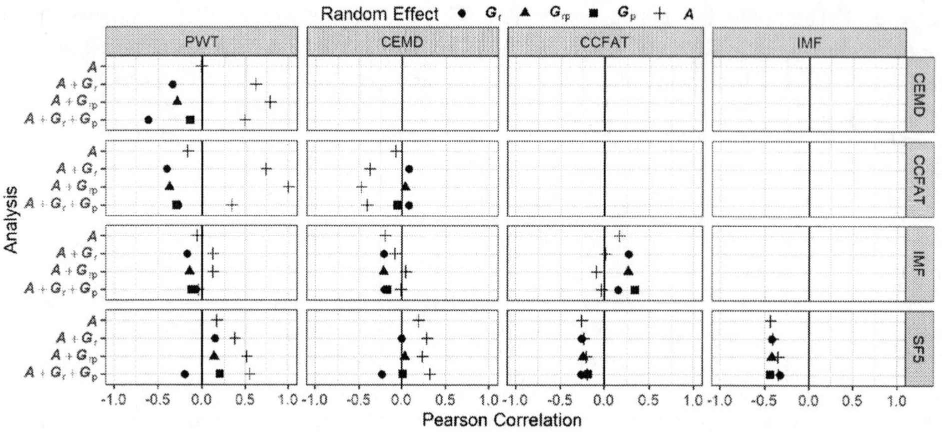


**Figure 2. Estimated correlations between traits for the genetic random effects for all four models. Abbreviations listed in Table 1. One correlation was estimated to be larger than one and was therefore modified to one for presentation**

## CONCLUSIONS

This study found that the current value of $\lambda = 0.5$ used in Australian sheep genetic evaluations was lower than that estimated for the carcase and eating quality traits examined. Higher log-likelihoods values were estimated for the models containing two GRMs, however, this often resulted in slightly lower heritabilities compared to a model that contained all SNPs in one GRM. Including GRMs in the analysis resulted in different genetic correlations for some trait pairs from different GRM/NRM combinations. These results suggest that not considering the GRM in variance component estimation for SS-GBLUP can result in variances incorrectly proportioned between NRM and GRM. Further work is required to examine these impacts in other populations with different genomic population structures and in different traits.

## ACKNOWLEDGEMENTS

This work was funded by MLA Project L.GEN. 1815.

## REFERENCES

Brown D.J., Swan A.A., Boerner V., Li L., Gurman P.M., McMillan A., Van der Werf J.H.J. (2018) *Proc. World Congr. Genet. Appl. Livest. Prod.* **Species-Ovine**:460.

Gurman P.M., Bunter K.L., Boerner V., Swan A.A., Brown D.J. (2019) *Proc. Assoc. Advmt. Anim. Breed. Genet.* **23**:254.

Lee S.H., Van der Werf J.H.J. (2016) *Bioinformatics* **32**:1420.

Li L., Gurman P.M., Swan A.A., Moghaddar N., Van der Werf J.H.J. (2021) *Proc. Assoc. Advmt. Anim. Breed. Genet.* **24**:*.These proceedings*

McMillan A.J., Swan A.A. (2017) *Proc. Assoc. Advmt. Anim. Breed. Genet.* **22**:557.

Moghaddar N., Khansefid M., Van der Werf J.H.J., Bolormaa S., Duijvesteijn N., Clark S.A., Swan A.A., Daetwyler H.D., MacLeod I.M. (2019) *Genet. Sel. Evol.* **51**:72.

VanRaden P.M. (2008) *J. Dairy Sci.* **91**:4414.

Van der Werf J.H.J., Kinghorn B.P., Banks R.G. (2010) *Anim. Prod. Sci.* **50**:998.

Yang J., Benyamin B., McEvoy B.P., Gordon S., Henders A.K., Nyholt D.R., Madden P.A., Heath A.C., Martin N.G., Montgomery G.W., Goddard M.E., Visscher P.M. (2010) *Nat. Genet.* **42**:565.

*Proc. Assoc. Advmt. Anim. Breed. Genet: 24: 139-142*

# CHARACTERISING THE QUANTITY AND QUALITY OF DATA USED IN MERINO SHEEP GENETIC EVALUATION SYSTEMS

**S.Z.Y. Guy[1] and D.J. Brown[1]**

[1] Animal Genetics Breeding Unit*, University of New England, Armidale, NSW, 2350 Australia

## SUMMARY

Estimated Breeding Values (EBVs) published by Sheep Genetics Australia have an accuracy estimated with them. While the EBVs, their accuracy, and errors of genetic parameter estimates are all influenced by both data quantity and quality, these calculations do not explicitly take into account all aspects of data quality. To encourage increased genetic gains, Sheep Genetics provides participating breeders with data quantity and quality metrics in a 'RAMping Up Genetic gains' report. This paper demonstrates the considerable variation in these metrics for Merino flocks, and proposes additional descriptors metrics to characterise the quantity and quality of sheep genetic evaluation data. Current results show that there are opportunities to improve the completeness of pedigree and reproduction trait recording. Flocks had on average 46.6 ± 36.1% (mean ± SD) of animals with full pedigree, and 4.1 ± 6.9% of animals within each flock with reproduction trait records. The average proportion of effective progeny was 64.3 ± 19.1%. Flocks had on average 40.2 ± 37.3% of animals in contemporary groups that had variation in birth date recording. Since variation in age within contemporary groups is expected, this highlights potential issues with accurate recording of birth dates. Additional metrics describing lambing date distributions and deviations from the expected dates were derived, and emphasise potential issues of birth date accuracy, with some flocks recording birth dates on a non-random proportion of days of the week. Feedback on the quantity and quality of their current data should help ram breeders target improvements on their recording program. However, the optimum or reasonable level of quantity and quality to maximise genetic gains is currently undefined.

## INTRODUCTION

The genetic evaluation systems available to the Australian sheep and beef industry through Sheep Genetics and BREEDPLAN, respectively, primarily rely on industry data submitted by seedstock producers. While there are standards and guidelines, there is wide variation in the data submitted. An accuracy figure is reported alongside estimated breeding values (EBVs). While the quality of data has been shown to influence the EBVs, their accuracy and the errors of genetic parameter estimates, accuracy is calculated using the amount and structure of information utilised (i.e. quantity), and not explicitly the quality of information. The difference between data quantity and quality is highlighted in the following example; a date of birth may be supplied for each animal (maximum data quantity), but a single generic date may be used for all animals irrespective of their actual date of birth within the lambing period (poor data quality). This will affect the ability to accurately correct for age and thus the accuracy of the EBVs. This highlights the need for additional metrics beyond EBV accuracy to characterise the quality of data.

Data Quality Grades, which reflect the level of recording for pedigree, scan and wool traits, were previously provided to LAMBPLAN clients as a practical approach to describing index accuracy (Banks, 1999). Currently, Sheep Genetics provides the 'RAMping Up Genetic gains' (RUGG) report to participating breeders, which includes metrics to describe the quantity and quality of pedigree and

---

* A joint venture of NSW Department of Primary Industries and the University of New England

performance recording, and data structure. These metrics have been shown to have an association with genetic gains for a flock (Stephen *et al.* 2018). This paper demonstrates the variation in the data quantity and quality metrics reported in RUGG reports for Merino flocks, and proposes additional metrics to characterise the quantity and quality of data being supplied to Sheep Genetics.

## MATERIALS AND METHODS

**Existing data metrics.** The metrics reported in RUGG reports were available for the 265 Merino flocks from the 12th December 2020 analysis. These flocks had a minimum of 100 animals per year and data available for the last 5 years. Unless stated, metrics were calculated as an average of the last 5 years and across contemporary groups. Metrics were classified as either quantity or quality metrics, although it must be acknowledged that some metrics can be placed in either category:

**1) Quantity:** the amount of data submitted and its completeness.

- **fullped** (%): proportion of animals from the flock in the analyses where both sire and dam are known (i.e. full pedigree).
- **avpedknown** (%): completeness of pedigree known from last 3 generations.
- **recorded** (%): proportion of animals with records submitted for any of the following: weight, fat, eye muscle depth, fleece weight and fibre diameter (all age stages) or number lambs weaned.
- **ngeno** (%): proportion of animals genotyped.

**2) Quality:** the appropriateness for its intended use, including accuracy and data structure.

- **synped** (%): proportion of animals with syndicate pedigree (i.e. where multiple rams are mated over a group of ewes, resulting in multiple potential parents for the progeny).
- **ages** (%): proportion of animals recorded that are in contemporary groups with variation in age. Variation in age within contemporary groups is expected with accurate birth date recording.
- **bt** (%): proportion of animals recorded that are in contemporary groups with variation in birth type recorded.
- **eff** (%): proportion of effective progeny (i.e. percentage of progeny from a given sire relative to all progeny within a group, as defined in Brown *et al.* 2001).

**Additional quantity metrics.** To take into account the different breeding objectives of each breed type, a 'recorded' metric was expanded to the proportion of animals recorded by trait groups:

- **rec_weights**: weight traits, ultrasound c-site fat depth, and ultrasound eye muscle depth.
- **rec_repro**: number of lambs weaned.
- **rec_wool**: greasy fleece weight and fibre diameter.

**Additional qualtity metrics.** These included genetic linkage metrics by trait group, as well as metrics to describe lambing date distributions and deviation from uniform distributions (inspired by DataAudit and StockTake; Johnston and Moore, 2005):

- Average proportion of animals recorded that are directly linked to external flocks, by trait groups: carcase scan traits (**link_carcase**), weight traits (**link_weights**), number lambs weaned (**link_repro**), wool traits (**link_wool**).
- **maxfreq_ywt** (%): the percentage of the most common single value appearing. Missing values were not included in this calculation. Only results for yearling weight (ywt) are reported in this paper as ywt was the most common weight trait recorded for the flocks examined.
- Chi-squared statistics: For a large sample size of data with sufficiently wide variation in values, the last digits are expected to have a uniform distribution (Dlugosz and Müller-Funk, 2009). Deviation from this expectation may be due to poor recording, equipment problems or non-randomisation of recordings. Since different traits are recorded in various increments (e.g. as whole number integers or various decimal places), chi-squared statistics were calculated for the last digits in the units (**chi_units_ywt**) and tenths (**chi_tenths_ywt**) place values for ywt:

$$\chi^2 = \sum_{i=0}^{10} \frac{(\text{expected\%} - \text{observed\%})^2}{\text{expected\%}}$$

where expected = 10%, and observed = % of records with the digit $i$

- **dayinweek:** mean square error of birth date for days in week. This metric is based on the same concept as the chi-squared statistics, where the likelihood of birth dates to occur on any given day of the work is expected to be equal. This was calculated as:

$$\text{dayinweek} = \sum_{i=1}^{7} (\text{expected\%-observed\%})^2$$

where expected = 1/7 × 100 % for each day of the week, and observed= % of animals born on the $i$th day of the week.

## RESULTS AND DISCUSSION

**Variation in data metrics.** Figures 1a and 1b demonstrate the considerable variation in the data quantity and data quality metrics across Merino flocks.

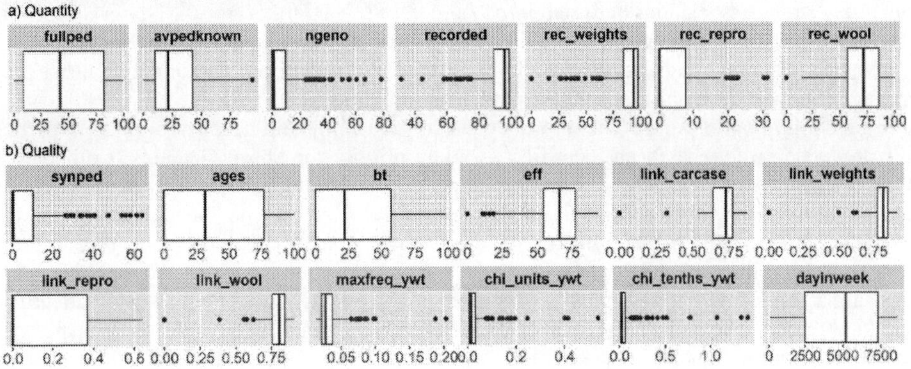


The data quantity metrics describe the amounts of pedigree and performance recording across Merino flocks. The average proportion of animals within a flock with full pedigree (*fullped*) was 46.6% ± 36.1% (mean ± SD), with 29.2± 36.1% of the pedigree complete over the last 3 generations (*avpedknown*). Flocks had an average of 7.4% of animals within the drop genotyped (*ngeno*). Recording by trait groups was more informative than a recording metric that included all traits. As expected, there was more recording for weight traits (*rec_weights*, 86.9 ± 19.2%) and wool traits (*rec_wool*, 69.0 ± 20.7%) than reproduction traits (*rec_repro*, 4.1 ± 6.9%). Since only a proportion of ewes enter the ewe flock, low values of *rec_repro* were as expected. However, it was also the most variable metric relative to the mean (range 0 to 31.3%, CV= 168.0%). These metrics highlight the opportunity for Merino breeders to improve recording for pedigree and reproduction traits.

The data quality metrics describe varying levels of pedigree accuracy and distribution of data. There was a low proportion of animals with syndicate pedigree (*synped*, 7.4 ± 12.3%). However this metric was also the most variable (range 0 to 63.9%, CV = 164.9%). The proportion of animals in a contemporary group that had variation in recording for birth dates (*ages*) was 40.2 ± 37.3% and 32.0 ± 32.1% for birth type recording (*bt*). That is, ~60% animals were in groups where there was no variation in birth date, and ~68% with no variation in birth type. This highlights potential issues with accurate recording of birth dates and birth types. The average proportion of effective progeny (eff) was 64.3 ± 19.1%. Since the eff metric can only be estimated if sire pedigree is known, this is expected to be an underestimate. The degree of linkage to other flocks reflected the level of recording by trait group and Merino breeding objectives, with the most linkage through weight, wool and carcase traits compared to reproduction traits (*link_weights*, 78.6 ± 15.8%; *link_wool*, 78.1 ± 13.6%; *link_carcase*, 58.2 ± 31.3%; *link_repro*, 14.3 ± 20.1%). An average of 3.3% of yearling weight records (*maxfreq_ywt*) were the same within each flock (range of 0.01% to 20.3%).

The quality metrics describing distributions of traits and deviations from expected distribution also varied across flocks. The chi-squared statistics, describing last digit distributions, were all less than the chi-squared critical value of 3.325, suggesting that the frequencies of last digits for ywt were as expected. Conversely, the average *dayinweek* was 4,841.5 ± 2777.53, and ranged from 20.3 to 8,571.0 (i.e. the maximum mean square error, with birth dates recorded on only one day of the week). Again, the required degree of accuracy for birth dates and what is considered a reasonable loss in age adjustment precision is currently unknown. Nevertheless, these distribution and deviation metrics can still be used as a way to highlight unusual data.

**Relationships between metrics.** The relationships between the quantity and quality metrics were quantified by Pearson's correlations (*r*). As expected, there were strong linear associations between *rec_repro* and *link_repro* (*r* = 0.82), and *fullped* and *avpedknown* (*r* = 0.77). There were moderately strong associations between *fullped* and *ages* (*r* = 0.60), *bt* (*r* = 0.61), *daysinweek* (*r* = -0.51), *link_repro* (*r* = 0.48) and *eff* (*r* = 0.40). There were also strong associations within categories (e.g. between *ages* and *daysinweek*, *r* = -0.85). Therefore, the quantity and quality metrics are not necessarily independent, and some metrics describe similar aspects.

**Industry implementation.** The improvement of the quality and quantity of data, in particular for reproduction traits, has been identified as a key priority for Sheep Genetics (Collison *et al.*, 2018). A framework to characterise genetic evaluation data, including a carefully developed overall 'data quality score, will benefit individual breeders, ram buyers and the industry as a whole. Feedback on the quantity and quality of their current data will allow ram breeders target improvements on their recording program, which support selection decisions and maximise genetic gains, and assess changes in recording across time. A data quality score could also help identify and highlight breeders who collect high quality data. In turn, this will provide increased transparency to ram buyers about the quality of data used to calculate EBVs. There is also potential to use these metrics to determine how data contributing to the reference population is valued and rewarded.

## CONCLUSIONS

This paper demonstrates the considerable variation in the quantity and quality of Merino sheep genetic evaluation data. While there are opportunities for Merino flocks to improve completeness and accuracy of pedigree recording, birth date and reproductive performance, the optimum or reasonable level of quantity and quality is currently undefined.

## ACKNOWLEDGEMENTS

This work was funded by MLA project L.GEN.2004. The authors also grateful for the AGBU and Sheep Genetics staff who provided valuable feedback and discussion on this topic.

## REFERENCES

Banks R.G. (1999) *ICAR Technical Series.* **3**: 521.

Brown D.J., Tier B. and Banks R.G. (2001) *Proc. Assoc. Advmt. Anim. Breed. Genet.* **14**: 441.

Collison C.E., Brown D.J., Gill J.S., Chandler H.R., Apps R., Swan A.A. and Banks R.G. (2018) *Proc World Congress on Genetics Applied to Livestock Production.* **11**: 661.

Dlugosz S. and Müller-Funk U. (2009*). Adv. Data Anal. Classif.* **3**: 281.

Johnston D.J. and Moore K.L. (2005) *Proc. Assoc. Advmt. Anim. Breed. Genet.* **16**: 161.

Stephen L.M., Brown D.J., Jones C.E. and Collison C.E. (2018) *Proc World Congress on Genetics Applied to Livestock Production.* **11**: 433.

*Proc. Assoc. Advmt. Anim. Breed. Genet: 24: 143-146*

# MAXIMISING GENETIC GAINS WITH DATA QUANTITY AND QUALITY IN MERINO FLOCKS

**S.Z.Y. Guy[1] and D.J. Brown[1]**

[1] Animal Genetics Breeding Unit*, University of New England, Armidale, NSW, 2350 Australia

## SUMMARY

Genetic gain can be maximised when selection is based on the most accurate breeding values and selection indices. To more explicitly take into account aspects pertaining to the quality of information used to estimate breeding values, metrics to characterise the quantity and quality of genetic evaluation data were previously proposed. This paper examines the relationships between these data quantity and quality metrics and genetic gains for Merino flocks. Stepwise regression analysis was used to analyse 3 metrics describing genetic gains: index accuracy, average index value and index trend. Index accuracy had the most number of significant predictors, with 4 quantity and 3 quality predictors explaining 85% of the observed variation. The most important metrics explaining index accuracy were level of genetic linkage for wool traits, average proportion of pedigree known in the last 3 years, and the level of wool and reproduction trait recording ($p < 0.0005$). Data characteristic metrics were also associated with average index and index trend, although to a lesser level (~24% variation explained). This study demonstrates that both data quantity and quality are associated with index accuracy and genetic gains in Merino flocks. This decomposition provides a basis for informing ram breeders on improvements in their data recording. Used in conjunction with optimum selection decisions, this will enable higher rates of genetic progress.

## INTRODUCTION

Genetic gain can be maximised when selection is based on the most accurate breeding values and selection indices. While the accuracy of estimated breeding values is calculated using the amount and structure of information utilised (i.e. quantity), some aspects pertaining to the quality of information can not explicitly taken into account in this calculation. Aspects of data quality, such as management group structure and accurate dates of birth, have been shown to affect the accuracy of estimation of genetic merit (Brown *et al.* 2001; Swan and Brown 2007). However, it is important to more specifically quantify the impact of data quality on the estimation of genetic merit due to the varying quality of data submitted by seedstock producers. Characterising both the quantity and quality of data will allow breeders to identify where their recording programs can be improved.

Sheep Genetics reports data quantity and quality metrics in their 'RAMping Up Genetic gains' (RUGG) report. Variation in these metrics has been shown to be associated with variation in rates of genetic progress (Stephen *et al.* 2018). Guy and Brown (these proceedings) reported considerable variation in key data metrics for Merino flocks, and proposed additional data quantity and quality metrics. This paper aims to demonstrate the value proposition of these metrics by examining their associations with genetic gains.

## MATERIALS AND METHODS

**Data quantity and quality metrics.** The metrics examined in this paper were available for the 265 Merino flocks in the 12$^{th}$ December 2020 analysis that had a minimum of 100 animals per year

---

* A joint venture of NSW Department of Primary Industries and the University of New England

and data available for the last 5 years. These metrics (defined in Guy and Brown, these proceedings) were calculated as an average of the last 5 years and across contemporary groups.
Briefly, the data quantity metrics were:

- **fullped** (%): overall proportion of animals in the analysis with full pedigree.
- **avpedknown** (%): completeness of pedigree known from last 3 generations.
- **count**: average flock size per year.
- **ngeno** (%): average proportion of animals genotyped.
- Average proportion of year drop recorded for weight (**rec_weights**), reproduction (**rec_repro**) and wool traits (**rec_wool**).

The data quality metrics were:

- **synped** (%): proportion of animals with syndicate pedigree.
- **ages** (%): proportion of animals in contemporary groups with variation in age recorded.
- **bt** (%): proportion of animals in contemporary groups that have variation in birth type recorded.
- **eff** (%): the average proportion of effective progeny (Brown *et al.* 2001).
- Average proportion of animals directly genetically linked to other flocks for weight (**link_weights**), reproduction (**link_repro**) and wool traits (**link_wool**).
- **Dayinweek:** mean square error of birth date for days in week from an expected uniform distribution.

**Data metrics and genetic gains.** Index accuracy (indexacc), average index value (avindex) and index trend (indextrend) were calculated for each Merino flocks for the Merino Production + index, and averaged over the last 5 years. A series of linear regression models was used to examine the relationships between each genetic gain metric and data quantity and quality metrics:

$$\text{GeneticGains\_perc}_i = \mu + \beta_1 x_i + \varepsilon$$

where GeneticGains_perc is the percentile of indexacc, avindex or indextrend of flock *i* (quintiles 1 to 5, with 5 being the highest index), and $x_i$ the flock's corresponding data quantity and quality metric refined above. Outliers, defined by 1.5 × Inter Quartile Range below the 1st quartile and above the 3rd quartile, were removed from data metrics due to potential leverage and influential points affecting results of this analysis.

Multiple linear regression was conducted using all data quantity and quality metrics as predictor variables. Identification of the strongest associations with index accuracy and genetic gains was via stepwise regression (combining both backward elimination and forward selection), based on Akaike Information Criterion (AIC). The final model only included significant data characteristic metrics:

$$\text{GeneticGains} = \mu + \beta_1 x_1 + \beta_2 x_2 + \beta_3 x_3 + \ldots + \beta_i x_i + \varepsilon$$

where $\beta_i$ is the partial regression coefficient of metric *i*. Flocks were included in the analysis if all data quantity and quality metrics were available ($n = 243$).

## RESULTS AND DISCUSSION

There were significant differences in metrics across percentile groups, although there was considerable variation in each metric across the genetic gain percentile groups, with distributions overlapping (strongest relationships shown in Figure 1. Flocks with higher indexacc, avindex and indextrend ($P < 0.005$) had more reproduction traits recorded, more genotyped animals and a higher degree of average pedigree known in last 3 years (with the exception of avindex). These flocks also had greater linkage with other flocks for reproduction, weight and wool traits, actual birth dates and birth types recorded (not for avindex) and birth dates recorded evenly across days of the week. Flocks with higher indexacc also had greater average effective progeny numbers for sires and less syndicate recording (not shown in Figure 1). Therefore, flocks with more records and better quality data were associated with higher index accuracies and greater index gains.

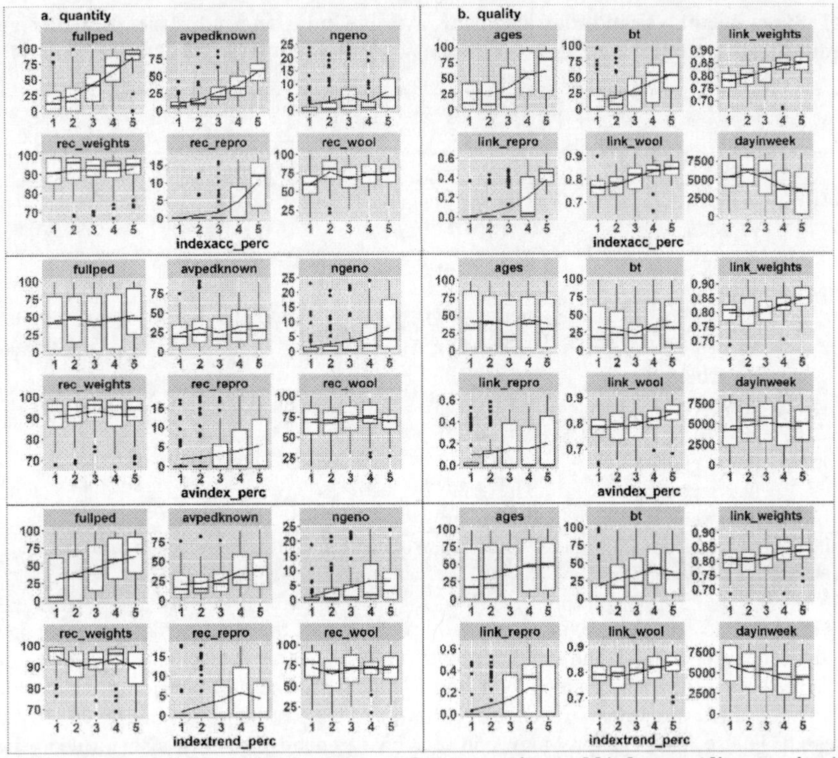


**Figure 1. Univariate relationships between a) data quantity and b) data quality metrics (*y* axes) and percentiles of index accuracy (indexacc_perc), average index value (avindex_perc) and annual index trend (indextrend_perc) (*x* axes) for the Merino Production Plus index, from 265 Merino flocks. Lines are the average metric values for each percentile group, which reflect the strength of each relationship**

Stepwise regression was used to analyse the indexacc and genetic gains, with all data quantity and quality metrics tested (Table 1). Indexacc had the most number of significant predictors, with 4 quantity and 3 quality in the final model. The most important descriptors of indexacc were level of linkage by wool traits, average pedigree known in the last 3 years, and level of wool and reproduction trait recording ($P < 0.0005$). This reflects the key priority areas of increasing complete pedigree, wool trait and reproduction trait recording identified by Sheep Genetics (Collison *et al.* 2018). The proportion of variation explained, taking into account number of predictor variables ($R^2_{adj}$), was also highest for indexacc ($R^2_{adj}$ = 0.85). Avindex had 3 significant quantity and 2 quality metrics ($R^2_{adj}$ = 0.25) and indextrend had 4 significant quantity and 1 quality metrics ($R^2_{adj}$ = 0.23). The metric that was a significant predictor in all models was avpedknown, while synped, bt and dayinweek were not significant predictors for all models.

These results have implications for the development of an overall 'data quality score'. Indexacc more appropriately reflects data characteristics than avindex and indextrend, which are influenced by breeder selection decisions. Along with a large proportion of variance explained by data quantity and quality metrics, indexacc may be the most appropriate measure to 'train' an overall score.

**Table 1. Stepwise analysis of index accuracy, average index and index trends, with significant data quantity and quality metrics as predictor variables ($p < 0.05$), for 243 Merino flocks**

| | | Index accuracy[+] ($R^2_{adj} = 0.85$) | Average index[+] ($R^2_{adj} = 0.25$) | Index trends[+] ($R^2_{adj} = 0.23$) |
|---|---|---|---|---|
| | Metrics[1] | | Coefficient estimates (SE)[*] | |
| Quantity | fullped | 0.03 (0.01) | -0.10 (0.05)[*] | - |
| | avpedknown | 0.11 (0.02) | 0.16 (0.08) | 0.03 (0.01) |
| | count | - | 0.007 (0.003) | - |
| | ngeno | - | 0.72 (0.17) | 0.07 (0.03) |
| | rec_weights | 0.06 (0.03)[*] | - | -0.09 (0.02) |
| | rec_repro | 0.18 (0.08) | - | -0.14 (0.06) |
| | rec_wool | 0.05 (0.01) | - | - |
| Quality | ages | - | -0.08 (0.04) | - |
| | eff | 0.04 (0.02) | - | - |
| | link_weights | - | 60.21 (23.98) | - |
| | link_repro | 8.02 (2.20) | - | 4.60 (1.70) |
| | link_wool | 40.27 (4.56) | - | - |

[+]using the Merino Production Plus index; [1] Description of metric acronyms provided in materials and methods section above; [*] P < 0.10

It is important to note that the most powerful predictors of the measures of genetic gain used in this study (indexacc, avindex and indextrend) are specific to this dataset and the index examined, and separately, that they may change over time. The effectiveness of providing feedback on data characteristics can be monitored by trends over time, and the cost-benefit of improved recording can be assessed. Future investigations may consider how genetic gains are also influenced by ram breeder selection decisions. This includes selection for traits not included in the index or use of outside genetics or selection differential, which has been identified as a key performance indicators of index gains across multiple beef cattle breeds (Johnston and Moore 2005).

## CONCLUSIONS

This study demonstrates some key components of data quantity and quality which are associated with index accuracy and metrics describing genetic gains in Merino flocks. This decomposition provides a basis for informing ram breeders on improvements in their data recording. Used in conjunction with optimum selection decisions, this will enable higher rates of genetic progress.

## ACKNOWLEDGEMENTS

This work was funded by MLA L.GEN.2004. The authors also grateful for the AGBU and Sheep Genetics staff who provided valuable feedback and discussion on this topic.

## REFERENCES

Brown D.J., Tier B. and Banks R. (2001) *Proc. Assoc. Advmt. Anim. Breed. Genet.* **14**: 441.

Collison C.E., Brown D.J., Gill J.S., Chandler H.R., Apps R., Swan A.A. and Banks R.G. (2018) *Proc World Congress on Genetics Applied to Livestock Production.* **11**: 661.

Guy S.Z.Y. and Brown D.J. (2021) *Proc. Assoc. Advmt. Anim. Breed. Genet.* **24**: In Press.

Johnston D.J. and Moore K.L. (2005) *Proc. Assoc. Advmt. Anim. Breed. Genet.* **16**: 161.

Stephen L.M., Brown D.J., Jones C.E. and Collison C.E. (2018) *Proc World Congreess on Genetics Applied to Livestock Production.* **11**: 433.

Swan A.A. and Brown D.J. (2007). *Proc. Assoc. Advmt. Anim. Breed. Genet.* **17**: 395.

*Proc. Assoc. Advmt. Anim. Breed. Genet: 24: 147-150*

# GENOMIC EVALUATION OF MALE FERTILITY OF AUSTRALIAN HOLSTEIN-FRIESIAN AND JERSEY BULLS

**M. Haile-Mariam[1], K. Konstantinov[1], E.G Jewell[1,2], G.J Nieuwhof[1,2] and J.E. Pryce[1,3]**

[1]Agriculture Victoria Research, AgriBio, 5 Ring Road, Bundoora, VIC, 3083, Australia
[2]DataGene Limited, AgriBio 5 Ring Road, Bundoora VIC 3083 Australia
[3]School of Applied Systems Biology, La Trobe University, Bundoora, VIC 3083, Australia

## SUMMARY

For a long time, the artificial insemination (AI) industry has provided high-quality semen for dairy cattle breeding. With the recent trend to widely use genomically selected bulls before adequate screening of their semen, predicting bull fertility early in life has become an important area of research. In this study we used 25-day non-return rate of about 3 million Australian cows that were inseminated using semen from 5943 Holstein (H) and 1258 Jersey (J) bulls that had high density SNP data (HD), to estimate the proportion of variance explained by SNP data and assess the accuracy of prediction of validation bulls. The proportion of variance explained by SNP data was about 1.2% in Jersey and 0.6% in Holstein bulls. The mean bull solution for both breeds was near zero (-0.05% for H and 0.43% for J). The standard deviation of the bull solutions of was 2.36% in H and 3.30% in J bulls. For both H and J bulls, the difference between the best and worst bulls was about 18% units. Genomic prediction (GP) accuracies were estimated using 5-fold cross validation and varied from 0.20 to 0.25 in H bulls and 0.08 to 0.36 in J bulls. For H bulls the GP accuracy for young bulls were lower (0.13) than average accuracies calculated from 5-fold cross validation. In the case of J bulls, the accuracy for young bulls were the same (0.22) as the average accuracy from 5-fold cross validation. The results show that despite the low heritability, GP of male fertility in Australian H and J breeds is possible and could be used for monitoring and making early decisions to avoid the use of semen from extremely poor fertility bulls.

## INTRODUCTION

Genetic improvement programs in dairy cattle have focused on female fertility but ignored male fertility assuming the artificial insemination (AI) industry is able to properly screen and standardize the quality of semen before it is widely distributed. Most studies have not found significant genetic difference in outcomes of insemination among bulls used for mating, possibly because of screening on semen parameters (e.g., Carrick *et al.* 2000; Kuhn and Hutchison 2008). As a result, bull fertility is a phenotypic evaluation used to rank bulls on AI success. Nevertheless, there is evidence that AI success varies among bulls and information on bull non-return rate (NRR) following insemination could be useful for improving overall herd fertility (e.g., Abdollahi-Arpanahi *et al.* 2017). The economic impact of even a small difference in semen fertility between bulls could be large because a single bull is mated to thousands of cows and the benefit of using bulls with good semen fertility is immediate and has a direct effect on the overall herd fertility.

With the recent shift in the dairy industry towards fast tracking of young genomically selected bulls for intensive use before adequate screening, exploring causes of variation in bull fertility early has become an emerging area of research (Taylor *et al.* 2018). The renewed interest to assess the extent of genetic variation in male fertility is partly due to the opportunity to carry out genomic-enabled screening of bulls before they are extensively used for semen collection (Abdollahi-Arpanahi *et al.* 2017; Rezende *et al.* 2019). The main aim of this study was to examine if the use of genomic evaluations can provide an opportunity for early culling of bulls based on 25-day non-return rate (success or failure of insemination outcomes) of their mates. For this study we used genotype and phenotype data of 5934 Holstein (H) and 1258 Jersey (J) bulls that mated to about 3 million cows. Accuracy of genomic predictions (GP) for both breeds were tested using a 5-fold cross validation and by predicting direct genomic values (DGVs) for younger bulls.

## MATERIALS AND METHODS

**Phenotype data**. Detailed description of the phenotype data used for this study is given by Carrick *et al.* (2000) and Haile-Mariam and Pryce (2021). Briefly the outcome of each insemination of AI bulls, called non-return rate (NRR), is derived by coding each insemination as successful (1) or failed (0) based on a minimum of interval of at least 25-days after insemination. In the first instance, any insemination performed at least 25-days before the end of the AI period was coded as successful and was changed to failed if it is followed by another insemination or mating at least 10 days after the previous insemination. Currently these data are used for calculating semen fertility values (SFV) of bulls by DataGene (https://datagene.com.au). In total there were 10941 bulls with 3.8 million inseminations between 1995 and 2020 in 3289 herds in Australia. AIs involving H and J bulls that mated to all breeds of cows (predominantly H and J, respectively) were selected for this study. The number of H and J bulls with phenotype and genotype data are given in Table 1.

**Table 1. The structure of Holstein and Jersey data used for genomic analyses**

| Reference set | Holstein bulls | Jersey bulls |
|---|---|---|
| No. of records | 2114529 | 300560 |
| No. of bulls with data | 4654 | 1057 |
| Year of birth of bulls | 1990-2014 | 1990-2012 |
| Mean NRR (%) | 51.77(49.97)[A] | 55.95(49.65)[A] |
| No. of inseminations per bull | 449(10-43221)[B] | 285(10-14147)[B] |
| Validation set | | |
| No. of records | 234401 | 61493 |
| No. of bulls with data | 799 | 201 |
| Year of birth of bulls | 2015-2019 | 2013-2019 |
| Mean NRR (%) | 49.69(49.99)[A] | 52.96(49.91)[A] |
| No. of inseminations per bull | 293(10-3313)[B] | 309(10-4497)[B] |

[A]Standard deviation; [B]Range in number of inseminations per bull.

**Genotype data.** Most bulls were genotyped using 50K SNP chips from various commercial providers, while about a quarter had HD genotypes. The first stage of the imputation was to a standard 50K SNP chip for all bulls followed by imputation to HD. Imputation of all 50K genotypes to HD was implemented using Fimpute v3 (Sargolzaei *et al.* 2014) with a reference set (RS) of 2700 HD genotypes. All 50K variants that passed quality control but did not overlap the HD set were then added back into the final imputed set which included the combined HD and 50K SNP sets. The 720,521 SNP set used for this study are located on all 30 chromosomes including the pseudo-autosomal region of the X Chromosome (Nguyen et al. 2021). The SNP data were used to create genomic relationship matrix (GRM) following Yang *et al.* (2011) separately for H and J bulls applying a minor allele frequency of 0.01 and 0.05 for H and J, respectively. To test if a joint RS of H and J bulls is beneficial, a third GRM using genotyped data of both breeds was also constructed.

**Statistical analyses.** This study used NRR coded as 100 (for successful) and 0 (for failed) as the response variable to evaluate male fertility compared to studies in the literature (Abdollahi-Arpanahi *et al.* 2017; Rezende *et al.* 2020) that used summarized bull solutions (e.g., sire conception rate or SFVs). The use of the raw NRR data jointly with important fixed and random effects and the GRM of bulls is expected to capture more of the variance and increase the accuracy

GP of bulls. Data analyses were carried out assuming a linear animal model using ASReml (Gilmour *et al.* 2015). Details of the fixed and random effects that were fitted are described by Haile-Mariam and Pryce (2021). Briefly a contemporary group effect that included herd-year-AI technician, mating number, cow breed, month of insemination, data processing centre, age of cow and bull at insemination, days in milk at insemination and days from insemination to the end of the AI period were fitted. The random effects fitted were the permanent environmental effect for the cow and the GRM for the bulls with insemination data. First, we used the genotype and phenotype data of all H and J bulls to quantify the proportion of variance captured by GRM. Then accuracies of GP were tested into 2 ways: Firstly, in a 5-fold cross validation scheme where the data were split into 5 parts of approximately equal size, by allocating the offspring of each sire to one of the 5 datasets. In this approach no bull in the validation set had paternal half sibs in the RS. This analysis was performed 5 times using each dataset in turn as a validation and the other 4 sets as the reference. Secondly, validation using young bulls (forward prediction) where bulls born after 2014 were used as a validation set and those born between 1990 and 2014 were used as RS in H. For J, bulls born after 2012 were included in validation set because the number born after 2014 were fewer (see Table 1). In both cases validation bulls were included in the GRM but had missing phenotypes when calculating their DGVs. Accuracy of prediction is calculated as the correlation between corrected phenotype (for effects considered in the model described above) and DGVs for bulls with at least 100 inseminations.

## RESULTS AND DISCUSSION

The mean NRR for both H and J bulls used in the reference and validation set are shown in Table 1. The mean NRR are lowest in H validation bulls and highest in J reference bulls. The proportion of variance explained by the GRM was lower in H (0.6%) than in J bulls (1.2%). In both cases the permanent environmental effect of the cow accounted for 3% of the total variance and more than 95% of the variation was not accounted for by the model. Despite this, the bull solutions for both breeds show considerable variation. The mean bull solutions for all Holstein bulls were close to zero (-0.05%) with a standard deviation (SD) of 2.36%. In the case of Jersey bulls, the mean was 0.43% with SD of 3.30%. The bull solutions for both breeds show an approximate normal distribution (-9.0 to +9.0%) with few extremely poor fertility bulls. There were 9 H and 7 J bulls with solutions of below -9.0%.

The accuracy of GP from the 5-fold cross validation are similar in both breeds despite the larger reference size of the H breed. The accuracy values for H bulls are lower than those reported by Abdollahi-Arpanahi *et al.* (2017) who used 7447 bulls with sire conception rate in the USA. Part of the reason for the difference could be the response variable used and the way the data were analysed in both studies. The difference in the RS between the two studies may also have contributed to the lower GP accuracy of the current study. For J bulls our estimates are slightly lower than those for J bulls from the USA (0.28-0.29) which was based on about 1500 bulls (Rezende *et al.* 2019). Interestingly for Australian J bulls, a bivariate model that used sire conception rate from the USA and SFV from Australia resulted in accuracy of 0.24 (Rezende *et al.* 2020), which is similar to our result in Table 3. The analyses by Rezende et al. (2020) used about half of J bulls used in the current study and about 1500 bulls from the USA.

To the best of our knowledge the accuracy of GP for young bulls for male fertility is not available in the literature. GP accuracy for young H bulls is lower than that the average from 5-fold cross validation (Table 3). This could be because the young bulls in H are less related to the RS set due to the fast turn-over of bulls in the post genomic era. Furthermore, the lower proportion of genetic variance explained by GRM and the higher genetic diversity of all H bulls relative to J bulls may have contributed to lower accuracy of prediction for the young bulls. Possibly also changes to the level of screening on semen parameters after the introduction of genomic selection may have contributed to low accuracy (Taylor *et al.* 2018). The use of joint H and J RS gave similar accuracy for young bulls (Table 3) suggesting a potential to have a single step genomic evaluation by including both genotyped and ungenotyped bulls of both breeds. This is appealing for the Australian dairy industry as the current evaluation for SFV uses data of all breeds.

**Table 2. Variance component estimates for semen fertility value and proportion of variance explained by the different random effects in Holstein and Jersey bulls**

| Random effects | Holstein bulls | | Jersey bulls | |
|---|---|---|---|---|
| | Variance | Proportion of total | Variance | Proportion of total |
| GRM | 13.60±0.68 | 0.006±0.000 | 27.74±2.24 | 0.012±0.001 |
| PE of cows | 70.08±1.49 | 0.031±0.001 | 73.05±4.10 | 0.033±0.002 |
| Residual | 2190.35±2.39 | 0.963±0.001 | 2139.93±6.37 | 0.955±0.002 |

**Table 3. Accuracy of genomic prediction for validation bulls for semen fertility value in Holstein and Jersey bulls with at least 100 inseminations**

| Breed | Five-fold cross validation | | Validation in young bulls | | |
|---|---|---|---|---|---|
| | No. | Accuracy | No. | Breed specific reference | Joint reference |
| Holstein | 717-898 | 0.197-0.252(0.220) | 482 | 0.128 | 0.123 |
| Jersey | 100-176 | 0.078-0.357(0.221) | 126 | 0.219 | 0.239 |

## CONCLUSIONS

The results of this study show that prediction of DGVs for H and J bulls using raw insemination data is feasible. At this stage the accuracies of GP particularly for young bulls are low. Nevertheless, there is a potential to use these results for monitoring and making early decisions to avoid using semen from extremely poor fertility bulls.

## ACKNOWLEDGEMENTS

We thank DataGene Ltd. for access to the data. Drs I. M. Macleod, Bolormaa S. and T. Nguyen performed the quality control and imputation of the genotype data. The research was funded by DairyBio (Dairy Australia, Agriculture Research Victoria and Gardiner Foundation) (Melbourne, Australia).

## REFERENCES

Abdollahi-Arpanahi R., Morota G. and Peñagaricano F. (2017) *J. Dairy Sci.* **100**: 9656.

Carrick M., Goddard M.E. and Bowman P.J. (2000) Pilot system for routine collation of non-return data for bulls. The University of Queensland, Australia pp. 9-24.

Haile-Mariam M. and Pryce J.E. (2021) *J. Dairy Sci.* In press.

Gilmour A.R., Gogel B.J., Cullis B.R. and Thompson R. (2015) ASReml User Guide Release 4.0. VSN International Ltd., Hemel Hempstead, UK.

Kuhn M.T. and J.L. Hutchison. (2008) *J. Dairy Sci.* **91**: 2481.

Nguyen T.V., Bolormaa S., Reich C. M., Chamberlain A. J., Medley A., Schrooten C., Daetwyler H. D. and MacLeod I. M. (2021) *Proc. Assoc. Advmt. Anim. Breed. Genet.* **24**:(in press).

Rezende F.M., Nani J. P. and Peñagaricano F. (2019) *J. Dairy Sci.* **102**: 3230.

Rezende F.M., Haile-Mariam M., Pryce J.E. and Peñagaricano F. (2020) *J. Dairy Sci.* **103**: 11618.

Sargolzaei M, Chesnais J.P. and Schenkel F.S. (2014) *BMC genomics* **15(1)**: 478.

Taylor J.F., Schnabel R.D. and Sutovsky P. (2018) *Animal* **12(S1)**: S172.

Yang J., Lee S.H., Goddard M.E. and Visscher P.M. (2011) *Am. J. Hum. Genet.* **88**: 76.

*Proc. Assoc. Advmt. Anim. Breed. Genet: 24: 151-154*

# QUANTIFYING GENETIC VARIATION IN URINATION TRAITS OF GRAZING DAIRY CATTLE

**R.C. Handcock[1], L.A. Box[2], C.M. Thomas[3] and D.J. Garrick[1]**

[1] A.L. Rae Centre for Genetics and Breeding, Massey University, Hamilton 3214, New Zealand
[2] AgResearch, Ruakura Research Centre, Hamilton 3240, New Zealand
[3] DairyNZ, 24 Millpond Lane, Lincoln 7608, New Zealand

## SUMMARY

This research quantified genetic variation in urination traits using sensors attached to grazing lactating dairy cattle that are designed to record timing, volume and nitrogen concentration of every urination event. The records from individual events were used to generate phenotypic traits that included daily urinary nitrogen (UN), urinary volume (Uvol), number of urination events (Unum) and average volume per urination event (VolEvent). Heritability estimates for these daily traits ranged from 0.20 to 0.37, showing these urination traits are heritable. Repeatability estimates ranged from 0.27 to 0.59, indicating there is considerable residual variation and sensor observations would need to be repeated over a number of days to get reliable phenotypic measures. Phenotypic and genetic correlations have been estimated, however due to the small number of animals in the current study, these preliminary estimates should only be viewed as indications. Overall, these results suggest there is potential for urination traits to be changed through selection however, these traits are difficult and expensive to measure and more cows need to be phenotyped in order to provide more reliable estimates of genetic parameters.

## INTRODUCTION

Pasture-based dairy cows in New Zealand predominately consume a sward containing perennial ryegrass (*Lolium perenne*) and white clover (*Trifolium repens*). These forages contain a high concentration of protein-nitrogen (N) relative to energy that is in excess of dairy cow requirements (Kolver and Muller 1998). Unless supplemented with a high-energy low-protein feed, these cows will excrete most of this surplus nitrogen in their urine (Selbie *et al.* 2015). Excreted N is concentrated in patches where it can be surplus to pasture requirements. Additionally the soil is unable to retain excess N which can lead to leaching through the soil and hence become a major source of N in waterways (Woods *et al.* 2016).

One method to reduce N losses at the farm level would be to reduce the total amount of urinary nitrogen (UN) excreted per cow per day. Another method would be to alter the urine patch dynamics, i.e. the UN load per urination event (Kennett *et al.* 2020). At a given daily UN load, a greater total urinary volume (Uvol) and smaller volume per urination event (VolEvent) would typically be favourable as it would lead to a more uniform spread of urine across the paddock and dilute the concentration of UN deposited in urine patches on pasture (Kennett *et al.* 2020).

The objective of this study was to quantify genetic variation in urination traits of grazing dairy cattle.

## MATERIALS AND METHODS

This experiment was conducted at Ashley Dene, situated near Burnham, Canterbury, New Zealand (43.6468° S 172.34679° E) between January and August 2020 with the approval of the Lincoln University Animal Ethics Committee (#2019-46). Six experimental runs were completed on a total of 180 Holstein-Friesian-Jersey crossbred cows milked twice-a-day. Each run was comprised of 30 cows split evenly into two grazing groups of 15 cows per group. Urine sensors (Mk II) developed by AgResearch (Betteridge *et al.* 2013; Shepherd *et al.* 2017) were attached to the cows between 8am and 9am on a Monday and removed around 6am on Friday. For data analysis, day was defined as the 24 hours from 9am to 8:59am the following day, run-group-day was defined as the contemporary group made up of experimental run (1-6), grazing group (1 or 2) and day of the experiment (1-4).

The urine sensors measure refractive index (RI), pressure, duration and time of each urination event. These data are used to estimate the urinary N concentration and volume of each urination event. The urinary nitrogen (UN) yield of each urination event (g N/event) was calculated as UN concentration (g/dL) x 10 x volume of the event (L).

Over the four days, there were periods from time-to-time where the urine sensor was not functioning for the measurement of N concentration or was not functioning at all. For each cow-day, the number of urination events (Unum), cumulative urinary volume (Uvol) and cumulative UN was known for the total elapsed time that the urine sensor was functioning. These cumulative measures were divided by their respective total elapsed time of valid observations to give a per minute value. The per minute value was multiplied by the number of minutes in the day to get the known cumulative measure representing daily Unum, Uvol and UN. For each cow-day the average volume per urination event (VolEvent) was calculated by dividing Uvol by Unum. Cow-days where the urine sensor was functioning for less than 50% of the day were not included in the final dataset (n=187 cow-days) for any of the urine traits. Similar edits to remove cow-days for UN were applied when the sensor recording N concentration was not functioning.

Two cows that were having extended lactations (>500 days) atypical of New Zealand pasture-based dairy cattle were removed from the dataset.

**Genotypes.** Cattle were genotyped by Weatherbys (www.WeatherbysScientific.com) on an Illumina 50,000 SNP bovine panel. The small proportion of SNPs that were missing on any animal were imputed using the methodology of FImpute (Sargolzaei *et al.* 2014). Only mapped SNPs assigned to chromosomes 1 through 29 were included in the analysis (n=42,231).

**Statistical analysis.** Bayesian univariate repeatability models for the urination traits were run using the Julia for Whole-genome Analyses Software (JWAS) package (Cheng *et al.* 2018) run in a Julia computing environment (julialang.org). Inference was based on MCMC chains of 90,000 samples, retaining every 10th sample, after a burn-in of 10,000 samples which had been discarded.

The repeatability model equation was:

$$y = \mathrm{RGD} + \mathrm{age} + \mathrm{DIM} + \mathrm{pJ} + \mathrm{het} + \mathrm{AnimPerm} + \mathrm{SNPs} + \boldsymbol{e}$$

where $y$ is the daily measurement on the trait of interest: UN (n=483 records on 164 cows), Uvol, Unum and VolEvent (n=517 records on 168 cows); RGD was the fixed class effect of run-group-day the animal was grazing; age was the fixed class effect of age of the cow in years from birth to most recent parturition; DIM was the fixed linear covariate of days in milk on day 1 of RGD; pJ was the linear covariate of Jersey breed proportion; het was the linear covariate of the specific heterosis coefficient between Holstein-Friesian and Jersey (Dickerson 1973); AnimPerm is the random permanent effect of animal assumed to be independently and identically normally distributed with variance $\sigma_c^2$; SNPs are additive covariates for all of the 42,231 autosomal loci with effects independently and identically normally distributed with variance $\sigma_a^2$; and $\boldsymbol{e}$ is the residual effects independently and identically normally distributed with variance $\sigma_e^2$.

Co(variance) components for UN, Uvol, Unum and VolEvent were estimated by fitting the model equation pairwise using six bivariate repeatability animal models.

The 95% credibility intervals were calculated by taking the 97.5th percentile of the MCMC samples as the upper bound and the 2.5th percentile as the lower bound.

## RESULTS AND DISCUSSION

Descriptive statistics for the final dataset are given in Table 1. Daily Uvol and UN (Table 1) were greater than that reported for Friesian-Jersey crossbred cows fitted with the same sensors as used in the current study (Bryant *et al.* 2018). Another study using similar urine sensors reported a daily Uvol of 42.2 L (Mangwe *et al.* 2019), comparable to the current study. The mean number of urinations per day was similar to that reported by Bryant *et al.* (2018) and Mangwe *et al.* (2019).

The estimates of heritability for the four urination traits were moderate (Table 1). Repeatability for VolEvent was greater than for UN (0.59 vs 0.27). Estimates of repeatability were similar to those observed for lactation test-day traits spread monthly or alternate monthly throughout a lactation. For example, estimates of repeatability were 0.52 for milk yield, 0.43 for fat yield and 0.44 for protein yield with approximately 80,000 multibreed cows and an average of two test-day records per cow (Lembeye *et al.* 2016).

Heritability and repeatability estimates of urination traits in cattle are scarce, although a Danish study reported a heritability of 0.12 for concentration of phosphorus in urine from random spot samples and a repeatability of 0.21 (Løvendahl and Sehested 2016). The same study reported a heritability of 0.05 and a repeatability of 0.38 for urinary creatinine, a nitrogen containing compound in urine.

**Table 1. Unadjusted phenotypic mean and standard deviation for daily urinary nitrogen (UN; g/d), urination volume (Uvol; L/d), urination number (Unum; count) and mean volume per urination event (VolEvent; L/event). Posterior means with lower and upper 95% credibility intervals (presented in brackets) of the genetic variance, heritability and repeatability**

| Trait | Mean | Standard deviation | Genetic Variance | Heritability | Repeatability |
|---|---|---|---|---|---|
| UN | 238 | 80 | 868 | 0.20 | 0.27 |
| | | | (441, 1,311) | (0.10, 0.30) | (0.18, 0.36) |
| Uvol | 36.8 | 12.4 | 50.2 | 0.36 | 0.50 |
| | | | (22.3, 78.9) | (0.17, 0.51) | (0.41, 0.58) |
| Unum | 13.0 | 4.4 | 3.9 | 0.24 | 0.46 |
| | | | (0.8, 7.5) | (0.05, 0.44) | (0.36, 0.55) |
| VolEvent | 2.9 | 0.7 | 0.16 | 0.37 | 0.59 |
| | | | (0.08, 0.26) | (0.18, 0.55) | (0.51, 0.66) |

Genetic and phenotypic correlations among the four urination traits are in Table 2. The phenotypic correlations among UN, Uvol and Unum were moderately high and positive, suggesting that cows that excreted a high volume of urine per day would be doing so with more urination events per day and at a greater daily UN load. The posterior means for the genetic correlations between Uvol and UN, Unum and VolEvent were moderate to moderately high and positive (Table 2). The genetic correlation between daily UN and VolEvent was near zero.

Due to the small numbers of animals in this study, the 95% credibility intervals around the posterior means of genetic parameters were wide, thus these preliminary estimates of genetic parameters should only be viewed as indications. Subsequent studies with larger cohorts of cattle are required to increase the reliability of the genetic parameters for urination traits. Nevertheless, based on this study, the genetic correlations of Uvol with Unum and VolEvent are likely to be positive.

Phenotyping cows for urination traits is expensive and logistically challenging, especially when cows are lactating and are outdoors grazing pasture. For this reason, there are few studies that have summarised whole day urination traits in grazing dairy cattle (Shepherd *et al.* 2017; Bryant *et al.* 2018; Mangwe *et al.* 2019), and none that have quantified genetic variation in the same traits. Comparing the square root of the estimated genetic variance to the raw mean shows that there is opportunity for urination traits to be included in the national breeding objective to ultimately reduce N losses to waterways, however, a cheaper and easier measurement to predict the urination traits would be advantageous to enable faster and more accurate selection over the national dairy herd.

**Table 2. Posterior means of the phenotypic (above the diagonal) and genetic (below the diagonal) correlations between daily urinary nitrogen (UN; g/d), urination volume (Uvol; L/d), urination number (Unum; count) and mean volume per urination event (VolEvent; L/event) with lower and upper 95% credibility intervals (presented in brackets)**

| Trait | UN | Uvol | Unum | VolEvent |
|---|---|---|---|---|
| UN | - | 0.68 | 0.60 | 0.09 |
| | - | (0.62, 0.73) | (0.53, 0.67) | (-0.01, 0.19) |
| Uvol | 0.59 | - | 0.75 | 0.31 |
| | (0.29, 0.78) | - | (0.70, 0.80) | (0.21, 0.41) |
| Unum | 0.58 | 0.75 | - | -0.29 |
| | (-0.03, 0.83) | (0.44, 0.91) | - | (-0.39, -0.18) |
| VolEvent | -0.09 | 0.47 | -0.17 | - |
| | (-0.49, 0.36) | (0.04, 0.76) | (-0.64, 0.41) | - |

## CONCLUSIONS

This study shows that there is genetic variation in the urination traits UN, Uvol, Unum and VolEvent. This suggests there is potential for urination traits to be changed through selection however, these traits are difficult and expensive to measure and more cows would need to be phenotyped in order to provide more reliable estimates of genetic parameters among the urination traits in addition to other important traits such as lactation and fertility.

## ACKNOWLEDGEMENTS

Low N Livestock is a DairyNZ-led collaborative research programme delivering science for better farming and environmental outcomes. The main funders are DairyNZ and the Ministry of Business, Innovation and Employment, with co-funding from CRV-Ambreed and Fonterra. Field work was conducted with the assistance of farm, technical and academic staff at Lincoln University.

## REFERENCES

Betteridge, K., Costall, D. A., Li, F. Y., Luo, D. and Ganesh, S. (2013) *Proc. New Zeal. Grassl. Assoc.* **75**: 33.

Bryant, R. H., Welten, B. G., Costall, D., Shorten, P. R. and Edwards, G. R. (2018) *Livest. Sci.* **209**: 46.

Cheng, H., Fernando, R. L. and Garrick, D. J. (2018) *Proc. World Congr. Genet. Appl. to Livest. Prod.* **11**: 859.

Dickerson G.E. (1973) *J. Anim Sci.* **(Suppl.)**: 54.

Kennett, J., Romera, A., Beukes, P. and Amer, P. (2020) *New Zeal. J. Anim. Sci. Prod.* **80**: 60.

Kolver, E. S. and Muller, L. D. (1998) *J. Dairy Sci.* **81**: 1403.

Lembeye, F., Lopez-Villalobos, N., Burke, J. L. and Davis, S. R. (2016) *Livest. Sci.* **185**: 142.

Løvendahl, P. and Sehested, J. (2016) *J. Dairy Sci.* **99**: 4580.

Mangwe, M. C., Bryant, R. H., Beck, M. R., Beale, N., Bunt, C. and Gregorini, P. (2019) *Anim. Feed Sci. Technol.* **252**: 11.

Sargolzaei, M., Chesnais, J. P. and Schenkel, F. S. (2014) *BMC Genomics* **15**: 478.

Selbie, D. R., Buckthought, L. E. and Shepherd, M. A. (2015) *Adv. Agron.* Elsevier Ltd.

Shepherd, M., Shorten, P., Costall, D. and Macdonald, K. A. (2017) *Agric. Ecosyst. Environ.* **236**: 285.

Woods, R. R., Cameron, K. C., Edwards, G. R., Di, H. J. and Clough, T. J. (2016) *Soil Use Manag.* **32**: 565.

*Proc. Assoc. Advmt. Anim. Breed. Genet: 24: 155-158*

# GROUP RECORDS WITH GENOMIC PREDICTON CONVERT ACCURACY INTO GENETIC GAIN MORE EFFICIENTLY THAN PEDIGREE PREDICTION

**M. Henryon[1,2], G. Su[3], A.C. Sørensen[1] and T. Ostersen[1]**

[1] Danish Pig Research Centre, SEGES, Denmark
[2] School of Agriculture and Environment, University of Western Australia, Australia
[3] Institute for Molecular Biology and Genetics, Aarhus University, Denmark

## SUMMARY

We tested the premise that genomic prediction (GBLUP) converts accuracy into genetic gain (ΔG) more efficiently than pedigree prediction (PBLUP) using group records at the same rate of true inbreeding (ΔF). We tested this premise by stochastic simulation. We estimated conversion efficiency (CE) of optimum-contribution selection (OCS) using individual and group records with PBLUP and GBLUP at 0.01 ΔF. We did this by allocating selection candidates to groups of 12 individuals. Animals in each group were measured as either individual or group records. Selection was for a single trait with heritability 0.2. The trait was controlled by 7702 biallelic quantitative-trait loci. We found that the CE of group records increased from 94 to 102% when we changed prediction from PBLUP to GBLUP. Group records generated EBV that were about 0.76 times as accurate as individual records with both PBLUP and GBLUP. However, group records realised only 0.70 times as much ΔG as individual records with PBLUP; they realised 0.79 times as much ΔG with GBLUP. Clearly, group records converted accuracy into ΔG more efficiently with GBLUP than they did with PBLUP. This makes group records a more attractive measure of phenotypic performance with GBLUP.

## INTRODUCTION

Group records measure the sum of phenotypic performances of animals reared in groups (e.g., feed intake of pigs in a pen). They can be particularly useful for traits that are difficult or expensive to measure as individual records (i.e., phenotypic performance of individual animals). Not only are group records often easier and cheaper to measure than individual records, estimated breeding values (EBV) predicted using group records are typically 50-90% as accurate as EBV using individual records (Olson *et al.* 2006, Su *et al.* 2018, Ma *et al.* 2020). This prompted a widely-held view that selection based on group records could realise most of the genetic gain (ΔG) realised by individual records at a fraction of the cost. However, Henryon *et al.* (*in prep.*) found that group records were only 82-90% as efficient in converting accuracy into ΔG as individual records – a parameter they referred to as conversion efficiency (CE). In their study, selection candidates were grouped and phenotyped, breeding values (BV) were predicted as BLUP of breeding values based on pedigree information (PBLUP), and selection was carried out by optimum-contribution selection (OCS) with rate of pedigree inbreeding constrained to 0.01. They found that group records had lower CE than individual records because OCS using group records reduced selection intensities. Selection intensities were reduced because EBV with group records expressed less within-family variation and candidates that ranked highest for EBV were more related. To realise the constrained rate of pedigree inbreeding, OCS using group records needed to select more candidates than OCS using individual records. This implies that if group records are to generate higher CE, we need EBV with more within-family variation. One way to do this is to replace PBLUP with genomic prediction of BV (GBLUP). With GBLUP, group records should generate higher selection intensities by enabling OCS to differentiate between candidates within full-sib families. Fewer candidates would need to be selected to realise the same rate of inbreeding as OCS using group records with PBLUP. This reasoning led us to believe that GBLUP results in higher CE than PBLUP when using group records at the same rate of inbreeding. We tested this premise by stochastic simulation.

## MATERIALS AND METHODS

**Procedure.** We used stochastic simulation of animal-breeding schemes to estimate CE generated by OCS using individual and group records with PBLUP and GBLUP at 0.01 rate of true inbreeding (ΔF), where the true inbreeding coefficient of an individual was defined as the observed proportion of loci in its genome with alleles that are identical-by-descent (IBD). We allocated selection candidates to groups of 12 individuals. Animals in each group were measured as either individual or group records. We also sampled relatives of the selection candidates. These animals were measured as individual records. They were included in the prediction models, but were not candidates for selection. Selection was for a single trait with heritability 0.2 (additive-genetic variance 1.0). The trait was controlled by 7702 biallelic quantitative-trait loci (QTL). It was also influenced by litter and group effects (litter and group variances 0.25). All animals were genotyped and phenotyped before selection in each generation. Breeding schemes were run for eight discrete generations ($t = 1 \ldots 8$) and replicated 120 times. Each replicate was initiated by sampling a unique base population from a founder population. Animals in the base populations were randomly selected in generation $t = 1$. In generations $t = 2 \ldots 8$, selection candidates were allocated matings by OCS.

**Breeding scheme.** A total of 600 matings were allocated to 3600 selection candidates by OCS in generations $t = 2 \ldots 10$. The number of matings that were allocated to each male could vary from 0, 1, 2 ... to 50 matings. Six-hundred females were allocated a single mating. The matings allocated to the sires and dams were paired randomly. Each dam produced seven offspring – four males and three females – resulting in 600 full-sib families and 4200 offspring (2400 males and 1800 females). Three males and three females from each full-sib family were randomly pre-selected as candidates for selection. These 3600 animals were allocated to groups of 12 and measured as individual or group records. The remaining male in each full-sib family was measured as an individual record but was not a candidate for selection. The BV of the selection candidates were predicted using their own phenotypes and their genetic relationships to the male in each full-sib family that was measured as an individual record.

**Grouping criterion.** Groups of 12 animals were established by dividing each full-sib family into two sub-families of three full-sibs. Four sub-families from four different full-sib families were randomly allocated to each group. Each full-sib family was represented in two groups. Selection candidates were allocated to a total of 300 groups in each generation.

**Genetic model.** The founder population was established using a Fisher-Wright inheritance model to generate linkage disequilibrium between QTL and markers. The genome was 30 M and consisted of 18 pairs of autosomal chromosomes. Each chromosome was 167 cM long. The genome contained 7702 QTL and 54218 biallelic markers. These markers were randomly distributed across the genome and in linkage disequilibrium with the QTL. They were used in GBLUP. An additional 6012 IBD loci were placed evenly across the genomes of animals in base populations. Unique alleles at these loci were used to calculate ΔF.

**Optimum-contribution selection.** OCS was carried out by maximising $\mathbf{U}_t(\mathbf{c}) = \mathbf{c}'\hat{\mathbf{a}} - \omega\mathbf{c}'\mathbf{A}\mathbf{c}$, where $\mathbf{c}$ is a vector of genetic contributions to the next generation, $\hat{\mathbf{a}}$ is a vector of BV predicted with PBLUP or GBLUP, $\omega$ is a penalty applied to the average-estimated relationship of the next generation, and $\mathbf{A}$ is a pedigree-relationship matrix (after Henryon *et al.* 2019). The penalty, $\omega$, was constant across generations. It was calibrated to realise 0.01 ΔF. The realisedΔF deviated from 0.01 by less than 0.0001.

**Statistical analyses.** We present CE, accuracy, ΔG, selection intensity, and additive-genetic standard deviation realised by OCS using individual and group records with PBLUP and GBLUP. CE measured the efficiency by which accuracy of EBV from group records was converted to ΔG relative to individual records: $CE = \frac{\Delta G_j/\Delta G_{ind}}{r_j/r_{ind}} \cdot 100$, where $\Delta G_j$ and $r_j$ are mean ΔG and accuracy of individual or group records (*j* = *ind*, *grp*). ΔG, accuracy, selection intensity, and additive-genetic standard deviation are presented as means (± sd) of the 120 replicates. ΔG in each replicate was calculated as the linear regression of $G_t$ on *t*, where $G_t$ is the average true breeding value of animals born at times *t* = 4 … 8. Accuracy, selection intensity, and additive-genetic standard deviation in each replicate were averaged over generations *t* = 4 … 8. Accuracy was calculated as the correlation between true breeding values and EBV of animals within generation. Selection intensity was calculated as the difference in average EBV of selected animals weighted by their contribution to the next generation and average EBV of selection candidates within generations divided by the standard deviation of the EBV. Additive-genetic standard deviation was calculated as the standard deviation of true breeding values of animals within generations. We present absolute and scaled ΔG, accuracy, selection intensity, and additive-genetic standard deviation. Scaling was carried out by setting values realised by individual records with PBLUP and GBLUP to 100. ΔF in each replicate was calculated as 1-exp(β), where β is the linear-regression coefficient of $\ln(1-F_t)$ on *t*, and $F_t$ is the average coefficient of true inbreeding for animals born at times *t* = 4 … 8 (after Sonesson *et al.* 2004).

**RESULTS AND DISCUSSION**

Our findings supported our premise that GBLUP results in higher CE than PBLUP when using group records at the same rate of inbreeding. We found that the CE of group records increased by eight percentage units – from 94 to 102% – when we changed prediction from PBLUP to GBLUP at 0.01 ΔF (Table 1). When prediction was changed from PBLUP to GBLUP, the accuracy of both individual and group records increased by about 1.4 times. That is, the relative difference in accuracy between individual and group records remained the same: group records generated EBV that were about 0.76 times as accurate as individual records with both PBLUP and GBLUP. However, group records realised only 0.70 times as much ΔG as individual records with PBLUP. They realised 0.79 times as much ΔG with GBLUP. Clearly, group records converted accuracy into ΔG more efficiently with GBLUP than they did with PBLUP. It suggests that the widely-held view that selection based on group records could realise most of the ΔG realised by individual records at a fraction of the cost is more applicable to GBLUP than it is to PBLUP. Of course, the ultimate decision of whether to invest in individual or groups records to measure difficult and expensive traits will be specific for each breeding scheme. It will depend on the relative cost and difficulty of gathering individual and group records and how managers of breeding schemes evaluate returns of investment. So, groups records are a more attractive measure of phenotypic performance with GBLUP than with PBLUP because they convert accuracy into ΔG more efficiently.

As we contented, OCS using group records generated higher CE with GBLUP than they did with PBLUP because selection intensity of OCS using group records relative to individual records was higher with GBLUP. We found that selection intensity using group records was only 0.89 times as high as individual records with PBLUP (Table 1). It was 0.95 times as high with GBLUP. The selection intensity of OCS using group records was higher with GBLUP presumably because genomic relationships generated more within-family variation for EBV. OCS using group records with GBLUP was able to differentiate between candidates within full-sib families. It could select fewer candidates to realise 0.01 ΔF than group records with PBLUP. Therefore, group records generate higher CE with GBLUP than PBLUP because they increase selection intensities by generating more within-family variation for EBV.

**Table 1. Conversion efficiency, accuracy, rate of genetic gain, selection intensity, and additive-genetic standard deviation realised by individual and group records at 0.01 ΔF with two predictions methods (PBLUP and GBLUP)**

| Prediction | Record | CE | r | ΔG | $i$ | $\sigma_a$ | r* | ΔG* | $i^*$ | $\sigma_a^*$ |
|---|---|---|---|---|---|---|---|---|---|---|
| PBLUP | Individual | 100 | 0.54 | 0.73 | 1.70 | 0.83 | 100.0 | 100.0 | 100.0 | 100.0 |
| | Group | 94 | 0.40 | 0.51 | 1.50 | 0.88 | 74.4 | 70.3 | 88.7 | 107.0 |
| GBLUP | Individual | 100 | 0.74 | 1.01 | 1.83 | 0.75 | 100.0 | 100.0 | 100.0 | 100.0 |
| | Group | 102 | 0.57 | 0.80 | 1.74 | 0.82 | 77.3 | 79.1 | 95.3 | 109.5 |

Absolute and scaled accuracies ($r$ and r*), rates of genetic gain (ΔG and ΔG*), selection intensities ($i$ and $i^*$), and additive-genetic standard deviation ($\sigma_a$ and $\sigma_a^*$) are means of 120 simulation replicates. r*, ΔG*, $i^*$, and $\sigma_a^*$ were calculated by setting $r$, ΔG, $i$, and $\sigma_a$ realised by individual records to 100 with PBLUP and GBLUP. SD between replicates ranged from 0.012 to 0.035 ($r$), 0.040 to 0.057 (ΔG), 0.030 to 0.091 ($i$), 0.144 to 0.180 ($\sigma_a$), 1.58 to 6.49 (r*), 3.99 to 7.81 (ΔG*), 1.66 to 4.96 ($i^*$), and 19.21 to 24.00 ($\sigma_a^*$).

We were surprised to find that CE was greater than 100 for group records with GBLUP. It was greater than 100 because there was more additive-genetic variation available for OCS using group records to convert accuracy into ΔG than OCS using individual records. Unlike selection intensity, the relative difference in additive-genetic variation between individual and group records remained the same with PBLUP and GBLUP: the additive-genetic standard deviation of OCS using group records was about 1.08 times higher than OCS using individual records (Table 1). More additive-genetic variation was available for OCS using groups records because selection was not as effective as individual records. It realised less ΔG, leading to less Bulmer effect and smaller changes in allele frequencies. So, CE of group records using GBLUP can be higher than 100 because OCS using group records results in more additive-genetic variation available to be converted into ΔG.

**ACKNOWLEDGEMENTS**

This study was financed by the Green Development and Demonstration Programme (GUDP Grant No. 34009-14-0849) under the Danish Ministry of Food, Agriculture and Fisheries; SEGES, Pig Research Centre; and Aarhus University.

**REFERENCES**

Henryon M., Liu H., Berg P., Su G., Nielsen H.M., Gebregiwergis, G.T. and Sørensen A.C. (2019) *Genet. Sel. Evol.* **51**: 39.

Ma X., Christensen O.F., Gao H., Huang R., Nielsen B., Madsen P., Jensen J., Ostersen T., Li P., Shirali M. and Su G. (2021) Heredity **126**: 206.

Olson K.M., Garrick D.J., Enns R.M. (2006) *J. Anim. Sci.* **84**: 88.

Sonesson A.K., Woolliams J. and Meuwissen T.H.E. (2004) In 'Selection and breeding programs in aquaculture', pp. 73-87, editors T. Gjedrem and K. Doordrecht, Springer, Dordrecht.

Su G., Madsen P., Nielsen B., Ostersen T., Shirali M., Jensen J. and Christensen O.F. (2018) *Genet. Sel. Evol.* **50**: 42.

*Proc. Assoc. Advmt. Anim. Breed. Genet: 24: 159-162*

# ECONOMIC VALUES FOR FARROWING RATE TO IMPROVE SEASONAL FERTILITY

**S. Hermesch**

Animal Genetics Breeding Unit*, University of New England, Armidale, NSW, 2350 Australia

## SUMMARY

Seasonal fertility is the variation in reproductive performance of sows across different seasons. A consistent fertility of sows across seasons is desirable. Seasonal fertility is reflected in farrowing rate because a reduction in farrowing rate is often observed during the summer-autumn period. An independent economic model was developed to derive economic values for farrowing rate. Economic values varied from $2.19 to $1.95 per 1% change in farrowing rate for mean farrowing rates of 72 to 85%. The economic value for farrowing rate predominately accounted for the costs of non-productive days of non-pregnant sows. The model and economic values presented in this study for farrowing rate can be used to extend existing maternal breeding objectives in pigs. Further, the variation in economic values for farrowing rates can be used to consider genotype by season interactions for farrowing rate in pig breeding programs.

## INTRODUCTION

Seasonal fertility is the variation in a fertility measure such as farrowing rate or litter size across different seasons and low seasonal fertility is desirable. Historically, research has focused on seasonal infertility which is characterised by poorer reproductive performance of sows during the summer and autumn period (e.g. Love *et al.* 1993; Auvigne *et al.* 2010). In contrast, a focus on seasonal fertility extends seasonal infertility because seasonal fertility aims to improve the consistency of high reproductive performance of sows across all seasons rather than focusing on reduced performance of sows in one season only.

Selection of sows across seasons for reproductive traits is expected to improve seasonal fertility somewhat. For example, a maternal line selected in hot and tropical environments across countries was better adapted to high temperatures than a line selected in one temperate environment only (Bloemhof *et al.* 2008). However, incorporating genetic variation in the response of sows to changes in seasonal conditions in breeding objectives enables more targeted selection for seasonal fertility.

A key trait to quantify seasonal fertility is farrowing rate which represents the proportion of sows served that farrow. Genetic variation in the response of sows to changes in photoperiod and ambient temperature has been found for farrowing rate (Sevillano *et al.* 2016). Further, farrowing rate was genetically a different trait in different temperature groupings in the Australian study by Bunz *et al.* (2019). These results support the inclusion of genotype by season interactions for farrowing rate in order to enhance genetic gain in seasonal fertility of sows. It was the aim of this study to derive economic values for farrowing rate taking into account differences in the level of performance for farrowing rate as they may be observed across seasons.

## MATERIALS AND METHODS

Farrowing rate is a binary trait and variance components may be based on the original scale (0 versus 1) or may be expressed as a percentage (0 versus 100). Production systems usually refer to changes in farrowing rate in 1% increments which was the basis of the model that was developed to

* A joint venture of NSW Department of Primary Industries and the University of New England

derive the economic value of farrowing rate. A higher farrowing rate improves profit by reducing costs of non-productive days of sows in each parity and reducing costs associated with each mating. Non-productive days arise for sows that fail to farrow by returning from a mating and for non-pregnant sows in general until they are removed from the herd.

The economic value for a 1% increase in farrowing rate (EV_FR) was calculated as: EV_FR = (ScreturnFR + ScmatingFR) – (ScreturnFR+1% + ScmatingFR+1%), where ScreturnFR are the costs to keep a returned sow (e.g. non-pregnant sow) in the herd until the sow is either mated again or removed from the herd assuming a base farrowing rate and ScmatingFR are the additional mating costs of returned sows ($ 30/mating) for the same assumed base farrowing rate. Both ScretrurnFR+1% and ScmatingFR+1% are the corresponding costs associated with a farrowing rate that is 1% higher than the assumed base farrowing rate.

The costs of keeping returned sows in the herd for the base farrowing rate depend on feed, housing and labour costs as well as the average number of days until a returned sow is either successfully remated or removed from the herd. Key assumptions made in the calculation of the economic value for farrowing rate were based on typical production parameters for Australia (Australian Pork Limited 2012a). These include production levels of sows, price assumptions for feed as well as those that relate to other aspects of the operation, including capital value of the buildings and facilities as assumed by Amer *et al.* (2014).

The daily costs per sow (dSc) were the sum of daily feed costs (dFc), daily housing costs (dHc) and daily labour costs (dLc) which were derived as: dFc = feed per day (kg) * costs of feed ($ / kg); dHc = (costs of sow place ($ / place) * annual interest rate (%) + costs of sow place ($ / place) * annual depreciation rate (%)) / 365; dLc = (labour costs per staff ($ / annum)/ number of sows per staff (n sows)) / 365.

The average number of days until a returned sow is either successfully remated or removed from the herd for the base farrowing rate depends on the proportion of sows that a) farrowed from each mating (n = 1 to 4) and b) were not pregnant or not-in-pig (NIPs) and subsequently removed from the herd. The NIPs were calculated as: NIPs = (1 – FR) * NIPs%, where FR is farrowing rate and NIPs% is the percentage of NIPs (12%) of returned sows from each mating (Australian Pork Limited 2012b).

The proportion of sows that farrowed from each mating (psow) was calculated as: $psow_n = (psow_{n-1} - NIPs)*FR$.

The costs of keeping returned sows in the herd were: ScreturnFR = NIPs * NIPsdays * dSc + $\sum_{n=2}^{4}(21 * psow_n * dSc + cmate)$, where cmate were costs of mating including semen costs and labour ($ 30 / mating). A mating interval of 21 days and removal of NIPs at 80 days after mating (NIPsdays) were assumed.

## RESULTS AND DISCUSSION

**Costs.** The daily costs to keep a sow in the herd were $4.73 $ per day (Table 1). Housing costs accounted for the largest cost component with $2.63 per day, reflecting the high capital costs of buildings in Australia. Comparison with costs structures outlined for other countries overseas (e.g. Krupa *et al.* 2017) are not possible because housing costs were not reported specifically and were part of other non-feed costs which were outlined for groups of animals of a full farrow-to-finish unit and not specifically defined for sows.

The proportion of sows that farrowed from the $2^{nd}$ to the $4^{th}$ mating for different farrowing rates are shown in Table 2. These percentages of sows farrowing from different matings and the corresponding NIPs corresponded to industry values (Australian Pork Limited 2012b).

**Economic values**. Economic values for farrowing rate were derived for different levels of farrowing rate using the base assumptions outlined above. The economic value for farrowing rate varied from $ 2.19 per 1% improvement for a low farrowing rate of 72% to $ 1.95 per 1%

improvement for a high farrowing rate of 85% (Table 3). The intermediate value of $ 2.06 may be appropriate for most farms as an overall average across the year, while the higher economic value may be more applicable for the summer-autumn period when farrowing rates are usually lower.

**Table 1. Daily costs per sow ($/day) due to feed, housing and labour**

| Cost component | Item | Input value | Costs per sow |
|---|---|---|---|
| | Daily feed per sow (kg) | 2.5 | |
| | Costs of feed ($/kg) | 0.4 | |
| Feed costs | | | 1.00 |
| | Costs of sow place ($) | 8,000 | |
| | Interest rate (%) | 7 | |
| | Depreciation rate (%) | 5 | |
| Housing costs | | | 2.63 |
| | Annual costs per staff ($) | 60,000 | |
| | Sows per staff | 150 | |
| Labour costs | | | 1.10 |
| **Total daily costs per sow** | | | **4.73** |

**Table 2. Percentage of sows farrowing from the second to fourth mating and percentage of non-pregnant sows (not-in-pig sows, NIPs) for different farrowing rates**

| Percentage of sows that farrow after | Farrowing rate (%) | | | | | |
|---|---|---|---|---|---|---|
| | 72.0 | 75.0 | 77.0 | 80.0 | 82.0 | 85.0 |
| 2$^{nd}$ mating (%) | 17.7 | 16.5 | 15.6 | 14.1 | 13.0 | 11.2 |
| 3$^{rd}$ mating (%) | 5.0 | 4.1 | 3.6 | 2.8 | 2.3 | 1.7 |
| 4$^{th}$ mating (%) | 1.4 | 1.0 | 0.8 | 0.6 | 0.4 | 0.3 |
| Percentage of NIPs (%) | 3.4 | 3.0 | 2.8 | 2.4 | 2.2 | 1.8 |

Sensitivity analyses showed the effect of modifying assumptions in housing costs and number of sows per person on economic values for farrowing rate (Table 3). Capital costs due to housing were the biggest cost component and changes in these costs affected economic values for farrowing rate most. The range of these economic values may be used to define the economic value for farrowing rate that is most appropriate for specific conditions observed on commercial farms.

This study extends the number of traits included in maternal breeding objectives for pigs outlined by Amer *et al.* (2014). The approach of using independent models for each trait improves the feasibility of extending breeding objectives. The economic value for farrowing rate mainly reflects costs of non-productive traits in sows complementing economic values for age at first oestrus and weaning to conception interval which also describes variation in non-productive days of gilts and sows as outlined by Amer *et al.* (2014). A longer farrowing interval, however, is also associated with higher culling rates of sows that ultimately result in poorer sow longevity. The economic value for sow longevity outlined by Amer *et al.* (2014) was derived from net returns and replacement costs of sows resulting from a 1% increase in survival of sows in each parity which was independent from the costs of non-productive days due to changes in farrowing rate.

Economic values for farrowing rate were not found in the literature. The model presented by de Vries (1989) has been widely used in pig breeding programs. The model considered culling rate of sows as breeding objective traits. Culling rates were defined for different stages of the reproductive cycle of sows including the stage from mating to farrowing. The number of non-productive days was constant in each stage, and culling rates effectively described sow longevity as illustrated by the author, who derived an economic value for sow longevity based on the association between culling rates and the number of farrowings per replacement gilt used in their model.

**Table 3. Economic values for farrowing rate ($ / 1%) assuming different levels of farrowing rate and alternative input values for housing and labour costs that vary from the base value by plus or minus 25%**

| | Farrowing rate (%) | | | | | |
|---|---|---|---|---|---|---|
| | 72.0 | 75.0 | 77.0 | 80.0 | 82.0 | 85.0 |
| Base assumptions | 2.19 | 2.15 | 2.12 | 2.06 | 2.02 | 1.95 |
| Base and $10,000 per sow place | 2.44 | 2.40 | 2.36 | 2.30 | 2.25 | 2.17 |
| Base and $6,000 per sow place | 1.94 | 1.91 | 1.88 | 1.83 | 1.79 | 1.73 |
| Base and 112.5 sows per person | 2.33 | 2.29 | 2.25 | 2.19 | 2.15 | 2.07 |
| Base and 187.5 sows per person | 2.11 | 2.07 | 2.04 | 1.98 | 1.94 | 1.88 |

**Breeding objective.** A breeding objective may consider farrowing rate as one trait, assuming that it is the same trait throughout the year. However, farrowing rate should be considered as a different trait in the hot summer-autumn period versus other seasons, given the genetic parameters estimated by Bunz *et al.* (2019). This can be accommodated in breeding objectives by defining farrowing rate as a separate trait for two separate seasons (hot summer-autumn versus other seasons) given the result by Bunz *et al.* (2019). Defining farrowing rate as a different trait for two seasons requires using appropriate economic values for each season taking differences in farrowing rate across seasons into account. The economic value for farrowing rate applicable to each season should then be weighted by the proportion of sows represented in each season. In the study by Bunz *et al.* (2019) about 24% of sows were mated in the hot summer-autumn period leaving 76% of sows for the other seasons.

## CONCLUSIONS

An independent economic model was developed and used to derive economic values for farrowing rate enabling the extension of maternal breeding objectives in pigs. Economic values for farrowing rate were higher for lower farrowing rates, which may be observed in the summer-autumn season. These higher economic values for lower farrowing rates may be used to consider genotype by season interactions for farrowing rate in pig breeding programs.

## ACKNOWLEDGEMENTS

This study was supported by Australian Pork Limited under project APL 2017/2208.

## REFERENCES

Amer P.R., Ludemann C.I. and Hermesch S. (2014) *J. Anim. Sci.* **92:** 5345.

Australian Pork Limited (2012a) 'Australian pig annual 2011 – 2012' Austr. Pork Ltd., Barton.

Australian Pork Limited (2013b) 'Fact Sheet Reproductive Health' Austr. Pork Ltd., Barton.

Auvigne V., Leneveu P., Jehannin C., Peltoniemi O. and Salle E. (2010) *Theriogenology* **74:** 60.

de Vries A.G. (1989) *Livest. Prod. Sci.* **21:** 49.

Bloemhof S., van der Waaij E.H., Merks J.W.M. and Knol E.F. (2008) *J. Anim. Sci.* **86:** 3330.

Bunz A.M.G., Bunter K.L., Morrison R., Luxford R.G. and Hermesch S. (2019) *Proc. Assoc. Advmt. Anim. Breed. Genet.* **23:** 155.

Krupa E., Krupova, Z. Wolfova, M. and Zakova E. (2017) *Livest. Sci.* **205:** 70.

Love R.J., Evans G. and Klupiec C. (1993) *J. Reprod. Fertil. Suppl.* **48:** 191.

Sevillano C.A., Mulder H.A., Rashidi H., Mathur P.K. and Knol E.F. (2016) *J. Anim. Sci.* **94:** 3185.

*Proc. Assoc. Advmt. Anim. Breed. Genet: 24: 163-166*

# ACROSS-COUNTRY PREDICTION OF METHANE EMISSIONS USING RUMEN MICROBIAL PROFILES

**M.K. Hess[1], A. Donaldson[2], H.M. Henry[1], D.L. Robinson[2], A.S. Hess[1], J.C. McEwan[1], V.H. Oddy[2] and S.J. Rowe[1]**

[1] AgResearch Ltd, Invermay, Mosgiel, New Zealand
[2] NSW Department of Primary Industries, University of New England, Armidale, NSW 2351

## SUMMARY

Rumen microbial profiles have been shown to be accurate predictors of methane emissions in a variety of species, however, it can be very costly and slow to generate a dataset with a sufficient number of individuals measured for methane who also have had rumen samples collected and processed into rumen microbial profiles for these benefits to be applied in industry. We evaluated the potential of combining datasets from New Zealand and Australian sheep to improve our ability to accurately predict methane emissions in Australian sheep. Prediction of Australian sheep methane emissions using rumen microbial profiles and phenotypes from New Zealand was possible, however, it was important to closely match the diets the sheep were fed to have confidence in the predictions. Prediction accuracies of Australian sheep methane emissions were higher when training on data collected on Australian sheep than training on New Zealand sheep; however augmentation of New Zealand data collected on a similar diet enabled more complex models to be run and an improvement in prediction accuracy.

## INTRODUCTION

The rumen microbiome has been shown to play an important role methane production and feed efficiency and improve prediction accuracy in these traits (Hess *et al.* Submitted-b). However, large sample numbers are typically required for accurate trait prediction. Over 3,000 New Zealand sheep rumen microbial profiles have been generated with associated methane emission phenotypes, representing a variety of breed compositions, ages and diets (Hess *et al.* Submitted-a). Robinson *et al.* (2020) describe a study in over 500 Australian merino sheep that have been measured for methane emissions with rumen samples collected during the study. This study predicted methane emissions in Australian merino sheep under two scenarios: 1) when Australian sheep had no methane data collected and models were trained using the New Zealand dataset, and 2) when some Australian sheep had methane data collected and added to the New Zealand training dataset. The models used in our study utilized genomic information, rumen microbial profiles or both.

## MATERIALS AND METHODS

**Australian Microbiomes.** Rumen samples were collected from 502 Information Nucleus Flock follower ewes on a chaffed lucerne and cereal hay diet at 1.5-1.6 times maintenance (Robinson *et al.* 2020). Restriction Enzyme-Reduced Representation Sequencing (Hess *et al.* 2020) was used to generate Reference Free Rumen Microbial Profiles, as described in Hess *et al.* (Submitted-a).

**New Zealand Microbiomes.** Reference Free Rumen Microbial Profiles were generated on 3,019 rumen samples from 1,200 dual purpose composite ewes (Hess *et al.* Submitted-a; Hess *et al.* Submitted-b). Rumen Microbial Profiles were separated into 3 groups based on diet (all fed ad lib) and age: lamb on ryegrass-based pasture/grass (GL, n = 1051), adult on ryegrass-based pasture/grass (GA, n = 1010), and lambs on a lucerne pellet diet (LL, n = 958).

**Methane Phenotypes.** Australian sheep had methane phenotypes collected in Respiration and Portable Accumulation Chambers (Robinson *et al.* 2020) during the same experiment in which rumen samples were collected. Methane emission phenotypes for the Australian sheep used in this study were the genetic plus permanent environmental effects for respiration chamber measurements based on the model without covariates for liveweight and feed intake of Robinson *et al.* (2020)

New Zealand methane phenotypes were the methane emission phenotypes from Portable Accumulation Chambers, adjusted for the fixed effects of birth rear rank, age of dam and birth date deviation (Hess *et al.* Submitted-b). Adjusted methane phenotypes were normalized within group, such that each group had a mean of zero and standard deviation of one to account for differences in measurement type (respiration chamber vs portable accumulation chamber), differences in methane emissions due to effects such as diet and age, and differences in the methane yield models.

**Genotypes.** High density genotypes were available on all New Zealand sheep and 322 of the Australian sheep. Sheep were genotyped on a variety of SNP chips, then imputed to a high density set of SNPs separately within each country. After imputation, the two datasets were combined and SNPs that were segregating in both populations (471,596 SNPs) were used to generate a genomic relationship matrix (GRM) using the first method of Van Raden (2008).

**Models.** Three models were run in ASReml v 4.1 (Gilmour *et al.* 2015), which explained variation in methane phenotype using genotypes, Microbial Profiles or both:

$$y = \mu + \mathrm{G} + e;$$
$$y = \mu + \mathrm{M} + e;$$
$$y = \mu + \mathrm{G} + \mathrm{M} + e$$

where $y$ is the adjusted methane phenotype; $\mu$ is the mean; G is the random animal genetic effect with relationships between animals represented by the GRM described above; M is the random microbial effect with relationships between samples represented by the cohort-adjusted microbial relationship matrix, calculated as described in Hess *et al.* (2020); and $e$ is the residual.

The above models were trained using GL, GA, LL or all NZ samples, and used to predict breeding values (BV) and microbial values (MV) in the Australian dataset. For models including both G and M, the BV and MV were summed to get the combined value (GMV). Accuracies were estimated as the correlation between the phenotype and the BV, MV or GMV. The accuracy of the microbial values were calculated using all Australian samples or just the samples associated with genotyped animals, and models containing G were only run for animals with genotype information available. Accuracies were estimated for each cohort separately and the standard errors of the accuracies estimated as the standard deviation across all cohorts. There were 10 validation cohorts with 50 ± 26 Australian sheep in the full dataset and 5 of these cohorts had 64 ± 29 genotyped sheep.

The three models above were also trained using Australian samples excluding the cohort that was being predicted, as well as these samples augmented with the LL or all NZ samples. Microbial relationship matrices used for each model were generated using tags that were present in all groups found in either the training or prediction set for that model. There were 79,328 tags present in both GA and AUS groups, 69,120 tags present in both GL and AUS groups, 39,502 tags present in both LL and AUS groups, 29,456 tags present in all groups (GA, GL, LL and AUS), and 150,687 tags present in the AUS group.

## RESULTS AND DISCUSSION

**Across-country prediction.** Our first analysis aimed to use various rumen microbial profiles from New Zealand sheep to predict methane emissions in Australian sheep. Microbial value estimates for either all Australian samples or samples associated with a genotyped animal were poor and tended to be negative when New Zealand samples were used as the training set, with the exception of the samples from lambs fed lucerne pellets (Table 1). The highest accuracy (0.23) was from BV estimated using the full NZ dataset and a model fitting both genomic and microbial effects. This model contains the most information, with up to three methane phenotypes collected on each individual (one each in GL, GA and LL), compared to one for each of the other groups.

Models trained on the LL data had low but positive accuracies (0.09-0.13) and the lowest standard errors (Table 1). Most training individuals were represented in all three NZ groups (GA, GL and LL), so the difference in BV accuracies from the model just fitting genomics is largely driven by differences in the methane phenotype. The diet fed to the Australian sheep (chaffed lucerne and cereal hay) is more similar to the lucerne pellet diet of the LL group than the ryegrass-based pasture of the other two New Zealand groups, therefore it is likely the drivers of methane emissions in these Australian sheep are most similar to those in the New Zealand LL group.

The model fitting both genomic and microbial effects and trained on the LL dataset showed the highest GMV accuracy, but this was no higher than the accuracy of the BV in the model just fitting genomic effects with the same training data (Table 1), suggesting that incorporating microbial information doesn't always improve accuracy beyond just fitting genotypes even when the microbial profiles had some predictive ability (e.g. LL). For the model fitting genomic and microbial effects and trained on the other NZ datasets (GL, GA and all NZ), there is some evidence that including the microbial component into the model can improve BV accuracy (0.08-0.23) compared to a model fitting only the genetic effect (-0.01-0.17).

**Table 1. Accuracy of predicting Australian methane emissions using Genotypes and/or Microbial Profiles from New Zealand sheep**

| Training set | All AUS | Genotyped AUS | | GRM+MRM | | |
|---|---|---|---|---|---|---|
| | MV | BV | MV | BV | MV | GMV |
| GL | -0.10 ± 0.14 | -0.01 ± 0.18 | -0.12 ± 0.09 | 0.08 ± 0.17 | -0.12 ± 0.10 | -0.11 ± 0.09 |
| GA | -0.20 ± 0.14 | 0.04 ± 0.15 | -0.22 ± 0.11 | 0.21 ± 0.21 | -0.23 ± 0.11 | -0.20 ± 0.11 |
| LL | 0.13 ± 0.08 | 0.13 ± 0.09 | 0.09 ± 0.07 | 0.12 ± 0.12 | 0.12 ± 0.08 | 0.13 ± 0.08 |
| NZ | -0.02 ± 0.12 | 0.17 ± 0.14 | -0.06 ± 0.04 | 0.23 ± 0.16 | -0.01 ± 0.06 | 0.01 ± 0.06 |

All AUS = Genotyped and non-genotyped Australian sheep, Genotyped AUS = genotyped subset of All AUS
GL = Grass lamb, GA = Grass adult, LL = Lucerne pellet lamb, NZ = All NZ samples (GL + GA + LL)
MV = Microbial value, BV = Breeding value, GMV = Genetic plus Microbial value

**Incorporating data from other countries.** Our second analysis aimed to evaluate whether including data from another country can improve prediction accuracy. All accuracies were higher when incorporating Australian data into the training set (Table 2) compared to training on different combinations of the New Zealand dataset (Table 1). BV and MV accuracies were high when using the training set of Australian samples despite the smaller size (Table 2). The highest accuracies were observed for GMV using the AUS+LL training set, followed by the MV estimated for genotyped animals when training on just the Australian dataset.

BV accuracy was not significantly impacted by adding LL or all NZ data to the Australian dataset (Table 2). This is likely driven by the different breed compositions between the two countries, leading to genomic relationships that were mostly negative between animals from NZ and Australia (Mean = -0.05; Range = -0.09 to 0.09); while those within Australia were mostly positive (Mean = 0.19; Range = -0.03 to 0.73).

The model fitting both genomic and microbial effects gave higher accuracies than the models fitting just genomic or just microbial relationships for the models trained on AUS+LL and AUS+NZ data (Table 2). The model fitting both genomic and microbial effects is more complex than the other two models used in our study, this led to singularity issues when used on the Australian dataset, likely driven by the smaller training set of 322 genotyped animals. Augmentation of the Australian dataset with New Zealand samples allows a more complicated model to successfully run and produces a higher prediction accuracy than using a model that just fits genomic or microbial effects.

**Table 2. Accuracy of predicting Australian methane emissions using Genotypes and/or Microbial Profiles from New Zealand and Australian sheep**

| Training set | All AUS | Genotyped AUS | | | | |
|---|---|---|---|---|---|---|
| | | | | GRM+MRM | | |
| | MV | BV | MV | BV | MV | GMV |
| AUS | 0.54 ± 0.12 | 0.45 ± 0.26 | 0.57 ± 0.16 | | Singularities | |
| AUS+LL | 0.47 ± 0.13 | 0.46 ± 0.24 | 0.49 ± 0.16 | 0.44 ± 0.25 | 0.48 ± 0.15 | 0.60 ± 0.17 |
| AUS+NZ | 0.40 ± 0.14 | 0.44 ± 0.25 | 0.39 ± 0.17 | 0.43 ± 0.25 | 0.37 ± 0.15 | 0.53 ± 0.15 |

All AUS = Genotyped and non-genotyped Australian sheep, Genotyped AUS = genotyped subset of All AUS AUS = Australian Samples, AUS+LL = Australian and Lucerne Lamb samples, AUS+NZ = AUS and All NZ samples; MV = Microbial value, BV = Breeding value, GMV = Genetic plus Microbial value

**Factors influencing these results.** Several factors will be influencing these results and their application to other datasets. The design of the Australian and New Zealand datasets were different in terms of sheep breed, the method for measuring methane (respiration chambers vs portable accumulation chambers), diet the sheep were on, and slightly different methods of rumen sample processing. Differences in rumen microbial profiles between New Zealand and Australian datasets were observed in Hess *et al.* (Submitted-a). These differences are likely largely driven by environmental factors, such as diet, but could also be partially due to differences in sample preparation. Cohort-adjusted rumen microbial profiles, as were used in this study, did not show the same differences between New Zealand and Australian samples (Hess *et al.* Submitted-a).

## CONCLUSIONS

This study shows that prediction of methane emissions across country using microbial profiles is possible even when genetic linkages are not strong, however, care needs to be taken in matching the diets as closely as possible to have some confidence in the predictions. Prediction accuracies of Australian sheep methane emissions were higher when training on data collected on Australian sheep than training on New Zealand sheep. Importantly, augmentation of the Australian dataset with data collected on New Zealand sheep that were on a similar diet enabled more complex models to be run and an improvement in prediction accuracy.

## REFERENCES

Gilmour A., Gogel B., Cullis B., Welham S. and Thompson R. (2015) VSN International Ltd.

Hess M.K., Hodgkinson H., Hess A.S., Zetouni L., Budel J., Henry H.M., Van Stijn T.C., Kirk M.R., Dodds K.G., Brauning R., McCulloch A.F., Hickey S.M., Johnson P.L., Jonker A., Morton N., Hendy S., Janssen P.H., McEwan J.C. and Rowe S.J. (Submitted-a) *BMC Genomics.*

Hess M.K., Johnson P.L., Brauning R., McCulloch A.F., Hickey S.M., Henry H.M., Stijn T.C.V., Dodds K.G., Hess A.S., Jonker A., Janssen P.H., McEwan J.C. and Rowe S.J. (Submitted-b) *Genet. Sel. Evol.*

Hess M.K., Rowe S.J., Van Stijn T.C., Henry H.M., Hickey S.M., Brauning R., McCulloch A.F., Hess A.S., Kirk M.R., Kumar S., Pinares-Patiño C., Kittelmann S., Wood G.R., Janssen P.H. and McEwan J.C. (2020) *PLOS One* 15, e0219882.

Robinson D., Dominik S., Donaldson A. and Oddy V. (2020) Animal Production Science 60, 880.

VanRaden P.M. (2008) *J. Dairy Sci.* **91**, 4414.

*Contributed paper*

# CHARACTERISATION OF SPERMATOZOAL TRANSCRIPTOMES IN SHEEP AND THE INFLUENCE OF BREED AND SEMEN QUALITY

**M.J. Hodge[1,2], S. de-las Heras-Saldana[3], S.J. Rindfleish[2], C.P. Stephen[1], and S.D. Pant[1]**

[1] Graham Centre for Agricultural Innovation, Charles Sturt University, Wagga Wagga, NSW, 2678 Australia
[2] Apiam Animal Health, Apiam Genetic Services, Dubbo NSW, 2830 Australia
[3] School of Environmental and Rural Science, University of New England, Armidale, NSW, 2351 Australia

## SUMMARY

Reproductive success, particularly after AI, is dependent on a number of contributing factors on both the ewe and ram sides. While there has been considerable emphasis on characterising ewe side contributions to reproductive success, relatively little emphasis has been placed on defining ram side contributors. In this context, the quality of semen used in AI is a crucial factor. Research details that spermatozoa contain around 14,000 mRNA transcripts (Selvaraju *et al.* 2017), which are transferred to the ova on fertilisation, conceivably influencing early embryonic development and successful conception. Therefore, this study aims to characterise the ovine spermatozoal transcriptome and investigate whether spermatozoal transcriptomes differ between breeds and between semen samples with high or low quality. Semen was collected (n=45) across three breeds; Merino, Dohne and Poll Dorset, and each ejaculate was subjected to computer assisted semen analysis (CASA) for assessment of quality parameters. RNA Sequencing and differential gene expression analysis identified 754 differentially expressed genes that were identified to play crucial roles in a variety of physiological functions, including fertilisation, embryonic development, and offspring production.

## INTRODUCTION

Artificial insemination (AI) is increasingly used in sheep breeding as it shortens the lambing period and allows for a single ejaculate to be used to inseminate a large number of ewes. While there are a number of contributing factors to conception success, a number are linked with seminal origin (Saacke 2008). Thus, it is crucial to characterise mechanisms underlying seminal factors which contribute to conception success. While it is generally accepted that semen quality influences conception outcomes, the magnitude of this influence has been difficult to characterize, primarily because visual assessment, frequently used to determine semen quality, can be highly subjective. However, computer assisted semen analysis (CASA), which enables repeatable assessment of semen quality parameters, now affords an objective alternative for determination of semen quality. Spermatozoa are known to contain a range of transcripts that can potentially influence fertilisation (Vijayalakshmy *et al.* 2018), and even offspring phenotype (Rando 2012). This study aims to characterise the transcriptome of three common sheep breeds in Australia, and determine whether spermatozoal transcriptome varies between breeds and between ejaculates of varying quality. These investigations could lead to the development of molecular markers and in vitro measures that could assist in predicting successful reproduction when specific ejaculates are used in AI programs.

## MATERIALS AND METHODS

**Animals and assessment.** Semen was collected from 3 sheep breeds; Merino (n=16), Dohne Merino (n=16), and Poll Dorset (n=13). Rams were closely matched for age (~18 months old), location and management conditions. Immediately following collection, each ejaculate was split into 2 aliquots; 250μL was diluted 1:10 (Nutrixcell, IMV Technologies) maintained at 37$^{O}$C for 4 hours then assessed utilising a CASA, with remaining semen snap frozen and stored in liquid nitrogen until RNA isolation. Each ejaculate was ranked (Gillian *et al.*, 2008) to identify ejaculates with high and low quality for RNA Sequencing analysis. Following RNA isolation (Kasimanickam and Kasimanickam 2019), the RNA integrity number was assessed and samples higher than 8 were kept for sequencing; Merino (n=12), Dohne (n=12), Poll Dorset (n=12).

**RNA sequencing.** Novogene (Singapore) utilised the NEBNext Ultra RNA Library Prep Kit for Illumina was used to fragment the RNA and synthesize the complementary DNA (cDNA) library. The sequencing of cDNA libraries was performed, obtaining 100 bp paired-end reads. Quality of reads was assessed, poor quality bases (Phred score Q<30), adaptors, and overrepresented sequences filtered out. Outliers and samples with low mapping rates to the ovine genome were also excluded.

**Differential expression analysis.** Quality control was performed and genes with low expression were removed. Differential gene expression analysis was performed within the R software environment to identify all genes that were either up or down regulated with a log fold change >1. A false discovery rate threshold of < 0.05 was applied to control type I error. Four contrasts were performed utilising Merino (n = 9), Dohne (n=10), and Poll Dorset (n = 12); three were between breeds (Dohne vs. Merino, Dohne vs. Poll Dorset, and Merino vs. Poll Dorset); and the fourth compared ejaculates of relatively high and low qualities fitting the breeds, to account for possible breed differences. The makeup per breed for the ejaculates ranked as being relatively low include Merino (n= 4), (Dohne (n= 5), and Poll Dorset (n = 6). Similarly, the ejaculates ranked as being relatively high include Merino (n= 5), (Dohne (n= 5), and Poll Dorset (n = 6).

## RESULTS AND DISCUSSION

Spermatozoal cells contain a range of RNA transcripts that are transferred to the ovum during fertilisation. However, the physiological role of spermatozoal RNA, particularly in relation to fertility and embryonic development, remain largely unknown. Therefore, the key objectives of this study were to characterise the ovine spermatozoal transcriptome, and determine whether transcriptomic profiles varied between breeds, and between semen ejaculates of varying quality.

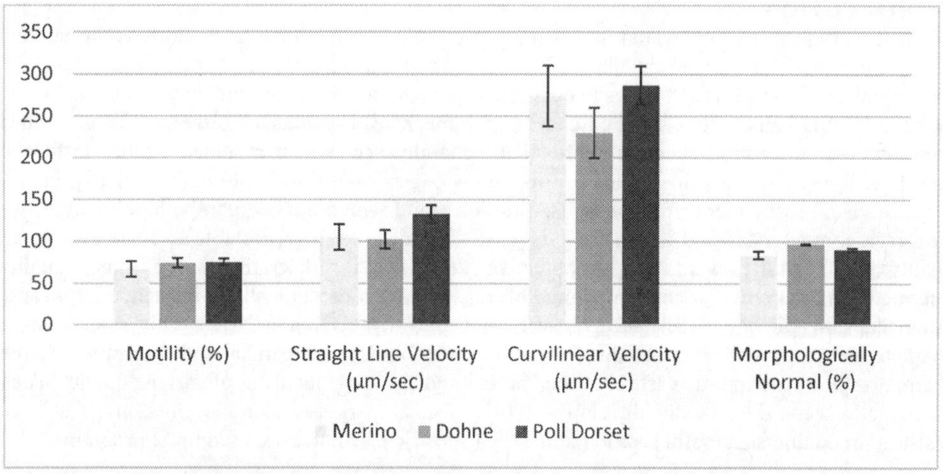


**Figure 1. Semen quality parameters (Mean ± SE) for Dohne, Merino, and Poll Dorset**

**Differentially expressed genes.** A total of 1,187,335,440 mapped reads across the three breeds sampled were mapped against the latest publicly available reference genome (Oar v.4.) at an average rate of 88%. According to the number of differentially expressed gene (DEGs) between contrasts, the transcriptomic profile of Merino and Dohne rams appear similar, in comparison to the transcriptomic profile of the Merino compared to the Poll Dorset. Respectively; 570, 72, 73, and 39 DEGs were found between the breed comparisons Merino vs. Poll Dorset, Dohne vs. Merino, Dohne vs. Poll Dorset, and ejaculate quality contrasts (descriptive statistics for quality parameters shown in Figure 1). Figure 2 displays the DEGs found for each comparison group.

Of the 39 DEGs found when contrasting ejaculates that were determined to be relatively high and low quality, 10 were found in literature to be associated with reproduction. Most noteworthy DEGs associated with reproduction included *PRM3*, *SPEM2*, and *OXCT2*. *Protamine 3* (*PRM3*) is significantly enriched for spermatogenesis, gonad development and hormone synthesis in sheep following next generation sequencing of sheep testis (Yang *et al.* 2018). Stafuzza *et al.* (2019) found *SPEM family member 2* (*SPEM2*) to be associated with embryonic development and number of piglets born alive following a GWAS. Georgiadis *et al.* (2015) used *3-oxoacid CoA-transferase* 2 (*OXCT2*) as a post-fertilisation and early embryonic marker using quantitative polymerase chain reaction (qPCR) when investigating high quality RNA in human semen.

Eleven of the 39 DEGs found in ejaculate quality contrast are associated with growth and production traits. Key genes linked with growth and production are *BRI3BP*, *LYRM4*, *KLK10*, and *MFSD9*. Following GWAS conducted in cattle, *BRI3 binding protein* (*BRI3BP*) has been shown to be associated with carcass traits (Lee *et al.* 2012), and *LYR motif containing 4* (*LYRM4*) is significantly associated with rib eye area (Wang *et al.* 2020). Kern *et al.* (2016) found *kallirein related peptidase 10* (*KLK10*) to be up regulated in a study looking feed intake and efficiency in cattle, suggesting that it could play a similar role for growth and development in sheep. Likewise, Perland *et al.* (2018) validated *major facilitator superfamily domain containing 9* (*MFSD9*) as a central solute carrier which is expressed in the food regulatory areas of the brain, resulting in increased feed intake, and increased growth.

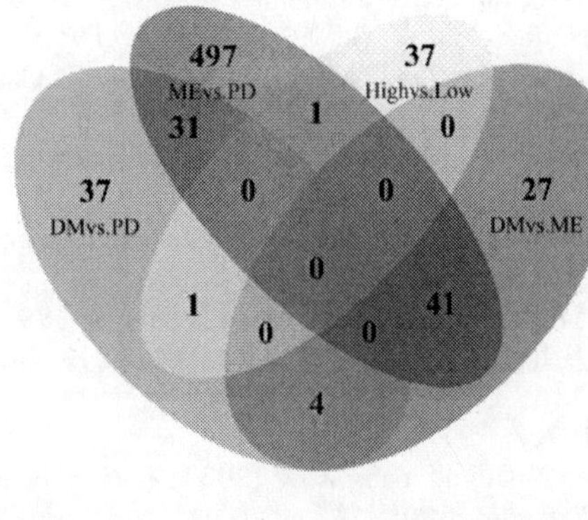


**Figure 2. Venn diagram of DEGs for each breed contrast and semen quality contrast; DM: Dohne, PD: Poll Dorset, ME: Merino**

Notable genes found when comparing three breeds sampled were subjected to a literature review. The *5'-aminolevulinate synthase 1* (*ALAS1*) is a gene of interest identified in the Merino vs. Dohne contrast, it is known to regulate circadian networks in cattle, which could influence the regulation of reproduction in seasonal breeding species like sheep (Wang *et al.* 2015). Likewise, Edwards *et al.* (2013) undertook a GWAS in cattle and found *capping protein regulator and myosin 1 linked* (*CARMIL1*) to be significantly associated with fertility.

In the Merino vs Poll Dorset contrast, *solute carrier family 35 member A5* (*SLC35A5*) and *integral membrane protein 2C* (*ITM2C*) were identified as key DEGs. In cattle, a GWAS found *SLC35A5* to be associated with fertility (Parker Gaddis *et al.* 2016). Similarly, expression of *ITM2C* is significantly enriched in the epididymis and vas deferens in both humans and mice during sexual maturation (Rengaraj *et al.* 2007).

Key DEGs identified in the Dohne vs. Poll Dorset contrast included; *DNA polymerase lappa* (*POLK*), which is developmentally regulated in testis of human and mice, and is hypothesised to play a crucial role in spermatogenesis (Ogi e*t al.* 2001); and *mannosidase alpha class 1A Member 1* (*MAN1A1*), which is associated with 6 month weight in sheep in a GWAS (Gholizadeh *et al.* 2015).

## CONCLUSION

The current study provides important insights into spermatozoal transcriptomes in sheep, and suggests that future investigations may target specific genes found to be differentially expressed in our study. Validation of our results in an independent population is also warranted. Furthermore, we have observed some of the differentially expressed genes are expressed stably within breeds, while others are expressed variably within breeds. This also deserves further scrutiny.

## ACKNOWLEDGEMENTS

This work was financially supported by the Graham Centre (Charles Sturt University) and Apiam Animal Health. We also kindly thank the producers who assisted with this research.

## REFERENCES

Edwards M., Liang Y., Kim T., and Cooper J.A. (2013) *Mol. Bio. Cell* **24**: 3047.

Georgiadis A.P., Kishore A., Zorilla M., Jaffe T.M., Sanfilippo J.S., Volk E., Rajkovic A. and Yatsenko A.N. (2015) *J. Urology* **193**: 352.

Gholizadeh M., Rahimi-Mianji G., and Nejati-Javaremi A. (2015) *J. Genet.* **94**: 143.

Gillian L., Kroetsh T., Maxwell W.M. and Evans G. (2008) *Anim. Reprod.* **103:** 201.

Kasimanickam V. and Kasimanickam R. (2019) *Bio-Protocol* **9:** 3284.

Kern R., Lindholm-Perry A.K., Freetly H.C., Kuehn L.A., Rule D.C. and Ludden P.A. (2016*) Livest. Sci.* **187**: 24.

Lee S. van der Werf J., Lee S.H., Lim D.J., Park E.W., Gondro C., ... and Thompson J. (2012) *Genes & genomics* **34**: 43.

Ogi T., Mimura J., Hikida M., Fukimoto H., Fukii-Kuriyama Y. and Ohmori H. (2001) *Genes to cells* **6**: 943.

Parker Gaddis K.L., Null D.J. and Cole J.B. (2016) *J. Dairy Sci.* **99**: 6420.

Perland E., Hellsten S.V., Schweizer N., Arapi V., Rezayee F., Bushra M. and Fredriksson R. (2018) *Plos one* **13**: e0197417.

Rando O. (2012) *Cell* **151**: 702.

Rengaraj D., Gao G., Liang X.H. and Yang Z.M. (2007) *Endocrine* **31**: 193.

Saacke R.G. (2008) *Theriogenology* **70**: 473.

Selvaraju S., Parthipan S., Somashekar L., Kolte A.P., Krishnan Binsila B., Arangasamy A. and Ravindra J.P. (2017) *Sci. Rep.* **7**: 42392.

Stafuzza N.B., Silva R.M.d.O., Gradomeni B.d.O,. Masuda Y., Huang Y., Gray K. and Lourenco D.A.L. (2019) *BMC Genom.* **20**: 321.

Vijayalakshmy K., Kumar D., Virmani M., Jacob N. and Kumar P. (2018) *Int. J. Curr. Microbiol. Appl. Sci.* **7**: 1188.

Wang M., Zhou Z., Khan M.J., Gao J. and Loor J.J. (2015) *J. Dairy Sci.* **98**: 4601.

Wang Y., Zhang F., Mukiibi R., Chen L., Vinsky M., Plastow G., ... and Li C. (2020) *BMC Genomics* **21**: 38.

Yang H., Wang F., Li F., Ren C., Pang J., Wan Y., ... and Zhang Y. (2018) *Biol. Reprod.* **99**: 650.

*Proc. Assoc. Advmt. Anim. Breed. Genet. 24: 171-174*

# THE PREDICTED RESPONSES TO GENOMIC SELECTION IN GROWING PIGS

**M. Shariful-Islam[1], J.H.J. van der Werf[2], V. Boerner[1] and S. Hermesch[1]**

[1]Animal Genetics and Breeding Unit, University of New England, Armidale, NSW 2351
[2]School of Environmental and Rural Science, University of New England, Armidale, NSW 2351

## SUMMARY

The responses to genomic selection in breeding programs for growing pigs were predicted using a selection index approach. Genomic selection increased overall predicted response by 2.6 (500 reference population) to 27.8% (5000 reference population) for a breeding objective consisting of backfat thickness (BFT), average daily gain (ADG), post-weaning survival (PWS) and feed conversion ratio (FCR) in growing pigs. Predicted response in PWS increased by 147% with genomic selection (5000 reference population) at the expense of the other traits like BFT, ADG, and FCR which had 14.5, 1.6, and 2.8% less genetic gain compared to the response in a conventional breeding program without genomic selection. The higher loss in genetic gain for BFT was due to a stronger genetic correlation with FCR in comparison to ADG. The predicted additional responses in the breeding objective is a guideline for the implementation of genomic selection in pig breeding programs.

## INTRODUCTION

Genomic selection is a method of predicting genetic merit of selection candidates utilizing dense marker genotyping covering the whole genome and basing predictions on a reference population that has both genotypes and phenotypes (Meuwissen *et al.* 2001). The impact of genomic information on response to selection is mostly determined by an increase in prediction accuracy and a decrease in generation interval. Since the generation interval of pigs is short, the genetic gain will largely be affected by the increased prediction accuracy with genomic information. Tribout *et al.* (2012) predicted 26% additional genetic gain from genomic selection compared to a breeding program without genomic selection. The breeding objective consisted of two genetically independent traits of growing pigs. However, breeding objectives consist of more than two traits in practial pig breeding programs and the prediction of response to genomic selection has not been reported for a broader breeding objective. Moreover, genomic selection is expected to benefit individual traits differently. Therefore, the objective of this study was to investigate how much genomic selection increases the predicted response in a breeding objective consisting of multiple correlated traits in growing pigs.

## MATERIALS AND METHODS

A selection index was used to predict the genetic gain in a conventional pig breeding program and compare that with ten different scenarios using genomic selection. The genomic breeding value (GBV) was included as an additional trait with a heritability of 0.99 and had zero economic value in the breeding objective (Dekkers 2007). Genomic selection (GS1 to GS10) was based on a reference population size that varied from 500 to 5000 pigs with an interval of 500. The accuracy of the genomic prediction was derived based on the trait heritability, size of reference population and effective population size according to the formula of Daetwyler *et al.* (2008). Effective population size was assumed to be 100 to provide an estimate of linkage disequilibrium in a historical population. Accordingly, increases in the size of the reference population increased the correlation

*A joint venture of NSW Department of Primary Industries and the University of New England

between the true breeding value and corresponding GBV (accuracy) of each trait. A deterministic simulation was used to predict the genetic gain per selection round, R = $i\ r_{iA}\sigma_A$, where R is the genetic gain, $i$ is the selection intensity of 1, $r_{iA}$ is the selection accuracy (i.e. correlation between the true and estimated breeding value) and $\sigma_A$ is the standard deviation of the breeding objective. Simulation was done using MTindex software (https://jvanderw.une.edu.au/).

**Breeding scenarios.** A terminal sire index for growing pigs included four breeding objective traits; back fat thickness (BFT, mm), average daily gain (ADG, g/d), feed conversion ratio (FCR, kg/kg), post-weaning survival (PWS, 0/1). These traits were also used as selection criteria, along with phenotype measurements of insulin-like growth factor 1 (IGF-1, ng/mL), muscle depth (MD, mm) and genomic breeding values for all traits. Base parameters for breeding objective traits and selection criteria are given in Table 1. Heritabilities and correlations were based on studies using Australian pig data. Economic values were taken from the study of Hermesch *et al.* (2014). The breeding objective is summarised in Table 2 including the source of information for each trait at the moment of selection (5 months of age). Sources of information for different traits varied depending on the availability of recording and recording cost. Genomic breeding values were available for the selection candidates before selection.

**Table 1: Genetic standard deviation (bold, on the diagonal), heritability ($h^2$), common litter effect ($c^2$), economic value (EV), genetic (below diagonal) and phenotypic (above diagonal) correlations of the selection criteria traits (BFT, mm; ADG, g/d; FCR, kg/kg; PWS, 0/1; IGF-1, ng/mL; MD, mm) in a terminal sire line index for growing pigs**

| Traits[1] | $h^2$ | $c^2$ | EV | Correlations | | | | | |
|---|---|---|---|---|---|---|---|---|---|
| BFT | 0.33 | 0.04 | -1.7 | **1.09** | 0.11 | 0.06 | 0 | 0.21 | -0.03 |
| ADG | 0.31 | 0.1 | 0.09 | 0.02 | **39.95** | -0.2 | 0 | 0.09 | -0.01 |
| FCR | 0.12 | 0.11 | -27.44 | 0.1 | -0.37 | **0.13** | 0 | 0.15 | 0 |
| PWS | 0.05 | 0 | 182.88 | 0 | 0 | 0 | **0.04** | 0 | 0 |
| IGF-1 | 0.22 | 0.19 | 0 | 0.57 | 0.06 | 0.65 | 0 | **11.63** | 0 |
| MD | 0.19 | 0 | 0 | -0.05 | 0.28 | 0 | 0 | 0.37 | **2.02** |

[1]BFT=back fat thickness, ADG= average daily gain, FCR= feed conversion ratio, PWS=post-weaning survival, IGF-1 = insulin like growth factor-1, MD = muscle depth

**Table 2: Relative emphasis on the breeding objective traits and the sources of information for the selection criteria traits**

| Traits | % Contribution to breeding objective | Sources of information | | | | |
|---|---|---|---|---|---|---|
| | | Own | Sire | Dam | Fullsibs | Halfsibs |
| BFT | 12.0 | 1 | 1 | 1 | 5 | 30 |
| ADG | 17.5 | 1 | 1 | 1 | 5 | 30 |
| FCR | 23.1 | 1 | 1 | 1 | 1 | 5 |
| PWS | 47.4 | 0 | 1 | 0 | 0 | 30 |
| IGF-1 | 0 | 1 | 1 | 1 | 2 | 12 |
| MD | 0 | 1 | 1 | 1 | 5 | 30 |

[1]BFT=back fat thickness, ADG= average daily gain, FCR= feed conversion ratio, PWS=post-weaning survival, IGF-1 = insulin like growth factor-1, MD = muscle depth

## RESULTS AND DISCUSSION

**Selection accuracy.** In comparison to a breeding program without genomic selection, overall accuracy of the breeding objective increased in genomic selection scenarios and showed an upward trend with the increase in the size of reference population (Table 3). The accuracy of PWS in the breeding objective increased by 81% in scenario GS10 whereas the accuracy of BFT, ADG and FCR increased by 14, 15, and 20%. Improvement of accuracy for different traits illustrates that traits with limited information prior to selection benefited more due to adding genomic information. Additional carcase and meat quality traits are also expected to benefit from genomic selection but were not considered in this breeding objective because they are not rewarded in most Australian markets.

**Table 3. Accuracy of breeding objective traits and the overall terminal sire line index in growing pigs in scenarios with no GS (conventional breeding program) and GS1 to GS10 (assuming different size of reference population starting from 500 to 5000 in increments of 500)**

| | Accuracy | | | | | | | | | | |
|---|---|---|---|---|---|---|---|---|---|---|---|
| Terminal Sire index | no GS | GS1 | GS2 | GS3 | GS4 | GS5 | GS6 | GS7 | GS8 | GS9 | GS10 |
| BFT | 0.71 | 0.72 | 0.74 | 0.75 | 0.76 | 0.77 | 0.78 | 0.79 | 0.80 | 0.81 | 0.81 |
| ADG | 0.69 | 0.70 | 0.72 | 0.73 | 0.75 | 0.76 | 0.77 | 0.78 | 0.79 | 0.80 | 0.80 |
| FCR | 0.56 | 0.57 | 0.58 | 0.60 | 0.61 | 0.62 | 0.63 | 0.64 | 0.65 | 0.66 | 0.67 |
| PWS | 0.27 | 0.31 | 0.34 | 0.37 | 0.39 | 0.42 | 0.43 | 0.45 | 0.46 | 0.47 | 0.49 |
| Overall Merit ($Index) | 0.47 | 0.48 | 0.50 | 0.51 | 0.53 | 0.55 | 0.56 | 0.57 | 0.58 | 0.59 | 0.60 |

[1]BFT=back fat thickness, ADG= average daily gain, FCR= feed conversion ratio, PWS=post-weaning survival.

**Predicted responses.** Genomic selection in this study showed the potential to improve overall response in the breeding objective (Table 4). The predicted genetic gain in PWS increased between 23 and 147% using genomic selection compared to the genetic gain in the conventional breeding program. On the other hand, BFT, ADG and FCR had 14.5, 1.65, and 2.89% lower gain in the most accurate genomic selection scenario (GS10). The genetic improvements in PWS were achieved at a diminishing rate from GS3 to GS10.

Relative improvement for the different breeding objective traits is explained by the relative emphasis on breeding objective trait, the accuracy of its estimated breeding values (EBV) and the correlation with EBVs from other objective traits. Back fat thickness had a relative economic value of 12% of the total breeding objective whereas FCR contributed 23.1% while having fewer records available before selection. As a result, FCR did not lose as much gain as BFT. A negative genetic correlation between FCR and ADG (-0.37) prevented ADG from losing as much gain as BFT. However, the different rate of predicted responses for different traits indicates a shift of genetic improvement towards the traits having the limited number of records, a feature of genomic selection that has not been well studied in pig breeding programs but has been reported in a sheep breeding study (van der Werf 2009). The current study illustrated the effects of genetic correlations between breeding objective traits on the magnitude of genetic improvement for different breeding objective traits due to genomic selection.

Genetic gain depends on the GBV prediction accuracy that ultimately depends on the size of the effective population (Daetwyler *et al.* 2008). In this study, effective population size was assumed to be 100 which is slightly higher than the value estimated by D'Augustin *et al.* (2017) that varied from 42 to 98 in three Australian pig breeds. However, genetic gain was predicted deterministically based on selection index theory. This approach provided an approximate figure of additional

1.16$/growing pig in genomic selection with a reference population of 5000 pigs. Further study should be conducted to investigate the long-term economic impact of using genomic selection in growing pigs.

**Table 4. Response per round of selection for the breeding objective traits in terminal sire line index in scenarios with no genomic selection (No GS, conventional breeding program) and GS1 to GS10 (assuming different size of reference population starting from 500 to 5000 in increments of 500)**

| | Responses in genetic standard deviation | | | | |
|---|---|---|---|---|---|
| Scenarios | BFT | ADG | FCR | PWS | Overall Merit ($Index) |
| No GS | -0.311 (100) | 0.462 (100) | -0.450 (100) | 0.119 (100) | 4.17 (100) |
| GS1 | -0.302 (97.1) | 0.461 (99.7) | -0.426 (94.6) | 0.147 (123.5) | 4.28 (102.6) |
| GS2 | -0.291 (93.5) | 0.458 (99.1) | -0.421 (93.5) | 0.176 (147.8) | 4.44 (106.4) |
| GS3 | -0.285 (91.6) | 0.456 (98.7) | -0.419 (93.1) | 0.199 (167.2) | 4.59 (110.0) |
| GS4 | -0.281 (90.3) | 0.457 (98.9) | -0.421 (93.5) | 0.214 (179.8) | 4.71 (112.8) |
| GS5 | -0.278 (89.3) | 0.454 (98.2) | -0.420 (93.3) | 0.236 (198.3) | 4.85 (116.2) |
| GS6 | -0.276 (88.7) | 0.458 (99.1) | -0.426 (94.6) | 0.246 (206.7) | 4.94 (118.4) |
| GS7 | -0.274 (88.1) | 0.456 (98.7) | -0.428 (95.1) | 0.262 (220.1) | 5.06 (121.3) |
| GS8 | -0.272 (87.4) | 0.459 (99.3) | -0.428 (95.1) | 0.273 (229.4) | 5.15 (123.5) |
| GS9 | -0.271 (87.1) | 0.461 (99.7) | -0.434 (96.4) | 0.284 (238.6) | 5.25 (125.8) |
| GS10 | -0.269 (86.5) | 0.459 (99.3) | -0.437 (97.1) | 0.295 (247.8) | 5.33 (127.8) |

[1]BFT=back fat thickness, ADG= average daily gain, FCR= feed conversion ratio, PWS=post-weaning survival. Values in parentheses indicate the percentage changes in the responses relative to the base scenario (conventional breeding program).

## CONCLUSIONS

This study predicted the additional overall response and additional response in individual breeding objective traits resulting from different scenarios of genomic selection. Overall genetic gain resulting from using GS is motivating for the implementation of GS in growing pigs. Before reaching a final conclusion, it is worthwhile to investigate the cost-benefit analysis of more realistic genomic selection scenarios.

## ACKNOWLEDGEMENT

The first author was supported by Animal Genetics and Breeding Unit (AGBU), Armidale, NSW through a scholarship for PhD study.

## REFERENCES

Daetwyler H.D., Villanueva B. and Woolliams J.A. (2008) *PloS one* **3(10):** e3395.

D'Augustin O., Boerner V. and Hermesch S. (2017) *Proc. Assoc. Advmt. Anim. Breed. Genet.* **22**: 241.

Dekkers J.C.M. (2007) *J Anim. Breed. Genet.* **124**: 331.

Meuwissen T.H.E., Hayes B.J. and Goddard M.E. (2001) *Genetics* **157**: 1819.

Hermesch S., Ludemann C.I. and Amer P.R. (2014) *J. Anim. Sci.* **92**:5358.

Tribout T., Larzul C and Phocas F. (2012) *J. Anim. Sci.* **90**: 4164.

Van der Werf J.H.J. (2009) *Proc. Assoc. Advmt. Anim. Breed. Genet.* **18**: 38.

*Proc. Assoc. Advmt. Anim. Breed. Genet: 24: 175-178*

# NEW ZEALAND NATIONAL DAIRY BREEDING OBJECTIVE REVIEW STAKEHOLDER SURVEY

**G.M. Jenkins[1], B.F.S. Santos[1], P.R. Amer[1], K. Stachowicz[1], A. Young[2] and B. Wickham[3]**

[1] AbacusBio Ltd, Dunedin, New Zealand
[2] DairyNZ, Hamilton, New Zealand
[3] NZAEL, Hamilton, New Zealand

## SUMMARY

The National Breeding Objective (NBO) for the New Zealand dairy industry is currently under review. As part of this process a stakeholder survey was carried out to help guide the direction of the review. The survey included questions on key issues related to the NBO, such as the direction of breeding worth (BW – the national breeding index which includes economically important traits), the role of important traits (particularly fertility, TOP traits - Traits Other than Production, liveweight and environmental traits, among others), and views on the dairy cow of the future. There were good levels of engagement with the survey and overall, the results show there is stakeholder support for prioritising the inclusion of a new and more accurate fertility breeding value in BW, applying a non-linear weighting to liveweight, and including key TOP traits (such as udder traits, capacity, feet and legs, and lameness) in BW.

## INTRODUCTION

The NBO for the New Zealand dairy industry had its last major review in 2012 (Amer *et al.* 2013). Since that time, it has been updated annually. Following discussions with breeding industry representatives along with the New Zealand Animal Evaluation Ltd (NZAEL) management and board, it was agreed that a major review of the NBO should take place. To guide this process a stakeholder survey was carried out to garner views from key stakeholders on the direction the NZ dairy herd is heading in and what the dairy cow of the future looks like for NZ. A farmer survey will also follow. In this paper, we will discuss key findings of the stakeholder survey and how this is shaping a plan for the future direction of the National Breeding Objective.

## MATERIALS AND METHODS

To gather stakeholder opinions on the NBO a survey was constructed in Alchemer (formerly Survey Gizmo). Questions in the survey covered the following points:

1. Fertility - After recognising the antagonistic effect that continued selection for milk production was having on NZ dairy herd fertility levels (Grosshans *et al.* 1997), fertility was added to the NBO in 2001. However, recent summaries show only a very small positive genetic trend in fertility (DairyNZ 2021). This is coupled with the opinion, expressed by many farmers, that fertility does not have a high enough weighting in Breeding Worth (BW). There is also some interest in changing the definition of the fertility trait breeding value (EBV) from CR42 (calving rate in the first 42 days after planned start of calving) to 6-week-in-calf rate (from planned start of mating). Questions in the survey asked whether respondents believed the fertility EBV was accurate enough, had high enough weighting in BW and whether a 6 week in-calf rate trait was more desirable definition for the fertility breeding value.
2. Environmental traits – environmental traits have become a major issue for NZ dairy farmers over the past decade. There is scope to increase focus on environmental traits in BW to achieve gains genetically as a low-cost approach to reaching on farm environmental goals. However, this would result in reductions in the rate of genetic progress in the existing BW traits. How much would farmers be willing to give up in profitability to make advances in environmental

traits?

3. Traits affecting survival - fertility and production traits play a large part in cow culling, however, there are a number of other traits that affect survival – these traits are currently encompassed in the residual survival breeding value. Functional survival is computed on the phenotypic level and is independent from fertility and production also. Functional survival is a newer alternative trait to residual survival and is a more accurate way of removing fertility and production related survival from the survival EBV. TOP traits, such as udder overall, can influence a cows functional survival. Is stakeholder opinion that it is sufficient to include udder overall indirectly in BW by including it as a predictor trait for survival or do stakeholders believe udder overall should be considered as a standalone trait in BW?
4. Liveweight – do stakeholders believe that the current liveweight penalty is appropriate?

There were also questions covering further TOP traits, optimal sire selection and decision support tools, gestation length, calf survival and calving difficulty, OAD (once a day) milking, high output systems, health traits, producing milk for specific markets based on genotype, milk price prediction, frequency of updates, as well as questions on traits that do not currently have breeding values (e.g. lameness). The survey was circulated among key stakeholders from DairyNZ, breeding companies (LIC, CRV), NZAEL Farmer Advisory Board, Massey University, and breed societies.

**RESULTS AND DISCUSSION**

In total, there were 459 responses - 280 complete and 179 partial responses. Of the total, approximately 50% of respondents were commercial farmers (211) and a further 25% breeders (109, produce milk and sell heifers, embryos, and occasionally semen or bulls). The remaining respondents (109) consisted of stakeholders from industry support and research groups, technical farmer support roles, bull breeding companies and milk processors. There were some differences in results between farmer and non-farmer survey respondents.

When respondents were asked which of the NZ indexes were most useful for genetic selection – 60% thought Breeding Worth (BW) was the most useful (Figure 1). The Australian Balanced Performance Index (BPI) was most commonly listed index under the 'Other' category.

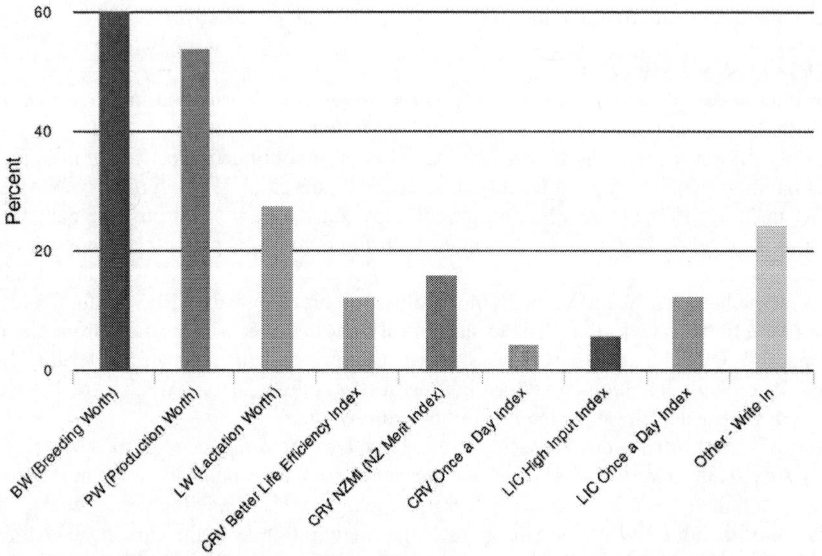


**Figure 1. Which of the NZ indexes are the most useful for genetic selection?**

Results from the survey show that overall stakeholders believe indexes are a useful tool for selection, however, support for BW was lower than expected because stakeholders believe there are some important traits missing in the national index. Some of the specific traits that stakeholders would like to see included in BW are udder traits, capacity, feet and legs, and lameness – along with further health traits (e.g. clinical mastitis and facial eczema). There was very strong support for the inclusion of lactation persistency in BW – with 71% of respondents either agreeing or strongly agreeing with this statement (Figure 2).

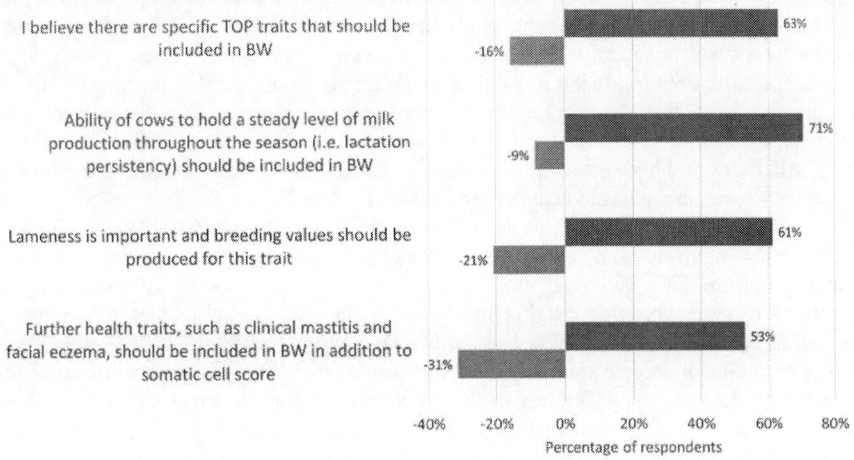


**Figure 2. Traits stakeholders would like to see included in BW (blue – strongly agree/agree, orange – strongly disagree/disagree)**

There were mixed opinions on the inclusion of environmental traits (low nitrogen and low methane production) in BW – however, a reasonable number of respondents (>60%) said they would sacrifice some profitability to achieve genetic gain in a Low Nitrogen (Low N) or Greenhouse Gas (GHG) trait. Currently, there are no direct environmental traits that have reached a level of proven integrity to justify inclusion in the NBO. Work on appropriate future weights for environmental traits is encompassed in research programs investigating the development and application of these traits.

Forty percent of respondents believed fertility is underweighted in BW. Of those that believe fertility is underweighted, a majority (strongly) agreed with the following: there is lost opportunity around sales of high value calves and heifers when fertility isn't high; the cost of artificial insemination is too high; the fertility EBV is not currently accurate enough; having to rear more replacement heifers (i.e. low fertility herd) is inconvenient; the economic impact of reduced fertility is very high for a dairy operation; and more weighting on fertility would result in faster improvements in herd fertility levels. The percentage of respondents who thought fertility was underweighted in BW was less than expected, which may be driven by a belief that the current fertility trait isn't accurate enough – in which case, increasing weighting on fertility still wouldn't achieve the desired gains. Respondents were in support of having a new fertility breeding value indicating likely change in 6-week-in-calf rate, rather than the current definition of 6-week re-calving rate (CR42). The rationale here is that 6-week-in-calf rate is the primary metric of fertility performance used by farmers at a phenotypic level, providing farmers with a more tangible means of understanding how changes in the fertility genetic merit of their herd is contributing to their fertility performance targets.

Most respondents either disagreed or strongly disagreed (43% vs 28% who agreed/strongly agreed) that the penalty for liveweight is appropriate for comparing across breeds. Holstein-Friesian farmers/breeders believed that liveweight (LW) is penalised too much within the Holstein-Friesian breed, these breeders also tended to think that LW is penalised too much within the Jersey breed (but not to the same magnitude). On the other hand, while Jersey farmers/breeders believed LW is penalised too heavily in Jerseys, they tended to disagree more than agree with the statement that LW is penalised too much within the Holstein-Friesian breed. Traits like BCS and LW have intermediate optimum levels in the minds of farmers and while Jerseys benefit as a breed from the current liveweight penalty, few farmers purchasing Jersey bulls wish them to produce very small sized cows in their herd. The current linear weighting applied to every breed equally creates a rigidity in deployment that causes mismatches between BW and farmer perception. Therefore, these survey responses support the thinking that a non-linear weighting on liveweight is more appropriate than the current weighting. Other key findings were that there is: strong support for custom selection tools along with mating allocation and inbreeding management tools; support for a NZAEL produced OAD index and high production index; and finally, there were mixed views on timelines of NBO reviews – annual updates versus every 3-5 years.

Full results from the stakeholder survey can be found via the DairyNZ website at the following address https://www.dairynz.co.nz/animal/animal-evaluation/national-breeding-objective-review/.

Following on from this stakeholder survey, a farmer survey is planned for later this year. Farmer trait prioritisation surveys provide insights into farmer preferences and traits that have an important influence over the cows they farm – over and above those traits which are associated with profitability by inclusion in selection indexes (Amer and Byrne 2019). Trait prioritisation surveys are an important way to engage farmers in the process of creating a selection index, influencing acceptance and adoption of the index (Axford 2018), and ensuring the industry is working together to create the NZ dairy cow of the future.

## CONCLUSIONS

Overall, results from the stakeholder survey suggest the following changes should be considered as priorities for the NBO: higher weighting on a new and improved fertility value; application of a non-linear weighting on liveweight; and direct weighting on udder, feet and legs. Following on from this stakeholder survey, a farmer survey including trait prioritisation is planned for 2021.

## ACKNOWLEDGEMENTS

We would like to thank all stakeholders who took the time to respond to the survey and to those who provided individual submissions.

## REFERENCES

Alchemer c2020. Retrieved from https://www.alchemer.com/ [acessed Feb 2021].

Amer P. and Bryne T. (2019) In 'Advances in breeding of dairy cattle', editors J. van der Werf and J. Pryce, Burleigh Dodds Science Publishing, Cambridge, UK.

Amer P.R., Santos B., Byrne T.J., Ludemann C., Visser B., Harris B.L. and Bryant J. (2013) *Proc. Assoc. Advmt. Anim. Breed. Genet.* **20**: 471.

Axford M.M. (2018) *Proceedings of the World Congress on Genetics Applied to Livestock Production* **11**: 791.

DairyNZ c2021. Genetic trends fertility. Retrieved from https://www.dairynz.co.nz/animal/animal-evaluation/animal-and-herd-averages/#category=sires&breed=all&status=ras [accessed Feb 2021].

Grosshans T., Xu Z.Z., Burton L.J., Johnson D.L., Macmillan K.L. (1997). *Livest. Prod. Sci.* **51(1-3)**: 41.

*Proc. Assoc. Advmt. Anim. Breed. Genet. 24: 179-182*

# EVALUATION OF AUSTRALIAN BREEDING VALUES FOR HEAT TOLERANCE UNDER US CONDITIONS

**L.M. Jensen[1], E.A. Jannaman[1], J.E. Pryce[2,3], A. De Vries[1] and P.J. Hansen[1]**

[1]Department of Animal Sciences, University of Florida, Gainesville FL, USA 32611-0910
[2]Agriculture Victoria Research, AgriBio, Centre for AgriBioscience, Bundoora, Victoria 3083, Australia
[3]School of Applied Systems Biology, La Trobe University, Bundoora, Victoria 3083, Australia

## SUMMARY

Reduced milk production and reproductive losses are common consequences of heat stress in dairy cattle and are likely to increase because of global climate change. The objective of this study was to compare body temperature regulation during heat stress between genetically heat-tolerant and heat-sensitive cows in peak summer (August 2020) on a California dairy farm. Genomic ABVs (ABVHT) were calculated for 12,487 cows from a single U.S. dairy farm. The herd had an average ABVHT of 102.5 with a standard deviation of 3.6. Rectal temperature was measured in 626 lactating cows with ABVHT $\geq$ 102 (heat tolerant) or <102 (heat sensitive) using a rectal thermometer. Vaginal temperature was measured in 118 cows with ABVHT $\geq$ 108 or <97 at 15-min intervals for five days in 118 cows using iButtons placed in blank CIDRs. Heat-tolerant cows had a 0.12°C (P=0.032) lower rectal temperature and a 0.07°C (P<0.001) lower vaginal temperature than heat-sensitive cows. The ABVHT can be used to select cows for resistance to heat stress under U.S. conditions.

## INTRODUCTION

Heat stress is one of the most significant environmental determinants of livestock productivity. In cattle, heat stress decreases milk production, reduces growth, diminishes sexual behavior, compromises female fertility, alters fetal development, and disrupts spermatogenesis (Hansen 2020). Global climate change means that limitations to sustainable livestock production caused by heat stress will become even more important than today (Battisti and Naylor 2009; Gauly and Ammer 2020). A breeding value for heat tolerance (ABVHT) in Holsteins and Jerseys based on the magnitude of the decline in milk, fat and protein yield per unit increase in temperature-humidity index for cattle is available for on-farm selection decisions in Australia (Nguyen *et al.* 2017). Inclusion of thermotolerance in selection indices for cattle present in hot climates may represent a useful approach for minimizing current and projected effects of heat stress on production and reproduction of dairy cows. The extent to which the breeding value is useful for other countries and systems, depends on whether the trait, measured in cattle under Australian conditions, also identifies dairy cattle in other countries that are genetically predisposed to be resistant or susceptible to heat stress. Given differences in management, including housing and feeding, this may or may not be the case. Accordingly, the purpose of this study was to evaluate the effectiveness of the genomic estimates of ABVHT for predicting resistance of lactating Holstein cows in the USA to heat stress. It is hypothesized that core body temperature is lower for females with high ABVHT than for females with low ABVHT.

## MATERIALS AND METHODS

The trial was approved by the University of Florida animal ethics committee. Data were collected from a commercial dairy farm in Riverdale, California. There were 3,613 lactating Holstein cows milked two times per day. The cows were housed in free-stall barns with shade cloth, fans, and sprinklers for heat abatement, and had access to dirt lots. In August, average daily milk yield was 38.1 kg.

**Climate data.** Dry bulb temperature and relative humidity were measured every 15 minutes for the duration of the study using HOBO U23 Pro v2 temperature and relative humidity data logger (Onset, Bourne, Massachusetts, USA) from three locations at the farm: exterior parking lot, the barn identified as the hottest by the farm manager and the barn identified as the coolest.

**Genotypes.** Genotypes (n=12,684) and pedigrees were sent to DataGene (Melbourne, Australia) and included in the August 2020 official genetic evaluation run. Standard procedures used by DataGene to edit and impute genotypes were applied (Nieuwhof *et al.* 2010). Samples with a call rate less than 0.9 or with more than 40% of markers heterozygous were removed. Animals with parentage or sex inconsistencies between the pedigree and genotype were also excluded. As many commercial providers provide genotypes to DataGene, a standard set of 45,685 SNP genotypes is used for routine evaluations (Nieuwhof *et al.* 2010) and missing genotypes were imputed by DataGene to satisfy this requirement. After quality control, genomic ABVHT were calculated for 12,487 cows following the methodology of Nguyen *et al.* (2016). The ABVHT is calculated to have a breed mean of 100 and a standard deviation of 5.

**Cow design and sampling.** Of the 12,487 cows with ABVHT, 2,925 cows were in the current herd. Rectal temperatures were measured daily using digital rectal thermometers across a random sample of cows (n=1078) once they had returned from milking in the afternoon (range of sampling time was 11:00 – 20:45 H). Of the 1,078 cows, 626 animals with ABVHTs ranging from 102 to 109 (termed HTR, n=354) or 95 to 101 (HSR; n=272) were used for statistical analysis. The most heat tolerant cows on the farm (ABVHT $\geq$ 108; termed HTV) and least heat-tolerant cows (ABVHT $\leq$ 97; termed HSV) were selected for vaginal temperature analysis. A blank CIDR containing an iButton 1922L (Maxim Integrated, San Jose, California, USA) was placed intravaginally for five days to record temperature every 15 minutes. The experiment was performed with 40 HTV and 23 HSV cows in week 1 and with a separate 26 HTV and 29 HSV cows in week 2. Herd records were obtained from the farm and accessed using DHI-Plus (Amelicor, Provo, Utah, USA).

**Statistical analysis.** Statistical analysis was performed using R (4.0.3). The model $y = \mu + X\beta + \varepsilon$ was used to analyze rectal temperature, where y is a vector of rectal temperature, $\beta$ is a vector of fixed effects including close barn temperature (barn closest to the location of the cows), test day milk yield, parity (primiparous or multiparous), day of calendar year, and ABVHT, and X is a design matrix of the fixed effects. Vaginal temperature was analyzed using the model, $y = \mu + X\beta + \varepsilon$, where y is a vector of vaginal temperature averaged across 5 days, $\beta$ is a vector of fixed effects including ABVHT, week, pen, milk yield, parity (primiparous or multiparous), and days in milk, and X is a design matrix for fixed effects. Week one and two were analyzed separately using the mixed model: $y = \mu + X\beta + Zg + \varepsilon$, where y is a vector of vaginal temperature, $\beta$ is a vector of fixed effects including ABVHT, date, time, date by ABVHT, time by ABVHT, milk yield, parity (primiparous of multiparous), and days in milk, g is a vector of random effects including cow nested within ABVHT, and X and Z are design matrices for fixed and random effects respectively.

## RESULTS AND DISCUSSION

Cows designated as HTR (ABVHT $\geq$102) had lower (P=0.032) rectal temperatures than cows designated as HSR (ABVHT <102) (Figure 1A). The mean rectal temperature was 38.46°C and 38.58°C for HTR and HSR cows, respectively (Figure 1B).

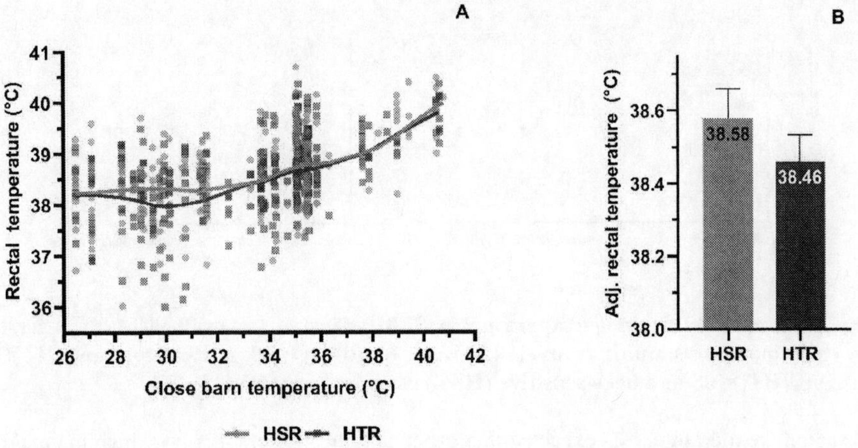


**Figure 1. Differences in rectal temperature between heat-sensitive and heat-tolerant cows as affected by dry bulb temperatures measured at the barn closest to the cow (n=626). Heat-tolerant (HTR) cows had an ABVHT ≥102 and heat-sensitive (HSR) cows had an ABVHT <102. (A) Data points are rectal temperatures for individual cows. (blue = HTR; orange = HSR). The line shows the group average at each barn temperature. (B) Mean rectal temperature ± SEM for each ABV class after adjustment for close barn temperature, milk yield, parity, and day**

Daily variation in vaginal temperature was characterized by lower vaginal temperatures in the morning and higher temperatures in the late afternoon. There was variation between weeks that is likely due to milking schedule (Figure 2A). Mean vaginal temperatures across both weeks were lower for extreme heat-tolerant cows (ABVHT ≥108; designated HTV) than for extreme heat-sensitive cows (ABVHT ≤97; designated HSV). The average vaginal temperature was 39.02°C for HTV and 39.09°C for HSV(P<0.001). These results are similar to those of Garner *et al.* (2016), working in environmental chambers in Australia, who found heat-tolerant cows (i.e. based on ABVHT) had significantly lower body temperatures during a heat challenge than heat-sensitive cows. There was a large effect of week on vaginal temperature. Analysis of vaginal temperature data to examine interactions between ABVHT and time of day were analyzed for week 1 separately from week 2. For week 1, there was a significant effect of ABVHT (P<0.001) but there was no interaction between time and ABVHT (Figure 2A). HTV cows had an average vaginal temperature of 39.11°C and heat-sensitive cows had an average vaginal temperature of 39.18°C. In week 2, there was an interaction between ABVHT and time of day (P <0.001) but no significant effect of ABVHT (Figure 2B). The difference between groups was more variable for week 2 throughout the day. In the evenings and early morning, there is a large difference in body temperature between HTV and HSV cows, while they become nearly identical between 12:00 AND 15:00. This result could indicate that HTV cows are more efficient at cooling their bodies when the environmental heat load is reduced than HSV cows. This idea is supported by the rectal temperature measurements where HTR cows maintained lower rectal temperatures than HSR cows until the dry bulb temperature reached ~33°C (Figure 1A). During both weeks, HTV cows maintained lower body temperatures than HSV cows throughout the day with the smallest difference in vaginal temperature from 12:00 to 15:00 when vaginal temperature declined for both groups. This time coincided with most cows returning from the parlor and drinking large amounts of water.

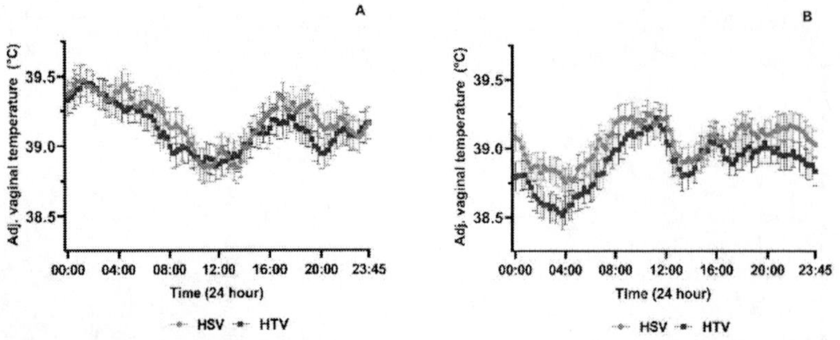


**Figure 2. Mean daily vaginal temperature ± SEM adjusted for milk yield, days fresh, and parity (primiparous vs multiparous) in (A) week 1 and (B) week 2. Heat-tolerant (HTV) cows had an ABVHT ≥108 and heat-sensitive (HSV) cows had an ABVHT ≤97**

Based on the literature, it is expected that genetically heat-tolerant cows which maintain lower body temperatures under heat stress conditions will sustain higher milk yields and better reproduction in the summer. The average difference in rectal temperature between heat-tolerant and heat-sensitive cows in the current experiment was 0.12°C and the average difference in vaginal temperature was 0.07°C. Future studies will determine whether differences in ABVHT are also reflected in genetically heat-tolerant cows having higher milk yield and conception rates under heat stress than their heat-sensitive counterparts.

## CONCLUSIONS

Cows with a high heat tolerance breeding value had lower body temperatures under heat stress than cows with a low heat tolerance value. Thus, the ABVHT can identify heat tolerant cows with superior ability to regulate body temperature under US conditions.

## ACKNOWLEDGEMENTS

The authors acknowledge Zoetis for providing blank CIDRs and the genotypes and Amelicor for providing access to DHI-Plus management software. Additionally, thanks are extended to DataGene for calculating the ABVHT. Special thanks are also extended to Steve Maddox, Juan Garcia, Daniella Demetrio, and all personnel at Maddox Dairy for providing the cows, genotypes, and assistance in data collection. Research was supported by the Southeast Milk Checkoff Program and the L.E. "Red" Larson Endowment. L.M. Jensen is supported by the Doak Graduate Fellowship from the National Association of Animal Breeders.

## REFERENCES

Battisti D.S., Naylor R.L. (2009) *Science* **323**: 240.

Garner J.B., Douglas M.L., Williams S.R.O., Wales W.J., Marett L.C., Nguyen T.T.T., Reich C.M., Hayes B.J. (2016) *Scientific Repts.* **6**: 34114.

Gauly M., Ammer S. (2020) *Animal* **14**: s196.

Hansen, P.J. (2020) *Theriogenology* **154**: 190.

Nieuwhof G., Beard K., Konstantinov K., Bowman P., Hayes B. (2010) *Interbull Bulletin* 42:35.

Nguyen T.T.T., Bowman P.J., Haile-Mariam M., Pryce J.E., Hayes B.J. (2016) *J. Dairy Sci.* **99**: 2849.

Nguyen T.T.T., Bowman P.J., Haile-Mariam M., Nieuwhof G.J., Hayes B.J., Pryce J.E. (2017) *J. Dairy Sci.* **100**: 7362.

*Proc. Assoc. Advmt. Anim. Breed. Genet: 24: 183-186*

# INVESTIGATION OF THE PATHOGENESIS OF SUSPECTED INHERITED NEUROLOGICAL DISEASES IN AUSTRALIAN SHEEP

**L.K. Manning[1,2], B.A. O'Rourke[1,3], M. Krockenberger[2], K.L.M. Eager[1,2] and I. Tammen[2,3]**

[1] Elizabeth Macarthur Agricultural Institute, Department of Primary Industries, Menangle, New South Wales 2568, Australia
[2] Sydney School of Veterinary Science, The University of Sydney, Sydney, NSW, 2006, Australia
[3] Anstee Hub for Inherited Diseases in Animals (AHIDA)

## SUMMARY

Several cases of neurological diseases in sheep were submitted opportunistically to a government veterinary diagnostic laboratory in Australia. Initial analysis suggested a possible genetic cause for segmental axonopathy, degenerative thoracic myelopathy, lissencephaly and cerebellar hypoplasia and cervicothoracic vertebral subluxation. Suitable case material is available and will be investigated further using in depth pathological investigation to establish diagnostic criteria, understand pathogenesis and propose candidate genes. Whole genome sequencing data will be used to identify likely causal variants with the aim to develop diagnostic tools for industry.

## INTRODUCTION

The Online Mendelian Inheritance in Animals database (OMIA, https://omia.org/home/) lists 194 inherited defects in sheep and likely causal variants have been identified for only 32 of these to date. A number of suspected inherited conditions have been reported recently in sheep in Australia and New Zealand. Our group has identified causal variants for neuronal ceroid lipofuscinosis (OMIA 001443-9940, Tammen *et al.* 2006), brachygnathia, cardiomegaly and renal hypoplasia syndrome (OMIA 001595-9940, Woolley *et al.* 2020) in Merino sheep and hydrops foetalis/pulmonary hypoplasia and anasarca (OMIA 000493-9940) in Persian sheep, but several suspected inherited neurological conditions are still under investigation (Woolley *et al.* 2019). Characterisation of inherited diseases and identification of causal variants is imperative to the development of diagnostic capabilities to identify affected and carrier animals and inform better management and breeding practices.

This study focussed on a detailed literature review of four diseases to inform further characterisation of the phenotype and pathogenesis to assist with selection of confirmed cases for genetic analysis, and identification of candidate genes associated with disease by comparison to similar genetic conditions in animals and humans.

## MATERIALS AND METHODS

**Review of previously submitted case material.** Selected suspected inherited neurologic diseases for which case material was available at the Elizabeth Macarthur Agricultural Institute, Department of Primary Industries, NSW (EMAI) were further characterised based on clinical presentation, clinical pathology, gross pathology, histopathology and where indicated, special stains, immunohistochemistry, transmission electron microscopy and additional diagnostics. Case material was both retrospective from previous submissions received at EMAI, and prospectively recruited from emergent disease cases submitted to EMAI during the study period. Upon identification of suspected heritable neurological conditions, ongoing investigation involves exclusion of differential diagnoses presenting with similar clinical and/or pathologic abnormalities. Existing SNP chip and whole genome sequencing (WGS) data from previous analysis were catalogued.

**Structured literature review.** A structured literature review of selected neurological conditions was conducted. PubMed, Web of Science and OMIA searches were used to identify relevant literature. Published literature was evaluated for each disease to identify new references and summarise current knowledge about disease phenotypes, mode of inheritance and candidate genes associated with disease. New information was used to update entries in OMIA. The literature was also reviewed to identify any additional suspected or known inherited neurologic phenes in sheep.

## RESULTS AND DISCUSSION

**Review of previously submitted case material.** Suspected inherited neurological diseases identified for initial investigation included ovine segmental axonopathy and degenerative thoracic myelopathy using retrospective analysis of case material, cervicothoracic vertebral subluxation from both retrospective and current investigations, and a recent investigation of lissencephaly and cerebellar hypoplasia diagnosed in a flock of crossbred sheep in NSW (Table 1).

**Table 1. Inherited ovine neurologic conditions investigated: available case material, whole genome sequencing (WGS) and SNPchip genotyping data**

| Disease | OMIA ID | Case material | WGS / SNP50* data |
|---|---|---|---|
| Segmental axonopathy | 001492-9940 | Multiple animals from several properties | 1 affected sheep (MGIseq System, Q30% = 88.05) |
| Degenerative thoracic myelopathy | 000079-9940 | Multiple animals, 1 property | None |
| Lissencephaly and cerebellar hypoplasia | 001867-9940 | Multiple animals, 1 property | 1 affected sheep (MGIseq System, Q30% = 88.28) |
| Cervicothoracic vertebral subluxation | 002313-9940 | Multiple animals from several properties | 2 affected sheep (Illumina HiSeqTM X Ten, Q30% = 92.16%); 9 affected & 2 obligate carriers (SNP50) |

*all WGS = 150bp paired-end reads, 30X coverage; SNP50 = Illumina® OvineSNP50 Genotyping BeadChip (CA, USA).

**Structured literature review.** A literature review did not yield any additional references for lissencephaly and cerebellar hypoplasia or cervicothoracic vertebral subluxation that were not already listed in OMIA. One review article referencing degenerative thoracic myelopathy and two review articles referencing segmental axonopathy were added to OMIA.

*Ovine Segmental axonopathy* ('Murrurundi disease') has been reported in Merino sheep of 1 to 5 years of age in Australia and New Zealand (Hartley and Loomis 1981; Harper *et al.* 1986). Clinically, sheep present with gradually progressive hindlimb ataxia (Hartley and Loomis 1981). Gross post-mortem lesions are absent or limited to hindlimb muscle atrophy (Harper *et al.* 1986; Jolly *et al.* 2006; Windsor 2006). Histologically, affected animals have vacuolation and spheroid formation throughout white matter in the brain and spinal cord (Hartley & Loomis 1981; Harper *et al.* 1986) with Wallerian degeneration of variable severity. Within the spinal cord, dorsal columns are more severely affected with spheroids compared to ventral and lateral columns (Harper *et al.* 1986). Axonal swellings associated with this condition ultra structurally contain membrane-bound vesicles (Jolly *et al.* 2006; Windsor 2006) and mitochondria (Windsor 2006). It has been postulated that the vesicles may originate from degenerating organelles (Jolly *et al.* 2006; Windsor 2006). Proteomic analysis has found cytoskeletal abnormalities in the trigeminal root, thought to be secondary changes (Jolly *et al.* 2006). It has been suggested that ovine segmental axonopathy may be an autosomal recessive inheritance (Jolly *et al.* 2006).

*Degenerative thoracic myelopathy* has been reported as a cause of hindlimb ataxia or paresis in Australian Merino sheep (Harper *et al.* 1991). Affected animals were between 5 and 24 months of age (Harper *et al.* 1991). Clinically, animals have slowly progressive hindlimb ataxia and paresis with neurologic examination consistent with a thoracolumbar lesion. Histologically this disease manifests as symmetrical Wallerian degeneration of variable severity, predominately affecting the ventromedial and dorsolateral tracts of the spinal cord (Harper *et al.* 1991). As this is a non-specific lesion, exclusion of differential diagnoses is essential in the investigative process, with potential differential diagnoses including plant toxicities, were excluded in this study (Harper *et al.* 1991). While a hereditary cause of degenerative thoracic myelopathy is suspected, it has not yet been proven.

*Lissencephaly and cerebellar hypoplasia* has been identified in mixed breed sheep on a property in New South Wales in 2019. Preliminary pathological investigation resulted in a diagnosis of LCH. LCH has previously been identified in Spanish Churra lambs (Pérez *et al.* 2013; Suárez-Vega *et al.* 2013), as well as humans, goats (Santos *et al.* 2013) and calves (Santos *et al.* 2016). In Churra lambs, affected animals were markedly ataxic and died within days of birth (Pérez *et al.* 2013). There was marked cerebellar hypoplasia, agyria and pachygyria with reduced and disorganised layers within the cerebral cortex, and disorganisation of the hippocampus histologically (Pérez *et al.* 2013). A monogenic autosomal recessive pattern of inheritance was suspected, with affected animals shown to have a 31-bp deletion in predicted exon 36 of the *RELN* gene and an absence of the protein reelin, a reported cause of LCH in humans (Pérez *et al.* 2013; Suárez-Vega *et al.* 2013). LCH in humans has also been associated with variants in genes including *DCX* and *LIS1*, although additional genetic variants are suspected (Ross *et al.* 2001).

*Cervicothoracic vertebral subluxation* has been reported in Poll Merino sheep in NSW (Hill *et al.* 1993; Cronin *et al.* 2019), Corriedale sheep in NSW (Hartley *et al.* 1994), Columbia lambs in the US (Lakritz *et al.* 1995) and Suffolk sheep in Scotland (Nisbet and Renwick 1961). Affected animals range in age and typically show a dropped or U shaped neck with hindlimb ataxia and variable neck rigidity or pain and inspiratory stridor (Hill *et al.* 1993; Hartley *et al.* 1994; Cronin *et al.* 2019). The consistent gross finding in this condition is subluxation or deviation at the junction of the cervical and thoracic vertebrae leading to spinal cord compression (Hill *et al.* 1993; Hartley *et al.* 1994; Cronin *et al.* 2019). Some reports have described gross changes in the paravertebral muscles including white streaks and pinpoint haemorrhages, characterised histologically by muscle degeneration, necrosis, regeneration, mineralisation, haemorrhage and fibrosis (Cronin *et al.* 2019). Cronin *et al.* (2019) postulated this was an inherited condition based on pedigree analysis; however, the responsible variant(s) remains to be determined. Woolley *et al.* (2019) reported on the findings of initial investigation of SNP genotyping and whole genome sequencing data, but a likely causal variant has not been identified.

**Ongoing research.** The literature review identified strong candidate genes for LCH and standard analysis of WGS of a single affected animal identified possible likely causal variants that are currently validated. Current investigation involves immunohistochemistry to describe the condition and compare the reported cases to the disease in Churra lambs and humans.

For the remaining three diseases the literature review did not identify strong candidate genes and the initial analysis of limited WGS data for cervicothoracic vertebral subluxation (Woolley et al 2019) and segmental axonopathy was inconclusive. Detailed analysis of the histopathology of all identified case material will be conducted with the aim to establish clear diagnostic criteria for each disease and to accurately characterise the phenotype and underlying pathogenesis. This information will assist in diagnosis of future cases as clinical presentation alone has resulted in misdiagnosis of neurological conditions in the past. Furthermore, possible candidate genes can be identified by comparison to similar genetic conditions in animals and humans.

DNA from six additional affected animals (2 each for segmental axonopathy, degenerative thoracic myelopathy and cervicothoracic vertebral subluxation) has been submitted for whole genome sequencing. Standard methods will be used to align reads to the ovine reference genome and to call and annotate variants. This new data, existing WGS data (Table 1) and publicly available control WGS data will be used to filter for private variants in identified candidate genes that segregate with the disease phenotype. Likely causal variants will be validated in additional samples.

## CONCLUSIONS

A number of neurological conditions recognised in sheep are suspected to be hereditary. Characterising the clinical aspects and pathology of these conditions will facilitate accurate diagnosis of affected animals, provides insight into possible pathogenesis, and can assist in guiding genetic investigations to identify causative genetic variants. This study is in the early stages and we aim to raise awareness and encourage submission of additional samples to be used in the validation stage of the genetic study.

## ACKNOWLEDGEMENTS

The authors thank farmers and veterinarians for submitting data and case material and the technical staff in the histology, molecular and genetic laboratories at EMAI. Dr. Cali Willet from The Sydney Informatics Hub Core Research Facility at The University of Sydney, Dr. Shernae Woolley and Monique Cauchi from The University of Sydney are acknowledged for initial analysis of WGS data. The authors acknowledge Dr. Pedro Pinczowski from DPI, Dr. Marina Gimeno and Emily Winkenwerder from The University of Sydney and district veterinarian Lou Baskind for their work on the recently diagnosed case of lissencephaly and cerebellar hypoplasia in sheep.

## REFERENCES

Cronin N, Loukopoulos P., Brown D., O'Rourke B.A., Morrice G. and Windsor P.A. (2019) **97**: 499.

Harper P.A.W., Duncan D.W., Plant J.W. and Smeal M.G. (1986) *Aust. Vet. J.* **63**:18.

Harper P.A.W., Plant J.W., Walker K.H. and Timmins K.G. (1991) *Aust. Vet. J.* **68**: 357.

Hartley W.J. and Loomis L.N. (1981) *Aust. Vet. J.* **57**: 399.

Hartley W.J., Stevenson J.M.K. and McCarthy P.H. (1994) *Aust. Vet. J.* **71**: 267.

Hill B.D., O'Dempsey N.D. and Carlisle C.H. (1993) *Aust. Vet. J.* **70**: 156

Jolly R.D., Johnstone A.C., Williams S.D., Zhang K. and Jordan T.W. (2006) *N.Z. Vet. J.* **54**: 210.

Lakritz J., Barr B.C., George L.W., Wisner E.R., Glenn J.S., East N.E. and Pool R.R. (1995) *J. Vet. Intern. Med.* **9**: 393.

Nisbet D.I. and Renwick C.C. *J. Comp. Pathol.* (1961) **71**: 177.

Pérez V., Suárez-Vega A., Fuertes M., Benavides J., Delgado L., Ferreras M.C. and Arranz J.J. (2013) *BMC Vet. Res.* **9**:156.

Ross M.E., Swanson K. snd Dobyns W.B. (2001) *Neuropediatrics*. **32**: 256.

Santos J.R.S., Dantas, A.F.M., Pessoa, C.R.M., Silva, T.R., Simões S.V.D., Correa, F.R. and Pedroa D. (2013) *Ciência Rural*. **43**: 1858.

Santos B.L., Damé M.C.F., Coelho A.C.B., Oliveira P.A., Marcolongo-Pereira C. and Schild A.L. (2016) *Ciência Rural*. **46**: 1622.

Suárez-Vega A., Gutiérrez-Gil B., Cuchillo-Ibáñez I., Sáez-Valero J., Pérez V., García-Gámez E, Benavides J. and Arranaz, J.J. (2013) *PLoS One*. **8**: e81072.

Tammen, I., Houweling, P.J., Frugier, T., Mitchell, N.L., Kay, G.W., Cavanagh, J.A., Cook, R.W., Raadsma, H.W. and Palmer, D.N. (2006) *Biochim. Biophys. Acta* **1762**: 898.

Windsor P.A. (2006). *Aust. Vet. J.* **84**: 169.

Woolley, S.A., Hayes, S.E., Shariflou, M.R., Nicholas, F.W., Willet, C.E., O'Rourke, B.A. and Tammen, I. (2020) *BMC Genet.* **21**: 106.

Woolley, S., Tsimnadis, E., Tulloch, R., Hughes, P., Hopkins, B., Hayes, S., Shariflou, M., Bauer, A., Hafliger, I., Jagannathan, V., Khatkar, M., Willet, C., and Tammen, I. (2019) *Proc. Assoc. Advmt. Anim. Breed. Genet.* **23**: 270.

*Proc. Assoc. Advmt. Anim. Breed. Genet: 24: 187-190*

# A GENOME-WIDE ASSOCIATION STUDY (GWAS) FOR CARCASS TRAITS IN HANWOO CATTLE USING IMPUTED WHOLE GENOME SEQUENCE DATA

**H. Kim, S. de las Heras-Saldana, N. Moghaddar and J.H.J van der Werf**

School of Environmental & Rural science, University of New England, Armidale, NSW, 2351 Australia

## SUMMARY

The identification of genomic region that are associated with phenotypic traits differences is important for improving genomic prediction accuracy. In this study, we aimed to find significant genomic regions for carcass traits in Hanwoo cattle using imputed whole genome sequence data on 13,715 animals. For carcass weight we found 285 SNPs in 7 QTL regions in which 54 candidate genes were identified on BTA4, BTA6 and BTA14. For back fat thickness we found 249 SNPs in 2 QTL regions containing 27 candidate genes on BTA17 and BTA19. The candidate genes from the top 5 significant SNPs were *ZFAT*, *TG* and *TOX* for carcass weight and *NOG* for back fat thickness. No significant SNPs for eye muscle area and marbling score were observed.

## INTRODUCTION

The fast development of the genomic technology enables the use of genomic information to improve the selection of animals in breeding programs. A genome-wide association study (GWAS) identifies associations between genetic variants along the genome and variation in phenotypes. These associations have been used to identify quantitative trait loci (QTL) and candidate genes for complex traits for humans and for diseases and production traits in livestock. QTL information can be used to prioritize genetic markers in order to improve the accuracy of genomic prediction of breeding value. Hanwoo is a Korean native beef cattle breed with the characteristic of high quality meat, mainly caused by high levels of intra muscular fat, also known as marbling. For finding significant QTLs for carcass traits, many GWAS studies have been reported on Hanwoo, However, many of those studies are limited due to their small sample size and low density of genetic markers. The objective of this study is to identify candidate genes for carcass traits using a larger number of samples with imputed sequence data.

## MATERIALS AND METHODS

In total, 13,715 animals with genotypes and phenotypes for carcass traits were used in this study. The four carcass and meat quality traits recorded were carcass weight (CWT), back fat thickness (BFT), eye muscle area (EMA) and marbling score (MS).

**Table 1. Summary statistics for carcass traits**

| Traits | Mean | Standard deviation | Min | Max | Coefficient of variation |
|---|---|---|---|---|---|
| CWT (kg) | 425.48 | 59.84 | 152 | 692 | 0.14 |
| BFT (mm) | 13.42 | 5.23 | 1 | 57 | 0.39 |
| EMA ($cm^2$) | 92.61 | 12.56 | 22 | 156 | 0.14 |
| MS (1-9) | 5.68 | 1.98 | 1 | 9 | 0.35 |

The phenotypic data were adjusted for fixed effects using a linear mixed model in ASReml v.4.1. (Gilmour *et al.* 2014):

$$y = CG + Sex + age + e$$

Where y is the observation vector, CG is the fixed effect of contemporary group, defined by birth season (4 season per year) and farm, Sex and age are covariates and e is the vector with residuals, which will be used as adjusted phenotypes in our GWAS.

Animals were genotype with the Illumina Bovine SNP 50K Bead Chip. After quality control, only 14K SNPs were remaining and these were used to first impute from 14K to 50k, then to high density, and finally to sequence with 203 reference animals using Beagle V5.1. The imputed SNPs with an accuracy of imputation ($R^2$) lower than 0.4 were removed. Finally, 17,549,506 SNPs and 13,715 animals were used in this study. A single SNP regression, GWAS was performed under a mixed linear model in (MLMA) in GCTA v.1.93 (Yang *et al.* 2011):

$$y^* = \mu + Xb + g + e$$

Where y* is a vector with adjusted phenotypes one for each of the four traits, $\mu$ is the overall mean, b is the allele substitution effects and X is the vector of genotype codes for SNP fitted. g is a vector of additive genetic effects with $g \sim N(0, G\sigma_g^2)$, where G is the genomic relationship matrix (GRM) calculate from 17,549,506 SNPs in PLINK v.1.9 and e is the residual effect. Manhattan plots were produced using ggplot2 packages in R. To reduce type-1 errors, the significance threshold was set at (P < 1.54-E08), derived as 0.05 divided by the number of independent variants, which in turn was calculated after not counting SNPs in linkage disequilibrium > 0.5 with other SNPs (Sham and Purcell 2014; Bedhane *et al.* 2019). We used the Ensamble database for *Bos taurus* UMD3.1 to identify the candidate genes that were located within 1Mb of the significant SNPs.

**RESULTS AND DISCUSSION**

We identified 9 QTL regions for carcass traits in Hanwoo.

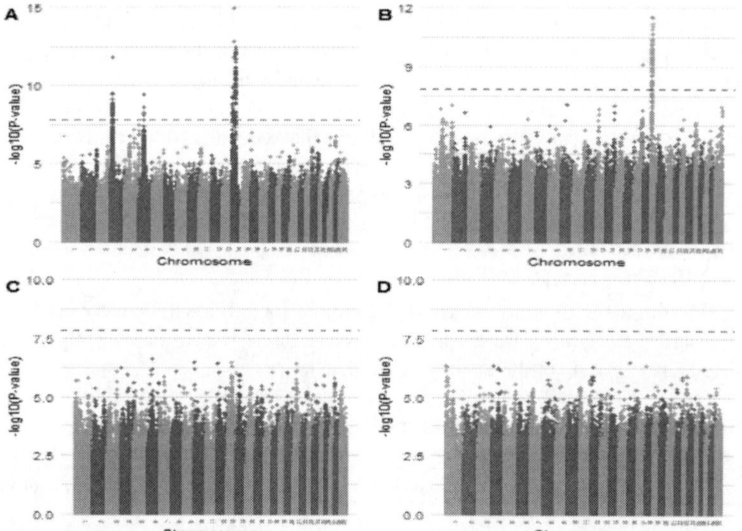


**Figure 1. Manhattan plot for carcass weight (A), back fat thickness (B), eye muscle area (C) and marbling (D)**

The Manhattan plot from the results of GWAS are shown in Figure 1 for all traits. In total, 285 SNPs in 7 QTL regions were detected for CWT on *Bos Taurus autosome* 4 (BTA4), BTA6 and BTA14, these chromosomes include 80, 8 and 197 SNPs respectively. The most significant QTLs for CWT were located on BTA14 which contained 32 candidate genes. The 22 candidate genes were found on BTA4 and nine candidate genes were found on BTA6 (Table 2). Previous GWAS also reported QTL for CWT on BTA4, BTA6 and BTA14 in Hanwoo (Lee *et al.* 2012; Srikanth *et al.* 2020) and BTA6 and BTA14 contains QTL regions in Chinese Simmental and multiple beef cattle breeds (Lu *et al.* 2013; Chang *et al.* 2018; Wang *et al.* 2020).

Significant SNPs associated with BFT were located on BTA17 and seven genes were found close to significant SNPs on BTA19. Significant QTL regions for BFT in Hanwoo were detected on BTA13 and BTA16 (Lee *et al.* 2012) and BTA19 (Srikanth *et al.* 2020) similar to the results obtained in this study.

**Table 2. QTL regions and candidate genes associates with carcass traits**

| Traits | Chr | Position (Mb) | Candidate genes |
|---|---|---|---|
| Carcass weight | 4 | 7.82 - 9.07 | FZD1 |
| | 4 | 9.35 - 12.4 | KRIT1, ANKLB1, TMBIM7, GATAD1, PEX1, RBM48, CDK6, SAMD9, CALCR, TFPI2, GNGT1, GNG11, BET1, VPS50, HEPACAM2, COLA2, CASD1, SGCE, PEG10, PPP1R9A, PON1 |
| | 6 | 38.52 - 39.52 | LAP3, MED28, FAM184B, LCORL, DCAF16, NCAPG |
| | 6 | 40.4 - 42.11 | SLIT2, PACRGL, KCNIP4 |
| | 14 | 4.91 - 6.33 | COL22A1, FAM135B |
| | 14 | 6.58 - 10.89 | KDHRBS3, ZFAT, ST3GAL1, NDRG1, WISP1, TG, SLA, PHF20L1, TMEM71, LRRC6, KCNQ3 |
| | 14 | 23.99 - 27.65 | RP1, KR4, TMEM68, TGS1, LYN, RPS20, MOS, PLAG1, SDR16C5, SDR16C6, PENK, IMPAD1, FAM110B, UBXN2B, SDCBP, CYP7A1, NSMAF, TOX, CA8 |
| Back fat thickness | 17 | 64.84 - 65.84 | SIRT4, MSI1, SRSF9, GATC, TRIAP1, COX6A1, COQ5, DYNLL1, RNF10, POP5, CABP1, MLEC, UNC119B, ACADS, SPPL3, HNF1A, OASL, C17H12orf43, ANKRD13A, GIT2, TCHP, GLTP, TRPV4 |
| | 19 | 7.02 - 8.15 | FAM222A, ANKFN1, NOG, C19H17orf67, DGKE, TRIP25, COIL, SCPEP1 |

No significant SNPs were detected for EMA and MS. In another recent study, no significant QTLs were detected for the MS in Hanwoo (Srikanth *et al.* 2020). MS appears to be mainly affected by many genes, each with a small effect.

**Table 3. Top 5 significant SNPs and candidate genes associated with carcass traits**

| Traits | Chromosome | Position | P-value | Candidate genes |
|---|---|---|---|---|
| Carcass weight | 14 | 8,160,456 | 1.20E-15 | ZFAT |
| | 14 | 9,518,339 | 1.41E-13 | TG |
| | 14 | 26,619,895 | 3.60E-13 | TOX |
| | 14 | 26,621,673 | 3.60E-13 | TOX |
| | 14 | 26,622,060 | 3.60E-13 | TOX |
| Back fat thickness | 19 | 7,645,081 | 3.02E-12 | NOG |
| | 19 | 7,620,479 | 3.31E-12 | NOG |
| | 19 | 7,646,102 | 6.87E-12 | NOG |
| | 19 | 7,620,249 | 8.93E-12 | NOG |
| | 19 | 7,618,889 | 1.21E-11 | NOG |

The *ZFAT* (zinc finger and AT-hook domain containing) gene which is located near to the most significant SNP for CWT has the potential for semi-lethality in Aberdeen Angus (Jenko *et al.* 2019) but was also associated with growth in humans and horses (Lango Allem *et al.* 2010; Makvandi-Nejad *et al.* 2012). The *TG* (thyroglobulin) gene plays a role in metabolism and has been associated with carcass and growth traits in cattle (Zhang et al. 2015). The *TOX* (thymocyte selection associated high mobility group box) gene has been associated with reproductive traits (de Camargo *et al.* 2015). All top five significant SNPs for BFT were located on BTA 19 were close to the *NOG* (Noggin) gene. *NOG* plays a role in inducing adipogenesis (Sawant *et al.* 2012) and was previously associated with BFT in Hanwoo (Srikanth *et al.* 2020).

## CONCLUSIONS

This study shows identified QTL regions and candidate genes associated with carcass traits in Hanwoo. Seven QTL regions with 63 candidate genes were found for carcass weight and two QTL regions with 31 candidate genes for back fat thickness. There were no significant genomic regions for eye muscle area and marbling score. This result can be helpful as genomic information to improve the accuracy of genomic prediction in Hanwoo breeding.

## REFERENCES

Bedhane M., van der Werf J.H.J., Gondro C., Duijvesteijn N., Lim D., Park B.H., Park M.N, Hee R.S and, Clark, S. (2019) *Front Genet.* **10**: 1235.

Chang C., Chow C., Tellier L.C., Vattikuti S., Purcell S.M. and Lee J.J. (2015) *Gigascience* .**4**: 7.

Chang T., Xia J., Xu L., Wang X., Zhu B., Zhang L., Gao X., Chen Y., Li J and Gao H. (2018) *Anim. Genet.* **49(4)**: 312.

de Camargo G.M., Costa R.B., de Albuquerque L.G., Regitano L.C., Baldi F. and Tonhati H. (2015) *Reprod. Fertil. Dev.* **27(3)**: 523.

Gilmour A.R., Gogel B.J., Cullis B.R., Welham S.J., Thompson R., Butler D., Cherry M., Collins D., Dutkowski G. and Harding SA. (2014) VSN international Ltd

Jenko J., McClure M.C., Matthews D., McClure J., Johnsson M., Gorjanc G. and Hickey, J. M. (2019) *Genet. Sel. Evol.* **51(1)**: 9.

Lango Allen H., Estrada K., Lettre G., Berndt S.I., Weedon M.N., Rivadeneira F., . . . Hirschhorn, J.N. (2010) *Nature* **467(7317)**: 832.

Lee S.H., van der Werf J., Lee S.H., Lim D.J., Park E.W., Gondro C., Yoon D., Oh S.J., Kim O.H, Gibson J. and Thompson J. (2012) *Genes & Genomics.* **34(1)**: 43.

Lu D., Sargolzaei M., Kelly M., Vander Voort G., Wang Z., Mandell I., Moore S., Plastow G. and Miller P. (2013) *BMC Genet.* **14**: 80.

Makvandi-Nejad S., Hoffman G.E., Allen J.J., Chu E., Gu E., Chandler A.M., Loredo A.I., Bellone R.R., Mezey J.G., Brooks S.A and Sutter N.B. (2012). *PLoS One* **7(7)**: e39929.

McClure M.C., Morsci N.S., Schnabel R.D., Kim J.W., Yao P., Rolf M.M., McKay S.D., Gregg S.J, Chapple R.H., Northcutt S.L. and Taylor, J. F. (2010) *Anim Genet.* **41(6)**: 597.

Sawant A., Chand D., Isayeva T., Tsuladze G., Garvey W.T., and Ponnazhagan S. (2012). *J Biol Chem.* **287(15)**: 12241.

Sham P.C. and Purcell S.M. (2014) *Nat. Rev. Genet.* **15(5)**: 335.

Srikanth K., Lee S.H., Chung K.Y., Park J.E., Jang G.W., Park M.R., Kim N.Y., Kim T.H., Chai H.H., Park W.C. and Lim D. (2020) *Genes (Basel)* **11(3)**.

Wang Y., Zhang F., Mukiibi R., Chen L., Vinsky M., Plastow G., Basarab J., Stothard P. and Li, C. (2020) *BMC Genomics* **21(1)**: 38.

Yang J., Lee S.H., Goddard M.E.and Visscher P.M. (2011) *Am. J Hum. Genet.* **88(1)**: 76.

Zhang L.P., Gan Q.F., Hou G.Y., Gao H.J., Li J.Y. and Xu, S. Z. (2015) *Genet. Mol. Res.* **14(2)**: 5320.

*Proc. Assoc. Advmt. Anim. Breed. Genet: 24: 191-194*

# MANAGEMENT OF INBREEDING AND COANCESTRY TO TARGET SHORT-TERM AND LONG-TERM GENETIC GAINS

**B.P. Kinghorn[1] and A.J. Kinghorn**

[1] University of New England, Armidale, NSW 2351, Australia

## SUMMARY

Inbreeding is not heritable. This means that breeding for low inbreeding generally has a transient and non-accumulating impact over generations. This is a bit like selecting for a trait that has a heritability of zero. However, coancestry (taken as the mean coancestry between an individual and the rest of the selected population) is heritable. Breeding for low coancestry in any one generation has a lasting impact over generations to increase genetic diversity and decrease the population mean inbreeding level. These issues have been poorly understood in some industries. Appropriate management of both coancestry and inbreeding is required to optimise the balance between short-term and long-term genetic gains, as well as to maintain genetic diversity. This paper tests and illustrates the implementation of such strategies. Management of coancestry is critical, whereas management of progeny inbreeding is of some transient value.

## INTRODUCTION

A high rate of genetic gain for the desired breeding objective(s) is central to most breeding programs. However, to sustain a high rate of gain in the long term, genetic diversity has to be maintained. Without genetic diversity, the better individuals are no better genetically than the worse individuals, and genetic gain stops.

Genetic diversity reduces more quickly in smaller breeding populations: In any one generation, few individuals contribute to the genetic mix in the long-term. Moreover, their contributions become less evenly distributed, giving more loss of diversity. This is because few parents in each future generation means that some individuals' descendant lineages die out, just by random chance. Selection on an index accelerates this loss of diversity, as less meritorious lineages die out more quickly than by chance.

This loss of diversity is essentially the same as the increase in the level of inbreeding. With a small population size, it becomes inevitable that relatives are mated with each other, and their progeny are thus inbred.

The inbreeding coefficient of an individual is the probability that the two genes that it inherits from its two parents are identical by descent – they are exact copies of a gene carried by an ancestor that is in the pedigree of both of its parents. However, an individual cannot pass its merit for inbreeding to its progeny – for that to happen we would have to squeeze two genes into one sperm or egg, and that is not how it works. Inbreeding is not heritable.

Highly inbred individuals have more identical genes, and thus have less genetic diversity or variation ***within themselves***, and this is a key reason that they tend to survive and perform poorly. However, within a small closed breeding population, there is also generally less genetic variation ***among*** individuals, because they share so many recent ancestors in common. The reduction in genetic diversity is simply related to the increase in average inbreeding coefficient:

*Genetic variation* = (1 – *average inbreeding coefficient*) x *genetic variation without inbreeding*.

So, inbreeding in a small closed population leads directly to loss of genetic diversity. We can reduce the rate of increase in inbreeding in three related ways:

1. Use more individuals as parents
2. Select individuals that are on average less related to each other across the group
3. Allocate more matings to individuals that are less related to the rest of the population

All of these are accommodated in correct balance with each other when we select parents and the numbers of matings to allocate to each by minimising ***the mean Parental Coancestry*** (**PC**) of the selected group. The coancestry between two individuals is the same as the inbreeding coefficient of the progeny they would produce. This means that Parental Coancestry would be the expected mean inbreeding in progeny if we had random mate allocation, including self-mating at random! We cannot achieve this in most species, so Parental Coancestry is in fact a measure of contribution to inbreeding in later generations – ***if we keep Parental Coancestry low, we keep long-term inbreeding low***. When combined with consideration of genetic gain, this constitutes Optimal Contributions Selection, whether coancestry and inbreeding are derived from pedigree or genomic information (Meuwissen *et.al.* 2021).

The three points above all relate to selection of individuals to be used as parents, with no attention to mate allocations. So, in addition, we can delay the expression of inbreeding by avoiding the mating of relatives, reducing inbreeding in the next generation. However, this only delays the inevitable mating of relatives in later generations. So, ***if we avoid the mating of relatives, we keep short-term inbreeding low***. This paper examines the interplay of these two inbreeding management strategies, together with the degree of emphasis on genetic gains.

## MATERIALS AND METHODS

A small population was simulated using PopSim (https://www.youtube.com/watch?v=5K4Q7SkBdMk&t) with the following properties: Discrete generations for simplicity, 25 breeding females each producing 4 offspring of random sex, maximum 10 females mated per male, BLUP selection on a single trait with heritability 25%.

Parental Coancestry was targeted to increase over $t$ = 20 generations at the same closely-controlled set of rates across treatments for Progeny Inbreeding. The rates chosen were dictated using different values for the balance between genetic merit ("Progeny Index") and Parental Coancestry, these being 0, 25, 45 and 75 Target Degrees (**TD**), where 0 degrees puts full emphasis on high Progeny Index, and 90 degrees puts full emphasis on low Parental Coancestry (Figure 1; and Kinghorn and Kinghorn, 2021).

However, policies on avoiding Progeny Inbreeding (and many other factors) can affect outcomes for Parental Coancestry. For example, strong emphasis on decreasing Progeny Inbreeding can result in decreased Progeny Index and increased Parental Coancestry in any one generation, as seen for the current solution in the Figure 1.

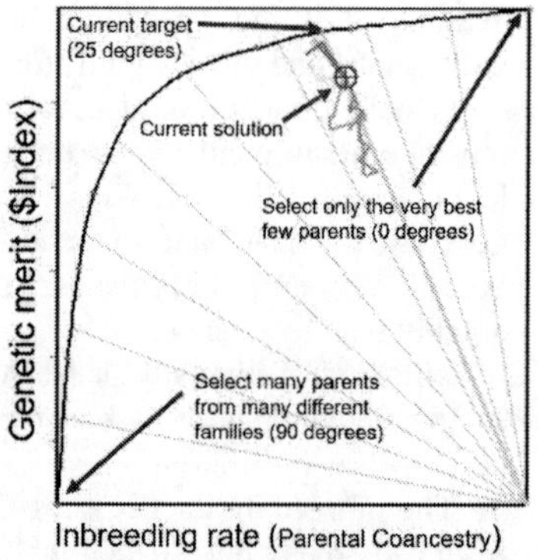


**Figure 1. Balancing Genetic merit and Genetic diversity using Target Degrees**

So rather than using TD, a hard limit was placed on maximum permissible Parental Coancestry, increasing each generation. This was done by first generating simulated populations at each of the three TDs > 0 (as 0 TD is unconstrained for Parental Coancestry), and the realised values for Parental Coancestry in each generation were then used to set a target maximum Parental Coancestry each generation in subsequent simulations. This approach was needed to give comparable results across treatments.

For each TD, three weightings for avoiding Progeny Inbreeding were applied (**W** = 0, -1, -100; See Kinghorn and Kinghorn 2021), plus a fourth treatment for which only close matings (full sibs, half sibs, parent-offspring) were avoided. All results are the means for 20 replicate simulations.

## RESULTS AND DISCUSSION

Figure 2 shows a wide range of results for Progeny Inbreeding and Progeny Index across TDs and treatments for Progeny Inbreeding (F). It can be seen that a strong weighting against inbred progeny (W=100) strongly reduces F, especially at low TDs. However, the need to select individuals able to give this outcome results in much lower Index response. For all TDs, the best strategy for genetic gain is a moderate weighting (W=1) against Progeny Inbreeding.

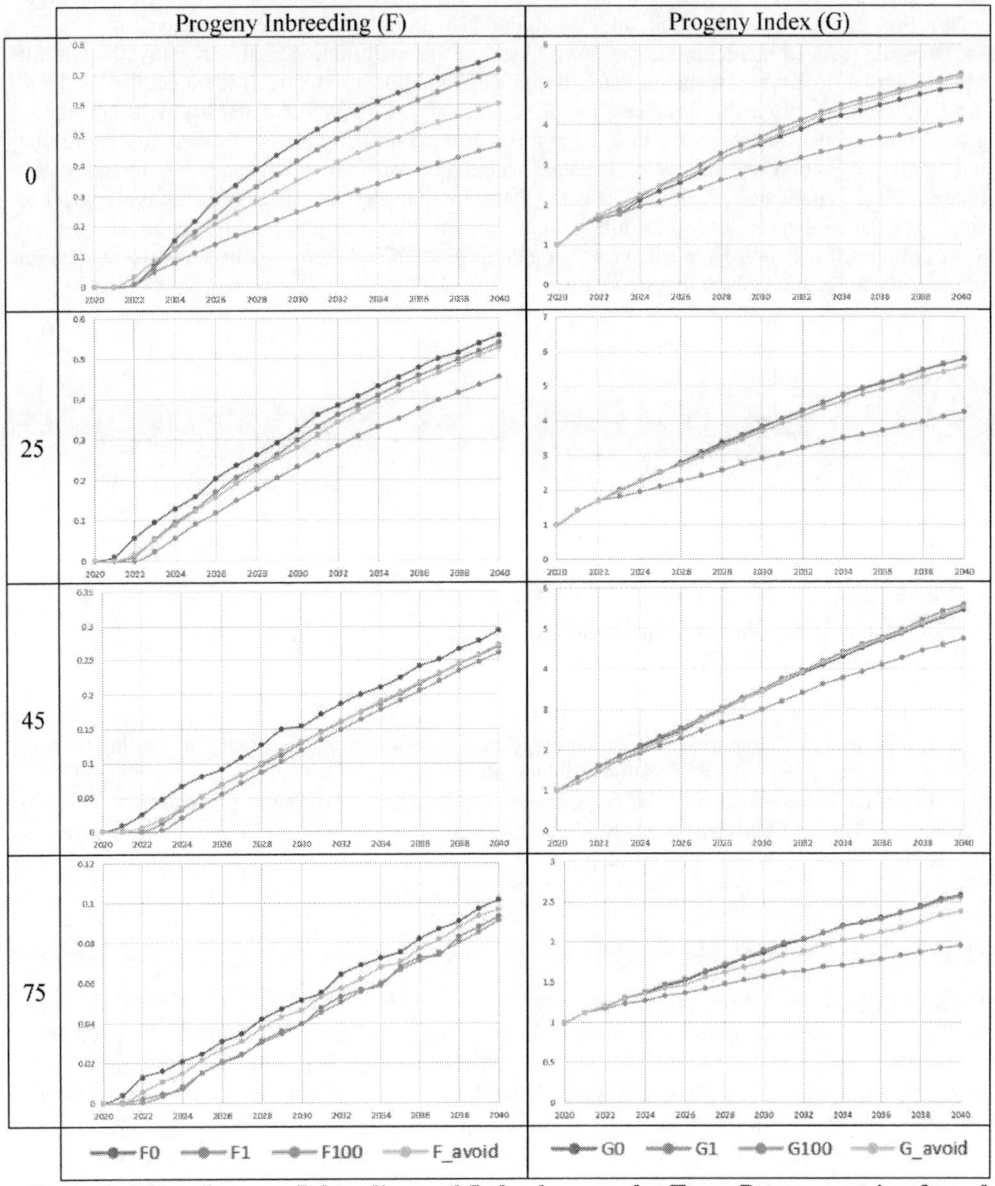


**Figure 2. Mean Progeny Inbreeding and Index by year for TargetDegrees ranging from 0 (aggressive) to 75 (conservative). Treatments are weightings of W = 0, 1, and 100 against Progeny Inbreeding, plus avoidance of mating close relatives, "avoid".**

The "avoid" strategy gives substantially reduced Parental Coancestry (Figure 2). Detailed observation shows that this is because avoiding the mating of close relatives has usually required selection of individuals that would not otherwise be selected. This causes increased diversity (reduced Parental Coancestry), but it also decreases selection response. However, this is due to the very small population size used here. In most breeding programs, use of this strategy will lead to re-shuffling of matings to avoid mating close relatives with little or no effect on selections.

Does the reduced inbreeding under W>0 mean we are increasing genetic diversity? Not at all! Mean Parental Coancestry is the measure of diversity, and this is essentially identical for W=0 and W=1 in Figure 3. Progeny Inbreeding for W=0 lags behind Parental Coancestry as follows: one generation, because the inbreeding in an animal equals the coancestry between its parents, plus about one more generation in this case, because calculation of Parental Coancestry includes self relationships, as indicated previously. It is this latter bit that accommodates the very real impact of small population size on long-term inbreeding. Progeny Inbreeding is usefully reduced by using W=1 in figure 3, with no cost to coancestry or genetic gain. In fact there may be some positive effect on gain due to reduced inbreeding depression.

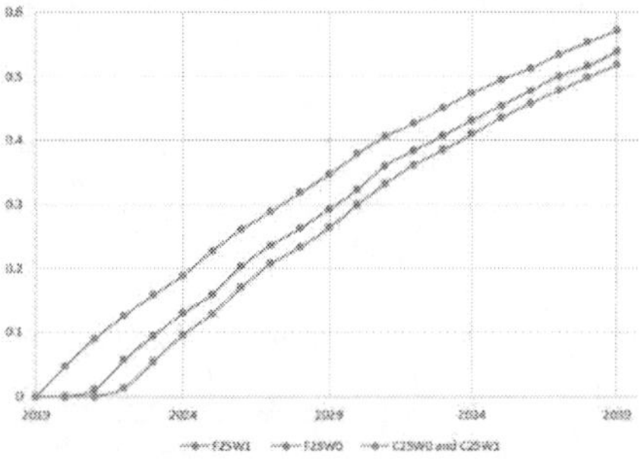


**Figure 3. Parental Coancestry and Progeny Inbreeding trends for W=0 and W=1 under TD=25**

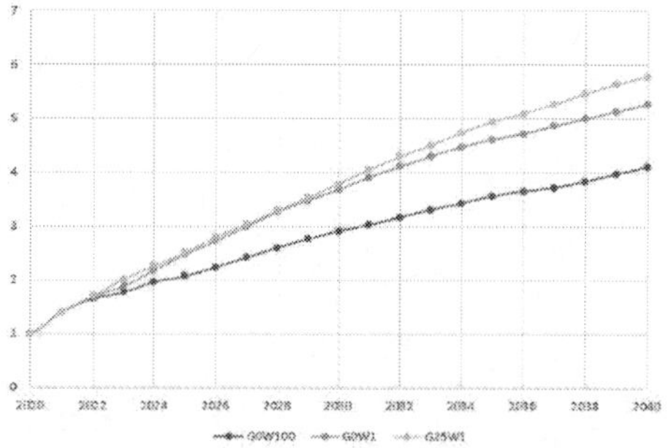


**Figure 4. Genetic gains for strategies to decrease progeny inbreeding, coancestry, and both**

As a "bottom line", Figure 4 shows that breeding to strongly reduce progeny inbreeding (TD=0, W=100) results in poor selection response. Moderate pressure to reduce progeny inbreeding (TD=0, W=1) is much better, but long-term response is reduced because of lack of attention to Parental Coancestry. The best strategy here is moderate attention to both Parental Coancestry and Progeny Inbreeding (TD=25, W=1). This strategy gave 12.2% more response in Index, and leaves the population with 61.5% more genetic variation after 20 generations. Overall, for genetic gain, coancestry management was 4.9 times more valuable than inbreeding management, and it was 7.7 times more valuable for genetic diversity.

## CONCLUSIONS

To manage inbreeding and genetic gain in breeding populations, attention to keep Parental Coancestry low (eg. choosing and invoking an appropriate Target Degrees) is generally very much more important than steps to reduce Progeny Inbreeding. However, both play a role and both should be attended to appropriately.

## REFERENCES

Kinghorn, B.P. and Kinghorn, A.J. (2021) https://matesel.com/content/documentation/MateSelInstructions.pdf

Meuwissen T.H.E., Sonesson A.K., Gebregiwergis G. and Woolliams J.A. (2020). *Front. Genet.*. https://www.frontiersin.org/article/10.3389/fgene.2020.00880

*Proc. Assoc. Advmt. Anim. Breed. Genet. 24:195-199*

# ACCURACY OF GENOMIC PREDICTION IN BRAHMAN CATTLE USING SIMULATED GENOTPYES FROM LOW-COVERAGE NANOPORE SEQUENCING

**H.J. Lamb[1], L.T. Nguyen[1], B.N. Engle[1], B.J. Hayes[1] and E.M. Ross[1]**

[1]Centre for Animal Science, Queensland Alliance for Agriculture and Food Innovation, The University of Queensland, St. Lucia, QLD 4067, Australia

## SUMMARY

Rapid, on-farm genotyping may be an alternative to SNP chip genotyping for genomic selection in certain agriculture industries. This study aimed to assess the accuracy of genomic breeding values, estimated from simulated Oxford Nanopore derived genotypes. Oxford Nanopore Technologies' (ONT) single nucleotide sequencing and genotyping accuracy was calculated from real sequencing runs of cattle DNA, and used to alter 50K SNP array genotypes in a population of 868 Brahman heifers. Genomic breeding values for age of first corpus luteum (an indicator of age of puberty, were estimated from the simulated ONT genotypes. The accuracies were compared to accuracies calculated using the original SNP array genotypes. Simulated ONT genotypes representing as little as 4 X sequencing coverage were able to generate accuracies not statistically different to SNP chip genotype accuracies.

## INTRODUCTION

Genomic selection (GS) first described by Meuwissen *et al.* (2001), is a technique widely used in agriculture, which uses genomic information to predict the genomic estimated breeding value (GEBV) of an individual for key traits. Typically, single nucleotide polymorphism (SNP) arrays are used to cost effectively genotype tens-of-thousands of SNPs, spread evenly across the genome, for genomic selection. Given a sufficiently large reference population of genotype and phenotype data, the GEBV can be accurately predicted from the SNP genotypes.

Turnaround time has limited the use of SNP genotyping and GS in Australia's northern beef industry, where cattle are often only handled once a year. With Queensland, the Northern Territory and Western Australia accounting for 62% of Australia's national beef herd, the difficulty of adopting GS in northern Australia represents a significant loss of potential productivity. We previously proposed a solution to this problem, namely crush-side genotyping (Lamb *et al.* 2020). Crush-side genotyping describes the use of ONT's MinION sequencer to rapidly, genotype cattle on-farm as they pass through the crush. A major limitation to the technology, is its high error rate. Improvements in flow cell chemistry and base calling algorithms has seen the error rate steadily decrease in recent years. However, the current error rate (between 5-8%) is still significantly higher than that of SNP array genotyping. The objective of this study was to ascertain the effect of ONT sequencing errors on the accuracy of genomic estimated breeding values in Brahman cattle.

## MATERIALS AND METHODS

**Ethics.** All analysis was performed using phenotypes and DNA samples previously collected with approval by the J.M. Rendel Laboratory Animal Experimental Ethics Committee (CSIRO, Queensland) as approvals TBC107 (1999 to 2009) and RH225-06 (2006 to 2010).

**Nanopore Sequencing Error Rates.** To determine ONT sequence error rates, ONT sequence data (approximately 8 X coverage) from a Brahman cow sequenced on MinION R9 flow cells was aligned against the Brahman genome (assembled from data from the same animal ; Ross 2019) using minimap2 (Li 2018) with the default settings for ONT alignment. Samtools mpileup (version 1.2, Li *et al.* 2009) was used to create a genome wide mpileup of the reads aligned to the reference genome. A maximum read depth of 50 was used to avoid chimeric repeats or ambiguously aligned regions of the genome. The number of single nucleotide mismatches for each locus across the genome was calculated from the mpileup using R. The error rates were reported as percentages of mismatches for each nucleic acid, given the total number of observations of nucleotides at all reference sites of a particular nucleic acid. For example, adenosine-guanine errors are the number of Guanine mismatches divided by all observations at reference adenosine sites.

**Nanopore Genotyping Error Rates.** A subset of reads, representing 4 X, 6 X, 8 X, 10 X and 18 X coverage from a second Brahman cow sequenced on the MinION, were then aligned using Minimap2 to the *Bos taurus* reference genome. Reference assembly UMD 3.1.1 was used to ensure reference loci and strand direction matched between sequencing and SNP chip genotypes. Samtools and BCFtools were used with a probability threshold (P value) of 1 for SNP discovery and a phred scaled base accuracy threshold (Q score) of 7, to genotype loci on the BovineSNP50 BeadChip (Illumina, San Diego, CA). Three methods (variable allele count, set ratio and minimum allele count) for assigning genotypes from the sequence were examined. The variable allele count method grouped loci by total coverage, and used a separate minimum allele count for each group to verify a genotype (Figure 1). This method was hypothesised to better distinguish between sequencing noise and heterozygous genotypes at higher coverages. The set ratio method called a particular observation as a likely true genotype if the allele was observed in greater than 10% of total observations at that loci. Finally, the minimum allele count method called a true genotype if a particular allele was observed more than twice no matter the total coverage. Any loci with more than two different alleles observed were considered incorrect genotype calls. All genotypes were then compared to the SNP chip genotypes to calculate genotyping accuracy, as well as the percentage of missing calls (loci with less than 2X coverage).

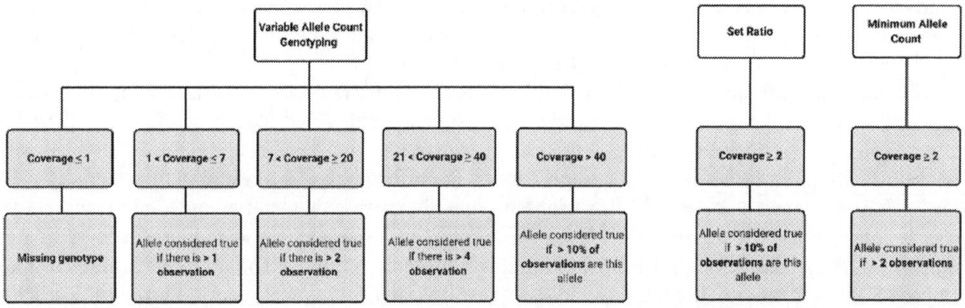


**Figure 1. Genotyping method. Three different SNP genotyping methods were used to call variable loci**

**Simulating Nanopore Genotypes and Genomic Breeding Value Prediction.** The cattle used in this experiment represent a subset of the Northern Breeding Project population, established by the Cooperative Research Centre for Beef Genetic Technologies. Phenotypes and management history for this herd have been extensively documented (Johnston *et al.* 2009; Engle *et al.* 2019). Records from a subset of 868 Brahman heifers was taken, including management history and age of first corpus luteum (AGECL), as determined using ultrasound scanning. The 868 heifers were also genotyped using the BovineSNP50 BeadChip (Illumina, San Diego, CA; Hawken *et al.* 2012).

Herd of origin, management cohort and birth month were concatenated into a single factor: contemporary group, which was modelled as a fixed effect (Engle *et al.* 2019). As only a Brahman subset was used in this study *Bos indicus* content was excluded as a covariate.

Genomic best linear unbiased prediction (GBLUP) was used to calculate GEBVs for AGECL using the univariate model:

$$y = XB + Zu + e$$

Where $y$ is the vector of phenotypes, $X$ is a design matrix allocating phenotypes to fixed effects, $B$ is a vector of the fixed effect contemporary group, $Z$ is a matrix of SNP genotypes and $u$ is a vector of additive SNP effects.

The genotyping error rate for each coverage was used to randomly select a number of SNP genotypes in $Z$ to alter. The calculated Nanopore sequencing error rate was then used to simulate errors at these loci consistent with the Nanopore error profile. The percentage of missing genotypes was also used to introduce missing SNPs.

To calculate the GEBV accuracy for AGECL 5-fold cross validation was used, with each validation population representing 20% of the total population (n = 868). Validation animals were included in the G matrix but coded with missing phenotypes. The package MTG2 (Lee and van der Werf 2016) was used for the predictions and the accuracy was calculated using $acc = r(GEBV, AGECL_{res})/\sqrt{h^2}$ where $h^2 = 0.55$. The 95% confidence interval was used to compare accuracies across the different simulations.

Two scenarios were simulated when calculating the accuracy of the GEBVs. The first simulation represented a scenario where, all animals, both reference and validation populations, were genotyped using ONT. This was simulated by simulating ONT errors in all animals. The second simulation represented, the more realistic situation where the reference population was SNP chip genotyped, while the validation population was genotyped using ONT. This was simulated by inducing errors into only animals in the validation population.

**RESULTS AND DISCUSSION**

Cytosine and thymine were found to have the lowest sequencing accuracies with 0.84% and 0.83% of bases at cytosine and thymine loci being inaccurately sequenced. The sequencing error rate revealed that for each nucleotide there was a single nucleic acid which was significantly more likely to be incorrectly called than the other nucleic acids (Table 1). For example, errors at adenosine loci were three time more likely to be called as guanine than either cytosine or thymine.

**Table 1. Nanopore sequencing error rates. The distribution of substitution errors observed in Nanopore sequencing data mapped to the reference genome built from the same animal**

| | | Reference Nucleotide[1] | | | |
|---|---|---|---|---|---|
| | | A | C | T | G |
| Alternate Nucleotide[2] | A | NA | 18.03% | 68.74% | 16.34% |
| | C | 17.85% | NA | 13.13% | 65.83% |
| | G | 65.70% | 13.27% | NA | 17.83% |
| | T | 16.46% | 68.70% | 18.13% | NA |

[1] The observed nucleotide in the reference genome

[2] The nucleotide observed in the mapped Nanopore reads

The minimum allele count method performed best at high coverages while the set ratio method had better genotype calling accuracies at lower coverages. Despite this the variable allele count method still outperformed the other two methods across all coverages (Table 2). At 18 X coverage the maximum genotyping accuracy achieved was 93.89%, in order to further increase the genotyping accuracy methods to disseminate between systematic sequencing errors, such as methylation, may still be required. Strand bias for example, could be used to filter out methylation signals to increase the accuracy of genotyping.

**Table 2. Nanopore genotyping accuracies and percentage of missing genotypes for various coverages**

| | Coverage | | | | |
|---|---|---|---|---|---|
| | **4** | **6** | **8** | **10** | **18** |
| **Percentage of loci not called**[1] | 41.2% | 9.5% | 4.4% | 4.1% | 0.6% |
| **Accuracy** (Variable allele count)[2] | 84.5% | 87.4% | 89.7% | 91.4% | 93.9% |
| **Accuracy** (Minimum allele count)[3] | 66.9% | 74.8% | 81.5% | 86.8% | 93.7% |
| **Accuracy** (Set ratio)[4] | 84.4% | 87.1% | 89.2% | 90.3% | 93.0% |

[1] Loci which did not meet the minimum depth criteria (>2 reads) for genotyping
[2] Variable SNP calling criteria were used based on the sequencing depth at each loci (See Figure 1)
[3] Alleles were called as present if observed more than 2 times
[4] Alleles were called as present if they comprised more than 10% of observed alleles at that locus

The genotyping errors observed (Table 2) also supported the ratios of nucleotide sequencing errors (Table 1), for example, at homozygous adenosine loci (AA) for 10 X coverage, 95.5% of loci were called correctly as AA or TT (the reverse compliment), while 3.3% of loci were called incorrectly as AG or GA. The other 12 genotype combinations shared the remaining 1.2% of AA loci evenly. This supports the earlier findings that A-G errors are more than three times more common in Nanopore sequencing than A-C or A-T. This pattern was observed in the results across all genotype combinations and could be leveraged to further increase the accuracy of Nanopore genotyping by incorporating a more stringent threshold for calling a genotype which corresponds to the most error prone nucleic acid given the reference loci. Using the AA example above, this would mean increasing the threshold for a guanine genotype call at an adenosine reference locus to decrease incorrect AG/GA genotype calls.

The GEBV accuracy of AGECL from the SNP chip genotypes was $0.39 \pm 0.03$ which is not statistically different to the accuracy reported by Engle *et al.* (2019), although removing tropical composites from the herd (effectively decreasing the reference population by 1,000 animals) likely describes the difference in average accuracy. At coverage as low as 4 X, there was no difference between the SNP chip accuracy and the simulated Nanopore genotype accuracies (Figure 2). Another study using Nanopore sequence data to predict genomic breeding values in cattle for three other traits: body condition score, hip height and body weight also reported accurate genomic predictions were possible from 4 X sequencing coverage without imputation (Lamb *et al.* 2021). This demonstrates accurate genomic prediction from Nanopore data is possible for a range of desirable traits.

A difference between the 95% confidence interval in the two different genotyping scenarios (reference and validation versus validation only) can be seen at 4 X coverage. However, this difference appears to decrease at higher coverages, likely due to the overall increase in genotyping accuracy.

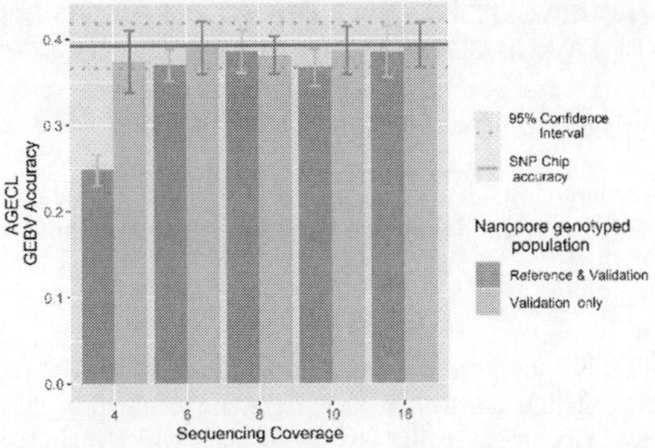


**Figure 2: GEBV accuracies for AGECL calculated from 33k genotyped loci. Genotypes were either directly observed in the SNP array data or had the error profile observed in SNP calling from ONT data simulated in the dataset. ONT errors were either simulated in both the reference and validation population or only in the validation population to represent two different sequencing scenarios**

## CONCLUSIONS

Here, we have demonstrated genotyping accuracies as high as 85% are achievable with just over 4 X Nanopore sequencing coverage. Using a SNP chip genotyped reference population, simulated Nanopore genotypes generated GEBV accuracies that were not significantly different ($P > 0.05$) from accuracies achieved using entirely SNP chip genotypes. This suggests ONT genotyping at low coverages can provide comparable GEBV accuracies to traditional SNP chip genotyping.

## ACKNOWLEDGMENTS

The authors would like to acknowledge the numerous contributions from the staff and students of the Hayes' lab group, in particular Mrs. Christie Warburton. We would also like to acknowledge the efforts of all individuals who have contributed to the collection and maintenance of the Beef CRC phenotype and genotype data. Financial contributions from Meat and Livestock Australia (project codes B.STU.2001 and L.GEN.1808) are also gratefully acknowledged.

## REFERENCES

Engle, B.N., Corbet, N.J., Allen, J.M., Laing, A.R., Fordyce, G., McGowan, M.R., Burns, B.M., Lyons, R.E., Hayes, B. (2019) *J. Anim. Sci.* **97**, 90-100.

Hawken, R.J., Zhang, Y.D., Fortes, M.R., Collis, E., Barris, W.C., Corbet, N.J., Williams, P.J., Fordyce, G., Holroyd, R.G., Walkley, J.R., Barendse, W., Johnston, D.J., Prayaga, K.C., Tier, B., Reverter, A., Lehnert, S.A. (2012) *J. Anim. Sci.* **90**, 1398-410.

Johnston, D.J., Barwick, S.A., Corbet, N.J., Fordyce, G., Holroyd, R.G., Williams, P.J., Burrow, H.M. (2009) *Anim. Prod. Sci.* **49**, 399-412.

Lamb, H.J., Hayes, B.J., Nguyen, L.T., Ross, E.M. (2020) *Genes (Basel)* **11**,

Lamb, H.J., Hayes, B.J., Randhawa, I.A.S, Nguyen, L.T., Ross, E.M. (2021) *BioRxiv*

Lee, S.H., van der Werf, J.H. (2016) *Bioinformatics* **32**, 1420-2.

Li, H. (2018) *Bioinformatics* **34**, 3094-3100.

Li, H., Handsaker, B., Wysoker, A., Fennell, T., Ruan, J., Homer, N., Marth, G., Abecasis, G., Durbin, R., Genome Project Data Processing, S (2009) *Bioinformatics* **25**, 2078-9.

Meuwissen, T.H.E., Hayes, B.J., Goddard, M.E. (2001) *Genetics* **157**, 1819-1829.

Ross, E. (2019) 'Characterisation of the Brahman Genome' Meat and Livestock Australia Limited.

*Contributed paper*

# EVALUATING THE BENEFITS OF INCLUDING PREDICTIVE SNP MARKERS IN SINGLE STEP EVALUATION IN SHEEP USING CROSS-VALIDATION

**L. Li[1], P.M. Gurman[1], A.A. Swan[1], N. Moghaddar[2] and J.H.J. van der Werf [2]**

[1] Animal Genetics Breeding Unit*, University of New England, Armidale, NSW, 2351 Australia
[2] School of Environmental & Rural Science, University of New England, Armidale, NSW, 2351, Australia

## SUMMARY

A SNP array of 50k SNP markers was used in single-step GBLUP (SS-GBLUP) models to estimate breeding values in the Australian sheep genetic evaluation system. In 2019, Neogen launched a new GeneSeek Genomic Profiler Ovine 50k chip, which included ~5000 SNPs that were identified based on Sheep CRC research as highly predictive for growth, carcass and eating quality traits. The objective of this work was to apply a five-fold cross-validation approach to compare different models for the use of predictive SNPs for post-weaning weight (PWT), carcass eye muscle depth (CEMD), carcass fat at C site (CCFAT), intramuscular fat (IMF) and shear force (SF5) based on the LAMBPLAN terminal sire genetic evaluation. Correlation and regression coefficients between adjusted phenotypes and SS-GBLUP EBVs for validation animals from the different models were calculated. The results indicated that adding predictive SNPs slightly improved the correlation and regression coefficient of EBVs, but there was no advantage in giving them more weight via a separate term in the model, confirming that the current industry evaluation model using a single genomic relationship matrix is the best of the tested models for these traits.

## INTRODUCTION

Single-step genomic BLUP (SS-GBLUP) procedures have been implemented in the Australian sheep genetic evaluation system since 2017 (Brown *et al.* 2018). Prior to 2020, the genomic relationship matrix (GRM) used in SS-GBLUP analyses was built using an ovine 50k panel of common SNPs. Recent genome-wide association studies have identified ~5000 predictive SNP markers for carcass and eating quality traits in sheep (Moghaddar *et al.* 2019). In 2019, Neogen launched a GeneSeek Genomic Profiler Ovine (GGP) 50k panel, which included these predictive SNPs. To accommodate these markers, the set of SNPs used in routine genetic evaluations was modified to be the union of all SNPs included on all panels used for sheep genotyping, resulting in a set of 60,410 SNPs. This set of SNPs was then implemented in the sheep SS-GBLUP analyses in a single genomic relationship matrix (GRM) from 2020. However, this method assumes equal weighting for all SNPs. An alternative approach is to use an additional term in the model, using a separate GRM based on predictive SNPs, effectively giving them more weight to those SNPs. In this study, models with one or two GRMs fitted in the SS-GBLUP model for the calculation of breeding values were investigated using a five-fold cross-validation approach. The correlation and regression of SS-GBLUP EBVs with adjusted phenotypes from the different models were compared.

## MATERIALS AND METHODS

**Phenotype data.** This study was conducted using data from the LAMBPLAN terminal sire industry evaluation, due to the new predictive SNPs targeting growth, carcass and eating quality traits. The data consisted of records from animals measured for the main slaughter traits in the Sheep

---

* A joint venture of NSW Department of Primary Industries and the University of New England

CRC Information Nucleus Flock (van der Werf *et al.* 2010) and the MLA Resource Flock databases which are used in the industry evaluation. Phenotypes were pre-adjusted for a combination of birth type, rearing type, age, and age of dam, depending on the trait. Five traits from two data sets were used in SS-GBLUP analyses to estimate breeding values for cross-validation (Table 1). The first data set included 9688 animals that had all five traits observed as well as SNP genotype information (the "small data set"). To investigate whether the extra phenotypes from ungenotyped animals affected the cross-validation results for those genotyped animals, the second data set extended the small data set by including all ungenotyped animals with at least one trait observed for any of the five traits in the analysis (the "large data set"). A summary of the two data sets is presented in Table 1. Pedigree information was extracted from the LAMBPLAN database and included 44,874 and 1,985,749 animals for the small and large data sets, respectively.

**Table 1. Traits (units), number of animals (N), mean and standard deviation (sd) for the small (animals with all phenotypes and genotypes) and large (all animals including ungenotyped animals with at least one phenotype) data sets in this study**

| Trait | Unit | Small data set | | | Large data set | | |
|---|---|---|---|---|---|---|---|
| | | N | mean | sd | N | mean | sd |
| Post-weaning weight (PWT) | kg | 9688 | 58.58 | 9.47 | 1,674,789 | 58.00 | 9.71 |
| Carcass eye muscle depth (CEMD) | mm | 9688 | 31.31 | 3.87 | 16,753 | 31.43 | 3.77 |
| Carcass fat at C site (CCFAT) | mm | 9688 | 4.13 | 1.96 | 16,560 | 4.63 | 2.48 |
| Intramuscular fat (IMF) | % | 9688 | 4.24 | 0.99 | 14,832 | 4.35 | 1.04 |
| Shear force (SF5) | Newtons | 9688 | 34.88 | 15.22 | 14,840 | 34.24 | 15.16 |

The five-folds subsets derived from the 9688 genotyped animals were used as the cross-validation data set for SS-GBLUP analyses. Animals were crosses between terminal sire breed rams and Merino ewes or Border Leicester x Merino ewes. The main ram breeds represented were White Suffolk (323 sires,3801 progeny), Poll Dorset (319 sires, 4080 progeny), Suffolk (40 sires, 499 progeny), White Dorper (35 sires, 309 progeny), Texel (31 sires, 413 progeny) and Dorper (29 sires, 235 progeny). Five-fold subsets were randomly allocated stratified by ram breeds and sire families with five replicates with the average number of sires and progeny ranging from 161 to 167 and from 1679 to 2043 for each subset, respectively.

**Genomic data.** Three sets of SNPs were used in this study: unselected (random) SNPs (55,709), the predictive SNPs (4,701) and the combined set (60,410). The first set was a combination of the original ISAG 50k sheep panel and the additional random SNPs from the Neogen GGP 50k, where the actual number of SNPs used is the set remaining after applying quality control measures. The predictive 4,701 SNPs (Moghaddar *et al.* 2019) were those originating from the CRC research that were then commercialised on the GGP 50k. Genomic relationship matrices (GRMs) were constructed based on these SNP sets, using the implementation of the breed-adjusted GRM as described by Gurman *et al.* (2019) and as implemented in the LAMBPLAN terminal sire SS-GBLUP analysis. Three genomic relationship matrices were calculated: $\mathbf{G}_r$, based on the random SNPs; $\mathbf{G}_p$, based on the predictive SNPs and $\mathbf{G}_{rp}$, based on the combined set.

**Models.** The multivariate linear mixed model used for estimating breeding values was $\mathbf{Y} = \mathbf{Xb} + \mathbf{ZQg} + \mathbf{Zt} + \mathbf{e}$, where **Y** is data in the multivariate form; **Xb** is the fixed contemporary group effects (defined as combinations of the management group, flock, year, sex, breed type and date of measurement); **ZQg** is the random genetic group effects; **Zt** represents combined effects of breeding values based on pedigree and genomic effects from different SNP sets, and **e** is residuals. Maternal effects were included as permanent environment effects for PWT. Four combinations of polygenic and genomic effects were compared to identify appropriate models: 1) A model: $\boldsymbol{Zt} = \boldsymbol{Za}$; 2) A+$G_r$ model: $\boldsymbol{Zt} = \boldsymbol{Za} + \boldsymbol{Zu_r}$; 3) A+$G_{rp}$ model: $\boldsymbol{Zt} = \boldsymbol{Za} + \boldsymbol{Zu_{rp}}$; and 4) A+$G_r$+$G_p$ model: $\boldsymbol{Zt} = \boldsymbol{Za} + \boldsymbol{Zu_r} + \boldsymbol{Zu_p}$, where **a**, $\mathbf{u}_r$, $\mathbf{u}_p$ and $\mathbf{u}_{rp}$ are N($\mathbf{0}$, $\boldsymbol{A}\otimes\boldsymbol{\Sigma}_a$), N($\mathbf{0}$, $\mathbf{H}_r\otimes\boldsymbol{\Sigma}_{g_r}$), N($\mathbf{0}$, $\mathbf{H}_p\otimes\boldsymbol{\Sigma}_{g_p}$) and N($\mathbf{0}$, $\mathbf{H}_{rp}\otimes\boldsymbol{\Sigma}_{g_{rp}}$) respectively, with $\mathbf{H}_r$, $\mathbf{H}_p$ and $\mathbf{H}_{rp}$ matrices derived from combining the genomic relationship matrixes $\mathbf{G}_r$, $\mathbf{G}_p$ and $\mathbf{G}_{rp}$ with pedigree relationship matrix **A**, respectively. $\boldsymbol{\Sigma}_a$, $\boldsymbol{\Sigma}_{g_r}$, $\boldsymbol{\Sigma}_{g_p}$, and $\boldsymbol{\Sigma}_{g_{rp}}$ are the multivariate genetic variance-covariance matrices due to those corresponding relationship matrices as estimated by Gurman *et al.* (2021).

The average accuracy of the different models was assessed by the correlation coefficient between EBVs and phenotypes adjusted for contemporary group effects (solutions from the same models with the full data set) for the animals in the test set which were removed from the analysis. Note that correlations were presented without scaling by heritability. The bias was evaluated based on the regression coefficient of adjusted phenotype on EBVs. This process was repeated for all five cross-validation sets.

## RESULTS AND DISCUSSION

The average correlation and regression coefficient for validation animals across the five cross replicates from cross-validation are shown in Table 2 for the small data set and in Table 3 for the large data set. Results from both data sets show that the average correlation increased by the largest amount when adding genomic information, from model A to model A+$G_r$, with much greater improvement for carcass and eating quality traits (17.6 ~ 43.5% increase) than growth traits (5.3 ~ 7.9 % increase for PWT). The correlation was also generally higher in the large data set compared to the small data set. There were small improvements in correlation when adding predictive SNPs in the combined GRM, from A+$G_r$ to A+$G_{rp}$, but no apparent benefit was observed in fitting predictive SNPs in a separate GRM in model A+$G_r$+$G_p$. The results confirm that the current LAMBPLAN model (A+$G_{rp}$), including predictive SNPs in a combined GRM is an appropriate solution to exploit the additional benefits of these SNPs.

**Table 2. Average correlation and regression coefficients for validation animals for post-weaning weight (PWT), carcass eye muscle depth (CEMD), carcass fat at C site (CCFAT), intramuscular fat (IMF), and shear force (SF5) for models A, A+$G_r$, A+$G_{rp}$ and A+$G_r$+$G_p$ across 5 replicates for the small data set**

| Models | PWT | CEMD | CCFAT | IMF | SF5 |
|---|---|---|---|---|---|
| | | | *Correlation* | | |
| A | 0.38 | 0.17 | 0.17 | 0.23 | 0.18 |
| A+$G_r$ | 0.40 | 0.20 | 0.23 | 0.33 | 0.23 |
| A+$G_{rp}$ | 0.41 | 0.21 | 0.24 | 0.36 | 0.25 |
| A+$G_r$+$G_p$ | 0.40 | 0.19 | 0.22 | 0.34 | 0.23 |
| | | | *Regression coefficient* | | |
| A | 0.97 | 1.01 | 0.90 | 0.92 | 0.92 |
| A+$G_r$ | 0.93 | 0.97 | 0.98 | 1.10 | 1.01 |
| A+$G_{rp}$ | 0.94 | 0.99 | 1.00 | 1.15 | 1.03 |
| A+$G_r$+$G_p$ | 0.93 | 0.83 | 0.88 | 1.08 | 0.91 |

[1] Standard deviation for correlation and regression coefficients ranged from 0.002 to 0.008

**Table 3. Average correlation and regression coefficients for validation animals for post-weaning weight (PWT), carcass eye muscle depth (CEMD), carcass fat at C site (CCFAT), intramuscular fat (IMF), and shear force (SF5) and for models A, A+$G_r$, A+$G_{rp}$ and A+$G_r$+$G_p$ across 5 replicates for the large data set**

| Models | PWT | CEMD | CCFAT | IMF | SF5 |
|---|---|---|---|---|---|
| | | | *Correlation* | | |
| A | 0.38 | 0.19 | 0.22 | 0.31 | 0.20 |
| A+$G_r$ | 0.41 | 0.23 | 0.27 | 0.39 | 0.25 |
| A+$G_{rp}$ | 0.41 | 0.24 | 0.28 | 0.41 | 0.26 |
| A+$G_r$+$G_p$ | 0.41 | 0.22 | 0.26 | 0.39 | 0.24 |
| | | | *Regression coefficient* | | |
| A | 0.87 | 0.91 | 0.95 | 0.93 | 0.96 |
| A+$G_r$ | 0.81 | 0.82 | 0.89 | 1.08 | 0.94 |
| A+$G_{rp}$ | 0.81 | 0.83 | 0.90 | 1.12 | 0.95 |
| A+$G_r$+$G_p$ | 0.82 | 0.74 | 0.83 | 1.08 | 0.84 |

[1] Standard deviation for correlation and regression coefficients ranged from 0.002 to 0.008

Regression coefficient estimates were generally within an acceptable range around the expected value of 1 in both data sets, although there was a greater degree of over-prediction (regression coefficient < 1) in the large data set relative to the small data set. This could be due to the variance components used in both data sets were estimated using the small data set. Over-prediction regression coefficient was also more remarkable for the weight trait, PWT. It is interesting to note that IMF is the only trait with under-prediction regression coefficient (regression coefficient >1), especially for the A+$G_{rp}$ model.

## CONCLUSIONS

Cross-validation analyses comparing the predictive ability of breeding values demonstrated the benefits of including genomic information, and that predictive SNPs do increase correlation by a small amount, and they can be included in a single genomic relationship matrix with all SNPs rather than used for an additional random term. This method is equivalent to the current industry evaluation model for these traits, highlighting that the current method is the more accurate of those investigated.

## ACKNOWLEDGEMENTS

This work was funded by MLA Project L.GEN.1815.

## REFERENCES

Brown D.J., Swan A.A., Boerner V., Li L., Gurman P.M., McMillan A., van der Werf J.H.J. (2018) *Proc. World Congr. Genet. Appl. Livest. Prod.* Species-Ovine:460.

Gurman P.M., Bunter K.L., Boerner V., Swan A.A., Brown D.J. (2019) *Proc. Assoc. Advmt. Anim. Breed. Genet.* **23**:254.

Gurman P.M., Li L., Swan A.A., Moghaddar N., and van der Werf J.H.J. (2021) *Proc. Assoc. Advmt. Anim. Breed. Genet.* **24**:.

Moghaddar N., Khansefid M., van der Werf J.H.J., Bolormaa S., Duijvesteijn N., Clark S.A., Swan A.A., Daetwyler H.D. and MacLeod I.M., 2019. *Genet. Sel. Evol.* **51**(1), 72.

*Contributed paper*

# RANKING BRAHMAN BULLS FOR FEMALE REPRODUCTIVE PERFORMANCE IN NORTHERN AUSTRALIAN COMMERCIAL ENVIRONMENTS USING DNA POOLING

**Y. Li[1], L. Porto-Neto[1], R. McCulloch[1], S. McWilliam[1], P. Alexandre[1], J. McDonald[2], A. Reverter[1] and S. Lehnert[1]**

[1] CSIRO Agriculture and Food, St Lucia, QLD, 4067 Australia
[2] MDH Pty Ltd, Cloncurry QLD 4824, Australia

## SUMMARY

Female fertility is one of the important reproductive traits that directly impact the profitability of commercial beef breeding herds. DNA pooling of cows with reproductive records can provide a cost-effective way for assessing and predicting the contribution of individual bulls to the fertility of their female offspring. However, panels of different SNP density exist and their impact on genomic prediction is unknown when DNA pooling is applied. In this study, using the genotype and phenotype (pregnancy test and lactation status) from two Brahman cattle populations in north Queensland, one containing 715 samples genotyped with 54,791 SNPs, the other consisting of 290 samples genotyped with 74,584 SNPs, we investigated genetic relationships between the two populations as well as rankings of individual bulls based on genomic prediction for pregnancy test outcome of their progeny. Our results show different outcomes obtained from using different density SNP panels in separating cow pooling samples, and estimating genomic breeding values for pregnancy test outcome of individual bull's progeny. The research highlights that extreme caution needs to be taken for choosing SNP panels of different densities to rank and select bulls for commercial beef production based on DNA pooling technology.

## INTRODUCTION

Genomic prediction of breeding values based on a genomic relationship matrix has revolutionized the ability to identify genetically superior livestock for improving traits that are difficult to measure (van der Werf 2009). However, in commercial herds, it is impractical to individually genotype all animals. DNA pooling of cows with reproductive records can provide a cost-effective way for assessing and predicting the contribution of individual bulls to the fertility of their female offspring (Reverter *et al.* 2016). A question that remains to be answered is what density SNP panel should be used to genotype DNA pooled cows to rank bulls to achieve accurate prediction of reproductive performance of their progeny? In this study, using two Brahman cattle populations in north Queensland, we aimed to investigate the impact of SNP panels of different density on the ranking of bulls.

## MATERIALS AND METHODS

**Animals.** Datasets from two Brahman cattle populations in north Queensland were used for the study. One (SmartF) consists of 290 samples from 2012-2014 herds (177 individual bulls and 113 pools representing 2,648 cows) genotyped with 74,584 SNPs (770K BovineHD BeadChip platform). The other (MDH2020) contains 715 samples from the 2020 herd (482 individual bulls and 233 pools representing 2,452 cows) genotyped with 54,791 SNPs (Neogen Australasia GGP TropBeef 50K chip). DNA pools were formed based on the pregnancy test (i.e. not pregnant or pregnant) and lactation status (dry or wet) of cows at 2$^{nd}$ joining. Details of the phenotype of pregnancy test outcome (PTO) and pooling techniques can be found in Reverter et al. (2016). In brief, animals were separated into 6 categories, that is, dry and empty (not pregnant, scored as 1), dry and early pregnant (scored 2), dry and mid pregnant (scored 3), dry and late pregnant (scored 4), wet and empty (not pregnant, scored as 5), and wet and pregnant (scored 6). DNA samples of animals with identical phenotypic scores were pooled together. The individual pool size ranged from 4-45 animals for SmartF (Reverter *et al.* 2016) and from 5-12 animals for MDH2020, depending on the number of animals available in each category. Details of the two datasets are presented in Table 1.

**Table 1. Composition of two genotyped populations**

| Population | Sex | Year | DNA samples | Total |
|---|---|---|---|---|
| SmartF | Cows | 2012 | 41 (pools) | |
| (74,584 SNPs) | | 2013 | 31 (pools) | |
| | | 2014 | 41 (pools) | 113 |
| | Bulls | 2013 | 27 | |
| | | 2014 | 150 | 177 |
| MDH2020 | Cows | 2020 | 233 (pools) | 233 |
| (54,791 SNPs) | Bulls | 2020 | 482 | 482 |

**Imputation of genotypic data.** Between the two populations, there were 19,089 SNP in common. The imputation from low to high-density SNP genotypes was conducted to both SmartF and MDH2020, using 730,000 SNPs from 5,040 Beef CRC Brahman cattle as the reference. PLINK (Change *et al.* 2015) and Eagle v2.4.1 (Loh *et al.* 2016) were applied for phasing and imputation, respectively. After quality checks with the threshold of R-square value >0.8 and removing SNPs on the sex chromosome, this resulted in 615,310 SNPs.

**Principal Component Analysis (PCA).** To visualize genetic relationships between two populations, we conducted a PCA using genotypes from either the low density (19,089 common SNP) or imputed high-density panel (615,310 SNP, HD).

**Genomic prediction**. Genomic estimated breeding values (GEBVs) of PTO of progeny for individually genotyped bulls were derived within each population. The conventional genomic prediction method was applied to derive GEBVs, that is, a mixed animal model was used by fitting a polygenic random effect with the GRM (genomic relationship matrix). The fixed effects included the size of pool (30 levels) and contemporary group (5 levels) for SmartF, and SNP chip row (3 levels) and column (24 levels) information for different pools in MDH2020, respectively. The GRM was constructed using the method described by Reverter *et al.* (2016). In brief, the B-allele frequencies from the genotypes of the pools of cows ($\leq$0.25, >0.25 and <0.75 or $\geq$0.75, best fitted the three genotypes based on the individual DNA samples and the genotype call algorithm employed by Illumina) were converted into the three possible genotypes (i.e. 0, 1 and 2 for AA, AB, and BB, respectively) and these were merged with the individual genotypes of each bull to generate a single GRM relating bulls with pools of cows. Then the Qxpak5 software program (Pérez-Enciso and Misztal 2011) was used to fit the GRM in a mixed animal model and obtain genomic estimates of variance components and genomic predictions (GEBVs) for PTO of the testing population. For comparison purposes of different density panels within populations, GEBVs were derived using four GRMs, either with 19,089, 54,791 (for MDH2020 only), 74,584 (for SmartF only), or high density (HD) SNP.

## RESULTS AND DISCUSSION

**Relationships between animals of two populations**. The results from the PCA on all 1,005 animals (290 from SmartF and 715 from MDH2020) are shown in Figure 1. When a low-density SNP panel data (19,089, Figure 1a) was used, 346 DNA pooled cow samples from both populations were clustered together with very small variation among them, suggesting high similarity in the number of alleles between pooled samples. For the 659 individually genotyped bulls (red and blue dots), there was a much wider range of variation than for cows. However, when the high-density SNP panel was applied (HD, Figure 1b), there was a clear separation of cow samples of within and across two populations. But bulls remained mixed up as low-density results show, with a much narrower range of variation. This indicates that the bulls in the two populations had some degree of relatedness among themselves, but not among the cows. Therefore, the separation of pooled cows would not have been detected if the HD was not used.

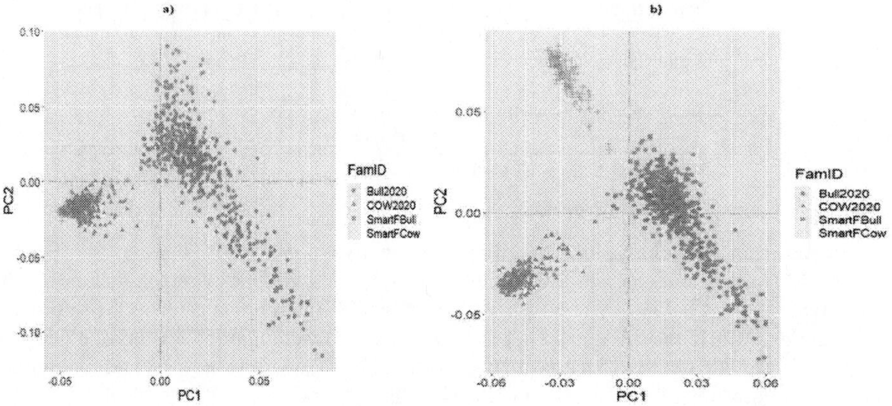


**individually genotyped bulls (Bull2020), 233 were pools of cow DNA samples (COW2020), 177 were individually genotyped SmartF bulls (SmartFBull) and 113 were pools of SmartF cows (SmartFCow). a) 19,089 common SNP; b) High density SNP**

**Genomic predictions of bull's PTO with different panels of SNP density.** Assuming the results from HD are true, Table 2 shows the Pearson's correlations among the PTO GEBVs from three SNP panels (19,089, 54,791 and HD) in the MDH2020 and SmartF respectively. Within MDH202, the correlations between GEBVs of PTO of 482 bulls were 0.74 between 19,089 and HD, and 0.82 between 54,791 and HD. The correlations were much lower (0.39 and 0.45 respectively) if only the top 25% bulls were considered (see Table 2 correlation for top 25%). Similar trends were observed in SmartF when the correlations of GEBVs for 177 bulls were compared (Table 2), despite slightly higher correlations between 19089 and 74584 with HD when the top 25% bulls were selected (0.54-0.59, Table 2). These suggest that if low-density panels were used to genotype pooled DNA cows for estimating the EBVs of PTO of bulls, at least 40-50% of the best bulls would not be selected.

When further investigating the bull GEBVs of PTO estimated using HD, Table 3 illustrates the profiles of the GEBVs of 482 MDH2020 bulls in different quartiles. The average GEBV difference between top and bottom 25% of bulls was 0.292, which is much larger than the difference obtained using low-density panels (0.120 from 19,089 or 0.158 from 54,791, results are not shown here). For animals being dry and empty (score 1) to become wet and pregnant (score 6), there could take conservatively up to 21-27 months to achieve. The GEBV difference of 0.292 from HD would translate into earlier conception by 1.31 months for the female progeny of the top 25% sires.

The study presents preliminary results for the comparison of different panels of SNP density in ranking commercial bulls in two populations. The phenotype score (1-6) of the $2^{nd}$ joining pregnancy test outcome was treated as a continuous trait in which wet and non-pregnant was scored as "5" instead of "2". Further research is underway to explore the impact of different score systems on ranking differences.

**Table 2. Pearson's correlations among GEBVs estimated from using 19,089, 54,791 and HD SNP panels within MDH2020 and SmartF populations, respectively**

| Population | | MDH2020 | | | SmartF | | |
|---|---|---|---|---|---|---|---|
| | SNP | 19089 | 54791 | HD | 19089 | 74584 | HD |
| All bulls | 19089 | 1 | 0.90 | 0.74 | 1 | 0.76 | 0.72 |
| | 54791 / 74584 | | 1 | 0.82 | | 1 | 0.81 |
| | HD | | | 1 | | | 1 |
| Top 25% | 19089 | 1 | 0.81 | 0.39 | 1 | 0.52 | 0.54 |
| | 54791 / 74584 | | 1 | 0.45 | | 1 | 0.59 |
| | HD | | | 1 | | | 1 |

**Table 3. Average genomic breeding values (GEBVs) of progeny pregnancy testing outcome (PTO) of the MDH2020 bulls in four quartiles using HD SNP panel**

| Quartile | # Bulls | Av. GEBV | Min | Max |
|---|---|---|---|---|
| 1 -Top 25% | 120 | 0.136 | 0.0833 | 0.323 |
| 2 | 120 | 0.055 | 0.0275 | 0.0831 |
| 3 | 121 | -0.004 | -0.0341 | 0.0261 |
| 4 – Bot. 25% | 121 | -0.156 | -0.2771 | -0.0345 |
| All | 482 | 0.023 | -0.277 | 0.323 |

## CONCLUSION

This research highlights the need for extreme caution to be taken when applying SNP panels of low or medium densities to study genetic relationships, and rank and select top bulls for commercial beef production based on DNA pooling technology.


## ACKNOWLEDGEMENTS

The authors wish to acknowledge Meat and Livestock Australia (MLA) for their financial support to the project P.PSH.1211 "Validation of pooled DNA gEBV for Brahman commercial cow fertility". We also like to thank Clint Smith, manager of Iffley Station of MDH Pty Ltd, for his great assistance in collecting samples and phenotypic information for the study.


## REFERENCES


Chang C.C., Chow C.C., Tellier L.C.A.M., Vattikuti S., Purcell S.M. and Lee J.J. (2015) *GigaScience*, **4**.

Loh P.-R., Palamara P.F. and Price A.L. (2016). *Nat. Genet.* **48**: 811.

Reverter A., Porto-Neto L.R., Fortes M.R., McCulloch R., Lyons R.E., Moore S., Nicol D., Henshall J., and Lehnert S.A. (2016). *J Anim Sci*. **94**: 4096.

van der Werf J. H. J. (2009). *Proc. Assoc. Advmt. Anim. Breed. Genet.* **18**:38.

*Contributed paper*

# A DETERMINISTIC ALGORITHM FOR OPTIMALITY OF THRESHOLD IN A GWAS EXPERIMENT

**Z. Loh, J. H. J. van der Werf and S. Clark**

School of Environmental and Rural Science, University of New England, Armidale, NSW, Australia

## SUMMARY
While genome-wide association study (GWAS) is an important tool for gene discovery for economic traits in livestock, its use of large numbers of genetic markers necessitates the use of multiple testing correction methods. Several of these methods have been suggested, but their optimality is not as well studied. The aim of this study is to present a deterministic algorithm to provide a framework for estimating the power and false positive rate (FPR) in a GWAS, and using these estimates to test the optimality of these correction method based on the Receiver Operating Characteristic (ROC) curve. This study suggests that both Bonferroni correction and Benjamini-Hochberg False Discovery Rate are overly conservative even if under the assumption of independence between markers.

## INTRODUCTION
Genome-wide association studies (GWAS) are commonly used to identify genes associated with quantitative traits. Due to the increasingly large number of markers used in GWAS however, it had been plagued by an unprecedented level of a multiple testing problem. To avoid the correspondingly increased number of false positives, a multiple testing method that increases the threshold for significance had been utilized in GWAS (Gondro 2015; Tam *et al.* 2019; Visscher *et al.* 2017).

The Bonferroni correction was originally proposed due to its effectiveness in controlling the false positives (Narum 2006), but has since been widely criticized for its conservativeness (de Smet *et al.*, 2004; Narum 2006; Tam *et al.* 2019). Alternative correction methods with reduced stringency in their threshold such as the frequently used Benjamini-Hochberg False Discovery Rate (BH-FDR) method have been suggested. A test on threshold optimality, defined as its ability to optimally balance the power and FPR of GWAS is lacking. Such an optimal threshold may depend on sample size, QTL effect distribution and marker allele frequencies.

The aim of this study is to test the degree of optimality of thresholds provided by Bonferroni and BH-FDR methods under varying relevant parameters. Optimality will be derived from an estimate of power and FPR of a GWAS using a deterministic algorithm, and using these estimates to establish the optimality of these thresholds.

## THEORY
In this study a threshold would be considered optimal if it could balance the power and FPR in a GWAS. Given a threshold $THR$, alongside with effect size of the marker $a$, phenotypic variance $Var(p)$, allele frequency $p$, sample size of GWAS $N$ and number of QTL $nqtl$, the power of GWAS can be defined as follow:

$$power = \frac{Number\ of\ true\ QTLs\ that\ \ exceed\ THR\ \left(-log_{10}(pvalue)\right)}{nqtl}$$

The expected $pvalue$ of a locus could in turn be calculated using the following equation:

$$pvalue = 2 - 2t_{CDF}\left(a\sqrt{\frac{2p(1-p)(N-2)}{Var(P) - 2p(1-p)a^2}}, N-2\right)$$

Where $t_{CDF}(t,n)$ is the cumulative density function (CDF) of Student's t-distribution with test statistic $t$ and degree of freedom $n$. While $nqtl$ is not estimated in this study, deterministic algorithms for this estimation are available with assumption on the distribution of QTL effect sizes, for example see Hall *et al.* (2016). With the same threshold $THR$, the FPR could be defined as follow:

$$FPR = \frac{Number\ of\ null\ marker\ that\ exceed\ THR}{Number\ of\ null\ marker}$$

As this model assumed independence between markers, linkage disequilibrium is not assumed, and null marker are modelled with effect size 0. Modelling of simulated null markers suggested that $FPR$ followed a 1-CDF of gamma distribution with shape and scale parameter of 1 and 0.4344 respectively, and FPR depends only on $THR$. Thus the equation of $FPR$ can be rewritten as follow:

$$FPR = (1 - gamma_{CDF}(THR;\ 1, 0.4344))$$

Where $gamma_{CDF}(x;\ k, \theta)$ is the CDF of gamma distribution at point $x$ with shape and scale parameter $k$ and $\theta$. To test the optimality of $THR$, a receiver operating characteristic (ROC) curve was used. The conventional ROC curve have its FPR and power plotted at x and y-axis, respectively, with optimal threshold being the point where the tangent of the curve equal to 1 (as described by de Smet *et al.* (2004) and mathematically proven by Kaivanto (2008)). Another interpretation which was used in this study, is the difference between number of true and false positives, which represent the numerator of $power$ and $FPR$ respectively. The optimal threshold can then be defined as the argument of the maxima of this differences, where the power is maximized and FPR minimized. This interpretation can also take into account the unequal chance between finding true QTLs and null markers. A sample of this reinterpreted ROC curve would be provided in Figure 1.

## VALIDATION OF THE MODEL

The model was validated through simulation using Python (Version 3.7.3), where the optimality of threshold calculated by Bonferroni and BH-FDR was compared under varying parameters.

A GWAS experiment with $N$ sample size was simulated with a genotype array with $M$ number of independent markers with their allele frequencies following a beta-distribution. A vector of effect sizes was assigned to $nqtl$ markers, which were considered QTL with their effect sizes following a gamma distribution while other markers had effect size of 0. Only markers with effect size of > 0.1 σ were considered in the calculation of power. For all simulations the heritability of the trait was set to 0.3. Using the genotype array, effect sizes and heritability, a vector of phenotypes was calculated, and a GWAS was conducted using Single SNP Linear Regression with the genotype array and phenotype vector. Using Bonferroni correction and BH-FDR at alpha = 0.05, the number of true and false positives were recorded. The ROC score was calculated by subtracting number of false positives from number of true positives. Correction methods with higher ROC score are deemed having its threshold more optimal and provide better balance between power and FPR. This simulation was repeated 200 times. When a parameter is under study the other parameters were kept at the Default Value. The parameters tested are presented in Table 1.

**Table 1. Parameter tested in this experiment**

| Parameter | Default Value | Alternative Value |
|---|---|---|
| Sample Size | 2000 | 800 |
| Number of Markers | 20k | 80k |
| Distribution of QTL Effect Size | Gamma(0.4, 1) | Gamma(0.8, 1) |
| Distribution of Allele Frequency | Beta(0.5, 0.5) | Beta(0.2, 0.2) |
| Number of QTLs | 100 | 2000 |

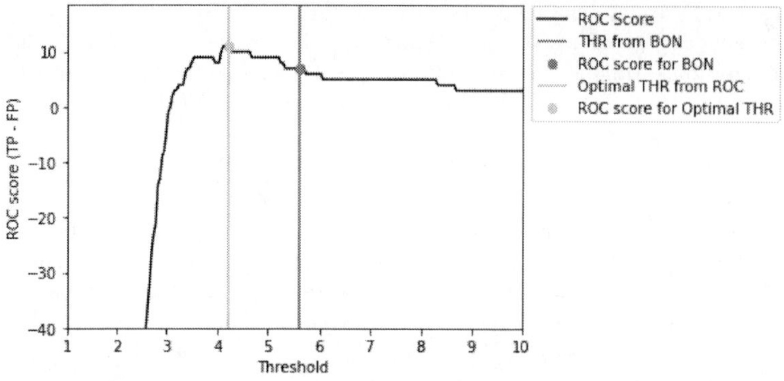


**Figure 1. The reinterpreted ROC curve under default scenario with Bonferroni correction, with TP and FP representing number of true and false positives respectively**

**RESULTS AND DISCUSSION**

The number of true and false positives from each correction method are provided in Table 2, and the ROC score and threshold of each correction methods were provided in Table 3.

**Table 2. The number of true positives (TP) and false positives (FP) for each correction methods under varying parameter values[1]**

| Parameter Tested | Values | Multiple Testing Correction Method | | | | | |
|---|---|---|---|---|---|---|---|
| | | Optimal Threshold from ROC | | Bonferroni Correction | | BH-FDR | |
| | | TP | FP | TP | FP | TP | FP |
| Sample Size (Default)[1] | 2000 | 11.36 | 0.86 | 7.82 | 0.09 | 9.77 | 0.63 |
| (Alternative) | 800 | 4.70 | 0.42 | 2.64 | 0.04 | 3.35 | 0.20 |
| Number of Markers | 80k | 9.53 | 0.72 | 6.92 | 0.05 | 8.30 | 0.44 |
| Distribution of QTL Effect Size | Gamma(0.8,1) | 11.70 | 1.01 | 7.52 | 0.02 | 9.72 | 0.50 |
| Distribution of Allele Frequency | Beta(0.2, 0.2) | 9.96 | 0.61 | 7.37 | 0.07 | 8.84 | 0.48 |
| Number of QTLs | 2000 | 6.77 | 2.31 | 1.08 | 0.02 | 1.47 | 0.07 |

[1] The default values are provided in Table 1.

**Table 3. The threshold (THR) and ROC score for each correction methods under varying parameter values[1]**

| Parameter Tested | Values | Multiple Testing Correction Method | | | | | |
|---|---|---|---|---|---|---|---|
| | | Optimal Threshold from ROC | | Bonferroni Correction | | BH-FDR | |
| | | THR | ROC | THR | ROC | THR | ROC |
| Sample Size (Default)[1] | 2000 | 4.29 | 11.18 | 5.60 | 7.49 | 4.62 | 9.10 |
| (Alternative) | 800 | 4.73 | 4.05 | 5.60 | 2.61 | 5.09 | 2.86 |
| Number of Markers | 80k | 5.06 | 8.82 | 6.20 | 6.88 | 5.28 | 7.86 |
| Distribution of QTL Effect Size | Gamma(0.8,1) | 4.32 | 10.84 | 5.60 | 7.79 | 4.58 | 9.41 |
| Distribution of Allele Frequency | Beta(0.2, 0.2) | 4.46 | 9.35 | 5.60 | 7.30 | 4.64 | 8.36 |
| Number of QTLs | 2000 | 3.93 | 4.58 | 5.60 | 0.93 | 5.34 | 1.24 |

[1] The default values are provided in Table 1.

Compared to both the Bonferroni and the BH-FDR methods, the threshold optimal to the ROC curve has a significantly higher number of false positives in all scenarios, which is associated with a significantly lower threshold. This suggests that the threshold optimal to ROC is less stringent compared to both correction method. Despite this, as suggested by the increased ROC score, the increment of power of GWAS due to the decreased threshold is more significant than the increment of FPR, which could suggest that both Bonferroni correction and BH-FDR are overconservative for all the scenarios in this study.

Between the two existing correction methods, BH-FDR provided a better balance between power and FOR when compared to the Bonferroni correction. While the number of false positives also increased in this correction method, as suggested by Huang *et al.* (2018), the increment in true positives is more significant than the increment of false positives. While with the Bonferroni correction, the power is significantly lower than with BH-FDR, it also had a significantly smaller proportion of false positives. Indeed, the Bonferroni correction had successfully maintained the number of false positives between 0.02 and 0.09 in all scenarios, whereas BH-FDR failed to maintained it in all the scenarios.

While this experiment has illustrated the optimality of threshold from the multiple correction methods, there were several assumptions being made. One of the main assumptions is the independence of the markers, which is unlikely to occur in actual GWAS. Huang *et al.* (2018) suggested threshold from correction methods that assumed independence between markers had increased conservativeness compared to those without such assumption. Despite this, even if this assumption is held, as in this experiment, both correction methods are still overconservative in respect with the optimal threshold. Further study on the effect of correlated markers on the optimality of thresholds from these correction methods would be required.

## CONCLUSION

This study had provided a framework for estimating the power and false positive rate of GWAS using a deterministic algorithm, and using these measures to test the optimality of threshold from two common multiple testing correction methods. This study had demonstrated the excessive conservativeness in both correction methods, especially in Bonferroni correction. The BH-FDR attained a better balance between true and false positives in the setting of independent markers and thus a more optimal threshold. Despite this the optimality of these threshold from correlated markers still warranted further study.

## REFERENCES


de Smet F., Moreau Y., Engelen K., Timmerman D., Vergote I. and de Moor B. (2004) *Br. J. Cancer* **91**: 1160.

Gondro C. (2015). Primer to analysis of genomic data using R. Springer International Publishing, Cham.

Hall D., Hallingbäck H.R. and Wu H.X. (2016) *Tree Genet. Genomes* **12**: 110.

Huang Q.Q., Ritchie S.C., Brozynska M. and Inouye M. (2018) *Nucleic Acids Res.* **46**: e133.

Kaivanto K. (2008) *J. Clin. Epidemiol.* **61**: 516.

Narum S.R. (2006) *Conserv. Genet.* **7**: 783.

Tam V., Patel N., Turcotte M., Bossé Y., Paré G. and Meyre, D. (2019) *Nat. Rev. Genet.* **20**: 467.

Visscher P. M., Wray N. R., Zhang Q., Sklar P., McCarthy M. I., Brown M. A. and Yang J. (2017) *Am. J. Hum. Genet.* **101**: 5

*Contributed paper*

# CURRENT CHALLENGES FOR IMPUTATION OF SNP CHIPS TO WHOLE- GENOME SEQUENCE IN CATTLE AND SHEEP

**I.M. MacLeod[1], S. Bolormaa[1], C.J. Vander Jagt[1], T.V. Nguyen[1], A.J. Chamberlain[1] and H.D. Daetwyler[1,2]**

[1] Agriculture Victoria, AgriBio, Centre for AgriBioscience, Bundoora, VIC, 3083 Australia
[2] School of Applied Systems Biology, La Trobe University, Bundoora, VIC, 3083 Australia

## SUMMARY

Imputation to whole-genome sequence data has been successfully exploited in livestock for fine-mapping causal variants, meta-GWAS and increasing the accuracy of genomic prediction. However, imputation of sequence variants from marker panel (SNP chip) genotypes involves several key challenges that do not generally cause issues for SNP chip level imputation. Here we consider the challenges and potential solutions for issues such as rare variants, sequencing errors, misalignment in regions with large segmental duplications and/or copy number variants.

## INTRODUCTION

Imputation of genotypes to sequence generally requires that target animals first have imputed or real marker panel (SNP chip) genotypes. Then the missing sequence variants between the markers are filled in using a reference set of real sequence genotypes. Imputation algorithms rely on the premise that animals sampled from a population will share a mosaic of haplotypes along the chromosome in common with one or more animals in the population. Even across breeds there are shared haplotypes due to their common ancestral origins. The observed length of the shared haplotypes depends on the marker density, local recombination rates, effective population size and importantly the level of relationships between the target individuals and the reference set. In livestock, it is commonplace to impute genotypes from lower density SNP chips to higher density chips. This imputation is highly accurate using a range of software (Calus *et al.* 2014) and has enabled genomic prediction of breeding values to become routine in the dairy, beef and sheep industries.

Imputation to whole-genome sequence from SNP panel genotypes is routinely undertaken for livestock research. The use of imputed sequence has been demonstrated to enable fine mapping of causal variants (e.g. Pausch *et al.* 2017), to facilitate meta Genome-Wide Association Studies (e.g. Bouwman *et al.* 2018) as well as increasing the accuracy of genomic prediction (e.g. Brøndum *et al.* 2015; Moghaddar *et al.* 2019; Xiang *et al.* 2021).

However, huge challenges remain compared to SNP chip level imputation for several reasons. First, 99% of the sequence variants are missing in high density SNP chip genotypes (HD: ~600k SNP) and the reference sequence data has higher error rates than SNP chip genotypes. This affects the accuracy of determining matching haplotypes between target and reference animals. Second, a large proportion of the sequence variants are less common (Minor Allele Frequency, MAF < 0.01) or rare compared to those selected for industry SNP chips and therefore may not be in strong linkage disequilibrium (LD) with the more common SNP on chips. This leads to inaccuracies for matching target to reference haplotypes. Third, it is costly to develop and maintain large representative sequence reference sets: a task that in addition to sequencing, requires considerable computational resources. Therefore, an attractive solution is for research groups to continue global collaborations to ensure that the databases continue to develop and grow by sharing costs/resources for sequence processing, storage and access.

The aim of this paper is to use examples from our own imputed and real sequence data to demonstrate the impact of some of the above challenges and briefly discuss potential solutions.

## MATERIALS AND METHODS

We imputed sequence data into over 46,000 sheep and over 200,000 cattle using Minimac3 and pre-phased with Eagle software following Pausch *et al.* (2017). The sheep in the target set represented a range of breeds and crosses common to the Australian sheep industry, while the target cattle were dairy breeds and their crosses (mainly Holstein, Jersey and Australian Reds). Both sheep and cattle target populations had been imputed first to HD genotypes (~600k SNP). The sheep sequence reference used for imputation included 726 animals from European breeds and crosses in SheepGenomesDB Run2 (Daetwyler *et al.* 2017). The reference cattle sequences were from *Bos taurus* Run 6 and Run 7 of the 1000 Bull Genomes project (Hayes & Daetwyler 2019) and included 2333 and 3090 animals representing > 50 breeds and crosses. There were several key differences in the Run 6 (Daetwyler *et al.* 2017) and Run 7 pipeline: Run 6 was aligned to the UMD3.1 reference genome, while Run 7 used the improved ARS-UCD1.2 reference genome (Rosen *et al.* 2020). Run7 used GATK v3.8 for variant calling instead of Samtools (Run 6).

Prior to imputation, the variants called in the sheep and cattle reference sequences were pre-filtered to retain only bi-allelic variants (most imputation algorithms do not impute multiallelic variants) with minor allele counts of 4 or more (to remove variants that may be sequencing errors or so rare they cannot be well imputed). Additional pre-filtering was applied in Run 7 where we retained variants with Beagle R2 >0.9 (from the imputation of missing genotypes) and variants in GATK Tranche 99.0 or better. We also identified chromosome segments of ≥ 0.5Mb with excessive heterozygosity among genotyped individuals: i.e. > 2% of variants with heterozygote frequency > 0.55 (maximum expected heterozygosity is 0.5 for neutral loci). These segments generally coincided with regions of large duplications (>1 kb) that generate alignment errors and false SNP calls, therefore variants in these regions with heterozygote frequency >0.5 were removed.

## RESULTS AND DISCUSSION

The pre-imputation filtering of variants in sheep Run 2 and cattle Run 6 reference sequences removed up to 25% of all variants called but this increased to 47% in Run 7, largely due to extra filters imposed. Table 1 compares the proportion of imputed variants above two Minimac R2 thresholds because the Minimac R2 statistic is a good proxy for empirical imputation accuracy (Bolormaa *et al.* 2019). The sheep imputation retained a larger number of imputed variants at Minimac R2 thresholds >0.4 and >0.8 compared to imputed cattle data. This is potentially due to the imputation target sheep having very recent relatives in the reference set compared to the cattle where relationships were more distant between the target and reference sets.

**Table 1. Numbers of variants (M=Millions) imputed into sheep and cattle**

| No. of Variants | Sheep Run2 | Cattle Run 6 | Cattle Run 7 |
|---|---|---|---|
| **Total Imputed** | 40 M (77% of total) | 34 M (75% of total) | 32 M (53% of total) |
| **Minimac R2 > 0.4** | 31 M (77% of imputed) | 18 M (53% of imputed) | 21 M (66% of imputed) |
| **Minimac R2 > 0.8** | 22 M (55% of imputed) | 14 M (41% of imputed) | 19 M (59% of imputed) |

Overall, only 40 to 60% of variants had a MinimacR2 >0.8. The main reason for this is due to the very high proportion of sequence variants with a MAF <0.01 (e.g. Figure 1) that are difficult to impute with accuracy above 0.8 (Pausch *et al.* 2017). Further, we hypothesise that due to purging selection, rare mutations with strong deleterious effects will tend to have arisen relatively recently, and therefore will be more difficult to impute accurately compared to rare variants that have been segregating in the population longer because they have small or neutral effects. Indeed, we found some evidence of this in both sheep (Bolormaa *et al.* 2019) and cattle where for example, missense and frameshift mutations (potentially damaging protein activity) showed a higher proportion of less accurately imputed variants compared to intergenic and intronic variants (Figure 1). In part, we may be able to improve the accuracy of imputation for rare variants by strategies such as skim whole genome sequencing (Daetwyler *et al*, these proceedings) but also by increasing the number of sequenced animals in the reference sets. An increase in the number of animals in cattle Run 7 may have helped increase the number of variants with Minimac R2 >0.8 compared to Run 6 (Table 1). However, other factors including the improved ARS-UCD1.2 reference genome map, different variant calling software and more stringent filtering of variants prior to imputation may also have contributed to the improvement and this will be further evaluated.

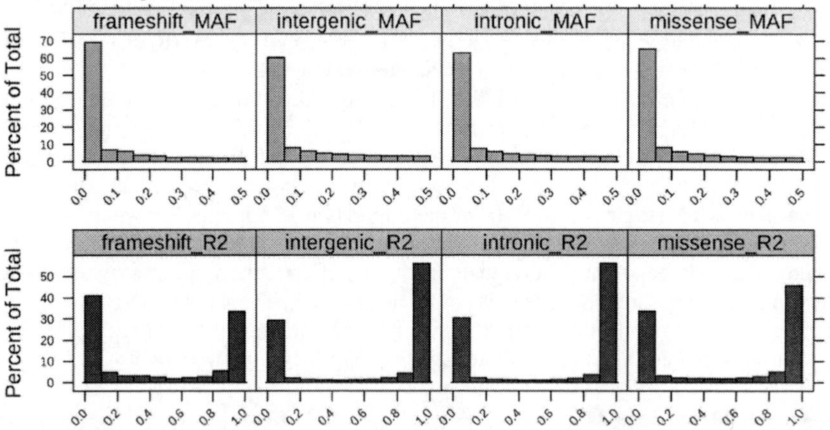


**Figure 1. MAF (Minor Allele Frequency) and Minimac R2 distribution in functional categories of variants from cattle Run7. Frameshift and missense variants show the highest frequency of variants with low imputation R2**

Another important factor causing low sequence imputation accuracy is an erroneous calling of SNP in the reference sequences, for example, due to alignment errors of short-read sequencing. Typically, this more frequently occurs in the many genome-wide regions of up to several Mb long that harbour large segmental repeats (each ≥ 5 kb in length) and/or large structural variants such as copy number variants (CNV) (Liu *et al.* 2010). For example, the major histocompatibility complex region has many segmental duplications and CNV (>86% synteny between cattle and sheep; Gao *et al.* 2010) and across this region the mean empirical accuracy within segments of 1 Mb length drops well below 0.8 in both sheep and cattle (Pausch *et al.* 2017; Bolormaa *et al.* 2019). In these regions, we typically observe excessive heterozygosity among reference sequence variants (i.e. heterozygosity >0.5) (Fig 2). Thus, in Run 7, prior to imputation we filtered out variants with heterozygosity >0.5 in these regions under the assumption that these are false SNP calls and may decrease the imputation accuracy of surrounding variants. As a result, on Chr X the Run 7 pre-imputation filtered variant set included only half the number of variants compared to Run 6 but the number of imputed variants in Run 7 with R2 >0.8 was almost double that of Run 6. Although stringent pre-filtering may be helpful, the low imputation accuracy of these regions (covering >3% of the genome) cannot be fully addressed with the current sequence reference sets because the short sequence reads (~150bp) cannot be accurately aligned, even though the reference genome map may be very accurate. A potential solution is to develop a reference resource where animals are sequenced using long-read technology as well as improved methods to impute large structural variants.

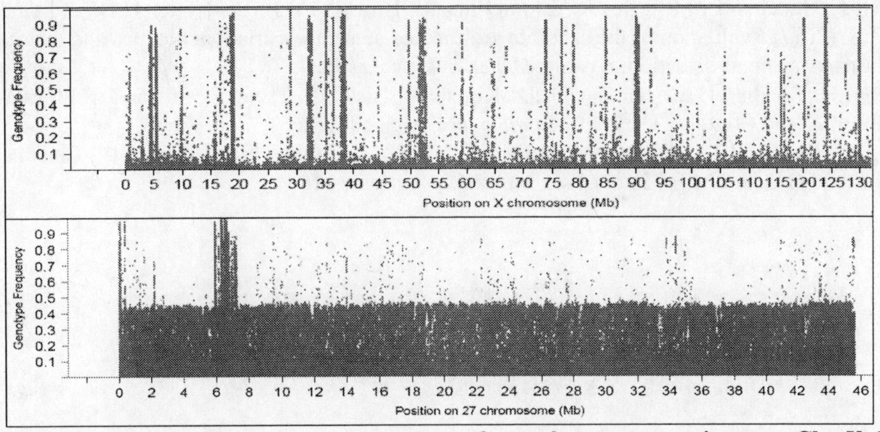


**Figure 2. Frequency of heterozygous genotypes for real sequence variants on Chr X (non-pseudo autosomal region) and Chr 27. The data was derived from 2470 bulls sequenced to > 10x average read depth). Banded regions of excessive heterozygosity (>0 on Chr X and >0.5 on Chr27) coincide with large segmental repeats and copy number variants. On Chr X in addition to bands of high heterozygosity, we also observe ubiquitous random errors across the genome: i.e. these were bull X chromosome sequences that should be haploid, with "homozygous" genotypes**

## CONCLUSIONS

Although imputed sequence has already advanced livestock genomics research there remain considerable challenges: including rare variant imputation and limitations of short-read sequencing.

## ACKNOWLEDGEMENTS

The authors acknowledge financial support from DairyBio, a joint venture between Agriculture Victoria, Dairy Australia and the Gardiner Foundation. We are grateful to the 1000 Bull Genomes Project, SheepGenomesDB, SheepCRC and DataGene for access to data.

## REFERENCES

Bolormaa S., Chamberlain A.J., Khansefid M., Stothard P., Swan A.A., Mason B., Prowse-Wilkins C.P., Duijvesteijn N., Moghaddar N., van der Werf J.H., Daetwyler H.D. and MacLeod I.M. (2019) *Genet. Sel. Evol.* **51**: 1.

Bouwman A.C., Daetwyler H.D., Chamberlain A.J., et al. (2018) *Nature Genet.* **50**, 362.

Brøndum R.F., Su G., Janss L., Sahana G., Guldbrandtsen B., Boichard D. and Lund M.S. (2015) *J. Dairy Sci.* **98**: 4107.

Calus M., Bouwman A., Hickey J., Veerkamp R. & Mulder H. (2014) *Animal* **8**: 1743.

Daetwyler H.D., Brauning R., Chamberlain A.J., McWilliam S., McCulloch A., Vander Jagt C.J., Bolormaa S., Hayes B.J. and Kijas J.W. (2017) *Proc. Assoc. Advmt. Anim. Breed. Genet.* **22**: 201.

Gao J., Liu K., Liu H., Blair H.T., Li G., Chen C., Tan P. and Ma R.Z. (2010) *BMC Gen.* **11**: 466.

Hayes B.J. and Daetwyler H.D. (2019) *Ann. Review Anim. Biosci.* **7**: 89.

Liu G.E., Hou Y., Zhu B., Cardone M.F., Jiang L., Cellamare A., Mitra A., Alexander L.J., Coutinho L.L. and Dell'Aquila M.E. (2010) *Genome Research* **20**: 693.

Moghaddar N., Khansefid M., van der Werf J.H., Bolormaa S., Duijvesteijn N., Clark S.A., Swan A.A., Daetwyler H.D. and MacLeod I.M. (2019) *Genet. Sel. Evol.* **51**: 72.

Pausch H., MacLeod I.M., Fries R., Emmerling R., Bowman P.J., Daetwyler H.D. and Goddard M.E. (2017) Evaluation of the accuracy of imputed sequence variant genotypes and their utility for causal variant detection in cattle. *Genet. Sel. Evol.* **49**: 24.
Rosen B.D., Bickhart D.M., Schnabel R.D., Koren S., Elsik C.G., Tseng E., Rowan T.N., Low W.Y., Zimin A. and Couldrey C. (2020) *GigaScience* **9**: giaa021.
Xiang R., MacLeod I.M., Daetwyler H.D., de Jong G., O'Connor E., Schrooten C., Chamberlain A.J. and Goddard M.E. (2021) *Nature Communications.* **12**: 860.

*Proc. Assoc. Advmt. Anim. Breed. Genet. 24:217-221*

# MACRO- AND MICRO-GENETIC ENVIRONMENTAL SENSITIVITY FOR 400 DAY WEIGHT IN AUSTRALIAN ANGUS

**M.D. Madsen[1], J.H.J. van der Werf[1], V. Börner[2] and S. Clark[1]**

[1] School of Environmental and Rural Science, University of New England
[2] Animal Genetics and Breeding Unit, University of New England

## SUMMARY

Genotype by environment interactions can be caused by both macro- and micro-genetic environmental sensitivity (GES). In the current study, 400 day weight (400DW) measured on Australian Angus was analysed using a variability model and a reaction norm model to obtain estimates for genetic variation due to macro- and micro-GES. The results showed additive genetic variance for both macro- and micro-GES. Over the range of contemporary group means the macro-GES impacted the genetic variance and ranking of sires across environments. The presence of micro-GES indicated the possibility of selecting to reduce the variability of phenotypes, but further investigation into the consequences is needed.

## INTRODUCTION

Genotype by environment interactions (G×E) occur when the phenotypes of different genotypes respond unequally to different environments. The genetic control of G×E is called genetic environmental sensitivity (GES). The environmental differences may be definable, such as temperature, location etc. These environments are termed macro-environments and are typically experienced by a cohort of animals (Falconer and Mackay 1996). Macro-environments are numerous in most livestock populations. Within macro-environments are micro-environments, which are experienced by individual animals and can be observed via differences in variation among progeny (Hill and Mulder 2010). Animals can exhibit GES in response to changes in both macro- and micro-environments, and GES is thus split into macro- and micro-GES.

The aim of this study was to estimate the levels of genetic variation due to macro- and micro-GES for 400 day weight in Australian Angus data.

## MATERIALS AND METHODS

**Data.** Angus Australia provided 400 day weight (400DW) measured in kg on live animals. Contemporary groups (CGs) were constructed by concatenating herd, year, observation date and breeder defined management group for each record (see Graser *et al.* (2005)). The records were then cleaned in four stages. Firstly, all records had to be measured at 301-500 days of age, from animals with known sex, sire and dam and the recorded weight could not be more than 3 standard deviations from the phenotypic mean of its CG. Secondly, repeated measurements were removed by keeping the record belonging to the largest CG out of the available records for that animal. Thirdly, records from animals born prior to 2015 were removed. Lastly, animals with less than 4 paternal half-sibs and animals belonging to CGs with less than 60 animals or to single sire CGs were removed in an iterative procedure, which ensured all 3 criteria were met in the final data set. The final data contained 52,446 400DW records (mean 393.15kg; SD 74.83kg) from 1370 sires (mean number of offspring 38.3; SD 79.2) and 33,201 dams (1.58; 1.43) distributed over 443 CGs (mean number of records 118.39; SD 81.67). The animals were reared across the temperate Australia. The pedigree spanned 13 generations.

**Statistical analysis.** Micro-GES was investigated using a two-step approach described in Mulder *et al.* (2009) where step 1 is a traditional animal model and step 2 is a variability model where the *ln*-transformed squared residual form step 1 was used as the phenotype (Mulder *et al.* 2009). The animal model (step 1) was also used to obtain the estimated environmental effect of CGs, which were used as environmental covariate in a linear reaction norm model to examine macro-GES (Falconer and Mackay 1996).

*Animal model.*

$$\mathbf{Y} = \mathbf{Xb} + \mathbf{Za} + \mathbf{Wc} + \mathbf{e} \quad (1)$$

where $\mathbf{Y}$ was a vector containing the 400DW records, $\mathbf{b}$, $\mathbf{a}$, $\mathbf{c}$ and $\mathbf{e}$ were vectors of fixed effects (age at observation and sex), additive genetic animal effects, random effect of CGs and random residuals, respectively. $\mathbf{X}$, $\mathbf{Z}$, and $\mathbf{W}$ were design matrices linking records to fixed effects, animals and CGs, respectively. The distribution assumptions were $\mathbf{a}\sim N(\mathbf{0}, \sigma_a^2 \otimes \mathbf{A})$, $\mathbf{c}\sim N(\mathbf{0}, \sigma_c^2\mathbf{I_c})$ and $\mathbf{e}\sim N(\mathbf{0}, \sigma_e^2\mathbf{I_e})$, where $\mathbf{A}$ was the numerator relationship matrix and $\mathbf{I_c}$ and $\mathbf{I_e}$ were identity matrices of appropriate dimensions.

*Variability model.*

$$\mathbf{ln(e^2)} = \mathbf{X_v b_v} + \mathbf{Z_v a_v} + \mathbf{e_v} \quad (2)$$

where $\mathbf{ln(e^2)}$ was the *ln*-transformed squared residuals from the animal model, $\mathbf{b_v}$ contained the fixed effects of age at observation and sex, $\mathbf{a_v}$ and $\mathbf{e_v}$ were the additive genetic variance and random residuals of the variability of 400DW. $\mathbf{X_v}$ and $\mathbf{Z_v}$ were design matrices linking records to fixed effects and animals, respectively. The distribution assumptions were $\mathbf{a_v}\sim N(\mathbf{0}, \sigma_{a_v}^2 \otimes \mathbf{A})$ and $\mathbf{e_v}\sim N(\mathbf{0}, \sigma_{e_v}^2\mathbf{I_e})$. The genetic variance estimated in this model ($\sigma_{a_v}^2$) was on the scale of the natural logarithm and thus a conversion was done to obtain the genetic variance of the additive genetic effect contributing to the residual variance $\sigma_{a_R}^2 = \sigma_{a_v}^2\left(\sigma_{a_v}^2 + \sigma_{e_v}^2\right)^{-1} 2(\sigma_e^2)^2$ (Mulder *et al.* 2009).

*Reaction norm model.*

$$\mathbf{Y} = \mathbf{Xb} + \mathbf{Za_{int}} + \mathbf{Ha_{sl}} + \mathbf{Wc} + \mathbf{e} \quad (3)$$

where $\mathbf{a_{int}}$ and $\mathbf{a_{sl}}$ were the additive genetic animal effects for the intercept and slope, respectively, of the reaction norm and $\mathbf{H}$ contained the estimated CG effects. The distribution assumption of the additive genetic effect was $\begin{bmatrix}\mathbf{a_{int}}\\ \mathbf{a_{sl}}\end{bmatrix}\sim N\left(\mathbf{0}, \begin{bmatrix}\sigma_{a_{int}}^2 & \sigma_{a_{int}a_{sl}}\\ \sigma_{a_{sl},a_{int}} & \sigma_{a_{sl}}^2\end{bmatrix} \otimes \mathbf{A}\right)$. The remaining effects and distribution assumptions were as in equation 1.

All analysis was performed in ASReml v4.1 (Gilmour et al., 2015).

**Heritabilities.** The heritability for the animal model was $h^2 = \frac{\sigma_a^2}{\sigma_a^2+\sigma_e^2}$. The heritability of the residual was $h_R^2 = \frac{\sigma_{a_R}^2}{3\sigma_{a_R}^2+2(\sigma_a^2+\sigma_e^2)^2}$ (Mulder *et al.* 2009). The heritability of the reaction norm model was only calculated for the average environment, i.e. replacing $\sigma_a^2$ with $\sigma_{a_{int}}^2$ in the formula given for the animal model.

## RESULTS AND DISCUSSION

Results in Table 1 show additive genetic variance due to both macro- and micro-GES. The variation due to macro-GES (slope of reaction norm) were relatively low when compared to the intercept. However, while it is often assumed that breeding stock is exposed to similar environmental conditions across cohorts, we found that the mean value of CGs ranged from -149 to 173kg. The variation due to CG ($\sigma_c^2$) was 2399kg$^2$ and thus the standardised estimated range of CG effects were -3.04–3.52$\sigma_c$. Over a given environmental range it is commonly assumed that the bulk of the data is present in non-extreme environments, resulting in low accuracy of estimated environmental effects. While the bulk of CGs have effects in non-extreme environments (Figure 1), the data filtration in the current study has resulted in a significant number of animals in all environments, ensuring accurate estimation of CG effects across the full range. Across a large range of environmental effects even a low genetic variance due to macro-GES can have significant impacts on the additive genetic variation across environments (Figure 2). The presence of macro-GES can result in scaling effects and/or re-ranking (Falconer and Mackay 1996). Scaling effects are differences in variance across macro-environments, which is of statistical concern and should be accounted for during analysis e.g. by using a reaction norm model. Re-ranking is of more practical concern since it occurs when animals are superior to others in one environment, but not in another. The estimated breeding values (EBVs) of the five most influential sires estimated with the reaction norm model show both scaling and re-ranking effects across environments (Figure 3). The sire represented by the grey line is the second poorest performer in the -3.0$\sigma_c$ environment and the best in the 3.5$\sigma_c$ environment, while the red sire performs consistently better than the black, blue, and green sires. If these sires were evaluated without consideration to macro-GES the red sire would be considered the best of the 5 sires (legend of Figure 3).

**Table 1. Additive genetic variance (SE) from the animal model ($\sigma^2_a$) and the variability model ($\sigma^2_{a_v}$) and the additive genetic variance of intercept ($\sigma^2_{a_{int}}$) and slope ($\sigma^2_{a_{sl}}$) from the reaction norm model**

| Model* | $\sigma^2_a$ | $\sigma^2_{a_v}$ | $\sigma^2_{a_R}$ | $\sigma^2_{a_{int}}$ | $\sigma^2_{a_{sl}}$ | $\sigma_{a_{sl},a_{int}}$ | $h^2$ | $h^2_v$ | $h^2_R$ |
|---|---|---|---|---|---|---|---|---|---|
| Animal | 509.07 (19.26) | - | - | - | - | - | 0.43 | - | - |
| Variability | - | 0.59 (0.05) | 96937.08 | - | - | - | - | 0.11 | 0.03 |
| Reaction norm | - | - | - | 473.65 | 0.12 (0.00) | 0.44 (0.13) | 0.45 | - | - |

*the units for $\sigma^2_a$, $\sigma^2_{a_{int}}$, and $\sigma^2_{a_{sl}}$ were kg$^2$, for $\sigma^2_{a_v}$ the unit was kg$^4$, and the unit for $\sigma_{a_{sl},a_{int}}$ was kg.

The genetic correlation between intercept and slope was only 0.06 meaning there was little association between the breeding value for the level and the macro-GES. It should thus be possible to select animals with high EBV for intercept and low EBV for slope. This would be relevant if breeders wish to breed for high producing, robust animals, i.e. animals that are less sensitive to changes in macro-environments and thus performs similarly in all environments. However, if a breeder is consistently providing a superior environment for their animals it may be relevant to select on environmental specific EBVs to ensure maximum profit.

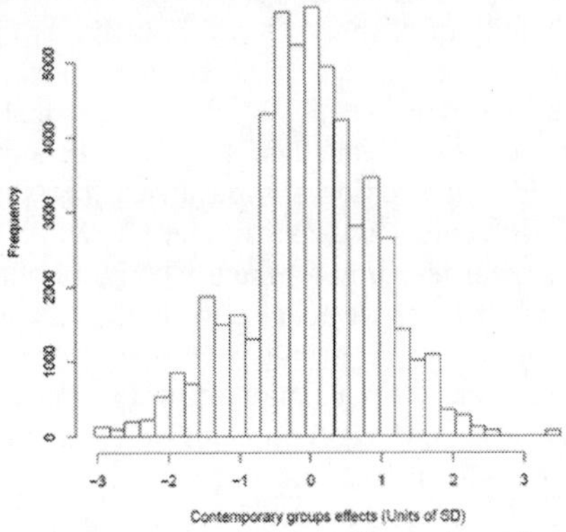


**Figure 1. Frequency of the contemporary group effects (standardised)**

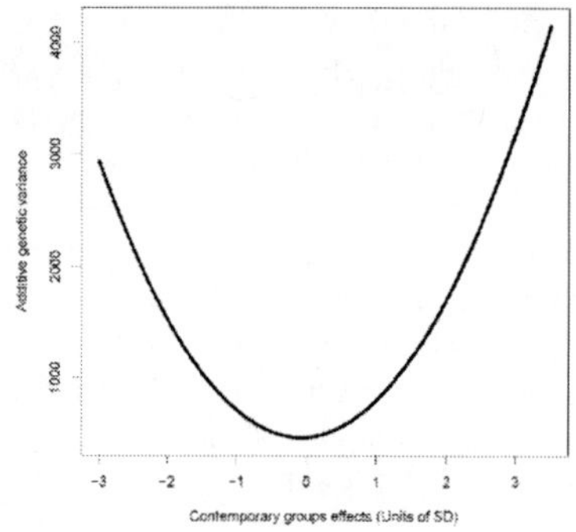


**Figure 2. Additive genetic variance across the contemporary group effects (standardised)**

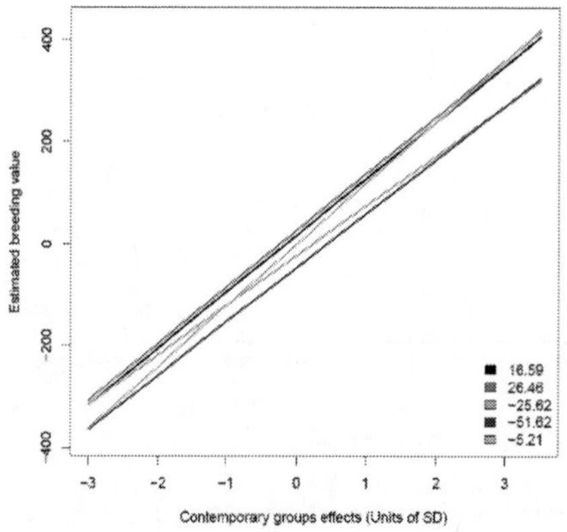


**Figure 3. Estimated breeding values (EBVs) of the 5 most influential sires. Lines represent EBVs from the reaction norm model plotted across the contemporary group effects (standardised). Legend shows the corresponding EBVs from the animal model**

Micro-GES affects the variability of phenotypes within macro-environments. A ten generation divergent selection experiment on litter size in rabbits have shown that selection to alter the variability of phenotypes is possible (Blasco et al., 2017). Thus, reducing micro-GES could reduce the variability and ensure more uniform production. This is especially relevant for traits, such as body weight in broilers, where the final product is penalised for falling outside a desired range (Mulder et al., 2009), i.e. traits with a non-linear profit margin. While 400DW itself does not have a non-linear profit margin it is an indicator trait for mature body weight and carcass weight, both of which may be penalised as slaughterhouses are not able to handle very small or overly large animals. The relatively high estimated variation due to micro-GES in 400DW showed that it should be possible to reduce the variation around the population mean for this trait, thus reducing the risk of the animals falling outside of the desired range for mature weight and carcass weight.

It has been shown that the variability model used in the current study has lower prediction ability than a double hierarchical generalised linear model (DHGLM) for estimation of micro-GES. Iung et al. (2017) observed lower accuracies of EBVs, partly because a DHGLM allows for estimation of the genetic correlation between $\sigma_a^2$ and $\sigma_{a_v}^2$. However, Iung et al. (2017) did not find significant differences between estimated variances. A DHGLM was not fitted in the current study due to the more stringent data structure requirements compared to variability models, but further research will be done to try and apply the DHGLM to the data and examine the difference between the two models.

## CONCLUSION

In conclusion, the analysis showed evidence of macro-GES in 400 d weight in Australian Angus causing re-ranking across environments amongst the five most influential sires. It would therefore be possible to select on macro-GES to either reduce the overall impacts of changes in macro-environments or to ensure high performance in specific macro-environments. Considerable levels of micro-GES were also present in 400 day weight, showing the potential to increase uniformity, but further research is needed to improve the analysis and investigate the outcomes of selection on micro-GES.

## REFERENCES

Blasco, A., M.-Á. Marina, G. Maria-Luz, I.-E. Noelia, and A. María-José. (2017) *Genet. Sel. Evol.* **49(1)**:1.

Falconer, D. S., and T. F. C. Mackay. (1996) Introduction to quantitative genetics. Pearson Education Limited, Essex, United Kingdom.

Gilmour, A. R., B. J. Gogel, B. R. Cullis, S. J. Welham, and R. Thompson. (2015) ASReml User Guide Release 4.1 Functional Specification,. VSN International Ltd, Hemel Hempstead, HP1 1ES, UK

Graser, H. U., B. Tier, D. J. Johnston, and S. A. Barwick. (2005) *Anim. Prod. Sci.* **45(8)**:913.

Hill, W. G., and H. A. Mulder. (2010) *Genet. Res.* **92(5-6)**:381.

Iung, L. H. S., H. H. R. Neves, H. A. Mulder, and R. Carvalheiro. (2017) *J Anim. Sci.* **95(4)**:1425.

Mulder, H. A., W. G. Hill, A. Vereijken, and R. F. Veerkamp. (2009). *Animal* **3(12)**:1673.

*Contributed paper*

# EFFECT OF BOVINE REFERENCE GENOME CHOICE IN RNA-SEQ ALIGNMENT AND DIFFERENTIAL GENE EXPRESSION ANALYSIS IN BRANGUS CATTLE

**E. Mantilla Valdivieso[1], E. Ross[1], A. Raza[1], B. Hayes[1], N. Jonsson[2], P. James[1] and A. Tabor[1]**

[1] Queensland Alliance for Agriculture and Food Innovation, University of Queensland, St. Lucia, QLD, 4072 Australia

[2] Institute of Biodiversity Animal Health and Comparative Medicine, University of Glasgow, Glasgow, United Kingdom

## SUMMARY

New and improved assemblies for bovine genomes have been released in the past two years, contributing to the growing field of livestock genomic information, but they still require to be more comprehensively evaluated in RNA-seq bioinformatic pipelines in terms of their reproducibility in mapping and differential gene expression analysis. The present study aimed to evaluate these parameters by mapping Brangus-derived leukocyte sequence data to three bovine reference genome assemblies (Hereford, Brahman, and Angus) in order to find differentially expressed genes related to ectoparasite host resistance. We observed similar mapping rates across the three genome assemblies and a similar number of differentially expressed genes (DEGs) detected with each genome (84-86 genes). However, using haplotype-resolved genomes (Angus and Brahman) was found to be important to discover an additional 45 DEGs that could not be identified with the non–haplotype-resolved Hereford reference genome.

## INTRODUCTION

High-throughput RNA sequencing technology (RNA-Seq) is currently the most powerful approach for profiling transcriptomes and identifying differentially expressed genes (DEGs) between experimental conditions (Wang *et al.* 2009). This technology is now extensively applied in the field of animal research, particularly to better understand the mechanisms responsible for genetic variation in complex phenotypes in livestock (Georges *et al.* 2019). In cattle, for instance, genetic improvement to enhance traits such as host resistance against parasites is highly desirable since the reduction of parasitic burden can improve animal welfare and increase productivity (Tabor *et al.* 2017). Ectoparasites such as the cattle tick (*Rhipicephalus microplus* species complex) represent a major animal health challenge for the cattle industry; thus, finding effective ways to control tick infestations is a priority for producers.

One of the most feasible options to protect cattle herds from ticks is through the use of tick-resistant breeds which have *Bos indicus* genetics, as *Bos taurus* breeds are mostly susceptible (Utech *et al.* 1978). Crossbred cattle (*B. indicus x B. taurus*), such as Brangus, have more desirable meat quality than purebred *Bos indicus* but exhibit a range of tick-resistant and susceptible phenotypes. On top of this, targeting host resistance for genetic improvement is challenging because the underlying biological mechanisms are not yet fully understood (Tabor *et al.* 2017). Previous work suggests that variation in immune gene expression can contribute to the variation in the phenotype (Piper *et al.* 2010). Therefore, it is hypothesised that biomarker discovery by differential gene expression analysis could provide feasible opportunities for selecting for tick-resistant hosts in cattle with *Bos taurus* content.

The accurate quantification of gene expression heavily relies on the availability of high-quality genomes and their corresponding annotations (Oshlack *et al.* 2010). Currently, the *Bos taurus* ARS_UCD1.2 assembly (Rosen *et al.* 2020), which originated from an inbred Hereford animal, is widely accepted as the reference genome for taurine and indicine cattle. However, Low *et al.* (2020) released two novel reference-quality assemblies UOA_Angus_1 and UOA_Brahman_1 from Angus and Brahman parental haplotypes of an F1 *B. taurus x B. indicus* hybrid (Brangus), which provides the opportunity to further study breed-specific gene expression patterns that could be related to the expression of host resistance. Therefore, this study aimed to investigate if the choice of bovine reference genome (Hereford, Angus, and Brahman) may affect the mapping rate of short-read sequencing data and produce substantial differences in downstream differential gene expression analysis in circulating leukocytes from Brangus cattle of high and low resistance to tick infestation.

## MATERIALS AND METHODS

**Animals.** 30 Brangus steers (~9 months old) without previous exposure to ticks were recruited for this study conducted under animal ethics approval (QAAFI/469/18). The animals were exposed to artificial infestation with approximately 10,000 tick larvae (*R. australis*) over 12 weeks, during which animals were ranked for their resistance to infestation and blood samples were collected. The number of developing adult ticks after an infestation cycle (21 days) was estimated with a tick scoring scale from 1 (<50 ticks = Resistant) to 5 (>300 ticks = Susceptible). The animals subsequently classified as the most resistant (R, n=3), and most susceptible (S, n=5) hosts were selected for RNA sequencing of leukocytes isolated from blood collected immediately before primary infestation.

**RNA extraction and sequencing.** RNA was extracted from frozen leukocytes in Qiazol reagent with the miRNeasy mini kit (QIAGEN, USA) as per manufacturer's instructions. RNA samples were treated with DNAse and RNA was quantified using the Nanodrop 2000 (Thermofisher, USA). The RNA RIN quality analysis was evaluated with the 2100 Bioanalyzer Instrument (Agilent Technologies, USA). The cDNA libraries were prepared with the TruSeq Stranded mRNA kit and sequenced as 100 bp single-end reads in one flow cell lane on the Illumina NovaSeq 6000 sequencer (Illumina, USA) through the Australian Genome Research Facility.

**Bioinformatics pipeline**. The RNA-Seq pipeline for this study is shown in Figure 1. Briefly, read quality control was performed with FastQC v.0.11.4 (Andrews 2015) and adapters were removed with Trimmomatic v.0.35 (Bolger *et al.* 2014). The reads were mapped with STAR .2.5.2b (Dobin *et al.* 2012) to the ARS-UCD1.2 (Rosen *et al.* 2020), UOA_Angus_1, and UOA_Brahman_1 (Low *et al.* 2020) assemblies. Genomes and annotations were sourced from the Ensembl Release 102 (https://asia.ensembl.org). The gene count matrices were processed in RStudio with the edgeR Bioconductor package (Robinson *et al.* 2009). A generalized linear model was fitted to test phenotype (S vs. R) as the main factor with sample RIN number as a covariate. Differentially expressed genes (DEGs) were considered significant based on a false discovery rate (FDR) < 0.05 and $|\log_2 \text{(fold change)}| > 1$.

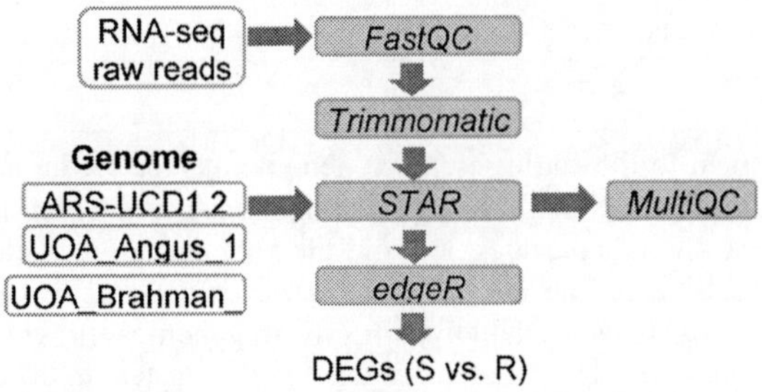


**Figure 1. RNA-Seq pipeline for differential gene expression in leukocytes of tick-susceptible (S) compared to resistant (R) Brangus cattle pre-infestation**

## RESULTS AND DISCUSSION

**RNA-Seq mapping.** The sequencing produced an average of 36.2 million raw single-end reads per sample. After the adapter trimming and QC steps, the average number of reads per sample was 35.5 million. The percentage of reads that uniquely mapped to the ARS_UCD1.2 genome was higher by approximately 4% and 6% compared to the UOA_Brahman_1 and UOA_Angus_1 genomes, respectively (Table 1). Additionally, the percentage of multi-mapped reads was between 5.1 and 7.1 across all three genomes, but a larger proportion (4%) of unmapped reads was obtained with the Angus genome. Therefore, for this Brangus-derived transcriptomic dataset, mapping rates were consistently high with all three genomes, but the performance of the STAR aligner improved slightly when using the Hereford assembly.

**Differential gene expression.** In total, 131 DEGs were identified in the circulatory leukocytes from tick-resistant compared to tick-susceptible Brangus with all three bovine reference genomes (Figure 2). Of these genes, 51 (38.9%) were commonly identified by all three genomes, 47 (35.9%) were common to the taurine genomes (ARS_UCD1.2 and UOA_Angus_1), and 19 (14.5%) were unique to the indicine genome (UOA_Brahman_1). Overall, mapping our sequencing data to the haplotype-resolved reference genomes was useful to identify an additional 45 DEGs that otherwise could not have been identified by the ARS_UCD1.2 genome alone; however, many of these genes did not have full annotations. This result further highlights the need for an improved gene annotation pipeline for both the UOA_Brahman_1 and UOA_Angus_1 assemblies, particularly to be able to characterise indicine-derived DEGs and their relevance in conferring host resistance against ticks.

Moreover, it was found that choice of reference genome did not significantly alter the total number of genes that were differentially expressed in the two phenotypes of host resistance (susceptible vs. resistant), but the number of up- and down-regulated genes varied slightly for each reference genome (Table 1).

**Table 1. RNA-seq mapping results (%) for three bovine reference genomes and the resulting number of differentially expressed genes (DEGs) in leukocytes from tick-susceptible compared to tick-resistant Brangus cattle**

| | Hereford<br>ARS_UCD1.2 | Angus<br>UOA_Angus_1 | Brahman<br>UOA_Brahman_1 |
|---|---|---|---|
| **Uniquely mapped reads** | 94.07 | 88.63 | 90.43 |
| **Multi-mapped reads** | 5.15 | 6.56 | 7.07 |
| **Unmapped reads** | 0.42 | 4.46 | 2.14 |
| **Total no. of DEGs** | 86 | 84 | 84 |
| Up-regulated | 33 | 26 | 20 |
| Down-regulated | 53 | 58 | 64 |

## CONCLUSIONS

Continuous improvement to the cattle reference genome has led to the latest release of the *B. taurus* ARS_UCD1.2 assembly. Although this is generally considered a high-quality assembly, it is based on an inbred taurine animal and does not hold the potential to characterise all the variation that exists in other cattle subspecies, i.e. *B.t. indicus, B.t. africanus*, and crosses thereof (Low *et al.* 2020). The UOA_Brahman/Angus_1 haplotype-resolved genomes provide an opportunity to address these concerns, but they have not been extensively tested in RNA-Seq bioinformatic pipelines. This study explored how the choice of reference genome input can influence short-read mapping and differential gene expression in leukocyte transcriptomic data from crossbred Brangus cattle.

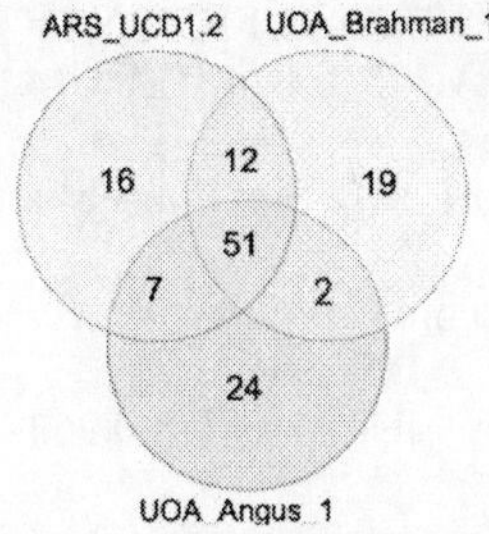


**Figure 2. Venn diagram showing the number of unique and overlapping DEGs (tick-susceptible vs. -resistant Brangus) detected from three bovine reference genomes.**

It was found that the choice of bovine reference genome had a mild effect on read mapping and different gene expression detection, likely reflecting on the very high-quality of all three genomes. Importantly, using haplotype-resolved genomes allowed the detection of additional DEGs that appeared to be specific to the indicine and taurine components of the Brangus breed (an Angus and Brahman cross). However, many of these genes are yet to be fully annotated, thus, further gene overlap could still be expected in addition to 51/131 DEGs discovered with all three genomes, once gene annotations pipelines improve. Further work on characterising which unannotated up- and down-regulated DEGs are homologous to other ARS_UCD1.2 sequences or orthologous to other species (human or rat) will be the first step towards elucidating these novel genes and potentially shed light on the biological mechanisms underlying tick host resistance. Ultimately, testing a variety of high-quality genome resources in well-established bioinformatic pipelines such as RNA-Seq can greatly improve interpretations from transcriptomic data, particularly if the end goal is discovering biomarkers that can assist for genetic improvement of a wider range of cattle breeds.

**ACKNOWLEDGMENTS**

Meat and livestock Australia (MLA) Donor Company project funding P.PSH.0798, and The University of Queensland RTP Scholarship for funding E. Mantilla's candidature.

**REFERENCES**

Andrews, S. (2015) *FastQC A quality control tool for high throughput sequence data.*; Available from: http://www.bioinformatics.babraham.ac.uk/projects/fastqc.

Bolger, A.M., Lohse, M., Usadel, B. (2014) *Bioinformatics* **30**(15): 2114.

Dobin A., Davis C.A., Schlesinger F., Drenkow J., Zaleski C., Jha S., Batut P., Chaisson M., and Gingeras T.R. (2012). *Bioinformatics* **29**:15.

Georges M., Charlier C., and Hayes B. (2019). *Nat Rev Genet* **20**: 135.

Low W.Y., Tearle R., Liu R., Koren S., Rhie A., Bickhart D.M., Rosen B.D., Kronenberg Z.N., Kingan S.B., Tseng E., *et al.* (2020) *Nat Commun* **11**: 2071.

Oshlack A., Robinson M.D., and Young M.D. (2010) *Genome Biol* **11**: 220.

Piper E.K., Jackson L.A., Bielefeldt-Ohmann H., Gondro C., Lew-Tabor A.E., and Jonsson N.N. (2010) *Int J Parasitol* **40**: 431.

Tabor A.E., Ali A., Rehman G., Garcia G., Zangirolamo A., Malardo T., and Jonsson N.N. (2017) *Front Cell Infect Microbiol* **7**: 1.

Robinson M.D., McCarthy D.J., and Smyth, G.K. (2009) *Bioinformatics* **26**: 139.

Rosen B.D., Bickhart D.M., Schnabel R.D., Koren S., Elsik C.G., Tseng E., Rowan T.N., Low W.Y., Zimin A., Couldrey C., *et al.* (2020) *Gigascience* **9**.

Utech K.B., Wharton R.H., and Kerr J.D. (1978) *Aus J Agric Res* **29**: 885.

Wang Z., Gerstein M., and Snyder M. (2009). *Nat Rev Genet* **10**: 57.

*Contributed paper*

# GENETIC PARAMETERS FOR FEED EFFICIENCY AND WEIGHT IN JERSEY COWS USING 3D CAMERAS IN COMMERCIAL DANISH FARMS

**C.I.V. Manzanilla-Pech[1], R.B. Stephansen[1] and J. Lassen[1, 2]**

[1] Center for Quantitative Genetics and Genomics, Aarhus University, Blichers Alle 20, 8830 Tjele, Denmark

[2] Viking Genetics, Ebeltoftvej 16, Assentoft, 8960 Randers, Denmark

## SUMMARY

A challenge of including feed intake in a breeding goal is to have sufficient phenotypic records of feed intake, given how difficult it is to measure on an individual cow basis. With new tools available, such as 3D cameras, this problem might be overcome. This is a preliminary study on estimating genetic parameters for dry matter intake (DMI) and body weight (BW) measured using 3D cameras to posteriorly calculate residual feed intake (RFI). A total of 24,746 weekly records of DMI and BW recorded from 3D cameras during 2019-2021 were available from 963 commercial Danish Jersey cows. These weekly records were complemented with milk and milk content records for the same period, and energy corrected milk (ECM) was calculated. Residual feed intake was calculated as the partial regression of dry matter intake on energy sinks (Tempelman et al., 2015). Estimated heritabilities were 0.08 (RFI), 0.18 (DMI), 0.35 (BW) and 0.29 (ECM). Genetic correlations between DMI with ECM (0.69) were highly positive and DMI with BW (0.37) were moderate positive. Genetic correlations of RFI and DMI were highly positive (0.90), whereas between RFI and BW (0.12) and ECM (0.39) were low to moderate with large standard errors. Phenotypic correlations of RFI with ECM and RFI with BW were close to zero as expected, whereas, between RFI and DMI were close to one. With these results, we conclude that feed efficiency (RFI) calculated using DMI and BW measured by 3D cameras is heritable. Given that DMI and BW were measured only on 963 animals in four commercial farms, adding more farms, animals and records may change the genetic parameters for DMI, BW and RFI.

## INTRODUCTION

In the last decade, several countries have included feed efficiency in their breeding goal (Veerkamp *et al.* 2014; Pryce *et al.* 2015). The Saved feed index in the Nordic Total Merit Index (NTM; NAV, 2020) now includes the breeding value for feed efficiency (also called metabolic efficiency) and maintenance (Lidauer *et al.* 2019). Residual feed intake (RFI) has been proposed as proxy trait for feed efficiency in several species including cattle, pig and poultry (Martin *et al.* 2021). Residual feed intake is commonly defined as the difference between the actual measured feed intake and the expected feed intake, and is a measure of how efficiently a cow utilizes the feed consumed. One way to calculate RFI is as the partial regression of dry matter intake on energy sinks (energy-corrected milk; ECM, metabolic body weight; BW, and BW change; Tempelman *et al.* 2015).

As dry matter intake (DMI) and BW records are required to calculate RFI, new technologies are being developed to fulfil the demand of individual records in an easy way. Using artificial intelligence and 3D cameras, the Cattle Feed InTake System (CFIT, Viking Genetics, 2020; Lassen *et al.* 2018) is one of the latest alternative tools to record feed intake and BW. Through CFIT 3D cameras located in barns can identify individual cows and using artificial intelligence algorithms, record individual DMI and BW for the entire herd. However, as with every new phenotype, the traits (DMI and BW) need to be quantified genetically and determine its associations with other traits (such as ECM). The first CFIT data are now available for Jersey cows. In this study, we estimated the genetic parameters including genetic correlations between DMI, BW (obtained by 3D cameras) and ECM in 963 Danish Jersey commercial cows, to posteriorly, calculate RFI.

## MATERIALS AND METHODS

**Data.** The data included 24,746 weekly records of DMI and BW from 963 Danish Jersey cows. Only data from 1st to 6th parity from the first 330 days in milk was utilized in the genetic evaluation of RFI. The Jersey cows were recorded for DMI and BW using 3D cameras technology on four commercial farms in Denmark during 2019-2021. The cows were fed with a total mixed ration diet that mainly consisted of maize silage, grass silage and concentrates. The cameras were located above the feeding area floor, and the cows were recorded when eating. An algorithm based on artificial intelligence identify the cows and translate their 3D images into phenotypes (DMI and BW). Lassen *et al.* (2018) have presented a complete description of the 3D cameras methodology to measure DMI. Body weight is also predicted using 3D images of the back of the cow (paper in preparation). From these images, couvertures of the back were obtained. Using a PLS method, a prediction model was developed based on scale measures of the cows. The prediction was done with high accuracy (0.9) and RMSE of 18 kilo. Posteriorly DMI and BW weekly averages were calculated. Weekly milk yield and content were available through the national milk recording system. Energy corrected milk was calculated using the following formula (Sjaunja *et al.* 1991), ECM (kg) = 0.25 Milk (kg) + 12.2 Fat content (kg) + 7.7 Protein content (kg). Residual feed intake (RFI) was the residual of the partial regression of DMI on metabolic BW (MBW), ECM and body weight change (ΔBW) (according to the two-steps RFI from Tempelman *et al.* 2015), along with fixed effects described posteriorly in the model. Metabolic BW (MBW) was defined as $BW^{0.75}$. Body weight change (ΔBW) is described as change in kg per day.

**RFI calculation.** The model used to calculate RFI is the one used to calculate the breeding values for RFI by the Nordic Cattle Genetic Evaluation (NAV; Stephansen *et al.* 2021):

$$y_{ijklmn} = \mu + HTYS_i + LW_j + ACC_k(P) + ACC^2{}_l(P) + \text{YSLACP}_m + ECM + MBW + \Delta BW + e_{ijklmn}$$

where $y_{ijklmn}$ is the phenotype for RFI; μ is the mean; HTYS is the fixed effect *i* for herd-trial-year-season; LW is the fixed effect *j* for week of lactation; ACC (P) is the fixed effect of the *k* age of cow at calving with parity nested, ACC2 (P) is the fixed effect of the *l* age of cow at calving squared with parity nested; YSLACP is the fixed effect *m* for year-season-lactation period, ECM is the regression on energy corrected milk, MBW is the regression of metabolic body weight, ΔBW is the regression of body weight change.

**Statistical analyses**. A univariate animal model for repeated measures was performed to estimate the variance and covariance components using DMU software (Madsen and Jensen 2014). The model used to estimate the variance components for DMI, BW, and ECM was:

$$y_{ijklmnop} = \mu + HTYS_i + LW_j + ACC_k(P) + ACC^2{}_l(P) + \text{YSLACP}_m + a_n + pe_o + e_{ijklmnop}$$

where $y_{ijklmnop}$ is the phenotype for DMI, BW, ECM; μ is the mean; HTYS is the fixed effect *i* for herd-trial-year-season; LW is the fixed effect *j* for week of lactation; ACC (P) is the fixed effect of the *k* age of cow at calving with parity nested, $ACC^2$ (P) is the fixed effect of the *l* age of cow at calving squared with parity nested; YSLACP is the fixed effect *m* for year-season-lactation period. Random effects are as follows: a is the additive genetic effect *n* distributed as N (0, $A\sigma^2 a$), in which A is the pedigree relationship matrix and $\sigma^2 a$ is the genetic variance, pe the permanent environmental effect o (within and across parities) distributed as N (0, $I\sigma^2 pe$), in which I is an identity matrix and $\sigma^2 pe$ is the permanent environmental variance and e is the residual effect *p* of $y_{ijklmno}$. To estimate the genetic correlations, pairwise bivariate models between all four traits were fitted. The pedigree included 6,903 animals up to 5 generations. The model to estimate variance components for RFI only included the additive genetic effect, the permanent environmental effect and the residual as RFI has been previously adjusted by fixed effects.

## RESULTS AND DISCUSSION

**Descriptive statistics.** Descriptive statistics for DMI, BW, ECM and RFI in Danish Jersey cows are presented in Table 1. The mean DMI was 21.88 kg with a phenotypic standard deviation of 3.87 kg, whereas, the mean BW was 467.98 kg with standard deviation of 41.73 kg. Both were slightly higher than averages reported previously in literature for Jersey cows by Li *et al.* (2018; primiparous cows) and Halachmi *et al.* (2011).

**Table 1. Descriptive statistics for dry matter intake (DMI), body weight (BW), energy corrected milk (ECM), residual feed intake (RFI) in Danish Jersey cows**

| Trait | No. of records | No. of animals | Mean | SD | Min | Max | CV (%) |
|---|---|---|---|---|---|---|---|
| DMI | 24,746 | 963 | 21.88 | 3.87 | 8.21 | 36.55 | 18 |
| BW | 24,746 | 963 | 467.98 | 41.73 | 312.0 | 603.0 | 9 |
| ECM | 24,746 | 963 | 33.99 | 6.97 | 4.05 | 55.88 | 20 |
| RFI | 24,746 | 963 | 0.00 | 3.10 | -14.64 | 14.45 | -- |

SD=standard deviation, CV= coefficient of variation.

**Genetic parameters.** There are few studies available reporting genetic parameters for feed intake and feed efficiency in Jersey cows. Genetic and permanent environmental variances, heritabilities, genetic correlations, and phenotypic correlations of DMI, BW, ECM and RFI in Danish Jersey cows are shown in Table 2. The genetic variance for DMI in this study was slightly higher (2.11) than previously reported by Li *et al.* (2016) who reported a range from 0.6 to 1.8 (depending on the lactation stage of Jersey cows), however DMI heritability (0.18) was within the range reported (0.17 to 0.52). Likewise, heritability for ECM was within the range of values (0.14-0.53) previously reported by Ulutas *et al.* (2008), Sabedo *et al.* (2018) and Li *et al.* (2018) in primiparous Jersey cows. Estimated heritability for BW was slightly lower than the values (0.46-0.61) reported by Li *et al.* (2018) across lactation stages. Genetic variance for RFI was in the range (0.4-1.4) reported by Li *et al.* (2017) in Holstein cows, same than the permanent environmental variance reported values (1-3.5). However, heritabilities were higher (0.10-0.23) than the reported in this study (0.08).

**Table 2. Genetic and phenotypic variances, heritabilities (diagonal), genetic correlations (lower diagonal) and phenotypic correlations (upper diagonal) of dry matter intake (DMI), body weight (BW), energy corrected milk (ECM), residual feed intake (RFI) in Danish Jersey cows**

| Trait (unit) | $\sigma^2_a$ | $\sigma^2_{pe}$ | DMI | BW | ECM | RFI |
|---|---|---|---|---|---|---|
| DMI (g/d) | 2.11 | 3.69 | **0.18 (0.05)** | 0.19 (0.02) | 0.35 (0.02) | 0.93 (0.00) |
| BW (kg) | 338.20 | 387.74 | 0.37 (0.15) | **0.35 (0.07)** | 0.02 (0.02) | 0.02 (0.02) |
| ECM (kg/d) | 9.55 | 12.12 | 0.68 (0.10) | 0.12 (0.15) | **0.29 (0.06)** | 0.03 (0.02) |
| RFI (g/d) | 0.76 | 2.70 | 0.90 (0.03) | 0.12 (0.18) | 0.39 (0.16) | **0.08 (0.03)** |

**Correlations between traits.** Moderate to high genetic correlations were estimated between DMI and BW, and DMI and ECM, these values were within the range of values previously reported in Danish Jersey cows across lactation stages (Li *et al.* 2018). Furthermore, Manzanilla *et al.* (2017)

reported lower genetic correlations (0.59) for DMI and ECM and higher (0.43) between DMI and BW in Dutch Holstein cows. Genetic and phenotypic correlations between RFI and DMI were large and positive (0.90 and 0.96, respectively) as expected given that RFI is the residual of DMI after been corrected by ECM, MBW and ΔBW. Phenotypic correlations between RFI and its regressors (BW and ECM) were close to zero as expected, whereas genetic correlations were low (0.12) for RFI-BW and moderate positive (0.39) for RFI-ECM, however, due to the large standard errors, the values between RFI and BW are not significantly different from zero. The correlations between DMI and BW-ECM show the importance of having a trait as RFI that is phenotypically independent of economically important traits as ECM and BW.

## CONCLUSIONS

This study shows that feed efficiency calculated using DMI and BW measured by 3D cameras is heritable. Despite the slightly low heritability of DMI, which could be influenced by the small number of farms, animals and records, the results of this study appear promising, endorsing a new technique of recording feed intake and weight that can be implemented in commercial farms. Measuring larger number of animals in more commercial farms, extending the period of measuring and making adjustments in the algorithm and the editing procedure might help to get better quality data and consequently more accurate estimates for genetic parameters.

## REFERENCES


Lassen J., Thomasen J.R., Hansen R.H., Nielsen G.G.B., Olsen E., Stentebjerg P.R.B., Hansen N.W., Borchersen S.. (2018) *Proceedings of the 11th WCGALP*. Auckland, New Zeland. 613.

Li, B., Fikse W.F., Løvendahl P., Lassen J., Lidauer M.H., Mäntysaari P., Berglund B. (2016) *J. Dairy Sci.* **99**: 7232.

Li, B., Fikse W.F., Løvendahl P., Lassen J., Lidauer M.H., Mäntysaari P., Berglund B.. (2018) *J. Dairy Sci.* **101**: 10011.

Lidauer, M.H., Leino A.M., Stephansen R.S., Poso J., Nielsen U.S., Fikse W.F., Aamand G.P.(2019) Proceedings of the 2019 Interbull meeting, Cincinnati, Ohio, USA June 22-24.

Madsen, P. and Jensen. (2014) *Aarhus Res Cent.* Foulum Box 50, 8830 Tjele Denmark.

Manzanilla-Pech, C.I.V., Veerkamp R., de Haas Y., Calus M. and J. tenNapel. (2017) *J. Dairy Sci.* **100**:9103.

Martin, P., Ducrocq V., Faverdin P., Friggens N.C. (2021) *J. Dairy Sci.* **106**: 6329.

NAV. (2020). https://www.nordicebv.info/wp-content/uploads/2020/11/Metabolic-efficiency-included-in-the-Saved-feed-index.pdf

Pryce J., Gonzalez-Recio O., Nieuwhof G., Wales W.J., Coffey M.P., Hayes B.J. and Goddard M.E. (2015) *J. Dairy Sci.* **98**: 7340.

Veerkamp, R.F., Calus M.P.L., de Jong G., van der Linde R., De Haas Y. (2014) *Proceedings of the 10th WCGALP,* Vancouver, Canada.

Sabedot, M., Romano G.S., Pedrosa V.B. and L.F.B. Pinto. (2018) *R. Bras. Zootec.* **47**:1806.

Sjaunja, L.O., Baevre L., Junkkarinen L., Pedersen J. and J. Setala. (1991) *Centre for Agricultural Publishing and Documentation (PUDOC),* Wageningen, the Netherlands.

Stephansen, R.V., Lidauer M. H., Nielsen U.S., Pösö J., Fikse F., Manzanilla-Pech C.I.V. and. Aamand G.P. (2021) Proceedings of the 2021 Interbull meeting.

Tempelman, R., Spurlock D., Coffey M. P., Veerkamp R. F., Armentano L., Weigel K., de Haas Y., Staples C., Connor E. E., Lu Y. and VandeHaar M. (2015) *J. Dairy Sci.* **98**: 2013.

Ulutas, Z., A. Sahin, and M. Saatci. (2008). *J. Appl. Anim. Res.* **34**:29.

Viking Genetics. (2020). https://www.vikinggenetics.com/about-us/innovative-breeding/ntm/saved-feed

*Contributed paper*

# REVEALING PHENOTYPIC AND GENETIC RELATIONSHIPS UNDERLYING THE THERMOTOLERANCE-PRODUCTION COMPLEX IN BEEF CATTLE

**R.G. Mateescu[1], F.M. Rezende[1], K.M. Sarlo Davila[1], S. Dikmen[2], A.N. Nunez Andrade[1], A. Hernandez[1] and M.A. Elzo[1]**

[1] Department of Animal Sciences, University of Florida, Gainesville, FL, USA
[2] Department of Animal Science, College of Veterinary Medicine, Bursa Uludağ University, Bursa, Turkey

## SUMMARY

Heat stress is a principal factor limiting production of animal protein in subtropical and tropical regions, and its impact is expected to increase dramatically. Development of effective strategies to improve the ability to cope with heat stress is imperative to enhance productivity of the livestock industry and secure global food supplies. However, selection focused on production and ignoring adaptability results in beef animals with higher metabolic heat production and increased sensitivity to heat stress. The heritabilities estimated in this study in an Angus-Brahman multibreed population demonstrate genetic variation, which supports the hypothesis that selection for improved thermal tolerance is possible. Moreover, the estimated genetic correlations are favorable and indicate the opportunity to develop genomic tools for simultaneous improvement of tolerance to heat stress as well as production.

## INTRODUCTION

In tropical and subtropical regions where more than half of the world cattle are maintained, climatic stress is a major limiting factor of production efficiency. This stress is expected to increase due to predicted changes in climate. Beef cattle when exposed to environmental high temperature and humidity, exhibit significant declines in feed intake, growth, fertility and welfare. Selection to increase productivity disregarding the genotype x environment interaction is likely to increase susceptibility to climatic stress. This makes the quest for heat-tolerant cattle with increased efficiency of production and reproduction increasingly important. *Bos indicus* cattle exhibit increased resistance to environmental stressors but they also have slower growth, are less fertile and have poorer meat quality relative to *Bos taurus* cattle. Beef producers in tropical and sub-tropical environments are incorporating a certain proportion of 'indicus' genes in their herds but, without knowledge of genes associated with thermotolerance, this also brings along negative aspects of indicus cattle. Research is needed to uncover the phenotypic and genetic relationships underlying this thermotolerance-production complex and subsequently identify the functional variants for thermotolerance without an antagonistic pleiotropy on production and reproduction. This will allow the incorporation of the GxE interaction in genomic selection programs for improvement of economically important traits in a predicted hotter world.

Animals vary in their ability to dissipate heat and, therefore, in their ability to cope with heat stress, and this variability has a genetic component. The goal of this research is to describe novel traits which can be used to characterize genetic pathways for thermotolerance which are independent or positively associated with production performance. This will allow the incorporation of the GxE interaction in genomic selection programs for improvement of economically important traits in a predicted hotter world.

## MATERIALS AND METHODS

**Animal population.** The University of Florida Institutional Care and Use Committee approved the research protocol (Approval no. 201203578). The population consisted of 330 heifers from the University of Florida multibreed herd (Elzo and Wakeman 1998; Elzo *et al.* 2016, 2017) over 2 years in 2017 and 2018. For mating purposes, animals in the multibreed herd are assigned to 6 breed groups based on breed composition: 100% Angus = 100% to 80% Angus; 75% Angus = 79% to 60% Angus; Brangus = 62.5% Angus; 50% Angus = 59% to 40% Angus; 25% Angus = 39% to 20% Angus; and 100% Brahman = 19% to 0% Angus. Heifers were managed similarly across both years. DNA was extracted from blood samples from all animals and genotyped with the Bovine GGP F250 array (Illumina Inc., San Diego, CA, United States).

**Skin biopsies.** Skin samples were taken during summer (July 17, 2017 and August 7, 2018) between 0700 and 1100 h. Skin samples were collected from the back, 4 inches down from spine and halfway along horizontal axis. The skin was cleaned and disinfected with 70% ethanol and chlorhexidine (Clorhexidine 2%; VetOne, Boise, ID). A skin biopsy sample was collected using a 0.6 cm diameter punch biopsy instrument (Biopsy Punch, Miltex Inc., PA) and fixed in 10% formalin for approximately 24 h. Samples were dehydrated in 70% ethanol and infiltrated in liquid paraffin and stored until sectioned and stained at the UF Molecular Pathology Core. Sections were cut on a microtome with a thickness of 7 um, and sections were placed on slides, then stained with Harros-Eosin Hematoxylin. All histological sections were analyzed from digitized images obtained from a Nikon T3000 inverted phase microscope equipped image capture equipment (DMZ1200F with NIS Image Elements software). Images were obtained with the microscope in 40 X, and analyzed with ImageJ software. Sweat gland area ($mm^2$) and sweat gland depth as the distance from the top of the sweat glands to the skin surface (mm) were determined from a constant 4.6 $mm^2$ cropped image area.

**Hair samples.** Hair samples were collected from the shoulder, 4 inches down from spine and halfway along horizontal axis of each animal, as described in Hamblen *et al.* (2018). Hair samples were measured for length using ImageJ software. Five long and 5 short hairs from each individual were measured to evaluate the length of the topcoat and undercoat, respectively. The averages of the 5 short and long hairs were used in the statistical analysis.

**Body temperature.** Core body temperature was measured as vaginal temperature at 15-min intervals for 5 d using an iButton data logger (Dikmen *et al.* 2014) inserted into a blank CIDR device and then into the vagina of each animal. Each iButton was calibrated before the study started and pre-programmed to record body temperature at 15-min intervals on a 24-h cycle. Ambient environmental conditions were monitored using HOBO data loggers which continuously record temperature, humidity dew-point temperature with HOBO-U23 data logger (Onset Computer Corp., Bourne, MA), and black globe temperature by using HOBO-U22 data logger. The temperature humidity index (THI) was calculated as:

$$THI = (1.8 \times T + 32) - [(0.55 - 0.0055 \times RH) \times (1.8 \times T - 26)],$$

where T = air temperature (°C) and RH = relative humidity (%). This equation has been shown to be a good indicator of heat stress (Dikmen and Hansen 2009). Only body temperatures from the 3 continuous days when cattle were on pasture undisturbed were analyzed, as described in Sarlo Davila *et al.* (2019). Based on the thresholds defined by the livestock weather hazard guide and the THI level encountered during our experiment, THI conditions between 84 and 86 were considered high THI. Body temperatures at high THI for each individual were calculated by averaging all the body temperature measurements collected during the time that the THI windows occurred. This was accomplished for each heifer by averaging the body temperature from all 15-minute windows when the heifer was exposed to a high THI interval.

**Carcass traits.** A certified technician recorded ultrasound images from yearling calves using an Aloka 500 ultrasound system (Hitachi Aloka Medical, Ltd., Wallingford, Connecticut, USA). Analysis of the ultrasonic images with UICS Scanning Software by Walter and Associates, LLC (Ames, 106 Iowa, USA) yielded yearling ultrasound backfat (UFAT, cm) and yearling ultrasound percent intramuscular fat (UPIMF, %) phenotypes.

**Statistical analyses.** Average information restricted maximum likelihood (AIREML) variance components, heritabilities, additive genetic correlations, and phenotypic correlations were estimated using single-trait and pairwise two-trait animal linear mixed models. The statistical model for both analyses included the direct additive genetic and residual as random effects, breed group (based on genomic breed composition) and group of data collection as class effect, except for short hair length and skin biopsy records, where group was not significant, and age at measurement as a covariate. The pedigree file consisted of 2,327 individuals, 715 sires and 1,286 dams. All analyses were performed using the airemlf90 package from BLUPF90 software (Misztal *et al.* 2002).

**RESULTS AND DISCUSSION**

Heritability estimates for skin histology characteristics, hair characteristics, body temperature under high THI conditions, and ultrasound carcass traits are provided in Table 1. A high heritability of 0.69 was estimated for the sweat gland area while the sweat gland depth had a low heritability estimate of 0.09. Heritability was estimated to be 0.33 for short hair length (undercoat) and 0.16 for long hair length (top coat). Heritability for coat score has been estimated to be 0.6, (Turner and Schleger 1960) and McEwan Jenkinson *et al.* (1975) estimated the heritability of hair follicle measurements to range from 0.15 to 0.76. The heritability for body temperature under high THI conditions was estimated to be 0.13 which is similar the heritability estimated reported for rectal temperature in a Brahman x Angus crossbred population (0.19; Riley *et al.* 2012) and dairy cattle (0.17; Dikmen *et al.* 2012). Both studies utilized cattle located in Florida. High heritability estimates were obtained for backfat (0.76) and intramuscular fat (0.37) ultrasound measures.

**Table 1. Additive genetic variance ($\sigma^2_a$), residual variance ($\sigma^2_e$), and heritability ($h^2$) estimates for skin histology characteristics (sweat gland area and depth), hair characteristics (short and long hair length), core body temperature under high THI conditions, and ultrasound carcass traits (backfat thickness and intramuscular fat) with approximate sampling errors (in parentheses)**

| Trait[1] | $\sigma^2_a$ | $\sigma^2_e$ | $h^2$ |
|---|---|---|---|
| Sweat gland area ($mm^2$) | 2.03 (0.62) | 0.89 (0.49) | 0.69 (0.18) |
| Sweat gland depth (mm) | 0.002 (0.004) | 0.02 (0.004) | 0.09 (0.15) |
| Short hair length (mm) | 1.95 (1.07) | 3.97 (0.99) | 0.33 (0.18) |
| Long hair length (mm) | 3.21 (3.39) | 16.82 (3.42) | 0.16 (0.17) |
| Temperature at high THI (°C) | 0.02 (0.02) | 0.10 (0.018) | 0.13 (0.15) |
| UFAT (cm) | 0.001 (0.0003) | 0.0003 (0.0002) | 0.76 (0.19) |
| UPIMF (%) | 0.22 (0.12) | 0.38495 (0.11) | 0.37 (0.19) |

[1]*UFAT, ultrasound backfat (cm); UPIMF, ultrasound intramuscular fat (%).*

Two-trait AIREML estimates of direct additive genetic and phenotypic correlations between skin histology characteristics, hair characteristics, body temperature under high THI conditions, and ultrasound carcass traits are presented in Table 2. Sweat gland area had a negative genetic correlation with sweat gland depth (-0.49), short and long hair length (-0.45 and -0.28, respectively), and body temperature under high THI conditions (-0.65). These negative correlations suggest a similarity in the genetic control underlying these traits which would allow for selection of animals with large sweat glands, short hair (both topcoat and under coat), and able to maintain a lower body temperature under high THI conditions. More importantly, although weak, the genetic correlations between sweat gland area and the two production traits (backfat and intramuscular fat) were favorable (0.22 and 0.20, respectively). Similarly, there was a medium negative genetic correlation between the body temperature under high THI and the two ultrasound carcass traits, suggesting animals able to maintain a lower body temperature would be more productive.

**Table 2. Two-trait AIREML estimates of phenotypic (above diagonal) and direct additive genetic (below diagonal) correlations between skin histology properties, hair characteristics, and carcass traits**

| Trait[1] | SWGA | SWGD | SHL | LHL | THighTHI | UFAT | UPIMF |
|---|---|---|---|---|---|---|---|
| SWGA | 0.69 | -0.18 | -0.22 | 0.02 | -0.23 | -0.05 | -0.13 |
| SWGD | -0.49 | 0.10 | 0.32 | 0.26 | 0.12 | 0.08 | 0.22 |
| SHL | -0.45 | 0.27 | 0.33 | 0.75 | 0.23 | 0.07 | 0.17 |
| LHL | -0.28 | 0.02 | 1.00 | 0.16 | 0.23 | 0.04 | 0.11 |
| THighTHI | -0.65 | -0.61 | -0.28 | -0.45 | 0.13 | -0.17 | 0.04 |
| UFAT | 0.22 | -0.57 | -0.34 | -0.60 | -0.38 | 0.76 | 0.23 |
| UPIMF | 0.20 | 0.49 | 0.08 | 0.09 | -0.33 | 0.42 | 0.37 |

[1]SWGA, sweat gland area (mm2); SWGD, sweat gland depth (mm); SHL, short hair length (mm); LHL, long hair length (mm); THighTHI, temperature at high THI (°C); UFAT, ultrasound backfat (cm); UPIMF, ultrasound intramuscular fat (%).

## CONCLUSIONS

The values of heritability estimated in this study indicate a large, exploitable genetic variance which can be used in selection programs to improve heat tolerance in cattle. Novel traits describing the thermotolerance phenotype such as sweat gland area, short hair length and body temperature under high THI conditions had medium to high heritabilities. More importantly, the genetic correlations estimated in this population are encouraging, indicating favorable relationships between the thermotolerance phenotypes and the production traits. This would suggest that genetic programs to improve resilience to environmental stress could be successful and opportunities exists for simultaneous improvement of production related traits.

## REFERENCES

Dikmen, S., and Hansen, P. J. (2009). *J. Dairy Sci.* **92**: 109.

Dikmen, S., Khan, F. A., Huson, H. J., Sonstegard, T. S., Moss, J. I., Dahl, G. E., et al. (2014). *J. Dairy Sci.* **97**

Elzo, M. A., Mateescu, R. G., Johnson, D. D., Scheffler, T. L., Scheffler, J. M., Carr, C., et al. (2017). *Livest. Sci.* **202**: 58.

Elzo, M. A., Mateescu, R., Thomas, M. G., Johnson, D. D., Martinez, C. A., Rae, D. O., et al. (2016). *Livest. Sci.* **190:** 104.

Elzo, M. A., and Wakeman, D. L. (1998). *J. Anim. Sci.* **76**: 1290.

Hamblen, H., Hansen, P. J., Zolini, A. M., Oltenacu, P. A., & Mateescu, R. G. (2018). *J. Anim. Sci.* ***96*(8)**: 3131.

Misztal, I., S. Tsuruta, T. Strabel, B. Auvray, T. Druet, and D. H. Lee. (2002) Proc. 7th World Congr. Genet. Appl. Livest. Prod. Montpellier, France. Communication No. 28–07.

Riley, D. G., Chase, J. C., Coleman, S. W., and Olson, T. A. (2012). *Livest. Sci.* 148, 109–118.

Sarlo Davila, Kaitlyn M., Hamblen, H., Hansen, P. J., Dikmen, S., Oltenacu, P. A., & Mateescu, R. G. (2019). *J. Anim. Sci.* ***97*(8)**: 3246.

Turner, H. G., and Schleger, A. V (1960). *Aust. J. Agric. Econ.* **11**: 645.

*Contributed paper*

# A NEW TOOL TO SELECT ANGUS BULLS TO BREED TO DAIRY COWS

**S.P. Miller[1], J.A. Archer[2], F. Hely[2], C. Quinton[2], K. Retallick[1], D. Moser[2] and P. Amer[2]**

[1]Angus Genetics Inc., 3201 Frederick Ave., Saint Joseph, MO, 64506 USA
[2] AbacusBio Limited, 442 Moray Place, PO Box 5585, Dunedin 9058, New Zealand

## SUMMARY

Widespread use of sexed semen among other factors has resulted in a dramatic increase in dairy cows being mated to beef bulls. Economic selection indexes were developed to identify the best Angus bulls, a predominant breed in this market, to use on Holstein and Jersey cows. Through interviews and site visits with key participants in the dairy beef supply chain, production and economic parameters were sourced to inform the modification of the standard American Angus terminal sire index $Beef. $Angus-on-Holstein Value ($AxH) and $Angus-on-Jersey Value ($AxJ) were developed and although highly correlated to each other (0.96), were considered to rank bulls different enough, especially at the top end, that both were needed. Generally, the dairy indexes identify bulls with the best $Beef but avoid three particular traits that are problematic in the dairy industry, with non-linear emphasis. Calving ease had relatively lower emphasis with a similar penalty in both the $AxH and $AxJ indexes, where muscling was heavily weighted with greatest emphasis in $AxJ and a penalty for excessive yearling height EPD was implemented in $AxH in an effort to reduce excessively long carcasses in that cross. These new indexes provide dairy farmers and players in the supply chain through to slaughter a tool to select Angus bulls to produce calves that are better suited to the requirements of this unique sector.

## INTRODUCTION

America's dairy cattle have always been one source in the beef supply chain. In recent years, this source of beef has been evolving due to a convergence of factors. Sexed semen has been revolutionary in dairy cattle, where breeders can target replacement heifers from their best cows and breed the remainder for beef production. Before sexed semen, the feeding of straight Holstein steers was common, and although this practice remains, there has been a movement towards less demand for these from processors, increasing incentives for breeding dairy cows to beef bulls. Jerseys are gaining market share in the USA but their straight Jersey male calves have very little value, making beef breedings even more attractive. Low milk prices and contraction in the dairy industry reduces the demand for surplus replacements, again pushing the incentive for more beef matings.

Angus has been the most common breed of sire for beef on dairy matings in the USA. $Angus-on-Holstein Value ($AxH) and $Angus-on-Jersey Value ($AxJ) were developed to help dairy farmers identify the most profitable Angus sires for those markets. Although these were the first indexes developed in the USA for beef bulls crossed on Dairy cattle, such indexes have been in place in other countries where beef from the dairy industry is significant such as in Ireland (Berry *et al.* 2019).

Described are the unique aspects that were considered in the development of the beef on dairy indexes. Differences in resulting selection choices and trait emphasis between $Values is described.

## MATERIALS AND METHODS

The development of the Angus on dairy $Values released in 2020 built on the Angus $Value indexes, released in 2019 which were the result of a complete rebuild of the bio-economic model at the time and included an industry wide survey described in Santos *et. al.* (2019). Beef Value or $Beef is a terminal index related to profitability differences on a per carcass basis when all progeny are fed through to slaughter. Differences in post-weaning performance and carcass revenue are considered. There is no emphasis on calving ease in $B as it is assumed that terminal matings involve mature cows only, which have a negligible dystocia incidence when bred to Angus bulls.

The biological models behind $Beef were adjusted to reflect differences in the production and performance of the Dairy cross calves. The dairy model considered the calves performance from birth, including calving ease and pre-weaning growth EPD. Differences in post-weaning performance, feed efficiency and mean carcass grading performance in Dairy cross calves were all considered. Information about the unique aspects and challenges, along with mean performance characteristics were obtained by visiting supply chain participants in the USA by way of in-person interviews and facility visits through the mid and Southwest. Included were dairy farms, calf raisers, feedlots and the processing sector. In addition to this production tour, interviews were also held with the USDA scientists behind dairy selection indexes (Drs. Paul VanRaden and John Cole pers comm). These interviews ascertained the importance of calving ease in the dairy production system as described in VanRaden *et al.* (2018).

The dairy cross calves were characterized with slower post-weaning growth and poorer feed conversion efficiency, lighter carcass weights, less back-fat but similar marbling compared to the straight beef animals as modelled in $Beef. Unique aspects of the dairy cross calves were also apparent, including problems with carcass length in the Holstein cross calves and lack of muscling in both Holstein and Jersey. Differences in mean growth and feed efficiency performance were relatively straight forward adjustments to the bio-economic model behind $B.

Lack of muscling in the dairy cross calves creates two problems. The term "sunken strips" was revealed through the packer interviews to describe the problem of some steak cuts, like strip loins, that are undesirable from a visual, "plate appeal" standpoint due primarily to shape. These poorly muscled animals also create a problem in the live animal as it is a way for the marketplace to visually distinguish animals from the dairy industry. To prevent discounts in the market, these beef animals resulting from the beef-on-dairy cross need to look like beef animals and not the stereotypical "narrow" dairy character.

Deficiencies due to muscling required the development of a genetic trait for use in the indexes, but is not reported. Standard American Angus carcass Expected Progeny Differences (EPD) are presented on an age constant basis (Miller *et al.* 2018). Muscling was determined via a genetic regression using the component traits of ultrasound rib-eye area in bulls adjusted for weight at time of ultrasound scanning (yearling age) using similar methodology employed for feed efficiency as described in MacNeil *et al.* (2011). To rank high for muscling an animal needs to have a large ribeye area relative to their weight.

The economic impact of deficient muscling was modelled in a similar manner to other carcass traits in the $B model, where different categories have different prices. With the thresholds where prices change being known, a new cumulative price can be determined based on a shift in the mean performance. Changes in the developed muscling trait was used to model the economic impact of a different proportion of animals being discounted for lack of muscling, which creates a non-linear relationship between muscling EPD and $Value, where increasing economic discounts are applied with decreasing muscling. Improved muscling was most important with the Jersey crosses.

The same approach was applied to yearling height EPD as a predictor of carcasses being out of specifications in $AxH with 20% of carcasses being over-length with 20 USD per 45.4 kg discount. This approach to modelling categorical traits resulting in a non-linear emphasis is described in Quinton *et al.* (2019).

Correlations were determined using Microsoft Excel between $Value indexes and related EPDs Sires in the analyses included 25,914 current sires with indexes reported for both $AxH and $AxJ.

**RESULTS AND DISCUSSION**

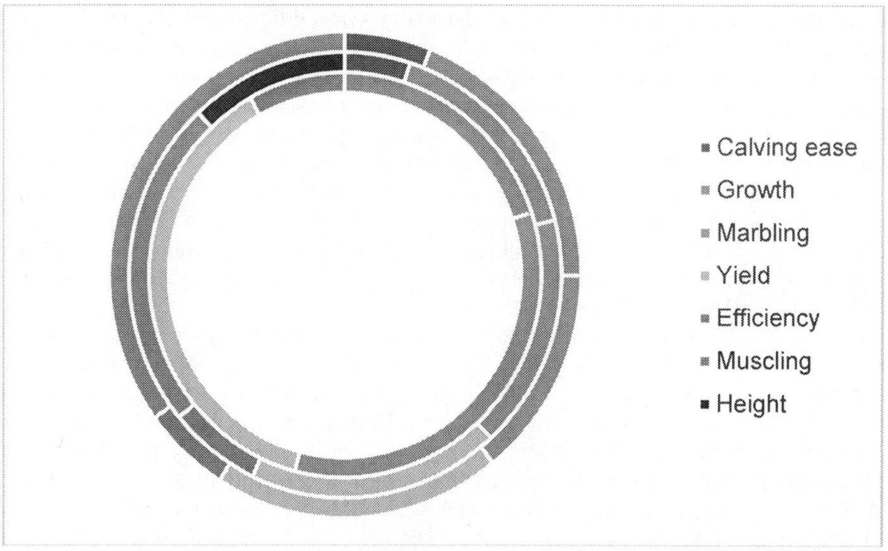


**Figure 1. Relative emphasis on each trait segment in $B, $AxH and $AxJ (from center)**

The dairy indexes are different in the traits that are added to $Beef and as a result, reduce emphasis on existing traits as illustrated in Figure 1. Although the emphasis on growth and efficiency remained similar, the additional traits of calving ease and muscling in both dairy indexes as well as yearling height in $AxH reduced the emphasis on marbling and yield compared to $Beef considerably. The heavy emphasis on muscling stands out as distinctly different with the greatest emphasis in $AxJ.

**Table 1. Correlations between $Value Indexes[1] and some important trait EPD[2] related to Angus on Dairy Indexes**

| | $Beef | $AxH | $AxJ |
|---|---|---|---|
| Calving Ease | 0.01 | 0.27 | 0.20 |
| Post-Weaning Gain | 0.75 | 0.53 | 0.63 |
| Marbling | 0.66 | 0.49 | 0.44 |
| Yearling Height | 0.54 | 0.09 | 0.28 |
| Muscling | 0.31 | 0.79 | 0.79 |
| $Beef | | 0.72 | 0.78 |
| $AxH | | | 0.96 |

[1]$Value indexes are economic selection indexes developed for American Angus including a standard terminal index $Beef when Angus bulls are bred to Angus cows along with newly developed indexes when Angus bulls are mated to Holstein ($AxH) and Jersey ($AxJ) dairy cows. Expected Progeny Differences are the genetic evaluation estimates from the American Angus weekly genetic evaluation.

Resulting $AxH and $AxJ were considerably different to $Beef with correlations of 0.72 and 0.78, respectively (Table 1). As most dairy matings are via artificial insemination, considerable re-ranking among top bulls on the $AxH and $AxJ indexes justified both indexes in the market place despite their high correlation to each other (0.96).

The emphasis on calving ease in both $AxH and $AxJ resulted in a low to moderate correlation with calving ease direct EPD of 0.27 and 0.20, respectively. This is different to the near zero correlation observed between $Beef and calving ease EPD, which can be surprising considering $Beef is a terminal index with considerable weight on growth and weight traits with a 0.75 correlation with post-weaning gain EPD. Amongst the current sires analysed the correlation between calving ease EPD and post-weaning gain EPD was only -0.04 and this combined with a positive correlation of 0.22 between calving ease direct and marbling, a heavily weighted trait in $Beef, are contributing to this neutral correlation observed.

There was a positive correlation between EPDs for yearling height and weight (0.59) and since traits like post weaning gain and carcass weight are positively weighted in $Beef, a positive relationship between $B and Yearling height (YH) EPD (.53 correlation) exists. High yearling height EPD sires then come to the top of $Beef rankings, which are not desirable sires for Holstein cows due to the problem with carcass length. This was addressed in $AxH with downward non-linear emphasis on yearling height resulting in a very small (0.09) correlation between YH EPD and $AxH. It is interesting that although there is no direct emphasis on YH in the $AxJ indexes, there was reduced correlation with YH EPD (0.28) compared to $Beef (0.54). This is partly due to the negative correlation (-0.10) between muscling and yearling height where the increased emphasis on muscling in $AxJ is putting downward pressure on height. Also, as traits are added to the dairy indexes, compared to $Beef, proportional emphasis on traits like growth are also reduced, which could also be influencing the relationship with height.

The indexes developed addressed a need from dairy farmers and participants in the dairy beef supply chain that was not being met with current tools. The main differences between the indexes developed and the standard $Beef index was the relationship with calving ease and the elimination of bulls at the top of the index rankings that are best described as "tall and narrow". Such genetics are not a favourable cross on dairy cattle as they tend to magnify the phenotype that buyers and processors identify as being less desirable. The resulting $AxH and $AxJ indexes were well received by semen companies, who are the primary marketers of genetics to dairy farmers.

## CONCLUSIONS

The $AxH and $AxJ indexes developed to select Angus sires to breed Holstein and Jersey cows address important and unique aspects relevant to these dairy crosses that don't exist in the straight beef supply chain, which the existing terminal index for American Angus, $Beef was designed for. The moderate correlation between $Beef with the $AxH and $AxJ indexes illustrates the major re-ranking that will exist with these new indexes compared to the standard terminal index for straight beef matings and hence their need in the marketplace.

## REFERENCES


Berry D.P., P.R. Amer, R.D. Evans, T. Byrne, A.R. Cromie and F. Hely. (2019) *J. Dairy. Sci.* **102**: 10056.

MacNeil M., N. Lopez-Villalobos, S. Northcutt. (2011) *J. Anim. Sci.* **89**:4124

Miller, S.P., L. Wang, K.J. Retallick and D.W. Moser. (2018) *Proc. 11th World Congr. Genet. Appl. Livest. Prod*

Quinton, C.D., P.R. Amer, T.J. Byrne, J.A. Archer, B. Santos, F. Hely (2019) *Proc. 23rd Conf. Assoc. Adv. Anim. Breed. Genet.*

Santos B., J.A. Archer, D. Martin-Collado, C.D. Quinton, J. Crowley, P.R. Amer and S.P. Miller (2019) *Proc. 23rd Conf. Assoc. Adv. Anim. Breed. Genet.*

VanRaden, PM, P.M., J.B. Cole and K.L. Parker Gaddis (2018). USDA,
< https://aipl.arsusda.gov/reference/nmcalc-2018.htm>

*Contributed paper*

# THE EFFECT OF GDF9 ON LITTER SIZE IN AUSTRALIAN SHEEP

**N. Moghaddar[1], K. L. Bunter[2], A. A. Swan[2] L. Li[2], P. M. Gurman[2] and J. H. J. van der Werf[1]**

[1] School of Environmental & Rural Science, University of New England, Armidale, NSW 2351, Australia

[2] Animal Genetics Breeding Unit, University of New England, Armidale, NSW, 2351 Australia

## SUMMARY

*Growth differentiation factor 9* (*GDF9*) is a known autosomal gene which regulates ovulation rate in mammals. In sheep, numerous polymorphisms have been reported in coding regions of *GDF9* with a significant impact on ovulation rate and hence litter size. To study the effect of *GDF9* on litter size in Australian sheep breeds, an association analysis was performed between 1,600,633 imputed sequence single nucleotide polymorphisms (SNPs) on OAR5 and litter size phenotypes in 8,850 Merino and 7,613 maternal sheep breed ewes (predominantly Border Leicester, Coopworth, Corriedale and composite maternal lines) respectively. Results showed a significant association between litter size and SNPs in the *GDF9* region in maternal breeds. After filtering for high linkage disequilibrium, a highly significant SNP (*p_value* = 9.09E-09) was found in an intron of the *GDF9* gene at OAR5:41841588, which accounted for a 0.22 increase in litter size and explained 4.75% of the total genetic variance. This SNP and the surrounding SNPs in the region of *GDF9* were not significantly associated with litter size in Merinos. Information on this SNP genotype could be useful for obtaining a more accurate estimate of genetic merit for reproduction traits in some breeds of sheep.

## INTRODUCTION

Sustainable livestock farming requires a constant increase in productivity and profitability of the enterprise. Reproductive performance is among the economically important traits in sheep breeding objectives as it directly affects the profitability of the enterprise. In sheep populations, the rate of genetic improvement in reproduction traits with conventional selection, based on breeding values derived from phenotypes and pedigree information, can be low. This is mainly due to low heritability of reproduction traits, incomplete recording in the industry, phenotypes being sex-limited and only available later in life, in particular for adult ewe performance.

Application of genomic information in breeding programs, including using information about polymorphisms affecting the genetic variation of a trait, can increase the accuracy of estimated breeding values (Moghaddar *et al.* 2019) and potentially can lead to significant improvements in genetic gain for reproduction traits. *GDF9* is a known autosomal gene with a significant impact on fertility traits in different mammals, including some sheep breeds. Literature shows some polymorphisms in *GDF9* are responsible for increased ovulation rate and higher litter size in both heterozygous and homozygous genotypes (e.g. Hanrahan *et al.* 2004; Silvia *et al.* 2011; Våge *et al.* 2013). However, sterility is reported for homozygous genotypes of some other mutation in *GDF9* gene in some sheep breeds, such as Belcare, Cambridge and Icelandic sheep breeds (Hanrahan *et al.* 2004; Davis 2005; Nicol *et al.* 2009; Pérez-Ruiz *et al.* 2020). The objective of this study was to perform a genetic analysis of segregating variants of the *GDF9* gene and estimate impact on litter size of adult ewes in Australian Merino and maternal sheep breeds using recently available whole genome sequence data.

## MATERIALS AND METHODS

**Phenotypes.** Adult litter size (LS) phenotypes for Merinos and maternal breeds, which were respectively derived from the national Sheep Genetics database MERINOSELECT and

LAMBPLAN were used in this study. Both data sets included research (Information Nucleus Flocks and MLA resource flock) and industry animals (Sheep Genetics). Merino population consisted of purebred animals and the maternal population were a multi-breed/admixture of maternal sheep breeds including predominantly Border Leicester, Coopworth, Corriedale and composite maternal lines. Litter size phenotypes reflected the number of lambs counted at birth or were derived from pregnancy scanning records (Bunter *et al.* 2019, 2021). The total number of genotyped animals with LS recorded were 8,850 and 7,613 respectively for Merinos and maternal breeds, and 82% and 43% of these ewes had repeated records for Merino and maternal data set respectively. These data belonged to ewes born between 2007 and 2018.

**Genotypes.** Imputed sequence data on OAR5 were used in this study. A description of the imputation procedure is provided in Bolormaa *et al.* (2019). Briefly, research and industry data with low-density genotypes (12k, 15k) were imputed to 50k genotypes based on a large 50k reference set, and then all the 50k genotypes were imputed to high-density genotypes (500k: HD) using a 2,266 multi-breed reference set. Subsequently, animals with HD genotypes were imputed to sequence level using 726 multi-breed animals as a reference set (with on average 10x coverage). The final set of sequence data provided 1,600,633 variants on OAR5 after quality control and filtration for variants with low imputation accuracy ($r < 0.63$). SNPs with minor allele frequency of greater than 0.005 and at least 0.95 call rate were used in this study.

**Statistical analysis.** Phenotypes used in the association study were first corrected for environmental effects separately for research and industry data and according to the following equation in ASReml 4.1 (Gilmour *et al.* 2009): $y = \boldsymbol{Xb} + \boldsymbol{Z_1a} + \boldsymbol{Z_2pe} + \boldsymbol{Z_1Qg} + \boldsymbol{e}$. In this equation, $y$ represents the phenotypes, $\boldsymbol{b}$ is a vector of fixed effects, consisting of mean, contemporary group (cohort of flock, birth year, management group) and age at lambing, $\boldsymbol{a}$ is the random direct additive genetic effect of the animal, fitted through the pedigree relationship matrix, $\boldsymbol{pe}$ is random permanent environmental effect of the animal, $\boldsymbol{g}$ is random effect of reed and **e** is random residual effect. $\boldsymbol{X}$, $\boldsymbol{Z_1}$ , and $\boldsymbol{Z_2}$ are corresponding incidence matrices and $\boldsymbol{Q}$ is a matrix of contributions of genetic groups for all animals in the pedigree. The pre-corrected phenotype for each individual was the sum of the within group genetic and residual effects ($y^* = Z_1\hat{a} + \hat{e}$ ).

Association analysis was performed according to the single SNP mixed model regression method based on the following equation $\boldsymbol{y^*} = \boldsymbol{Xb} + \boldsymbol{Zu} + \boldsymbol{e}$ in the Gemma V0.96 program (Zhou *et al.* 2014). In this equation $y^*$ refers to the pre-corrected phenotypes, $\boldsymbol{b}$ refers to mean and allele substitution effect of the investigated SNP, $\boldsymbol{u}$ refers to the random additive genetic effect of the animal fitted by genomic relationship matrix (G), and $\boldsymbol{e}$ is the residual effect. G was calculated using 50k genotypes based on Yang *et al.* 2011, and $X$ and $Z$ are incidence matrix relating fixed and random effects to phenotypes.

## RESULTS AND DISCUSSION

The litter size results showed the maternal breeds on average were more prolific (LS = 1.73) than Merinos (1.34) (Table 1). However, the heritability of litter size was higher in Merinos (0.13) compared to maternal breeds (0.08).

**Table 1. Summary statistics of phenotypes and pedigree-based heritability of litter size in Merino and maternal sheep populations**

| Population | No. of Records | No. of Animals | Average | sd | range | $h^2$(se) |
|---|---|---|---|---|---|---|
| Merinos | 547,807 | 295,748 | 1.34 | 0.53 | 1- 4 | 0.13 (0.02) |
| Maternal breeds | 703,503 | 305,916 | 1.74 | 0.63 | 1- 5 | 0.08 (0.01) |

The association results showed that SNPs in the *GDF9* region significantly affected LS in maternal breeds (41,841,034 to 41,843,517 bp, Oar_V3.1, Ensembl Genome Browser; www.ensembl.org) (Figure 1). However, SNPs in this region did not have a significant effect on LS in Merinos. The significant region in maternal breeds was within the *GDF9* gene as well as both upstream and downstream of the gene. A total of 298 SNPs in this region were significantly associated initially (-log p value ≥ 6). However, the number of significant SNPs retained in this region after pruning for high LD (LD≥0.95) was 34 and spanned from position 40,685,116 to 45,175,518 bp on OAR5. Conditional and joint analysis of these remaining SNPs in stepwise multiple regression (Yang *et al.* 2012) identified that the most significant SNP within the *GDF9* coding region to be located at OAR5:41841588, which is in intron location of the *GDF9* gene. This SNP was associated with a 0.22 increase in LS and explained 4.75% of the total genetic variance. The frequency of this SNP was 1.2% in the maternal breed population. However, this SNP was not segregating in Merinos and other SNPs in the *GDF9* region were also not significantly associated with LS in Merinos.

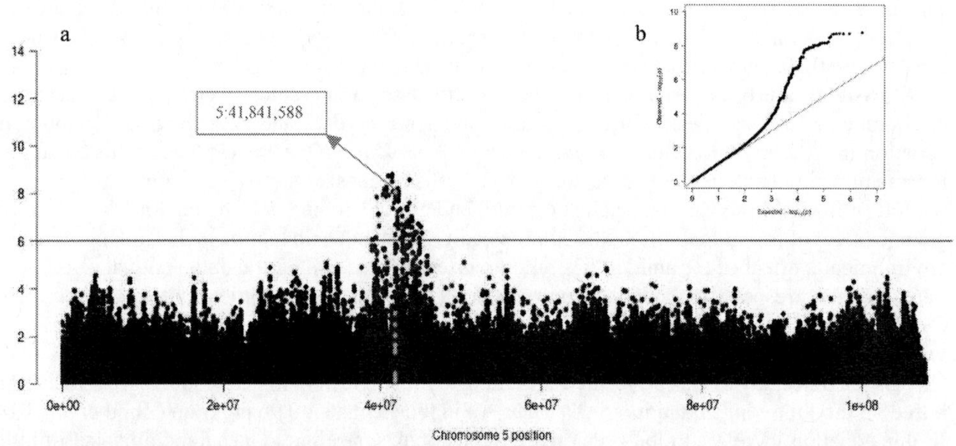


**Figure 1. Manhattan plot of p_value of association between genetic markers on OAR5 and litter size in maternal breeds (a) and the associated QQ plot (b). Green dots show SNPs located within *GDF9* coding region**

The *GDF9* gene, located on OAR5 in sheep, is important for normal folliculogenesis. Some polymorphisms including missense mutations in *GDF9* have been associated with between 0.2 to 0.7 increase in litter size (Davis, 2005) in various sheep breeds. In this study, we observed a highly significant region in *GDF9* and prioritized a SNP located within intron of *GDF9* in the maternal breeds only. This significant SNP was 243 base pairs apart from the causative mutation reported in Norwegian White Sheep (Våge *et al.* 2013), which was introduced to this breed by crossing with Finnish Landrace sheep. It is highly possible that the significant region found in maternal breeds here also originated from Finnish Landrace, due to historical introductions. The increase in litter size (0.22 lambs) in maternal breeds observed herein was within the range of increase in litter size reported for Finnish landrace (Våge *et al.* 2013).

Sterility associated with homozygous genotypes for some mutations in *GDF9* has been reported in some breeds (e.g. Nicol *et al.* 2009). However, in other breeds, such as Finnish Landrace, Norwegian White Sheep and Santa Ines, and for other mutations in *GDF9*, homozygous genotypes have been reported to be fertile (Våge *et al.* 2013; Silvia *et al.* 2011). Herein, three animals homozygous for the significant SNP in the *GDF9* region were all fertile and showed higher LS than population average. However, due to low frequency of the SNP markers in the significant region and the small number of homozygous genotypes in this study, further investigation is required to confirm the fertility status in homozygous animals.

## CONCLUSIONS

This study showed sequence variants located in *GDF9* were significantly associated with litter size in maternal breeds. The allele frequency of the favourable allele was 1.2% in the maternal population and explained 4.75% of the total genetic variance in LS. No such association was observed in Merinos. Further work is required to investigate the relationship between the significant region in *GDF9* with other reproductive traits and also the impact of homozygous genotypes on fertility and litter size. Information about *GDF9* could be useful for more accurate prediction of the genomic merits of selection candidates for LS in some breeds.

## ACKNOWLEDGEMENTS

Phenotypic and data and genotypes used in this study were provided by Sheep Genetics from the MERINOSELECT and LAMBPLAN maternal evaluations, and the research was funded by MLA projects L.GEN.1815 and L.GEN.1704.

## REFERENCES

Bolormaa S., Chamberlain A.J., Khansefid M., Stothard P., Swan A.A., Mason B., and MacLeod I.M. (2019) *Genet. Sel. Evol.* **51**: 1.

Bunter K.L., Swan A.A., Gurman P.M., and Brown D.J. (2021) *Anim. Prod. Sci.* **61** 333.

Bunter K.L., Swan A.A., Gurman P.M., Boerner V., McMillan A.J., and Brown D.J. (2019) *Proc. Assoc. Advmt Anim. Breed. Genet.***23***:* 560.

Davis G.H. (2005). *Genet. Sel. Evol.* **37:** 11.

Gilmour A.R., Gogel B.J., Cullis B.R., Welham S., and Thompson R. (2015) Hemel hempstead: VSN international ltd.

Hanrahan J.P., Gregan S.M., Mulsant P., Mullen M., Davis H.H., Powell R., and Galloway S.M. (2004) *Biol. Reprod.* **70**: 900.

Moghaddar N., Khansefid M., Van der Werf J.H.J., Bolormaa S., Duijvesteijn N., Clark S.A., Swan A.A., Daetwyler H.D., MacLeod I.M. (2019) *Genet. Sel. Evol.* **51**:72.

Nicol L., Bishop S.C., Pong-Wong R., Bendixen C., Holm L.E., Rhind S.M., McNeilly A.S. (2009). *Reproduction* **138**: 921.

Pérez-Ruiz E., Gallegos-Sánchez J., Cortez-Romero C., Segura-León O.l., Salinas-Ruíz J., and Salazar-Ortiz J (2020). *Anim. Genet.* **51**: 341.

Silva B.D.M., Castro E.A., Souza C.J.H., Paiva S.R., Sartori R., Franco M.M. and Melo E.O. (2011) *Anim. Genet.* **42:** 89.

Våge DI,. Husdal M,. Kent MP,. Klemetsdal G,. Boman IA (2013). *BMC Genet.* **14**: 1.

Yang J,. Ferreira T., Morris AP., Medland S.E. (2012). *Nat. Genet.* **44(4)**: 369.

Zhou X., and Stephens M., (2014). *Nat. Methods* **11:** 407.

*Contributed paper*

# DAIRYFARMER PERCEPTIONS AND ATTITUDES TO FEMALE GENOMIC TESTING

**L.C. Monks[1], B. A. Scott[2,3,4], S.L. Bullen[2], J. J. Andony[2] and M.M. Axford[3,4,5]**

[1] Monks Communication, PO Box 193, Bli Bli Qld 4560
[2] Dairy Australia, Level 3, HWT Tower, 40 City Road, Southbank,
[3]School of Applied Systems Biology, La Trobe University, Bundoora, Victoria 3083, Australia
[4]Agriculture Victoria, AgriBio, Centre for AgriBioscience, Bundoora, Victoria 3083, Australia
[5]DataGene, 5 Ring Rd, Bundoora Vic 3083

## SUMMARY

Increasing the adoption of female genomic testing, as a driver to accelerate the rate of genetic gain for net profit is a key priority for the Australian dairy industry. The aim of this research was to understand the motivations and barriers to adoption of female genomic testing through semi-structured interviews and self-administered questionnaires. The results showed that farmer awareness of genomics was high, but many had limited understanding of the practicalities of testing. An awareness-building campaign should therefore focus on building understanding of how genomics fits within a farm business. Ensuring farmers have the necessary support to make use of their results will be critical in achieving sustainable adoption. These findings provide the 'people perspective' to inform research, development and extension strategies to increase the rate of adoption of female genomic testing by farmers.

## INTRODUCTION

Commercial genomic testing of bulls began in Australia in 2011 and has played a significant role in increasing the rate of genetic gain in the Australian dairy herd and the range of traits (Pryce *et al.* 2018; Newton *et al.* 2021). Adoption of genomic testing of bulls has been rapid. Of the Holstein and Jersey bulls registered for artificial breeding and born in the past 5 years, 91% and 84%, respectively, have genotypes included in DataGene's genetic evaluation.

Genomic testing of females has the potential to enable a quantum leap in genetic gain in the Australian dairy herd. It offers significant benefits to individual dairy herds (Newton *et al.* 2021). Genomic testing of females at a young age gives dairy farmers the ability to identify high, medium and low genetic merit animals and the opportunity to manage them differently (DataGene 2019b). The ImProving Herds project determined that the direct benefits of genomics outweighed the testing costs in more than half Australian herds (Newton *et al.* 2018) with the greatest benefits being in herds with low replacement rates and high reproductive performance. DataGene reports that around 20,000 females are tested annually which is less than 1% of heifers born each year. As a proportion of the number of herd recorded cows, the animals genotyped in Australia is 4% (DataGene 2019a, 2020) compared to 22% reported by the Council on Dairy Cattle Breeding in the USA (2020).

Previous studies have shown that farmer breeding choices and attitudes towards genetics vary (Nettle *et al.* 2010; Martin-Collado *et al.* 2015; Ooi *et al.* 2021). The aim of this study was to better understand farmers' motivations to undertake routine female genomic testing and the barriers to adoption in order to advise the development of industry communication and extension activities directed at increasing the rate of genomic testing.

## MATERIALS AND METHODS

We conducted semi-structured telephone interviews with 17 dairy farmers and two managers of commercial GSP (GSP) businesses. The interviews involved a semi-structured conversation process that captures what people think and enables participants to reflect on why they hold these views (Stanfield 1997). Interviewee selection was based on purposeful sampling (Patton 2002). We aimed to gather a full range of perspectives while understanding that it is more likely that ideas are repeated as the number of interviewees increases (Ooi *et al.* 2021). We interviewed two managers from GSP first, to gain insight from their broad experience in discussing genomics with farmers.

Discussion topics for farmer interviews were modified slightly according to their level of interest in genomics which we described as: 'genomics convert', 'genomics is on my radar', and 'non-converts'. Number of interviewees for each of these categories were 7, 8 and 2; respectively.

The self-administered survey of commercial GSPs was conducted through email with follow up phone calls to prompt responses. The survey asked GSPs what they thought were motivations for, and barriers to, the adoption of genomic testing by farmers, based on experience. Eight people from GSPs were invited to participate, with seven responses received by the deadline.

Responses from both interviews and survey were collated and similar ideas were grouped into themes by the research team.

## RESULTS AND DISCUSSION

The initial attractions of genomics were similar across all farmer interviewees, with the two biggest attractions being parentage verification (especially for crossbred or large herds with intensive calving blocks) and to identify heifers to keep as replacements and not having the expense of rearing those animals unlikely to perform in the herd (Table 1). With experience, converts had discovered additional benefits of female genomics, e.g. identifying suitable candidates for sexed semen to breed replacements and beef semen as a terminal cross which are more sophisticated applications to their business.

**Table 1. Reasons why farmers genomic test females**

| |
|---|
| Parent (and pedigree) verification, especially in herds with crossbreeding programs or large herds with intensive calving batches. |
| Heifer rearing decisions<br>• Select the right heifers to rear as replacements<br>• Sell heifers that don't have a future in the milking herd *("Identify the tail end of the herd")* |
| Breeding decisions<br>• Matching different types of straws to animals of high, medium, and low genetic merit (e.g. sexed over high; conventional over medium, beef over low)<br>• Monitor impact of breeding decisions |
| Business decisions: developing alternative income streams e.g. elite genetics, heifer exports. |

Overwhelmingly, the non-convert farmer interviewees had heard of genomics, and the ability to test young females, however their understanding was limited in terms of how the test worked, the costs involved, practicalities of sampling, the application of the results to decision making and the benefits/value to their business. This presents a communication challenge, as one GSP pointed out: "*Nobody wants to look silly and admit they don't know about genomics.*"

There appeared to be regional variation in understanding, with it being greater in areas where peers had tried genomics. For example, in Western Australia, 'everyone around us is testing' so hearing peer success had given them confidence in genomics.

The lack of commercial genomic service for crossbred and minority breeds was a barrier for some farmers in the 'on my radar' category.

Barriers cited by 'non-convert' interviewees fell into the following themes: the herd/business was not yet ready for genomics, other priorities, logistics, costs and confidence in the technology. The barrier of 'other priorities' has been previously recognised in the context of animal breeding and genetics extension programs (Dodd *et al.* 2015) and highlights the need for repeated activities to maintain farmers' awareness of, and interest in female genomics.

One interviewee summed it up as: "*Genomics is currently in my too hard basket and it's easier to justify spending money elsewhere.*"

Converts reported similar concerns before deciding to test. While the cost was initially off-putting, converts' focus changed from cost to 'value' of genomics once they began using the results. This is reflected in the very high repeat testing rate (at least 80%) reported by GSPs interviewed and surveyed. Both converts and GSPs confirmed that the logistics of sampling, especially the first time, was a genuine challenge for many farmers. This could be overcome with practical support and farm protocols to incorporate the testing process into routine activities such as vaccination and disbudding or calf feeding. GSPs also highlighted the importance of providing follow up support, tools and reports to help farmers interpret the results and make breeding and management decisions. They saw it as vital to take people through the results the first time with some farmers needing ongoing support (with each new set of results).

Based on these insights, the research team developed a four-phase adoption pathway for female genomic testing in Australian dairy herds: awareness and understanding, consideration and overcoming barriers, deciding and sampling and interpreting and applying the results (Figure ).

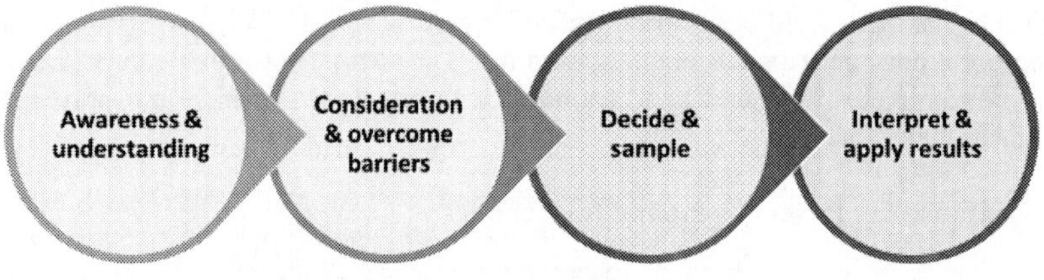


**Figure 1: Adoption pathway for female genomics by farmers**

This pathway has some similarities with the Transtheoretical model of behaviour change (Prochaska *et al.* 1992). It has formed the basis of developing and delivering a communication and extension program to fast track the uptake of female genomics by the Australian dairy industry.

The communication and extension program requires tailored communication formats for the different stages of the adoption pathway. Nettle *et al.* (2010) and Ooi *et al.* (2021) have previously reported that farmers' decisions are influenced by a range of advisors. Therefore, a collaborative approach across the herd improvement industry is expected to be the most effective way to support farmers in their journey along the adoption pathway for female genomics. Different actors in the industry are better equipped to deliver via different communication formats (Table 2). Online delivery formats offer the opportunity to allow farmers to engage with communication and extension resources when the time is right for their individual circumstances.

This study found that Australian dairy farmers have heard of genomic testing but understanding and application of the test results is variable. Those who have not previously tested have limited understanding of the costs, practicalities of sampling, the application of the results and the benefits/value to their business. One-way communication involving mass media is essential in maintaining awareness and can help build understanding. However, fast-tracking adoption will also require interactive communication such as group activities, learning resources and, in some cases, individual one-on-one support.

**Table 2: Communication formats suited to stages in the adoption pathway**

| Stage in adoption pathway | Communication/extension formats | Potential delivery partners |
|---|---|---|
| Awareness and understanding | Mass media, including industry/trade media, print, digital, social media. | Dairy Australia (DA) DataGene, GSPs |
| Consideration and overcoming barriers | Group activities e.g. such as pub nights, field days, discussion groups, local industry events.<br>Peer testimonials and case studies. | Regional Programs DataGene, GSPs |
| Decide and sample | Special interest discussion groups (virtual)<br>One-on-one conversations<br>Practical (on-farm) support with sampling<br>Tools / resources e.g. how-to videos, checklists | DataGene<br>GSPs<br>Dairy Australia |
| Interpret results and apply to decisions | Practical (one-on-one) support<br>One-on-one conversations, individual support<br>Learning resources (including online)<br>Report demonstrations (including online)<br>Special interest discussion groups (virtual) | GSPs, DataGene<br>Breeding Advisers<br>Semen resellers<br>Dairy Australia |

## CONCLUSIONS

Understanding the hurdles and motivations for on-farm adoption of female genomic testing adds to the existing knowledge of genetic trends. These insights provide the 'people perspective' to inform research, development and extension strategies designed to increase the rate of adoption of female genomic testing by farmers. The proposed adoption pathway can inform the development of a communication and extension program to promote the uptake of female genomics in dairy herds.

## ACKNOWLEDGEMENTS

This work was conducted as part of the Accelerating Heifer Genomics Project which is a joint initiative of Dairy Australia and DataGene. The authors thank the farmers who generously shared views and experiences and Heather McLaren, Dairy Australia for her advice.

## REFERENCES

CDCB (2020) Council on Dairy Cattle Breeding, Bowie, MD. Available at https://www.uscdcb.com/wp-content/uploads/2020/10/2020-CDCB-Activity-Report_103020_lowres.pdf [Accessed 26/02/2021].

DataGene (2019a) DataGene, Bundoora. Available at https://datagene.com.au/0/8b2c39cd71015148ca257b010083d5ad [Accessed 26/02/2021].

DataGene (2019b) DataGene, Bundoora, Vic. Available at https://datagene.com.au/node/1367 [Accessed 26/02/2021].

DataGene (2020) DataGene, Bundoora. Available at https://datagene.com.au/node/1190 [Accessed 26/02/2021].

Dodd J, Thomas M and Barber S (2015) Ipsos Available at https://www.mla.com.au/research-and-development/search-rd-reports/final-report-details/Drivers-and-Barriers-to-Genetic-and-Genomic-Tool-Adoption-Market-Research-Qualitative-Phase/4203 [Accessed 26/02/2021].

Martin-Collado D, Byrne TJ, Amer PR, Santos BFS, Axford M and Pryce JE (2015) *J Dairy Sci* **98**, 4148.

Nettle R, Paine M and Penry J (2010) *Anim. Prod. Sci* **50**, 429.

Newton JE, Axford MM, Ho PN and Pryce JE (2021) *Anim. Prod. Sci* **61**, 220..

Newton JE, Hayes BJ and Pryce JE (2018) *J Dairy Sci* **101**, 6159..

Ooi E, Stevenson MA, Beggs DS, Mansell PD, Pryce JE, Murray A and Pyman MF (2021) *J Dairy Sci*
Patton MQ (Ed. MQ Patton (2002) 'Qualitative research and evaluation methods.' (Thousand Oaks, Calif. : Sage Publications: Thousand Oaks, Calif.)
Prochaska JO, DiClemente CC and Norcross JC (1992) *Am Psychol* **47**, 1102.
Pryce JE, Nguyen TTT, Axford M, Nieuwhof G and Shaffer M (2018) *J Dairy Sci* **101**, 3702.

*Proc. Assoc. Advmt. Anim. Breed. Genet: 24: 247-250*

# GENETIC ANALYSIS OF BODY CONDITION AND GROWTH TRAITS IN BEEF FEMALES WITHIN AND ACROSS AGES AND PHYSIOLOGICAL STATES

**K.L. Moore[1], T.P. Grant[2] and D.J. Johnston[1]**

[1]Animal Genetics Breeding Unit*, University of New England, Armidale, NSW, 2350 Australia
[2]Queensland Department of Agriculture and Fisheries, Toowoomba, QLD, 4350 Australia

## SUMMARY

This study estimated variance components of body condition and growth traits and the genetic relationships across time and traits for approximately 2,200 females from three tropically adapted northern Australian beef breeds. Body condition score, measured in yearling heifers and subsequently at the commencement of their annual mating seasons (1st and 2nd), was estimated to be heritable ($h^2$: 0.32to 0.36) and with high genetic correlations ($r_g$) over time, ranging from 0.76 to 0.85. Hip height was also estimated to be strongly heritable at the three time points ($h^2$: 0.59 to 0.67) and was genetically the same trait across the time points ($r_g$: 0.94 to 0.99). Similar results were found for live weight, with heritability estimates ranging between 0.61 and 0.65 and weight being strongly correlated across the different time points ($r_g$: 0.81 to 0.95). Genetic correlations between traits within the same time point showed that when cows were undergoing the fastest growth (commencement of mating 1) the genetic relationships varied compared to times points with slower growth. As yearling heifers and into mating 2 the genetic relationship between hip height and body condition score was small to moderately negative. However, at commencement of mating 1, a strong negative genetic correlation was observed. Likewise, the genetic correlation between live weight and body condition score was moderately positive, except for the commencement of mating 1, when it was not significantly different from zero. Body composition is moderately heritable but the physiological state impacts on the genetic relationships between traits, so having a clearly defined time of measurement will be essential in the trait definition.

## INTRODUCTION

Cow body condition score is an important trait in beef production. It describes body reserves of fat and is potentially an indirect measure of both fertility and survival. Wolcott *et al.* (2014b) demonstrated in Australian Brahman females a positive moderate genetic correlation between body condition score and pregnancy success from the first mating. Overall body condition is affected by both environmental conditions and the physiological state of the cow. A thorough understanding of the genetics of body condition score (reflecting composition) is required to be included into a genetic evaluation programme. This requires evaluating different physiological states and the genetic relationships to other traits.

The aim of this study was to estimate the genetic parameters for body condition score, hip height and live weight, within and across ages in tropically adapted northern Australian beef breeds.

## MATERIALS AND METHODS

Female reproduction traits have been extensively recorded as part of the Repronomics™ project in northern Australia (Johnston *et al.* 2017). Briefly, three breeds (Brahman, Droughtmaster and Santa Gertrudis) were managed together at two sites in Queensland (QLD). Approximately 2,200 measurements were considered from females measured as yearling heifers (~14 months) and at the

* A joint venture of NSW Department of Primary Industries and the University of New England

commencement of the annual mating seasons1 (~26 months) and 2 (~38 months). Cows not producing a calf were culled and data at mating 2 is censored with only lactating cows included.

Cow body condition score was assessed based on a 1 (poor) to 5 (fat) scoring system, with plus/minus amendments to scores allowed. Where a plus/minus score is recorded the body condition score is adjusted by 0.33 increments, for example 3-, 3 and 3+ are analysed as 2.66, 3.0 and 3.33, respectively. Within each cohort a single experienced assessor scored all animals. Hip height (cm) was recorded in the crush as the distance from the ground to the top of the hip. Live weight (kg) was recorded using electronic scales.

Statistical models were developed for each trait using PROC MIXED in SAS (SAS Institute, 2007), sire was fitted as random and model terms were tested with step-wise elimination until only significant terms remained. All traits were analysed as linear, with breed (Brahman, Droughtmaster and Santa Gertrudis) and cohort fitted as significant fixed effects. A cohort was defined as purebred females born together at the same site in the same year. All breeds at each site were managed together. Mating outcomes were recorded and included the date of birth and sex of calves. Animals with unknown parentage, calf sex, date of birth or dam age were removed from the dataset, as were multiple births and animals that were not purebred.

For body condition score at matings 1 and 2 and all hip height measures, birth month and dam group (a concatenation of dam's project herd, breed type, herd of origin and age group) nested within cohort was also significant. Except formating 2 live weights, age at measurement was a significant linear covariate fitted for all traits, with the quadratic age term also significant for mating 1body condition score. The first order interaction between birth month and measurement age was included in the final model for all body condition score traits and mating 1 hip height. Calf birth month,dam group nested within birth cohort and the interaction between calf birth month and cohort were significant for mating 2 hip height, and calf birth month and age of calf at foot were significant for live weight at the start of mating 2.

To estimate genetic parameters, mixed linear animal models including significant fixed effects were fitted using ASReml (Gilmour *et al.* 2009). Univariate models were fitted to estimate variance components, with genetic relationships estimated fitting tri-variate models that grouped traits by stages (i.e. all mating 1 traits) or by trait type (i.e. all body condition score traits). Fitting a tri-variate model accounts for data censorship at the later time points. A combined breed pedigree was used including up to 3 generations where available.

**RESULTS AND DISCUSSION**

The number of records by time and trait are presented in Table 1. There are fewer animals at the later time points due to data censoring from culling cows that do not calve from mating 1 and recent cohorts not yet being mated a 2nd time. The largest increase in skeletal size and live weight occurred between the yearling heifer and commencement of mating 1 times. Skeletal structure and live weight still increased (at a slower rate) between matings 1 and 2 as the cows grew, but body condition score decreased as cows were rearing calves and losing condition. Cows were leanest at the commencement of mating 2 and there was more raw variation in body condition score at mating 2 compared to body condition scores at the earlier ages. Averagehip height and live weight increased with each subsequent measurement, but the raw variation increased initially but was then similar for mating 1 and 2.

Estimated variance components (Table 1) showed that all traits were heritable, and that heritability was similar across the different time points considered in this study. Body condition score heritability estimates were moderate ($h^2$=0.32 to 0.36) with large estimates for hip height ($h^2$=0.59 to 0.67) and live weight ($h^2$=0.61 to 0.65). After adjusting for significant fixed effects, including age and reproduction status, the phenotypic variance of body condition score at the commencement of mating 2 was approximately twice the size compared with the variance at the commencement of mating 1 and the variance of live weight also increased over time. The greater variation in body condition score and live weight observed at mating 2 reflects that at this stage cows are meeting the energy demands associated with both lactation and growth. The variance of hip height also increased initially but was then similar for mating 1 and 2.

**Table 1. Data summary statistics, estimated additive variance and heritability for body condition score (1 – 5 score), hip height (cm) and live weight (kg) measured in yearling heifers and at the commencement of mating seasons 1 and 2 for pooled breeds (Brahman, Droughtmaster and Santa Gertrudis)**

| | N | Mean | Standard deviation | Range | Additive Variance | Heritability |
|---|---|---|---|---|---|---|
| | | | | **Body condition score** | | |
| Yearling Heifer | 2,253 | 3.02 | 0.41 | 1.66 – 4.00 | 0.019 | 0.32 (0.05) |
| Mating 1 | 2,219 | 3.13 | 0.41 | 1.66 – 4.66 | 0.033 | 0.36 (0.05) |
| Mating 2 | 1,585 | 2.66 | 0.59 | 1.00 – 4.33 | 0.064 | 0.36 (0.06) |
| | | | | **Hip height** | | |
| Yearling Heifer | 2,167 | 124.4 | 4.5 | 109 -141 | 8.27 | 0.61 (0.05) |
| Mating 1 | 2,213 | 135.7 | 5.4 | 115 - 156 | 10.5 | 0.67 (0.05) |
| Mating 2 | 1,548 | 140.3 | 5.1 | 121 - 162 | 9.24 | 0.59 (0.08) |
| | | | | **Live weight** | | |
| Yearling Heifer | 2,391 | 258 | 37.3 | 142 - 381 | 317.7 | 0.64 (0.05) |
| Mating 1 | 1,979 | 381 | 60.8 | 176.5 - 572 | 612.1 | 0.61 (0.06) |
| Mating 2 | 1,486 | 431 | 61.9 | 244 - 648 | 1111.6 | 0.65 (0.06) |

**Table 2. Genetic (above diagonal) and phenotypic (below diagonal) correlations (standard errors) between body condition score (BCS, 1 – 5 score), hip height (HH, cm) and live weight (LW, kg)measured in yearling heifers and at the commencement of mating season 1 and 2 for pooled breeds (Brahman, Droughtmaster and Santa Gertrudis)**

| | | **Yearling heifers** | | | **Mating 1** | | | **Mating 2** | | |
|---|---|---|---|---|---|---|---|---|---|---|
| | | **BCS** | **HH** | **LW** | **BCS** | **HH** | **LW** | **BCS** | **HH** | **LW** |
| **Yearling heifers** | **BCS** | | -0.19 (0.09) | 0.38 (0.08) | 0.79 (0.07) | | | 0.85 (0.08) | | |
| | **HH** | -0.02 (0.03) | | 0.66 (0.04) | | 0.99 (0.01) | | | 0.94 (0.03) | |
| | **LW** | 0.30 (0.02) | 0.62 (0.02) | | | | 0.95 (0.02) | | | 0.81 (0.04) |
| **Mating 1** | **BCS** | 0.39 (0.02) | | | | -0.50 (0.08) | -0.11 (0.10) | 0.76 (0.08) | | |
| | **HH** | | 0.75 (0.01) | | -0.25 (0.02) | | 0.67 (0.05) | | 0.98 (0.01) | |
| | **LW** | | | 0.79 (0.01) | 0.10 (0.03) | 0.56 (0.02) | | | | 0.91 (0.02) |
| **Mating 2** | **BCS** | 0.26 (0.03) | | | 0.33 (0.03) | | | | -0.17 (0.12) | 0.27 (0.10) |
| | **HH** | | 0.71 (0.02) | | | 0.80 (0.01) | | -0.12 (0.03) | | 0.71 (0.06) |
| | **LW** | | | 0.64 (0.02) | | | 0.73 (0.01) | 0.36 (0.03) | 0.55 (0.02) | |

Table 2 presents estimates of genetic correlation between traits recorded at the same time point and for individual traits across time points. All traits were strongly correlated across the different time points. Body condition score was estimated to have very strong genetic correlations (between 0.76 and 0.85) across time points. Likewise, genetic correlations for live weight across time were very strong ranging between 0.81 and 0.95. Genetic correlation estimates for hip height indicated that the trait was genetically the same at each time point with all correlations greater than 0.94. Wolcott *et al.* (2014a) reported similar relationships between pre-calving and mating 2 measurements for all three traits considered in this study. These correlations indicate that selection based on records of younger animals will have a consistent impact on the trait genetically through mating 1 and 2.

Hip height and live weight at all stages were strongly correlated ($r_g$: 0.66 to 0.71). In contrast, the correlation between hip height and body condition score varied depending on the physiological state. As a growing yearling heifer and at the commencement of mating 2 (i.e. lactating heifer) there were small negative genetic correlations that were not significantly different from zero. At the commencement of mating 1, a moderate negative correlation ($r_g$=-0.50) was estimated between hip height and body condition score. This suggests that in periods of high growth, genetically taller cattle put energy towards structural growth before laying down body condition. Similarly, genetic correlations between body condition score and live weight varied at the different stages. At the commencement of mating 1, the genetic relationship was not significantly different from zero, but at the other times in the study, a moderate positive genetic correlation was estimated. Wolcott *et al.* (2014a) from Brahman and Tropical Composite animals reported similar results for traits measured pre-calving and at mating 2.

## CONCLUSIONS

This study confirms that body condition score, hip height and live weight were heritable when recorded in yearling heifers and at the commencement of mating 1 and 2. Further, within trait estimates of genetic correlations across time showed that selection of animals at one physiological state can be effective when selecting to improve the same trait at another physiological state, however it is not genetically the same trait. The between trait genetic correlations illustrate how animals with more rapid growth (i.e. at the start of mating 1) tend not to lay down body condition whilst they are growing in skeletal size. Once the active growth slows, these animals partition more resources to body condition and there is no longer a significant genetic relationship between body condition score and hip height. Genetic selection of body composition is achievable but having a clearly defined time of measurement will be essential in the trait definition.

## ACKNOWLEDGEMENTS

The authors acknowledge funding from MLA (project B.NBP.0759). Thank you to the collaborating organisations and the large team involved in designing, running and collecting data. We acknowledge the three breed societies and their members for access to their database.

## REFERENCES

Gilmour A.R., Gogel B.J., Cullis B.R. and Thompson R (2009) ASReml user guide, release 3.0. VSN International Ltd, Hemel Hempstead, UK.

Johnston D.J., Grant T.P., Schatz T.J., Burns B.M., Fordyce B.M. and Lyons R.E. (2017) *Proc. Assoc. Advmt. Anim. Breed. Genet.* **22**:385.

SAS Institute (2007) SAS user's guide in statistics, 9th edition. SAS Institute Inc., Cary, NC, USA.

Wolcott M.L., Johnston D.J., Barwick S.A., Corbet N.J. and Williams P.J. (2014a) *Anim. Prod. Sci.***54(1)**, 37.

Wolcott M.L., Johnston D.J., and Barwick S.A. (2014b) *Anim. Prod. Sci.***54(1)**, 60.

*Proc. Assoc. Advmt. Anim. Breed. Genet: 24: 251-254*

# EWE REPRODUCTION STATUS AND ITS IMPACT ON GREASY FLEECE WEIGHT BREEDING VALUES

**S.I. Mortimer[1], A.A. Swan[2] and K.L. Bunter[2]**

[1] NSW Department of Primary Industries, Armidale, NSW, 2351 Australia
[2] Animal Genetics and Breeding Unit*, University of New England, NSW, 2351 Australia

## SUMMARY

Using industry records (n=12,912), the effect of ewe reproductive status (defined as combined lambing outcomes during previous and current production cycles) on adult greasy fleece weight (AGFW) was estimated for pre-joining and pre-lambing shearing systems. Increasing ewe reproductive output significantly decreased AGFW, by up to 26% over 2 cycles of shearing. Differences in adjustments between shearing systems reflected that AGFW was most affected by the reproductive cycle completed before shearing. Estimated breeding values for AGFW of sires and ewes were little affected by bias due to reproductive status (< 1% for the highly reproductive ewes comprising 36% of the data) and re-ranking of animals was limited. Nevertheless, adjusting AGFW for reproductive status is proposed, but this would be difficult under the current low recording levels.

## INTRODUCTION

The recording of lifetime productivity traits in Merino ewes is encouraged by both Sheep Genetics and, more recently, through the activities of the Merino Lifetime Productivity project (Ramsay *et al.* 2019). This follows studies (Brown *et al.* 2013; Swan and Brown 2013) which confirmed that recording of at least 1 measurement of adult greasy fleece weight (AGFW) would increase genetic gains in AGFW and overall selection accuracy for lifetime wool production. These studies used available expressions of AGFW recorded on both males and females from MERINOSELECT flocks. Reproductive level of ewes was not included as a fixed effect in models fitted to AGFW due in part to constraints with the genetic evaluation software at that time.

For a production system where ewes were shorn with a lamb at foot, Waters *et al.* (2000) reported that ewes rearing multiple lambs during the current production cycle had 0.12 kg lighter AGFW than ewes rearing single lambs, but effects due to rearing performance during the previous production cycle were not significant. Richards *et al.* (2018) examined effects of cumulative lifetime reproductive performance on clean fleece weight of Merino ewes, finding generally no significant differences in fleece weight between ewes with higher and lower number of lambs scanned over 3 consecutive reproductive records. As their study used data from 2 commercial flocks, genetic and environmental influences could not be separated, and it was not clear when ewes were shorn in relation to stage of the reproduction cycle.

Using data from the MERINOSELECT database, this study aimed to evaluate the effects on AGFW of ewe reproductive status, defined as the combined lambing outcomes from its previous and current production cycle. The effect of ewe reproductive status was estimated for these consecutive cycles within 2 shearing systems.

## MATERIALS AND METHODS

Greasy fleece weights of ewes with known and consecutive reproductive outcomes recorded between their second and fifth adult shearings were extracted from the MERINOSELECT database.

---

* A joint venture of NSW Department of Primary Industries and the University of New England

The fleece weights were collected under 2 annual shearing systems, where ewes either were shorn pre-joining (dry period) or pre-lambing (mid-gestation). For each shearing system, initially reproductive performance at each ewe's previous and current cycle was described as not pregnant (DRY), single lamb born and lost (S_L), single lamb born and reared (S_S), multiple lambs born and lost (M_L), multiple lambs born but a single lamb reared (M_S) and multiple lambs born and reared (M_M). Ewe reproductive status then was defined by concatenating reproductive performance during the previous and current production cycles (36 levels per shearing system). A total of 12,912 AGFW records from 9,934 ewes across shearing systems were available for previous and current reproductive performances relevant to each fleece weight's production cycle (Table 1).

Analysis of the effect of ewe reproductive status on AGFW was conducted using ASReml (Gilmour *et al.* 2015). The fixed effects fitted included the birth-rearing type of the ewe (3 levels), age of dam (10 levels), age in days at measurement (fitted as a linear covariate), contemporary group (defined as combinations of flock, birth year, date of measurement and management group) and its reproductive status, described above. Random effects included sire and ewe permanent environment, to accommodate repeated records for ewes. Predicted means for greasy fleece weight for the reproduction effects were estimated using solutions from the full model. The impact of fitting reproductive status on BLUP estimated breeding values (EBVs) for AGFW was evaluated using single trait models with and without the effects of reproductive status within shearing system.

**Table 1. Descriptive statistics for adult greasy fleece weight (kg) in each shearing system**

| | Number of records | Number of ewes | Mean (SD) | Minimum | Maximum | Mean ewe age (days) |
|---|---|---|---|---|---|---|
| Pre-joining | 6636 | 5226 | 5.6 (1.32) | 2.2 | 12.8 | 1436 |
| Pre-lambing | 6276 | 4708 | 4.9 (1.58) | 1.2 | 13.8 | 1267 |

## RESULTS AND DISCUSSION

The predicted means for AGFW were higher for the pre-joining system (5.3 kg for DRY ewes, Table 2), but this difference at least partially reflected a change in the predominant type of ewe: stronger wool ewes tended to be shorn pre-joining, whereas fine-ultra fine wool ewes were mainly shorn during gestation. Within a single shearing cycle, reductions in AGFW from DRY to M_M were 14% within pre-joining and 20% within pre-lambing shearing systems. The high reproductive output of twice M_M ewes (M_M-M_M category) reduced AGFW by 22% and 26% when compared to twice DRY ewes under pre-joining and pre-lambing shearings respectively. For twice S_S ewes, the reduction was 18% and 21% respectively. These highly reproductive ewes (S_S-S_S and M_M-M_M categories combined, Table 2) contributed 36% and 37% of records to the pre-joining and pre-lambing shearing data, respectively. In agreement with the finding of Bunter and Swan (2021), of small unfavourable genetic correlations between reproduction and AGFW, having accurate reproduction records would assist Merino breeders to avoid culling of ewes with lower fleece weight but higher reproductive output, and so improve both fleece weight and reproduction.

Reproductive status had a significant detrimental impact on AGFW, increasing in magnitude with number of lambs reared over more than one annual cycle (Table 2). Ewes with persistently high reproductive performance will have reduced AGFW as a result of competition for limiting nutritional resources during pregnancy and lactation, with lactation taking priority (Corbett 1979). This result was consistent across both shearing systems. However, a difference in timing of shearing relative to lambing and lactation altered which of current or previous reproductive outcome (i.e. lambing was several months before or shortly after the shearing event) influenced AGFW more. The largest effects were evident where the full reproductive cycle (including lactation) was completed prior to shearing. Therefore, current reproduction effects were larger under the pre-joining system

**Table 2. Predicted means of adult greasy fleece weight (kg) for ewe reproductive status categories of ewes shorn under pre-joining and pre-lambing shearing systems**

| Previous reproduction | | | Current reproduction[1] | | | | | |
|---|---|---|---|---|---|---|---|---|
| | | | DRY | S_L | S_S | M_L | M_S | M_M |
| Pre-joining shearing | | | | | | | | |
| | | N | 297 | 347 | 2901 | 164 | 97 | 2010 |
| Previous reproduction | DRY | 468 | 5.29 | 4.86*** | 4.54*** | 4.68** | 4.56*** | 4.54*** |
| | S_L | 441 | 5.02 | 4.68*** | 4.45*** | 4.45*** | 4.42*** | 4.12*** |
| | S_S | 3357 | 5.07* | 4.64*** | 4.36*** | 4.65*** | 4.34*** | 4.21*** |
| | M_L | 150 | 4.89‡ | 4.70** | 4.39*** | 4.32*** | 4.34*** | 4.24*** |
| | M_S | 815 | 5.29 | 4.73*** | 4.44*** | 4.61*** | 4.34*** | 4.15*** |
| | M_M | 1405 | 5.22 | 4.58*** | 4.36*** | 4.53*** | 4.27*** | 4.10*** |
| Pre-lambing shearing | | | | | | | | |
| | | N | 421 | 509 | 2933 | 235 | 744 | 1434 |
| | DRY | 302 | 3.74 | 3.40** | 3.37*** | 3.96 | 3.46 | 3.68 |
| | S_L | 520 | 2.88*** | 3.25*** | 3.14*** | 3.28** | 3.17*** | 3.25*** |
| | S_S | 3725 | 3.02*** | 2.96*** | 2.95*** | 2.99*** | 2.98*** | 3.00*** |
| | M_L | 105 | 3.45 | 2.83*** | 3.06*** | 2.94** | 3.15** | 3.09*** |
| | M_S | 608 | 2.80*** | 2.80*** | 2.81*** | 2.93*** | 2.81*** | 2.96*** |
| | M_M | 1016 | 2.98*** | 2.89*** | 2.80*** | 2.87*** | 2.84*** | 2.77*** |

[1] DRY: not pregnant; S_L: single lamb born, lost; S_S: single lamb born, reared; M_L: multiple lambs born, lost; M_S: multiple lambs born, single lamb reared; M_M, multiple lambs born, reared. ***, $P < 0.001$, **, $P < 0.01$, *, $P < 0.05$ and ‡, and $P \leq 0.10$ tested within each shearing system as a contrast to DRY_DRY predicted mean.

(range of 12-22% reduction in AGFW of M_M ewes within each previous reproduction category), while previous reproduction effects were larger under the pre-lambing system (range of 14-29% reduction in AGFW of M_M ewes within each current reproduction category).

The difference between unadjusted and adjusted EBVs for AGFW of ewes (i.e. bias) was around 2.5% for twice dry ewes, but much less in ewes consecutively bearing and raising singles (S_S) and twins (M_M) (Figure 1A). The large positive bias in AGFW EBVs for twice DRY ewes indicated that their EBVs were overestimated when reproductive status was not accounted for during genetic evaluation. However, very few records for AGFW were available for twice dry ewes (1% of records for both shearing systems), similar to industry flocks where ewes are usually culled if dry once, and so few ewes would have EBVs affected. Across the other reproductive categories, industry recording of AGFW is also low, e.g. industry data used by Bunter and Swan (2021) had 6% of AGFW with known previous reproductive status. For both ewes and sires, EBVs based on models where AGFW was unadjusted and adjusted for ewe reproductive status were highly correlated (correlations > 0.98; Figure 1B, D), indicating that little re-ranking of animals on AGFW would occur when reproductive effects are ignored. Independently, the relationship between an EBV for litter size and bias in AGFW for sires was negative and not strong (Figure 1C). While sires with lower EBVs for litter size had higher EBVs for AGFW after adjusting for reproduction status, the overall effect was small as daughters of sires were spread across all reproductive categories.

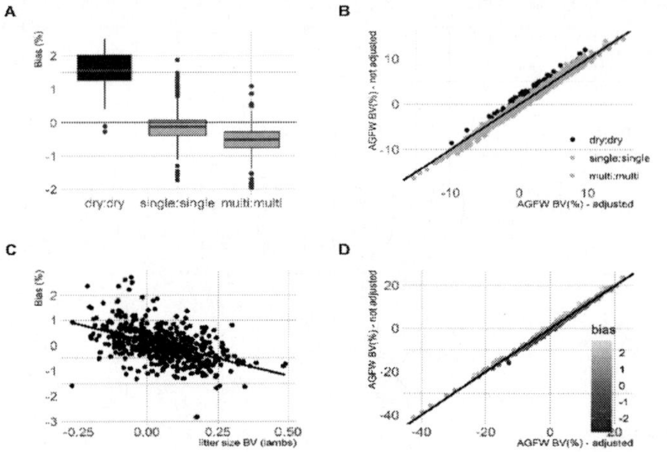


**Figure 1. Bias in and impact on estimated breeding values (BV) for adult greasy fleece weight (AGFW) in ewes (A and B respectively) and sires (C and D respectively)**

## CONCLUSIONS

Ewe reproductive status significantly influenced AGFW, with timing of shearing relative to the reproductive cycle influencing size of the effects. Sire and ewe EBVs for AGFW, though, were little affected by bias due to reproductive status and re-ranking of animals was limited, largely because reproductive status was mainly unknown in industry data. Current low levels of recording for ewe reproductive status make it difficult to apply such adjustments to AGFW. However, it is proposed that applying these adjustments would increase confidence in using EBVs for AGFW by ram breeders and producers. Well-recorded reproductive information is needed to avoid culling of more reproductive ewes with lower fleece weights and for increased selection accuracy of young animals.

## ACKNOWLEDGMENTS

This research was made possible through the efforts and support of the Australian sheep industry in submitting data to MERINOSELECT, as well as funding provided by MLA and AWI.

## REFERENCES

Brown D.J., Swan A.A. and Gill J.S. (2013) *Proc. Assoc. Advmt. Anim. Breed. Genet.* **23**: 110.

Bunter K.L. and Swan A.A. (2021) *Proc. Assoc. Advmt. Anim. Breed. Genet.* These proceedings.

Corbett J. L. (1979) In 'Physiological and Environmental Limitations to Wool Growth', pp. 79-98 editor J. L. Black and P. J. Reis., University of New England Publishing Unit, Armidale.

Gilmour A.R., Gogel B.J., Cullis B.R., Welham S.J. and Thompson R. (2015) 'ASReml User Guide Release 4.1 Functional Specification'. VSN International Ltd, Hemel Hempstead, UK.

Ramsay A.M.M., Swan A.A. and Swain B.C. (2019) *Proc. Assoc. Advmt. Anim. Breed. Genet.* **23**: 512.

Richards J.S., Sladek M.A. and Lee G.J. (2018) *Anim. Prod. Sci.* **58:** 1470.

Swan A.A. and Brown D.J. (2013) *Proc. Assoc. Advmt. Anim. Breed. Genet.* **20:** 232.

Waters C.M., Coelli K.A., Lee G.J. and Atkins K.D. (2000) *Aust. J. Exp. Agric.* **40**: 931.

*Proc. Assoc. Advmt. Anim. Breed. Genet: 24: 255-258*

# VARIATION BETWEEN MERINO SIRES IN LAMB CARCASS VALUE

**S.I. Mortimer[1], J. L. Smith[2], B.C. Hine[2], S.M. Fowler[3], B.W.B. Holman[3], D.L. Hopkins[3], K.L. Egerton-Warburton[4] and A.A. Swan[5]**

[1] NSW Department of Primary Industries, Armidale, NSW, 2351 Australia
[2] CSIRO, Agriculture and Food, F.D. McMaster Laboratory, Armidale, NSW, Australia
[3] NSW Department of Primary Industries, Cowra, NSW, 2794 Australia
[4] NSW Department of Primary Industries, Orange, NSW, 2800 Australia
[5] Animal Genetics and Breeding Unit*, University of New England, NSW, 2351 Australia

## SUMMARY

Carcass value and its components were evaluated for a range of Merino sires based on progeny performance in diverse climates and production systems. Sire adjusted means for carcass value had a range of $31.33 under a mixed farming system while the range under a fine wool production system was $62.48. This preliminary analysis shows that considerable variation exists in carcass value of individual Merino sires when based on a simple economic model.

## INTRODUCTION

Incorporating meat production traits together with traditional fine wool traits into sheep breeding programs is driven by an increased demand for sheep meat and changes in relative prices paid for wool and meat. Gross margin analyses of 10 model sheep enterprises by the NSW Department of Primary Industries (NSW DPI) have shown over time that sheep enterprise performance over a 10-year period has been steadily improving, driven by increases in both returns from wool and sheep sales (G.C. Casburn, personal communication).

Genetic benchmarking of Merinos through central test sire evaluation has been focussed on the performance recording of measured and visually assessed traits that are relevant to wool production, such as yearling, hogget and early adult fleece traits. Establishment of the Merino Lifetime Productivity (MLP) project, through a partnership between Australian Wool Innovation Limited and the Australian Merino Sire Evaluation Association combined with committees and hosts at 5 sites (Ramsay *et al.* 2019), has enabled the design of a sire evaluation scheme, which will assess lifetime productivity of ewe progeny across a range of environments. At 2 sites, an additional project has recorded the performance of sire progeny for a range of carcass composition and meat quality traits, providing comparative information on Merino sires for these traits. Previously, Clarke *et al.* (2019) reported variation between sires in value of production (wool and meat), based on live animal data being used to assign animals to market segments and therefore estimate sale value. For Merino producers looking to diversify their businesses and take advantage of the potential higher returns from running self-replacing flocks and selling trade wether lambs, such information would assist in identifying sires more suited to their commercial enterprises.

A preliminary study of carcass value and its components for a range of Merino sires based on their wether progeny born and raised in diverse climates and production systems is presented in this paper.

## MATERIALS AND METHODS

Data were recorded on the carcasses produced from F1 wethers at 2 MLP sites managed within

* A joint venture of NSW Department of Primary Industries and the University of New England

mixed farming (Macquarie, MCQ) and fine wool production (New England, NE) systems. The design of the MLP project has been described by Ramsay *et al.* (2019), with the protocols that produced the F1 progeny at the MCQ site described by Egerton-Warburton *et al.* (2019). These protocols were implemented also at the NE site. The data were recorded on the F1 wethers born in 2017 and 2018 following AI mating of industry sires in each of two years to foundation ewes. The MLP project web site (https://merinosuperiorsires.com.au/mlp-project) provides details on the sources of sires and foundation ewes at each site. The wethers were the progeny of 30 (MCQ) and 28 (NE) sires, with 2 sires used across both sites. From weaning to slaughter, the wethers were production fed to achieve a target liveweight of 48 kg due to drought conditions affecting both sites. Following supplementation on pasture, the NE wethers were finished in a commercial feedlot, whilst MCQ wethers were finished on-site. The wethers were weighed and transported to a commercial abattoir, and held overnight in lairage with water before slaughter the next day.

The MCQ wethers were slaughtered during mid-March to late-May in 2018 and mid-February to early-May in 2019, whereas the NE wethers were slaughtered in early-August and mid-September in 2018 and early-July and mid-August in 2019. MCQ wethers were slaughtered at an average age of 11.6 months and 11 months while NE wethers were slaughtered at an average age of 11.7 and 10.8 months in 2018 and 2019, respectively.

Assessments on each carcass included: hot carcass weight (HCWT, kg), dressing percentage (DP, %), and tissue (fat) depth at the 12th rib, 110 mm from the backbone and measured using a GR knife (GRFAT, mm). The carcass data were collected from 462 and 499 MCQ wethers and 242 and 244 NE wethers in 2018 and 2019, respectively. Carcass value (CVAL, AUD$ per head) was calculated for each carcass using its HCWT, GRFAT and over the hook (OTH) price information from the abattoir feedback reports for the slaughters. GRFAT was used to adjust for carcasses being outside specifications i.e. deductions of $0.30 per kg for carcasses with fat score 1 (≤5 mm) or of fat score 5 (≥21 mm). Summary statistics for the sites are shown in Table 1.

**Table 1. Descriptive statistics for carcass traits in MCQ and NE data**

| **Birth year** | | **Macquarie** | | **New England** | |
|---|---|---|---|---|---|
| | | Mean (SD) | Range | Mean (SD) | Range |
| 2017 | HCWT | 24.4 (1.87) | 19.3 - 33.7 | 23.8 (2.64) | 16.7 - 33.0 |
| | DP | 45.0 (2.31) | 38.5 - 61.9 | 46.7 (2.11) | 35.6 - 51.8 |
| | GRFAT | 10.8 (4.04) | 2 - 25 | 13.8 (3.81) | 3 - 25 |
| | CVAL | 153.21 (15.16) | 110.01 - 215.68 | 174.90 (26.24) | 113.56 - 254.10 |
| 2018 | HCWT | 25.9 (1.86) | 21.4 - 32.3 | 29.5 (3.38) | 21.4 - 40.1 |
| | DP | 46.7 (1.94) | 38.8 - 52.0 | 47.9 (2.18) | 35.3 - 54.5 |
| | GRFAT | 14.4 (3.32) | 7 - 31 | 21.7 (5.13) | 11 - 42 |
| | CVAL | 164.64 (12.76) | 128.40 - 206.72 | 224.20 (29.11) | 149.80 - 308.77 |

For each data source, separate analyses to calculate adjusted sire means for each trait were performed using ASReml (Gilmour *et al.* 2015). The fixed effects of sire, ewe bloodline and their interaction were first tested, with non-significant effects then excluded from the model. Random effects fitted in the model were birth type (single, twin, triplet (MCQ data only)), rearing type (single, twin) and dam age (2 (NE data only), 3, 4, 5, 6 and 7 year old at mating), as well as a contemporary group effect (accounting for management and slaughter group effects).

## RESULTS AND DISCUSSION

Sire was significant for all traits at both sites ($P < 0.001$), while ewe bloodline was significant for HCWT ($P < 0.05$) and CVAL ($P < 0.05$) at the MCQ site only. The interaction between sire and ewe bloodline was not significant for any trait at either site.

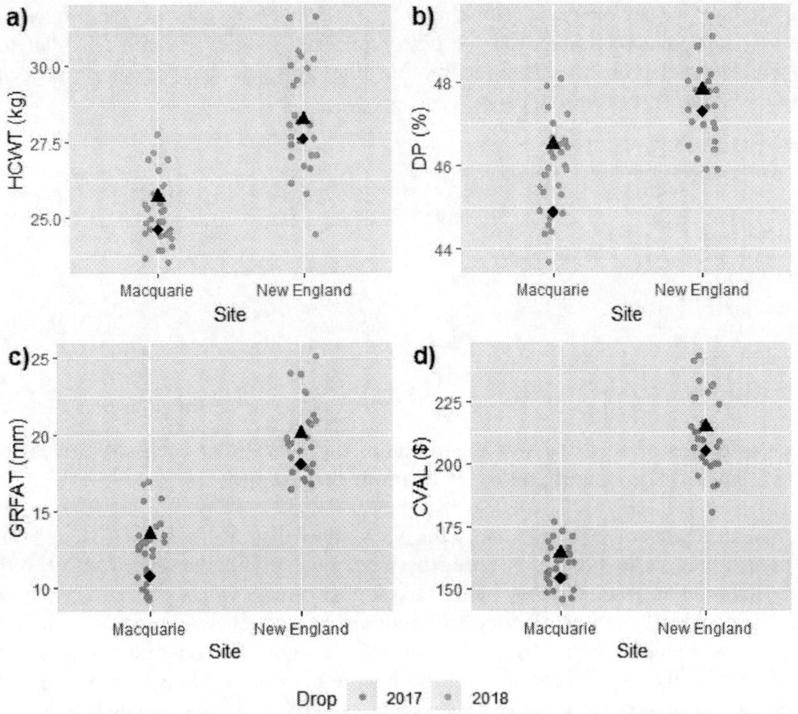


**Figure 1. Adjusted sire means for a) HCWT, b) DP, c) GRFAT and d) CVAL at MCQ and NE sites, with black diamonds and triangles representing median values within site for 2017 and 2018 birth years respectively**

At the MCQ site, sire adjusted means for HCWT, DP, GRFAT and CVAL ranged between 23.5±0.30 and 27.8±0.30 kg, 43.7±0.35 and 48.1±0.35%, 9.2±0.67 and 17.0±0.57 mm and $145.78±2.24 and $177.11±2.20 per head, respectively (Figure 1). Across both birth years, the averages were 25.2 kg, 45.9%, 12.5 mm and $159.06 per head for these traits, respectively. The ranges in sire adjusted means at the NE site, where the wethers were finished in a commercial feedlot, were 24.4±0.85 to 31.6±1.00 kg, 45.9±0.50 to 49.6±0.53%, 16.4±1.25 to 25.1±1.20 mm and $180.82±7.91 to $243.30±9.21 for HCWT, DP, GRFAT and CVAL, respectively (Figure 1). The average values across birth years for these traits were 28.2 kg, 47.5%, 19.7 mm and $212.49 per head, respectively. Among the sires of the 2017 born progeny at the MCQ and NE sites, the ranges in CVAL means were $17.99 and $51.87, respectively. The ranges in CVAL for the 2018 born progeny were $23.46 and $60.23, respectively.

Due to the assumptions used in this study, carcass value was essentially determined by carcass weight (Figure 2a; unity correlation between CVAL and HCWT at both sites). However, for sires with similar adjusted means for carcass value, a range in mean carcass fat levels was evident (Figure 2b, correlations of HCWT with GRFAT of 0.81 at the MCQ site and 0.69 at the NE site).

Sires were only compared within site and consequently within their own finishing system, where both systems had a target liveweight of 48 kg at slaughter. This, together with the NE progeny being finished under feedlot conditions, produced fatter carcasses from the NE wethers at the same weight and similar ages to carcasses from the MCQ wethers. Adjusted sire means for GRFAT were 21 mm and over for 29% of NE sires, versus none for MCQ sires. This contrasts with the perception that fine wool Merinos are late maturing (Hopkins *et al.* 2005) and suggests that the progeny of certain NE sires may not have been managed for best expression of their genetic potential for growth balanced with fat level, probably due to feedlot finishing. This was not apparent for progeny of sires at the MCQ site that were pasture finished.

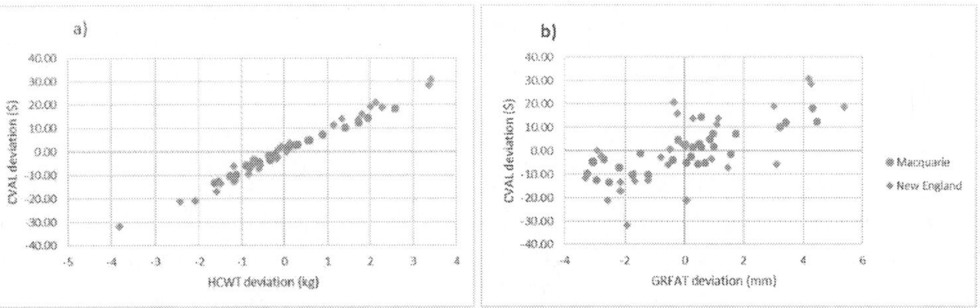


**Figure 2. Deviations of adjusted sire means from the average at each site for carcass value relative to a) hot carcass weight and b) fat depth at the GR site**

MLA market reports of OTH indicators for NSW show that during both 2018 and 2019 prices received at the time of slaughters of NE progeny were much higher than when MCQ wethers were processed (as for the feedback reports), hence their higher carcass values. Also, information was not readily available on the impact of price differentials for fat levels on carcass prices to use in predicting carcass value. Future work will involve more rigorous economic analyses, where both returns and costs are considered for both finishing systems, wool value is included, the impacts of reproduction are evaluated and relationships with breeding values are estimated (following Hall *et al.* 1997). Furthermore, rather than relying on actual prices received at one point in time, the analyses will evaluate the sensitivity of income from carcasses to changes in the relative value of component traits. In conclusion, this preliminary study has shown that considerable variation exists in carcass value of individual Merino sires when based on a simple economic model.

**ACKNOWLEDGEMENTS**

We acknowledge the co-investment by Meat & Livestock Australia in the conduct of this study through its funding (P.PSH.1032). Australian Wool Innovation Limited is the owner of pedigree and fixed effect data from the MLP Project used by this study. Staff are thanked at: NSW DPI Cowra, for the recording of data in the abattoir; Fletcher International Exports, for their support; and NSW DPI Trangie and CSIRO Armidale, for their contributions to sheep management.

**REFERENCES**

Clarke, B.E., Young, J.M., Hancock, S. and Thompson, A.N. (2019) *Proc. Assoc. Advmt. Anim. Breed. Genet.* **23**: 516.

Egerton-Warburton, K.L., Mortimer, S.I. and Swan, A.A. (2019) *Proc. Assoc. Advmt. Anim. Breed. Genet.* **23**: 520.

Gilmour, A.R., Gogel, B.J., Cullis, B.R., Welham, S.J. and Thompson, R. (2015) 'ASReml User Guide Release 4.1 Functional Specification'. VSN International Ltd, Hemel Hempstead, UK.

Hall, D.G., Fogarty, N.M. and Holst, P.J. (1997) *Proc. Assoc. Advmt. Anim. Breed. Genet.* **12**: 355.

Hopkins, D.L., Hatcher, S., Pethick, D.W., and Thornberry, K.J. (2005) *Aust. J. Exp. Agric.* **45**: 1225.

Ramsay, A.M.M., Swan, A.A. and Swain, B.C. (2019) *Proc. Assoc. Advmt. Anim. Breed. Genet.* **23**: 512.

*Proc. Assoc. Advmt. Anim. Breed. Genet: 24: 259-262*

# THE IMPORTANCE OF EARLY ENVIRONMENTAL EFFECTS ON MERINO FLEECE TRAITS ACROSS TWO SHEARINGS

**K.L. Egerton-Warburton[1], J.L. Smith[2], S.I. Mortimer[3], A.A. Swan[4], and B.C. Hine[2]**

[1] NSW Department of Primary Industries, Orange, NSW, 2800 Australia
[2] CSIRO, Agriculture and Food, F.D. McMaster Laboratory, Armidale, NSW, Australia
[3] NSW Department of Primary Industries, Armidale, NSW, 2351 Australia
[4] Animal Genetics and Breeding Unit*, University of New England, NSW, 2351 Australia

## SUMMARY

The importance of early environmental effects, and their estimates, on yearling and adult fleece traits recorded in 2 flocks managed under Lifetime Ewe Management (LTEM) guidelines were evaluated. Significance and overall influence of the effects of birth type and rear type were generally consistent with previous reports. However, estimates of the size of effects were generally larger than those previously reported, with the specific context of LTEM and impacts of management on early environmental effect estimates requiring further investigation.

## INTRODUCTION

In breeding programs, we seek to disentangle the effects of genes from environment to obtain accurate estimates of the genetic merit of individuals and maximise genetic progress. To do so, adjustment factors are applied in the MERINOSELECT genetic evaluation system (Brown *et al.* 2007) to account for environmental influences such as age of dam, birth type, rear type, and date of birth or age at trait assessment. For example, classer grades are known to be influenced by birth/rearing type (Clarke and Thompson 2021; Mortimer *et al.* 2009) and age of dam (Mortimer *et al.* 2009). From the earliest genetic studies of Merino sheep, summarised by Turner and Young (1969), it is established that most fleece traits are influenced by early environmental effects, with later work showing these effects to be important across a range of ages (e.g. Huisman *et al.* 2008). Genetic evaluation systems in Australia generally use 'multiplicative' adjustments to account for different levels of performance across breeds and sites when accounting for fixed effects (Graser *et al.* 2005; Gilmour 1993).

The Lifetimewool project established that improving the nutritional management of Merino ewes during pregnancy and lactation resulted in their progeny having heavier, finer fleeces across several shearings (Thompson *et al.* 2011). Guidelines from this project underpin the Lifetime Ewe Management (LTEM) program (http://www.lifetimewool.com.au/), providing recommendations for base ewe management in sire evaluation flocks (AMSEA 2018). This paper reports the importance and persistence of early environmental effects (birth/rearing type, dam age) on measured fleece traits assessed at 2 ages (yearling, and first adult shearings) and at 2 locations in progeny from dams managed to LTEM targets.

## MATERIALS AND METHODS

Data used in these analyses were from 2 shearings of the F1 ewe and wether progeny of the Merino Lifetime Productivity (MLP) project conducted at the Macquarie (MCQ) and New England (NE) sites. The design of the MLP Project has been described previously by Ramsay *et al.* (2019). Assessment data were collected according to the AMSEA guidelines (AMSEA 2018). Dams of the

---

* A joint venture of NSW Department of Primary Industries and the University of New England

progeny were managed in line with Lifetimewool regional guidelines, with multiple bearing ewes managed differentially to single bearing ewes (http://www.lifetimewool.com.au/guidelines.aspx).

F1 progeny at the MCQ site were born in May-June, whilst F1 progeny at the NE site were born in late August/early September of each year. At each site, a yearling (Y, 300 to 400 days age) and an adult (A, 540 days or older) fleece assessment were completed. Shearing at the MCQ site occurred in late February (Y) and mid October (A), whilst shearing at NE occurred in August (Y) and July (A). Greasy fleece weight (GFW, kg), clean fleece weight (CFW, kg) and fibre diameter (FD, µm) data were analysed.

For each data source (MCQ or NE site), analyses were performed using R (R Core Team 2020). Significance of early environmental effects was tested in models that fitted a random effect of sire. For both sites, the fixed effects examined included birth type (BT: single, twin, triplet), rearing type (RT: single, twin, triplet) and age of dam (DAGE: 2, 3, 4, 5, 6 and 7 years old at mating), as well as a contemporary group effect defined by year of birth, management group, ewe bloodline source and sex. Interactions among the BT, RT and DAGE were tested but were found to be not significant for these traits and therefore were not fitted in the final models. Age at observation was fitted as a linear covariate. Table 1 summarises the data available at each site for each trait.

**Table 1. Descriptive statistics for selected fleece traits (including number of records, n)**

| | | YGFW | AGFW | YCFW | ACFW | YFD | AFD |
|---|---|---|---|---|---|---|---|
| Macquarie | Mean[1] | 3.8 | 7.6 | 2.4 | 4.7 | 17.2 | 18.8 |
| | | (0.75) | (1.30) | (0.51) | (0.89) | (1.38) | (1.46) |
| | n | 2013 | 961 | 2013 | 961 | 2015 | 961 |
| | Range | 1.6 - 6.2 | 3.4 - 12.0 | 1.0 - 4.2 | 1.4 - 7.8 | 13.1 - 22.1 | 15.0 – 24.4 |
| New England | Mean | 2.7 | 4.5 | 2.0 | 3.3 | 15.2 | 16.1 |
| | | (0.66) | (0.68) | (0.47) | (0.53) | (1.05) | (1.26) |
| | n | 2170 | 2151 | 2170 | 2151 | 2175 | 2152 |
| | Range | 1.2 - 5.6 | 2.7 - 7.3 | 0.8 - 4.2 | 1.9 - 5.5 | 11.4 - 19.9 | 12.5 - 21.0 |

[1]Standard deviations shown in brackets below the mean

## RESULTS AND DISCUSSION

All Y and A fleece weight traits examined were influenced significantly by BT (Table 2). The lighter fleeces of multiple-born animals relative to single-born animals were still evident at their first adult shearing, consistent with effects on fleece weights of yearlings, hoggets and adults reported previously by Huisman *et al.* (2008). Rearing type (RT) was significant for Y and A fleece weight traits at the NE site, but only significant for the Y fleece weight traits at the MCQ site. Early environmental effects were not significant for FD at either Y or A stage at the MCQ site; only BT was significant for YFD and AFD at the NE site. In general, the significance of the effects of RT and dam age declined with stage. The significance of the BT effect on fleece weight traits was maintained across the two shearings, although the size of the effect generally tended to decrease. Age at shearing was only significant for the Y fleece weights at the NE site ($P < 0.0001$).

The results of this analysis contrast with those of Huisman *et al.* (2008) who reported that the RT effects on Y fleece weights were approximately half of those of BT effects. Additionally, when effects are converted to a 'multiplicative' basis (Table 2), the adjustments for BT≥2 are larger than published estimates for sheep (Gilmour 1993), but do decline with stage. The size of the estimate of the BT and RT effects on fleece weight at these MLP sites are large in comparison with those reported by Thompson *et al.* (2011), on a site in Victoria across annual shearings from 15 months of age. Our results are consistent with Thompson *et al.* (2011) with respect to fleece weight (e.g. their estimate had twin-born animals producing 0.19 kg less wool than single-born animals), but do not in general support their effects reported for FD (twin born animals had 0.26 µm broader fibres).

**Table 2. Significance of fixed effects[1] and their estimates[2] and multiplicative adjustments[3] for greasy fleece weight (GFW, kg) clean fleece weight (CFW, kg) and fibre diameter (FD, µm) at yearling (Y) and adult (A) assessments**

| | | Macquarie | | | | New England | | | | | |
|---|---|---|---|---|---|---|---|---|---|---|---|
| Effect | Level | YGFW | AGFW | YCFW | ACFW | YGFW | AGFW | YCFW | ACFW | YFD | AFD |
| Birth type | Twin | -0.46*** | -0.45** | -0.37*** | -0.17* | -0.23*** | -0.22*** | -0.18*** | -0.20*** | 0.11** | 0.14*** |
| | Triplet | -0.61*** | -0.49** | -0.47*** | -0.26* | -0.42*** | -0.37*** | -0.32*** | -0.29*** | 0.24* | 0.62*** |
| Rear type | Twin | -0.34*** | | -0.22*** | | -0.33*** | -0.11** | -0.27*** | -0.11*** | | |
| | Triplet | -0.41** | | -0.32** | | -0.18 | -0.03 | -0.18* | -0.13 | | |
| Age of dam | 3 | | | | | 0.08* | 0.07‡ | 0.07** | 0.09* | | |
| (years) | 4 | 0.13‡ | | | | 0.10** | 0.09* | 0.09*** | 0.10** | | |
| | 5 | 0.03 | | | | 0.14*** | 0.18*** | 0.12*** | 0.16*** | | |
| | 6 | 0.03 | | | | 0.11‡ | 0.07 | 0.10* | 0.10‡ | | |
| | 7 | -0.07 | | | | 0.09 | -0.01 | 0.14 | -0.003 | | |
| Age | Linear | | | | | 0.018*** | | 0.013*** | | | |
| Multiplicative | 2:1 | 1.12 | | 1.16 | | 1.08 | 1.05 | 1.09 | 1.06 | | |
| (BT:RT) | 2:2 | 1.23 | 1.06 | 1.28 | 1.04 | 1.22 | 1.08 | 1.26 | 1.10 | 0.99 | 0.99 |
| | 3:1 | 1.17 | | 1.21 | | 1.15 | 1.09 | 1.18 | 1.09 | | |
| | 3:2 | 1.29 | | 1.34 | | 1.32 | 1.12 | 1.38 | 1.14 | | |
| | 3:3 | 1.32 | 1.07 | 1.42 | 1.06 | 1.24 | 1.09 | 1.30 | 1.14 | 0.98 | 0.96 |

[1] ‡, $P < 0.10$; *, $P < 0.05$; **, $P < 0.01$; ***, $P < 0.001$; level of significance based on t-test between estimates
[2] Estimates expressed as: birth type as the deviation from estimates for single animals of estimates for twin and triplet animals; rearing type as the deviation from estimates for single animals of estimates for twin and triplet animals; age of dam as the deviation from estimates for youngest dams (2 year olds) of estimates for older dams. Where cells are blank the effect was not significant.
[3] Multiplicative adjustments: calculated from marginal means of BT and RT expressed relative to BT:RT 1:1 (single born, single reared)

This is an important observation considering that the intent of the differential management of multiple bearing ewes under LTEM is to provide optimal nutrition to offset any lifetime effects of BT and RT. Further investigation of this finding could be achieved by extending this analysis to other MLP sites, especially the Balmoral site whose environment and sheep type are most similar to the Victorian site at which data were collected and reported by Thompson *et al.* (2011). In addition to MLP sites where LTEM management is known to be applied, identifying the size of these effects in the wider MERINOSELECT database is also necessary.

This exploratory study will also be expanded to examine other measured and visual wool traits at the two sites reported herein, and later stages of assessment (third adult and later shearings).

**CONCLUSIONS**

This study of the importance of early environmental effects, and their estimates, on yearling and adult fleece traits recorded in two flocks managed under Lifetimewool guidelines found the significance and overall influence of the effects of birth type and rear type were generally consistent with previous reports. Estimates of effects were generally larger than those previously reported, but the specific context of LTEM and impacts of management on early environmental effect estimates requires further investigation.

**ACKNOWLEDGEMENTS**

The Merino Lifetime Productivity Project is being undertaken in partnership between the Australian Merino Sire Evaluation Association Incorporated (AMSEA) and Australian Wool Innovation (AWI). AMSEA and AWI acknowledge those entities who also contribute funding: woolgrowers through sire evaluation entry fees, site hosts, site committee in-kind contributions, and sponsors of AMSEA. The MLP Macquarie site gratefully acknowledge the generous financial support provided by the site host NSW Department of Primary Industries. The MLP New England site gratefully acknowledges the generous financial support provided by the site host, CSIRO.

Projects ON 536 (MCQ site) and ON 511 (NE site) also provided data for this study, and we gratefully acknowledge AWI for their co-investment. Staff and Site Committee volunteers at the respective MLP sites are also thanked for their contributions to data collection.

**REFERENCES**

AMSEA (2018) 'Merino Sire Evaluation Site Requirements'. Version 180831.

Brown D.J., Huisman A.E., Swan A.A., Graser H-U., Woolaston R.R., Ball A.J., Atkins K.D. and Banks R.G. (2007) *Proc. Assoc. Advmt. Anim. Breed. Genet.* **17**: 187.

Clarke B.E. and Thompson A.N. (2021) *Anim. Prod. in Aust.* **33:** clii

Gilmour A.R. (1993) 'BVEST: A multiple trait animal model Best Linear Unbiased Predictor breeding value estimation program'. NSW Agriculture, Orange, Australia.

Graser H-U., Tier B., Johnston D.J. and Banks R.G. (2005) *Aust. J. Exp. Ag.* **45**: 913.

Huisman A. E., Brown D.J., Ball A.J. and Graser H.-U. (2008) *Aust. J. Exp. Ag.* **48**: 1177.

Mortimer S.I., Robinson D.L., Atkins K.D., Brien F.D., Swan A.A., Taylor P.J. and Fogarty N.M. (2009) *Anim. Prod. Sci.* **49**: 32.

R Core Team (2020) R Foundation for Statistical Computing, Vienna, Austria. https://www.R-project.org/.

Ramsay A.M.M., Swan A.A. and Swain B.C. (2019) *Proc. Assoc. Advmt. Anim. Breed. Genet.* **23**: 512.

Thompson A.N., Ferguson M.B., Gordon D.J., Kearney G.A., Oldham C.M. and Paganoni B.L. (2011) *Anim. Prod. Sci.* **51**: 794.

Turner H.N. and Young S.S.Y. (1969) 'Quantitative Genetics in Sheep Breeding'. Macmillan of Australia, Melbourne.

*Proc. Assoc. Advmt. Anim. Breed. Genet: 24: 263-266*

# USING MID-INFRARED SPECTROSCOPY PREDICTIONS OF FERTILITY TO OPTIMISE SEMEN ALLOCATION IN DAIRY HERDS

**J.E. Newton[1], P.N. Ho[1] and J.E. Pryce[1,2]**

[1] Agriculture Victoria, AgriBio, Centre for AgriBioscience, Bundoora, Victoria 3083, Australia
[2] School of Applied Systems Biology, La Trobe University, Bundoora, Victoria 3083, Australia

## SUMMARY

As most dairy cows require more than one artificial insemination (AI) to fall pregnant, prioritising more fertile cows for insemination with expensive semen could support optimised semen usage. In this study, we explored two approaches for identifying high (H) and low (L) fertility cows in dairy herds; the use of calving dates (CD) and probability of conception to first service (pMIR) predicted from milk mid-infrared (MIR) spectra and other early lactation data. We found cows classified as H fertility by pMIR had 1$^{st}$ service and overall conception rates (CR_1, CR_O) 2.0 to 6.2% higher compared to those classified H by CD. Both H subgroups had higher CR_1 and CR_O than herd average with the reverse also true for L subgroups. Differences in CR_1 between H and L cows were approximately 50% greater (up to 17.7%) when pMIR was used to classify cows compared to CD (up to 10.8%). This shows pMIR was better than CD at identifying cows most and least likely to conceive at first service. However, total number of AI events between cows classified using pMIR or CDs were similar. A preliminary case study exploring three strategies for assigning sexed dairy semen (SS), conventional dairy semen (DS) and beef semen (BS) in dairy herds found that the net benefit (calf values minus semen costs) was greatest when pMIR was used to assign SS and BS to H and L fertility cows, respectively, followed by CD and random semen allocation, \$70.26 ± 3.05, \$68.68 ± 3.05 and \$66.73 ± 3.11/cow, respectively. Differences in net benefit were largely due to the higher number of heifer replacements generated in the pMIR strategy. Therefore, pMIR has promise as a tool for identifying the most and least fertile dairy cows. The pMIR predictions could be used alone, or in conjunction with other fertility indicators to support optimised allocation of semen, including sexed semen, in dairy herds and offer the next generation of breeding tool.

## INTRODUCTION

Average CR_1 in dairy herds in Australia is 39% but ranges from 22 to 61% (Dairy Australia 2011). This indicates there are differences in the ability to conceive at first service. Being able to identify cows that are more likely to conceive at first service could support optimised semen usage, particularly the incorporation of more expensive semen, such as sexed semen, into breeding programs. Cows with well-managed transition periods are more likely to fall pregnant again (Roche *et al.* 2013), conversely cows that calve later in the calving period are less likely to conceive in the subsequent joining period (Dennis *et al.* 2018). New phenotyping technologies and computational approaches offer additional opportunities to identify more fertile cows. Ho and Pryce (2020) have previously demonstrated that mid-infrared (MIR) spectroscopy and other data collected on-farm in early lactation can be used to rank cows on the probability of conception to first service (pMIR) with accuracies of up to 76% achieved in identifying cows that are least likely to fall pregnant. The hypothesis of Ho and Pryce (2020) is that cows that are phenotypically divergent in fertility give clearer biological signals for training MIR prediction models. Our aims were two-fold: firstly, to compare reproductive performance of cows identified as high or low likelihood of conceiving to first service based on pMIR or CD and second, to explore the net benefit of three strategies for allocating semen in dairy herds: random allocation, allocation based on earliest to latest CD and allocation based on highest to lowest pMIR.

## MATERIALS AND METHODS

Lactation parameters including MIR data from first herd test after calving, subsequent AI records and calving details were available for 11,369 dairy cows (13,379 records) across 76 herd-year-calving seasons (HYC). The pMIR was generated for each cow using a model which combines MIR spectral data and other on-farm parameters (milk production, milk SCC, days from calving to insemination, calving season, days in milk and age at calving) previously described in Ho and Pryce (2020). Briefly, a training population of cows with good (conceived to first service) and poor (did not conceive within the mating season and had only one insemination) fertility was created and used to train a prediction model using partial least squares discriminant analysis. This model was then used to derive pMIR (0 to 1) of cows in a new herd that had not been included in the training set. To test the ability of pMIR to identify cows with higher and lower fertility compared to other approaches, each HYC was divided into H and L subgroups using one of the 2 classification approaches and reproductive parameters calculated for each subgroup. The H and L fraction (increments from 5 to 50%) of each HYC was selected based on pMIR in the first instance, and in the second approach were selected based on earliest (H subgroup) and latest (L subgroup) CD. Then CR_1, CR_O, average number of inseminations overall and to achieve a pregnancy were calculated and compared between H and L subgroups and to HYC average performance.

As similar trends in performance of H and L subgroups were seen irrespective of fraction selected, a case study was used to compare 3 strategies for allocating SS, DS and BS. The case study assumed 20% of cows (H subgroup) were assigned to SS at first service and DS for subsequent services, 20% of cows (L subgroup) received BS only, and the remaining 60% of cows received DS within each HYC. The strategies were as follows:

**Strategy 1 (pMIR)**: cows were assigned to H and L subgroups based on pMIR.

**Strategy 2 (CD)**: cows were assigned to H and L subgroups based on earliest to latest CD.

**Strategy 3 (random)**: cows within a HYC were randomly sorted using a random number generator then split into subgroups representing 20%, 60% and 20% of cows and assigned SS, DS and BS, respectively. The results were averaged over 100 replicates of random sampling.

Net benefit was calculated as calf values minus semen costs, assuming number of AI events and calves born remained static across strategies. National average semen prices of $50, $20 and $10 were assumed for SS, DS and BS, respectively, while dairy heifers, male dairy calves and dairy-beef crossbred calves were valued at $275, $54.30 and $100, respectively (Byrne *et al.* 2016). All analyses were conducted within HYC with overall averages presented here.

## RESULTS AND DISCUSSION

High subgroups selected by CD and pMIR both identified cows with higher conception rates than HYC averages, regardless of the fraction of the herd chosen (Figure 1a,b). However, cows with higher pMIR had higher CR_1 and CR_O than cows that calved earliest in the calving period. For example, the average CR_1 in the data set was 38.8%. When 20% of cows were selected as H using pMIR, CR_1 was 45.7% compared to 41.5% when H cows were selected using CD, thus showing a 4.2% advantage of using pMIR. Conversely, CR_1 was on average 2.8% lower in L sub-herds selected on pMIR, compared to CD. Both L subgroups had lower conception rates than average HYC, regardless of fraction of herd compared. The difference in CR_1 between H and L subgroups ranged from 9.7% to 17.7% when cows were chosen based on pMIR, while the difference was only 5.6 to 10.8% when cows were chosen based on CD. This suggests that pMIR may be better than CD at identifying cows with highest and lowest likelihood of conceiving.

When more than 15% of cows were selected (Figure 1c), H subgroups had slightly fewer total AI events that L subgroups ranging from 0.05 to 0.08 fewer AI events/cow. No clear difference was seen between classification based on pMIR or CD. When less than 20% of HYC was selected, H cows had up to 0.11 and 0.09 more AI events to conceive than L cows, for CD and pMIR, respectively (Figure 1d). In scenarios above 20%, L cows had more AI events to achieve pregnancy though differences were small (<0.03). These small observed differences in AI records could be influenced by the strategy used for classifying L cows using pMIR. Ho and Pryce (2020) defined a poor fertility as a cow with only 1 recorded AI event who did not fall pregnant. A failure to show return of oestrus after first insemination could indicate physiologically different reasons for not conceiving than a cow who fails to conceive after multiple AI events.

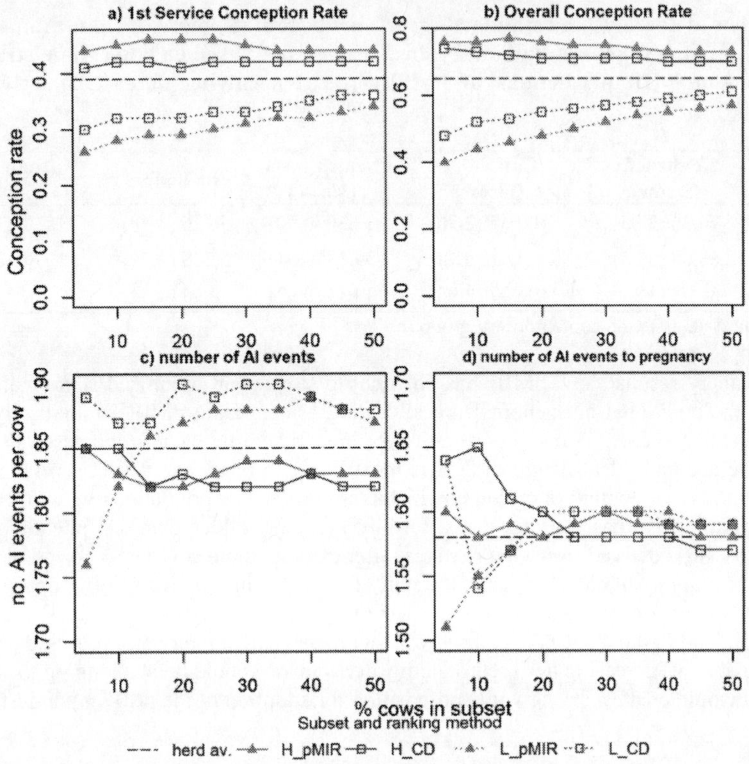


**Figure 1. Comparison of average (av.) 1st service conception rate (CR), overall CR, number of artificial insemination (AI) events and number of AI events to achieve pregnancy in a variable (5-50%) proportion of cows within a contemporary group, selected as having highest (H) or lowest (L) probability of conceiving to 1st service (pMIR) or earliest (H) or latest (L) calving date (CD). Dotted line shows herd av.**

In our case study, semen allocation using pMIR resulted in a higher net benefit than CD or randomly assigning semen, though all strategies showed a high level of variability (Table 1). Average net benefit per cow for pMIR was \$70.26 ± 3.05 compared to \$68.68 ± 3.05 for CD and \$66.73 ± 3.11 for random semen allocation. Average semen costs only differed by ~\$0.10/cow across the 3 strategies. Differences in net benefit were mainly driven by differences in the number of each calf type, with 1.0 and 3.28 more dairy heifer calves in the pMIR strategy compared to CD and random semen allocation strategies, respectively. This was largely driven by higher CR_1 to SS in H cows selected by pMIR. Conversely, lower CR_1 in L subgroups saw fewest dairy-beef calves in the pMIR strategy, followed by CD while random allocation resulted in most dairy beef calves. As a preliminary case study of potential applications of pMIR applications on farm, the net benefit calculations solely considered semen prices and calf value. Potential effects of reduced CR_1 with SS were ignored. Although SS conception rates are approaching parity with DS, this area warrants further study. Future analyses could also consider a wider range of semen prices, calf genetic merit and the impact of calving date on a calf's future value on-farm.

**Table 1. Overview of average (and standard error[1]) net benefit ($/cow, calf value minus semen costs), number of dairy heifers and dairy beef calves for 3 semen allocation strategies, allocation based on MIR predictions of fertility (pMIR), calving date (CD) or random allocation**

| Semen strategy | Net benefit ($/cow) | Calf value ($/cow) | Semen costs ($/cow) | No. heifers | No. dairy beef calves |
|---|---|---|---|---|---|
| 1. pMIR | 70.26 (3.05) | 109.55 (2.90) | 39.29 (0.90) | 56.26 (4.78) | 18.38 (1.85) |
| 2. CD | 68.68 (3.05) | 115 (3.18) | 39.23 (0.91) | 55.26 (4.73) | 19.80 (2.01) |
| 3. random | 66.73 (3.11) | 106.07 (2.96) | 39.33 (0.92) | 52.98 (4.64) | 23.52 (2.07) |

[1]reported across herd-year-season contemporary group (n=76)

These preliminary results show pMIR has potential to support an optimised semen allocation strategy. While the additional net benefit from allocating semen using pMIR is small, given the importance of fertility to dairy farms opportunities for incremental net benefit increases should be considered. More accurate identification of cows most likely to conceive at first service may be possible through the development of an index which combines pMIR predictions with other easily accessible information like past calving dates, fertility breeding values or novel phenotypes like sensor data. As pMIR is derived from first herd test after calving, there could be a period of up to 8 weeks between availability of pMIR data and the start of joining. This could also offer opportunities to provide management interventions for cows, especially those identified as least likely to conceive at first service (Ho and Pryce, 2020). Validating whether management interventions based on pMIR data are then capable of increasing herd reproductive performance could be challenging to achieve, but if successful would create a strong value proposition for adoption of the pMIR by industry.

## CONCLUSION

This study shows that pMIR identifies the most and least fertile cows in the milking herd better than CD and has potential as a next generation breeding tool. The pMIR predictions could be used alone, or in conjunction with other fertility indicators to support optimised allocation of semen, including sexed semen, to increase the number of dairy heifer replacements on farm or to support additional income streams like dairy beef.

## REFERENCES

Dairy Australia (2011) The InCalf Fertility Data Project 2011.

Byrne T.J., Santos B.F.S., Amer P.R., Martin-Collado D., Pryce J.E., Axford M. (2016) *J. Dairy Sci* **99**: 8146.

Dennis N.A., Stachowicz K., Visser B., Hely F.S., Berg D.K., Friggens N.C., Amer P.R., Meier S., Burke, C.R. (2018) *J. Dairy Sci* **101**: 3176.

Ho P.N and Pryce J.E. (2020) *J. Dairy Sci* **103**: 11535.

Roche J.R., Bell A.W., Overton T.R., Loor J.J. (2013) *Anim. Prod Sci.* **53**: 1000.

*Proc. Assoc. Advmt. Anim. Breed. Genet: 24: 267-271*

# EXPLORING IMPUTATION ACCURACY ACROSS THE BOVINE X CHROMOSOME

**T.V. Nguyen[1], S. Bolormaa[1], C.M. Reich[1], A.J. Chamberlain[1], A. Medley[2], C. Schrooten[3], H.D. Daetwyler[1,4] and I.M. MacLeod[1]**

[1] Agriculture Victoria, AgriBio, Centre for AgriBiosciences, Bundoora, Victoria, 3083, Australia
[2] CRV, Hamilton, New Zealand
[3] CRV, Arnhem, Netherlands
[4] School of Applied Systems Biology, La Trobe University, Bundoora, Victoria, 3083, Australia

## SUMMARY

Many of the current imputation benchmarking studies are performed on autosomes with limited studies addressing the X chromosome. Furthermore, the X chromosome genome map has recently been updated in the new ARS-UCD1.2 bovine reference genome. In this study, we evaluated the empirical accuracy of imputation from a low-density SNP array (LD) to 50K and then high-density (HD) for the pseudo-autosomal region (PAR), non-PAR, and autosomes across several scenarios using multiple dairy breed groups. Overall, imputation accuracies for the PAR were very low when imputing from LD to 50K, while accuracy for the non-PAR was comparable with autosomes. We demonstrated that imputation accuracies for the PAR increased when the PAR & non-PAR were merged for imputation. However, while this strategy performed well for imputing LD to 50K, there was no advantage when imputing from real 50K genotypes to HD. In addition, when imputing all chromosomes to HD level, imputing from real 50K to HD resulted in an overall higher accuracy than imputing from LD to 50K to HD, with the PAR region showing the most improvement. By separately imputing only the end segment of five autosomes and comparing accuracy with the PAR region, we demonstrated that the PAR region is more difficult to impute accurately, perhaps due to higher recombination rates. Therefore, future SNP genotyping panels should have SNP density in the PAR at least equivalent to that of the 50K SNP panel to achieve a good imputation result.

## INTRODUCTION

Genomic selection (Meuwissen *et al.* 2001) has created a dramatic breakthrough in the dairy industry during the last two decades. Accurate prediction of breeding values requires medium to high-density genome-wide markers, but many of the dairy genomic reference populations have been genotyped on a range of lower density platforms (6,000 to 25,000 markers) to reduce costs. Genotype imputation is considered an effective approach to provide the marker density required by the industry. To date, most studies that examined the empirical accuracy of imputation from low-density (LD) to medium (50K) or high-density (HD) SNP genotypes investigated imputation of autosomes only and this is generally highly accurate (Calus *et al.* 2014). The X chromosome generally requires modifications to the imputation pipeline because it has a 5.7 Mb region of homology between chromosome X & Y called the pseudo-autosomal region (PAR) and a larger non-PAR that is haploid in males. Two studies investigated the accuracy of imputation on the X chromosome (LD to 50K) and found it was much less accurately imputed compared to autosomes in cattle (Su *et al.* 2014; Mao *et al.* 2016). However, the imputation of the X chromosome warrants further study for three key reasons. First, these studies used the UMD-3.1 reference genome map, while recently the X chromosome map has been extensively updated on the ARS-UCD1.2 bovine reference, in particular the PAR region (Figure 1A). Second, these authors tested imputation to 50K density only and did not investigate strategies to improve the PAR imputation accuracy. Third, there may be important genetic variation on the X chromosome for economically important traits as reported for fertility (Pacheco *et al.* 2020). In this study, we evaluated the empirical accuracy of imputation from a LD SNP array to 50K and then to HD for the pseudo-autosomal region (PAR), non-PAR and autosomes across several scenarios using multiple dairy breed groups.

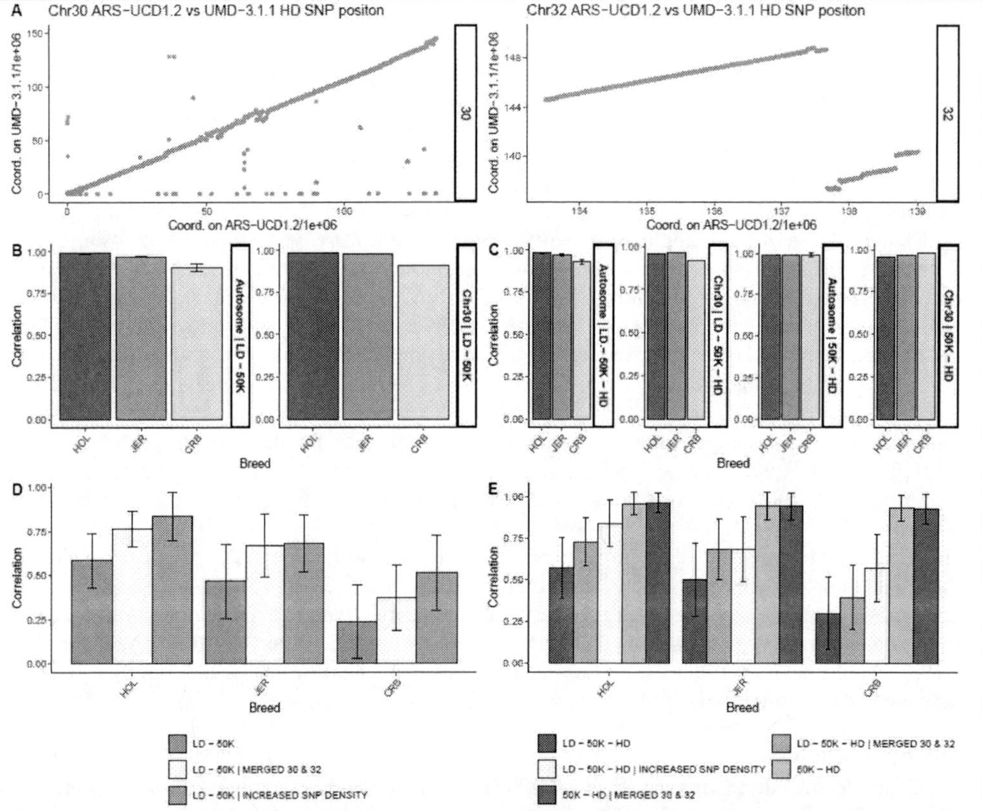


**Figure 1. (A) Bovine HD SNP coordinates on Chr30 (Left) and Chr32 (Right) on ARS-UCD1.2 and UMD-3.1.1 reference genome. Red dots represent SNPs that were relocated from other chromosomes. (B) Average imputation accuracy for autosomes and Chr30 imputed from LD to 50K, error bars represent Standard deviation (SD) of accuracies across autosomes. (C) Average imputation accuracy for autosomes and Chr30 imputed to HD level from either LD (via 50K) or direct from real 50K. Error bars represent SD of accuracies across autosomes. (D) Imputation scenario accuracy for Chr32 at 50K level. Error bars represent SD of accuracy across SNP. (E) Imputation scenario accuracy for Chr32 imputed to HD either from LD (via 50K) or direct from real 50K**. Error bars represent SD of accuracy across SNP.

## MATERIALS AND METHODS

The target animals used for this study included 35 Jersey (JER), 35 Holstein (HOL), and 35 crossbred (HOL, JER) bulls (CRB) and were genotyped using the Illumina® BovineHD chip. GenCall threshold score was set at 0.6: animals and SNP were removed if >10% of genotypes fell below this threshold. The marker map positions were based on the ARS-UCD1.2 reference genome (Rosen *et al.* 2020). The boundary between the non-PAR and PAR (hereby noted as Chr30 and Chr32 respectively) was set to 133,300,518bp (Johnson *et al.* 2019). Chr30 and Chr32 were imputed separately unless otherwise stated. We masked the HD genotypes (714,452 SNPs) to simulate either a LD SNP-chip of 7,135 markers or the 50K chip (40,397 markers). Two sub-experiments were conducted: (1) All autosomal (Chr1 to 29), non-PAR (Chr30), and PAR (Chr32) LD genotypes were either imputed to 50K and then to HD level or from real 50K genotypes to HD. (2) For comparison between accuracy of imputation on the PAR and autosomes, we selected the last 5,708,563 bp segment (equivalent to the length of Chr32) on Chr 1,2,3,4 and 5 and re-imputed only these short segments. On these autosomal segments the LD SNP density (N≈30) was double that of the PAR, therefore we compared imputation at two SNP densities: first we reduced every other marker of the autosome sets to mimic the density on Chr32 (N=15), and second, we doubled the density on Chr32 by including several 50K variants to mimic the LD marker density on autosomal segments (N=30). Imputation was performed using FImpute V3.0 (Sargolzaei *et al.* 2014). We estimated the accuracy of imputation as Pearson's correlation coefficient (r) between imputed and real genotypes and results are reported based on the mean per-SNP accuracy. Imputation to 50K was performed with a reference set of 14,000 animals that included HOL and JER, and imputation to HD was conducted with a similar mixed breed reference of 2,700 animals.

**RESULTS AND DISCUSSION**

In this study, we tested several imputation strategies for the PAR & non-PAR on the X chromosome and compared the accuracy to that of the autosomes. At 50K level, we found that pooling all samples (HOL, JER, and CRB) and using a mix breed reference gave similar imputation accuracy compared to imputing HOL or JER target sets separately with only the same breed in the reference. Therefore, we present results using pooled target and reference sets but show average accuracies for each breed group. We found some differences in accuracies between the breed groups: the CRB were lowest for LD to 50K (Fig. 1B) but as high as HOL and JER when imputing from real 50K to HD (Fig. 1C). However, the CRB were more related to the smaller HD reference than the 50K reference, implying that this caused the variation in imputation accuracies, to confirm this, we masked the HD reference down to 50K level to act as a new 50K reference and found similar imputation accuracy for all three breed groups (~0.96) .

We found that Chr30 imputation accuracy was high (>0.97) and comparable to autosomes for both 50K (Figure 1B) and HD level across target breed sets (Figure 1C), indicating that it is useful to include imputed genotypes from the non-PAR for downstream analysis. Conversely, Chr32 imputation accuracy was very low when imputing from LD to 50K (Figure 1D). Although it is recommended that Chr30 and Chr32 are imputed separately, by combining Chr32 and Chr30 (and re-extracting Chr32 genotypes) the accuracy increased by at least 15% for all breed groups when imputing to 50K (Figure 1D). Nonetheless, the accuracy is still rather low for downstream analysis. It should be noted that this strategy slightly reduced the imputation accuracy on Chr30 (results not shown), so markers on Chr30 should be imputed separately. Per SNP statistics for Chr32 showed that accuracy was improved in the borderline region between Chr30 & Chr32. This strategy of merging Chr30 and Chr32 provides a practical approach for historical datasets with low-density genotypes because increasing SNP density is not an option but should also be tested in females because our target animals were all males. When the SNP density was doubled on Chr32 (15 to 30) to mimic the number of SNPs in the last 5.7 Mb segment of Chr 1,2,3,4 and 5 the accuracy increased further (Figure 1D). At HD level, imputation of Chr32 from real 50K genotypes was always more accurate (0.92-0.97) than imputation from LD regardless of scenario (0.29-0.84). Although this was a little lower than the accuracy for Chr30 and autosomes, it was of high enough quality for downstream analyses. Notably, there was no longer an advantage in merging Chr32 with Chr30 for imputation from real 50K to HD (Figure 1E). This contradicts the result observed for LD to 50K level, suggesting that the denser markers available on the 50K SNP chip on Chr32 (99 SNPs) enable good resolution of Chr32 haplotypes.

One potential reason for the low accuracy on Chr32 may be simply that it is a very short segment to impute, and typically on all chromosomes the imputation accuracy tends to drop at the ends of the chromosomes. However, the results of our autosomal segment imputation test, demonstrated that when the marker density was made equivalent (either reducing density on the autosomal segments 1-5 or increasing density on Chr32), the accuracy was always better for the autosomal segments relative to Chr32 (Figure 2). We believe that higher recombination frequencies on Chr32 compared to the autosomes (Van Laere et al. 2008) might be responsible for increased haplotype complexity of this region. Therefore, when designing SNP panels for genotyping, it is perhaps critical to use SNP densities on Chr32 that are at least equivalent to those on the 50K chip.

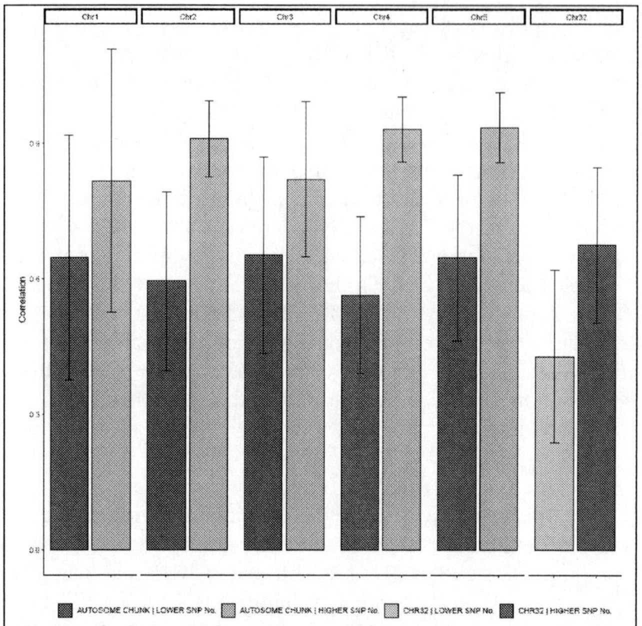


**Figure 2. Imputation accuracy for markers at two LD densities imputed to 50K on a 5.7 Mb end segment of Chromosome 1,2,3,4 and 5, compared with markers on the PAR region (Chr32)** Results were pooled across three breed groups & error bars show SD across SNPs.

## CONCLUSIONS

This study compares accuracy of imputation for autosomes and the X chromosome including several imputation scenarios for the PAR on bovine genome ARS-UCD1.2. We demonstrated that the accuracy of PAR imputation can be improved from LD to 50K by imputing the PAR & non-PAR together and re-extracting the PAR markers. However, if designing new SNP genotyping panels, we recommend SNP density in the PAR should be equivalent to that of the 50K SNP panel because this can greatly increase imputation accuracy.

## ACKNOWLEDGEMENTS

DairyBio, a joint venture project between Agriculture Victoria (Melbourne, Australia), Dairy Australia (Melbourne, Australia) and the Gardiner Foundation (Melbourne, Australia), funded resources used in the analysis.

## REFERENCES

Calus MPL, Bouwman AC, Hickey JM, Veerkamp RF, Mulder HA. (2014) *Animal* **8**: 1743.

Johnson T, Keehan M, Harland C, Lopdell T, Spelman RJ, Davis SR, Rosen BD, Smith TPL, Couldrey C. (2019) *J. Dairy Sci.* **102**: 3254.

Mao X, Johansson AM, Sahana G, Guldbrandtsen B, De Koning D-J. 2016. *Genetics* **157**: 1819.

Pacheco HA, Rezende FM, Peñagaricano F. (2020). *J. Dairy Sci.* **103**: 3304.

Rosen BD, Bickhart DM, Schnabel RD, Koren S, Elsik CG, Tseng E, Rowan TN, Low WY, Zimin

A, Couldrey C et al. (2020) GigaScience 9, doi: 10.1093/gigascience/giaa021
Sargolzaei M, Chesnais JP, Schenkel FS. (2014). *BMC Genomics* **15**: 478.
Su G, Guldbrandtsen B, Aamand GP, Strandén I, Lund MS. (2014) *Genet. Sel. Evol.* **46**: 47.
Van Laere A-S, Coppieters W, Georges M. (2008) *Genome Res.* **18**: 1884.

*Contributed paper*

# GENETIC IMPROVEMENT OF GOATS OWNED BY SMALLHOLDER GOAT KEEPER WOMEN IN BIHAR, INDIA, WITH THE HELP OF A DATABASE TOOL

**C. Nimbkar[1], P. Amer[2], A. Schurink[2], T. Sawhney[3] and R. Azam[3]**

[1] Nimbkar Agricultural Research Institute, Phaltan 415523, Maharashtra, India
[2] AbacusBio Limited, Dunedin 9016, New Zealand
[3] Aga Khan Foundation (India), New Delhi 110001, India

## SUMMARY

A community based pilot goat breeding program (CBBP) is being implemented under Project Mesha which seeks to improve goat production in Muzaffarpur district of Bihar state in India. Goat performance recording and selection of male kids for breeding have been carried out since 2018. The breeding goal established through consultation with rearers, targets improved kid growth rate and twinning. Birth weight, weight at 3 and 6 months, average daily gain (ADG) up to 120 days and adult doe weights were analysed with fixed models. There is wide variation in each trait, indicating potential for selecting animals with very high values for breeding goal traits. The progress of the CBBP is encouraging.

## INTRODUCTION

Bihar is one of the poorer states of India. The per capita net state domestic product of Bihar was US$663 in 2019, 31% of India's per capita gross domestic product (MOSPI 2019). Eighty nine per cent of Bihar's population lives in rural areas, and about 34% of this rural population live below the poverty line. The literacy rate is 64%, but women's literacy is only 54%. In Muzaffarpur district, 36% rural houses are temporary structures (Census of India 2011). As some villages get inundated by flood waters during the monsoon, residents have to move with their livestock to higher ground for varying periods every year.

Bihar state has a goat population of 12.8 million out of 149 million goats in India (BAHS, 2019). The predominant goat breed type is the highly prolific Black Bengal (BB) reared for meat production (Dey *et al.* 2007). Small size and low weight of these goats are constraints on goat production (BLSA 2019). The importance of goats to support livelihoods of socio-economically marginalized households through income generation and enhancing financial resilience is well recognized (Barooah *et al.* 2016). The Aga Khan Foundation is therefore implementing a community based program called Project Mesha since 2016 for about 50,000 households in 240 villages of 4 out of 16 blocks of Muzaffarpur district of Bihar to improve goat production, transform the lives of the rural poor and bring about rural women's empowerment. Any increase in income from goat rearing is expected to lead to an improvement in the well-being and status of women as they primarily care for goats.

Project Mesha's approach is improvements in goat nutrition, health, shelter, genetics and marketing through community institutions. As a part of Project Mesha, a community based pilot goat breeding program (CBBP) is being implemented with participation of the goat rearer communities since 2018.

This paper describes the participatory processes of the CBBP including the effective use of the database tool Dtreo (https://abacusbio.com/ventures/dtreo/) and findings from an analysis of the records collected.

## MATERIALS AND METHODS

**Participatory processes.** A cadre of trained women community based small ruminant health workers (*pashu sakhis*) has been established by Project Mesha. *Pashu sakhis* provide a range of fee-based preventive health services for goats and also castration of male kids to be raised for meat production. Each step in the CBBP was taken in consultation with the goat rearer women by holding several meetings with the 'producer groups' established under Project Mesha.

**Production system.** The average number of adult does per household in the project area is 2 and goats are mostly grazed or tethered in harvested fields, fallows or other common property grazing grounds (Barooah 2016). Goat nutrition and other management vary greatly among households. Before the start of the CBBP, does were mostly mated by roaming bucks let lose as a customary ritual or by young male kids that go grazing together with does. Before Project Mesha and the advent of *pashu sakhis*, mortality rates of up to 40% in adult goats and 50% in kids were reported (Population Council 2018).

**Selection of villages to establish recording.** Individual goat identification with numbered tags (with a unique number for each goat) and performance recording were started in 4 villages in 2018 and then increased to 8 after 2 years. The criteria used for village selection were partly external such as reasonable availability of goat feed resources. Community-related criteria were also important, such as willingness of the community to participate, at least 200 breeding does with reasonably even ownership, average or above average performance of goats compared to the general goat population in the Project Mesha area, indicating reasonable proficiency in goat rearing and substantial income being obtained from goat rearing.

**Determination of breeding goals and selection criteria.** The breeding goals for Project Mesha were determined in consultation with goat rearers. These are: increased size and weight, faster growth up to 90 days, twinning but not litter sizes larger than twins (although at this stage only a small number of kids have an identified sire), increased adaptation to local conditions and kid rearing ability of the dams. A scoring system to calculate an overall index score for buck kids was devised in consultation with the field team. The criteria used in the scoring system are measured by trained enumerators and include the predicted weight of each buck kid at 100 days and four traits of the kid's dam which are, the dam's chest girth, its condition at the time of assessment, its litter size history and kid survival history.

**Recording and evaluation system.** The Dtreo (dtreo.io) software application has been customized to capture performance data (online and offline), store data while ensuring its integrity and convert it into information based on the needs of goat rearers, the CBBP and genetic analysts. Data recording can be done in English or Hindi which is the language used in the CBBP area. Dtreo has been set up to calculate the index score for each buck kid and make it available in a report for the field team to use for primary selection. Data entries are usually made by the veterinarians in the Project Mesha field team. They also monitor the data, assess the buck kids attaining the requisite index score and maintain a continuous dialogue with the goat rearing community.

**Buck purchase and rotation.** After the primary selection, the buck kid's owner's consent has to be obtained to keep the buck uncastrated until the second selection point. The buck is weighed and its soundness for breeding assessed every month up to the age of 6 to 8 months. If approved, the buck is purchased by Project Mesha at a price premium over its market value for meat. It is then transported to a village, at least 20 km away where it is placed with a willing buck-keeper household. Thirteen bucks have so far been placed for mating does in different villages. The bucks are moved again to another village at the end of a year to control inbreeding in accordance with rules specially formulated by genetics advisers to Project Mesha.

**Data and models.** Average daily gains (ADG) of kids that were weighed $\geq$3 times up to the age of 120 days, were estimated with a regression of weight on age for each kid. Kid weight at birth (BWT), 3 months (3WT), 6 months (6WT), ADG and adult doe weights (DOEWT) were analysed. The number of records was 93 for BWT, 148 for 3WT, 151 for 6WT, 353 for ADG and 301 for DOEWT. Fixed models were fitted with the effects of village (6 classes), year of birth (2018, 2019, 2020), season of birth (rainy, summer, winter), kid birth type (single, twin, triplet, quadruplet) and kid sex (male, female). The effect of dam parity (first and 'later or unknown') could be fitted only for ADG as almost all the records for the other traits were for the 'later or unknown' parity. Only the effect of village was fitted for DOEWT as data for other effects was not available. No interactions were fitted because of the limited number of records. Least squares means (LSM) were estimated with only significant effects in the model.

## RESULTS AND DISCUSSION

Twenty per cent of the does had singles, 51% had twins, 26% had triplets and 3% had quadruplets, yielding an average litter size of 2.14, similar to Dey *et al.* (2017) who reported an average litter size of 2.1 under field conditions in Bihar. The BB field unit of the All India Coordinated Research Project on Goat Improvement (AICRP 2017) has reported an average litter size of about 1.8 in BB goats in West Bengal state.

There was substantial variation in each trait. BWT ranged from 0.5 to 3 kg, 3WT from 2.5 to 11.1 kg, 6WT from 6 to 20 kg and ADG from 10 to 140 g. There is thus a good chance of identifying candidates superior for the breeding goal traits and using them for further breeding.

**Table 1. Significance of fixed effects for the traits analysed**

| | Fixed effects | | | | | |
|---|---|---|---|---|---|---|
| Traits | Village | Year of birth | Season of birth | Kid birth type | Kid sex | Dam parity |
| BWT | * | * | not significant | * | not significant | not fitted |
| 3WT | * | * | * | * | not significant | not fitted |
| 6WT | * | * | not significant | * | * | not fitted |
| ADG | * | * | * | * | not significant | * |
| DOEWT | * | not fitted | not fitted | not fitted | not fitted | not fitted |

*Significant

6WT of bucks was higher by about 14% than that of does. BWT, 3WT and ADG declined from 2019 to 2020. This could be because of the addition of new villages for data collection.

The LSM (kg) for BWT, 3WT, 6WT and DOEWT were 1.5±0.2, 6.3±0.4, 9.1±0.5 and 23.5±0.7 respectively. The LSM for ADG was 57.5±4.5 g. Dey *et al.* (2017) report BB goat ADG of 30 g and adult doe weight of 12 kg. AICRP (2017) has reported BWT, 3WT and 6WT to be 1.23 kg, 5.30 kg and 7.50 kg respectively for BB goats. It is likely that weights in this study are higher because of the way villages were selected for recording. There may also have been selection of more cooperative households or of larger does for recording.

Thirteen bucks have been selected based on their index scores being above the set threshold and used for breeding in different villages. About 500 does have so far been mated with these bucks. Observations of their progeny have shown excellent vigour and growth, indicating that the improvement may be due to reduction in inbreeding.

The CBBP has created awareness among the community about the basic principles of genetic improvement, inbreeding and its impact. Before the CBBP, it was thought here that for genetic improvement, 'good' breeding bucks had to be brought from outside the state. It was also felt that no one would be willing to maintain a breeding buck. The Project Mesha team has now compiled a list of households in several villages ready to maintain breeding bucks. There are still challenges for the CBBP but the progress is encouraging. As data accumulates over time and pedigree records build up, more accurate genetic evaluations will be possible, leading to more progress. The success of Project Mesha's CBBP is likely to lead to expansion of the program in more blocks of Muzaffarpur district and then many more districts of Bihar.

## CONCLUSIONS

The CBBP reported here is likely to be the first systematic CBBP for goats in India. Many difficulties have been overcome and the goat rearer community has cooperated well with the CBBP implementing team. There is good opportunity to exploit hitherto untapped genetic variation in the highly adapted local breeds to improve goat productivity genetically.

## ACKNOWLEDGEMENTS

We acknowledge gratefully the funding to Project Mesha by the Bill and Melinda Gates Foundation.

## REFERENCES

AICRP (2017) West Bengal University of Animal and Fishery Sciences, Kolkata, India. <https://pcgoatcirg.icar.gov.in/black_bengal_kolkata/pdf_download/Breed%20Information.pdf > Accessed on 26 Feb. 2021.

BAHS-2019. (2019) Government of India, Ministry of Animal Husbandry and Dairying, New Delhi.

BLSA (2019) Animal and Fisheries Resources Department, Government of Bihar. https://cgspace.cgiar.org/bitstream/handle/10568/100523/bihar_livestock_sector_analysis.pdf?sequence=2&isAllowed=y Accessed 24 Feb. 2021.

Barooah B., Sarkar R., Dubey P., Bagai A. (2016) 'Project Mesha formative evaluation report' International Initiative for Impact Evaluation.

Census of India (2011) <https://censusindia.gov.in/2011-common/censusdata2011.html> Accessed 16 Feb. 2021.

Dey A., Barari S. K. and Yadav B. P. S. (2007) *Livest. Res. for Rural Dev* 19 Retrieved 26 Feb. 2021. <http://www.lrrd.org/lrrd19/9/dey19123.htm>

MOSPI (2019) <http://mospi.nic.in/data> Accessed 16 Feb. 2021.

Population Council (2018) Population Council, New Delhi, India.

*Contributed paper*

# IDENTIFICATION OF GENETIC VARIANTS LINKING DAIRY FERTILITY AND MILK PRODUCTION TRAITS

**E. Ooi[1,2], J.E. Pryce[2,3] and M.E. Goddard[1,2]**

[1] Faculty of Veterinary and Agricultural Sciences, University of Melbourne, Parkville, VIC, 3052 Australia
[2] AgriBio, Department of Environment & Primary Industries, Bundoora, VIC, 3083 Australia
[3] Biosciences Research Centre, La Trobe University, Bundoora, VIC, 3083 Australia

## SUMMARY

Fertility in dairy cattle has declined as an unintended consequence of selection for high milk yield. The negative genetic correlation between milk yield and fertility is now well-documented, however, the underlying biological causes are still uncertain. The objective of this study was to examine this problem from a genomic perspective by first identifying the variants that link dairy fertility and milk production traits, and then using an archetypal clustering method to group variants with similar patterns of effects. Each cluster was finally subjected to over-representation analysis to identify the biological processes underpinning variants with similar effects. Nine groups with distinct effects on production, fertility and conformation traits were identified. Initial results from over-representation analysis suggest that the clusters formed are consistent with prior knowledge about the associated genes, but also suggest new areas of interest for further research.

## INTRODUCTION

Fertility in dairy cattle has declined over the last 50 years as an unintended consequence of selection for high milk yield. Lactation is obviously contingent on parturition, making fertility a key driver of profitability, particularly on pasture-based dairy farms. The ideal cow does not only conceive – she does it at the right time, on the first attempt, and achieves and maintains pregnancy despite producing 60+ litres of milk per day.

The exact physiological mechanisms linking fertility and milk production are still uncertain, despite significant research investment. Results from observational studies and in vivo experimentation have been equivocal – largely because nutrition, health, management interventions and environmental factors all combine to confound analysis of herd reproductive performance.

Advances in genomics allow a direct approach to testing hypotheses. However, from a genetic perspective, fertility is a complex trait composed of successive biological events, with phenotypes that are difficult to measure. In this study, the use of a genome-wide association study incorporating large multi-breed reference population and a subset of variants which have been pre-selected for significance gives us significantly more power to identify variants of interest. It also allows us to identify variant clusters that have similar effects on multiple traits possibly indicating a common physiological pathway.

This study aims to uncover the physiological mechanisms underlying milk production and fertility, which may assist herd managers in uncoupling these traits to breed cattle that are both productive and highly fertile.

## MATERIALS AND METHODS

**Data preparation.** Genotype and phenotype data for a total of 5,123 bulls and 29,081 cows from DataGene, Australia were used for this study. This data included a mix of Holstein-Friesians (4,061 bulls/22,899 cows) and Jerseys (1,062 bulls/6,174 cows).

Genotypes included a total of 46,771 sequence variants, which were selected from a total of 17,669,372 imputed variants prepared according to a multi-phase method which includes regression involving FAETH scores, variant clustering and pruning, and Bayesian approaches (Xiang *et al.* 2021). Two hundred and forty-seven variants thought to be informative for milk fat and protein percentage from an analysis performed by van den Berg *et al.* (2020) were also included.

Phenotype data included trait deviations and daughter trait deviations for cows and bulls, which were calculated using a model that corrects for fixed effects including herd, season, and year. Twelve traits were selected which are thought to have effects on production and/or fertility, including protein yield, fat yield, protein percentage, fat percentage, milk yield, fertility, direct survival, stature, angularity, bone quality, udder texture and body condition score.

**Single-trait GWAS.** Each trait was analysed one at a time in each sex with linear mixed models using GCTA (Yang *et al.* 2011). Results for both genders were then combined using a weighted meta-analysis based on a method described in (Xiang *et al.* 2018). This allowed us to fully utilize GWAS summary data and thereby expand the power of our analysis.

Although most of the initial 46,771 variants were the result of LD pruning in the set of 1.7 million variants identified by (Xiang *et al.* 2021), we found that for known QTL with large effects such as DGAT1, some variants remained in high LD. To remove these, further post-processing was undertaken using the snp_clumping function within R package bigsnpr (Privé *et al.* 2018). This function is analogous to the –clump function implemented in PLINK 1.9 but has been adapted for memory-efficient usage within the R environment. For our study, as we were most interested in the relationship between milk production and fertility traits, we used fertility t-values as our ranking statistic. This reduced the starting set of 46,771 variants to 15,220 variants.

**Archetype-based clustering.** We then clustered the sequence variants according to their pattern of effects on the 12 traits of interest. This was done by first ranking the variants in descending order according to the magnitude of their effect size on these traits, and then completing iterative pairwise comparisons of their cosine similarity. Whenever a variant was identified which had < 0.8 cosine similarity with the index variant, it was considered a new archetype. Subsequent variants were considered to represent new archetypes only if this held true for all preceding archetypal variants.

Using this method, we identified 9 archetypal sequence variants with unique patterns of effects on the traits of interest. The remaining 15,211 variants were then assigned to the archetype with which they had the highest measure of cosine similarity, forming 9 variant clusters. The direction of effects was standardised across variants.

**Enrichment analysis.** To better understand the underlying biology for each of the 9 clusters, pathway analysis was performed on each cluster using the over-representation analysis (ORA) function provided by a gene-set analysis toolkit, WebGestalt (Liao *et al.* 2019).

## RESULTS AND DISCUSSION

It is important to note in Figure 1 that, as the fertility trait is measured by calving interval, positive effects represent infertility. With this in mind, we can distinguish 4 broad groups amongst the 9 variant clusters. One primarily affects fertility (i.e., clusters 3 and 8), one affects production traits with a negative effect on fertility (i.e., cluster 9), and one affects production traits without impacting fertility (i.e., clusters 1, 5). Another group could be considered to include clusters which have varying effects on conformation, particularly in clusters 4 and 8.

Cluster 1 includes genes such as DGAT, FASN and MGST1, which have all be implicated in fat synthesis. The pattern of effects is consistent with this, with fat, fat percentage and protein percentage traits in the opposite direction to milk yield and protein. There is little impact on other traits. The most represented GO terms reported by ORA include fat cell differentiation, carbohydrate derivative biosynthetic process, response to toxic substance, lipid biosynthetic process and endocrine system development.

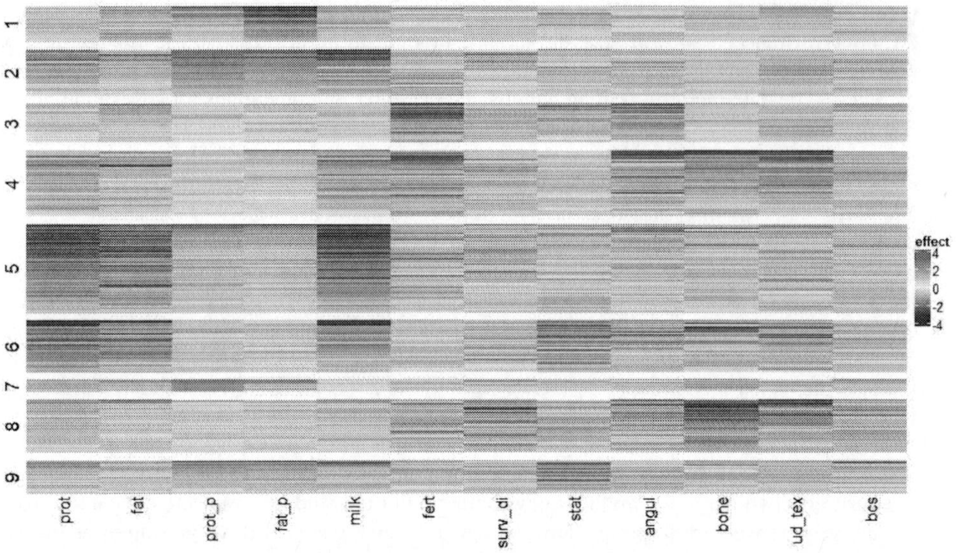


**Figure 1. Nine clusters exhibiting shared patterns of effect for 15,220 variants on 12 traits**

Cluster 2 has strong effects on protein and fat percentage, which is likely due to an antagonistic effect on milk volume. Notable genes include RORA, LAMA4 and PROX1. The most represented GO terms included cardiovascular system development, tube morphogenesis, regulation of cellular response to stress, and cell fate commitment.

Cluster 3 displays strong effects on fertility, direct survival, and angularity. Notable genes include NOG, ASCL1 and GDNF. The most represented GO terms included neuron death, neuron development, appendage development, regulation of system process, and regulation of cell development.

Cluster 4 also has strong effects on fertility and direct survival, with some interaction with conformation traits and weaker but consistent effects on production traits. Notable genes include LRRK2, DHX36 and BMP7. The most represented GO terms included regulation of nervous system development, regulation of cell development, regulation of cell projection organisation, regulation of secretion, and response to inorganic substance.

Cluster 5 represents very strong production effects, without impacting conformation or fertility. Notable genes include ADCYAP1, EDN1, and TGFBR1. The most represented GO terms included carbohydrate derivative transport, multicellular organismal response to stress, circulatory system process, anion transport, and response to growth factor.

Cluster 6 comprises variants with effects on fat, protein and milk yield that do not affect fat and protein percentage. Notable genes include BMP4, TP63 and WNT5A. The most represented GO terms included negative regulation of developmental process, signal transduction by p53 class mediator, cranial skeletal system development, positive regulation of cell proliferation, and epithelial cell proliferation.

Cluster 7 affects protein percentage and not much else. Notable genes include BCL2, IL6 and ISL1. The most represented GO terms included peptidyl-threonine modification, peptidyl-serine modification, tricarboxylic acid metabolic process, negative regulation of transcription, and regulation of ion transport.

Cluster 8 represents conformation traits, along with body condition score and possibly fertility.

Notable genes include BMP7, MEF2C and SYK. The most represented GO terms included connective tissue development, cardiovascular system development, appendage development, tube morphogenesis, and integrin-mediated signaling pathway.

Cluster 9 is similar to cluster 2 in that it primarily affects protein and fat percentage. However, unlike cluster 2 it also has effects on fertility, direct survival, and stature. Notable genes include ARRDC3, LGR4 and CIB1. The most represented GO terms included second-messenger-mediated signaling, G protein-coupled receptor signaling pathway, coupled to cyclic nucleotide second messenger, secretion by cell, cellular component disassembly, and cell-cell adhesion.

Care must be taken when interpreting these preliminary results, particularly when pathway analysis has been performed. Pathway analysis is still a developing area in computational biology, with no current consensus as to the best tool, method, or annotation database to utilise. Pathway analysis also requires a gene to be linked to each variant, which is a complex problem. Although GWAS can identify genetic loci associated with complex traits, the causal gene associated with each locus is often difficult to determine. This is because firstly, LD between loci can mask the identity of the causal variant and secondly, the causal variants at most associated loci are not coding, instead acting through gene regulatory mechanisms which are difficult to determine (Weeks et al. 2020). Validation of our results is still ongoing, through the development of new statistical methods and the cross-validation of our findings against experimental datasets comprising expression QTL results.

## CONCLUSIONS

This study shows that clustering variants by their patterns of effects and combining the results with pathway analysis may help to elucidate the underlying genes and biological processes which link genetically associated traits.

## ACKNOWLEDGEMENTS

The ideas and assistance of colleagues, including Ruidong Xiang and Ed Breen, have been invaluable for this project. We would also like to thank DataGene and the herd managers who have provided data for our work. This project was supported through an Australian Government Research Training Program Scholarship, as well as financial contributions from the University of Melbourne and DairyBio (a joint initiative of Agriculture Victoria, Gardiner, and Dairy Australia.

## REFERENCES

van den Berg, I., Xiang, R., Jenko, J., Pausch, H., Boussaha, M., Schrooten, C., Tribout, T., Gjuvsland, A.B., Boichard, D., Nordbø, Ø., Sanchez, M.-P., Goddard, M.E. (2020) *Genet. Sel. Evol.* **52**, 37.

Liao, Y., Wang, J., Jaehnig, E.J., Shi, Z., Zhang, B. (2019) *Nucleic Acids Research* **47**, 199.

Privé, F., Aschard, H., Ziyatdinov, A., Blum, M.G.B. (2018) *Bioinformatics* **34**, 2781.

Weeks, E.M., Ulirsch, J.C., Cheng, N.Y., Trippe, B.L., Fine, *et al.* (2020) medRxiv. doi:10.1101/2020.09.08.20190561

Xiang, R., Hayes, B.J., Vander Jagt, C.J., MacLeod, I.M., Khansefid, M., Bowman, P.J., Yuan, Z., Prowse-Wilkins, C.P., Reich, C.M., Mason, B.A., Garner, J.B., Marett, L.C., Chen, Y., Bolormaa, S., Daetwyler, H.D., Chamberlain, A.J., Goddard, M.E. (2018) *BMC Genomics* **19**, 521.

Xiang, R., MacLeod, I.M., Daetwyler, H.D., de Jong, G., O'Connor, E., Schrooten, C., Chamberlain, A.J., Goddard, M.E. (2021) *Nat Commun* **12**, 860.

Yang, J., Lee, S.H., Goddard, M.E., Visscher, P.M. (2011) *The American Journal of Human Genetics* **88**, 76.

*Contributed paper*

# INVESTIGATION OF METHODS FOR INCLUSION OF FIXED EFFECTS FOR ULTRASOUND SCAN CARCASS TRAITS IN LARGE SCALE SHEEP GENETIC EVALUATION

**U. Paneru[1], D. J. Brown[2], N. Moghaddar[1] and J.H.J. van der Werf[1]**

[1] School of ERS, University of New England, Armidale, NSW, 2350 Australia
[2]Animal Genetics and Breeding Unit, University of New England, Armidale, NSW, 2350 Australia

## SUMMARY

Australian sheep genetic evaluation is conducted routinely for millions of animals for many traits. In the current analysis implemented by the OVIS software, phenotypes are pre-adjusted for systematic fixed effects to make fair genetic comparison between animals. This study assessed whether correction factors used in OVIS remain valid, and to explore whether the pre-adjustment method is still suitable and is comparable with a linear model. Furthermore, importance of interactions between body weight and sex, or body weight and flock were estimated. Regression slopes were calculated from forward prediction, using eye muscle depth data on 234,810 White Suffolk and 249,136 Poll Dorset sheep and fat depth data on 246,149 White Suffolk and 268,002 Poll Dorset sheep. Updated pre-adjustment factors produced regression slopes of progeny performance on their sire's estimated breeding values (EBVs) equal to 0.67 and 0.62 (averaged over breeds) for eye muscle depth and fat depth, respectively. Regression slopes were same for eye muscle depth and slightly better for fat depth than OVIS (0.66 and 0.64 respectively). A linear model produced significant improvements in regression slopes (0.60 and 0.50 respectively). Including interaction effects between fixed effects did not significantly influence the accuracy of prediction of progeny performance. A linear model will be implemented in future OVIS evaluation for ultrasound scan carcass traits.

## INTRODUCTION

Genetic evaluation is conducted to provide information to breeders about the genetic merit of their animals in the form of estimated breeding values (EBVs) and selection index values. EBVs are calculated by correcting observed phenotypes for systematic environmental effects to allow fair genetic comparisons between animals. There are two common approaches to correct for the environmental effects: 1) pre-adjustment of phenotypes for environmental effects before genetic evaluation (Brown and Reverter 2002; Schaeffer 2019) or 2) fitting environmental effects in the mixed model equations to estimate them jointly with the breeding values (Laird and Ware 1982; Meyer 2004). The analytical software that implements the Australian genetic evaluation for sheep (OVIS) uses a pre-correction method, including correction of scanned carcass traits for the animal weight at scanning animal via linear and quadratic regression coefficients. The only fixed effect that is directly fitted in an animal model in OVIS is the contemporary group (CG) which includes breed, flock, management group, sex, and year of measurement subclass (Brown *et al.* 2016).

Theoretically, fixed effects such as the weight of animals and interaction effects between fixed effects should be included directly into the mixed model equation because the linear model corrects for the systematic environmental effects and gives an unbiased estimate of breeding value directly from the model (Laird and Ware 1982; Henderson 1984; Meyer 1998). However, estimating all effects jointly in the routine analysis increases the computational burden which can be prohibitive for large-scale genetic evaluation with millions of animals for many traits. With increasing computing power and further advances in analysis algorithms, this is becoming less problematic. Another consideration in potentially changing adjustment methods is that pre-adjustment factors are multiplicative and hence non-linear, and such corrections cannot always be implemented in a linear mixed model in the same manner. The current adjustment factors were estimated many years ago and they may need to be updated.

Given these considerations, this paper aims to determine whether the adjustment factors currently used by OVIS are still appropriate, and to propose updated adjustment factors if required. Furthermore, we examined whether correcting scanned carcass traits for body weight differs significantly between sexes or flocks. Finally, we compared the effectiveness of a linear model in the evaluation compared to using pre-correction factors.

## MATERIALS AND METHODS

Data were retrieved from the LAMBPLAN database, comprising ultrasound measurements of eye muscle depth (EMD) and fat depths (FD) and associated body weight recorded at post-weaning in Australian and New Zealand sheep. A subset of the terminal dataset was extracted including animals born from 2009 onwards. Data were filtered according to the guidelines of OVIS (Brown *et al.* 2000). There were 234,810 and 249,136 animals for eye muscle depth and 246,149 and 268,002 animals for fat depth for the White Suffolk (WS) and Poll Dorset (PD) breeds, respectively. Estimated variance components were estimated for the scanned traits using the following mixed model equation:

$$y = \boldsymbol{X_1}b + \boldsymbol{Z_1}\,a + \boldsymbol{Z_2}m + \boldsymbol{Z_3}mp + \boldsymbol{Z_4}sfy + e$$

Where y is the vector of observations, *b* is a vector of fixed effects, $a$ is a vector of breeding values of animals, $m$ is a vector of maternal breeding values, $mp$ is a vector of maternal permanent environmental effects, *sfy* is a vector of sire by flock year interaction effects, and $e$ is a vector of random residuals. $\mathbf{X_1}$ is an incidence matrix relating b to y and $\mathbf{Z_1}$, $\mathbf{Z_2}$, $\mathbf{Z_3}$ and $\mathbf{Z_4}$ are incidence matrices relating *a, m, mp* and *sfy* to y. Then, variance components were used to estimate BLUP EBVs using the above mixed model equation. Contemporary group was only fitted as the fixed effect component, b, when EBVs were estimated from pre-adjustment because phenotypes were already adjusted for other fixed effects.

**Estimating fixed effects and their interactions.** The fixed effects currently included in OVIS for scanned carcass traits are contemporary group and a linear and quadratic regression on body weight of the animal and these effects were fitted in a complete linear mixed model that was used as a reference model. The reference linear model was expanded by adding interaction effects, one at a time, including sex by body weight, year of birth by body weight, flock by body weight and flock by sex by body weight. A complete mixed model was fitted and the significance of extra interaction effects was evaluated. Significant interaction effects were then tested for their effect on the EBVs through forward prediction(Huisman *et al.* 2015; Legarra and Reverter 2017).

**Regression of progeny performance on sire EBVs.** Forward prediction was conducted to test the predictive ability of the EBVs from the various models and their effectiveness in predicting progeny performance. The breeding values of sires for post-weaning body weight were estimated from the training data by different mixed models and by pre-adjustment of the phenotype. The training data included animals born before 2017. EBVs of sires were validated only if they had progeny born after 2016. Progeny performance was corrected for all of the fixed effects using solutions from a linear model, and were regressed on their sire's EBV. The expectation of the regression coefficient is 0.50. A lower value indicates an over-dispersion of sire EBVs relative to the variance observed in the progeny performance data, while a higher value reflects under-dispersion.

## RESULTS AND DISCUSSION

**Genetic parameters.** Variance components estimated from the current data are presented for eye muscle depth and fat depth in Table 1. Heritability estimated for post-weaning eye muscle depth and fat depth were 0.25 and 0.18 respectively, averaged over breeds. These heritability values were smaller than previous estimates of 0.32 and 0.26 for post-weaning EMD and FD respectively (Brown *et al.* 2016). The difference might be due to that previous study not including a sire by flock-year interactions in the model, this study having more recent records and heritability estimates in this paper are breed specific while the latter are across-breeds estimates.

**Table 1. Variance components used to estimate BLUP solutions**

| Traits(Breed) | $V_a$ | $V_m$ | $V_{mp}$ | $V_{sfy}$ | $V_e$ | $h^2$ |
|---|---|---|---|---|---|---|
| EMD (WS) | 1.17 | 0.050 | 0.03 | 0.028 | 3.24 | 0.25 |
| EMD (PD) | 1.10 | 0.064 | 0.08 | 0.073 | 3.37 | 0.23 |
| FD (WS) | 0.08 | 0.005 | 0.01 | 0.009 | 0.35 | 0.18 |
| FD (PD) | 0.09 | 0.005 | 0.01 | 0.009 | 0.35 | 0.19 |

EMD: Eye Muscle Depth, FD: Fat Depth, WS: White Suffolk, PD: Poll Dorset

$V_a$, additive genetic variance, $V_m$, maternal genetic variance, $V_{mp}$, permanent environment effect of the dam, $V_{sfy}$, sire by flock year variance, $h_2$, direct heritability

**Comparison between pre-adjustment factors.** The average linear regression of eye muscle depth on body weight is higher (0.38; Table 2) than the OVIS assumption (0.31) indicating that the eye muscle depth of animals, relative to the body weight, has increased over the years. On the other hand, the average linear component for fat depth (0.08) was lower than the current OVIS factor (0.09) (Brown and Reverter 2002), indicating that fat depth of the animals relative to the body weight has decreased over the years.

**Table 2. Adjustment factors currently used in OVIS and updated estimates**

| Fixed effect | Level | OVIS EMD | Updated EMD | | OVIS FD | Updated FD | |
|---|---|---|---|---|---|---|---|
| | | | WS | PD | | WS | PD |
| weight | Intercept | 27.44 | 28.48 | 29.01 | 3.03 | 3.39 | 2.88 |
| | Linear | 0.31 | 0.38 | 0.38 | 0.09 | 0.07 | 0.08 |
| | Quadratic | -0.001 | -0.003 | -0.003 | -0.004 | -0.0001 | -0.0004 |

EMD: Eye Muscle Depth, FD: Fat Depth, WS: White Suffolk, PD: Poll Dorset

**Comparison between the linear model and pre-adjustment of data.** Updated pre-adjustment factors produced a slightly better regression slope (0.62) than pre-adjustment factors that are currently used in OVIS (0.64) for fat depth but identical prediction for eye muscle depth (Table 3). The complete linear model produced significantly better regression slopes of progeny performance on sire EBV (0.60 and 0.50) than with the EBVs based on pre-adjustment (0.67 and 0.62), comparing values averaged across breeds for eye muscle depth and fat depth, respectively. The regression slope for eye muscle depth was higher than 0.50, indicating under-dispersion of EBVs. Regression slopes for fat depth were close to 0.50, indicating that sire EBVs were able to predict progeny performance reliably. Further, regression slopes are closer to expectation in White Suffolk than in the Poll Dorset breed. Moreover, the regression slopes obtained from models with interactions did not give significantly different estimates of slope. Models with extra interaction effects, use significantly more computation time, and require more degrees of freedom. Based on these results, including interaction effects in routine evaluation may not be necessary.

**Table 3. Regression slopes of progeny performance on sire EBVs for ultrasound scan traits**

| Models | Eye muscle depth | | Fat depth | |
|---|---|---|---|---|
| | White Suffolk | Poll Dorset | White Suffolk | Poll Dorset |
| Pre- adjustment (OVIS) | 0.63±0.01 | 0.70±0.01 | 0.61±0.02 | 0.67±0.02 |
| Pre-adjustment (updated) | 0.63±0.01 | 0.71±0.01 | 0.58±0.02 | 0.66±0.02 |
| | Linear models | | | |
| 1 = (CG + Wt + $Wt^2$) | 0.57±0.01 | 0.64±0.01 | 0.49±0.02 | 0.50±0.02 |
| 1 + sex*Wt + sex*$Wt^2$ | 0.56±0.01 | 0.64±0.01 | 0.49±0.02 | 0.50±0.02 |
| 1 + YOB*Wt + YOB*$Wt^2$ | 0.56±0.01 | 0.64±0.01 | 0.49±0.02 | 0.50±0.02 |
| 1 + flock*Wt + flock*$Wt^2$ | 0.58±0.01 | 0.64±0.01 | 0.50±0.02 | 0.52±0.02 |
| 1 + flock*sex*Wt + F*S*$Wt^2$ | 0.58±0.01 | 0.63±0.01 | 0.49±0.02 | 0.52±0.02 |

CG: Contemporary Group, Wt: Weight, S: Sex, YOB: Year of Birth, F: Flock,

## CONCLUSIONS

The predictive ability of a model can be improved marginally by using updated pre-adjustment factors for ultrasound scanned carcass traits, and is not recommended. A complete linear model brings more improvement in the capability of EBVs to predict future progeny performance and is recommended for use in future OVIS evaluations if it is computationally feasible. Interaction effects between body weights with other fixed effects did not significantly increase the predictive capability of a model and can be ignored to simplify computation.

## ACKNOWLEDGEMENTS

We would like to thank Sheep Genetics for the data. The first author is supported by UNE via International Postgraduate Research Award (UNE IPRA).

## REFERENCES

Brown D., Tier B., Reverter A., Banks R. and Graser H. (2000) *Wool Tech. Sheep Breed.* **48:** 285.
Brown D. J. and Reverter A. (2002) *Liv. Prod. Sci.***75:** 281.
Brown D. J., Swan A. A., Gill J. S., Ball A. J. and Banks R. G. (2016) *Anim. Prod. Sci.* **56:** 1442.
Henderson C. R. (1984) 'Application of Linear Model in Animal Breeding. University of Guelph', Guelph, Ontario.5.
Huisman A. E., Brown D. J. and Fogarty N. M. (2015) *Ani. Prod. Sci.* **56:** 95.
Laird N. M. and Ware J. H. (1982) *Biom.* **38:** 963.
Legarra A. and Reverter A. (2017) *Proc. Assoc. Advmt. Anim. Breed. Genet***.** 22.
Meyer K. (1998) *Gen. Sel. Evol.* **30:** 221.
Meyer K. (2004) *Liv. Prod. Sci.* **86:** 69.
Schaeffer L. R. (2019) 'Animal Model'. Self publication.33.

*Contributed paper*

# IDENTIFYING THE BREEDING PREFERENCES AND ATTITUDES OF THE AUSTRALIAN BEEF CATTLE PRODUCER

**J.E. Peart[1], A.I. Byrne[1], M.J. Reynolds[1] and C.J. Duff[1]**

[1]Angus Australia, Armidale, NSW, Australia

## SUMMARY

This study, through the method of quantitative survey, investigates bull selection criteria preferences and understanding of genetic technologies of Australian beef producers and breed utilisation within their operation. The survey captured 1,023 producer responses from a representative proportion of beef cattle businesses in each state. Participants were asked to value bull selection criteria preferences on a 1 (lowest value) to 10 (highest value) scale. Respondents were also asked to rate their knowledge of genetics and nominate their breed of choice utilized in their operations. Nationally, temperament was ranked the most valued bull selection criteria, followed by polledness, visual appraisal and BullCHECK. The results were relatively consistent between states. Angus was the dominant breed in the female breeding population, with 5.6 million head (48%) of the Australian breeding female herd influenced by Angus genetics. Members of breed societies, particularly Angus Australia members, rated their knowledge of genetics more highly than their non-member counterparts.

## INTRODUCTION

The development and commercialisation of genetic selection tools have provided an accurate and objective description of genetic merit upon which producers can select breeding candidates and achieve breeding objectives (Johnson 2007). It is commonly recognised that Angus genetics and associated genetic technologies (e.g. Estimated Breeding Values, genomics) have made a significant contribution to the wider beef industry in terms of lifting productivity through gene introgression and genetic gain for commercially relevant traits (Parnell 2015). However there have been few wide scale studies that have been formally undertaken to understand producer perception and utilisation of these technologies.

Quantifying producer knowledge in genetics and the emphasis that they attribute to the objective information available for selection identifies extension and development opportunities for applicable genetic tools and technology (Bell *et al.* 2019). To provide this knowledge, Angus Australia facilitated a study by way of quantitative survey methodology via an independent market research group. The study aimed to determine the level of penetration of Angus and Angus influenced genetics throughout Australia, in addition to gauging beef producer's knowledge and attitudes towards the available genetic technologies, the latter being the focus of this paper. The broader findings of this study have been extensively reported in the Australian Beef Breeding Insights report (Angus Australia 2020).

## MATERIALS AND METHODS

The independent research market group Chi Squared was engaged due to their experience in agricultural market research and primary producer focus. A quantitative survey process was conducted over a 50-day period (11$^{th}$ May to 30$^{th}$ June 2019) and gathered 1,278 responses through four streams;

1. Telephone interviews conducted by an Australian based call centre, consisting of retired producers and agricultural students;
2. Online survey promoted via email correspondence to the Crackerjack Farming database;

3. Online survey promoted via email correspondence to the Angus Australia membership;
4. Online survey promoted via the Angus Australia website and Facebook page

To ensure the survey captured responses that were representative of viable beef breeding enterprises across the wider beef industry there were disqualifying parameters put in place. These included;

- Herd size less than 20 head of breeding females;
- Participant younger than 18 years of age;
- Less than 3 years of experience;
- Participant wasn't actively involved in the management decision making process of the operation;
- Main enterprise did not involve breeding or trading;
- Participant didn't intend to be breeding cattle in 5 years' time

This ultimately resulted in 1,023 eligible, unique responses. Sample size was monitored to ensure that the proportion of responses was comparable to the proportion of beef producing business entities in each state, as reported by the Australian Bureau of Statistics (Australia Bureau of Statistics, 2020). Due to the limited sample size of Northern Territory respondents, no values have been reported in this paper for this state.

Participants were asked a series of questions regarding their operations including knowledge of genetics (1 being poor, 10 being excellent), breed of choice and perceived value (1 of least value and 10 of greatest value) of selection criteria available when selecting bulls, such as EBVs.

In order to gain survey results that reflected the Australian beef industry, the bias of Angus members participating in the survey was corrected. This was achieved by removing those respondents who were contacted through the Angus Australia membership streams and focussing on the randomized data collection of the Chi Squared and Crackerjack farming databases. Overall, 781 responses formed the 'adjusted' data on which the breed influence findings in this study were based. Where findings are reported for the selection criteria preference and rating of genetic knowledge, respondents from all four streams were included.

## RESULTS AND DISCUSSION

**Bull selection criteria preferences.** The results of the survey suggest that, overall, producers prioritise bull selection criteria related to fitness for purpose (e.g. temperament, polledness, visual appraisal and BullCHECK (Australian Veterinary Association (2007))) before criteria associated with genetic progress (EBVs, Pedigree, DNA enhanced EBVs, Selection Indexes) (Table 1).

These priorities were generally consistent across most states however there were some variations reflecting the difference between past experiences and education, production systems, profit drivers and climate. For example, producers in NSW placed higher importance on EBVs compared to raw data (e.g. weight, ultrasound scans) for bull selection, while this was opposite in Queensland.

The bull selection criteria related to DNA factors (e.g. sire/parent verification, enhanced EBVs) generally rated at the lower end of importance. This may be a result of the relatively recent availability of these selection criteria, particularly DNA enhanced EBVs for bull selection.

Selection indexes were consistently ranked the lowest importance criteria for bull selection. Further research is warranted to understand this outcome and determine strategies to increase the importance placed on selection indexes for bull selection.

**Table 1. Importance rating of bull selection criteria nationally and by state**

| Selection Criteria | National | NSW | Qld | SA | Tas | Vic | WA |
|---|---|---|---|---|---|---|---|
| Temperament | 9.3 | 9.3 | 9.3 | 9.3 | 9.7 | 9.3 | 9.3 |
| Polledness | 8.7 | 8.5 | 8.5 | 9.3 | 9.4 | 9.0 | 8.9 |
| Visual Appraisal | 8.7 | 8.7 | 8.6 | 8.7 | 8.7 | 8.9 | 8.7 |
| BullCHECK | 8.1 | 8.1 | 8.4 | 8.0 | 7.2 | 7.8 | 8.3 |
| Information on genetic conditions | 7.9 | 7.9 | 7.7 | 7.9 | 7.9 | 8.0 | 8.0 |
| EBVs | 7.6 | 7.7 | 7.1 | 7.6 | 7.1 | 7.7 | 8.0 |
| Coat Colour | 7.5 | 7.7 | 7.1 | 7.7 | 6.9 | 7.8 | 7.1 |
| Pedigree | 7.3 | 7.4 | 7.1 | 7.2 | 7.5 | 7.5 | 7.5 |
| Raw data | 7.2 | 7.1 | 7.4 | 7.4 | 7.2 | 7.2 | 7.0 |
| Sire/Dam DNA verification | 6.7 | 6.8 | 6.4 | 6.5 | 6.8 | 7.0 | 6.8 |
| DNA enhanced EBVs | 6.5 | 6.6 | 6.3 | 6.3 | 6.2 | 6.8 | 6.7 |
| Selection Indexes | 6.5 | 6.5 | 6.3 | 6.6 | 5.8 | 6.4 | 7.0 |

*Ratings are an average value score of a 1 (of least value) to 10 (greatest value) scale*

The value that Angus Australia members put on each selection criteria was generally higher than their non-member and other breed society member counterparts (Table 2). Their priorities generally reflected the national results however information of genetic conditions was rated more highly, resulting from exposure to some of the genetic conditions identified in the Australian Angus herd. Also of an elevated priority was coat colour, reflecting their breed preference. Selection indexes were also of the lowest value to this group of respondents.

Participants belonging to breed societies other than Angus Australia, placed the lowest value on polledness of the groups. Meanwhile, non-members, both Angus users and other breed users alike, placed least value on DNA enhanced EBVs and sire/dam DNA verification, reflecting the commercial nature of their operations.

**Table 2. Importance rating of bull selection criteria by breed society membership**

| Selection Criteria | Angus Australia Members | Non-members | | Other Societies Members |
|---|---|---|---|---|
| | | Angus users | Other breeds | |
| Temperament | 9.3 | 9.2 | 9.3 | 9.4 |
| Visual Appraisal | 8.9 | 8.5 | 8.5 | 8.8 |
| Polledness | 8.9 | 9.0 | 8.6 | 8.1 |
| Information on genetic conditions | 8.5 | 7.4 | 7.2 | 8.1 |
| BullCHECK | 8.3 | 7.9 | 7.8 | 8.3 |
| Coat Colour | 8.1 | 7.4 | 6.6 | 7.1 |
| EBVs | 7.9 | 7.5 | 7.1 | 7.4 |
| Pedigree | 7.8 | 6.8 | 6.8 | 7.6 |
| Sire/dam DNA verification | 7.7 | 5.9 | 5.7 | 7.1 |
| DNA enhanced EBVs | 7.2 | 6.0 | 5.9 | 6.7 |
| Raw data | 7.2 | 7.1 | 7.0 | 7.5 |
| Selection Indexes | 6.6 | 6.4 | 6.3 | 6.5 |

*Ratings are an average value score of a 1 (of least value) to 10 (greatest value) scale*

**Genetic knowledge.** Producers associated with a breed society rated their knowledge of genetics more highly than their non-member counterparts, with Angus Australia members having the greatest confidence in their knowledge of genetics (7.9), by comparison to members of other societies (7.4). Non-members of breed societies reported an average score of 6.4. When observed on a state basis,

Victoria and New South Wales had the highest averages (7.3 and 7.2, respectively), reflecting the greater Angus Australia membership base in those states.

**Breed influence.** Nationally, a total of 48% of females had some percentage of Angus influence in their breeding (Table 3). Angus was the most utilized breed in all states except Queensland.

The female beef cattle population figures for each state from the ABS Agricultural Commodities report for 2018-19 (Australia Bureau of Statistics, 2020) were used to extrapolate the breed findings of the survey. This resulted in an estimated population of 5.6 million head influenced by Angus genetics in Australia – with the largest populations of Angus females in Queensland (1.8 million head) and New South Wales (1.5 million head).

**Table 3. Estimated proportion of Angus influenced females and extrapolated herd size by state**

| | National | NSW | Qld | SA | Tas | Vic | WA |
|---|---|---|---|---|---|---|---|
| Influence | 48% | 78% | 32% | 78% | 53% | 77% | 40% |
| No. of head | 5,606,199 | 1,461,977 | 1,824,097 | 311,002 | 104,382 | 768,429 | 425,927 |

## CONCLUSIONS

The survey approach implemented in this study proved to be an effective method of identifying the breed and selection criteria preferences of Australian beef breeders. The representative nature of surveys is an obvious limitation however the robust number of participants lends credibility to the finding. The results suggest that producers value a bull's contribution to the current herd, such as their ability to join and produce a viable calf, alongside safety and welfare considerations, above selection criteria associated with genetic progress. The number of Angus influenced cattle in the Australian breeding herd, as well as the higher confidence of Angus Australia members in their knowledge of genetics, illustrate the magnitude that any advances in technology, performance and research can be amplified through engagement and extension with Angus breeders. It further illustrates the benefits that could be gained through similar extension activities in the wider beef industry.

## ACKNOWLEDGEMENTS

We would like to acknowledge Meat & Livestock Australia, who, through a co-funded project, contributed financially to this report. We would also like to extend our gratitude to the producers who spent time and effort participating in this survey.

## REFERENCES

Angus Australia (2020), https://www.angusaustralia.com.au/research/australian-beef-breeding-insights-survey/ (last viewed March 2021)

Australia Bureau of Statistics (2020), https://www.abs.gov.au/statistics/industry/agriculture/agricultural-commodities-australia/2018-19#data-download (last viewed March 2021)

Australian Veterinary Association (2007), https://www.ava.com.au/about-us/ava-groups/cattle/resources/schemes/bullcheck/ (last viewed on February 2021)

Bell AM, Byrne AI. Duff CJ, Dominik S. (2019) *Proc. Assoc. Advmt. Anim. Breed. Genet.* **23:** 508

Johnson D.J., (2007) *Proc. Assoc. Advmt. Anim. Breed. Genet.* **17**: 8.

Parnell PF (2015) *Proc. Assoc. Advmt. Anim. Breed. Genet.* **21**: 221.

*Contributed paper*

# HOW ARE YOU FEELING, GIRLS? – BEHAVIOURAL TRAITS AS EMERGENT PROPERTIES OF THE COMMUNITY

**G.E.L. Petersen[1], A.C.L. Walters[1], P.F. Fennessy[1] and P.K. Dearden[2,3]**

[1] AbacusBio Ltd, Dunedin 9016, Otago, New Zealand
[2] University of Otago, Department of Biochemistry, Dunedin 9016, Otago, New Zealand
[3] Genomics Aotearoa, Dunedin 9016, Otago, New Zealand

## SUMMARY

While individual properties arising from interactions between individuals can have a significant impact on trait expression across a vast range of species in both animals and plants, nowhere does this become more immediately apparent than in trait definition and assessment of honeybees.

Due to the eusocial nature of these insects, phenotypic observations are, for the most part, limited to the superorganism, a complex community of up to 80,000 honeybees. Individual assessment of the behaviour and physiology of both worker and queen bees is possible (e.g. aggression in workers, egg-laying in queens) but of limited usefulness since the opportunity for their expression in the context of the honeybee colony might be limited or favoured via their interaction. As a result, a different overall phenotype can be expressed by the colony than would be expected based on the individual assessment.

By including a "community" component in our understanding of trait expression, we can explore the many layers of individual trait expression which contribute to worker brood viability in honeybee colonies and their intersection with both genetic and environmental factors. The inclusion of community aspects allows us to include fundamentally separate aspects that cannot be explained or captured by traditional models defining phenotypes as a result of G, E and GxE.

## INTRODUCTION

Great strides have been made in genetic improvement of plants and animals over the past decades, but some challenges in genetic evaluations remain, especially around behavioural traits (Chang *et al.* 2020). Complex phenotypes and a limited understanding of all the factors influencing the expression of traits hinder the adequate parcelling and attribution of variation, both genetic (G) and non-genetic or environmental (E). In the classic partitioning of variation into components of genetic and environmental variation, variation due to interactions of the individuals with the external world is often lost in the environmental component or masked (Foris *et al.* 2018). This was partially overcome by the introduction of an interaction component between the genetics of an individual and the environment, thereafter referred to as GxE (Falconer 1952). However, the G+E+GxE framework neglects the role that social interactions as well as underlying factors such as population density and population structure play for the realisation of genetic potential.

Phenotypic variation is not fully explained by current methods, although additional aspects like epigenetics can contribute to our understanding of the occurrence of variation (Triantaphyllopoulos *et al.* 2016). Contributors to phenotypic variation that can be considered both "environmental" and "social" have been recognised in the field of animal breeding since the 1970s (Willham 1972) and have found entry into genetic evaluations in some species in the form of maternal effects (Solé *et al.* 2021). However, a large proportion of community-driven factors that contribute to variation (e.g. competition, genetic makeup of the population) are still often either considered completely environmental or entirely due to individual genetics.

## METHODS AND DISCUSSION

**Understanding individual contributions to the superorganism.** As eusocial insects, honeybees are limited to a small number of reproductively active females, the queens, who are supported by functionally sterile female worker bees at a ratio from 1:5,000 to 1:80,000. This puts them in a unique situation where phenotypic observations are, for the most part, limited to the superorganism, a complex community of tens of thousands of honeybees, while selection can only act on the core individual, the queen.

While some traits, such as the reaction to *Varroa* mite infestations, can be assessed in individual workers (e.g. Currie and Tahmasbi 2008), these observations are of limited usefulness since they might not find an opportunity to be expressed by the individual in the context of the entire colony.

Honeybee traits are often the result of multiple populations within a colony working together in cohorts of sisters of a similar age, which means that the exact expression of any trait relies on the performance of hundreds of individuals, each of them with an individual response threshold that triggers behaviours which contribute to the observed trait (Beshers and Fewell 2001). An approach trying to integrate studies of the behavioural, physiological and neurobiological aspects of division of labour developed a push-pull model explaining the relationship between different castes and their respective ages under natural conditions (Johnson 2010).

**The GCE model in theory.** Assuming that division of labour impacts on trait expression, a new approach is needed to interpret observations of honeybee performance before these can be used as the basis of genetic evaluations. The required model must allow for the consideration of modifiers that contribute to the outcomes for an individual worker and her life history. These modifying factors can be split into two classes: *effectors* and *responses*.

To fully explain the expression of the genetics of an individual bee, *effectors* are both the community that the individual partakes in and the environment they live in. These two effectors are necessary for the expression of an individual's genetics but exist largely independently from an individual bee's life or are only slightly influenced by her individual contribution (see circles in Figure 1).

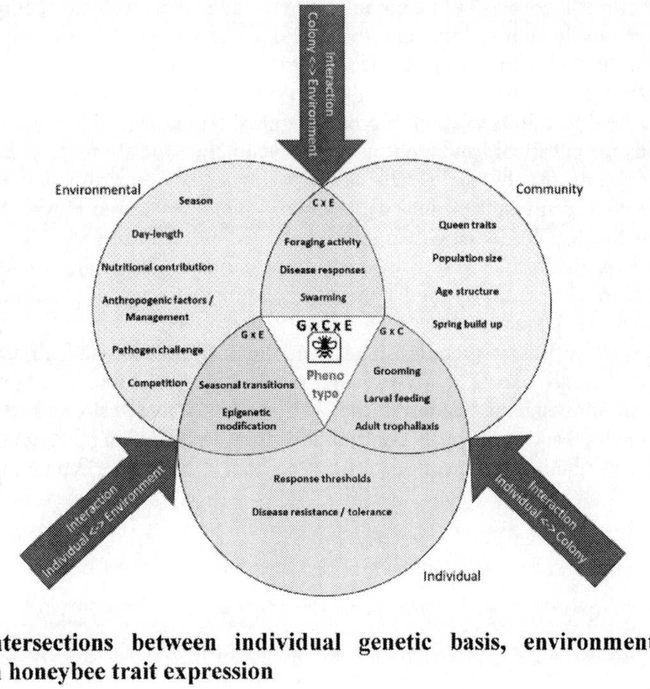


**Figure 1. Intersections between individual genetic basis, environmental aspects and community in honeybee trait expression**

The *responses* that an individual show as a result of their interaction with the effectors are adaptable outcomes of said interactions which are specific to the individual, since they are based on their individual response thresholds. These responses serve as measurable phenotypic outcomes that add to the individual's life history and in turn contribute to the phenotype of the superorganism. Responses arise from the interaction of individual genotypes with the environment (GxE), and the other members of the community (GxC) which is influenced by environmental effects on the community through individual interactions with the environment (CxE).

**Applying the GCE model to worker brood viability.** In applying the GCE framework to a particular trait, contributors to trait expression can be broken down and attributed directly to their source, rather than having to rely on a theoretical ability to control for these contributors.

Worker brood viability, the percentage of brood cells that are capped over in a patch of brood of similar age, is a complex phenotype driven by numerous factors.

Queen egg laying rate (see Figure 2, "Community") establishes an upper limit for the number of worker brood cells that can be capped at any time, since it determines how many eggs are developing within the roughly 10-day time frame that covers larval and pupal development under wax caps, the only time when brood viability can be readily observed. While queen laying traits are hard to observe without somehow limiting their expression (e.g. by supplying a limited number of cells in a confined area of the hive for a set number of hours), they can be measured. However, they do not present an adequate representation of the colony's ability to live up to the upper limit they present.

Brood care is a core part of the inner workings of a honeybee hive, and its success depends on the availability of capable nurse bees (who need to be of the correct age to be able to produce larval food) as well as the availability of food. Food can be stored as nectar, honey and bee bread, which can be assessed by the beekeeper and would reduce its function as an effector to an environmental effect. However, at times of high food availability outside the hive and depleted stores, the standard situation in spring, food resources can be directly distributed by worker bees foraging outside of the colony. This both brings in an aspect of a community dependency as well as a complication for the observation and quantification of the environmental effectors by the beekeeper since these transitional food sources are almost impossible to assess.

The individual genetic factors which contribute to observed brood viability apply to the larvae in development at the time of observation of brood viability, and include genetic disease resistance, response thresholds for nutritional and environmental factors like the ability to tolerate variation in brood comb temperature, and the allele status at the honeybee sex determination locus *csd*, which homozygosity at which can result in non-viable diploid male larvae see Figure 2, "Individual").

**Limitations of the model.** While the GCE model presented here can serve to identify contributors to honeybee trait expression that were previously impossible to determine in the context of the superorganism, it cannot be readily applied in the interpretation of honeybee performance to improve on genetic evaluations with the use of traditional observational data collected by beekeepers, as this information is not of sufficient granularity to generate insights into the community effector. However, with the increased use of beehive telemetry in routine beekeeping, datasets are becoming available which will allow the application of the GCE model to hive performance data in order to tease apart environmental and community contributions to superorganism phenotypes and thus fully define the genetic contribution and increase accuracy in honeybee genetic evaluations.

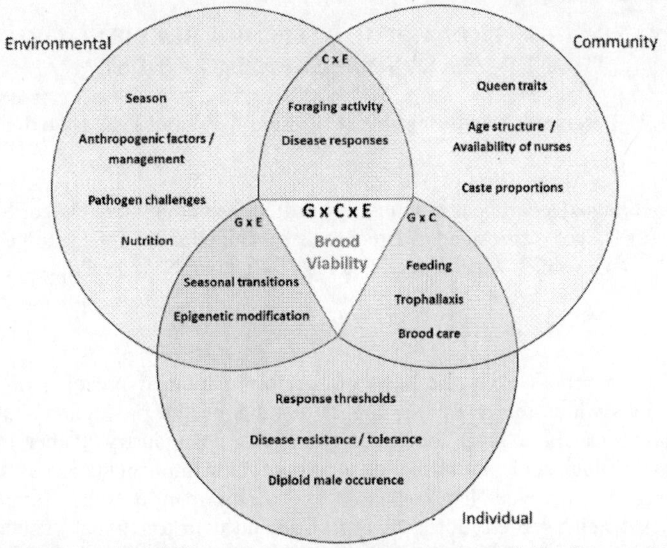


**Figure 2. Worker brood viability as an emergent property of the interaction of Genotype, Community and Environment**

## CONCLUSIONS

Treating phenotypes of the honeybee superorganism as emergent properties of the interaction between genetics, environment, and the community within the colony can help define contributors to observed variation and strip away variation which has previously clouded our understanding of the genetic effect on honeybee performance.

While the framework can be most readily applied in eusocial insects, it is likely to have applications in other livestock species, e.g. in defining the effects of competition on performance and survival, as well as in plant production systems where competition for natural resources and space cannot be avoided via translocation.

## ACKNOWLEDGEMENTS

This work was funded by the New Zealand Ministry for Business, Innovation and Employment, under the grant 'Selecting Future Bees' UOOX1610 to PKD.

## REFERENCES

Beshers S. N. and Fewell J. H. (2001) *Annu. Rev. Entomol.* **46**: 413.
Chang Y., Brito L.F., Alvarenga A.B., Wang Y. (2020) *Anim. Front.* **10**: 29.
Currie R.W., Tahmasbi G.H. (2008) *Canadian J. Zool.* **86**: 1059.
Falconer D.S. (1952) *The American Naturalist* **86**: 293
Foris B., Zebunke M., Langbein J., Melzer N. (2018) *PLoS ONE* **13**: 1.
Solé E., Ros-Freixedes R., Tor M., Reixach J., Pena R.N., Estany J. (2021) *PLoS ONE* **16**: 1.
Triantaphyllopoulos K.A., Ikonomopoulos I., Bannister A.J. (2016) *Epigenetics Chromatin* **9**: 1.
Willham R.L. (1972) *J. Anim. Sci.* **35**: 1288.

*Contributed paper*

# INDUSTRY CONSULTATION AS THE BASIS OF A BREEDING OBJECTIVE FOR THE NEW ZEALAND BEEKEEPING INDUSTRY

**G.E.L. Petersen[1], I.A. Balogun[1,2], P.F. Fennessy[1] and P.K. Dearden[3,4]**

[1] AbacusBio Ltd, Dunedin 9016, Otago, New Zealand
[2] University of Otago, Department of Human Nutrition, Dunedin 9016, Otago, New Zealand
[3] University of Otago, Department of Biochemistry, Dunedin 9016, Otago, New Zealand
[4] Genomics Aotearoa, Dunedin 9016, Otago, New Zealand

## SUMMARY

Trait prioritisation processes as the basis of the formulation of breeding objectives can be difficult in situations where the economic impact of traits on the production system are poorly understood. Surveys can be a great tool to interact with the industry, gather information and ultimately generate enough context information to allow for the implementation of stringent genetic evaluation systems. We surveyed New Zealand beekeepers to identify traits of importance from a list of 9 preselected honeybee characteristics to be included in a national genetic improvement scheme. Trait preferences were found to vary between groups within the industry (e.g., commercial beekeepers vs. queen breeders), but emphasis on *varroa* mite resistance, honey yield and gentle temperament leading to better workability was put on by all groups.

## INTRODUCTION

Despite being an important agricultural species, the Western honeybee, *Apis mellifera*, has received considerably less attention in animal breeding than more traditional livestock species with more accessible life histories. Selection is often performed *ad hoc* and based mainly on beekeeper intuition and experience (Cauia *et al.* 2011), and the adoption of structured breeding programs applying genetic evaluation tools has generally been low among commercial beekeepers. Linguistic discrepancies between beekeepers and other livestock producers around the use of the term "breeding" (which in industry jargon is used almost exclusively to refer to the multiplication of queens, both from selected and unselected dams) and honeybee mating strategies complicate the direct transfer of animal breeding methodologies from other industries.

For the formulation of a clear honeybee breeding objective, an instrumental tool in making beneficial livestock selection decisions (Dickerson 1970), beekeepers from different sectors of the industry (honey- and pollination fee-driven) need to be part of the process, both to improve the understanding of the profitability of commercial beekeeping operations and to disseminate the fundamental concepts of modern animal breeding strategies before making the corresponding tools available to the wider industry.

Industry consultation through surveys has been found to increase adoption rates of genetic evaluation services by aligning the breeding objective with the requirements of breeders and end-users of improved genetics across multiple industries and species such as pasture crops (Smith and Fennessy 2011, 2014), sheep (Byrne *et al.* 2012) and dairy cattle (Martin-Collado *et al.* 2015). Involving beekeepers directly in the formulation of a breeding objective will hopefully result in similar improvements in the adoption both of genetic evaluation services while lifting the understanding of both the promise and the limitations of genetic evaluation and mate selection tools.

## MATERIALS AND METHODS

**Trait pre-selection.** 9 honeybee traits were selected based on literature research and preliminary beekeeper consultation for relevance, measurability, presumed heritability and observed variation in the field (for details on this see Petersen 2019). An overview of the traits included in the survey, their unit (or observation, where units are hard to define) as well as the levels addressed as part of the prioritisation process can be found in Table 1.

**Table 1. Honeybee traits included in a survey to determine trait prioritisation in the New Zealand beekeeping industry**

| Trait | Unit or Observation | Levels |
|---|---|---|
| Honey production | kg / hive / season | Unchanged / +1kg / +2kg |
| Worker brood viability | percentage of viable brood | 90% / 100% viable |
| Workability (Gentleness) | Likelihood of bees stinging | Less likely / unchanged / more likely |
| Defensive behaviour | Ability of bees to defend the hive | Less able / unchanged / more able |
| Swarming | Swarming urge | Management needed / not needed |
| Queen longevity | Queen survival | 1 season / 2 seasons |
| *Varroa destructor* mite resistance | Ability of bees to control mites | Treatment needed / not needed |
| Body colour | Colour of drones produced | Drones are the same colour / different |
| Wintering ability | % surviving bees | Current winter survival / 4% better |

**Survey design and beekeeper recruitment.** The survey was entirely designed and distributed online. It consisted of a demographics part built in SurveyGizmo (Alchemer Inc., Boulder CO, USA) and the core trait prioritisation using multi-criteria decision-making tool 1000minds® (1000minds Ltd, Dunedin, New Zealand). Beekeepers were streamed into 3 distinct sets of demographic and management questions (commercial operator, designated queen breeder and hobbyist) based on their response to the first question and asked questions about their operation (e.g., size in hives, location, staff), hive management strategies, beliefs, and preferences around queen selection. After completing the demographic survey, they were directed to 1000minds®, where they were asked to make a number of trade-off decisions to determine their personal priorities (for details see Hansen and Ombler 2008).

The survey was made available to the public via a link on the website of a national honeybee genetic improvement research project, and beekeepers were encouraged to participate throughout the 2019 Apiculture NZ conference. When participation continued to be low throughout the beekeeping season 2019/20, a priority set of around 50 beekeepers was identified and contacted directly, with the survey being conducted interview style.

**Data analysis.** Data analysis was carried out on the combined dataset of the demographic survey as well as 1000minds® in R. The Kruskal-Wallis test was used to determine the difference in traits preference ranks for different beekeeper demographics. Principal component analysis (PCA) was employed to reduce the dimensionality of the data and to investigate patterns of preferences in trait rankings. PCA was followed by Correspondence Analysis (CA) of the principal components. Hierarchical clustering was performed using Ward's criterion on the selected principal components. K-means clustering was used to improve the initial partition obtained from hierarchical clustering and to determine the final number of clusters.

## RESULTS AND DISCUSSION

A total of 61 responses were recorded. Survey responses were excluded from the sample if they did not complete the 1000minds® survey. The final sample used in the analysis was 41 responses made up of 24 commercial operators, 11 queen breeders and 6 hobbyists.

Results from the combined dataset of all 3 beekeepers demographics showed a strong preference for *varroa* mite resistance and workability, two traits primarily associated with operational costs (e.g. mite treatments, labour), as well as honey yield, which was stated to be the source of at least 50% of income for all beekeepers. Defensive behaviour against wasps and other intruders, and body colour were found to be the least preferred traits (Figure 1, left).

While the general trend seen in the whole dataset was also found in the preferences of commercial beekeepers only (Figure 1, right), the ranking of honey yield was significantly higher among commercial beekeepers, while body colour was considered irrelevant.

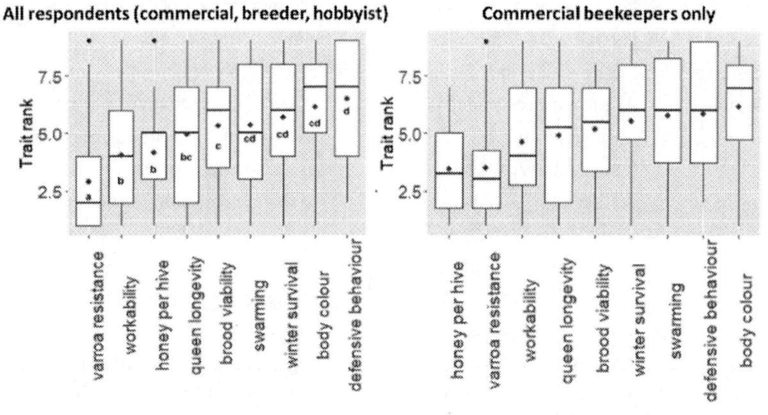


**Figure 1. Ranking of trait preferences for all respondents compared to commercial beekeepers**
Boxplots represent mean (blue), median (solid lines), first and third quartiles (contained in the boxes), and outliers (open points) of the distribution of the ranks of each trait improvement. Order of preferences for trait improvements is from most preferred (left) to least preferred (right). Different letters indicate significant (P-value<0.05) differences between the traits.

PCA revealed underlying patterns in the trait preferences, the most surprising of which was that although *varroa* mite resistance ranked highly in the results overall, the preference for mite resistance showed a high level of variation within the principal component (Figure 2, left). Honey yield in contrast was found to have almost no variation, due to having been given high emphasis by all respondents. Further analysis of the trait preferences showed the existence of heterogeneity even among players in the same value chain i.e. commercial operator, where 3 clusters spearheaded by queen longevity, *varroa* mite resistance and winter survival respectively could be identified (data not shown).

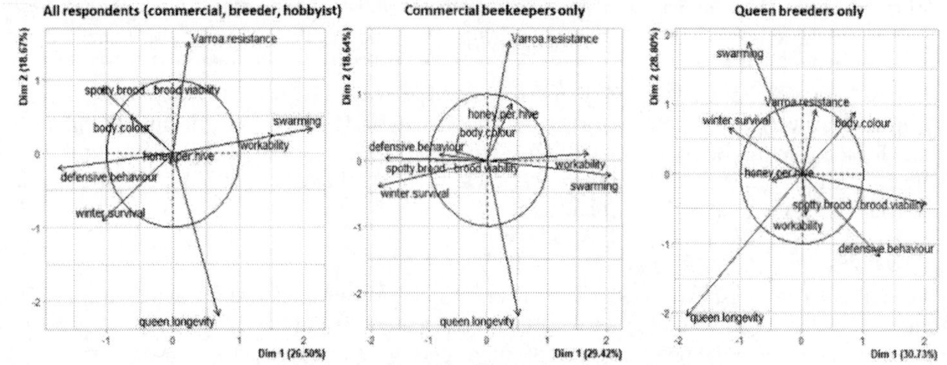


**Figure 2. Patterns of trait preferences in varying respondent groups**

One of the questions central to both commercial beekeepers who purchase queens from breeders and queen breeders themselves is whether queen breeder selection goals match the perceived needs in the industry. Comparing the patterns of preferences for all respondents with commercial beekeepers and queen breeders (Figure 2) reveals that this might not always be the case, since queen breeders more consistently put emphasis on mite resistance and traits that are low priority for commercial beekeepers, such as body colour, while showing variation in their emphasis on honey yield. Honey yield only placed fifth overall out of the 9 traits in the preferences of queen breeders (data not shown), indicating that they are either not able to observe honey production due to the constant "interference" with hives that is required during the queen rearing process, or that they do not consider honey yield a trait that can primarily be manipulated by selection.

An obvious limitation of this study is the number of responses from beekeepers, which limits its ability to identify e.g. clusters of preferences that could form the basis of different selection indices. However, representation of certain industry groups is strong; New Zealand currently has around a dozen specialised queen breeders out of which 11 responded to the survey or were interviewed. Within the group of 24 commercial operators, 10 fell into the range of >3,000 hives or "mega commercial" operators, representing 20.5% of these businesses which currently manage around 50% of the country's honeybee population (New Zealand Ministry for Primary Industries 2020).

Based on these rates of representation, our results can be considered meaningful despite their small sample size.

## CONCLUSIONS

The presented study showed that there is considerable heterogeneity in the trait preferences of different groups within the beekeeping industry, but that surveys present a valuable tool in ranking traits with no direct monetary value attached to them (such as bee behaviour traits) to allow scaling them to production traits (e.g. honey yield) with a set value or to potentially verify a calculated value based on a set of vague assumptions against their perceived value based on ranking.

## ACKNOWLEDGEMENTS

This work was funded by the New Zealand Ministry for Business, Innovation and Employment, under the grant 'Selecting Future Bees' UOOX1610 to PKD.

## REFERENCES

Byrne T. J., Amer, P. R., Fennessy P. F., Hansen P., and Wickham B. W. (2012) *Animal.* **6**: 778.

Cauia E., Siceanu A., Patruica S., Bura M., Sapcaliu A., Magdici M. (2011) *Econ. Manag. Financ. Mark.* **6**: 1146.

Dickerson G. (1970) *Journal of Animal Science* **30**: 849.

Hansen P. and Ombler F. (2008). *J. Multi-Crit Decis Anal* 2008; **15**: 87.

Martin-Collado D., Byrne T.J., Amer P.R., Santos B.F.S., Axford M., and Pryce J.E. (2015) *J. Dairy Sci.* **98**: 4148.

Nielsen H.M., Amer P.R.. (2007) *Animal.* **1**:1254.

New Zealand Ministry for Primary Industries (2020) Farm Monitoring Report NZ - Apiculture Report 2019, NZ MPI.

Petersen G.E.L. (2019) PhD Thesis, University of Otago.

Smith K.F. and Fennessy P.F. (2011). *Crop and Pasture Science.* **62**: 355.

*Contributed paper*

# AN EVALUATION OF THE EFFECT OF THE BOOROOLA GENE, FEC B, ON PRODUCTIVITY IN A BORDER LEICESTER X MERINO PRIME LAMB PRODUCTION SYSTEM

**L.R. Piper[1], A.A. Swan[2], J.M. Elsen[3], L. Bodin[3], H.G. Brewer[1] and B.M. Bindon[1]**

[1]CSIRO Agriculture and Food, F D McMaster Laboratory, Armidale, NSW 2350,
[2]Animal Genetics and Breeding Unit, University of New England, Armidale, NSW 2351,
[3]INRAe, GenPhySe, 31326 Castanet-Tolosan, France

## SUMMARY

The direct effect of the Booroola gene, Fec B, on ewe reproduction rates and ewe productivity traits has been evaluated in a typical Border Leicester x Merino prime lamb production system. The ovulation rate and prolificacy of the Border Leicester x Booroola Fec $B^{b+}$ ewes were significantly higher than those of traditional Border Leicester x Merino ewes. This advantage was offset by significantly lower lamb rearing ability with the result that there was no advantage in terms of lambs weaned/ewe joined or in $ returned/ewe joined. Targeted supplementary management strategies (ultra-sound scanning, supplementary feeding) appear to show promise in realising the gains from the increased prolificacy.

## INTRODUCTION

In a previous report (Bindon and Piper 1990), the role of the Booroola Merino in the Australian prime lamb industry was evaluated in a series of experiments conducted at the Armidale, NSW, CSIRO research stations, Longford and Arding during the period 1982-1992. Bindon and Piper (1990) reported results from a seven-year period (1982-1988) of a typical autumn joining system where Border Leicester x Booroola (BLxB) or Border Leicester x Merino (BLxM) ewes were joined with Suffolk, Polled Dorset or SIROMT (Bindon *et al.* 1984) rams. Over that period, (Bindon and Piper 1990, Table 7), the BLxB ewes " had a 56% higher ovulation rate, a 43% higher prolificacy (lambs born per ewe lambing, LB/EL), a 22 % lower lamb survival (lambs weaned per lamb born, LW/LB) and a 15 % higher lambs weaned per ewe joined (LW/EJ) than the BLxM ewes, This resulted in an advantage of 7 % in revenue returned per ewe joined in favour of the BLxB ewes)".

In the Bindon and Piper (1990) study, the BLxB ewes were generated by crossing Border Leicester rams with Booroola Merino ewes maintained in an auxiliary flock independent of the main Booroola Merino breeding flock. This flock contained a mixture of the three Fec B genotypes (bb; b+; ++) and the resulting BLxB ewes were therefore also a mixture of Fec B genotypes (b+; ++). As a result, the effect of the Booroola gene, Fec B, on productivity in the Border Leicester x Merino prime lamb production system, was not clearly established.

This paper extends the scope of the original study by reporting new analyses on the estimated direct effect of the Booroola gene, Fec B, on ewe reproduction rate and productivity in a typical autumn joined Border Leicester x Merino prime lamb production system.

## MATERIALS AND METHODS

**Sheep.** The ewes in this study were generated by joining Border Leicester rams with Booroola Merino or Control Merino ewes each year from 1976 to until 1990 except for 1981. The CSIRO Booroola Merino and the randomly bred Control Merino flocks have been described in detail elsewhere (Turner 1978; Piper and Bindon 1982). In all, 2243 records from 560-587 (depending on the trait being analysed) Border Leicester x ewes comprised the data analysed in this study. Over the period of the study (1982-1992) the ewes were joined with rams from the Suffolk, Poll Dorset or SIROMT (synthetic line derived from crosses of the Dorset Horn, Corriedale and Cheviot) breeds.

**Traits Measured.** For each of the ewes in each of the lambing years (1982-1992) the traits recorded were: Ovulation Rate (OR, assessed by laparoscopy), Fertility (ewes lambing/ewe joined, FERT), Prolificacy (or litter size, lambs born/ewe lambing, LS), Ewe Rearing Ability (lambs weaned/lamb born, ERA), Reproduction Rate (no. lambs weaned/ewe joined, NLW) and three ewe productivity traits, Lambs Sold/ewe joined (LSO/EJ), Lamb Weight Sold/ewe joined (LWS/EJ) and Revenue Returned/ewe joined ($ ret/EJ). The management of the ewe flocks and the reproduction rate and productivity traits observed have been described by Bindon *et al.* (1984).

**Statistical Methods.** For the analyses to estimate ewe genotype at the Fec B locus, the lifetime prolificacy and ovulation rate records of the BLxB (B) ewes and of their Booroola Merino dams, were analysed using segregation analysis methodology developed by Elsen *et al.* (1988) and by Foulley and Elsen (1988). The software returns a probability that each of the B ewes is either b+ or ++. The data set analysed comprised 2243 records (1181 records from BLxC (C)) ewes and 1062 records from B (b+, ++) ewes with probabilities of being b+ (631 records) or ++ (431 records)) ranging between 0.9 and 1.

The reproduction traits were analysed using repeated record, mixed linear models adjusting for fixed effects using ASReml (Gilmour *et. al*. 2014). The model employed was:

$y \sim mu + \text{ewe type}_i + \text{ewe age}_j + \text{lambing year}_k + \text{sire breed of lamb}_l + \text{ewe}_{im} + \text{within ewe}_{imp}$

Ewe type (B b+, B ++, C) ewe age (2-7), lambing year (1982-1992) and Ram type (sire breed of lamb - Suffolk, Poll Dorset, SIROMT) were fitted as fixed effects while the between ewe effects ($\text{ewe}_{im}$) were fitted as random effects.

## RESULTS

The means and standard errors (se) for the ewe reproduction and ewe productivity traits are given in Table 1. For the reproduction traits, and by comparison with the C ewes, the B ewes had 47% higher OR (P<0.001), 5% lower FERT (P=0.047), 35% higher LS (P<0.001), 16% lower ERA (P<0.001) and 2% lower NLW (n.s.). For the same reproduction traits, and by comparison with the ++ ewes, the b+ ewes had 100% higher OR (P<0.001), 11% lower FERT (P<0.001), 68% higher LS (P<0.001), 31% lower ERA (P<0.001) and 3% lower NLW (n.s.).

**Table 1. Reproduction trait and ewe productivity trait means for the Border Leicester x Control (C), BorderLeicester x Booroola (B) and for the B++ and Bb+ ewes**

| Trait | No. ewes | C | B | ++ | b+ |
|---|---|---|---|---|---|
| OR | 584 | 2.09 ± 0.04 | 3.08 ± 0.04 | 2.04 ± 0.06 | 4.07 ± 0.05 |
| FERT (EL/EJ) | 587 | 0.82 ± 0.02 | 0.78 ± 0.02 | 0.83 ± 0.02 | 0.74 ± 0.02 |
| LS (LB/EL) | 560 | 1.72 ± 0.04 | 2.33 ± 0.04 | 1.74 ± 0.05 | 2.93 ± 0.05 |
| ERA (LW/LB) | 560 | 0.85 ± 0.02 | 0.71 ± 0.02 | 0.83 ± 0.02 | 0.57 ± 0.02 |
| NLW (LW/EJ) | 587 | 1.16 ± 0.05 | 1.14 ± 0.04 | 1.15 ± 0.06 | 1.12 ± 0.05 |
| Lambs sold/EJ (LSO/EJ) | 587 | 1.13 ± 0.05 | 1.09 ± 0.04 | 1.12 ± 0.06 | 1.06 ± 0.05 |
| Lamb weight sold (LWS/EJ) | 587 | 37.85 ± 1.49 | 35.51 ± 1.37 | 36.81 ± 1.97 | 33.17 ± 1.74 |
| Revenue ($ ret/EJ) | 587 | 22.2 ± 1.23 | 21.75 ± 0.87 | 23.1 ± 0.94 | 20.4 ± 1.09 |

For the Ewe productivity traits, and by comparison with the C ewes, the B ewes had 4% lower LSO/EJ (n.s.), 6% lower LWS/EJ (n.s.) and 2% lower $ ret/EJ (n.s.) For the same Ewe Productivity traits, and by comparison with the B++ ewes, the Bb+ ewes had 5% lower LSO/EJ (n.s.), 10% lower LWS/EJ (P=0.007) and 12% lower $ ret/EJ (P=0.011).

The analyses of variance for the ewe reproduction and productivity traits are given in Table 2.

**Table 2. Probabilities from the Analyses of variance of the ewe reproduction and productivity traits.**

| Source | Mu | B v C | ++ v b+ | Lambing Year | Ewe Age | Ram Type |
|---|---|---|---|---|---|---|
| df | 1 | 1 | 1 | 10 | 6 | 2 |
| OR | <0.001 | <0.001 | <0.001 | <0.001 | <0.001 | 0.904 |
| FERT (EL/EJ) | <0.001 | 0.047 | <0.001 | <0.001 | 0.002 | <0.001 |
| LS (LB/EL) | <0.001 | <0.001 | <0.001 | 0.015 | <0.001 | 0.687 |
| ERA (LW/LB) | <0.001 | <0.001 | <0.001 | <0.001 | 0.059 | 0.627 |
| NLW (LW/EJ) | <0.001 | 0.907 | 0.606 | <0.001 | <0.001 | 0.024 |
| LSO (LSO/EJ) | <0.001 | 0.677 | 0.204 | <0.001 | <0.001 | 0.022 |
| LWS (LWS/EJ) | <0.001 | 0.495 | 0.007 | <0.001 | <0.001 | 0.050 |
| $ ret/EJ ($/EJ) | <0.001 | 0.704 | 0.011 | <0.001 | <0.001 | 0.118 |

For the fixed effects in the model, the differences between lambing years and between ewe ages were almost always significant. The effect of Ram Type varied and was not significant for OR, LS, ERA, or $ ret/EJ. There were significant differences between Ram Types for FERT (P<0.001), NLW (P=0.024), LSO (P=0.022) and for LWS (P=0.05).

**DISCUSSION**

At the time of this experiment, genotyping at the Fec B locus was not available because the causative mutation on the Bmpr1b gene was not discovered until about 10 years later (Souza *et al.* 2001; Wilson *et al.* 2001; Mulsant *et al.* 2001) and no blood samples were preserved. Utilisation of the segregation analysis software developed by Elsen *et al.* (1988) and Foulley and Elsen (1988) was therefore needed to estimate Fec B genotype, and to estimate the direct effect of the Fec B gene on reproduction rate and ewe productivity in a typical Border Leicester x Merino prime lamb production system.

The analyses reported above demonstrate that the increased ovulation rate and litter size of the B ewes is a direct effect of the Fec B gene. However, the lower fertility and ewe rearing ability of the B b+ ewes compared with the C ewes results in there being no advantage to the B ewes in NLW or any of the ewe productivity traits. These results agree with those of Southey *et al.* (2001) who reported that Merino–Rambouillet crossbred ewes introgressed with the Fec B allele do not produce more total weight of lamb at 30, 60 or 120 days postpartum than purebred Rambouillet ewes in spite of their higher reproductive performances.

These results contrast with those from Bindon *et al.* (1984) where the BLxB ewes had 21% higher NLW (LW/EJ) and 18% higher $ ret/EJ than the BLxC ewes. They also contrast with the results reported by Bindon and Piper (1990) where the BLxB ewes had 15% higher NLW (LW/EJ and 7% higher $ ret/EJ then the BLxC ewes. It is not clear why the results of the present analyses differ from those obtained from the smaller and earlier sub-sets of the data analysed by Bindon *et al.* (1984) and by Bindon and Piper (1990). These results also contrast with results from studies of the productivity of heterozygous Booroola Merino d'Arles ewes (MAb+) which is higher than that of the MA++ and pure-breed Merinos d'Arles (MA) under the pastoral management of the Merinos d'Arles breed in south-eastern France. The weight of 70-day lamb produced per ewe joined increased by 41% compared to non-carrier MA++ and MA, despite a lower lamb body weight at 70 days (Teyssier *et al.* 1998).

However, lower ERA is the main contributor to the lack of difference between the B and the C ewes in NLW, and a study by Hinch *et al.* (1996) demonstrates that ERA can be significantly improved (from 58% to 73%) by a combination of targeted supplementary feeding and pregnancy scanning strategies. These results suggest that optimum utilisation of the Fec B gene in prime lamb production systems will require additional management inputs to capitalise on the increased LS of ewes carrying one copy of the Fec B gene.

The lack of difference between the C and the ++ ewes in all of the ewe reproduction and productivity traits is perhaps surprising given that the CSIRO Booroola Merino flock was under continuous selection for increased reproduction rate from 1965 to 1990. However, this outcome may have been a consequence of the selection process focussing on recruiting Fec B genotypes (bb and b+) resulting in the development of a negative linkage disequilibrium between the Fec B locus and reproduction trait genes of small effect. In these situations, the polygenic values are negatively selected (e.g. Gibson 1994).

Utilisation of the Fec B gene in prime lamb production systems has been facilitated by the transfer of the Fec B gene into the Border Leicester breed creating a new sheep breed named the Booroola Leicester (Bindon *et al.* 1997). Rams homozygous for Fec B can be utilised to create BLxMerino ewes heterozygous for Fec B with prolificacy (LB/EL) 20-40 % higher than traditional BLxM ewes. And, as outlined above, the increased prolificacy, may be converted into increased $ ret/EJ by appropriate additional management inputs.

**REFERENCES**

Bindon B.M. and Piper L.R. (1990) The role of the Booroola Merino in the Australian prime lamb industry. Armidale, NSW: CSIRO Division of Animal Production; 1990.

Bindon B.M., Nethery R. D. and Piper L.R. (1997) Final report Meat Research Corporation Project CS.141. Armidale, NSW: CSIRO Division of Animal Production; 1997.

Bindon B.M., Piper L.R. and Ch'ang T.S. (1984) In "Reproduction in Sheep", AWC Tech. Publication (1984) Proc. of Conference, Lorne, Victoria. Eds. D.R. Lindsay and D. T Pearce.

Elsen J.M., Vu Tien Khang J. and Le Roy P. (1988) *Génét. Sél. Evol.*, **20**: 211.

Foulley J.L. and Elsen J.M. (1988). *Génét. Sél. Evol.*, **20**: 227.

Gilmour A. R., Gogel B. J., Cullis B. R., Welham S. J. and Thompson R. (2014) ASReml User Guide Release 4.1 Functional Specification. VSN International Ltd, Hemel Hempstead, HP1 1ES, UK.

Gibson J. (1994) In Proc. 5th WCGALP: 7-12 August 1994. Guelph:1994

Hinch G.N., Lynch J.J., Nolan J.V., Leng R.A., Bindon B.M. and Piper L.R. (1996) *Aust. J Exp. Agric.,* **36**: 129.

Mulsant P., Lecerf F., Fabre S., Schibler L., Monget P., Lanneluc I., Pisselet C., Riquet J., Monniaux D., Callebaut I., Cribiu E., Thimonier J.,Teyssier J., Bodin L., Cognie Y., Elsen J.M., (2001) *Proc. Natl. Acad. Sci.* U.S.A., **98**: 5104.

Piper L.R. and Bindon B. M. (1982) In L.R. Piper, B.M. Bindon and Nethery, R.D. (eds). The Booroola Merino, CSIRO, Melbourne, 9-19.

Southey B.R., Thomas E.R., Gottfredson R.G. and Zelinsky R.D. (2001) *Livest. Prod. Sci.* **75**: 33.

Souza C.J., MacDougall C., Campbell B. K., McNeilly A.S. and Baird D.T. (2001) *J. Endocrinol* **169**: R1.

Teyssier J., Elsen J.M., Bodin L., Bosc P., Lefevre C. and Thimonier J. (1998) In: Proc. VI World Congress on Genetics Applied to Livestock Production, Armidale, pp. 117–120, Vol. 24.

Turner H.N. (1978) *Aust. J Agric. Res.,***29**: 327.

Wilson T., Wu X.Y., Juengel J.L., *et al.* (2001) *Biol. Reprod.*, **64**: 1225.

*Contributed paper*

# MULTIBREED GENOMIC PREDICTION FOR MALE FERTILITY IN TROPICAL BEEF CATTLE

**L.R. Porto-Neto[1], J. Bertram[2], S. McWilliam[1], M.R.S. Fortes[3], P. Alexandre[1], M.R. McGowan[4], B.J. Hayes[5] and A. Reverter[1]**

[1] CSIRO Agriculture & Food, St Lucia, QLD 4067 Australia
[2] Agriculture Consultant, Livestock management and breeding, Toowoomba QLD 4350 Australia
[3] The University of Queensland, School of Chemistry and Molecular Bioscience, St Lucia, QLD 4072 Australia
[4] The University of Queensland, School of Veterinary Sciences, St Lucia, QLD 4343 Australia
[5] The University of Queensland, Queensland Alliance for Agriculture and Food Innovation, St Lucia, QLD 4072 Australia

## SUMMARY

Regardless of the mating method (natural or artificial insemination), bull fertility impacts the reproductive outcomes of any breeding herd. There is a need to improve our ability to genetically select fertile bulls, and genomic selection approaches could assist this process. Aiming at this gap in genomic approaches, we collected phenotypes and SNP genotypes on more than 6,000 bulls across six tropically-adapted breeds. Phenotypes related to male fertility were measured during Bull Breeding Soundness Examinations. The genomic correlations of the same trait observed in different breeds were positive for scrotal circumference and sheath score in most breed comparisons but close to zero for percentage normal sperm, suggesting a divergent genetic background for this trait. We confirmed the importance of breeds being part of the reference population while estimating breeding values in an across-breed scenario. Using this dataset, multibreed genomic predictions were obtained with useful accuracies.

## INTRODUCTION

Fertility is a key driver of profitability for beef breeding herds in tropical and semi-arid environments. The standardized bull breeding soundness examination (BBSE) involves a general physical examination, a detailed examination of the external and internal genitalia, and a microscopic examination of semen cells (Entwistle and Fordyce 2003). Quantitative traits of the BBSE are heritable (Corbet *et al.* 2013) and can be improved by selection. However, the BBSE is labor intensive resulting in a limited number of animals being tested every year, which hinders the assembly of a reference population. By combining information across breeds, we were able to generate a reference population of reasonable size (>6,000 animals,) and we postulate that the use of multibreed genomic selection approaches could allow the estimation of breeding values with useful accuracy to assist the improvement of commercially relevant male traits.

## MATERIALS AND METHODS

**Animals and phenotypes**. Phenotypic data was sourced on bulls from six different populations varying in number from 535 to 1,093 (Table 1). These were Brahman (BB) and Tropical Composite (TR) from the Beef CRC (Barwick *et al.* 2009), and cattle from four performance recorded breeding herds in Queensland, a Santa Gertrudis (SG), a Droughtmaster (DM), a Belmont Tropical Composite (BT) and an Ultra Black (UB) herd. The observed phenotypes included scrotal circumference (SC, cm), sheath score (Sheath, score 1-5), and the percentage of morphologically normal spermatozoa (PNS, %). The age at which the phenotype was observed varied across the populations; for the CRC cattle, the mean age at SC was around 360 d, and for Sheath and PNS around 700 d. For SG and DM all phenotypes were observed at around 600 d of age, while for UB and BT were around 440 d and 390 d, respectively.

**Genotypes.** Most animals were genotyped using a commercial SNP chip with ~50K markers. Genotypes were imputed to ~720K SNP using a reference population that combined Beef CRC and industry cattle genotyped on the higher density platform. Genotypes were first phased using Eagle (Loh *et al.* 2016) and then imputed using Minimac3 (autosomes) or Minimac4 (BTAX) (Das *et al.* 2016). SNP with imputation $r^2 > 0.8$ were kept for further analyses. To visualise the genetic relationship between animals a principal components analyses were calculated using PLINK1.9 (Chang *et al.* 2015).

**Table 1. Number of records and descriptive statistics of the observed traits***

| | Number of records | | | Mean (SD) of measurements | | |
|---|---|---|---|---|---|---|
| **Population**** | **SC** | **Sheath** | **PNS** | **SC** | **Sheath** | **PNS** |
| **BB** | 1,089 | 1,093 | 947 | 21.26 (2.69) | 3.79 (0.92) | 73.70 (21.95) |
| **TR** | 985 | 985 | 985 | 26.55 (3.17) | 3.12 (1.54) | 73.01 (20.59) |
| **SG** | 918 | 928 | 896 | 34.46 (3.10) | 2.95 (0.78) | 73.28 (21.57) |
| **DM** | 568 | 722 | 680 | 33.68 (3.13) | 3.14 (0.68) | 63.55 (26.28) |
| **UB** | 836 | 841 | 771 | 33.80 (3.38) | 1.78 (0.80) | 68.77 (25.30) |
| **BT** | 527 | 535 | 429 | 28.11 (3.29) | 1.64 (0.59) | 54.65 (29.70) |

* SC scrotal circumference (cm), Sheath score (1-5), PNS percentage of normal sperm (%).
** BB Brahman, TR Tropical Composite, SG Santa Gertrudis, DM Droughtmaster, UB Ultra Black, BT Belmont Tropical Composite.

**Statistical analyses.** The phenotypes were adjusted using SAS 9.4 (www.sas.com) before the genomic analyses. The model for adjustment included the fixed effects of population (one per farm), year of birth and management group (within farm). The covariates of age and the first two principal components were also used. The genomic relationship matrices (GRM) were constructed following method 1 of VanRaden *et al.* (2008). Univariate, and the GBLUP analyses were run using QXPAK (Perez-Enciso and Misztal 2011).(Porto-Neto *et al.* 2015) The accuracies of the genomic predictions were calculated as the correlation of adjusted phenotypes divided by the square root of heritability and by the method LR (Legarra and Reverter 2019) that compares the predictions based on the whole and partial datasets to estimate accuracies and biases.

## RESULTS AND DISCUSSION

The estimates of heritability for SC, Sheath and PNS across-breeds were moderate, with mean heritabilities, estimated using across-breed bivariate models, of 0.45, 0.59, and 0.33, respectively. These were at the lower end of the reported estimates for SC, but similar to values reported in the literature for the other traits (Corbet *et al.* 2013; Fortes *et al.* 2020). The mean genomic correlation between these traits calculated using the same across-breed bivariate analyses were close to zero, apart from a modest 0.11 between SC and Sheath (results not shown in Tables).

Using bivariate models, we also estimated the genomic correlation of the same trait observed in different breeds. The mean correlation estimate for all pair-wise combinations of populations were 0.34, 0.40 and 0.00 for SC, Sheath and PNS, respectively (Table 2). There is very low genomic correlation between all pair-wise combinations for PNS, suggesting different genetic architecture of the trait in the different breeds, except for BB and TR with a moderate -0.30. For SC, the relative lower genomic correlation between BB and the other breeds suggests that this trait is more genetically different when comparing BB to other breeds. The strong genomic correlations between breeds for SC and for Sheath might hint at the presence of common haplotypes affecting the traits in both populations.

**Table 2. Genomic correlation for a given trait in two separate populations*, ****

| Pop 1 | Pop 2 | SC | Sheath | PNS |
|---|---|---|---|---|
| **BB** | **TR** | 0.2694 | 0.7217 | -0.3052 |
| **BB** | **SG** | 0.1248 | 0.5781 | 0.0133 |
| **BB** | **DM** | 0.1619 | 0.5123 | 0.0289 |
| **BB** | **UB** | 0.1036 | 0.5498 | -0.0191 |
| **BB** | **BT** | 0.0151 | 0.2347 | -0.0124 |
| **TR** | **SG** | 0.5370 | 0.4773 | -0.0017 |
| **TR** | **DM** | 0.6504 | 0.4785 | -0.0024 |
| **TR** | **UB** | 0.5445 | 0.7920 | 0.0431 |
| **TR** | **BT** | 0.4803 | 0.2636 | 0.1332 |
| **SG** | **DM** | 0.8174 | 0.0303 | -0.0003 |
| **SG** | **UB** | 0.5693 | 0.7512 | -0.0093 |
| **SG** | **BT** | 0.0627 | 0.0301 | 0.0209 |
| **DM** | **UB** | 0.2470 | 0.2925 | 0.0006 |
| **DM** | **BT** | 0.0263 | -0.0051 | 0.0038 |
| **UB** | **BT** | 0.5031 | 0.2610 | 0.0126 |
| | Mean | 0.3408 | 0.3979 | -0.0063 |

* Analyses performed using a bi-population GRM (ie. for the two populations under comparison). ** Traits and populations as described in Table 1.

GEBV accuracy estimates for a breed, when the breed was not represented in the reference, were lower than those when some animals of the breed were included in the reference (comparison between scheme 1 vs 2, Table 3), with the largest impact on BB. This observation was expected given the known relationship between accuracy and genetic distance to the reference population for a given test animal (de Roos *et al.* 2009). Moreover, BB is the most divergent breed among the six populations, even though it was used during the formation of some of the other breeds.

## CONCLUSIONS

There are some genomic correlations between the same trait observed in different breeds, implying there exists at least some similarities in the genetic background across breeds; however, this was not observed across all traits. We confirmed that higher accuracies are obtained by including the targeted breed in the reference population. Finally, it was possible to estimate GEBVs with useful accuracies, for fertility-related traits in bulls, in a multibreed scenario. This approach could be further developed in the future, aiming at a broader adoption of the technology by the industry.

## ACKNOWLEDGMENTS

This work was supported by the Bull Fertility update project, an initiative from CSIRO, the University of Queensland, and Meat & Livestock Australia. The authors are thankful to the beef producers who took part in this collaborative effort and Meat & Livestock Australia that co-funded the project (L.GEN.1818).

**Table 3. Multibreed genomic prediction accuracies calculated using the method LR****

| Population | SC | Sheath | PNS | Mean |
|---|---|---|---|---|
| **Scheme #1: From a given population, all records missing in the reference** | | | | |
| BB | 0.217 | 0.217 | 0.217 | 0.217 |
| TR | 0.479 | 0.696 | 0.211 | 0.462 |
| SG | 0.367 | 0.366 | 0.233 | 0.322 |
| DM | 0.497 | 0.358 | 0.251 | 0.368 |
| UB | 0.381 | 0.512 | 0.176 | 0.356 |
| BT | 0.263 | 0.323 | 0.227 | 0.271 |
| Mean | 0.367 | 0.412 | 0.219 | |
| **Scheme #2: From a given population, a random 20% records missing in the reference (mean across five 80/20 cross-validation splits)** | | | | |
| BB | 0.513 | 0.399 | 0.319 | 0.410 |
| TR | 0.648 | 0.812 | 0.402 | 0.621 |
| SG | 0.501 | 0.412 | 0.341 | 0.418 |
| DM | 0.593 | 0.402 | 0.473 | 0.489 |
| UB | 0.629 | 0.573 | 0.406 | 0.536 |
| BT | 0.610 | 0.343 | 0.510 | 0.488 |
| Mean | 0.582 | 0.490 | 0.408 | |

* Traits and populations as described in Table 1. ** Legarra, and Reverter (2019)

## REFERENCES

Barwick S.A., Johnston D.J., Burrow H.M., Holroyd R.G., Fordyce G., Wolcott M.L., Sim W.D., and Sullivan M.T. (2009) *Anim. Prod. Sci.* **49**: 367.

Chang C.C., Chow C.C., Tellier L., Vattikuti S., Purcell S.M., and Lee J.J. (2015) *Gigascience* **4**.

Corbet N.J., Burns B.M., Johnston D.J., Wolcott M.L., Corbet D.H., Venus B.K., Li Y., McGowan M.R., and Holroyd R.G. (2013) *Anim. Prod. Sci.* **53**: 101.

Das S., Forer L., Schonherr S., Sidore C., Locke A.E., Kwong A., Vrieze S.I., Chew E.Y., Levy S., McGue M., Schlessinger D., Stambolian D., Loh P.R., Iacono W.G., Swaroop A., Scott L.J., Cucca F., Kronenberg F., Boehnke M., Abecasis G.R., and Fuchsberger C. (2016) *Nat. Genet.* **48**: 1284.

de Roos A.P., Hayes B.J., and Goddard M.E. (2009) *Genetics* **183**: 1545.

Entwistle K., and Fordyce G. (2003). Australian Association of Cattle Veterinarians, Australia.

Fortes M.R.S., Porto-Neto L.R., Satake N., Nguyen L.T., Freitas A.C., Melo T.P., Scalez D.C.B., Hayes B., Raidan F.S.S., Reverter A., and Boe-Hansen G.B. (2020) *Genet. Sel. Evol.* **52**.

Legarra A., and Reverter A. (2019) *Genet. Sel. Evol.* **51**.

Loh P.R., Danecek P., Palamara P.F., Fuchsberger C., Reshef Y.A., Finucane H.K., Schoenherr S., Forer L., McCarthy S., Abecasis G.R., Durbin R., and Price A.L. (2016) *Nat. Genet.* **48**: 1443.

Perez-Enciso M., and Misztal I. (2011) *BMC Bioinformatics* **12**: 202.

Porto-Neto L.R., Barendse W., Henshall J.M., McWilliam S.M., Lehnert S.A., and Reverter A. (2015) *Genet. Sel. Evol.* **47**.

VanRaden P.M. (2008) *J. Dairy Sci.* **91**: 4414.

*Contributed paper*

# NARROWING THE SEARCH SPACE: PUTATIVE CAUSAL VARIANTS ARE ENRICHED IN ANNOTATED FUNCTIONAL REGIONS FROM 6 BOVINE TISSUES

**C.P. Prowse-Wilkins[1,2], J. Wang[2], M.E. Goddard[1,2], R. Xiang[1,2], J.B. Garner[2] and A.J. Chamberlain[2]**

[1] Faculty of Veterinary & Agricultural Science, The University of Melbourne, Parkville, Victoria, Australia

[2] Agriculture Victoria, AgriBio, Centre for AgriBiosciences, Bundoora, Victoria, Australia

## SUMMARY

Identifying causal variants in the bovine genome is difficult as there are millions of variants. Work in humans shows that most variants affecting complex traits lie in non-coding functional regions. However, functional regions are generally species specific and not well annotated in non-model organisms. This project annotated functional regions directly in dairy cows using a laboratory technique called ChIP-seq (Chromatin Immunoprecipitation followed by sequencing).

We generated 86 functional datasets across 6 tissues from 3 lactating Holstein dairy cows. This represents millions of putative functional regions in the bovine genome including, for the first time, in the mammary gland of lactating dairy cows. These regions were highly enriched for putative causal variants (eg milk trait QTL and eQTL). The results represent the largest database of functional regions in the bovine genome to date and can be used to narrow the search space for causal variants and improve genomic predictions.

## INTRODUCTION

Genomic prediction aims to predict the phenotypes of animals based on their genotypes. It does this by finding genotypes which associate with the phenotype in a training population. However, this association could be based on linkage disequilibrium (LD) and not a direct causal relationship between the trait and the genotype. This means the accuracy of genomic predictions can break down over time as LD breaks down and is not useful in breeds which have different LD to the training population. If we could use the genetic variant which is directly affecting the phenotype (the causal variant) in our predictions, this would not occur (Hayes *et al.* 2016).

Work in other species has found that causal variants are enriched in functional regions (Schaub *et al.* 2012). Until recently, these were not well annotated in the bovine genome (Fang *et al.* 2019). Functional regions can be identified with Chromatin Immunoprecipitation followed by sequencing (ChIP-seq) to identify functional marks which pinpoint these regions in the genome. Examples of functional marks include histone modifications and transcription factors. Histone modifications are alterations to the histone proteins which DNA is wrapped around in the cell. Four histone modifications of interest are H3K4Me3-found at promoters, H3K4Me1-found at enhancers, H3K27ac-found in active regions and H3K27Me3-found in inactive regions (Kimura 2013). Another marker of interest is the binding site for the transcription factor CTCF which is found at insulators and other regions of importance (Kim *et al.* 2015). This study annotated these functional markers in 6 tissues (mammary, liver, kidney, spleen, lung and heart) in Holstein dairy cows and tested whether these regions are enriched for causal variants.

## MATERIALS AND METHODS

**Chromatin Immunoprecipitation and Sequencing.** Heart, kidney, liver, lung, mammary gland, and spleen were sampled from 3 Holstein dairy cows post-mortem and snap frozen in liquid nitrogen before being stored at -80°C until use. At sampling animals were at 5$^{th}$, 7$^{th}$, and 1$^{st}$ parity and 208, 173 and 65 days of lactation respectively. Ethics approval for 2 of the cows were obtained from Department of Jobs, Precincts and Regions Ethics Committee (Application No. 2014-23). The 3rd cow was not euthanised for this study but culled as a result of injury. Frozen tissue was ground for 3 minutes in the Geno/Grinder (SPEX SamplePrep) and fixed for 10 minutes with 10% formaldehyde. Chromatin was prepared using the Magnify Chromatin Immunoprecipitation kit (ThermoFisher) as per the manufacturer's instructions. Fixed chromatin was sheared to 200-500bp using the Covaris S2 (Covaris) for three minutes, duty cycle five, % intensity four and 200 cycles per burst. Chromatin immunoprecipitation was performed using the Magnify Chromatin immunoprecipitation kit (ThermoFisher) with some modifications. Sheared chromatin was immunoprecipitated with 0.25-0.5µg of antibody for the histone modifications (H3K4Me3, H3k4Me1, H3K27ac and H3K27Me3) or 10µl of antibody for CTCF. Sequence libraries were prepared for each ChIP sample and a control for each chromatin preparation (input sample) using the NEBNext Ultra II DNA Library Prep Kit for Illumina (New England Biolabs) as per the manufacturer's instructions and run on the Hiseq 3000 (Illumina) in a 150 cycle paired end run. Each library was sequenced with 20-300 million reads. Raw sequence reads were trimmed of adapters and poor-quality bases at the ends (quality less than 20) using Trimmomatic (Bolger *et al.* 2014). Trimmed reads with length less than 50 were removed. Trimmed reads were mapped to UMD3.1 bovine genome using BWA mem with default settings (Li 2013). Poor-quality reads with q>15 were removed with Samtools (Li *et al.* 2009) and marked duplicate reads were also removed. MACS2 with default settings was used to call peaks from mapped ChIP reads with input reads as control (Zhang *et al.* 2008). The quality of peaks was checked with deepTools plotFingerprint (Ramirez *et al.* 2016) and SPP (Kharchenko *et al.* 2008).

**Enrichment of Causal SNP in Functional Regions.** Enrichment of putative causal SNP in functional regions was calculated using the formula described in (Ernst & Kellis 2010) as outlined below. A variety of SNP datasets were used as putative causal SNP (Table 1). Statistical significance of enrichment or depletion was calculated in R using a hypergeometric test.

Enrichment=(C/A)/(B/D) where:
A= number of positions under peaks
B=number of positions under peaks and also a putative causal SNP
C=number of positions that were putative causal SNP
D=number of positions in the genome

## RESULTS AND DISCUSSION

In total we sequenced 86 ChIP-seq samples, with three biological replicates in 6 tissues assayed for 5 marks (four samples were excluded due to low quality). There was an average of 480,000 peaks per sample covering an average of 13% of the genome. All samples were high quality. These data represent millions of putative functional regions in the bovine genome.

Peaks were significantly enriched for putative causal variants (P<0.001) as expected (Table 2). The QTL for milk traits were particularly strongly enriched within peaks and particularly enriched within peaks found in the mammary gland (Figure 1). This is consistent with studies in other species which show that trait QTL are particularly enriched within histone markers specific to tissues relevant to the trait (Trynka *et al.* 2013). The 80k SNP dataset was the least enriched although these were still significantly enriched within peaks. It is possible that this is because these SNPs are contributing to multiple traits which may not be relevant to the tissues represented in this study.

## CONCLUSION

This work substantially increases the number of putative functional regions found in different tissues in the bovine genome, including the mammary gland of lactating dairy cows. As seen in other species, these regions are substantially enriched for putative causal variants for important traits suggesting SNP within these regions should be prioritised for genomic selection.

**Table 1. Details of putative causal SNP tested for enrichment within functional regions**

| Dataset | Number of SNP | Description | Reference |
|---|---|---|---|
| Allele specific eQTL | 1,100,446 | Allele specific expression QTL from white blood cells and milk cells in 112 holstein cows ($P$<1e-4) | (Chamberlain *et al.* 2018) |
| Exon eQTL | 945,832 | Exon expression QTL from white blood cells, milk cells, liver and muscle in 209 holstein cows ($P$<1e-4) | (Xiang *et al.* 2018, Xiang *et al.* 2019) |
| Gene eQTL | 110,200 | Gene expression QTL from white blood cells, milk cells, liver and muscle in 209 holstein cows ($P$<1e-4) | (Xiang *et al.* 2018, Xiang *et al.* 2019) |
| Conserved regions | 378,472 | SNP conserved in 100 species lifted over from human to bovine genome | (Xiang *et al* 2019) |
| SNP 80k | 83,454 | Top 80,000 sequence variants ranked for their contributions to 34 traits | (Xiang *et al.* 2021) |
| Splice QTL | 1,112,324 | Splice QTL from blood, milk cells, liver and muscle in 209 holstein cows ($P$<1e-4) | (Xiang *et al.* 2018, Xiang *et al.* 2019) |
| QTL Protein Yield | 3,317 | GWAS in 32347 cows for protein yield with $P$<1e-7 | (Xiang *et al.* 2020) |
| QTL Fat yield | 4,815 | GWAS in 32347 cows for fat yield with $P$<1e-7 | Xiang *et al.* 2020) |
| QTL Milk Yield | 6,883 | GWAS in 32347 cows for milk yield with $P$<1e-7 | Xiang *et al.* 2020) |
| QTL Fat percentage | 12,373 | GWAS in 32347 cows for fat percentage with $P$<1e-7 | Xiang *et al.* 2020) |
| QTL Protein percentage | 17,012 | GWAS in 32347 cows for protein percentage with $P$<1e-7 | Xiang *et al.* 2020) |

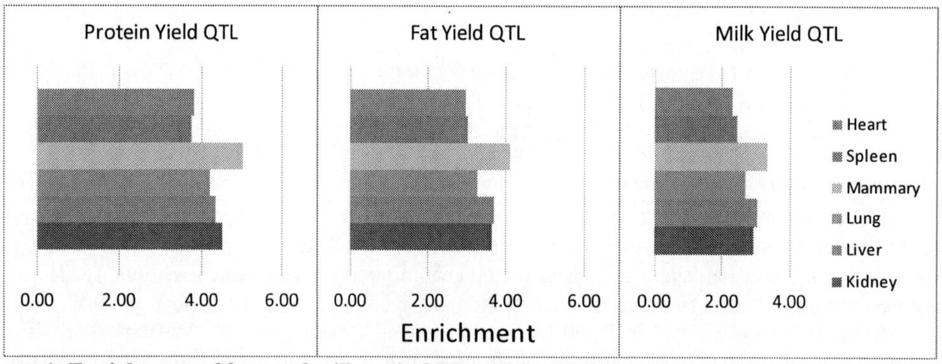


**Figure 1. Enrichment of 3 sets of milk trait QTL within H3K27ac peaks. Peaks in mammary gland have the highest enrichment for these milk trait QTL**

**ACKNOWLEDGEMENTS**

The authors would like to thank the Agriculture Victoria staff at Ellinbank and Bundoora who helped with sample collection. We also thank DataGene for access to data used in this study.

**Table 2. Enrichment of causal SNP in ChIP-seq peaks. Enrichment of each SNP dataset within each histone modification or CTCF averaged across tissues**

| | H3K4Me3 | H3K27ac | CTCF | H3K4Me1 | H3K27Me3 |
|---|---|---|---|---|---|
| **Allele specific eQTL** | 1.86 | 1.96 | 1.93 | 1.76 | 1.69 |
| **Exon eQTL** | 1.68 | 2.21 | 1.73 | 1.61 | 1.33 |
| **Gene eQTL** | 2.24 | 2.37 | 2.27 | 1.97 | 1.82 |
| **Conserved regions** | 1.66 | 1.46 | 1.42 | 1.21 | 1.14 |
| **SNP 80k** | 1.20 | 1.16 | 1.18 | 1.16 | 1.15 |
| **Splice QTL** | 1.70 | 1.77 | 1.75 | 1.63 | 1.58 |
| **QTL Protein Yield** | 4.46 | 4.27 | 4.06 | 3.21 | 2.93 |
| **QTL Fat yield** | 3.72 | 3.46 | 3.43 | 2.82 | 2.60 |
| **QTL Milk Yield** | 3.09 | 2.79 | 2.85 | 2.35 | 2.24 |
| **QTL Fat percentage** | 2.78 | 2.51 | 2.58 | 2.19 | 2.16 |
| **QTL Protein percentage** | 1.85 | 1.91 | 1.80 | 1.58 | 1.40 |

## REFERENCES


Bolger A.M., Lohse M. and Usadel B.J.B. (2014) *Bioinformatics* **30**: 2114.

Chamberlain A., Hayes B., Xiang, R., Vander Jagt C., Reich C., Macleod I., Prowse-Wilkins C., Mason B., Daetwyler H. and Goddard M. (2018) *Proc. Of. WCGALP*. **Vol Mol Gen 1**: 254

Ernst J & Kellis M. (2010) *Nature Biotech.* **28**: 817.

Fang L., Liu M., Kang X., Lin S., Li B., Connor E. E., Baldwin R.L., Tenesa A., Ma L., Liu G.E. and Li C. (2019) *BMC Biology* **17**: 68.

Hayes B., Chamberlain A., Daetwyler H., VanderJagt C. & Goddard M. (2016) *J.An. Sci.* **94**: 3.

Kharchenko P. V., Tolstorukov M. Y. and Park P. (2008) *Nature Biotech.* **26**: 1351.

Kim S., Yu N.-K. & Kaang B.-K. (2015) *Exp. & Mol. Med.* **47**: e166.

Kimura H. (2013) *J. Hum. Gen.* **58**: 439.

Li H., Handsaker B., Wysoker A., Fennell T., Ruan J., Homer N., Marth G., Abecasis G. and Durbin R. (2009) *Bioinformatics* **25**: 2078.

Li H. (2013) *arXiv*. 1303.3997.

Ramirez F., Ryan D.P., Gruning B., Bhardwaj V., Kilpert F., Richter A. S., Heyne S., Dundar F. and Manke T. (2016) *Nucleic Acids Res.* **44**:W160.

Schaub M.A., Boyle A.P., Kundaje A., Batzoglou S. and Snyder M. (2012) *Gen. Res.* **22**:1748.

Trynka G., Sandor C., Han B., Xu H., Stranger B.E., Liu X.S. and Raychaudhuri S. (2013) *Nature Genetics.* **45**:124.

Xiang R., Hayes B.J., Vander Jagt C.J., Macleod I.M., Khansefid M., Bowman P.J., Yuan Z., Prowse-Wilkins C.P., Reich C.M. and Mason B.A. (2018) *BMC Genomics.* **19**:521.

Xiang R., Van Den Berg I., Macleod I.M., Hayes B.J., Prowse-Wilkins C.P., Wang M., Bolormaa S., Liu Z., Rochfort S.J and Reich C.M. (2019). *PNAS.* **116**: 19398.

Xiang R., Van Den Berg I., Macleod I.M., Daetwyler H.D. and Goddard M.E. (2020). *Comm. Bio.* **3**: 1.

Xiang R., Macleod I.M., Daetwyler H.D., de Jong G., O'Connor E., Schrooten C., Chamberlain A.J. and Goddard M.E. (2021) *Nature Comms.* **12**: 860.

Zhang Y., Liu T., Meyer C.A., Eeckhoute J., Johnson D.S., Bernstein B.E., Nusbaum C., Myers R.M., Brown, M. and Li W. (2008) *Genome Biology.* **9**: 1.

*Contributed paper*

# INDEXES SUPPORTING GENOMIC TOOLS FOR SELECTING COMMERCIAL ANGUS HEIFER REPLACEMENTS AND IDENTIFYING STEERS FOR LONG-FED PROGRAMMES IN AUSTRALIA

**C.D. Quinton[1], J.A. Archer[1,2], P.R. Amer[1], S. Harburg[1], G. Petersen[1], A. Byrne[3], C. Duff[3] and P. Parnell[3]**

[1] AbacusBio Limited, Dunedin 9016, New Zealand
[2] *Current address:* Beef + Lamb New Zealand Genetics, Dunedin 9054, New Zealand
[3] Angus Australia, Armidale, NSW 2350, Australia

## SUMMARY

Angus Australia, in collaboration with the Commonwealth Scientific and Industrial Research Organisation (CSIRO), have developed new genomic tools for early life evaluation of commercial straightbred Angus heifers and steers. To aid producers to make optimal multi-trait selection decisions, two new commercial economic indexes have been developed. These indexes are based on economic value models for core GEBVs calculated with the new genomic products.

The heifer index is designed to aid selection of replacement heifers in commercial herds and is based on costs and revenues from cows and their offspring in Australian short/mid-fed and long-fed production systems. This index contains maternal (birth weight, weaning weight, milk, mature cow weight) and terminal (post-wean growth, feedlot growth and intake, rib fat and marbling) traits. Non-linear functions are applied to value birth weight as it relates to calving ease, milk, and marbling. This index should identify more efficient heifers with genetic potential to produce progeny with improved post-wean growth, feed efficiency and carcase merit.

The long-fed steer index is designed to identify steers best suited to Australian long-fed production systems. This index focuses on feedlot growth and intake, and carcase traits rib fat and marbling. This index should identify efficient steers with high marbling.

## INTRODUCTION

Angus Australia, in collaboration with the Commonwealth Scientific and Industrial Research Organisation, have developed new genomic products to evaluate commercial straightbred Angus beef heifers and steers. The products include GEBVs for birth weight (BW), weaning weight (WW), yearling weight (YW), mature cow weight (MCW), milk, average daily gain (ADG), dry matter intake (DMI), carcase weight (CWT), eye muscle area (EMA), MSA marbling (MSA), rib fat (RIB), ossification, antibody, cell-mediated antibody and ImmuneDEX (Hine *et al.* 2021 submitted). These evaluations are designed to aid producers to make early life selection and management decisions on commercial animals.

For commercial cow herds, a major decision is selection of heifer calves to be retained as replacements. A commercial heifer selection index was needed to identify heifers with genetic potential for maternal traits they (and possibly their daughters) will express, as well as beef traits expressed by their calves.

For commercial market animals, producers have a management decision to direct steers to either short/mid fed (SF) production, or to long-fed (LF) production. A LF steer index was needed to identify steers with genetic potential for marbling as well as feedlot growth and feed efficiency.

The objective of this work was to develop two new commercial economic indexes to aid the above decisions. These indexes are based on economic value models for core GEBVs calculated with the new genomic products and consider amount and timing of gene expression in animals' lifetimes.

## MATERIALS AND METHODS

**Economic index model.** An economic index model was built to calculate index traits' economic values (EVs) and discounted genetic expression coefficients (DGEs). This contains models for Australian Angus commercial breeder cow daily growth and feed requirements, and separate models for steer and heifer daily growth, feed requirements and carcase value in short/mid fed (SF) and long fed (LF) production systems. The SF model assumed 70% steers in feedlot from 420 to 520 days old with exit live weight (LW) 640 kg, and 30% heifers in feedlot from 360 to 430 days old with exit LW 490 kg. The LF model assumed only steers in feedlot from 420 to 620 days with exit LW 760 kg. Both models assumed age-constant slaughter endpoints. Feed requirements were based on Primary Industries Standing Committee (2007) metabolizable energy. Carcase value and feed costs were based on Australia industry averages in last 5 years. System-specific trait EVs were calculated as described below, as well as trait DGEs. Ossification and immune traits GEBVs were not included in these indexes.

**Heifer Index.** This index contains maternal and terminal traits. Calculations assume 75% of surplus calves will enter SF systems and 25% will enter LF systems.

Birth weight (kg) is valued with a non-linear EV function based on its relationship with heifer calving ease and associated effects on labour, heifer survival and calf survival. Angus Australia data of bull BW EBVs and their daughters' calving ease phenotypes were applied to fit an exponential function that related BW EBVs to calving costs (based on observed calving difficulty scores). This approach has been applied in other beef indexes to value dystocia (Quinton *et al.* 2019).

Weaning weight (kg) linear EV was calculated as change in profit expected from a 1 kg increase in WW, assuming animals grow at the same rate post-weaning to reach 1 kg heavier live weight at slaughter age. This increases carcase weight and revenue, with increased feed requirements and costs.

Milk (kg, defined as maternal genetic WW) is valued with a non-linear EV function as differences in milk genotype have the greatest economic impact, compared to other traits, at low GEBVs, but less relative economic impact at higher GEBVs. This approach has been applied in other beef maternal indexes (Quinton *et al.* 2019). The EV function incrementally decreases EV of milk up to an optimum GEBV, above which all individuals receive the same value. The optimum GEBV was defined as midpoint of $10^{th}$ and $90^{th}$ percentile of bull population GEBVs.

Mature cow weight (kg) linear economic weight was calculated from 3 component EVs. Replacement heifer MCW EV was calculated from the increase in feed costs associated with 1 kg additional growth from yearling to maturity ($2^{nd}$ calving), expressed in heifers. Annual cow MCW EV was calculated from the increase in maintenance feed costs for a 1 kg heavier cow, expressed annually from maturity over the cow's lifetime. Cull cow MCW EV was calculated from the increase in carcase revenue from a 1 kg LW heavier cow, expressed at average culling age. The MCW index economic weight was calculated as the sum of each component EV multiplied by the component DGE.

Post-wean gain (PWG, kg) was defined as a proxy trait where $GEBV_{PWG} = GEBV_{YW} - GEBV_{WW}$. This was done because YW is composed of two phenotypes WW + post-wean growth; but the EVs for WW and growth between weaning and 1 year need to be independent. The PWG linear EV was calculated as change in profit expected from 1 kg increase in PWG, assuming animals grow at same rate pre-weaning and post-yearling to reach 1 kg heavier live weight at slaughter age. This increases carcase weight and revenue, with increased feed requirements and costs.

Feedlot gain (FG, kg) was also defined as a proxy trait where $GEBV_{FG} = (GEBV_{CWT}$ / dressing %) – $GEBV_{YW}$. This was done because CWT is composed of YWT=WW+PWG at fixed dressing % and post-yearling growth, but these EVs need to be independent. Note FG differs from the ADG GEBV which is defined differently. The FG linear EV was calculated from change in revenue expected from 1 kg increase in FG, assuming animals grow at same rate pre-feedlot. This increases carcase weight and revenue. This EV is independent of feedlot feed costs valued via DMI.

Feedlot DMI (kg/d) linear EV was calculated from increased feed costs from 1 kg/d increased intake during fixed feedlot time.

Rib fat (RIB, mm) linear EV was based on industry rib fat pricing categories which penalize under- and over-fat carcases. Assuming RIB has an underlying standard normal distribution which is expressed as percentages of animals that occur in the rib fat price categories, the EV is calculated from the change in carcase revenue that results from shifting the distribution of RIB by 1 mm with according changes in proportions of animals in the rib fat price categories.

MSA marbling (MSA, score) is valued as a non-linear economic value based on the shift in marbling distribution expected for an individual GEBV. Marbling is assumed to have a normal distribution, with thresholds determining the value paid for an animal within a proportion of the distribution. A change to the proportion of animals falling within each marbling price category occurs in response to a shift in the distribution mean. The function also considers the different SF and LF industry marbling pricing categories and weights the value according to the proportions of animals in each system.

For WW, PWG, FG, DMI and RIB, separate EVs were calculated for SF and LF systems and a weighted average EV was calculated based on Australian industry proportions. Other traits' EV calculations incorporated SF and LF parameters. For each trait, EVs were multiplied by DGE coefficients that incorporate timing and frequency of expression in heifers and their calves.

The structure of this index is as follows, where *f(GEBV)* represent non-linear functions and *b* are linear index economic weights, to calculate an index value in units \$/heifer at selection:

$$I_{Heifer} = f(GEBV_{BW}) + (b_{WW}GEBV_{WW}) + f(GEBV_{Milk}) + (b_{MCW}GEBV_{MCW}) + (b_{PWG}GEBV_{PWG}) + (b_{FG}GEBV_{FG}) + (b_{DMI}GEBV_{DMI}) + (b_{RIB}GEBV_{RIB}) + f(GEBV_{MSA})$$

**Long-fed Steer Index.** This index contains only terminal beef traits FG, DMI, RIB and MSA. Economic value calculations for these traits followed the same methods as described for the heifer index, but incorporated only LF system parameters. These traits were assumed to be expressed at steer slaughter and therefore DGEs were set to 1.

This structure of this index is as follows to calculate an index value in units \$/steer fed:

$$I_{Steer} = (b_{FG}GEBV_{FG}) + (b_{DMI}GEBV_{DMI}) + (b_{RIB}GEBV_{RIB}) + f(GEBV_{MSA})$$

**Preliminary index selection predictions.** At this time, the pipeline for routinely calculating GEBVs for commercial heifers and steers is under development and the availability of GEBVs for large numbers of individuals is limited. For this study, the effectiveness of each selection index was assessed using a set of GEBVs from 333 bulls that represent the range of genotypes in the population (Table 1). Because the bull GEBVs are based on DNA only, we expect very similar outcomes from heifer or steer GEBVs. Mean GEBVs were calculated and compared for all bulls in the set and for the top 20% of bulls according to each index.

## RESULTS AND DISCUSSION

Mean GEBVs of the top 20% of bulls selected according to preliminary versions of the new indexes are shown in Table 1.

Top bulls with the Heifer Index had substantially higher mean GEBVs for growth (WW, YW, PWG, FG), CWT, EMA and MSA, with only slightly higher DMI, as well as lower RIB and MCW. Mean GEBVs for BW, WW and Milk were similar to the population average. Therefore, this index should identify heifers that are on average more efficient at maintaining similar mature weight and milk production, but with genetic potential to produce progeny with improved post-wean growth, feed efficiency and carcase merit.

Top bulls with the LF Steer Index had substantially higher mean GEBVs for FG, CWT, EMA and MSA than population average, but lower mean DMI and RIB. Therefore, this index should identify efficient steers with high marbling suited to long-fed production systems. The LF steer index contains growth only in terms of feedlot gain and therefore does not differentiate between steers that have different feedlot entry weight but the same growth rate in feedlot. This index assumes that the user accounts for pre-feedlot growth value by selling/purchasing steers on a per kg basis.

**Table 1. GEBV means, SD, minimum and maximum values for all bulls in data set and mean GEBVs of top 20% selected according to the Heifer Index, and Long-fed Steer Index**

| | All bulls (N=333) | | | | Heifer Index Top 20% bulls | LF Steer Index Top 20% bulls |
|---|---|---|---|---|---|---|
| GEBV, unit | mean | SD | min | max | mean | mean |
| BW, kg | -1.24 | 1.80 | -6.76 | 3.51 | -1.29 | - |
| WW, kg | -2.94 | 6.35 | -24.30 | 16.71 | -0.25 | - |
| YW, kg | 2.48 | 10.65 | -36.45 | 31.08 | 10.24 | - |
| PWG, kg | 5.43 | 6.92 | -22.46 | 26.81 | 10.49 | - |
| FG, kg | 12.23 | 28.94 | -78.44 | 78.81 | 37.05 | 26.03 |
| MCW, kg | -5.45 | 14.88 | -44.63 | 39.35 | -7.36 | - |
| Milk, kg | -5.39 | 3.70 | -15.51 | 6.93 | -4.77 | - |
| ADG, kg/day | 0.03 | 0.09 | -0.21 | 0.33 | 0.07 | 0.03 |
| DMI, kg/day | 0.51 | 0.71 | -1.46 | 2.51 | 0.69 | 0.38 |
| CWT, kg | 8.09 | 19.01 | -51.87 | 51.71 | 26.01 | 15.17 |
| EMA, $cm^2$ | 0.73 | 4.35 | -11.19 | 12.94 | 4.18 | 3.29 |
| MSA, score | 90.57 | 59.03 | -76.19 | 262.34 | 136.79 | 157.55 |
| RIB, mm | 0.08 | 1.66 | -4.69 | 5.37 | -0.08 | -0.08 |

## CONCLUSIONS

The recent development of commercial genomic tools for cost-effective evaluation of commercial Angus heifers and steers provide producers with new information to access an animals' genetic potential for performance in different sectors of the Australian beef industry. The new commercial replacement heifer and long-fed steer indexes offer tools to aid producers in multi-trait selection and management decisions.

## REFERENCES

Hine, B.C., Duff, C.J., Byrne, A., Parnell, P., Porto-Neto, L., Li, Y., Ingham, A.B., Reverter, A. (2021 submitted conference paper) *J. Anim. Prod. Sci.*

Primary Industries Standing Committee. (2007) 'Nutrient Requirements of Domesticated Ruminants'. CSIRO Publishing.

Quinton, C.D., Amer, P.R., Byrne, T.J., Archer, J.A., Santos, B. and Hely, F. (2019). *Proc. Assoc. Advmt. Anim. Breed. Genet.* **23:** 47.

*Contributed paper*

# DEFINING LONGEVITY AND ESTIMATING GENETIC PARAMETERS IN AUSTRALIAN MERINO EWES

**M.O.A. Rahman[1], D.J. Brown[1] and S.F. Walkom[1]**

[1] Animal Genetics Breeding Unit*, University of New England, Armidale, NSW, 2350 Australia

## SUMMARY

Currently, there are no measures of ewe longevity recorded by Australian sheep breeders for utilisation as part of their breeding objective. In the absence of disposal codes, this study explored the potential to use production records to define ewe longevity in Merino sheep and estimate genetic parameters for the resulting trait. The longevity trait was defined as the ewe's total life (TL) in the flock from birth to their last available production record. To identify suitable non-censored data, cohorts were selected based on the amount of pedigree and consistent annual production recording. Under these assumptions, the MERINOSELECT database provided 267,517 longevity records from 143 flocks. The heritability of TL was 0.22 ± 01. Adjusting TL for the ewe's lifetime reproductive performance, accounted for 94% of the variation in TL, reduced the heritability to 0.11 ± 01. The results herein indicate that it is possible to describe longevity in the Merino ewes using production records from the MERINOLSELECT database. TL was found to be heritable but further exploration is required before incorporation in industry breeding objectives.

## INTRODUCTION

Longevity can be defined as the duration of a ewe's productive life in the flock. In Australian Merino, ewes are usually first mated at 1.5 years old with most ewe's final mating at 4.5 to 6.5 years of age, after which they are culled as cast for age ewes (Kleemann *et al.* 2016). Age based culling is extensively used in Australian commercial sheep flocks (Hatcher *et al.* 2009), most commonly at 6 years of age. Longevity is a composite trait describing aspects of production, health, and reproduction and it is considered a trait of high economic importance for sheep production systems (McLaren *et al.* 2020). Greater longevity in sheep production leads to an increased overall mean age of the flock, more lambs available for sale, and higher reproductive performance (Conington *et al.* 2004). Conington *et al.* (2001) defined longevity as the period from birth to culling or death (days). According to previous research, the heritability estimates of longevity commonly ranges between 0.05 to 0.08 with a range from 0 to 0.33 depending on the species, production system and trait definition (Conington *et al.* 2001; El-Saied *et al.* 2005). In the MERINOSELECT database (Brown *et al.* 2007), culling date and reason are sparsely recorded. Therefore, building on the proposition by McLaren *et al.* (2020), we explored the potential of using the ewe's last production record as a proxy for culling age. The objective of this study was to define ewe longevity using production records and estimate genetic parameters for the resulting trait.

## MATERIALS AND METHODS

**Describing Longevity**. At the time of the analysis, there was no standard recording practice for capturing ewe longevity and limited recording of disposal date in the MERINOSELECT database. The MERINOSELECT database, described by Brown *et al.* (2007), currently includes pedigree and phenotypic records for 3,078,163 animals from 1,759 flocks submitted by Australian and New Zealand Merino breeders. A longevity trait was built based on the birth date of an individual and the

* A joint venture of NSW Department of Primary Industries and the University of New England

date of their last submitted production record, which was suggested by McLaren *et al.* (2020) as an alternative to a specified culling date. Total life of the ewes was referred to as the time between these two dates in years and aligns with the longevity traits presented by Conington *et al.* (2001); El-Saied *et al.* (2005). For TL to reflect the lifetime of the ewe it was assumed that the ewe's cohort (site of birth, flock and year of birth) is routinely recorded and so the absence of the ewe record reflects her departure from the flock and not that the cohort was not recorded. Cohorts considered to have suitable data were characterized as; 1) born since 2000, 2) had a minimum of 3 years of production records, 3) had an annual record for wool and reproduction recorded up to 6 years of age, and 4) contained at least 30 ewes and a minimum of 70% of the animals were assigned a sire. Approximately 20% of ewes in the database were from a cohort with sufficient recording (Table 1).

**Table 1. Number of flocks, cohorts and animals represented in the cleaned data set after implementing each of the data assumptions to ensure the eventual phenotype will reflect the animal's longevity in the flock are instated**

| Assumptions | Flocks | Cohorts | Animals |
|---|---|---|---|
| Ewes | 1,086 | 7,923 | 1,574,855 |
| Born since 2000 | 771 | 5358 | 1,371,799 |
| Minimum lifespan of cohort 3+ years | 676 | 4,410 | 1,187,133 |
| Cohort recorded annually up to 6 years of age | 447 | 3,249 | 1,147,498 |
| Cohorts contain at least 30 ewes and > 70% of animals assigned sire | 285 | 1,451 | 473,698 |
| Cohorts had annual reproduction and wool production records | 143 | 746 | 267,517 |

**Statistical Analysis.** Genetic parameters for TL were estimated from a series of univariate analyses using an animal model in ASReml version 4.1 (Gilmour *et al.* 2015). A pedigree spanning 2 generations back from the phenotyped animals, due to computational restraints, was extracted from the MERINOSELECT database (Brown *et al.* 2007) and incorporated 335,704 animals.. The phenotyped individuals descended from 6,030 sires and 103,730 dams. The base animal model (Model 1) used in the analysis can be described by the following equation:

$$Y = Xb + Z_1a + e$$

Where $Y$ is the vector of TL records, $b$ is the vector of fixed effects that include the birth type (1, 2, 3, 4+), rear type (1, 2, 3+), age of dam (linear) and cohort. Where cohort was defined by the animals' site of birth, flock and year of birth. $a$ is the vector of animal genetic effects with $X$ and $Z$ the incidence matrices that relate the respective effects to $Y$ and $e$ is the vector of the random residual effect. The phenotypic variance was calculated as the sum of the additive and residual variance.

TL reflects the lifetime of the ewe in the flock, which describes both the ewe's fitness and survivability but also the ewe's production performance and merit relative to the flock's breeding objective. To create a trait that more closely reflects the ewe's fitness and survivability and understand the underlying factors that may impact the ewe's time in the flock, TL was adjusted by fitting a series of co-variates, nested within cohort.

The covariates included; 1) the ewe's Merino Production Plus (MPP) index value (Swan *et al.* 2017), 2) the ewe's annual wool production across their lifetime (AWP, total lifetime greasy fleece weight (kg) / TL), 3) proportion of successful lambing opportunities (SLO, number of successful lambing opportunities / TL), and 4) proportion of lambs born (ALP, sum of lambs given birth to / TL). Animals culled from the flock prior to producing a lamb or fleece were assigned covariate values of zero. In Model 2, the covariates were fitted as fixed effect terms where they were nested within cohort. To better understand the proportion of variation in TL described by each of the covariates and which of these is having the greatest impact on the phenotypic and additive variance, the terms were fitted as random effects (Model 3).

**RESULTS AND DISCUSSION**

The result of the final assumption filters left 267,517 ewes with TL records from 143 flocks across 746 cohorts (Table 1). Only 76% of ewes born stayed in the flock to the yearling stage (42% to 2 years of age). This includes 203,350 ewes with at least one record describing lambing outcomes or fleece production (missing covariates were given a value of 0). The average TL for the Merinos was 2.37 (SD = 1.89) years with a maximum of 13.06 years. The results are much lower than 4.5 to 6.5 years reported by Kleemann *et al.* (2016) in Merino breeding flocks. In another study in Merino commercial flocks, Hatcher *et al.* (2009) stated that animals are usually culled at 6 years of age. This is in part likely due to the greater selection pressure placed on the ewe flock to achieve desired genetic gains within the seed stock sector. The mean SLO and ALP were 0.30 (SD = 0.30) and 0.41 (0.45), respectively.

The phenotypic variance and heritability of TL using Model 1 were 3.34 ± 0.02 and 0.22 ± 0.01, respectively (Table 2). After adjusting for the overall genetic merit, wool production, and reproductive performance (Model 2) the phenotypic variance and heritability were reduced to 1.22 ± 0.01 and 0.11 ± 0.01. The heritabilities herein for TL align with the low heritability estimates reported in the literature of 0.08 ± 0.01 in Scottish Blackface ewes by Conington *et al.* (2001) and 0.02 ± 0.01 to 0.06 ± 0.02 in Spanish Churra ewes by El-Saied *et al.* (2005). In an Australian Merino research flock, Hatcher *et al.* (2009) reported the heritability of ewe survival at 2$^{nd}$, 3$^{rd}$, 4$^{th}$ and 5$^{th}$ year in a range from 0 to 0.12. Brash *et al.* (1994) also described a heritability in an Australian Dorset sheep population of 0.06. As mentioned earlier, the breeders submitting data to the database tend to apply selection pressure and maintain heavily selected ewe flocks, which leads to censored data, which in part might explain the higher heritability of Model 1.

**Table 2: Variance components estimates (± se.) for total life with Model 1 (base), Model 2 (base + covariates nested within cohort) and Model 3 (base + covariates nested within cohort and fitted as random effects). The percent column represents the proportion of total variation accounted for by each random effect fitted in Model 3**

| | Model 1 | Model 2 | Model 3 | Percent |
|---|---|---|---|---|
| **Heritability ($h^2$)** | **0.22 ± 0.01** | **0.11 ± 0.01** | **0.11 ± 0.01** | |
| **Phenotypic ($\sigma^2_p$)** | **3.34 ± 0.02** | **1.22 ± 0.01** | **1.22 ± 0.01** | |
| Residual ($\sigma^2_e$) | 2.61 ± 0.01 | 1.08 ± 0.01 | 1.08 ± 0.01 | 4 % |
| Direct additive ($\sigma^2_a$) | 0.73 ± 0.02 | 0.14 ± 0.01 | 0.14 ± 0.01 | 1 % |
| Cohort | | | 0.16 ± 0.01 | 1 % |
| Success rate of lambing opportunities (SLO)* | | | 14.94 ± 0.83 | 63 % |
| Annual lamb production (ALP)* | | | 7.25 ± 0.42 | 31 % |
| Annual wool production (AWP)* | | | 0.03 ± 0.01 | 0 % |
| Merino production plus index (MPP)* | | | 0.00 ± 0.01 | 0 % |

*Covariates were fitted as nested terms within cohort

The majority of the variation in ewe longevity was described by the ewe's reproductive performance (94%) with the more fertile (SLO = 63%) and larger litter producing (ALP = 31%) ewes retained in the flocks longer (Table 2). This suggests that the underlying factor determining the length of time ewes are retained in the flock is associated with breeders' selection decisions around reproduction and not necessarily because of the ewe's inherent fitness to survive. Reproduction is one of the most important traits in the profitability of sheep farming enterprises, and ewes with high reproductive performance are most likely to perform better in the longer term (Zishiri *et al.* 2013). In the current study, AWP explained only a small proportion of the variation in ewe longevity, and it is hypothesised that variation in wool produced across her lifetime seems to have had little impact on the breeders' selection decisions to keep the ewe in the flock.. However, genetic merit for wool production is likely to have a significant influence on the ewe's value to the flock and this could be explored further by estimating the genetic association between longevity and wool production and quality traits. The MPP index explained no variation in the TL of the ewe. This may be in part because the index values used in this study was based on the information available at time of analysis not when the animal was being dispersed from the flock.As the flock's breeding objectives are likely to differ, the MPP index whilst moderately correlated with most flock's breeding objectives may not entirely reflect all the selection decisions placed on the ewe flock at the individual breeder level.

## CONCLUSIONS

The results of this study indicated that it is possible to capture and define longevity in Australian Merino ewes by utilizing performance records available within the MERINOSELECT database. Reproductive performance is the largest factor behind ewe longevity and should be accounted for if the desired trait is to more closely reflect the ewe's fitness to survive and not just her ability to produce a lamb. Correlations between longevity and key production traits as well as the estimation of its economic value require exploration before determining the value of TL within industry breeding objectives.


## ACKNOWLEDGEMENTS

Sheep Genetics is funded by Meat and Livestock Australia. The authors acknowledge the data contribution from the Australian sheep breeders.


## REFERENCES


Brash L., Fogarty N. and Gilmour A. (1994) *Aust. J. Agri. Res.* **45:** 427.

Brown D., Huisman A., Swan A., Graser H., Woolaston R., Ball A., Atkins K. and Banks R. (2007) *Proc. Assoc. Advmt. Anim. Breed. Genet* **17:**187.

Conington J., Bishop S., Grundy B., Waterhouse A. and Simm G. (2001) *Anim. Sci.* **73:** 413.

Conington J., Bishop S., Waterhouse A. and Simm G. (2004) *J. of Anim. Sci.* **82:** 1290.

El-Saied U., De La Fuente L., Carriedo J. and San Primitivo F. (2005) *J. Dairy Sci.* **88:** 3265.

Gilmour A., Gogel B., Cullis B., Welham S. and Thompson R. (2015) *Hemel hempstead: VSN international ltd*

Hatcher S., Atkins K. and Thornberry K. (2009) *Proc. Assoc. Advmt. Anim. Breed. Genet* **18:**580.

Kleemann D. O., Walker S. K., Ponzoni R. W., Gifford D. R., Walkley J. R., Smith D. H., Grimson R. J., Jaensch K. S., Walkom S. F. and Brien F. D. (2016) *Anim. Prod. Sci.* **56:** 716.

McLaren A., McHugh N., Lambe N. R., Pabiou T., Wall E. and Boman I. A. (2020) *Small Rumin. Res.* **189:** 106145.

Swan A., Banks R., Brown D. and Chandler H. (2017) *Proc. Assoc. Advmt. Anim. Breed. Genet* **22**

*Contributed paper*

# CIRCULAR GENOMIC PERMUTATIONS CAN LIMIT THE CONFOUNDING EFFECTS OF THE REFERENCE POPULATION IN THE ANALYSES OF SELECTION SIGNATURES

**I.A.S. Randhawa[1,2], M.S. Khatkar[1], P.C. Thomson[1] and H.W. Raadsma[1]**

[1] The University of Sydney, Sydney School of Veterinary Science, Camden, NSW, Australia
[2] The University of Queensland, School of Veterinary Sciences, Gatton, QLD, Australia

## SUMMARY

Analyses of selection signature are extensively used to detect chromosomal regions underlying phenotypic diversity which have been subjected to selective pressure. However, breeds with common origin, recent divergence, or similar production types are often confounded because the candidate and reference populations, compared in such analyses, exhibit similar patterns in genomic data. This study has applied a circular genome permutation method to generate a reference population to investigate selection signatures in Angus cattle (n=29) by using 1.6 million SNPs and applying the composite selection signals (CSS) method. Significant CSS were compared in two sets of analyses based on different reference populations, i.e., CSS-1: using circular genome permutation to form the breed neutral reference population (n=29) for Angus, and CSS-2: using five beef breeds as a reference population (n=36). Notably, several genomic regions were detected using CSS-1 (e.g., on chromosome 14, 16, 21) in Angus underlying commonly known genes of major effects on beef traits which were not detected by CSS-2 because of the confounding genetic background of Angus with the reference beef breeds. The results highlight the importance of selecting an appropriate reference population to circumvent the confounding breed effects.

## INTRODUCTION

In livestock species, genomic data of various breeds are compared for genotypic and haplotypic distributions against each other, individually or as groups, to characterize the historic selective pressures for breed specific traits of production, health and adaptation. These signatures of selection can be used to discover genetic variants and genes to understand the biological control of agricultural and health traits (Kemper and Goddard 2012). Genomic investigations are frequently resource-intensive; however, detection of selection signatures can provide insights into the genetic architecture underlying breed-specific traits in a relatively cost-effective manner (Gibbs *et al.* 2009). Ubiquitous cattle breeds, such as Angus have been selectively improved for economic traits by increasing the frequency of beneficial alleles throughout the genome.

A review of recently published signatures of selection showed that a few regions are commonly found in multiple breeds, suggesting genomic hotspots underlying genes of major effect, e.g., *PLAG1* on bovine autosome 14 (BTA14) (Randhawa *et al.* 2016). It was noted that the locations of selection signatures generally varied in those studies due to differences in sample size, SNP density and reference population. Rapidly lowering costs have allowed a larger number of samples to be assayed for high-throughput genotyping and genome-wide sequences. However, selection of an appropriate reference panel and thus avoiding confounding effects of the reference population remains a challenge. Inclusion of related breeds in the reference population can mask the detection of common selection signatures. This study has applied a new approach of circular chromosomal permutations (Cabrera *et al.* 2012) to generate a breed neutral reference population though permutation of the genome of the same breed used in the comparison and thus is likely to be free from breed bias. This approach was used to detect selection signatures in Angus and the results were compared with conventional breed-vs-breed approach.

## MATERIALS AND METHODS

Briefly, the circular genomic permutation approach considered SNPs along each chromosome as a circular fragment, which was then rotated for each sample (animal) to pick a starting location randomly. The approach shuffles the chromosomal fragments while keeping intra-sample haplotypic and linkage structures. All permuted samples were assembled and considered as a reference population for the same breed used in the comparison, expecting genome-wide neutral and uniform genetic diversities. The composite selection signal (CSS) method (Randhawa *et al.* 2014) was used to detect across-breed selection signatures in two data sets, i.e., CSS-1: Angus vs Permuted reference genome (using the 29 Angus samples), and CSS-2: Angus vs 5 beef breeds as reference. A total of 65 beef cattle samples (Table 1) and ultra-high density genotypic data (1,583,288 SNPs) were used.

**Table 1. Cattle breeds, their geographic origin, country of sampling and DNA samples**

| Breeds | Type | Geographic origin | Country of sampling | Samples |
|---|---|---|---|---|
| Angus | Beef | Scotland | USA, New Zealand, Australia | 29 |
| Brahman | Beef | India | India, Brazil, USA, Australia | 8 |
| Hanwoo | Beef | Korea | South Korea | 11 |
| Murray Grey | Beef | Australia | Australia | 1 |
| Simmental | Beef | Switzerland | USA | 6 |
| Wagyu | Beef | Japan | USA | 10 |
| **Total** | - | - | - | **65** |

## RESULTS AND DISCUSSION

Genome-wide analyses detected a number of expected genomic regions in Angus by CSS-1 (within breed), while CSS-2 (between breeds) missed most of the regions, as shown by three example chromosomes (BTA14, BTA16 and BTA21) in Figures 1, 2 and 3.

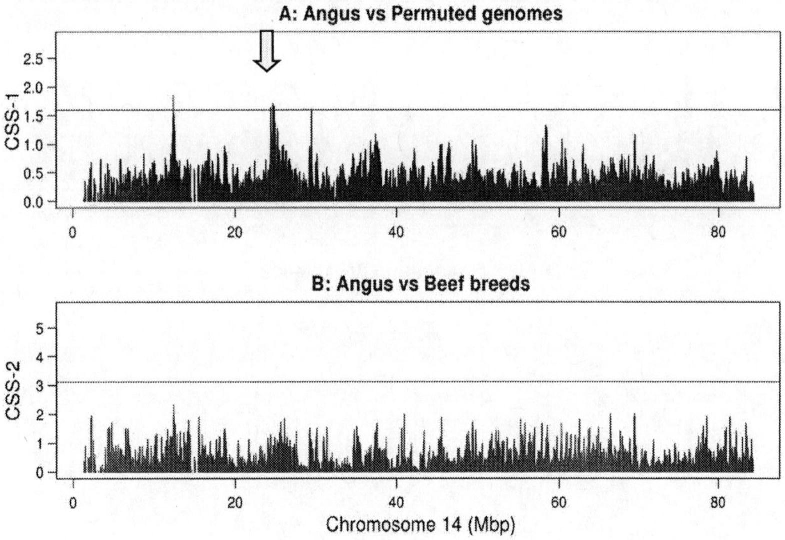


**Figure 1. Composite selection signals on chromosome 14 in Angus cattle, computed by using two different reference populations; A) Permuted genomes, B) Beef breeds. Blue lines at genome-wide top 0.1%. Arrow at the top shows position of the expected signatures of selection in Angus**

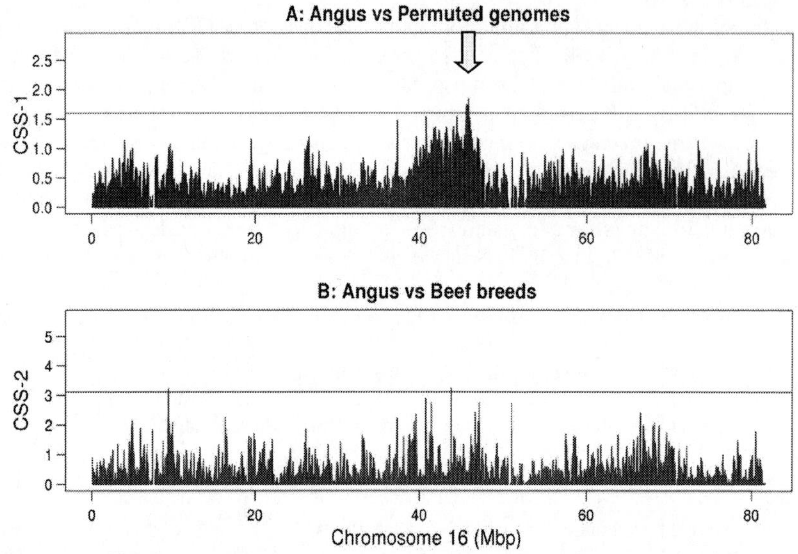


**Figure 2. Composite selection signals on chromosome 16 in Angus cattle, computed by using two different reference populations; A) Permuted genomes, B) Beef breeds. Blue lines at genome-wide top 0.1%. Arrow at the top shows position of the expected signatures of selection in Angus**

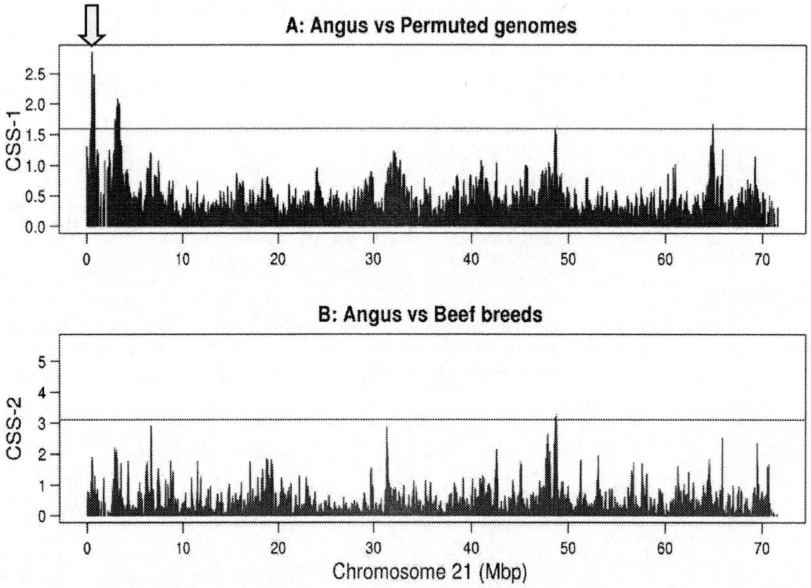


**Figure 3. Composite selection signals on chromosome 21 in Angus cattle, computed by using two different reference populations: A) Permuted genomes, B) Beef breeds. Blue lines at genome-wide top 0.1%. Arrow at the top shows position of the expected signatures of selection in Angus**

On BTA14 (24-27 Mbp of UMD3.1 bovine genome assembly), a strong selective sweep has been established due to selection targeting beneficial variants in *PLAG1-CHCHD7* in Angus and many other breeds (Randhawa *et al.* 2016). On BTA16 (42-48 Mbp), an extensive region under selection has been found underlying several genes important for both dairy and beef production. At the start of BTA21 (0.3-4 Mbp), many breeds (Angus, Belgian Blue, Brahman, Holstein) have shown strong selection signatures (Randhawa *et al.* 2016). Therefore, this study expected to find significant CSS on the target regions of BTA14, BTA16 and BTA21. CSS-1 succeeded in detecting these regions, while the genetic composition of reference population composed of beef breeds in CCS-2 showed a major confounding influence on detection of selection signatures in Angus (Figures 1-3). Interestingly, several genomic regions, e.g., BTA1, BTA4 and BTA13, also strongly selected in Angus were captured by both CSS-1 and CSS-2 (results not shown). This suggested that only the genomic patterns of allele frequencies and haplotype structures which are relatively more similar in a candidate breed and reference population can neutralise the across-breed statistics of selection signatures. Moreover, the magnitude of CSS values also varied between CSS-1 (max: 2.85, top 0.1%: 1.6) and CSS-2 (max: 5.67, top 0.1%: 3.1). The circular permutation approach was initially proposed to decide on a significance threshold for the empirical genome-wide association (Cabrera *et al.* 2012) and selection signatures (Stainton *et al.* 2015). The genome-wide results suggest that using the maximum value of CSS-1 as a significance threshold for conventional CSS-2 approach may have recovered a few confounded regions. However, most of the expected regions did not show a cluster of high values in CSS-2 analyses. Thus, our new approach of using permuted genomes as the reference population has been proved advantageous. This approach can be used to detect selection signatures from single breed data by permuting from its own samples where data are limited, or a unique set of variants is available through whole-genome sequencing.

## CONCLUSIONS

Molecular data can provide insights into historical natural or artificial selection events and the genetic architecture underlying breed-specific traits. This study examined the impact of reference population to validate known genomic regions under selection in Angus cattle. The results provide evidence that a reference population of closely related or phenotypically similar to the candidate breeds affects the power to detect selection signatures. Our new approach of circular genomic permutations can potentially limit such confounding effects and resource-limited data can be efficiently analysed to detect historic selection pressures in any species.

## ACKNOWLEDGEMENTS

We thank Robert D. Schnabel and Jerry F. Taylor from University of Missouri, Columbia, USA for providing the genotypic data.

## REFERENCES

Cabrera C.P., Navarro P., Huffman J.E., Wright A.F., Hayward C., Campbell H., Wilson J.F., Rudan I., Hastie N.D., *et al.* (2012) *G3-Genes Genom. Genet.* **2**:1067.

Gibbs R.A., Taylor J.F., Van Tassell C.P., Barendse W., Eversole K.A., Gill C.A., Green R.D., Hamernik D.L., Kappes S.M., *et al.* (2009) *Science* **324**:528.

Kemper K.E. and Goddard M.E. (2012) *Hum. Mol. Genet.* **21**:R45.

Randhawa I.A.S., Khatkar M.S., Thomson P.C. and Raadsma H.W. (2014) *BMC Genet.* **15**:34.

Randhawa I.A.S., Khatkar M.S., Thomson P.C. and Raadsma H.W. (2016) *PLoS ONE* **11**:e0153013.

Stainton J.J., Haley C.S., Charlesworth B., Kranis A., Watson K. and Wiener P. (2015) *Anim. Genet.* **46**:37.

*Contributed paper*

# GENOME-WIDE ANALYSES OF SCUR GENETICS IN CATTLE

**I.A.S. Randhawa[1], M.R. McGowan[1] and R.E. Lyons[2]**

[1] The University of Queensland, School of Veterinary Science, Gatton, QLD 4343, Australia
[2] Agri-Genetics Consulting, QLD 4074, Australia

## SUMMARY

This research has investigated genomic data (770K SNP genotypes) on 197 animals of Brahman, Droughtmaster and Hereford. Analyses of genome-wise association (GWAS) and composite selection signals (CSS) were conducted to find genomic regions underlying polledness (test) and scurs (discovery) in individual and combined (multi-breed) cohorts of horned, polled and scurred animals. Both GWAS and CSS successfully detected the poll-locus, whereas the GWAS results failed to localize any novel or previously proposed scur regions. CSS results coincided with 4 out of 7 previously detected regions as well as found several novel genomic regions. However, none of the significant regions harbour genes of profound effect on scur development. Overall, the results suggest that scur genetics has complex inheritance patterns and we discuss that many genetic factors and non-genetic effects interact variably to control development of scurs in cattle.

## INTRODUCTION

Scurs are horn-like-structures, grow slower than horns and remain unattached from the cranium in cattle. Scurs appear in genetically heterozygous (HP) animals at the POLL locus – genomic region on bovine autosome (BTA) 1 associated with complete absence of horns and scurs called polledness (PP), which is caused by either Celtic (Pc) or Friesian (Pf) types of insertion-deletions (mutations). However, scurs are seen in a relatively smaller proportion of a cattle population, given that many heterozygous animals remain polled due to conditional factors, such as sex of the animal (Aldersey *et al.* 2020). Eradication of scurs in cattle is as vital as horns due to the related economic and welfare impact because scurred animals often undergo dehorning and certainly transmit a horn (H) allele to their offspring. Inheritance of scurs has been alleged to be under single gene control (White and Ibsen 1936). However, the "scur gene" in cattle remains to be discovered. To date, investigations to identify a causal gene have pointed to different genomic regions across different cattle breeds. A well-known region on BTA19 (26-29 Mbp of ARS-UCD1.2 bovine assembly location) was initially discovered in Canadian beef cattle and recently affirmed an epistatic interaction with the POLL locus (Asai *et al.* 2004; Ketel and Asai-Coakwell 2020). Those findings were not sustained in Hereford, Angus and French Charolais cattle (Capitan *et al.* 2009). A dominant inheritance pattern was later reported in the French Charolais and their proposed type-2 scurs development has been linked to *TWIST1* gene on BTA4 (Capitan *et al.* 2011). Notably, appearance of type-2 scurs was specific to heterozygous French Charolais and the underlying homozygous frameshift mutation has embryonic lethality. The scurs locus in Simmental cattle was mapped on BTA19 (48-49 Mbp), which is a different location from those reported earlier (Tetens *et al.* 2015). The study also pointed to a multi-locus involvement regarding the development of scurs. Recently, genic and non-genic regions on BTA5 (44-45 Mbp), BTA12 (7.5-8.5 Mbp), BTA16 (40-41 Mbp, *SUCO* gene) and BTA18 (46-47.5 Mbp, *ARHGAP33* gene) were found in a multi-locus association in Holstein (Gehrke *et al.* 2020). As yet, the reported studies have not converged and elucidation for inheritance patterns and genetic control of scur development is an ongoing task. Investigation of scurs in other breeds and by using high-density genotyping dataset can be valuable in understanding the scurs genetics. This study has performed the analyses of genome-wide association and signatures of selection to localize scur genomic regions within Brahman, Droughtmaster and Hereford breeds and in a multi-breed framework.

## MATERIALS AND METHODS

A total of 197 animals of 3 breeds (Table 1) were sampled (tail-hair or blood), phenotyped (horned, polled and scurred), and diagnosed for the POLL genotype by the optimized poll testing (OPT) assay (Randhawa *et al.* 2020). Genomic DNA were used for Illumina BovineHD Genotyping BeedChip array. The 770K genotypes were quality control filtered to remove SNPs with MAF < 5% and call rate < 90%, and 657,543 SNPs were retained. Imputation of missing genotypes and haplotype phasing was performed with BEAGLE 3.3 (Browning and Browning 2009). The genome-wise association (GWAS) analyses were performed using the *qtscore* function (trait = "binomial") in R-package: GenABEL (Aulchenko *et al.* 2007). A significance threshold of $p < 1.0^{-7}$ was used to detect putative SNPs underlying the phenotypes. The composite selection signal (CSS) analyses were performed in R program for pairwise contrasting phenotypes (Randhawa et al. 2014). The smoothed CSS scores were used to capture the putative genomic regions using the top 0.1% threshold. Analyses of GWAS and CSS were conducted to find genomic regions underlying polledness (control: HH-horned vs PP-polled) and scurs (HP-scurred vs HP-polled) in individual and combined (multi-breed) cohorts (Table 1).

**Table 1. Phenotyped and genotyped animals for genome-wide analyses**

| Breeds | Control (Horn vs Poll) | | Discovery (Poll vs Scur) | | Total |
|---|---|---|---|---|---|
| | **Horned** (HH) | **Polled** (PcPc) | **Polled** (HPc) | **Scurred** (HPc) | |
| Brahman | 8 | 9 | 25 | 24 | 66 |
| Droughtmaster | 8 | 8 | 25 | 25 | 66 |
| Hereford | 8 | 8 | 24 | 25 | 65 |
| **Combined** | **24** | **25** | **74** | **74** | **197** |

HH: homozygous horned animals, PcPc: homozygous polled animals, HPc: heterozygous animals

## RESULTS AND DISCUSSION

Analyses to localize POLL region (control) were successful by CSS for all datasets. However, GWAS showed sensitivity to sample size with significant peak for only combined data (Figure 1).

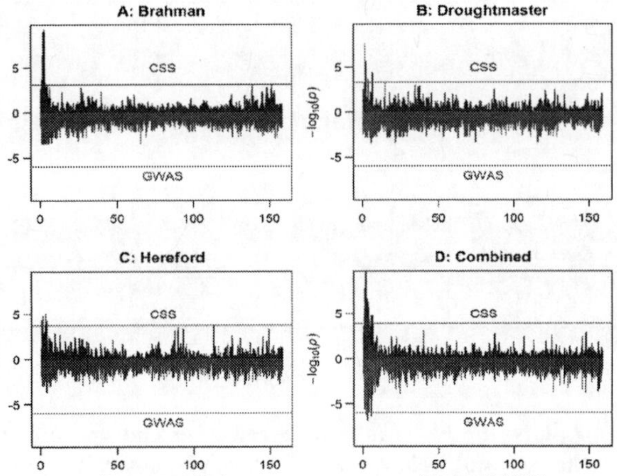


**Figure 1. CSS (blue, positive) and GWAS (red, negative) results to detect the POLL region on chromosome 1 in A) Brahman, B) Droughtmaster, C) Hereford and D) Combined data. Red and blue lines show genome-wide significance at top 0.1% (CSS) and p=10$^{-7}$ (GWAS), respectively**

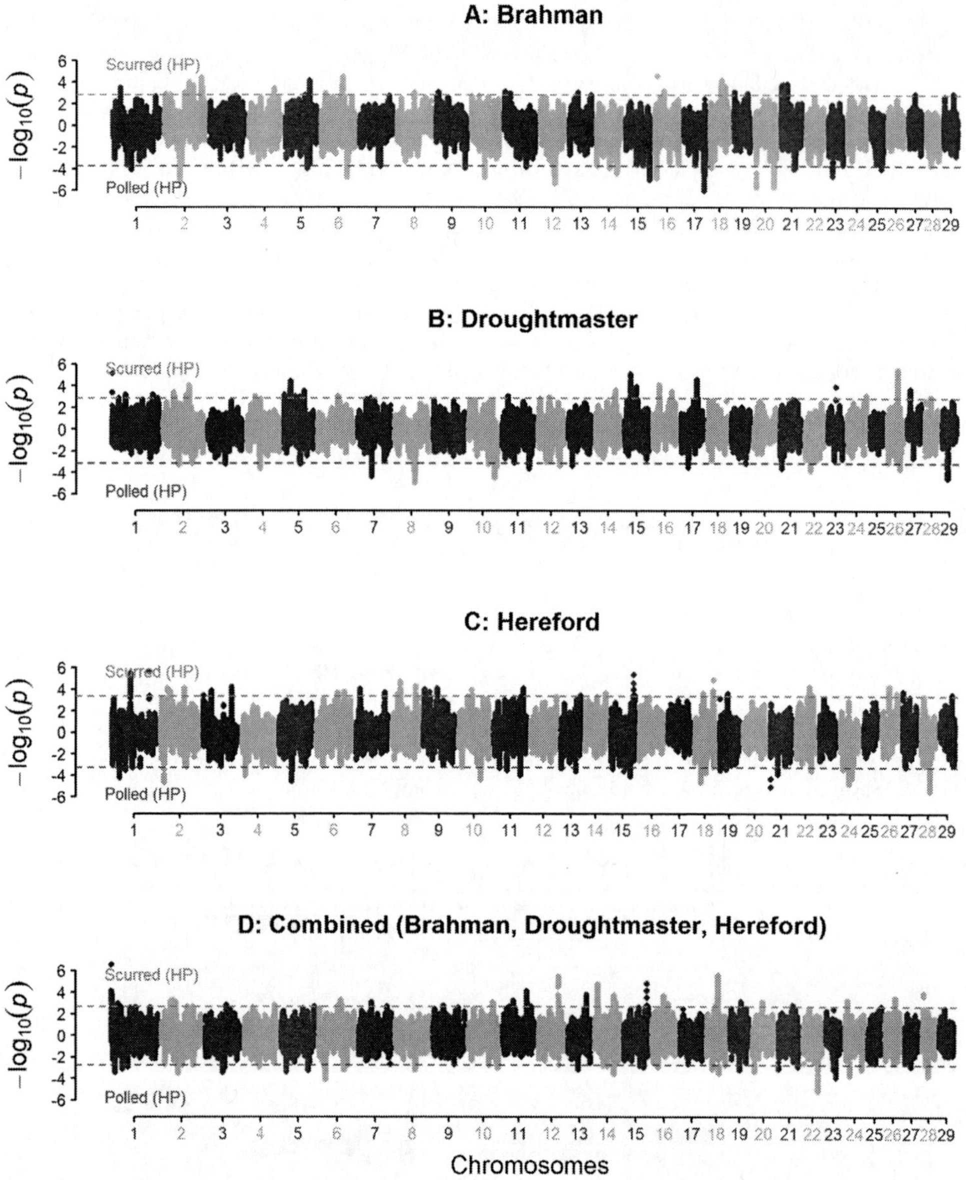


**Figure 2. Manhattan plots of CSS results in breed-wise and combined data using 657,543 SNPs. Peaks above the red and below the blue dashed lines capture candidate regions for scurs and polledness, respectively at the significance thresholds (0.1%)**

In the discovery analyses, CSS detected multiple regions within Brahman, Droughtmaster, Hereford and the combined dataset (Figure 2). However, GWAS analyses failed to reach above the significance thresholds, although many regions were found above a suggestive threshold of $-\log_{10}$

$(p)10^{-5}$ (results not shown). Note that CSS found some genes – *SUCO* (BTA16) and *ARHGAP33* (BTA18) – and non-genic regions (BTA5, BTA19) that overlapped with previous research (Tetens *et al.* 2015; Gehrke *et al.* 2020). Multiple genomic regions identified by CSS (Figure 2) suggest that many genes interacting through complex polygenic networks may control scurs development. Hence, the finding does not support a simple mono-genic inheritance model as initially proposed by White and Ibsen (1936). However, the results are non-conclusive to identify regions or genes with strong association with the scurs phenotype. The GWAS analyses might have limited power for successful association mapping for the complex trait of scurs due to lower sample size. Therefore, further investigations by including higher sample sizes and improved phenotyping accuracy will increase the power of genomic association mapping of scur controlling variants. Nonetheless, other factors, such as POLL locus heterogeneity (Pc, Pf) and sex of the animal further complicate the understanding of scurs genetics (Gehrke *et al.* 2020), thus extensive research designs are suggested. Although scurs are substantially less common than horns (Randhawa *et al.* 2020), and are less damaging and easily manageable, they will continue to appear as the beef breeds transition from horned to polled cattle. With increased frequencies of polled alleles (Pc and Pf), and as more homozygous polled breeding stock become available, the incidence of scurs is expected to decrease.

## CONCLUSIONS

Generally scurs develop in genetically heterozygous cattle, which carry an allele of either Celtic or Friesian mutations at the POLL locus. However, understanding of scurs genetics remains limited because it is unclear why some heterozygous animals remain polled. Interestingly, previous studies have associated 7 genomic regions on 6 chromosomes and none of them coincided in independent populations. This study has found 4 out of 7 previously detected regions as well as identifying several novel genomic regions, suggesting a polygenic inheritance model. None of the significant regions harbour genes of profound effect on scur development. Several factors including sex and mutation type have also been found to effect scur development, but the discovery of "scur genes" remains a challenge.

## ACKNOWLEDGEMENTS

Meat and Livestock Australia financially supported the project (L.GEN.1713).

## REFERENCES

Aldersey J.E., Sonstegard T.S., Williams J.L. and Bottema C.D.K. (2020) *Anim. Genet.*

Asai M., Berryere T.G. and Schmutz S.M. (2004) *Anim. Genet.* **35**:34.

Aulchenko Y.S., Ripke S., Isaacs A. and van Duijn C.M. (2007) *Bioinformatics* **23**:1294.

Browning B.L. and Browning S.R. (2009) *Am. J. Hum. Genet.* **84**:210.

Capitan A., Grohs C., Gautier M. and Eggen A. (2009) *BMC Genet.* **10**:33.

Capitan A., Grohs C., Weiss B., Rossignol M.-N., Reversé P. and Eggen A. (2011) *PLOS ONE* **6**:e22242.

Gehrke L.J., Capitan A., Scheper C., König S., Upadhyay M., Heidrich K., Russ I., Seichter D., Tetens J., *et al.* (2020) *Genet. Sel. Evol.* **52**:6.

Ketel C. and Asai-Coakwell M. (2020) *Can. J. Anim. Sci.* **100**:479.

Randhawa I.A.S., Burns B.M., McGowan M.R., Porto-Neto L.R., Hayes B.J., Ferretti R., Schutt K.M. and Lyons R.E. (2020) *G3-Genes Genom. Genet.* **10**:539.

Randhawa I.A.S., Khatkar M.S., Thomson P.C. and Raadsma H.W. (2014) *BMC Genet.* **15**:34.

Tetens J., Wiedemar N., Menoud A., Thaller G. and Drögemüller C. (2015) *Anim. Genet.* **46**:224.

White W.T. and Ibsen H.L. (1936) *J. Genet.* **32**:33.

*Contributed paper*

# EFFICIENCY OF OPTIMIZED POLL TESTING ASSAY IN AUSTRALIAN BEEF CATTLE

**I.A.S. Randhawa[1], K.M. Schutt[2], M.R. McGowan[1] and R.E. Lyons[3]**

[1] The University of Queensland, School of Veterinary Science, Gatton, QLD 4343, Australia
[2] Neogen Australasia, University of Queensland, Gatton, QLD 4343, Australia
[3] Agri-Genetics Consulting, QLD 4074, Australia

## SUMMARY

Genetic testing for the presence of POLL gene in cattle has been proposed in Australia because it helps avoid dehorning and disbudding in young calves. Animals can be true polled if they carry two copies of either Celtic (PcPc) or Friesian (PfPf) mutations, or one of each (PcPf). Optimized poll testing (OPT) – a 5 SNPs based assay to detect both type of mutations – was developed to improve efficiency of commercial tests, which are used in selective breeding for rapidly increasing the poll gene frequency in herds. This study evaluates the efficiency of OPT assay across various breeds by using a high number of commercial test results (n=98,744). Overall, OPT consistently showed high success rate of 99.56% in commercial tests, which is consistent with previous results (99.60%) based on experimental data. The OPT has been rapidly adopted across the industry leading to greater accuracy and more confidence. OPT has been equally efficient for the taurine (99.50%) and indicine (99.63%), Zebu and other indicus-influenced composite breeds.

## INTRODUCTION

The genetics of horns and polledness (absence of horns) is complex in the bovidae family. In cattle, inheritance of polledness (P) is dominant to horns (H); however, the underlying genes and causal mutations display an array of genetic heterogeneity and phenotypic diversity (Medugorac *et al.* 2012; Wiedemar *et al.* 2014). Genetic control of the polledness – the so called "POLL gene" – has been mapped on the starting end of bovine chromosome 1 (BTA1) (Long and Gregory 1978). To date, four different genetic mutations that can cause polledness have been identified in cattle worldwide, all of which are physically located in a narrow neighbourhood on BTA1 (Figure 1, (Aldersey *et al.* 2020)). The known mutations are named according to their geographic origin in cattle (Capitan *et al.* 2011; Tetens *et al.* 2015; Medugorac *et al.* 2017; Utsunomiya *et al.* 2019) e.g., Celtic (Pc), Friesian (Pf), Mongolian (Pm) and Guarani (Pg). Of those, only Pc and Pf have been found prevalent in Australian cattle herds. Animals can be true polled if they carry two copies of either Celtic (PcPc) or Friesian (PfPf) mutations, or one copy of each (PcPf). Due to the genetic complexity the heterozygous animals (HP: HPc or HPf) which carry one copy of horn (H) and one copy of a poll (Pc or Pf) can be polled or may develop small size and unattached horn-like-structures called scurs (Aldersey *et al.* 2020; Gehrke *et al.* 2020).

As a consequence of rising concerns about animal welfare and the costs of bruising and dehorning (Huertas *et al.* 2015), increasing the polled cattle population is a way forward for a sustainable beef industry. Identification of true polled cattle has been a challenge (Connors *et al.* 2018), given that some HP animals can be polled but can potentially pass on an H (horn) allele to its offspring. Therefore, two phenotypically polled animals can produce a horned offspring. Genetic testing for the presence of POLL gene in cattle has been proposed in Australia because it helps avoid dehorning and disbudding in young calves (Prayaga 2007). Poll gene testing has been in practice since 2012 and has evolved through the use of different types of genetic markers, initially based on microsatellites and more recently based on single nucleotide polymorphisms (SNPs). Optimized poll testing (OPT) – a 5 SNPs based assay to detect both type of mutations – was developed to improve efficiency of commercial tests (Randhawa *et al.* 2020), which are used in selective breeding for rapidly increasing the poll gene frequency in Australian herds. This study evaluates the efficiency of OPT assay across various breeds by using a high number of test results.

## MATERIALS AND METHODS

Genetic markers for the prediction of Celtic (Pc) and Friesian (Pf) types of poll associated SNP alleles (Table 1) are available on commercial bovine BeadChip assays (Illumina) including Neogen's proprietary GGP Bovine 100Kand GGP Indicus 50K assays (Neogen Corporation, Lincoln, NE). The Pc genotype is predicted by translating a single SNP marker rs383143898 (ARS-UCD1.2 position on BTA1: 2,429,319) based on its horn or poll allele (Table 1). The Pf genotype is predicted based upon four markers associated with Pf (Table 1, Figure 1). Pf associated markers include: rs801127025 (BTA1: 2,372,456), rs799403053 (BTA1: 2,486,811), rs210350155 (BTA1: 2,491,161) and rs797088784 (BTA1: 2,578,598). Results of OPT represent reconciled outcomes from both Pc and Pf predictions to generate genotypes such as HH, HPc, HPf PcPc, PcPf or PfPf. However, if the Pc-associated SNP or more than one Pf-associated SNPs fail during genotyping, or one or more SNPs differ in predicted genotype (H versus Pf) then the result is considered ambiguous and termed as a "No Result". For this study, OPT results on commercial samples (n=98,744) were obtained to check the efficiency of mutation predictions. In addition, call rate, genotyping error and prediction efficiency of the OPT and an additional SNP: rs799920960 (BTA1: 2,748,715), which is also available on the above mentioned commercial genotyping assays, were investigated by using a subset of the commercial tests and previous data (Randhawa *et al.* 2020).

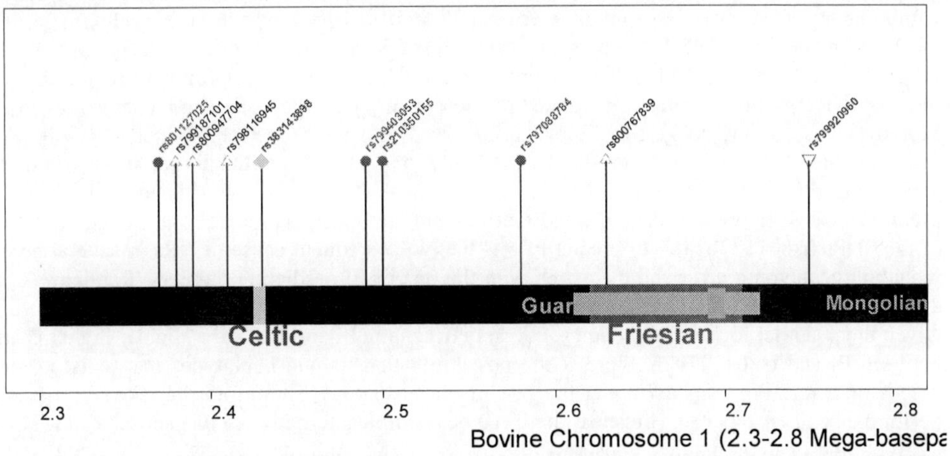


**Figure 1. POLL region on chromosome 1 (Bovine assembly: ARS-UCD1.2) showing locations of four known insertion-deletions (Celtic, Friesian, Mongolian and Guarani) associated with polledness across various worldwide breeds of cattle. The Optimized Poll Test (OPT) is based on the 5 coloured SNPs (1-green to predict Celtic and 4-purple to predict Friesian mutations). The blue SNP is localized close to Friesian and have shown strong linkage with Pf. The remaining SNPs have been used in poll testing assays previously and are available on most SNP chip assays**

**Table 1. Single nucleotide polymorphisms (SNP) on BTA1 for predicting the Celtic (Pc) and Friesian (Pf) mutations**

| SNPs | Positions* | Mutations | Poll alleles | Predicting mutation |
|---|---|---|---|---|
| rs801127025 | 2,372,456 | $P_{5ID}$ | T | Friesian (Pf) |
| rs383143898 | 2,429,319 | $P_{202ID}$ | T | Celtic (Pc) |
| rs799403053 | 2,486,811 | T>C | C | Friesian (Pf) |
| rs210350155 | 2,491,161 | C>A | A | Friesian (Pf) |
| rs797088784 | 2,578,598 | G>A | A | Friesian (Pf) |

* Genomic positions based on bovine genome assembly ARS-UCD1.2 (GCA_002263795.2).

## RESULTS AND DISCUSSION

Table 2 shows results from the obtained commercial tests performed using OPT based predictions. The available data were combined into two groups: Taurine (*Bos taurus*) and Zebu (*Bos indicus* and composite), based on the breed information about each sample. A total of 53,310 Taurine and 45,434 Zebu results show that OPT was generally successful with 99.56% efficiency. The remaining 0.44% (438) samples providing "No results (NRs)" are more likely be due to a failure to amplify one or more markers during the genotyping process. Previously, Zebu cattle showed very high number of failure rate with over 10% of NRs by using previously available POLL gene testing assays (Randhawa *et al.* 2020). Hence, we compared the NRs from OPT between the Taurine and Zebu, and respectively found that 0.50% and 0.37% of their samples returned an NR (Table 2). As such, these results are markedly lower than the previous tests and within the expected range of genotyping errors (Wu *et al.* 2019). However, the results suggest that the Zebu (n=170 out of 45,434) had significantly less (Fisher's Exact test, $p<0.001$) NRs than the Taurine (n=268 out of 53,310). This suggests that OPT test has successfully overcome the high rates of NRs in the commercial application, especially for the Zebu and composite breeds. Moreover, results may suggest that the Taurine breeds have an undetectable lack of LD between some of the SNP markers and the POLL mutations or an additional variation within the genotyping probe regions, either of which is causing the decrease in POLL gene prediction. A preliminary investigation of the collected samples and previously available results suggest that one of the SNPs (rs801127025) – to predict Pf – is likely the frequent cause of NRs in several breeds, including genotyping error and mismatch with the rest of Pf predicting markers (Randhawa *et al.* 2020). Note that rs801127025 is located farthest from Pf, rather upstream of the Pc (Figure 1). Given the potential for recombination (~0.2%) between the Pf and Pc, and a slightly higher chance between Pf and rs801127025, there is the possibility that unique haplotypes may exist in some breeds or herds. We emphasise that the rate of NRs (~0.44%) should not be taken as a lack of performance of OPT per se. However, there can be a simple alternative to further reduce the NRs.

We investigated another SNP (rs799920960) in a small dataset, which has not shown genotyping error or discordance with Pf. Hence, rs799920960 can be used either to substitute rs801127025 or as an additional marker for a leverage to accept two mismatches in OPT. The proposed marker is strongly linked to Pf because it is closely localized than any other SNPs being used to predict Pf (Figure 1). However, we suggest that additional research is required to evaluate the utility of SNP marker (rs801127025) of the OPT assay and the proposed increase in the efficiency in some breeds by including the additional marker (rs799920960). Overall, the OPT is performing as expected by providing commercial efficiency (99.56%) concordant to previously reported experimental results (99.6%) used to design the OPT (Randhawa *et al.* 2020). OPT has shown greater accuracy of head phenotype predictions, but phenotyping and sampling errors may deflect. All in all, the OPT has been rapidly adopted – replacing the poll haplotype test (Connors *et al.* 2018) – across the industry for poll breeding to achieve sustainable beef production.

**Table 2. Performance efficiency of OPT in the Taurine and Zebu breeds**

| OPT output | Taurine | Zebu | Total | Percentage |
|---|---|---|---|---|
| HH | 24,011 (45.04%) | 19,908 (43.81%) | 43,919 | 44.48% |
| HPc | 14,968 (28.07%) | 17,147 (37.74%) | 32,115 | 32.52% |
| HPf | 796 (1.49%) | 445 (0.97%) | 1241 | 1.26% |
| PcPf | 1,224 (2.29%) | 333 (0.73%) | 1557 | 1.58% |
| PfPf | 185 (0.34%) | 12 (0.02%) | 197 | 0.20% |
| PcPc | 11,858 (22.24%) | 7,419 (16.32%) | 19,277 | 19.52% |
| No results (NR) | 268 (0.5%) | 170 (0.37%) | 438 | 0.44% |
| **Total** | **53,310** | **45,434** | **98,744** | - |

## CONCLUSIONS

This study shows that OPT has been very successful (99.56%) for commercial testing of POLL gene in Australian beef cattle, both Taurine (99.50%) and Zebu (99.63%) breeds. Being compatible with genomic products, the test is also available at lower cost than the previous stand-alone tests. The OPT is performing as expected and it has been rapidly adopted across the industry leading to greater accuracy and more confidence to achieve a more sustainable beef industry.

## ACKNOWLEDGEMENTS

Meat and Livestock Australia financially supported the project (L.GEN.1713). Provision of test results by Neogen Australasia is duly acknowledged.

## REFERENCES

Aldersey J.E., Sonstegard T.S., Williams J.L. and Bottema C.D.K. (2020) *Anim. Genet.*

Capitan A., Grohs C., Weiss B., Rossignol M.-N., Reversé P. and Eggen A. (2011) *PLOS ONE* **6**:e22242.

Connors N.K., Tier B. and Johnston D.J. (2018) Current status of Australia's diagnostic poll haplotype test. In: *World Congress on Genetics Applied to Livestock Production*, p. 344.

Gehrke L.J., Capitan A., Scheper C., König S., Upadhyay M., Heidrich K., Russ I., Seichter D., Tetens J., *et al.* (2020) *Genet. Sel. Evol.* **52**:6.

Huertas S.M., van Eerdenburg F., Gil A. and Piaggio J. (2015) *Vet. Med. Sci* **1**:9.

Long C.R. and Gregory K.E. (1978) *J. Hered.* **69**:395.

Medugorac I., Graf A., Grohs C., Rothammer S., Zagdsuren Y., Gladyr E., Zinovieva N., Barbieri J., Seichter D., *et al.* (2017) *Nat. Genet.* **49**:470.

Medugorac I., Seichter D., Graf A., Russ I., Blum H., Göpel K.H., Rothammer S., Förster M. and Krebs S. (2012) *PLoS ONE* **7**:e39477.

Prayaga K. (2007) *Aust. J. Agric. Res.* **58**:1.

Randhawa I.A.S., Burns B.M., McGowan M.R., Porto-Neto L.R., Hayes B.J., Ferretti R., Schutt K.M. and Lyons R.E. (2020) *G3-Genes Genom. Genet.* **10**:539.

Tetens J., Wiedemar N., Menoud A., Thaller G. and Drögemüller C. (2015) *Anim. Genet.* **46**:224.

Utsunomiya Y.T., Torrecilha R.B.P., Milanesi M., Paulan S.d.C., Utsunomiya A.T.H. and Garcia J.F. (2019) *Anim. Genet.* **50**:187.

Wiedemar N., Tetens J., Jagannathan V., Menoud A., Neuenschwander S., Bruggmann R., Thaller G. and Drögemüller C. (2014) *PLOS ONE* **9**:e93435.

Wu X.L., Xu J., Li H., Ferretti R., He J., Qiu J., Xiao Q., Simpson B., Michell T., *et al.* (2019) *Anim. Genet.* **50**:367.

*Contributed paper*

# EFFECTS OF POLL BREEDING ON REPRODUCTIVE TRAITS IN BEEF CATTLE

**I.A.S. Randhawa[1], R.E. Lyons[2] and M.R. McGowan[1]**

[1] The University of Queensland, School of Veterinary Science, Gatton, QLD 4343, Australia
[2] Agri-Genetics Consulting, QLD 4074, Australia

## SUMMARY

Efficient reproduction is considered the backbone of sustainable livestock production. This study has evaluated the estimated breeding values (EBVs) of seven beef breeds (Charolais, Hereford, Limousin, Shorthorn, Brahman, Droughtmaster and Santa Gertrudis). Intra-breed genetic merits (EBVs) were compared between the polled and horned cohorts using 548,775 animals born between 2000 and 2018 for five traits (scrotal size, gestation length, days to calving, calving ease direct and calving ease daughters). All breeds have shown genetic gain in the reproductive traits. Moreover, more traits in polled cohorts were found to have higher genetic merit as compared to horned cohorts. For example, scrotal size were found significantly higher in polled cohorts of Charolais, Hereford, Limousin, Brahman, Droughtmaster and Santa Gertrudis, and in horned cohort of Shorthorn. EBVs of gestation length were significantly lower (desirable) in polled cohorts of all breeds. All in all, this research concludes that polledness has no detrimental effects on the genetic merit of reproductive traits in beef cattle.

## INTRODUCTION

Cattle breeding programs require reproductively sound animals of superior genetics. Genetic merits of the nucleus herds are routinely computed as estimated breeding values (EBVs) of recorded production and reproduction traits to rank animals and select them for various breeding programs. Reproductive traits have been generally found with low-to-moderate heritability (Meyer *et al.* 1990), and genetically favourable to neutral correlations with the production traits have been reported in beef cattle (Wolcott *et al.* 2013). Therefore, several reproductive traits, both in male and female, are measured and genetically evaluated (EBVs) for selective breeding (Barwick *et al.* 2013). In male, scrotal size (SS) is measure from scrotal circumference (cm) of bulls at 300-700 days (adjusted for 400 days of age). Higher EBVs of SS are favourable because the larger scrotal circumference is associated with more semen production and earlier age at puberty in bulls. Furthermore, heifer progeny of Brahman and Tropical Composite bulls with larger SS reached puberty earlier and had shorter days to calving (Johnston *et al.* 2013). Heifers and cows are measured for several performance traits, including gestation length (GL), days to calving (DTC), calving ease direct (CEdr) and calving ease of daughters (CEdt). EBVs of GL (days) calculated based on the number of days from the date of conception to the date of calf birth. Lower EBVs are favourable because shorter GL is generally associated with lighter birth weight, improved calving ease and improved re-breeding performance among dams. In addition, calves born with a shorter GL are often heavier at weaning due to more days of growth. For DTC, lower values are favourable for EBVs estimated from the date when the female is introduced to a bull (joining period) until subsequent calving. Note that the time taken by cows to conceive after the commencement of the joining period primarily cause variation in DTC. Moreover, cows that had early puberty as heifers and return to oestrous earlier after calving will have lower DTC EBVs. Both CEdr and CEdt are favourable at lower EBVs, which are based on the ability of a sire's calves to be born unassisted from 2-year-old heifers and ability of a sire's daughters to calve at 2 years of age without assistance, respectively. Recently, due to increased awareness of animal welfare, consumer choices and costs and risks associated with physical dehorning, commercial beef producers and feedlots have emphasized on poll breeding. Polledness has been perceived by some farmers to have negative effect on some beef traits, including reproduction. Therefore, genetic merit of reproductive traits between the polled and horned cohorts of beef cattle are compared in this study.

## MATERIALS AND METHODS

Phenotypes for horn status (polled or horned) and EBVs (accuracy > 50%) of five reproductive traits (SS, GL, DTC, CEdr and CEdt) were obtained on a total of 548,775 animals (birth years: 2000 to 2018) of seven beef breeds (Charolais = 14,219, Hereford = 25,2837, Limousin = 43,351 Shorthorn = 58,603, Brahman = 81,617 Droughtmaster = 17,686 and Santa Gertrudis = 80,462) from BREEDPLAN database (https://breedplan.une.edu.au). Within each breed, dataset analyses were performed for the poll-*vs*-horn cohorts (Table 1) by using the R program (R Core Team 2021) to compute the summary statistics of Mean ± Standard Deviation (SD). Descriptive statistics for pairwise comparisons between the means were performed by the t-tests with pooled SD, and p-values were obtained by using the *t.test* function in R-package "*stats*". Effect sizes on each trait due to polledness within breeds were computed using the Cohen's d (Cohen 1977; Lakens 2013).

## RESULTS AND DISCUSSION

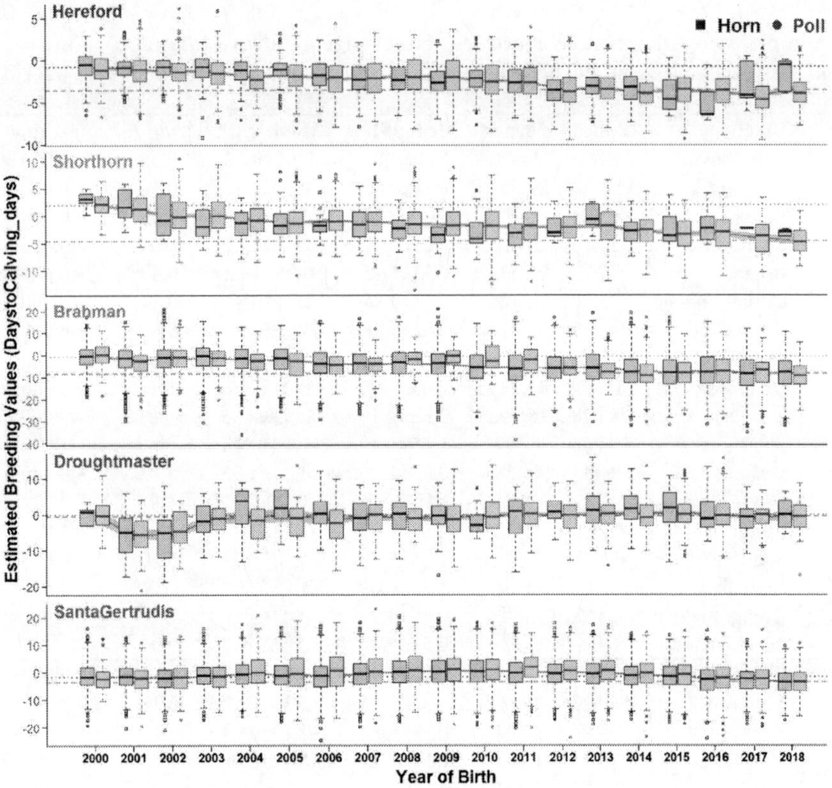


**Figure 1. Boxplots of EBVs of Days to Calving for horn and poll cohorts. Red lines in the background show overall annual averages (pink: 95% confidence intervals). Dotted (....) and dashed (----) lines refer to breed-averages at the start (2000) and end (2018) of the selected period**

Genetic merit for reproductive traits of the seven beef breeds have been consistently improving since 2000 though at variable rates. For example, Figure 1 shows EBVs of DTC for five breeds with decreasing trends depicting that all breeds have achieved shorter calving intervals. Intra-breed comparisons showed that genetic merit of DTC in the polled animals improved at significantly ($p < 0.05$) higher rates in Hereford and Brahman than horned cohorts, and vice versa in Shorthorn and Santa Gertrudis (Table 1). EBVs of SS were significantly higher with high effect sizes ($d$ = 0.14-0.65) in polled cohorts of all breeds except Shorthorn ($d$ = -0.1). The favourably decreasing trends for GL of polled animals in five breeds showed significantly better genetic gains ($d$ = -0.11 to -0.77) except for Brahman ($d$=0.17), while Shorthorn were non-significant. Both CEdr and CEdt have genetically improved as measured in polled cattle. However, Hereford and Limousin have shown significantly lower EBVs for calving difficulty in their polled cohorts. It is also evident that the number of polled animals were higher in five breeds (Charolais, Hereford, Limousin, Shorthorn and Droughtmaster) as compared to Brahman and Santa Gertrudis (Table 1). While the polled and horned cohorts were generally represented by higher sample sizes, Shorthorn (horned = 5-7%) and Brahman (polled = 7-13%) had uneven representation for horned and polled respectively, and therefore both breeds may have shown discordant trends of EBVs from 2000 to 2018 born animals.

**Table 1. Sample sizes, descriptive statistics, effect size (Cohen's *d*)and p-values of the comparison between polled and horned cohorts for five reproduction traits in beef breeds**

| Trait | Breed | Polled | Horned | Mean±SD$^P$ | Mean±SD$^H$ | $d$ * | $p$-value |
|---|---|---|---|---|---|---|---|
| SS (cm) | Charolais | 8,742 | 4,461 | 1.20±0.87 | 0.65±0.83 | 0.65 | <0.0001 |
| | Hereford | 159,151 | 86,357 | 1.71±0.85 | 1.43±0.72 | 0.36 | <0.0001 |
| | Limousin | 16,424 | 5,851 | 1.06±0.70 | 0.73±0.67 | 0.48 | <0.0001 |
| | Shorthorn | 55,626 | 2,761 | 1.26±0.70 | 1.33±0.67 | -0.10 | <0.0001 |
| | Brahman | 5,783 | 71,526 | 1.20±1.11 | 0.75±1.22 | 0.39 | <0.0001 |
| | Droughtmaster | 12,258 | 3,592 | 1.53±0.86 | 1.41±0.81 | 0.14 | <0.0001 |
| | Santa Gertrudis | 20,875 | 57,317 | 0.69±0.88 | 0.41±0.93 | 0.31 | <0.0001 |
| GL (days) | Charolais | 9,211 | 4,635 | -3.36±2.11 | -1.77±2.04 | -0.77 | <0.0001 |
| | Hereford | 115,106 | 41,739 | -0.45±1.77 | -0.27±1.56 | -0.11 | <0.0001 |
| | Limousin | 26,928 | 15,292 | -2.66±2.19 | -0.93±2.05 | -0.81 | <0.0001 |
| | Shorthorn | 28,806 | 1,497 | -1.45±1.49 | -1.40±1.44 | -0.03 | 0.22 |
| | Brahman | 600 | 4,458 | -0.03±1.06 | -0.24±1.35 | 0.17 | <0.0001 |
| | Droughtmaster | 538 | 401 | 0.32±1.78 | 0.67±1.63 | -0.20 | 0.002 |
| DTC (days) | Hereford | 8,064 | 1,602 | -2.53±2.27 | -1.78±1.93 | -0.36 | <0.0001 |
| | Shorthorn | 6,077 | 421 | -1.26±3.29 | -1.90±3.11 | 0.20 | <0.0001 |
| | Brahman | 2,541 | 34,293 | -5.87±6.89 | -4.45±7.28 | -0.20 | <0.0001 |
| | Droughtmaster | 1,885 | 840 | -0.65±4.76 | -0.47±5.60 | -0.03 | 0.43 |
| | Santa Gertrudis | 8,975 | 23,820 | -0.66±5.74 | -0.87±5.47 | 0.04 | 0.0028 |
| CEdr (%) | Charolais | 4,426 | 2,272 | 2.65±7.32 | -0.90±6.18 | 0.52 | <0.0001 |
| | Hereford | 56,827 | 18,933 | -0.32±5.85 | -2.34±6.42 | 0.33 | <0.0001 |
| | Limousin | 7,759 | 3,541 | 1.59±3.61 | -0.16±4.03 | 0.46 | <0.0001 |
| | Shorthorn | 21,511 | 1,152 | 0.23±5.93 | 0.10±6.48 | 0.02 | 0.51 |
| CEdt (%) | Charolais | 2,028 | 1,333 | -0.26±6.73 | -0.30±6.73 | 0.01 | 0.89 |
| | Hereford | 17,689 | 3,516 | 0.74±4.34 | -1.39±4.74 | 0.47 | <0.0001 |
| | Limousin | 4,588 | 2,484 | 1.28±3.37 | 0.31±4.50 | 0.24 | <0.0001 |
| | Shorthorn | 7,500 | 459 | -0.33±5.22 | -0.33±5.29 | 0.01 | 0.99 |

* Cohen's $d$ (effect sizes) are interpreted as; $d$ 0.01: very small, $d$ 0.20: small, $d$ 0.50: medium, $d$ 0.80: large, $d$ 1.20: very large, $d$ 2.0: huge (Sawilowsky 2009).

Overall, the results suggest that genetic improvements in reproductive traits and selection for polledness have been favourably in action in the nucleus herds of beef cattle during the last two decades. Given the positive associations between polledness, production (Randhawa *et al.* 2021) and studied traits, selection could be undertaken to improve them simultaneously to achieve sustainable beef production. Breeding polled animals can continually improve fertility and pregnancy traits at a rate governed by the respective trait's heritability. However, generalization of genetic potentials for some breeds with significantly unequal samples represented in their polled and horned cohorts may be substantially biased. In addition, phenotyping accuracy of head-status and subsequent recording in the BREEDPLAN database may have confounded the comparisons of this study (Connors *et al.* 2018). Poll gene testing assays can not only eliminate the impacts of phenotyping bias, but can also exclude genetically heterozygous animals (i.e., carry a horn allele but phenotypically polled (Randhawa *et al.* 2020). With widespread gene diagnostics tools and high-density genotyping being implemented into nucleus and commercial herds, larger proportion of genomic evaluated breeding animals will become available for future investigations based on genotype-phenotype concordant head-status to account for the perceived bias.

## CONCLUSIONS

This study shows that reproductive traits in beef cattle have generally improved along with the proportion of polled animals and their genetic merits in most of the studied breeds. Selection for polledness and reproductive traits could be undertaken simultaneously to achieve sustainable beef production. However, the findings require caution, as bias may be introduced through limited sampling, phenotyping inaccuracy and underlying genetic heterogeneity in the polled phenotypes. Further investigations by using recently developed poll diagnostic assays in genome-evaluated larger populations will enhance our understanding about the true genetic merit of polled cattle.

## ACKNOWLEDGEMENTS

Meat and Livestock Australia financially supported the project (L.GEN.1713).

## REFERENCES

Barwick S.A., Johnston D.J., Holroyd R.G., Walkley J.R.W. and Burrow H.M. (2013) *Anim. Prod. Sci.* **54**:97.

Cohen J. (1977) *Statistical power analysis for the behavioral sciences.* . Routledge, New York.

Connors N.K., Tier B. and Johnston D.J. (2018) Current status of Australia's diagnostic poll haplotype test. In: *World Congress on Genetics Applied to Livestock Production*, p. 344.

Johnston D.J., Corbet N.J., Barwick S.A., Wolcott M.L. and Holroyd R.G. (2013) *Anim. Prod. Sci.* **54**:74.

Lakens D. (2013) *Frontiers in psychology* **4**:863.

Meyer K., Hammond K., Parnell P.F., MacKinnon M.J. and Sivarajasingam S. (1990) *Livest. Prod. Sci.* **25**:15.

R Core Team (2021) R: A language and environment for statistical computing. R Foundation for Statistical Computing, Vienna, Austria.

Randhawa I.A.S., Burns B.M., McGowan M.R., Porto-Neto L.R., Hayes B.J., Ferretti R., Schutt K.M. and Lyons R.E. (2020) *G3-Genes Genom. Genet.* **10**:539.

Randhawa I.A.S., McGowan M.R., Porto-Neto L.R., Hayes B.J. and Lyons R.E. (2021) *Animals* **11**:870.

Sawilowsky S.S. (2009) *Journal of Modern Applied Statistical Methods* **8**:467.

Wolcott M.L., Johnston D.J. and Barwick S.A. (2013) *Anim. Prod. Sci.* **54**:60.

*Contributed paper*

# EBVS PREDICT PROGENY PERFORMANCE DIFFERENCES

**M.J Reynolds[1], C.J Duff[1], P.F Parnell[1] and A.I Byrne[1],**

[1]Angus Australia, 86 Glen Innes Rd, Armidale, NSW, 2350, Australia

## SUMMARY

Estimated Breeding Values (EBVs) form a key component of modern cattle breeding programs and are the foundation for genetic improvement within the Angus breed in Australia. Demonstrating the ability of EBVs to predict differences in progeny performance in a practical, real world scenario is seen as vital to ensure the continued growth in industry acceptance of EBVs.

This work explores the ability of EBVs predicted differences in progeny performance of sires entered in cohorts 5, 6 and 7 of the Angus Sire Benchmarking Program (ASBP) by comparing the expected differences in progeny performance based on EBVs with the observed differences in average progeny performance.

The study demonstrated that EBVs predicted differences in the breeding value of sires in the ASBP for birth, growth and carcase traits, and reinforces the merits of focussed adoption strategies pertaining to EBVs within the Angus genetic improvement pipeline.

## INTRODUCTION

EBVs function as both a breeding and marketing tool within modern cattle breeding programs with strong adoption by Angus seedstock and commercial breeders (Angus Australia 2020). The continued availability of evidence-based support of the technology is seen as vital to ensure confidence and increased use of the technology, enabling the industry to benefit from improvements in the rate of genetic gain.

## MATERIALS AND METHODS

The TransTasman Angus Cattle Evaluation includes pedigree, performance and genomic information from the Angus Australia and Angus NZ databases to evaluate Angus and Angus-influenced animals. The major components of this analysis rely on BREEDPLAN analytical software developed by the Animal Genetics and Breeding Unit (Graser *et al.* 2005).

To evaluate the ability of EBVs to predict differences in average progeny performance the EBVs of sires prior to entering the ASBP were collated as follows;

- Cohort 5 – 46 sires – March 2015 TransTasman Angus Cattle Evaluation
- Cohort 6 – 41 sires – March 2016 TransTasman Angus Cattle Evaluation
- Cohort 7 – 34 sires – March 2017 TransTasman Angus Cattle Evaluation

The average EBV difference between the highest 10 and lowest 10 sires within each cohort were calculated for each trait to determine the expected difference in average progeny performance between the two sire groups. With the expected difference, half the EBV difference, reflecting the contribution of the sire genetics to the average performance of progeny.

Performance data from the sires progeny was collected as part of the ASBP program. Calves are produced in co-operator herds, which involves the joining of approximately 2,500 females each year via fixed time artificial insemination to 40 sires. Performance measurements for birth and early growth traits are then collected on all calves, with male progeny castrated. Male progeny are grown to feedlot entry weight, before being measured for feed intake over a 70-day test period, at which point they enter a commercial feedlot finishing program. Meat quality traits were assessed in the steer carcases following slaughter, with samples taken for meat science assessment (e.g. IMF%, shear force) (Parnell *et al.* 2019).

The range of progeny number per sire for birth weight was 12 to 47, and 8 to 47 for the growth traits. The progeny number per sire for carcase traits was lower (from 4 to 27 progeny), as only male progeny were measured for these traits.

The progeny performance data for each trait was analysed through the Statistical Analysis System (SAS) to generate Least Squares Means (LSMs) for each sire, within their cohort. The LSMs were estimated using adjusted data and accounting for contemporary group as explained by Graser *et al*. (2005).

The LSMs are used to determine the observed differences between the mean progeny performance of the highest and lowest EBV sire groups. The expected differences were then compared to the observed differences to determine the reliability of the EBVs in predicting differences in progeny performance.

## RESULTS AND DISCUSSION

A comparison of the average EBV differences and observed differences in progeny performance for birth and growth traits is shown in Table 1 and Table 2. When the average expected difference is compared to the average actual difference, the results demonstrate that EBVs predict differences in the genetic merit of animal's for birth weight and the growth traits. As an example, for Birth Weight, the average EBV difference between the highest 10 and lowest 10 EBV sires was 3.7kg. The EBV difference was then halved to determine the average expected difference, of 1.9kg, and compared to the average actual difference of 1.5kg.

**Table 1. Comparison between average EBV difference and progeny performance for Birth Weight of the highest 10 and lowest 10 EBV sires for this trait**

| | Cohort 5 | Cohort 6 | Cohort 7 | Average |
|---|---|---|---|---|
| Average High EBV (kg) | 6.1 | 6.7 | 6.0 | 6.3 |
| Average Low EBV (kg) | 2.6 | 2.7 | 2.4 | 2.6 |
| Difference in EBV (kg) | 3.5 | 4.0 | 3.6 | 3.7 |
| Expected Difference (kg) | 1.8 | 2.0 | 1.8 | 1.9 |
| Average High LSM (kg) | 38.5 | 38.3 | 38.4 | 38.4 |
| Average Low LSM (kg) | 37.3 | 36.3 | 37.1 | 36.9 |
| Actual Difference (kg) | 1.2 | 2.0 | 1.3 | 1.5 |

The results for the carcase composition traits show alignment with those seen in birth and growth traits, when the average expected difference is compared to the average actual difference. This was most evident in Carcase Weight (Table 3), Eye Muscle Area (Table 4) and Intramuscular Fat (Table 7). The results show that the expected differences in performance based on EBVs was observed in the differences in average progeny performance, apart from carcase rump fat (Table 6). The discrepancy observed for carcase rump fat may be a function of unintended abattoir effects, such as hide puller damage, leading to lower precision in measuring this trait.

The methodology enabled the ability of EBVs to predict differences between progeny performance to be assessed, but did not fully account for the effect of the low number of progeny per sire. To fully account for this effect and to fully understand the statistical significance a much larger study would need to be undertaken.

**Table 2. Comparison between average EBV difference and progeny performance for 200 Day Weight of the highest 10 and lowest 10 EBV sires for this trait**

| | 200 Day Weight | | | | 400 Day Weight | | | | 600 Day Weight | | | |
|---|---|---|---|---|---|---|---|---|---|---|---|---|
| Cohort | 5 | 6 | 7 | Avg. | 5 | 6 | 7 | Avg. | 5 | 6 | 7 | Avg. |
| Average High EBV (kg) | 55.4 | 56.4 | 58.4 | 56.7 | 102.5 | 101.8 | 105.7 | 103.3 | 137.8 | 137.2 | 138.2 | 137.7 |
| Average Low EBV (kg) | 33.4 | 40.5 | 44.2 | 39.4 | 63.8 | 76.5 | 82.3 | 74.2 | 81.6 | 99.0 | 106.1 | 95.6 |
| Difference in EBV (kg) | 22.0 | 15.9 | 14.2 | 17.3 | 38.7 | 25.3 | 23.4 | 29.1 | 56.2 | 38.2 | 32.1 | 42.1 |
| Expected Difference (kg) | 11.0 | 8.0 | 7.1 | 8.7 | 19.3 | 12.7 | 11.7 | 14.6 | 28.1 | 19.1 | 16.0 | 21.1 |
| Average High LSM (kg) | 251.0 | 217.6 | 231.6 | 233.4 | 375.7 | 360.4 | 362.9 | 366.3 | 571.3 | 623.2 | 586.8 | 593.8 |
| Average Low LSM (kg) | 237.4 | 209.2 | 227.9 | 224.8 | 359.2 | 347.0 | 350.2 | 352.1 | 545.7 | 603.2 | 572.6 | 573.8 |
| Actual Difference (kg) | 13.6 | 8.4 | 3.7 | 8.6 | 16.5 | 13.4 | 12.7 | 14.2 | 25.6 | 19.9 | 14.2 | 19.9 |

**Table 3. Comparison between average EBV differences and progeny performance for Carcase Weight of the highest 10 and lowest 10 EBV sires for this trait**

| | Cohort 5 | Cohort 6 | Cohort 7 | Average |
|---|---|---|---|---|
| Average High EBV (kg) | 77.3 | 83.2 | 86.3 | 82.3 |
| Average Low EBV (kg) | 40.6 | 52.9 | 60.8 | 51.4 |
| Difference in EBV (kg) | 36.7 | 30.3 | 25.5 | 30.8 |
| Expected Difference (kg) | 18.4 | 15.1 | 12.7 | 15.4 |
| Average High LSM (kg) | 429.3 | 435.2 | 429.9 | 431.5 |
| Average Low LSM (kg) | 411.2 | 423.4 | 419.8 | 418.1 |
| Actual Difference (kg) | 18.1 | 11.9 | 10.1 | 13.4 |

**Table 4. Comparison between average EBV difference and progeny performance for Carcase Eye Muscle Area of the highest 10 and lowest 10 EBV sires for this trait**

| | Cohort 5 | Cohort 6 | Cohort 7 | Average ($cm^2$) |
|---|---|---|---|---|
| Average High EBV ($cm^2$) | 10.6 | 11.1 | 8.4 | 10.0 |
| Average Low EBV ($cm^2$) | 2.8 | 3.6 | 3.6 | 3.3 |
| Difference in EBV ($cm^2$) | 7.8 | 7.5 | 4.8 | 6.7 |
| Expected Difference ($cm^2$) | 3.9 | 3.8 | 2.4 | 3.3 |
| Average High LSM ($cm^2$) | 89.2 | 94.1 | 90.3 | 91.2 |
| Average Low LSM ($cm^2$) | 87.3 | 89.7 | 88.8 | 88.6 |
| Actual Difference ($cm^2$) | 1.9 | 4.4 | 1.6 | 2.6 |

**Table 5. Comparison between average EBV difference and progeny performance for Carcase Rib Fat of the highest 10 and lowest 10 EBV sires for this trait**

| | Cohort 5 | Cohort 6 | Cohort 7 | Average |
|---|---|---|---|---|
| Average High EBV (mm) | 1.9 | 2.1 | 1.8 | 1.9 |
| Average Low EBV (mm) | -2.2 | -1.5 | -1.6 | -1.8 |
| Difference in EBV (mm) | 4.1 | 3.6 | 3.4 | 3.7 |
| Expected Difference (mm) | 2.0 | 1.8 | 1.7 | 1.8 |
| Average High LSM (mm) | 18.2 | 14.7 | 15.3 | 16.1 |
| Average Low LSM (mm) | 15.6 | 14.6 | 12.8 | 14.3 |
| Actual Difference (mm) | 2.6 | 0.1 | 2.5 | 1.8 |

**Table 6. Comparison between average EBV difference and progeny performance for Carcase Rump Fat of the highest 10 and lowest 10 EBV sires for this trait**

| | Cohort 5 | Cohort 6 | Cohort 7 | Average |
|---|---|---|---|---|
| Average High EBV (mm) | 2.2 | 19.9 | 1.3 | 1.8 |
| Average Low EBV (mm) | -2.6 | -1.9 | -2.2 | -2.2 |
| Difference in EBV (mm) | 4.8 | 3.8 | 3.5 | 4.0 |
| Expected Difference (mm) | 2.4 | 1.9 | 1.7 | 2.0 |
| Average High LSM (mm) | 19.6 | 19.6 | 22.9 | 20.7 |
| Average Low LSM (mm) | 19.5 | 19.6 | 20.3 | 19.8 |
| Actual Difference (mm) | 0.1 | 0.0 | 2.6 | 0.9 |

**Table 7. Comparison between average EBV difference and progeny performance for Carcase Intramuscular Fat of the highest 10 and lowest 10 EBV sires for this trait**

| | Cohort 5 | Cohort 6 | Cohort 7 | Average |
|---|---|---|---|---|
| Average High EBV (%) | 2.8 | 3.9 | 4.0 | 3.6 |
| Average Low EBV (%) | 0.5 | 0.9 | 1.4 | 0.9 |
| Difference in EBV (%) | 2.3 | 3.0 | 2.6 | 2.6 |
| Expected Difference (%) | 1.2 | 1.5 | 1.3 | 1.3 |
| Average High LSM (%) | 9.9 | 9.3 | 9.4 | 9.5 |
| Average Low LSM (%) | 8.4 | 7.8 | 7.8 | 8.0 |
| Actual Difference (%) | 1.5 | 1.5 | 1.6 | 1.5 |

## CONCLUSIONS

The work has demonstrated that EBV differences are a predictor of differences in progeny performance for birth, growth and carcase traits. The expected difference in progeny performance calculated from the difference between the average initial EBV of the highest 10 and lowest 10 sires provided a prediction of the observed difference in progeny of the two groups of sires, assessed through the ASBP program.

Breeders should have confidence in using EBVs to identify genetics that are most aligned with their breeding objectives, as EBVs provide an indication of the genetics that sires are delivering to a breeding herd.

## ACKNOWLEDGEMENTS

Angus Australia gratefully acknowledges the co-funding contribution from the Meat and Livestock Australia Donor Company in the conduct of the Angus Sire Benchmarking Program

(P.PSH.0528). The results of this study are part of the "Lessons from the Angus Sire Benchmarking Program" resources, the full suite of resources can be found by visiting the Angus Australia website.

## REFERENCES

Angus Australia (2020) https://www.angusaustralia.com.au/content/uploads/2020/10/FINAL-Australian-Beef-Breeding-Insights-Report.pdf (accessed March 2021)

Graser H-U., Tier B., Johnston D.J. and Barwick S.A. (2005) *Aust. J. Exp. Agric.* **45**: 913.

Parnell P.F., Duff C.J., Byrne A.I., Butcher N.M. (2019) *Proc. Assoc. Advmt. Anim. Breed. Genet.* **23**:492.

*Proc. Assoc. Advmt. Anim. Breed. Genet: 24: 337-340*

# ECONOMIC ANALYSIS OF MERINO EWE PERFORMANCE FROM DIVERSE INDUSTRY SIRES USING GRASSGRO™

**J.S. Wallace[1], K.L. Egerton-Warburton[1], and A.A. Swan[2]**

[1] NSW Department of Primary Industries, Orange, NSW, 2800, Australia
[2] Animal Genetics and Breeding Unit*, University of New England, NSW, 2351 Australia

## SUMMARY

Differences in profitability between genetically different sire groups at the Macquarie site of the Merino Lifetime Productivity Project (MLP) were compared using GrassGro™ to simulate animal performance based on the interactions between animal production and pasture growth determined by historical climate data. Mean gross margin (GM) differences of $13/Dry Sheep Equivalent (DSE) and $42/head (hd) were found between sire groups for wool and meat median prices. In median to high markets wool income had a higher influence on GM/DSE than meat income, with fibre diameter being the trait of most influence. In low markets meat income had a greater influence on GM/DSE with weaning rate the most influential single trait. Utilising a combination of traits, through either of three different selection indexes, showed the strongest correlations with GM/DSE.

## INTRODUCTION

Previous studies (Clarke *et al.* 2019) have reported large (over $50) differences in production value of wether progeny on a per head basis between sire groups. Analysis of combined wether trial data using GrassGro™ (Merino Bloodline Performance, www.merinobloodlines.com.au) also reveals differences among bloodlines of up to $13/hd. However, one limitation of using wether data is an inability to account for differences in reproductive performance in the financial analysis.

GrassGro™ (Moore *et al.* 1997) is a decision support tool that enables the economic performance of livestock enterprises to be simulated using animal production data and their interactions with seasonal variation in pasture, historical weather data and specified market scenarios.

The MLP project was designed to evaluate 134 diverse industry sires based on the lifetime performance of their ewe progeny for a wide range of wool, growth, carcase, reproduction and disease resistance traits. This paper presents a preliminary analysis using GrassGro™ to investigate differences in economic performance between sire groups using a production dataset generated from the first one and two reproductive opportunities of the 2018 and 2017 drop ewes, respectively, at the Macquarie site of the MLP project.

## MATERIALS AND METHODS

The design of the MLP project has been described by Ramsay *et al.* (2019). The specific design of the Macquarie site has been described by Egerton-Warburton *et al.* (2019). Data from ewes born at the Macquarie site in 2017 (n=425) and 2018 (n=536) were available for analysis. Ewes were joined by artificial insemination in December to lamb the following May, with shearing occurring in October after the weaning of progeny in August. Traits evaluated included greasy fleece weight (GFW, kg), clean wool yield (YLD, %), mean fibre diameter (FD, µm), bodyweight (WT, kg) and reproduction (conception and number of lambs weaned per ewe joined, LW/EJ) in adult ewes.

A representative model farm was set up in GrassGro™ for the Macquarie site. Historical climate and rainfall data for Trangie Agricultural Research Centre (TARC, Lat -31.99, Long 147.95) was

* A joint venture of NSW Department of Primary Industries and the University of New England

sourced from SILO (Jeffrey *et al.* 2001) for the period 1970 to 2019, commencing when the required data set became available. The base parameters of the model were set to reflect the annual calendar of operations, and livestock management policies (feeding, selling) at TARC. A conservative fixed stocking rate (1.5 animals/ha) was used to ensure heavier sire groups were not unreasonably restricted in average seasons and was based on the regional estimate of 3-7 DSE/ha (Hassall and Associates 2006).

An across-year analysis was undertaken using the MERINOSELECT OVIS software (Brown *et al.* 2007) to estimate sire breeding values (BVs), accounting for fixed effects such as birth and rear type, dam age, dam source and management group. Predicted sire progeny group means were then calculated as mean flock production level + 0.5 × sire group BV, with assumed production level means shown in Table 1. Body weight sire group means were then adjusted to reflect animals in condition score 3 (standard reference weight). These predicted sire group means were used as inputs to the GrassGro™ simulation, and the range between sire group means are shown in Table 1. Sire differences in survival rate cannot be expressed (in combination with conception rates) in GrassGro™. Therefore, adjustments to conception rates were used to achieve desired weaning rate differences. The proportion of empty ewes was entered and then conception rates for singles and twins were adjusted in GrassGro™ to reflect desired weaning rate outcomes in alignment with weaning rate BVs for each sire group.

**Table 1. Mean, minimum and maximum predicted sire group means for production inputs to the GrassGro™ decision support tool**

| | FD (µm) | GFW (kg) | YLD (%) | Standard reference WT (kg) | LW/EJ |
|---|---|---|---|---|---|
| Mean | 19.5 | 7.0 | 71.4 | 60.0 | 0.98 |
| Minimum | 18.4 | 6.5 | 69.8 | 57.9 | 0.86 |
| maximum | 20.6 | 7.3 | 72.7 | 62.3 | 1.11 |

Three wool and meat price scenarios (30, 50 and 70$^{th}$ percentile, denoted as low, median and high) were used from weekly Australian Wool Exchange (AWEX) and Meat and Livestock Australia (MLA) market reports between January 2015 to December 2019 and supplementary feed costs (barley) were averaged over the same timeframe (ABARES 2020). Husbandry costs were calculated from NSW DPI Farm Enterprise budgets (18µm Merino) in 2019 (https://www.dpi.nsw.gov.au/agriculture/budgets/livestock).

Mean GM/DSE was simulated for each sire group in response to historic seasonal conditions over the period from 1970 to 2019 using median prices for 2015 to 2019. GM/DSE was plotted against four adult production traits (GFW, FD, WT and LW/EJ) as a deviation from the mean of all sire groups. Three selection indexes based on the Dual Purpose Plus (DPP), Merino Production Plus (MPP) and Fibre Production Plus (FPP) MERINOSELECT standard indexes (Brown and Swan 2016), were used to combine the production traits into index values, which were then correlated with GM/DSE under the 3 wool and meat price scenarios. Each of these indexes were modified to only include yearling and adult sire BVs for the traits in Table 1 and excluded any additional traits.

## RESULTS AND DISCUSSION

On average, GM/DSE ranged $13 between the top and bottom sire groups under median and high market scenarios and $10 per DSE with low prices. Mean GM/DSE may undervalue higher weaning rates and mean GM/hd provides an alternative comparison for properties that are understocked and can accommodate additional lambs without increasing supplementary feed or reducing ewe numbers. There was a difference in GM/hd of $35, $42 and $45 in low, median and high price scenarios respectively between the highest and lowest sire groups. There was a range of 3.8 to 4.5 mean DSE/ha between the highest and lowest sire groups in the scenario examined. Mean GM/DSE will be used for all other comparisons in this paper to account for changes in resource requirements, such as increases in feed.

Figure 1 shows the resulting distributions of mean GM/DSE at median prices against FD, CFW, WT and LW/EJ for each sire group as a deviation from the mean of all sire groups. The correlations between GM/DSE and these traits were -0.65, 0.32, -0.06 and 0.42, respectively. Wool income had a larger effect on GM/DSE (r = 0.62) than meat (r = 0.42) with FD the main trait of influence, followed by LW/EJ. Higher wool and meat prices resulted in similar trends in the relationship between traits of influence and the GM/DSE as those of median prices. Interestingly, when wool and meat prices were lower, meat income had a greater influence (r = 0.69) than wool (r = 0.17) on GM/DSE and the trait of largest influence was LW/EJ (r = 0.68) then CFW (r = 0.47).

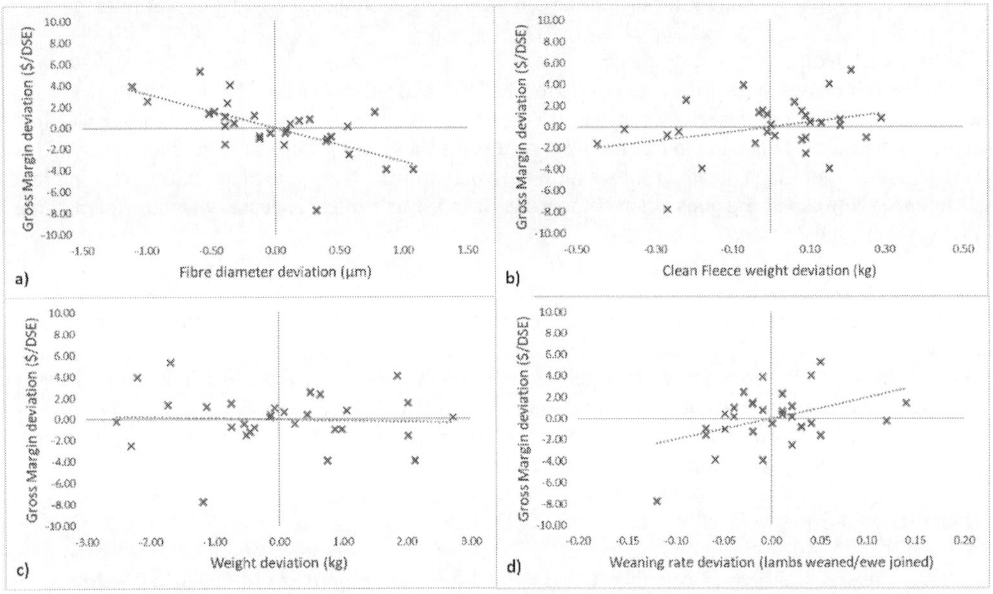


**Figure 1. Gross margin per DSE for each sire group against adult ewe a) FD, b) CFW, c) WT and d) LW/EJ as deviations from the mean of all sire groups at Macquarie (median prices)**

Correlations of GM/DSE for high and low prices with median prices were 0.99 and 0.84 respectively. This suggests that ranking of sires on GM/DSE will be similar in median and high markets. However, due to a higher emphasis on meat prices when wool and meat markets decline, a stronger influence of LW/EJ and CFW may lead to a re-ranking of sires during these periods.

**Table 2. Correlations between DPP, MPP and FPP MERINOSELECT indexes and GM/DSE in low, median and high market prices**

| | DPP | MPP | FPP |
|---|---|---|---|
| Low | 0.93 | 0.92 | 0.79 |
| Median | 0.72 | 0.82 | 0.92 |
| High | 0.76 | 0.87 | 0.95 |

Combining traits, using a selection index, resulted in higher correlations with GM/DSE than single traits, as shown in Table 2. These results show stronger correlations for the DPP and MPP indexes when markets were low and the FPP index when markets were high. This reflects the earlier findings where wool income had a stronger influence on GM/DSE in higher markets and meat income in lower markets.

The various seasonal conditions between 1970 and 2019 resulted in simulated mean production values that differed from the predicted sire means that were entered as breed references in GrassGro™, but these inputs and outputs were highly correlated (r>0.98). Variation between sire progeny group weaning rates had a large influence on GM/DSE, highlighting the importance of accounting for differences in reproductive performance between genotypes. However, it was difficult to model these directly in GrassGro™, and more accurate results may be achieved if genetic differences in survival rates could be included as inputs. Different resources (eg. nutrition) are required for similar weaning rates depending on variations in conception and survival rates.

## CONCLUSIONS

This study shows large differences of up to $13/DSE and $42/hd mean GMs between sire groups under a median price scenario when based on simulated environmental impacts across multiple seasons. It would be valuable to extend these results for these sire groups in different environments and across a larger range of sire groups within the MLP project. The value in including weaning rate differences between sire groups when comparing GM/DSE is evident especially when experiencing low wool and meat markets.

## ACKNOWLEDGEMENTS

The MLP Project is being undertaken in partnership between the Australian Merino Sire Evaluation Association Incorporated (AMSEA) and Australian Wool Innovation (AWI). AMSEA and AWI would like to acknowledge those entities who also contribute funding: woolgrowers through sire evaluation entry fees, site hosts, site committee in-kind contributions, and sponsors of AMSEA. The MLP Macquarie site gratefully acknowledge the generous financial support provided by the site host NSW Department of Primary Industries.

## REFERENCES

ABARES 2020, Australian crop report, Australian Bureau of Agricultural and Resource Economics and Sciences, Canberra, December. CC BY 4.0. https://doi.org/10.25814/5e3a014fa7e2d

Brown D.J., Huisman A.E., Swan A.A., Graser H-U., Banks R.B., Ball A.J., Atkins K.D. and Woolaston R.R. (2007) *Proc. Assoc. Advmt. Anim. Breed. Genet.* **17**: 187.

Brown D.J and Swan A.A. (2016) *Anim. Prod. Sci.***56:** 690

Clarke B.E., Young J.M., Hancock S. and Thompson A.N. (2019) *Proc. Assoc. Advmt. Anim. Breed. Genet.* **23**: 516.

Egerton-Warburton K.L., Mortimer S.I. and Swan A.A. (2019) *Proc. Assoc. Advmt. Anim. Breed. Genet.* **23**: 520.

Hassall and Associates Pty Ltd (2006) The Structure and Dynamics of Australia's sheep population - Report prepared for Australian Government. Available at https://www.agriculture.gov.au/sites/default/files/sitecollectiondocuments/animal-plant/animal-health/livestock-movement/sheep-movement-ead.pdf.

Jeffrey S.J., Carter J.O., Moodie K.B. and Beswick A.R. (2001) *Environ. Model. Softw.* **16:** 309.

Moore A.D., Donnelly J.R. and Freer M. (1997) *Agric. Syst.* **55**: 535.

Ramsay A.M.M., Swan A.A. and Swain B.C. (2019) *Proc. Assoc. Advmt. Anim. Breed. Genet.* **23**: 512.

*Proc. Assoc. Advmt. Anim. Breed. Genet. 24:341-344*

# ASSESSING THE POTENTIAL OF PARENTAGE TESTING USING PORTABLE LONG READ SEQUENCING TECHNOLOGIES

**E.M. Ross[1], H.J. Lamb[1], B.N. Engle[1], L.T. Nguyen[1] and B.J. Hayes[1]**

[1] Queensland Alliance for Agriculture and Food Innovation, University of Queensland, St. Lucia, QLD, 4072 Australia

## SUMMARY

Parentage testing in cattle based on DNA markers generates pedigree information for predicting breeding potential, as well as for the inclusion of animals in breed specific stud books. Here we present a method that could potentially be used for parentage assignment of calves. We propose the use of the portable minION (Oxford Nanopore Technologies; ONT) sequencer to achieve the parental assignments. The method uses ultra-low-coverage sequence data and a combination of genetic distances taken from a genomic relationship matrix, and simple Mendelian inheritance patterns - if both parents are homozygous for the same allele at a loci, then the offspring must also be homozygous at that loci.

The method was tested by simulating base calls based of the read length, loci distribution, and error profile observed in a real ONT sequenced sample of Brahman (*Bos indicus*) tail hair. These variables were used to simulate the expected data that would be obtained from sequencing the genome of 1500 Brahman calves with 100,000 mapped reads each. The algorithm assigned 98.7% of calves to the same sire/dam pair as the DNA based pedigree information. This study suggests that ONT could be successfully used to perform parentage testing in cattle.

## INTRODUCTION

Genotype based parental assignment in cattle is the most accurate way of pairing up bulls and cows with their progeny (e.g. McClure et al 2018). For cattle that are handled often, there is ample opportunity to sample tissue for DNA analysis and later make management decisions based on the results. For cattle in regions such as Northern Australia, cattle are rarely handled and so the time between taking a sample for DNA analysis and getting the parental assignment results that are needed to make management decisions is a limiting factor in the technologies use.

Portable sequencers are now available that can sequence long stretches of DNA in a single read. These have been applied in the field to address circumstances that require rapid turnaround of results, such as disease outbreaks (e.g O'Donnel *et al.* 2020). Here we suggest a new application: parentage testing of cattle. The implementation of parentage testing has the potential to be the first step towards crush-side-genotyping, where animals are genotyped and have genomic estimated breeding values calculated on farm, allowing for rapid management decisions to be implemented.

## MATERIALS AND METHODS

**SNP data.** 2675 Brahman heifers, cows and bulls from a single property were genotyped with the Neogen TropBeef V2 array, with 50045 SNP (after quality control, with genotypes with QC score <0.6 set to missing, monomorphic SNP excluded and SNP with all heterozygous calls excluded). All animals genotypes were imputed to 611,000 SNP on the Bovine HD array (following further QC) using Eagle (Loh *et al.* 2016) for phasing and Minimac3 (Das *et al.* 2016) for imputation.

All of the dams and bulls (1175 in total) were used as the potential parental population against which each calf was tested.

**ONT data.** DNA was extracted from the tail hair of a Brahman heifer from Queensland Australia that was collected as part of routine industry genomic evaluation by a commercial supplier. DNA was extracted using the PureGene (Qiagen) DNA extraction kit. The DNA was quantified on a Qubit

4.0. A sequencing library was prepared from the DNA using the SQK-LSK109 kit from ONT. The library was sequenced on a flow cell (FLO-MIN106; ONT) for 72 hours. Guppy v4.2.2 was used to convert the raw data to fastq format. The fastq data was aligned to the ARS_UCD1.2 *Bos taurus* genome using minimap2 version 2.17 (Li 2018). Alignment positions for each read were extracted from the .sam output file. The observed sequence lengths and alignment locations were then used to simulate the expected SNP coverage in the test population of animals.

The error rates of the ONT data were taken from Lamb *et al.* (2021), where ONT data was mapped back to a reference assembly generated from the same animal. The frequency of each substitution error was then calculated for each of the four possible reference nucleotides.

**ONT Simulation algorithm.** For each read the start position of the alignment to the reference genome and the read length were used to calculate if that read is expected to overlap a SNP location in the 700K SNP data. If the read did not overlay a SNP the algorithm moved onto the next read. If the read did overlay a SNP location, the "true" genotype of the test animal at that location was taken. If the animal was heterozygous at that location one of the two alleles was randomly chosen as the allele that was sequenced. An error was then induced into the base call using the specific ONT error rates for each base, such that the error profile of the final base call reflected the error profile of the real ONT data. The SNP location and basecall was then output to be tested by the parentage calling algorithm. Additionally, an error rate of ten times the observed rate was also tested, by increasing the probability of each error by a factor of 10.

**Parentage assignment algorithm.** Each calf was individually assigned to a parental pair. To reduce the search space a two-stage parental algorithm was used. First the simulated ONT genotype calls for the test calf were merged with the imputed SNP array genotypes of all potential sires and dams (N=1175). For each test calf loci without simulated ONT coverage were removed from the matrix, approximately 30K SNP remained in the matrix. The genotype matrix was used to calculate a genomic relationship matrix (GRM) using the *A.mat* command (default settings) of the rrBLUP package v4.6.1 in R version 4.0.0. The relationship values between the test calf and each bull were extracted, and the 50 bulls with the highest relationship value to the calf were used in the next step of the parentage assignment. The same approach was taken to highly related cows (n = 50).

The 50 bulls and 50 cows that were selected for further parentage testing were combined into 2500 possible parental pairs (50 x 50). A minimum minor allele frequency (MAF) cutoff was used to avoid large numbers of loci being homozygous in the parental pair and calf by chance. Unless otherwise stated, a MAF of 0.4 was used to filter SNP, and SNP on unplaced scaffolds and the X chromosome were removed. Then, for each pair of potential parents, loci where both the bull and the cow were homozygous were identified. If the test calf simulated ONT data had coverage in this location then the loci was used to create a score. The score was initialized at 0. For every loci that was homozygous in the bull and the cow, and where the calf had matching simulated ONT data, a +1 was added to the 'match' score. Alternatively, for every loci that was homozygous in the bull and the cow, and where the calf had a different simulated ONT genotype, +1 was added to the 'non-match' score. The final parental score for the bull cow pair was returned as M/(M+Nx10), where M is the match score and N is the non-match score. The unmatched loci were weighted more highly than the matched loci because the likelihood of them appearing by chance in the true sire-dam-calf trio is expected to be very low, proportionate to the error rate of the sequence data. This was repeated for all 2500 bull/cow combinations. The highest score was used to identify the most likely parental pair. If another pair of animals was greater than 90% of the highest score, that parental pair was also reported. The most likely parental pair was then compared to the pedigree data of the animals.

## RESULTS AND DISCUSSION

A known trio of animals consisting of a cow-bull-calf families were used to test the hypothesis that the highest level of concordance in homozygous sites would be between the true parents and their offspring. An additional 18 animals from the same herd were used to assess the method. At each minimum MAF tested the highest proportion of concordant calls was the correct parental pair (Table 1). The second highest level of concordance at all MAF contained at least one of the true parents. The difference between the true parental pair and the next highest match increased as the minimum MAF increased (Table 1). Interestingly, when the GRM from the same calf was examined, the highest relationship to any bull was not to the true sire, but to one of his sons (a half sibling of the test calf). The difference between the top (correct) match and the next highest match increased with an increase in the MAF.

**Table 1. Details of loci used to assign calves to sire and dam**

| Min MAF | N loci | Correct parental match[1] | Highest incorrect match[2] | Highest incorrect match (No parents)[3] |
|---|---|---|---|---|
| **0** | 676430 | 97.84% | 95.08% | 93.97% |
| **0.1** | 420021 | 95.48% | 89.02% | 86.65% |
| **0.2** | 289094 | 93.82% | 84.22% | 80.35% |
| **0.3** | 183804 | 92.45% | 80.08% | 75.02% |
| **0.4** | 89926 | 91.42% | 76.63% | 70.72% |

[1] Matches between the test calf and the true parents
[2] Matches between the test calf and the highest incorrect sire/dam pair (one parent can be correct)
[3] Matches between the test calf and the highest sire/dam pair where neither sire or dam is correct

The parental assignment algorithm described, which includes identifying the 50 most likely bulls and cows that for the parental pair of the calf being tested, was applied to simulated ONT base calls of 1500 calves. For the simulated reads only 11 (0.73%) calves had two or more potential sire/dam matches. When the error rate in the simulated reads was increased 10 fold, this number increased to 19 (1.26%).

When the parentage assignment results were compared to the pedigree information and historical mating records, the concordance was 98.7% and 98.6% for the low and high simulated error rates (Figure 1). After consultation with the producer, all but 5 of the calves were found to have errors in the pedigree data including misreporting of animal ID numbers. A number of the calves had been assigned to one of two potential bulls in the pedigree – with the DNA identifying the other as the true sire. Overall, after consultation with the producer, the level of agreement between the algorithm presented here and the pedigree information was 99.7% and 99.6%.

The best algorithm for parentage testing needs to play to the strengths of the technology being used. Here we opted for low pass sequencing, which allows many samples to be processed simultaneously. At the depth presented here approximately 100 samples could be processed on a flow cell, with an overall cost of under AU$20 per sample. This results in approximately 1 hour of sequencing per animal, which is a reasonable time to hold animal in yards while the parentage is determined, and clearly much shorter than the several weeks that other DNA parentage assignment tests take. Read alignment of the data can be performed in parallel with the sequencing on the device, and so does not add time to the test.

While using a GRM based approach with this data is possible, we observed that in at least some animals the highest relationship score of the potential bulls tested was not to the true sire, but rather to a half sibling of the test animal. Although the GRM is calculated on a reasonable number of loci (~30,000 loci), the genotype calls were constrained to information from one read, and hence heterozygous loci were randomly genotyped as homozygous. Consequently, the GRM alone was not sensitive enough to accurately differentiate between relationship levels of highly related individuals.

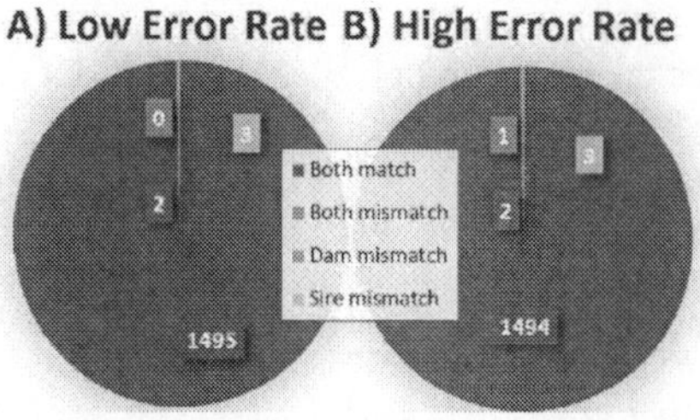


**Figure 1. Concordance between parental pairs assigned based on simulated ONT sequencing data and producer curated pedigree information. $N_{Calves}$ =1500; $N_{Dams}$=1034; $N_{Sires}$=141**

One consideration is that the number of usable loci is directly affected by the genetic distance between the cow and the bull. Animals that are highly related are expected to share alleles by descent, therefore have a higher number of matching homozygous loci. In this population we observed 600-700 matching homozygous loci per bull/cow pair. The required amount of sequence data may differ based on the genetic diversity of the test animals.

The assignment of each calf took approximately 25 seconds of computational time, which includes the calculation of the GRM. The GRM calculation constitutes > 95% of the computational time. Potentially, the GRM based potential parents selection could be removed from the algorithm, and the costs in terms of computational time should be considered based on the size of the test population, which would typically be much smaller than what was tested here. One important consideration is that the number of scores that the homozygosity based test must calculate is equal to the product of the potential sires times potential dams. Hence, with 50 bulls and 50 cows the number of tests is 2500, while with 100 of each the number of tests is 10,000 (a four fold increase even though the population has only doubled). Hence, where the size of the potential sire/dam herd is large, some reduction in the number of animals being tested is likely to save significant computation time. While the computational time for the test is minimal, and the sequencing and bioinformatics analysis can be completed in ~ 1 hour per animal. Laboratory methods (DNA extraction, library preparation) have not been examined here, research into the optimisation of those approaches is being undertaken with promising results (Mason and Botella 2020; Gowers *et al.* 2019).

## CONCLUSIONS

Here we present an algorithm for parentage testing from data obtained from ONT sequencing. When tested on simulated ONT data of 1500 calves, the accuracy of parental assignment (compared to curated pedigree information) was 99.7%. This work is the first step towards using ONT data to perform on-farm parentage assignment of cattle.

## REFERENCES

Das S., Forer L., Schönherr S., Sidore C., Locke A.E. et al. (2016) *Nat. Genet.* **48:** 1284.

Gowers G., Vince O.; Charles J., Klarenberg I.; Ellis T., Edwards A. (2019) *Genes*. **10**: 902.

Lamb H.J., Hayes B.J., Nguyen L.T. and Ross E.M. (2020) *Genes* **11**:1478.

Lamb H.J., Hayes B.J., Nguyen L.T., Engle B.N. and Ross E.M. (2021) *Proc. Assoc. Advmt. Anim. Breed. Genet:* **24**: *submitted.*

Li, H (2018) *Bioinformatics* **34**, 3094-3100.

Loh P.R., Palamara P.F. and Price AL. (2016) *Nat Genet.* **48**:811.

Mason, M.G., Botella, J.R. (2020) *Nat Protoc* **15**: 3663.

O'Donnell V., Grau F., Mayr G., Samayoa T., Dodd K., *et al.* (2020) *J. Clin. Microbiol.* **58**:e01104

*Proc. Assoc. Advmt. Anim. Breed. Genet: 24: 345-348*

# ESTIMATION OF OPTIMUM POLYGENIC AND GENOMIC WEIGHTS IN SINGLE-STEP GENETIC EVALUATION OF CARCASS TRAITS IN AUSTRALIAN ANGUS BEEF CATTLE

**A.M. Samaraweera[1], J.A. Torres-Vázquez[1], M G. Jeyaruban[1], D.J. Johnston[1] and V. Boerner[1]**

[1]Animal Genetics Breeding Unit*, University of New England, Armidale, NSW, 2350 Australia

## SUMMARY

Optimum polygenic and genomic weights enhance the accuracy of breeding value estimates in single-step genomic evaluations. This study estimated the contribution from marker information to total additive genetic variation referred as λ using an extended single-step model in a multi-trait variance component estimation based procedure using data for six Australian Angus carcase traits. The λ for these traits ranged from 0.54 (for carcass intramuscular fat) to 0.79 (for carcass eye muscle area). Heritabilities were similar between the pedigree only and the extended single-step multi-trait model when using the total genetic variance, and ranged from 0.37 (for carcass rib fat) to 0.53 (for carcass weight), suggesting that the single-step model did not explain more genetic variance than pedigree based models. Results suggest that the scalar λ in the current single-step routine evaluation could be replaced by an extended single-step model allowing for different proportions of the additive genetic co-variance explained by markers for all elements of the genetic co-variance matrix.

## INTRODUCTION

Increasing availability of genomic information requires ongoing modification to incorporate genotypes efficiently in routine genetic evaluation of Australian beef cattle. Single-step genomic evaluation developed by Legarra *et al.* (2009) and Christensen and Lund (2010) combines both pedigree and genotypes in a unified analysis. This method integrates numerator relationship matrix ($\boldsymbol{A}$) and genomic relationship matrix ($\boldsymbol{G}$) into a single $\boldsymbol{H}$ matrix, depicting co-variance between both genotyped and non-genotyped animals in the analysis. An improved $\boldsymbol{G}$ matrix ($\boldsymbol{G_w}$) that can be obtained as $\lambda\boldsymbol{G} + (1-\lambda)\boldsymbol{C}$ was suggested by Christensen and Lund (2012), where $\boldsymbol{C}$ is often the numerator relationship matrix among the genotyped animals, and λ is a non-zero weight with $0 < \lambda < 1$. λ is usually referred to as the proportion of additive genetic variance explained by the marker effects. For current BREEDPLAN single-step multi-trait breeding value estimation λ is set to a scalar value of 0.5, implying that for all genetic co-variances in the model, the same proportion is explained by markers.

Previous studies aimed at estimating λ by a cross-validation grid-search procedure to maximise the accuracy of predicted breeding values expressed as $\left(cov(\hat{u}, y)/\sigma_{\hat{u}}\sigma_y\right) * \sqrt{1/h^2}$, with the cross-validations performed on single trait data sets using a genetic variance $H\sigma$ (McMillan *et al.*, 2017; Zhang *et al.* 2017). However, the problem with the cross-validation approach is that contradicting values for λ in two single-trait analysis are difficult to accommodate when both traits are included in a multi-trait evaluation. Further, a multi-trait cross-validation grid-search would have to evaluate a high dimensional grid, which makes the approach computationally infeasible.

It can be shown that a model using $\boldsymbol{G_w} = \lambda\boldsymbol{G} + (1-\lambda)\boldsymbol{C}$ is simply the condensed form of the extended single-step model containing two genetic factors, one using $\boldsymbol{A}\otimes\boldsymbol{\Sigma_A}$ and the other using

---

* A joint venture of NSW Department of Primary Industries and the University of New England

$\boldsymbol{H}\otimes\boldsymbol{\Sigma}_H$ with $\boldsymbol{G}_w = \boldsymbol{G} + \boldsymbol{I}0.001$ where $\boldsymbol{\Sigma}_A$ and $\boldsymbol{\Sigma}_H$ are co-variance matrices and $\boldsymbol{I}$ is an identity matrix. The total genetic variance $\boldsymbol{\Sigma}_G = \boldsymbol{\Sigma}_H + \boldsymbol{\Sigma}_A$ with a scalar λ only being obtainable if $\boldsymbol{\Sigma}_H \oslash \boldsymbol{\Sigma}_G \equiv \boldsymbol{i}\boldsymbol{i}'k$ where $k$ is a scalar and $\boldsymbol{i}$ is an identity vector. Therefore, the partitioning of the genetic variance implicit in λ can be obtained by variance component estimation using the general model with two genetic factors, where the results may not support a scalar λ and in turn may require the use of the general model in genetic evaluation. However, the estimation of variance components for such a model via restricted maximum likelihood (REML) is challenging due to the mixed model coefficient matrix containing large non-zero blocks, and REML algorithms using the phenotypic co-variance matrix are severely limited with regard to the number of observations that can be accommodated.

This study investigated methods to optimally partition the genetic variances in Australian Angus carcass data. To overcome REML limitations, Bayesian methods were used.

## MATERIALS AND METHODS

A total of 59,616 pre-corrected records (Graser *et al.* 2005) for Australian Angus carcass traits were analysed consisting of carcass weight (CWT), carcass rib fat (CRF), carcass P8 fat (CP8), carcass eye muscle area (CEA), carcass retail beef yield percentage (CMY), and carcass intramuscular fat (CIM). Numbers of phenotypes, genotypes, and pedigree information available for each trait are given in Table 1. The pedigree consisted of 2.6 million animals, 110,000 of which were genotyped with 56,009 markers per genotype.

**Table 1. Number of phenotypic records, number of genotyped animals, and descriptive statistics for carcass traits, weight (CWT (kg)), rib fat (CRF (mm)), P8 fat (CP8 (mm)), eye muscle area (CEA (cm$^2$)), retail beef yield (CMY (%)), and intramuscular fat (CIM (%))**

| Trait | Records | Genotyped | Mean | Stddev | Minimum | Maximum |
|---|---|---|---|---|---|---|
| CWT | 16875 | 3340 | 422.9 | 60.2 | 186.8 | 636.0 |
| CRF | 5319 | 1059 | 15.5 | 5.1 | 1.6 | 36.8 |
| CP8 | 14793 | 3054 | 19.7 | 5.6 | 1.6 | 42.7 |
| CEA | 7392 | 839 | 83.9 | 9.1 | 41.8 | 120.7 |
| CMY | 2140 | 505 | 69.0 | 4.6 | 55.8 | 77.9 |
| CIM | 13097 | 2630 | 8.8 | 3.4 | 1.7 | 30.9 |

**Models.** A multi-trait linear mixed model (model 1) was fitted as follows:

$$\begin{pmatrix} y_1 \\ \cdot \\ y_6 \end{pmatrix} = \begin{pmatrix} X_1 & \cdot & \cdot \\ \cdot & \cdot & \cdot \\ \cdot & \cdot & X_6 \end{pmatrix} \begin{pmatrix} b_1 \\ \cdot \\ b_6 \end{pmatrix} + \begin{pmatrix} Z_1 & \cdot & \cdot \\ \cdot & \cdot & \cdot \\ \cdot & \cdot & Z_6 \end{pmatrix} \begin{pmatrix} u_1 \\ \cdot \\ u_6 \end{pmatrix} + \begin{pmatrix} e_1 \\ \cdot \\ e_6 \end{pmatrix}$$

where $[y_1,.,y_6]$ is a vector of phenotypic observations for traits 1 to 6, matrices $[X_1,.,X_6]$ and $[Z_1,.,Z_6]$ link fixed effects of contemporary group and random additive genetic effects, respectively to their respective observations, and $[e_1,.,e_6]$ is a vector of residuals. $[u_1,.,u_6]{\sim}N([0,.,0],\boldsymbol{A}\otimes\boldsymbol{\Sigma})$ where $\boldsymbol{\Sigma}$ is the co-variance matrix between genetic factors and **A** is the pedigree derived co-variance matrix between animals.

The single-step multi-trait linear model (model 2) was fitted as follows:

$$\begin{pmatrix} y_1 \\ \cdot \\ y_6 \end{pmatrix} = \begin{pmatrix} X_1 & \cdot & \cdot \\ \cdot & \cdot & \cdot \\ \cdot & \cdot & X_6 \end{pmatrix} \begin{pmatrix} b_1 \\ \cdot \\ b_6 \end{pmatrix} + \begin{pmatrix} Z_1 & \cdot & \cdot \\ \cdot & \cdot & \cdot \\ \cdot & \cdot & Z_6 \end{pmatrix} \begin{pmatrix} u_1 \\ \cdot \\ u_6 \end{pmatrix} + \begin{pmatrix} Z_1 & \cdot & \cdot \\ \cdot & \cdot & \cdot \\ \cdot & \cdot & Z_6 \end{pmatrix} \begin{pmatrix} g_1 \\ \cdot \\ g_6 \end{pmatrix} + \begin{pmatrix} e_1 \\ \cdot \\ e_6 \end{pmatrix}$$

where $[u_1,.,u_6]{\sim}N([0,.,0],\boldsymbol{A}\otimes\boldsymbol{\Sigma}_A)$ is a vector of polygenic effects and $[g_1,.,g_6]{\sim}N([0,.,0],\boldsymbol{H}\otimes\boldsymbol{\Sigma}_H)$ is a vector of genomic effects. Matrix $\boldsymbol{H}$ contains a genomic relationship matrix $\boldsymbol{G} = \boldsymbol{M}\boldsymbol{M}' + \boldsymbol{C}$, where $\boldsymbol{M}$ is centred and scaled marker genotypes matrix and $\boldsymbol{C}$ is a diagonal matrix of small values (e.g. 0.0001) ensuring invertability of $\boldsymbol{G}$. The total additive genetic variance $(\boldsymbol{\Sigma}_G)$ is equal to $\boldsymbol{\Sigma}_A + \boldsymbol{\Sigma}_H$, and when there are no genotyped animals model 2 essentially becomes model 1. Therefore, an underlying assumption about λ is that $\boldsymbol{u}$ and $\boldsymbol{g}$ are vectors of orthogonal random effects.

Variance components based on model 1 and model 2 were estimated using Gibbs sampling. The analysis were conducted with model 1 using $\boldsymbol{A}$ and with model 2 using $\boldsymbol{A}$ and $\boldsymbol{H}$, where in both cases the blocks of $\boldsymbol{A}$ and $\boldsymbol{H}$ related to the union of all phenotyped individuals were extracted from $\boldsymbol{A}$ and $\boldsymbol{H}$ built using all animals in the pedigree and all available genotypes. For the pedigree model prior variances were calculated from the phenotypic variances. For the extended single-step model prior variances were those obtained from the pedigree model, with a prior variance partitioning equal to $\boldsymbol{\Sigma}_H = 0.1\boldsymbol{\Sigma}_G$ and $\boldsymbol{\Sigma}_A = 0.9\boldsymbol{\Sigma}_G$. However, for both models the prior weight was zero. Variance components and genomic weights were obtained by discarding the first 30,000 samples as burn-in and averaging the sum of every 100th sample from a total of 200,000 samples.

## RESULTS AND DISCUSSION

The heritabilities for six carcass traits for model 1 which used $\boldsymbol{A}$ as the between animals co-variance matrix, and for model 2 where the variances were partitioned between the genomic and polygenic factor are presented in Table 2. The total heritabilities for six carcass traits in model 2 ranged from 0.37 for CRF to 0.53 for CWT , and they were almost identical to those derived from model 1 (Table 2).

The proportion of additive genetic variation explained by markers (λ) is greater for almost all carcass traits than the λ assumed in the current BREEDPLAN evaluations of 0.5, and ranged from 0.54 for CIM to 0.79 for CEA (Table 2). Therefore, future genetic evaluations should allow higher and different λ in BREEDPLAN routine genetic evaluation of carcass traits. These results suggest that the BREEDPLAN genetic evaluation model would have to allow for two genetic factors where the implications for model dimensionality, solver convergence rate, and breeding value accuracy must be investigated.

**Table 2. Pedigree based heritability ($h^2$) when using model 1 and matrix $A$, and polygenic ($h_A^2$), genomic ($h_H^2$) and total heritability ($h_G^2$), genomic weights (λ) and phenotypic variances ($\sigma_p^2$) when using model 2 for 6 Australian Angus carcase traits**

| Parameter | CWT[1] | CRF[2] | CP8[3] | CEA[4] | CMY[5] | CIM[6] |
|---|---|---|---|---|---|---|
| $h^2$ | 0.51 (0.03)[a] | 0.38 (0.05) | 0.45 (0.03) | 0.47 (0.04) | 0.51 (0.07) | 0.46 (0.03) |
| $h_A^2$ | 0.17 (0.03) | 0.12 (0.04) | 0.13 (0.03) | 0.10 (0.04) | 0.23 (0.07) | 0.22 (0.04) |
| $h_H^2$ | 0.35 (0.02) | 0.25 (0.03) | 0.34 (0.03) | 0.37 (0.03) | 0.29 (0.07) | 0.25 (0.02) |
| $h_G^2$ | 0.52 (0.03) | 0.37 (0.04) | 0.47 (0.03) | 0.47 (0.04) | 0.52 (0.05) | 0.47 (0.03) |
| λ[†] | 0.67 (0.04) | 0.68 (0.08) | 0.73 (0.06) | 0.79 (0.08) | 0.56 (0.13) | 0.54 (0.05) |
| $\sigma_p^2$ | 844.40 (13) | 16.80 (0.40) | 21.89 (0.36) | 46.51 (0.92) | 2.72 (0.10) | 5.50 (0.09) |

[1]weight, [2]rib fat, [3]P8 fat, [4]eye muscle area, [5]retail beef yield, [6]intramuscular fat; [a]standard deviation from 1700 samples in parenthesis; [†]$\lambda = diag(\boldsymbol{\Sigma}_H)/diag(\boldsymbol{\Sigma}_G)$

Directions (and values) of between trait total genetic correlations from model 2 were similar to those from model 1 (Table 3). However, comparison of trait correlations between polygenic and genomic factor in model 2 shows that for many traits this correlation is in the in opposite direction (Table 4). One notable example is CEA where positive genetic correlations were observed for polygenic factor whereas those correlations were negative in genomic factor (Table 4). Global correlations between CEA and fat traits (CRF and CP8) were negative regardless of whether model 1 or model 2 was used. However, for model 2 genomic correlations remained negative whereas polygenic correlations turned positive. The opposite pattern was observed for correlations between CEA and CMY, where the global correlation remained positive, but was larger, and the polygenic correlation turned negative. It needs to be confirmed whether these findings have a biological foundation or were caused by insufficient variance partitioning due to the low number of genotyped and phenotyped animals for CEA and CMY.

**Table 3. Genetic correlation (lower triangle) when using model 1 and total genetic correlation (upper triangle) when using model 2 for 6 Australian Angus carcass traits**

| Trait[1] | CWT | CRF | CP8 | CEA | CMY | CIM |
|---|---|---|---|---|---|---|
| CWT | | -0.03 (0.02)[a] | -0.12 (0.02) | 0.01 (0.02) | 0.10 (0.02) | 0.04 (0.02) |
| CRF | -0.02 (0.02) | | 0.55 (0.01) | -0.14 (0.02) | -0.43 (0.02) | 0.01 (0.02) |
| CP8 | -0.12 (0.02) | 0.55 (0.01) | | -0.19 (0.02) | -0.25 (0.02) | -0.02 (0.02) |
| CEA | 0.08 (0.02) | -0.19 (0.02) | -0.23 (0.02) | | 0.39 (0.01) | 0.05 (0.02) |
| CMY | 0.11 (0.02) | -0.54 (0.03) | -0.29 (0.02) | 0.43 (0.01) | | -0.08 (0.02) |
| CIM | 0.07 (0.02) | 0.02 (0.02) | -0.07 (0.02) | 0.03 (0.02) | -0.02 (0.02) | |

[1]CWT, weight; CRF, rib fat; CP8, P8 fat; CEA, eye muscle area; CMY, retail beef yield; CIM, intramuscular fat; [a]standard deviation from 1700 samples in parenthesis

**Table 4. Polygenic factor correlation (upper triangle) and genomic factor correlation (lower triangle) matrix when using model 2 for 6 Australian Angus carcass traits**

| Trait[1] | CWT | CRF | CP8 | CEA | CMY | CIM |
|---|---|---|---|---|---|---|
| CWT | | 0.31(0.02)[a] | -0.13 (0.03) | 0.22 (0.02) | -0.09 (0.03) | 0.33 (0.02) |
| CRF | -0.20 (0.03) | | 0.49 (0.01) | 0.27 (0.02) | -0.50 (0.04) | 0.04 (0.02) |
| CP8 | -0.12 (0.03) | 0.57 (0.01) | | 0.06 (0.02) | -0.22 (0.03) | -0.15 (0.03) |
| CEA | -0.07 (0.03) | -0.30 (0.03) | -0.27 (0.03) | | -0.21 (0.03) | 0.10 (0.02) |
| CMY | 0.22 (0.02) | -0.40 (0.04) | -0.27 (0.03) | 0.69 (0.01) | | -0.06 (0.03) |
| CIM | -0.15 (0.03) | 0.00 (0.03) | 0.05 (0.02) | 0.03 (0.02) | -0.10 (0.03) | |

[1]CWT, weight; CRF, rib fat; CP8, P8 fat; CEA, eye muscle area; CMY, retail beef yield; CIM, intramuscular fat; [a]standard deviation from 1700 samples in parenthesis

## CONCLUSIONS

The proportion of additive genetic variation explained by markers (λ) ranged from 0.54 to 0.79 for the six carcass traits in Australian Angus beef cattle. This finding is significant because the current BREEDPLAN single-step evaluation uses a single λ for all traits, 0.5. The results of this study do not support the use of the same λ for all traits. However, accounting for these findings requires a change in the BREEDPLAN model which must be preceded by further investigations into computational feasibility and breeding value accuracy.

## ACKNOWLEDGEMENTS

The authors would like to thank the Meat and Livestock Australia (MLA) for their financial support through L.GEN.1704, and Angus Australia and New South Wales Department of Primary Industries (NSW DPI) for providing data used in this study.

## REFERENCES

Christensen O.F., and Lund M.S. (2010) *Genet. Sel. Evol.* **42**: 2.
Christensen O.F., Madsen P., Nielsen B., Ostersen T., and Su G. (2012) *Animal.* **6**: 1565.
Graser H.-U., Tier B., Johnston D.J., and Barwick S.A. (2005) *Aust. J. Exp. Agric.* **45**: 913.
Legarra A., Aguilar I., and Misztal I. (2009) *J. Dairy Sci.* **92**: 4656.
McMillan A., and Swan A. (2017) *Proc. Assoc. Advmt. Anim. Breed. Genet.* **22**:557.
Zhang Y., Swan A., Johnston D., and Girard C. (2017) *Proc. Assoc. Advmt. Anim. Breed. Genet.* **22**:309.

*Proc. Assoc. Advmt. Anim. Breed. Genet: 24: 349-352*

# SUPPORTING DATA-DRIVEN SUSTAINABLE LIVESTOCK INDUSTRIES IN DEVELOPING COUNTRIES

**B. Santos[1], G. Jenkins[1], T. Byrne[2], M. Teviotdale[1], A. Haile[3], C. Nimbkar[4] and P. Amer[1]**

[1]AbacusBio Limited, PO Box 5585, Dunedin 9058, New Zealand
[2]AbacusBio Int. Limited, Roslin Innovation Centre, Edinburgh EH25 9RG, United Kingdom
[3]Int. Cent. for Agric. Res. in the Dry Areas (ICARDA), Addis Ababa, Ethiopia, PO Box 5689
[4]Nimbkar Agricultural Research Institute (NARI), Phaltan 415 523, Maharashtra, India

## SUMMARY

The absence of data capture systems and structures to produce information is a chronic problem of many livestock industries, particularly in developing countries. Information can be a key driver of transformation for these industries and could be created through data collection and integration across livestock supply chains, resulting in more efficient, profitable, and sustainable production systems. A technology platform, Dtreo, has been developed to transform livestock performance data, initially captured to promote breeding initiatives, into actionable information, supporting farmers and connecting producers to markets. The availability of data also has the potential to address a major issue for establishing breeding programs in unstructured livestock industries – namely the lack of phenotyping systems to support genetic evaluation models including genomic selection initiatives. The Dtreo platform has now been deployed in community-based breeding programs in Ethiopia, India, and Uganda with an initial focus on genetic improvement. Our current objective is to expand its implementation to facilitate better coordination of decision making across the supply chain.

## INTRODUCTION

The livestock sector in many countries is affected by the absence of basic structures to capture data and enable flows of information, impacting farmers, traders, and processors. In most developing countries around the world there is limited ability to provide support, advisory services, and to create functional information systems, thus weakening the market leverage of small holder farmers, and creating many inefficiencies in supply chain function.

Genetic improvement initiatives are a critical component of developed livestock industries. The infrastructure required to establish these initiatives rely on data and information systems. These same information systems are frequently used to inform other areas of the livestock production sector, including the market. Unsurprisingly the livestock sector of most developing countries can be unstructured, lacking availability of data and information which limits their ability to provide support to their farmers in key areas, such as extension and animal health. Similarly, information on volume and quality of supply to the market is frequently absent in such unstructured industries.

In many developing countries livestock markets have been operational for centuries. In most cases however, such markets provide minimal returns to smallholder farmers. In these cases, it is common that traders and intermediaries capture most of the value that could be otherwise destined to farmers, especially if there were supporting data and information systems in place.

Prediction of livestock performance can be reasonably accurate when phenotypic measures of performance, production system (including traditional practices and environmental data), and parentage information datasets are available. However, when none of these exist, predicting and understanding productivity in each population is challenging.

A technology platform has been developed to support improvements in livestock production outcomes and boost self-sufficiency in rural communities. The Dtreo platform combines breeding principles with livestock production initiatives through data and information, enabling farmers globally to generate genetic insights for tangible improvements. The system was developed to have a low fixed cost, to be presented in local languages, and to be easy to use.

A subset of case-studies is presented in this paper to illustrate the mechanisms through which principles of genetic improvement can be used to support development of livestock industries in countries which lack the basic infrastructure to support their farmers.

## MATERIALS AND METHODS

The platform enables online and offline data collection and flow of information to support decision making at multiple levels in specific production systems, or across one or multiple supply chains. It enables capturing animal characteristics and performance recording of quality data. The data collected in remote locations, where connectivity is often limited, is transferred into a designed Microsoft Azure table storage and Cosmos DB SQL API which uses entity (e.g. location, flock/herd, animal, etc.) and event (e.g. birth, weaning, sales, etc.) associations to produce information based on analysis of the relationships within and across multiple entities and events.

Dtreo is now deployed in several countries (including Ethiopia, India, and Uganda), where most users are smallholder low-input farmers and the technicians (enumerators) that provide support. Farmers, technicians, and breeders use the platform and its analytics to make decisions in support of traditional practices that have been established for many years, whilst the flow of information supports other stakeholders. The platform facilitates the digitalization process, enabling users to record, store and analyse animal performance, reproductive management (including artificial insemination), and animal health interventions. It delivers analytic reports for decision making associated with these events, as well as data integration for genetic evaluation and access to market.

Examples of development projects in which Dtreo is currently used are:

1) Community-based breeding programs, CBBP (Ethiopia), supporting smallholder farmers organized in breeding cooperatives to address market demand for small ruminants by tailoring genetic improvement programs to local pastoral systems (Getachew *et al.* 2018).
2) Project Mesha (India), supporting village production of goats and establishing a genetic improvement strategy in Bihar state in collaboration with Nimbkar Agricultural Research Institute (NARI) and the Aga Khan foundation. Project Mesha aims at improving the quality of life for marginalized landless people and empowering and raising the incomes of women goat keepers through improving productivity of their goats.
3) PigBoost (Uganda), bridging the gap between pig farmers, veterinarians, and extension service-providers, providing a platform to capture data and produce information. Pig production has been largely influenced by increased demand for pork in Uganda. Different research and development initiatives have been associated with PigBoost to leverage the value of the data and information produced.

## RESULTS AND DISCUSSION

By February 2021, 70,753 animals have been recorded in the Dtreo platform (Table 1). A total of 4,091 households have been impacted in the referred programs - CBBP, Project Mesha, and PigBoost. Outcomes are still early stage, and these results are based on projects that have only recently been established, or initiated usage of Dtreo as their main data platform.

Since its inception the CBBP in Ethiopia has directly benefited more than 3,200 households in more than 40 villages and over 18,000 people across the supply chain. Increased productivity (more births, better growth, and reduced mortality), as well as increased income from small ruminant production has been achieved. Breeder cooperatives have been formed to commercialize rams and bucks from the program, building on the initial revolving funds supported by the project (Haile *et al.* 2020). The CBBP communities have a sufficiently large and equally distributed sheep/goat flock/herd, frequently with more than 500 ewes/does in the combined flock/herd of the village. More than 68,000 animals have been included in the Dtreo platform so far. The flock is composed of local breeds, mainly Bonga, Doyogena, Horro, Konso, Menz and Pare. The data recorded has been used to inform within-flock (village) selection of young sire candidates based on a set of agreed selection traits. The retained individuals are then further judged based on weights (at birth, weaning, six months and one year), functional conformation, and body scores reflecting carcass value, relative to contemporaries, all adjusted for maternal information. The aim is that selection decisions align with traditional practices (i.e. where selection was based only on size and colour), improving acceptance within the community, and eventually enabling new market channels and/or supply agreements to be established.

Project Mesha has established individual goat identification and initiated performance recording in a limited number of villages. An overall index score for buck kids was created including the criteria of 100-day weight, dam chest girth, condition at the time of assessment, litter size and kid survival history. Dtreo calculates and reports the index score for each buck kid as one of the selection criteria. So far, thirteen bucks have been selected based on their index scores and about 500 mated with these bucks. Further details on the achievements and context of Project Mesha can be found in Nimbkar *et al.* (2021).

PigBoost has developed a package of services to support pig-farmers in Uganda, focusing on animal identification and data collection. Since early in 2021, approximately 200 pig owners were recruited, with over 1,000 pigs included in the database. The most common breeds are Camborough (46%), Landrace (32%), and Large White (17%), all crossed with local breeds. Data on inseminations and matings, farrowing, weights and animal health interventions have been collected. There is high awareness of biosecurity, given the incidence of serious diseases in some of the herds, and the impact of pig mortality to the overall farm profitability. Major challenges reported by farmers are incidence of diseases (68% of farmers), finances (57%), poor growth rates (30%) and poor fertility/abortions/still births (16%). Most striking was that African Swine Fever was reported to have affected 25% of herds in the last 12 months. Other diseases of major concern are skin lesions, gastrointestinal, and nervous syndromes. Main services required by farmers are disease treatments (85% of farmers), management support (75%), diagnostic services (72%), artificial insemination (38%), and improved breeding (26%). Almost all farmers intend to expand their activities in the next 12 months (96%), with strong emphasis on better genetics (98%), and market access (98%).

**Table 1. Numbers of households and livestock impacted by development projects as of February 2021 using the data platform Dtreo for data capturing and information flow**

| Program | Country | Species | Households included | Animals included | Target households | Target animals |
|---|---|---|---|---|---|---|
| CBBP | Ethiopia | Sheep/Goats | 2,073 | 68,365 | 11,000 | 250,000 |
| Project Mesha | India | Goats | 1,812 | 1,319 | 50,000 | 500,000 |
| Pigboost | Uganda | Pigs | 206 | 1,069 | 10,000 | 50,000 |
| Total | - | - | 4,091 | 70,753 | 71,000 | 80,000 |

There are important challenges to overcome before wider adoption of data recording is observed, and before delivery of genetic improvement impact is realized. These challenges are:

1. Extension & education – support and training to smallholder farmers is required, such thatfarm management practices, feeding strategies, and animal health interventions are provided, optimising traditional production systems, whilst maintaining sustainable interaction with local environmental conditions.

2. Infrastructure – unstructured livestock industries frequently lack the required data and information systems, genetic improvement pipeline, and market access. Basic systems for developed industries (such as animal identification, phenotyping tools, breeding organizations, and market information) are inexistent in most developing countries.

An important limitation in establishing structured smallholder systems is defining breeding schemes that are suitable for the low-input, smallholder farming (Gizaw *et al.* 2014). The small size and weak genetic connectedness amongst flocks/herds is challenging to support approaches that allow separation of genetic and environmental effects in these conditions (Selle *et al.* 2020). A centralized dataset that combines animal performance records, and parentage information, from smallholder farms introduces a potential alternative to the paradigm of dealing with traditional breeding approaches in these systems. For instance, genetic evaluation models in which contemporary groups or herd effects are defined using production levels derived from data might be a feasible solution for smallholder farming systems (Ojango *et al.* 2019). These examples assume that a minimal structure to mitigate the absence of routine phenotyping exists, something that could be a reality with the use of a data platform for data capture and collation.

According to (Ojango *et al.* 2019), the need for a simple approach to data capture from smallholder systems is critical to support estimation of genetic parameters for these systems. This should be coupled with tangible incentives for continued recording and longer-term monitoring of the livestock population as a basis for implementation of sustainable genetic improvement programs. These systems would also support establishing a new paradigm of market integration through data and information, using digital technology.

## CONCLUSIONS

Digitalisation of performance records and information is a requirement that should be met if livestock industries of most developing countries intend to improve output, efficiency, and sustainability. Data can influence farm management, application of principles of genetic improvement, and access to market. There are several challenges involved in this process, particularly given smallholder farmers normally maintain a low number of animals in their herds and flocks. Nevertheless, the large number of smallholder farmers operating under specific groups, sharing similar environmental conditions, provide a sufficiently large population to be used as a source of information for breeding programs, and volume/scale for commercialisation.

The case-studies presented in this short summary demonstrate that smallholder farmers in developing countries, when assisted by proper infrastructure and digital technology, can be stimulated to participate in initiatives that will ultimately result in improved livelihoods through better structured livestock supply chains.

## REFERENCES

Getachew T., Haile A. and Rischkowsky B. (2018) *Proceedings of the World Congress on Genetics Applied to Livestock Production, 11, p.858.*

Gizaw S., Rischkowsky B., Valle-Zarate A., Haile A., van Arendonk J.A.M., Mwai A.O. and Dessie T. (2014) *J. Anim. Breed. Genet.* **131:** 341.

Haile A., Wurzinger M., Mueller J., Mirkena T., Duguma G., Rekik M., Mwacharo J., Mwai O., Sölkner, J. and Rischkowsky B. (2020). *ICARDA - Tools and guidelines*, No.1.

Nimbkar C., Amer P., Schurink A., Sawhney T. and Azam R. (2021) *Proc. Assoc. Advmt. Anim. Breed. Genet* (In submission).

Ojango J.M.K., Mrode R., Rege J.E.O., Mujibi D., Strucken E.M., Gibson J. and Mwai O. (2019) *J. Dairy Sci.* **102** No. 6.

Selle M.L., Steinsland I., Powell O., Hickey J.M. and Gorjanc G. (2020). *Genet Sel Evol,* **52**: 69.

*Proc. Assoc. Advmt. Anim. Breed. Genet: 24: 353-356*

# IMPROVING CARCASE VALUE BY INCORPORATING PRIMAL WEIGHTS INTO PIG BREEDING OBJECTIVES

**N.R. Sarker, B.J. Walmsley and S. Hermesch**

Animal Genetics & Breeding Unit (AGBU[1]), University of New England, Armidale, NSW, 2351

## SUMMARY

This study aimed to evaluate the benefits of including primal cut measures of pig carcases in the breeding objective by comparing the efficacy of two different approaches; a detailed approach and a simpler approach. The detailed approach included economic values for the loin and belly primals separately ($1.54 and $2.24/pig respectively), where the simpler approach included an economic value for the combination of these (the 'middle primal' at $1.89/pig). Each approach was evaluated in two different scenarios by adding information on the primal cut(s). Inclusion of primal traits in the breeding objective increased the predicted response to selection by 2.47% and 3.20% for both approaches (I and II) and primals contributed 15% and 12% to the new breeding objectives. The predicted response to selection was greater for the approach that included the middle primal, which was consistent with moderate to high genetic correlations with other traits in breeding objectives.

## INTRODUCTION

The current pig carcase payment system in Australia focuses on fat depth at a given carcase weight without giving any consideration to variation in primal cut weights. Moreover, selection pressure to change back fat has been substantially reduced over the last decade, and further reducing genetic carcase fatness presents little opportunity for economic benefit (Hermesch 2005). However, there are significant premiums in wholesale markets for the belly and the loin primals, which combined make up the middle portion of a pig carcase. According to Mérour et al. (2009), substantial economic differences in carcase value are connected to the variability in individual primal cut weights at a fixed carcase weight and fat depth. The aim of this study was to compare two different approaches for incorporating primal cuts into pig breeding objectives that either included economic values for the loin and belly primals or placed economic emphasis only on the middle portion of the carcase, which combined the loin and belly primals to produce a single trait.

## MATERIALS AND METHODS

**Data sets.** Data describing primal traits were available for 2,198 pig carcases collected from March to September 2012. Pigs came from three different terminal sire lines, which were nested within five different grow-out facilities. Variance and covariance components were estimated with an animal model using ASReml 4.1 (Gilmour *et al.* 2015).

**Estimation of primal price at the farm gate level.** Firstly, the wholesale prices of different primals obtained from the Australian Pork Limited (APL) weekly reports were converted back to farm-gate-level price (FP). The farm-gate prices for loin ($FP_{Ln}$), belly ($FP_B$) and middle ($FP_M$) were obtained as follows,

$$FP_{Ln} = WP_{Ln}.\ k,\ FP_B = WP_B.\ k \text{ and } FP_M = (FP_{Ln} + FP_B)/2,$$

where k was a constant derived as $k = P_p / (WP_{Sh}{*}pCwt_{Sh} + WP_L{*}pCwt_L + WP_{Ln}{*}pCwt_{Ln} + WP_B{*}pCwt_B)$, where, $P_p$ was the total price of the carcase adjusted to a fixed weight of 80 kg and a

---

[1] AGBU is a joint venture of NSW Department of Primary Industries and University of New England

price of \$AU3.05/kg carcase weight. The wholesale prices for different primal cuts (shoulder, leg, loin and belly) were denoted by $WP_{Sh}$, $WP_L$, $WP_{Ln}$ and $WP_B$, and $pCwt_{Sh}$, $pCwt_L$, $pCwt_{Ln}$ and $pCwt_B$ were the proportion that each primal represented of an 80kg carcase, respectively.

**Estimation of the economic value of primals.** The economic values for the loin and belly primals were considered to improve as a percentage basis rather than weight to look at the effect of changing proportions of the carcase to improve profitability. The economic values for either loin, belly or middle were the price difference between each of these primals and the average price of the shoulder and leg primals. This price difference was multiplied by 0.80, since a 1% increase of either loin, belly or middle represented 0.8 kg of a carcase (Table 1).

**Index description.** The existing pig breeding objective ($I_0$) includes six traits: average daily gain (ADG, g/d), back fat thickness (BFT, mm), feed conversion ratio (FCR, kg/kg), post-weaning survival (PWS, %), belly fat percentage (BFP, %), drip loss percentage (DLP, %) and two selection criteria traits; juvenile insulin-like growth factor-I (IGF1, ng/ml) and muscle depth (MD, mm). This study expanded the breeding objective to include primal cut traits (loin percentages, LnP and belly percentages, BP or middle percentages, MP), as shown in Table 1. The genetic parameters and economic values outlined in Table 1 were based on Hermesch *et al.* (2015) and Hermesch and Jones (2010). The heritabilities for primal percentages were estimated in this study and were corresponded to heritabilities for weights of primal cuts presented by Sarker *et al.* (2019) based on these data.

**Table 1. Heritabilities ($h^2$), genetic standard deviations (GSD), economic values (EV), genetic (below diagonal) and phenotypic (above diagonal) correlations between traits**

| Trait[1] | $h^2$ | GSD | EV | ADG | BFT | FCR | PWS | BFP | DLP | IGF1 | MD | LnP | BP | MP |
|---|---|---|---|---|---|---|---|---|---|---|---|---|---|---|
| *ADG* | 0.31 | 30.00 | 0.09 | | 11 | -2 | 0 | -6 | 6 | 9 | 12 | -1 | 27 | 21 |
| *BFT* | 0.33 | 1.00 | -1.70 | 2 | | 6 | 0 | 45 | -8 | 6 | 8 | -11 | -7 | -15 |
| *FCR* | 0.12 | 0.15 | -27.4 | -20 | 27 | | 0 | 20 | -8 | 15 | 1 | -14 | 2 | 0 |
| *PWS* | 0.05 | 0.04 | 1.82 | 0 | 0 | 0 | | 0 | 0 | 0 | 0 | 0 | 0 | 0 |
| *BFP* | 0.34 | 10.96 | -0.20 | 16 | 63 | 21 | 0 | | -4 | 0 | 1 | -20 | 20 | 0 |
| *DLP* | 0.23 | 0.84 | -2.25 | 11 | -18 | -16 | 0 | -4 | | 0 | -7 | 3 | -10 | 0 |
| *IGF1* | 0.21 | 13.07 | 0.00 | 6 | 20 | 57 | 0 | 0 | 0 | | 0 | 0 | 0 | 0 |
| *MD* | 0.30 | 1.93 | 0.00 | 35 | 16 | -8 | 0 | 1 | 20 | 0 | | -1 | -5 | -5 |
| LnP | 0.13 | 0.57 | 1.54 | -3 | -4 | -13 | 0 | -37 | 1 | 0 | -12 | | -27 | 70 |
| BP | 0.21 | 0.62 | 2.24 | 59 | 65 | 1 | 0 | 32 | -8 | 0 | 19 | 6 | | 56 |
| MP | 0.24 | 0.94 | 1.89 | 36 | 44 | -17 | 0 | 0 | -1 | 0 | 8 | 64 | 76 | |

*Traits in Italics represent referenced results. Genetic and phenotypic correlations were multiplied by 100. [1] Trait[1] abbreviations: ADG: average daily gain (g/day), BFT: back fat thickness (mm), FCR: feed conversion ratio (kg/kg), PWS: post-weaning survivability (%), BFP: belly fat percentage (%), DLP: drip loss percentage (%), IGF1: juvenile insulin-like growth factor-I (ng/ml), MD: muscle depth (mm), LnP: loin percentage (%), BP: belly percentage (%), MP: middle percentage. $h^2$: heritability, GSD: genetic standard deviation and EV: economic value

The information for ADG, BFT and MD was available for the selection candidates, sires, dams, 5 full sibs and 30 half sibs. For FCR, records were only available for the candidates, sires, 1 full sib and 5 half sibs because this trait is expensive and difficult to measure. For IGF1, the candidates, sires, dams, 2 full sibs and 10 half-sibs had records. For PWS, it was assumed that only the sires had information available. Carcase traits BFP, DLP, LnP, BP and MP were assumed to be available for 2 full sibs and 10 half sibs. Index calculations were performed using the MTIndex program of van der Werf (https://jvanderw.une.edu.au/software.htm).

**Approaches to include primal cut percentages in pig breeding objectives.** Approaches that included the loin and belly primals or fitted the combined (middle) term (Approaches I and II, respectively) were evaluated under two different scenarios (A and B). In the first scenario for both approaches ($I_A$ and $II_A$), the economic values of different primal cuts were included in the breeding objective, while the second scenario ($I_B$ and $II_B$) included information for different primals from relatives.

## RESULTS AND DISCUSSION

**Genetic correlations between primal cuts and breeding objective traits.** Results presented in Table 1 show that the loin primal (LnP) had negative genetic correlations with BFT and BFP which were favourable for selection ($r_g$ = -0.04 and -0.37), but the genetic association of LnP with FCR and MD were unfavourable ($r_g$ = -0.13 and -0.12). In comparison, BP had favourable genetic correlations with ADG and MD and was highly positively correlated with BFT and BFP ($r_g$ = 0.59, 0.19, 0.65, and 0.32 respectively). A favourable correlation existed between MP and ADG ($r_g$ = 0.36), but MP had unfavourable correlations with FCR and BFT ($r_g$ = -0.17 and 0.44). The genetic and phenotypic correlations between MP and LnP or BP were high because of the part-whole relationship between these traits. Overall, MP had favourable correlations with most other traits.

**Comparisons of different scenarios for adding approaches.** Adding primal cut traits to existing pig breeding objectives without including additional data in the analyses produced a positive correlated response in LnP, while BP had a negative response ($I_0$) (Table 2). In comparison, MP showed a negative response due to the genetic correlations with other breeding objective traits in the index (Table1).

Current breeding objectives do not usually take these correlated responses with primal weights into account. Adding economic values for different primal cuts to existing pig breeding objectives in the first scenario increased the overall response in the breeding objective. The individual responses of LnP, BP and MP were positive for both approaches (I and II). The economic implications can be calculated by multiplying each primal response by its economic weight (0.046 and 0.067 for loin and belly, respectively and 0.094 for middle cut). The overall response was $4.07 per pig in first scenario for both approaches ($I_A$ and $II_A$) and an increase of 0.74% relative to the base index $I_0$. The addition of information describing loin and belly or middle primal cuts to the breeding objective both produced changes in the relative contribution of index components. The overall responses in the second scenario were highest for approach II (3.22%) compared to approach I (2.47%). The differences in response for the two approaches could be associated with inconsistencies in butchering practices of the loin and belly primals that affects genetic parameters as well.

In the first scenario, ($I_A$ and $II_A$) index accuracies were lower for both approaches when compared to the base index ($I_0$) when adding primal cuts as breeding objectives trait without including any additional information. However, accuracies were increased in the second scenario where information from relatives for different primal cuts was added to the index.

**Impact of primal cuts to other breeding objective traits.** The responses for ADG and MD increased for the first approach in both scenarios when compared to the base index by including primals in breeding objectives. This change was due to the positive genetic correlation between muscle depth and growth rate. However, in the second approach, for ADG, responses were slightly lower when compared to first approach for both scenarios. In comparison, the MD responses were similar in both scenarios for approach I and II. The responses for BFT and BFP were reduced for both approaches and all scenarios when information on the middle primal (or loin and belly) were included in the breeding objectives due to the high genetic correlation between these fatness and primal traits.

The relative contribution of different traits to both existing and extended breeding objectives indicates that including information about the loin and belly or middle primals led to a decreased percent contribution of the original index traits. The relative contribution of LnP, BP and MP cuts were 6% and 9% or 12% of the relative emphasis in their respective approaches (I and II), respectively, demonstrating the importance of these traits in pig breeding objectives.

Based on the conducted research, it is worth mentioning that the inclusion of additional trait(s) is important in the estimation of the commercial value of pig carcases. These results show that the approach which fitted information about the combined middle primal was superior as it produced the highest genetic response in comparison to the approach which fitted the loin and belly primals separately. Moreover, the proposed approach effectively improves the total carcase value of pigs and reduces the costs as well as relative errors or biasness for the individual measurements of the loin and belly.

**Table 2. Traits measured in indexes, the response in individual traits for different indexes, and resulting index value and accuracy**

| Traits | Indices | | | | | Relative contribution of different traits in breeding objectives (%) | | |
|---|---|---|---|---|---|---|---|---|
| | $I_0$ | $I_A$ | $I_B$ | $II_A$ | $II_B$ | Base index $I_0$ | Index $I_B$ | Index $II_B$ |
| ADG | 10.33 | 12.67 | 12.76 | 12.16 | 12.13 | 21 | 18 | 19 |
| BFT | -0.50 | -0.40 | -0.38 | -0.41 | -0.38 | 13 | 11 | 12 |
| FCR | -0.06 | -0.06 | -0.06 | -0.06 | -0.06 | 32 | 28 | 28 |
| PWS | 0.00 | 0.00 | 0.00 | 0.00 | 0.00 | 57 | 49 | 50 |
| BFP | -3.51 | -2.84 | -2.93 | -2.98 | -3.06 | 17 | 15 | 15 |
| DLP | 0.07 | 0.05 | 0.05 | 0.06 | 0.06 | 15 | 13 | 13 |
| IGF1 | -3.79 | -3.59 | -3.58 | -3.64 | -3.60 | 0 | 0 | 0 |
| MD | 0.19 | 0.25 | 0.25 | 0.24 | 0.24 | 0 | 0 | 0 |
| LnP | 0.03* | 0.03 | 0.06 | | | | 6 | |
| BP | -0.05* | 0.03 | 0.05 | | | | 9 | |
| MP | -0.02* | | | 0.05 | 0.11 | | | 12 |
| $ Index | 4.04 | 4.07 | 4.14 | 4.07 | 4.17 | | | |
| Accuracy | 0.60 | 0.58 | 0.59 | 0.57 | 0.58 | | | |

For trait[1] abbreviations and index descriptions see Table 1. Extended breeding objective includes both LnP and BP or MP.

*Correlated responses due to selection using the current breeding objective $I_0$

**ACKNOWLEDGEMENTS**

Data recording was funded by Australian Pork Limited (Project 1025). The first author was supported by an Australian Government RTP International Scholarship.

**REFERENCES**

Gilmour A., Gogel B., Cullis B., Welham S. and Thompson R. 2015. ASReml user guide release 4.1 structural specification. VSN international ltd, Hemel hempstead, UK.

Hermesch S. (2005) *Australasian Pig Science Association (APSA)* 99.

Hermesch S., Arnal M., Börner V. and Dominik S. (2015) *Proc. Assoc. Advmt. Anim. Breed. Genet.* **21**:1.

Hermesch S. and Jones R. M. (2010) *In Pig Genetics Workshop Notes* Armidale, Australia:39.

Mérour I., Hermesch S., Jones R. and Tribout T. (2009) *Proc. Assoc. Advmt. Anim. Breed. Genet* **18**:195.

Sarker N., Walmsley B. and Hermesch S. (2019) *Proc. Assoc. Advmt. Anim. Breed. Genet.* **23**:318.

*Proc. Assoc. Advmt. Anim. Breed. Genet: 24: 357-360*

# INTEGRATION OF GWAS, NETWORK AND PATHWAY ANALYSIS REVEALS NOVEL INSIGHTS INTO THERMOTOLERANCE IN BEEF CATTLE

**K.M. Sarlo Davila[1], E.E. Rodriguez[1], A.N. Nunez Andrade[1], G. Zayas[1], D. Rodriguez[1] and R.G. Mateescu[1]**

[1] Department of Animal Sciences, University of Florida, Gainesville, FL, USA

## SUMMARY

Thermotolerance, the ability to maintain production under heat stress conditions, is a complex trait determined by many component traits. Recent approaches combining traditional genome wide associations studies (GWAS) with gene network interactions theory could be more efficient in dissecting the genetic architecture of complex traits such as thermotolerance. Genes in common between several different gene ontology (GO) term groups might point towards key regulator genes with a greater impact on the thermotolerance complex. Highly connected genes identified in this analysis include *SYK, NOS2,* and *CD36.* While these genes have not been previously associated with thermotolerance, they have been associated with adaptation to other extreme environments including cold climates and high altitudes. These results indicate that there may be crucial genetic architecture responsible for environmental adaptation regardless of the nature of the challenging environment.

## INTRODUCTION

Thermal stress in hot and humid conditions limits beef cattle production. Over 65% of the world's cattle (beef and dairy) reside in tropical or subtropical climates known for their hot and humid conditions. Thermotolerance, the ability to maintain production under heat stress conditions, is a complex trait determined by many component traits. Component traits related to heat loss are particularly of interest as there is a strong correlation between production level and metabolic heat production (Renaudeau *et al.* 2012). Greater capacity for heat loss rather than a lower metabolic heat level may allow cattle to maintain elevated production levels in the presence of heat stress (O'Brien *et al.* 2010). Many of the component traits that impact an animal's ability to lose heat are found at the hair-skin interface. Sweating capacity is of great importance as animals lose a majority of their heat through sweating when heat stress conditions become severe (Finch 1986). However, hair characteristics impact the effectiveness of sweating. Short, sleek hair coats allow for effective evaporative cooling during sweating, as well as reflect a greater proportion of solar radiation and facilitate conductive and convective heat flow (Hansen 2004). Recent approaches combining traditional GWAS with gene network interactions theory could be more efficient in dissecting the genetic architecture of complex traits such as thermotolerance. One advantage of association weight matrix/partial correlation information theory (AWM/PCIT) methodology is the ability to include SNP with relatively small effects that do not reach genome-wide statistical significance but are potentially linked to elements controlling the trait of interest. It is well recognized that many elements with minor effects are usually not able to reach significance at the genome level but will be uncovered through a gene network when multiple correlated traits are used in the analysis. GO term analysis of significant genes can be used to explore the functional mechanisms underlying thermotolerance.

## MATERIALS AND METHODS

The University of Florida Institutional Care and Use Committee approved the research protocol used in this study (Approval no. 201203578). This study utilized 2,409 commercial Brangus heifers from the Seminole Tribe of Florida, Inc. Samples were collected from 12 groups of 200 animals: 4 groups over 4 consecutive weeks in each (August 15- September 12), 4 groups over 4 consecutive weeks in 2017 (July 31 – August 28), and 4 groups over 4 consecutive weeks in 2018(July 26- August 23). Heifers within a year were from the same cohort and approximately the same age (about 2 years old).

The length and diameter of the undercoat (shorter coat closer to the body of the animal) and topcoat (longer coat that covers the undercoat) measured as described by Sarlo Davila *et al.* (2019). Coat score was measured for each heifer while in the chute and scored as 1 = very smooth, 2 = smooth, 3 = long, and 4 = woolly, as described by Hamblen *et al.* (2018). Sweating rate was measured using a calibrated, digital moisture sensor (Vapometer, Delphin Tech. Ltd, Kupio, Finland) that determines trans-epidermal water loss. The Vapometer uses a closed system approach, free of ambient airflow, to measure ambient relative humidity and temperature. The average body temperature of each heifer for each THI class from 0600 to 2000 hour was used in a random regression mixed model to estimate the reaction norm parameters for each individual: an intercept (RN intercept) and a slope (RN slope), as described in (Mateescu *et al.* 2020). The RN intercept describes the body temperature when animals are exposed to low heat stress (THI of 74 to 76), and the RN slope describes the change in body temperature in response to an increase of 5 THI units.

DNA was extracted from blood samples and genotyped with the Bovine GGP F250 array (Illumina Inc., San Diego, CA, United States). GWAS was performed as described by Sarlo Davila *et al.* 2020). The p-values and additive genetic values for each SNP were obtained for each phenotype and used to construct the association weight matrix (AWM) (Reverter and Fortes 2013). The AWM approach was used to synthesize the results from the GWAS. Topcoat length was chosen as the key phenotype to describe the complex of traits related to both thermotolerance and production. An initial set of 620 SNP with additive effects for topcoat length were selected based on their raw $P < 0.005$. To build the AWM, a vector of posterior mean estimates of the 620 SNP effects from topcoat was enhanced with the vectors of effects of all the other 7 phenotypes. This 620 x 8 matrix of posterior mean estimates of SNP effects was used as the input for PCIT to detect similar effects for any SNP across multiple phenotypes. All SNP pairs within the matrix were tested for association with at least one other SNP in order to establish network connections. SNP pairs without a significant partial correlation to at least one other SNP were removed from the dataset and discarded from subsequent network association analysis as they would appear isolated. SNP were then replaced with the gene the SNP were located in (within 2.5 kb), resulting in a network of 363 genes.

GO term enrichment and clustering were performed on all annotated genes from the AWM as well as the top genes from the GWAS for each trait. Functional grouping based on kappa score and visualization in a functionally grouped network was performed using the ClueGO plugin (Bindea *et al.* 2009) in Cytoscape. A kappa coefficient of 0.4 was used as a threshold value.

**Table 1. GWAS results for the eight thermotolerance traits**

| Trait | n | Mean | Min | Max | $h^2$ | Top Genes |
|---|---|---|---|---|---|---|
| Topcoat length (mm) | 2163 | 14.95 | 3.84 | 33.53 | 0.36 | *PRLR* |
| Undercoat length (mm) | 2163 | 6.98 | 2.19 | 23.11 | 0.22 | *PRLR, PCCA* |
| Topcoat diameter (mm) | 2163 | 0.16 | 0.039 | 0.70 | 0.11 | *MYC, RUNDC3A* |
| Undercoat diameter (mm) | 2163 | 0.14 | 0.039 | 0.795 | 0.08 | *RUND3CA, MYC* |
| Coat Score | 2397 | 1.35 | 1 | 4 | 0.23 | *PRLR, MMP19* |
| Sweating Rate | 1319 | 60.68 | 2.00 | 218.0 | 0.11 | *UBE2D2, CLIC3* |
| RN Intercept | 2067 | 38.75 | 36.54 | 39.74 | 0.18 | *ANKH, HELB, MGAT4B, PLA2G4* |
| RN Slope | 2067 | 0.22 | -0.001 | 0.442 | 0.10 | *SLC22A10, RUNDC3A, LRRC49* |

**RESULTS AND DISCUSSION**

A total of 363 annotated genes were found to be associated with at least one other gene and had significant direct and partial correlations. This correlation network generated a gene network consisting of 363 genes (nodes) and 22,928 relationships (edges). The top connected genes from the AWM were *RUNDC3A, TIGD7, OR4F73, YIPF1*and *PTPN21.* A functionally grouped annotation network was created from a list of 374 genes, combining the AWM results with the top GWAS results (Table 1). A network (Figure 1) was developed and visualized using the ClueGO plug-in for Cytoscape. 233 genes were associated with 103 biological function GO terms and pathways, forming 16 functional groups. The most representative group of terms was "anion transport" (18.52%), followed by "positive regulation of myeloid cell differentiation" (16.67%), and "organic acid transport" (11.11%). Higher connectivity between GO terms with similar molecular functions is to be expected, however, a high priority in terms of future research will be placed on genes in common between several different GO term groups as these might point towards key regulator genes with a greater impact on the thermotolerance complex. Highly connected genes include *SYK(spleen associated tyrosine kinase), NOS2 (Nitic Oxide Synthase 2),* and *CD36 (thrombospondin receptor). SYK* and *NOS2* were both connected to seven different functional groups while was *CD36* connected to six.

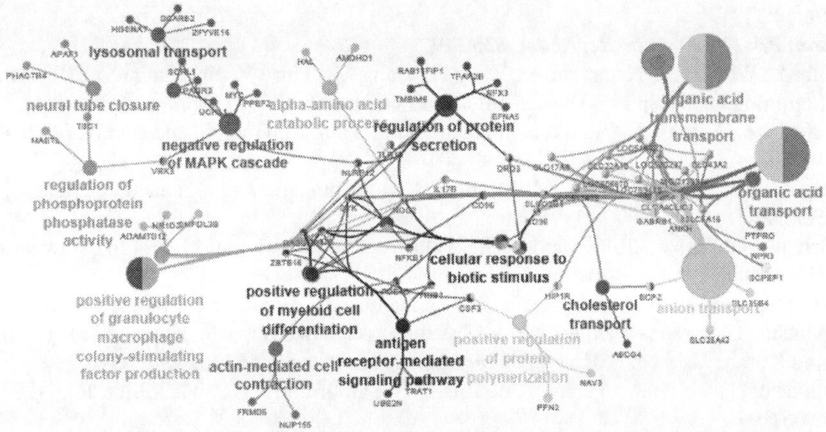


**Figure 1. Functionally grouped network for thermotolerance. Biological process terms and genes (in red) as nodes. The node size represents the enrichment significance of the term**

While none of these genes have been previously associated with thermotolerance, they have all been previously associated with adaptation to other harsh environments. *SYK* and *CD36* regulate brown adipose tissue and have been identified as candidate genes for adaptation to extreme cold. *SYK* was identified in a selection signature in two Russian breeds of cattle (Yurchenko *et al.* 2018) and *CD36* in Yanbian cattle (Shen *et al.* 2020). *SYK* is involved in brown adipocyte differentiation and *SYK* inhibition has been demonstrated to impair thermogenesis in mice (Knoll *et al.* 2017). *CD36* facilitates the uptake of energy substrates by brown adipose tissue and is essential for thermogenesis during cold exposure. *CD36* KO mice have been shown to have a drastically reduced body temperature after cold exposure (Putri *et al.* 2015). Brown adipose tissue is a key organ in non-shivering thermogenesis and helps cattle to conserve body heat in extremely cold environments. It is possible the absence of brown adipose tissue may help cattle lose excess body heat in hot environments.

*NOS2* has been identified as a candidate gene for high altitude adaptation in cattle native to the

Ladakh region of India (Verma *et al.* 2018) as well as Zhangmu cattle native to China (Liu *et al.* 2020). *NOS2* upregulation was found to prevent hypoxia and is related to vasodilation. Enhanced expression of *NOS2* may increase the production of NO, resulting in vasodilation and increased blood flow to increase the $O_2$ supply (Verma *et al.* 2018). Increased blood flow to the skin also allows for effective heat dissipation via sweating (Finch 1986).

## CONCLUSIONS

These results indicate that there may be crucial genetic architecture such as fat content and blood responsible for environmental adaptation regardless of the nature of the challenging environment, although the direction of selection for these traits changes with the environment. However, fat content, in particular, can also impact the production value of beef cattle as it affects meat quality. Further investigation of the impact of these traits on beef production is warranted.

## REFERENCES


Bindea G., Mlecnik B., Hackl H., Charoentong P., Tosolini M., Kirilovsky A., Fridman W.H., Pagès F., Trajanoski, Z., and Galon, J. (2009) *Bioinformatics.* **25**(8): 1091.

Finch V.A. (1986) *J Anim Sci.* **62**(2):531.

Hamblen H., Hansen P.J., Zolini A.M., Oltenacu P.A., and Mateescu, R.G. (2018) *J Anim Sci.* **96**(8):3131.

Hansen P.J. (2004) *Anim Repro Sci.* **82:** 349.

Knoll M., Winther S., Natarajan A., Yang H., Jiang M., Thiru P., Shahsafaei A., Chavarria T.E., Lamming D.W., Sun L., Hansen J.B., and Lodish H.F. (2017). *Nature Comm.* **8**(1).

Liu X., Li Z., Yan Y., Li Y., Wu H., Pei J., Yan P. Yang R., Guo X., and Lan X. (2020) *Evol App.* **14.**

Mateescu R.G., Sarlo Davila K.M., Dikmen S., Rodriguez E., and Oltenacu P.A. (2020). *J Anim Sci.* **98**: 1.

O'Brien M.D., Rhoads R.P., Sanders S.R., Duff G.C. and Baumgard L.H. (2010) *Domestic Anim Endo* **38**: 86.

Putri M., Syamsunarno M.R.A.A., Iso T., Yamaguchi A., Hanaoka H., Sunaga H., Koitabashi N., Matsui H., Yamazaki C., Kameo S., Tsushima Y., Yokoyama T., Koyama H., Abumrad N. A., and Kurabayashi M. (2015) *Biochem Biophys Res Comm.* **457**: 520.

Renaudeau D., Collin A., Yahav S., de Basilio V., Gourdine J. L., and Collier, R. J. (2012) *Animal.* **6**(05): 07.

Reverter A. and Fortes M.R.S. (2013) *J Anim Sci.* **91**(2): 530.

Sarlo Davila K M, Howell A., Nunez A., Orelien A., Roe V., Rodriguez E., Dikmen S. and Mateescu R. G. (2020) *Anim Gen.* **51**

Sarlo Davila K.M., Hamblen H., Hansen P.J., Dikmen S., Oltenacu P.A. and Mateescu R. G. (2019) *J Anim Sci.* **97**(8): 3246.

Shen J., Hanif Q., Cao Y., Yu Y., Lei C., Zhang G. and Zhao Y. (2020) *Frontiers Gen.* **11**: 1.

Verma P., Sharma A., Sodhi M., Thakur K., Kataria R.S., Niranjan S.K., Bharti V.K., Kumar P., Giri A., Kalia S. and Mukesh M. (2018) *SciReports.* **8**(1): 1.

Yurchenko A., Daetwyler H., Yudin N., Schnabel R., Vander Jagt C., Soloshenko V., Lhasaranov B., Popov R., Taylor J. and Larkin D. (2018) *Sci Reports.* **8**(1): 1.

*Proc. Assoc. Advmt. Anim. Breed. Genet: 24: 361-364*

# DOES SELECTING FOR THE A2 Β-CASEIN ALLELE INCREASE INBREEDING?

**B.A. Scott[1,2], M. Haile-Mariam[1], I.M. MacLeod[1] and J.E. Pryce[1,2]**

[1]Agriculture Victoria Research, AgriBio, 5 Ring Rd, Bundoora, Victoria 3083, Australia
[2] School of Applied Systems Biology, La Trobe University, Bundoora, Victoria 3083, Australia

## SUMMARY

Milk products from cows that are homozygous for the A2 β-casein allele are marketed in several countries by the A2 Milk Company. The alleles present at the β-casein locus of genotyped sires is published by DataGene and available to farmers when making selection decisions, alongside the estimated breeding values and selection indices. We hypothesised that intense selection for the A2 allele may have resulted in increased inbreeding. In this study we compared differences in genome wide and regional homozygosity between the two homozygotes of the β-casein alleles (A1 and A2) using medium density genotypes (50K SNP chips) of Holsteins cows. The A2 mutation was imputed into study animals, having first validated this approach in a group of bulls with known or certified genotypes for the β-casein locus. This study shows that the frequency of the A2 homozygote has increased by 20% since 2000 in Holstein cows. Our results suggest that selection for the β-casein A2 allele has increased inbreeding both across the genome and on chromosome 6 in Holstein cows that are homozygous for the A2 allele. Animals that had two A2 alleles were twice more likely to have a run of homozygosity of at least 35 SNP or 1000 kb long across the β-casein locus compared to animals that were homozygous for A1.

## INTRODUCTION

Seventy-five percent of milk protein content and composition can be linked to four casein genes CSN1S1, CSN2, CSN1S2, and CSN3, encoding the casein proteins alpha S1 (αS1), beta (β), alpha S2 (αS2), and kappa (κ), respectively (Ferretti *et al.* 1990, Threadgill and Womack 1990). This casein cluster spans ~250kb on BTA 6 (Boettcher *et al.* 2004). Selection for the A2 β-casein allele, has increased rapidly due to the commercialisation of milk products by the a2 Corporation in 2000. Milk products from the A2 Corporation are produced from cows that are homozygous for the A2 β-casein protein where herds need to have cows that are exclusively A2 homozygous. The alleles present at the β-casein locus of genotyped sires is published by DataGene and available to farmers when making selection decisions alongside the estimated breeding values and selection indices. Female genotyping at the β-casein locus can also be pursued by farmers who wish to build an A2 homozygous herd.

An implication of this is that intense selection for homozygosity at a given locus (A2) may result in increased inbreeding. Inbreeding can result in a loss of genetic diversity, decreased response to selection, reduced animal performance and ultimately, decreased farm profitability. Traditional pedigree methods to calculate inbreeding often underestimate the level of inbreeding due to incomplete pedigree and errors. With the availability of genotype information, we can now calculate inbreeding coefficients more accurately and are able to distinguish between recent and ancient inbreeding using runs of homozygosity. With the introduction of genomic selection the rate of inbreeding per year has increased (e.g. Doekes *et al.* 2018, Doublet *et al.* 2019, CDCB 2020), irrespective of the selection for specific alleles.

In this study we determined the frequency of homozygotes of the β-casein alleles (A1 and A2) over an 18-year period (2000-2017) in Australian Holstein cows and compared differences in genome wide and regional homozygosity between the two homozygotes of the β-casein allele using 50K SNP chip genotypes.

## MATERIALS AND METHODS

**Data.** A total of 139,898 genotyped individuals were available for Holsteins, Jerseys and their crosses from DataGene. The genotyping was carried out by various commercial providers. DataGene imputed the genotypes to a standard set of 45,685 SNP genotypes for routine evaluations (Nieuwhof *et al.* 2010). The breed of genotyped cows (Holstein) was validated using the ADMIXTURE program (Alexander *et al.* 2009). After correcting the breed information we had 114,567 Holsteins cows for subsequent analysis. Of these 73,003 cows born between 2000 and 2017 were used .

We imputed the A2 alleles while imputing animals to whole genome sequence. Genotypes were imputed in a stepwise fashion by imputing any low-density genotypes to 50k, then to high density and finally full sequence. The sequenced reference population used for imputation was Run 7 of the 1000 Bull Genomes project that includes 3090 *Bos taurus* animals after QC (Hayes and Daetwyler 2019). Only homozygote animals for the A1 and A2 alleles were selected for analysis. This imputation approach for the A1 and A2 alleles was validated in a group of 443 bulls with known or certified genotypes for the β-casein locus and showed 98.4% concordance.

**Inbreeding coefficients.** Genomic inbreeding coefficients were calculated from runs of homozygosity (ROH), identified across autosomes. A ROH was defined as a homozygous segment of at least 35 SNPs or 1000 kb long, with at least one SNP per 75 kb. Two consecutive SNP could not be included if they were more than 300 kb apart. ROH were identified using the PLINK "*homozyg*" function (Purcell et al., 2007) (command link: plink --cow --bfile genotyping_data_filename --homozyg --homozyg-kb 1000 --homozyg-snp 35 --homozyg-window-snp 50 –homozyg-window-density 75 --homozyg-gap 300 --out output_filename). ROH-based inbreeding estimates, $F_{ROH,i}$, were computed as the proportion of the genome included in the ROH as follows:

$$F_{ROH,i} = \frac{\sum L_{ROH,i}}{L_{auto}},$$

where $\sum L_{ROH,i}$ is the total length of ROH for individual i, and $L_{auto}$ the length of the autosome genome covered by SNPs after withholding gaps longer than 300 kb between two SNPs, corresponding to the length of the autosomal genome on which ROH can be detected. This parameter allowed for the detection of ROH on 92.2% of the autosomal genome.

For each individual, we also calculated the mean ROH length which is defined as:

$$L_{ROH,mean,i} = \frac{\Sigma L_{ROH,i}}{N_{ROH,i}},$$

where $\Sigma L_{ROH,i}$ is the total length of ROH for individual *i* in kb, and $N_{ROH,i}$the total number of ROH for individual *i.*

Additionally, we also compared genomic inbreeding (ROH) for Holstein cows that were homozygous for A2 and A1 alleles specifically on chromosome 6 (Chr6:87181619).

Wilcoxon-Mann Whitney tests were used to determine if there were significant differences between the two homozygote groups (A1/A1 versus A2/A2) in inbreeding levels for: all animals (born between 2000-2017) as well as between the same groups but only for young animals (born after 2013). A Chi-squared tested was used to determine if there was a difference between the observed versus expected number of animals with ROH over the A2 position. A total of 39,157 cows had both a known homozygosity over the β-casein locus and inbreeding coefficient and were used for the analysis.

## RESULTS AND DISCUSSION

**Frequency.** There were more than four times the number of A2/A2 cows (31,814) compared to A1/A1 (7,335) in the dataset. Figure 1 demonstrates how the frequency of the β-casein allele has changed in Holstein females since 2000. The A2/A2 frequency has increased from 32% in 2000 to 52% in 2017, suggesting growth of interest in β-casein possibly sparked by the a2 Corporation.

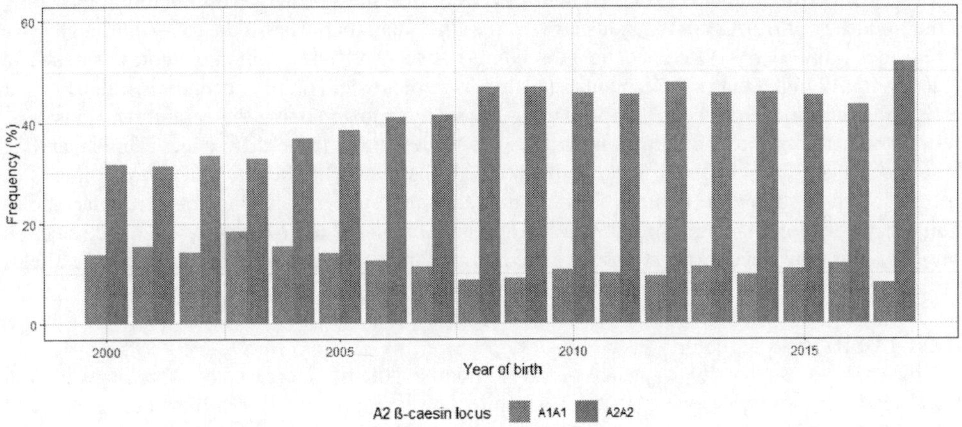


**Figure 1 Frequency of the A1/A1 and A2/A2 homozygous Holstein cows born in 2000 to 2017**

**Inbreeding.** The median $F_{ROH}$ was consistently higher in A2 than A1 homozygotes (Table 1). Over the whole genome, larger differences were observed when all animals were included (difference of 0.43%) than only young animals (animals born after 2013; difference 0.15, $p = 0.002$).

We observed significantly more regional inbreeding on Chromosome 6 for A2/A2 animals than for A1/A1, with median $F_{ROH}$ of 9.44% and 8.07%, respectively.

**Table 1 The median genomic inbreeding values ($F_{ROH}$*) for all animals (born in 2000-2017) or young animals (born in 2014-2017) homozygous for the β-casein A1 or A2 allele and the Wilcoxon-Mann Whitney significance test p-values**

| | No. A1/A1 | No. A2/A2 | Median $F_{ROH}$ A1/A1 | Median $F_{ROH}$ A2/A2 | p-value |
|---|---|---|---|---|---|
| *Whole Genome* | | | | | |
| All | 7,335 | 31,814 | 7.89 | 8.32 | < 2.2e-16 |
| Young Animals | 3,907 | 18,893 | 8.63 | 8.78 | 0.002 |
| *Chromosome 6* | | | | | |
| All | 5407 | 25301 | 8.07 | 9.44 | < 2.2e-16 |
| Young Animals | 3055 | 14977 | 8.89 | 9.60 | 8.211e-05 |

*ROH - Runs of homozygosity

**Inbreeding over the β-casein locus.** We found that while A2/A2 animals were more likely to have a ROH over the β-casein locus (chi-square statistic 262, $p < 0.00001$), the length of the ROH was longer for the A1/A1 animals (median ROH length 6,136 kb vs. 3,706 kb). When comparing this subset of animals to the median inbreeding observed over the whole genome, we found that animals with an ROH over the β-casein locus had higher overall level of inbreeding (median $F_{ROH}$ 10.0% subset vs. 7.89% population and 9.2% subset vs. 8.2% population for A1/A1 and A2/A2 animals, respectively).

In this study, we did not determine if the genotyped cows were representative of the entire population. It is possible that farmers who breed strictly for A2/A2 individuals are more likely to genotype their cows than those not breeding for A2/A2, resulting in an overestimation in this study of the frequency of A2/A2 homozygous cows across the Australian dairy industry. Additionally, we did not determine if the differences in inbreeding between A1 and A2 homozygotes was due to selection for A2 rather than simultaneous selection for protein yield or other economic traits. Further work using imputed sequence genotypes across the region encompassing the casein gene cluster as well as the heterozygote individuals may allow us to determine these differences. Understanding these mechanisms could have a wider implication for assessing the benefits and shortcomings of narrow selection strategies. Perhaps farmers that are interested in selecting for particular alleles should pay attention to monitoring inbreeding and its through the use of appropriate mate selection methods. This may have implications for breeders that may consider selection for specific alleles that are currently at low frequency, such as, the polled region.

## CONCLUSIONS

This study shows that the frequency of the A2 homozygote has increased by 20% since 2000 in Holstein cows. Our results suggest that A2/A2 animals were more inbred over the whole genome as well as on chromosome 6 and were more likely to have a ROH over the β-casein locus.

## REFERENCES

Alexander D.H., Novembre J. and Lange K. (2009) *Genome Res* **19**(9): 1655.

Boettcher P.J., Caroli A., Stella A., Chessa S., Budelli E., Canavesi F., Ghiroldi S. and Pagnacco G. (2004) *J Dairy Sci* **87**(12): 4311.

CDCB. 2020. Retrieved June, 2020, from https://queries.uscdcb.com/eval/summary/inbrd.cfm.

Doekes H.P., Veerkamp R. F., Bijma P., Hiemstra S. J. and Windig J. J. (2018) *GSE* **50(1)**: 15.

Doublet A.-C., Croiseau P., Fritz S., Michenet A., Hozé C., Danchin-Burge C., Laloë D. and Restoux G. (2019) *GSE* **51**(1): 52.

Ferretti L., Leone P. and Sgaramella V. (1990) *Nucleic acids research* **18**(23): 6829.

Hayes B. J. and Daetwyler H. D. (2019) *Annu Rev Anim Biosci* **7**: 89.

Nieuwhof G., Beard K., Konstantinov K., Bowman P. and Hayes B. (2010) *Interbull Bulletin* **42**: 35.

Purcell S., Neale B., Todd-Brown K., Thomas L., Ferreira M. A., Bender D., Maller J., Sklar P., de Bakker P. I, Daly M. J. and Sham P. C. (2007). *Am J Hum Genet* **81**(3): 559.

Threadgill, D. W. and Womack J. E. (1990) *Nucleic acids research* **18**(23): 6935.

*Proc. Assoc. Advmt. Anim. Breed. Genet: 24: 365-368*

# GENOMIC BREEDING VALUES FOR RESIDUAL FEED INTAKE IN AUSTRALIAN MATERNAL COMPOSITE EWES

**B.J. Sepulveda[1,2], S.K. Muir[3], S. Bolormaa[1], M.I. Knight[3], R. Behrendt[3], I.M. MacLeod[1], J.E. Pryce[1, 2] and H.D. Daetwyler[1, 2]**

[1] Agriculture Victoria, AgriBio, Centre for AgriBioscience, Bundoora, VIC, 3083 Australia
[2] School of Applied Systems Biology, La Trobe University, Bundoora, VIC, 3083 Australia
[3] Agriculture Victoria, Hamilton Centre, Hamilton, VIC, 3300, Australia

## SUMMARY

Residual feed intake (RFI) is difficult to measure, involving either labour-intensive measurements of feed intake and liveweight or specialised equipment, which makes genomic methods ideal for industry-wide selection. We estimated genetic parameters and investigated the accuracy of genomic prediction using five-fold cross-validation, using RFI phenotypes obtained from 465 Maternal Composite ewes measured at post-weaning, hogget and adult ages as a reference population. A genomic relationship matrix was constructed from 37,035 imputed markers. Uni- and multivariate GBLUP for RFI was performed, where records at different ages were either included as repeated measures or treated as separate traits. The first five principal components of the genomic relationship matrix were fitted as fixed effects to account for breed composition. The $h^2$ of RFI was estimated at 0.19 (±0.04) when all ages were fitted together, and genetic correlations between PW-hogget, PW-adult, and hogget-adult were estimated as 0.29 (±0.28), 0.24 (±0.43), and 0.50 (±0.37), respectively. The accuracy of genomic prediction across all ages was 0.22 (±0.03), and the bias was 1.00 (±0.19). The results suggest that after increasing the training set, breeding values for RFI in the Maternal Composite ewes could be developed.

## INTRODUCTION

Feed is the highest cost of sheep production. Subsequently, the sheep industry could potentially increase its profits by selecting for improved feed conversion efficiency (FCE). Residual feed intake (RFI) is the difference between actual and predicted dry matter intake (DMI) required for maintenance, growth, and production (Koch *et al.* 1963) and can be considered to be an indicator of FCE. However, an accurate measurement of DMI of animals at pasture is difficult, and thus RFI testing relies on the measurement of DMI and liveweight gain of intensively housed animals. The process is time-consuming expensive and may require specialised equipment, making it difficult to record at scale on commercial farms. For that reason, RFI is a good candidate trait for improvement through genomic selection.

Genomic selection predicts genomic estimated breeding values (GEBVs) for selection candidates based on their genotype even when their phenotype is unknown (Meuwissen *et al.* 2001). As a reference population, 445 crossed ewes from a population that included Coopworth, East Friesian, Finn, Border Leicester, South African Meat Merino, Texel, Poll Dorset, White Suffolk, Merino, Corridale, NZ Romney and Perendale were recorded for feed intake. Using a common training population usually leads to higher accuracies than using breed-specific reference populations, especially for crossbreeds (Bolormaa *et al.* 2013).

This study estimated genetic parameters and investigated the accuracy of genomic prediction via 5-fold cross-validation for RFI in a reference population of Maternal Composite crossbreds.

## MATERIALS AND METHODS

**Phenotypes and genotypes.** The feed intake and growth rate of 445 Maternal Composite ewes, 251 born in 2013 and 194 in 2014, was obtained using the automated feed intake facility validated by Muir *et al.* (2020a) at Agriculture Victoria, Hamilton, Victoria (Australia). From the 2013 born ewes, 81 were measured at post-weaning (PW), 195 at hogget, and 218 at adult ages. From the 2014 born ewes, 193 were measured at PW and 189 at hogget ages. At start of the tests, the ewes were 313 (±14), 534 (±19), and 858 (±23) days old at PW, hogget and adult ages, respectively. Animals were adapted to the pelleted diet for 10-14 days before the *ad libitum* feed intake was recorded individually in a group pen with automated feeders. Ewes were distributed across 10 group pens considering a balanced distribution of sires and ewes' liveweights across pens. All sheep were offered hay-based pellets for the duration of feed intake measurements. Pellets had 65% (±2.4) digestibility, 9.8% (±1.6) crude protein, 48% (±3.18) neutral detergent fiber, and 9.6 (±0.58) MJ/kg of dry matter. Feed intake measurements lasted 53 (±3), 42 (±3) and 32 (±0) days for PW, hogget, and adult age ewes, respectively. Live weights were measured three times weekly for the duration of the feed intake measurements. Details of the phenotypes and measurements were reported by Muir *et al.* (2020b).

The animals were genotyped with 12,785, 15,000, or 54,241 single nucleotide polymorphism (SNP) chips and imputed to 38,379 SNPs using Fimpute (Sargolzaei *et al.* 2014). Then, SNPs with minor allele frequency <0.05 were removed, and 37,035 remained for downstream analysis. A genomic relationship matrix (G) was constructed using the function Gmatrix of the R package AGHmatrix (Amadeu *et al.* 2016) using the method of Yang *et al.* (2010).

**Residual feed intake.** RFI was calculated as the residual DMI after energy sinks and corrected by fixed effects with the expression

Observed DMI = $\mu$ + b1ADG + b2MMWT + b3YOB + b4PEN + b5 STAGE + b6AGE + RFI,

where observed DMI is average daily dry matter intake over the measurement period, $\mu$ is the overall mean, b1-b6 are partial regression coefficients, ADG is average daily gain (kg/day), MMWT is metabolic mid-weight (kg), YOB is the year of birth, PEN is the pen, STAGE correspond to PW, hogget, or adult, and AGE is the age (days) at the start of the experiment. MMWT was calculated as the average between the liveweight at the start and the final of the test to the power of 0.75. RFI is the residual error of the equation. This model was used to estimate the RFI for the different life stages for all stages together, as preliminary analyses showed that a higher correlation between observed and predicted DMI was obtained when the three life stages were fitted together.

**Genomic prediction analysis.** Uni- and multivariate genomic best linear prediction (GBLUP) for RFI was performed with the R package ASReml-R (Butler *et al.* 2009). The univariate model fitted the trait at combined PW, hogget, and adult ages as a single trait with repeated measures. Additionally, univariate models for RFI at PW, hogget, and adult as different traits were also conducted. Those distinct traits were also included independently in a multivariate model. The number of records in the models was 876, 274, 384, and 218 for all ages as a single trait, PW, hogget, and adult, respectively. The first five principal components of the genomic relationship matrix were fitted as fixed effects to account for breed composition in all models.

**Measurement of accuracy and bias.** The accuracy and bias of genomic prediction were estimated for each univariate model using five-fold cross-validation. Initially, the animals were randomly grouped into five cohorts. One of the cohorts was used as a validation cohort by removing its RFI records from the dataset and training with the remaining four cohorts' RFI data and their genotypes. Prediction accuracy was calculated as the Pearson correlation between the GEBVs and the RFI phenotypes. The model bias was assessed as the regression slope of RFI on the GEBV. This was repeated with every validation cohort and averaged across cohorts.

## RESULTS AND DISCUSSION

**Genetic and phenotypic parameters.** The univariate model across all ages resulted in an RFI heritability of 0.19 (±0.04, Table 1). The heritabilities ($h^2$) at PW was higher in the univariate (0.72 ± 0.21) and multivariate (0.69 ± 0.22) models. The heritabilities of RFI at the hogget age were 0.40

± 0.16 in the univariate and 0.40 ± 0.15 in the multivariate models, consistent with the literature reported for growing animals. Most reported heritabilities of RFI in growing sheep are between 0.17 ±0.07, and 0.45 ±0.08 (François *et al.* 2002; Snowder and Van Vleck 2003; Paganoni *et al.* 2017; Hess *et al.* 2019; Tortereau *et al.* 2020). A lower heritability of 0.11 ±0.05 in growing sheep was also reported by Cammack *et al.* (2005), possibly because their RFI estimation did not adjust for metabolic weight.

**Table 1. Heritabilities and genomic prediction accuracies of RFI at PW, hogget, and adult ages**

| Age | Univariate models | | | Multivariate model *** | | |
|---|---|---|---|---|---|---|
| | $h^2$ ** | Accuracy | Bias | PW | Hogget | Adult |
| All ages * | 0.19 ± 0.04 | 0.22 ± 0.03 | 1.00 ± 0.19 | | | |
| PW | 0.72 ± 0.21 | 0.20 ± 0.19 | 0.69 ± 1.09 | 0.69 ± 0.22 | 0.29 ± 0.28 | 0.24 ± 0.43 |
| Hogget | 0.40 ± 0.16 | 0.24 ± 0.07 | 1.05 ± 0.71 | 0.18 ± 0.06 | 0.40 ± 0.15 | 0.50 ± 0.37 |
| Adult | 0.35 ± 0.27 | 0.11 ± 0.16 | 0.46 ± 1.12 | 0.00 ± 0.13 | 0.24 ± 0.07 | 0.37 ± 0.21 |

SE in brackets. * Univariate model fitted with RFI at all ages as a single trait with repeated measures. ** $h^2$ = narrow-sense heritability. *** $h^2$ in the main diagonal, genetic correlations (upper triangle), and phenotypic correlations (lower triangle).

The heritabilities in adults were 0.35 ± 0.27 and 0.37 ± 0.21 in the univariate and multivariate models, respectively. The heritability in adults being lower than in growing animals is consistent with the literature. In Australia, Merino sheep were measured at PW (n = 1,866), hogget (n = 1,010), and adult (n = 444) ages, with their heritabilities estimated at 0.29 ± 0.08, 0.17 ± 0.07, 0.07 ± 0.08, respectively (Paganoni *et al.* 2017). The heritabilities and their standard error in our study were larger than estimated by Paganoni *et al.* (2017), probably due to the smaller data set in our work.

The correlations had a large standard error, which suggests that the dataset was not large enough to obtain accurate correlations. The phenotypic correlations were 0.18 ± 0.06, 0.00 ± 0.13, and 0.24 ± 0.07 for PW-hogget, PW-adult, and hogget-adult, respectively. Similar correlations between the same ages in Australian Merino were reported, being 0.15 ± 0.03, 0.04 ± 0.05 and 0.33 ± 0.04, respectively (Paganoni *et al.* 2017). With the same population as used in our study, the phenotypic correlation between PW and hoggets born in 2014 was 0.20, and between hogget and adults born in 2013 was 0.17 (Muir *et al.* 2020b). These phenotypic correlations are similar, but not equal to the estimates reported here probably due to the differences in RFI estimation methods and the number of animals included in the analyses derived from the restriction per year Muir *et al.* (2020b) implemented.

The genetic correlation of RFI measured between PW and hogget, PW and adults, and hogget and adults were 0.29 ±0.28, 0.24 ±0.43, and 0.50 ±0.37, respectively. The genetic correlations for the same ages in Australian Merino were 0.36 ± 0.22, 0.00 ±0.53, and 0.75 ±0.74, respectively (Paganoni *et al.* 2017). The higher genetic correlation between hoggets and adults in Paganoni *et al.* (2017) agree with our results. It remains to be confirmed with a larger population whether RFI measurements in lambs are genetically correlated with measurements as adults.

**Genomic prediction accuracy and bias.** The all-ages single trait model (n = 876) achieved an accuracy of genomic prediction of 0.22 ±0.03, and the bias was 1.00 ±0.19. This model had a lower standard error and was less biased than the univariate models per age. The accuracy of RFI at PW (n = 274), hogget (n = 384), and adult (n = 218) were 0.20 ±0.19, 0.24 ±0.07, and 0.11 ±0.16, respectively and the bias values of those models were 0.69 ±1.09, 1.05 ±0.71, and 0.46 ±1.12, respectively. Adult RFI was less accurate than the other models, which is expected as RFI in adults had higher variance than RFI at PW and hogget ages.

In cattle, genomic selection for RFI is expected to improve feed efficiency, although the reference population size and its relationship to the predicted population are still factors that limit high reliabilities (Li *et al.* 2020). The reliability (~accuracy$^2$) of genomic predictions obtained in 3,947 Holstein cows in the USA with a 5-fold cross-validation was 0.34 (Li *et al.* 2020). The same study reported a high RFI reliability in the top 10 sires with most RFI daughters (0.85), but the reliability dropped to < 0.17 in the remaining animals (Li *et al.* 2020). In Australia, GEBVs for RFI were estimated in 4,106 unphenotyped Holstein sires from a reference population of 2,036 individuals obtained with a multi-trait GBLUP model and the reliability was 0.06 ±0.07. (Pryce *et al.* 2015).

## CONCLUSIONS

Our results suggest that it is feasible to develop genomic breeding values for RFI in the Maternal Composite ewes. However, expanding the training set is required in order to achieve higher accuracies and to confirm whether RFI in lambs and adults is the same trait.

## REFERENCES


Amadeu, R.R., Cellon, C., Olmstead, J.W., Garcia, A.A., Resende, Jr. M.F., and Muñoz, P.R. (2016) *Plant Genome - US* **9**: 1.

Bolormaa, S., Pryce, J., Kemper, K., Savin, K., Hayes, B., Barendse, W., Zhang, Y., Reich, C., Mason, B., and Bunch, R. (2013) *J. Anim. Sci.* **91**: 3088.

Butler D., Cullis B.R., Gilmour A. and Gogel B. (2009) ASReml-R reference manual. The State of Queensland, Department of Primary Industries and Fisheries, Brisbane.

Cammack, K.M., Leymaster, K.A., Jenkins, T.G., and Nielsen, M.K. (2005) *J. Anim. Sci.* **83:** 777.

François, D., Bibé, B., Bouix, J., Brunel, J., Weisbecker, J., and Ricard, E. (2002) *7th Proc. WCGALP.*, Montpellier, France.

Hess, M.K., Johnson, P., Knowler, K., Hickey, S., Hess, A., McEwan, J, and Rowe, S.J. (2019) *Proc. Assoc. Advmt. Anim. Breed. Genet.* **23**: 302.

Koch, R.M., Swiger, L.M., Chambers, D., and Gregory, K.E. (1963) *J. Anim. Sci.* **22**: 486.

Li, B., VanRaden, P., Guduk, E., O'Connell, J., Null, D., Connor, E., VandeHaar, M., Tempelman R., Weigel K., and Cole, J.B. (2020) *J. Dairy Sci.* **103**: 2477.

Meuwissen, T.H.E., Hayes, B.J., and Goddard, M.E. (2001) *Genetics* **157**: 1819.

Muir, S.K., Linden, N.P., Kennedy, A., Calder G., Kearney, G., Roberts, R., Knight, M.I., and Behrendt, R. (2020a). *Translational Animal Science.* **4**: 1006.

Muir, S.K., Linden, N., Kennedy, A., Knight, M., Paganoni, B., Kearney, G., Thompson, A., and Behrendt, R. (2020b). *Small. Ruminant Res.* **192**: 106241.

Paganoni, B., Rose G., Macleay C., Jones C., Brown D., Kearney G., Ferguson M., and Thompson A. N. (2017) *J. Anim. Sci.* **95**: 3839.

Pryce, J.E., Gonzalez-Recio O., Nieuwhof G., Wales W., Coffey M., Hayes B., and Goddard, M.E. (2015) *J. Dairy Sci.* **98**: 7340.

Pryce, J.E., Wales, W.J., de Haas, Y., Veerkamp, R.F., and Hayes, B.J. (2014) *Animal* **8**: 1.

Sargolzaei, M., Chesnais, J.P., and Schenkel, F.S. (2014) *BMC genomics* **15**: 478.

Snowder, G.D., and Van Vleck, L.D. (2003) *J. Anim. Sci.*, **81**: 2704.

Tortereau, F., Marie-Etancelin C., Weisbecker J.-L., Marcon D., Bouvier F., Moreno-Romieux C., and François, D. (2020) *Animal* **14**: 681.

Yang, J., Benyamin, B., McEvoy, B.P., Gordon, S., Henders, A.K., Nyholt, D.R., Madden, P.A., Heath, A.C., Martin, N.G., Montgomery, G.W., Goddard, M.E. & Visscher, P.M. (2010). *Nat. Genet.,* 42: 565.

*Proc. Assoc. Advmt. Anim. Breed. Genet: 24: 369-372*

# EXPLORING MACHINE LEARNING APPROACHES TO PREDICT THE INCIDENCE OF LAMENESS IN DAIRY COWS

**S. Shahinfar[1*], M. Khansefid[1], M. Haile-Mariam[1] and J.E. Pryce[1,2]**

[1]Agriculture Victoria Research, AgriBio, Centre Bundoora, Victoria 3083, Australia
[2]School of Applied Systems Biology, La Trobe University, Bundoora, Victoria 3083, Australia

## ABSRTACT

In this study, we investigated the ability of three machine learning algorithms, Naïve Bayes (NB), Random Forest (RF) and Multilayer Perceptron (MLP), in the prediction of cases of lameness. Performance of these algorithms were compared with logistic regression (LR) as the gold standard approach for binary classification. There were negligible differences between LR, NB and RF, while MLP underperformed the other three methods. However, the F1-score in NB (22%) outperformed LR (11%), suggesting NB potentially could be a more reliable method for prediction of lameness in practice if there is enough relevant data available for proper training.

## INTRODUCTION

Lameness along with mastitis and fertility problems are the most prevalent health issues in dairy cattle which have detrimental effects on the welfare and economic performance of the cows (Bruijnis *et al.* 2010). The direct economic impact of lameness which includes the costs of treatment and early culling are evident. However, the effects of lameness on reduced milk yield and impaired fertility are less obvious but have large contribution in total economic loss due to lameness incidence (Green *et al.* 2002; Huxley 2013).

Genetic improvement to reduce lameness is difficult because the accuracy of lameness predictions is often low. Considering the complexity of prediction of lameness incidence, machine learning (ML) was shown to have promise to detect the risk level of lameness at the herd level according to 20 routinely pre-collected farm based records related to management, housing, production, reproduction, longevity and genetics merits (Warner *et al.* 2020).

Predicting lameness incidence at the cow-level can help farmers detect susceptible cows (high risk category). Hence, the objective of this study was to evaluate the usefulness of ML approaches in the prediction of lameness incidence and compare it with classic binary classification method.

## MATERIAL AND METHODS

**Data.** Lameness scores, milk production and conformation traits data were collected from 11 Australian dairy farms in spring 2018. The lameness scoring was performed by trained classifiers after morning milking according to Dairy Australia guidelines[1]; where 0=walking evenly, 1=walking unevenly, 2=moderate difficulty in walking and 3=severe lameness. In this study, cows were classified to either sound (score 0) or unsound (score 1-3) group because there were a limited number of cows with non-zero scores. The milk production traits were test-day milk yield, fat, protein and lactose percentage as well as somatic cell count (SCC) measured within a week before the lameness scoring visit. Further, we also investigated the following potential predictors in our study; breed, parity, age at calving (in months), age at lameness scoring visit (in months), days in milk (DIM) at lameness scoring and test-day visit.

Any column or row with more than 50% missing values was excluded. The remaining data comprised 2,640 cows in 11 herds with records of lameness and 42 predictor features. Missing

[1] https://www.dairyaustralia.com.au/dairytas/animal-management-and-milk-quality/animal-health/lameness

values for about 30% of lactose percentage and parity number were imputed using rfImpute procedure from randomForest packages in R (Liaw and Wiener 2002). Feature selection was performed using combination of mean reduction in Gini index and mean decrease in accuracy from randomForest Package combined with the potential predictor traits reported in previous literature (Solano *et al.* 2015; Ranjbar *et al.* 2016; O'Connor *et al.* 2020). In total 31 features were selected as predictors of lameness incidence. Four of these features were categorical; breed (Holstein, Jersey, Holstein × Jersey and Holstein × non-Jersey crossbreds); herd (11 levels); parity (1, 2, 3, 4, and 4+); and month of calving (MOC; 12 levels). The summary statistics of the rest of the predictors used in this study is provided in Table 1.

**Lameness prediction.** Three machine learning methods were used in this study and their performance was compared with the classic binary prediction method, logistic regression (LR). *Multilayer Perceptron (MLP)* is a feedforward artificial neural network that calculates a sequential linear combination of inputs into a set of appropriate outputs via its hidden layers and activation functions (Mitchell 1997). Package 'h2o' in R was used for this purpose (LeDell *et al.* 2020). *Naïve Bayes (NB)* is one of the most efficient and effective inductive learning algorithms for machine learning and data mining. It is a statistical classifier based on Bayes rule (Domingos and Pazzani 1997). Package 'e1071' in R was used for this purpose (Meyer *et al.* 2019). *Random Forest (RF)* is one of the ensemble prediction methods in which predictor trees are trained on bootstrap samples drawn from the training data (Ho 1995; Breiman 2001). Package 'randomForest' in R was used for this purpose (Liaw and Wiener 2002).

Hyper-parameter tuning was conducted via a grid search on 50% of randomly selected data. Training and testing of models were performed using 10-fold cross validation and repeated 10 times. Performance metrics were aggregated. The entire training and validation process was conducted in R v4.0.2 programming language (R-Core-Team 2020).

## RESULTS AND DISCUSSION

Table 2 shows model performance metrics for algorithms used in this study to predict incidence of lameness. There was not a consistent best performer among algorithms used to predict lameness. In terms of accuracy (ACC) and precision (PRE) LR outperformed the ML algorithms at 0.86 and 0.28 respectively. Among the ML algorithms, MLP had the lowest false positive rate (FPR) at 0.04, however, it had a high standard deviation in performance. Considering true positive rate (TPR), it was NB that outperformed the other methods with a relatively low standard deviation (0.26). As the current study encountered an unbalanced classification problem (unbalanced numbers of lame to sound cows), using F1 score (harmonic average of precision and recall) was a more suitable metrics for comparing different classification algorithm. The naïve Bays classifier had the highest TPR and F1 score (0.22) and moderate precision relative to other tested algorithms. In real life different types of misclassification error varies in cost, without considering those costs, identifying the optimum classifier is not possible (Shahinfar *et al.* 2015). In the absence of misclassification cost, we base our classifier selection on F1-score.

The Area under ROC curve (AUC) indicates the overall performance of classifier asymptotically. In the current study LR had the highest AUC at 0.65 followed by NB (AUC= 0.63). Warner *et al.* (2020), reported AUC = 0.73-0.75 for risk prediction of lameness at the herd level.

Considering all the performance criteria, NB had significantly higher F1 Score compare to LR, therefore NB would be the recommended algorithm to predict incidence of lameness. Nevertheless, NB still misclassified a large proportion of animals (i.e. high FPR and low PRE). This sub-optimal performance can be firstly due to the fact that the training data set was limited in size and highly imbalanced; and secondly, lameness is indeed a very complex trait affected by genetics, environment and management factors such as nutrition, production level, bedding, weather, walking track, laneway quality and pasture condition (Ranjbar *et al.* 2016; O'Connor *et al.* 2020). Thus, for an accurate prediction of lameness incidence, a very comprehensive dataset of management factors affecting lameness (both at farm and animal level) is needed, which is often not accurately and consistently collected in dairy farms (O'Connor *et al.* 2020).

**Table 1. Summary of data used in this study**

| trait | mean | sd | min | max | mean decrease accuracy | mean decrease Gini |
|---|---|---|---|---|---|---|
| Age at calving | 47.71 | 22.33 | 22 | 161 | 6.44 | 22.4 |
| Age at lameness scoring | 52.33 | 22.72 | 23 | 162 | 7.36 | 21.9 |
| BCS | 3.59 | 0.76 | 1.0 | 8.0 | 3.28 | 15.44 |
| Dairy strength | 11.02 | 1.67 | 3 | 16 | 3.68 | 14.3 |
| Feet & legs | 10.52 | 1.55 | 3 | 15 | 2.35 | 12.83 |
| Mammary system | 10.28 | 1.35 | 5 | 14 | 1.77 | 11.02 |
| Overall type | 9.88 | 1.32 | 1 | 13 | 2.98 | 10.81 |
| Rump | 10.88 | 2.12 | 1 | 16 | 2.1 | 17.57 |
| DIM at lameness scoring | 138.68 | 145.25 | 1 | 485 | 8.37 | 22.45 |
| DIM at milk test-day | 115.75 | 105.56 | 2 | 314 | 8.67 | 27.06 |
| Fat % | 3.85 | 0.97 | 1.13 | 9.84 | 4.3 | 27.72 |
| Lactose% | 5.05 | 0.26 | 3.61 | 5.84 | 7.26 | 29.04 |
| Angularity | 5.55 | 0.98 | 2 | 8 | 4.62 | 10.43 |
| Body depth | 6.03 | 1.09 | 2 | 9 | 2.36 | 9.98 |
| Bone quality | 6.8 | 1.11 | 1 | 9 | 0.22 | 12.24 |
| Median suspensory | 6.4 | 1.07 | 2 | 9 | 3.19 | 9.86 |
| Foot angle | 5.36 | 0.92 | 2 | 9 | 1.41 | 10.29 |
| Heel depth | 5.64 | 0.83 | 2 | 9 | 1.5 | 11 |
| Loin strength | 6.34 | 0.91 | 2 | 9 | 2.52 | 10.77 |
| Pin width | 6.24 | 1.32 | 1 | 9 | 2.11 | 14.5 |
| Rear attachment width | 5.63 | 1.31 | 1 | 9 | 5.1 | 12.38 |
| Rear legs - rear view | 5.92 | 1.03 | 1 | 9 | 1.3 | 13.99 |
| Stature | 6.29 | 1.51 | 1 | 9 | 1.46 | 12.19 |
| Udder depth | 5.31 | 1.39 | 1 | 9 | 2.13 | 13.36 |
| Milk yield | 27.25 | 8.93 | 32 | 606 | 8.23 | 27.08 |
| Protein % | 3.44 | 0.38 | 2.00 | 5.86 | 3.34 | 27.41 |
| SCC | 129.84 | 477.43 | 1 | 9590 | 2.88 | 26.81 |
| Breed | * | * | * | * | 1.21 | 2.1 |
| Herd | * | * | * | * | 12.18 | 27.86 |
| Parity | * | * | * | * | 3.59 | 9.15 |
| MOC | * | * | * | * | 5.46 | 19.39 |

'*'these features were considered as factor

**Table 2. Model performance metrics for algorithms used in prediction of incidence of lameness in dairy cows. ACC=Accuracy; PRE=Precision; TPR=True Positive Rate; FPR=False Positive Rate; F1 = F1 scores**

| algorithm | ACC | PRE | TPR | FPR | F1-score |
|---|---|---|---|---|---|
| LR | **0.86(0.032)[ab]** | **0.28(0.072)[a]** | 0.09(0.086)[b] | **0.04(0.046)[ab]** | 0.11(0.072)[b] |
| MLP | **0.88(0.022)[a]** | 0.17(0.173)[b] | 0.03(0.038)[c] | **0.02(0.029)[a]** | 0.04(0.045)[c] |
| NB | 0.80(0.036)[c] | 0.20(0.015)[b] | **0.26(0.070)[a]** | 0.14(0.048)[c] | **0.22(0.020)[a]** |
| RF | 0.84(0.057)[b] | **0.22(0.097)[ab]** | 0.13(0.130)[b] | 0.07(0.080)[b] | 0.12(0.084)[b] |

* The values with different superscript letters in each column are significantly different ($p<0.05$) according to Tukey-HSD multiple comparison test.

## CONCLUSION

Prediction of incidence of lameness in dairy cattle is a difficult task. Multiple environmental effects influence lameness and their interactions and causal-effect pathways are often not considered in lameness prediction. Prediction of incidence of lameness on the cow level is possible with Naive Bayes classifier and logistic regression. Lack of a comprehensive dataset was the main limitation of this study. Although the classification performance was suboptimal in our study, we expect additional information on the herd level such as bedding, nutrition, and weather will improve prediction accuracy. Nevertheless, this study provided proof of concept for prediction of lameness at the cow level.

## REFERENCES

Breiman L. (2001) *Mach. Learn.* **45:** 5.

Bruijnis M.R.N., Hogeveen H., Stassen E.N. (2010) *J. Dairy Sci.* **93:** 2419.

Domingos P., Pazzani M.J. (1997) *Machine Learning* **29:** 103.

Green L.E., Hedges V.J., Schukken Y.H., Blowey R.W., Packington A.J. (2002) *J. Dairy Sci.* **85:** 2250.

Ho T.K. (1995) *in Proc. 3rd ICDAR, Montreal, QC, Canada* 278.

Huxley J.N. (2013) *Livestock Science* **156:** 64.

LeDell E., Gill N., Aiello S., Fu A., Candel A., Click C., Kraljevic T., Nykodym T., Aboyoun P., Kurka M., Malonhava, M. (2020)

Liaw A., Wiener M. (2002) *R News* **2:** 18.

Meyer D., Dimitriadou E., Hornik K., Weingessel A., Leisch F. (2019)

Mitchell T.M., McGraw-Hill: New York. (1997)

O'Connor A.H., Bokkers E.A.M., de Boer I.J.M., Hogeveen H., Sayers R., Byrne N., Ruelle E., Shalloo L. (2020) *J. Dairy Sci.* **103:** 9238.

O'Connor A.H., Bokkers E.A.M., de Boer I.J.M., Hogeveen H., Sayers R., Byrne N., Ruelle E., Engel B., Shalloo L. (2020) *Preventive Veterinary Medicine* **181:** 105077.

R-Core-Team (2020) 'R: A language and environment for statistical computing.' (R Foundation for Statistical Computing: Vienna, Austria)

Ranjbar S., Rabiee A.R., Gunn A., House J.K. (2016) *J. Dairy Sci.* **99:** 7495.

Shahinfar S., Guenther J.N., Page C.D., Kalantari A.S., Cabrera V.E., Fricke P.M., Weigel K.A. (2015) *J. Dairy Sci.* **98:** 3717.

Solano L., Barkema H.W., Pajor E.A., Mason S., LeBlanc S.J., Zaffino Heyerhoff J.C., Nash C.G.R., Haley D.B., Vasseur E., Pellerin D., Rushen J., de Passillé A.M., Orsel K. (2015) *J. Dairy Sci.* **98:** 6978.

Warner D., Vasseur E., Lefebvre D.M., Lacroix R. (2020) *Comput. Electron. Agric.* **169:** 105193.

*Proc. Assoc. Advmt. Anim. Breed. Genet: 24: 373-376*

# NEW MODEL FOR GENETIC EVALUATION OF FERTILITY IN NEW ZEALAND DAIRY CATTLE

**K. Stachowicz[1], P.R. Amer[1], B. Falconer[2] and D.P. Garrick[3,4]**

[1]AbacusBio Limited, PO Box 5585, Dunedin 9058, New Zealand
[2]DairyNZ Limited, Private Bag 3221, Hamilton 3240, New Zealand
[3]Theta Solutions LLC, Lacey, WA 98516, USA
[4]A.L. Rae Centre for Genetics and Breeding, Massey University, New Zealand

## SUMMARY

The current genetic evaluation of female fertility in New Zealand (NZAEL2.0) is based on the binary phenotype of calving rate in the first 6 weeks of the calving season. Recent research suggests that using a continuous phenotype and including a heifer calving trait would increase the accuracy of fertility breeding values. This paper describes and investigates these factors in the context of a new fertility model proposed for New Zealand (NZAEL3.0). Industry implementation steps and results and comparisons of the validation work undertaken on NZAEL2.0 and NZAEL3.0 are presented. The results of the validation undertaken show that NZAEL3.0 is substantially better at predicting future fertility phenotypes compared to NZAEL2.0.

## INTRODUCTION

In New Zealand, fertility is currently evaluated in NZAEL2.0 using two binary traits measured in first parity: calving rate within 6 weeks (CR) from planned start of calving (PSC) and percentage mated within 3 weeks (PM) from planned start of mating (PSM). This model has been substantially reduced compared to the previous model (NZAEL1.0) which used data across parities 1 to 3 (Harris *et al.* 2005). The reduction was to enable computational feasibility of the model when used in conjunction with genomic data. Research undertaken since 2013 indicated that redefining fertility as a continuous calving season day trait (CSD, Table 1) instead of using binary scores as per the current trait definition would increase the accuracy of the fertility evaluation (Bowley *et al.* 2015; Stachowicz *et al.* 2014a; Stachowicz *et al.* 2014b). This research has been summarised by Stachowicz *et al.* (2015), where rationale and recommendations for the proposed changes have been provided. The proposed model would consist of seven traits: four CSD and three PM traits. However, with application of genomic methodology such a model would not be computationally feasible. Therefore, a number of reduced models were tested (data not shown) and a seven-trait three breeding value model, where CSD or PM traits in parities 1 to 3 are modelled with the same breeding value, was selected for further assessment. Variance components estimated for this selected model as well as validation of the NZAEL2.0 and NZAEL3.0 models are presented in this paper.

## MATERIALS AND METHODS

The proposed model for genetic evaluation of fertility consists of seven traits: CSD0, CSD1, CSD2, CSD3, PM1, PM2, and PM3, and three breeding values (EBVs): CSD0, CSD123, and PM123 (Table 1), where CSD1-3 and PM1-3 are records for 2$^{nd}$ to 4$^{th}$ calvings and matings leading to those calvings. For the purposes of variance component estimation, the data and the model described by Amer *et al.* (2016) and Stachowicz *et al.* (2015) was employed but fit with a repeated records model for CSD123 and PM123. Variance component estimation was carried out using ASReml software (Gilmour *et al.* 2009). NZAEL3.0 evaluations were carried out using BOLT software (Garrick *et al.* 2018) where the required 7x7 residual covariance matrix was derived from the estimated residual and permanent environment variance components estimated by the repeatability model fit in ASReml. The 3x3 additive genetic covariance matrix remained unchanged.

Current testing has focused on a lifetime fertility estimated breeding value:

Fert EBV = -1*(0.28*CSD0 + 1.32*CSD123)

where the coefficients applied to CSD0 and CSD123 are based on a relativity determined by the number of discounted expressions, and a scale based on conversion of CSD to CR scale.

The following validation approach was used. The genetic evaluation model was run on national data but with the last four seasons of phenotypic data removed (2016-2019). The full pedigree of ~26.7 million animals including those for cows with removed phenotypes was retained. Fertility records were extracted for 200 herds (approximately 19,000 cows). These cows were all born in 2014 and were sourced from 100 sire proving scheme herds of two breeding companies (CRV & LIC) and a further 100 herds identified as having high data quality scores. Table 1 describes in detail the phenotypes that were derived for these animals.

The validation approach involved classifying these ~19,000 cows into quintiles (5 groups of equal size) based on a specific genetic evaluation option where records from 2016 onwards are excluded (set to missing). Thus, we have 5 groups of animals ranging from high parent average prediction for fertility to low parent average prediction for fertility. We then fit each of the validation phenotypes (Table 1) from the 19,000 cows as a dependent variable in a model with quintile, contemporary group associated with the validation phenotype, and age at corresponding calving also fitted as independent variables. Least Squares Means (LSM) for quintile groups are then compared. A higher separation between the 1st and the 5th quintile group LSMs of validation phenotype values indicates a better performing genetic evaluation system for the target traits of interest.

**Table 1. Abbreviations and trait definitions of validation phenotypes**

| Abbreviation | Trait definition |
|---|---|
| CR0 | Calving rate for heifers, =1 (success) if calved in the first 6 weeks from PSC (binary) |
| CR1-3 | As above but for cows' 2nd to 4th calving |
| TCD1-3 | Timing of conception, days from PSM until the last recorded (successful) mating date after 1st to 3rd calving |
| 3wICR1-3 | 3 week in-calf rate, =1 (success) if conceived within 3 weeks from PSM after 1st to 3rd calving (binary) |
| 6wICR1-3 | 6 week in calf rate, =1 (success) if conceived within 6 weeks from PSM after 1st to 3rd calving (binary) |
| CSD0 | Calving season day for heifers, days from PSC to 1st calving |
| CSD1-3 | Calving season day for cows, days from PSC to calving for 2nd to 4th calving |
| PM1-3 | Presented for mating in the 1st 3 weeks from PSM (binary, 1=success) after calvings 1 to 3 |
| GL1-3 | Gestation length culminating at 2nd to 4th calving |

## RESULTS AND DISCUSSION

The estimated variance components for the proposed model are presented in Table 2. Heritabilities and genetic correlations from the repeated records model tended to be higher than corresponding values from the multiple trait model (Stachowicz *et al.* 2015).

**Table 2. Heritabilities (repeatabilities; on diagonal) and genetic correlations (off diagonal) for proposed repeated records model for genetic evaluation of fertility**

| | CSD0 | CSD123 | PM123 |
|---|---|---|---|
| CSD0 | 0.034 | | |
| CSD123 | 0.686 | 0.047 (0.272) | |
| PM123 | -0.525 | -0.828 | 0.069 (0.143) |

The results of the validation work are presented in Tables 3-5. Table 3 shows the phenotypic performance contrast between the 1st and 5th quintiles when validation heifers were assigned to these based on alternative parent average genetic predictions. The NZAEL3.0 fertility EBV outperformed the NZAEL2.0 fertility EBV for predicting 6-week in-calf rate, with the difference being substantial (i.e. roughly twice as good) for later parities. Typically, second calving cows constitute a little under 20% of the herd, and so when it comes to predicting whole herd fertility, a paradigm shift in accuracy of fertility genetic merit prediction should be expected with the new approach. There was less difference between the methods for predicting submission rate than there was for in-calf rate after three weeks (Table 4). The NZAEL3.0 fertility EBV was substantially better than the NZAEL2.0 EBV for predicting the ability of cows to calve earlier in the season (Table 5), and also to conceive earlier in the season.

**Table 3. Phenotypic performance difference between animals assigned to their 1st versus 5th quintile based on alternative parent average (PA) EBVs taken from alternative evaluation validation runs – Six week in calf rate (6wICR) and gestation length (GL)**

| Fertility EBV | PA Spread[1] Proposed | PA Spread[1] Current | 6wICR1 | 6wICR2 | 6wICR3 | GL1 | GL2 | GL3 |
|---|---|---|---|---|---|---|---|---|
| NZAEL 3.0 | 10.63 | 2.92 | 0.076 | 0.070 | 0.070 | -0.9 | -1.2 | -1.2 |
| NZAEL 2.0 | 6.41 | 4.57 | 0.068 | 0.048 | 0.031 | -0.7 | -0.6 | -0.5 |
| CSD0 | -7.24 | -1.81 | -0.048 | -0.041 | -0.035 | 1.3 | 1.1 | 1.3 |

[1]PA spread for current and proposed EBV indicate the predicted differences in parent average based spread between the first and 5th quintiles of validation animals for the proposed new fertility breeding value, and for the current fertility breeding value, respectively.

The NZAEL3.0 fertility EBV will be expected to bring heifer calving forward (e.g. 1.5 day difference and 1.5% more calving in the first 6 weeks between the 1st and 5th quintiles when compared with the NZAEL2.0 EBV in Table 4), with about 1/3 of the difference (i.e. about 0.5 days when weighted across parities) attributable to shorter gestation length (Table 3). In general, the relationship between fertility EBV and gestation length is only slightly stronger for NZAEL3.0 than for NZAEL2.0 EBV. For example, NZAEL3.0 gives a 66% increase in separation for 6-week in calf rate weighted across parities, and this corresponds to a 100% increase in separation for gestation length (Table 3). In general, it seems likely that farmers would accept a 1.2 day shorter average gestation length as part of a genetic package that gives a 7% gain in 6-week in calf rate. Further, a comprehensive analysis of NZ data by Jenkins *et al.* (2016) indicated that the disadvantages at the extreme short end of the gestation length scale from shortening population average gestation length are more than offset by the gains from reduction in calvings at the extreme long gestation length end of the scale. While earlier born heifer calves have one less day in utero, they have one more day post-partum before the commencement of the heifer mating season.

**Table 4. Phenotypic performance difference between animals assigned to their 1st versus 5th quintile based on alternative parent average EBVs taken from alternative evaluation validation runs – Three week in calf (3wICR) and three week submission rates (PM)**

| Fertility EBV | 3wICR1 | 3wICR2 | 3wICR3 | PM1 | PM2 | PM3 |
|---|---|---|---|---|---|---|
| NZAEL 3.0 | 0.114 | 0.134 | 0.103 | 0.107 | 0.119 | 0.102 |
| NZAEL 2.0 | 0.074 | 0.091 | 0.063 | 0.092 | 0.101 | 0.094 |
| CSD0 | -0.069 | -0.071 | -0.058 | -0.042 | -0.058 | -0.043 |

The alignment between the difference in EBVs between quintiles, and the difference in phenotypic performance for calving by 6 weeks was approximately 1:1 for the NZAEL3.0 EBV (compare 10.63% difference in PA in Table 3 with .104 to .125 proportional differences observed in CR1, CR2 and CR3 in Table 4). In comparison, the 6.41% difference in PA from the NZAEL2.0 EBV in Table 3 under predicted the .07 to .09 differences in CR1 observed phenotypically).

Results for a univariate prediction of breeding values for calving season day for first calving heifers have been included in Tables 3 to 5. Low values (earlier calving) are better for this trait, so the signs of the phenotypic differences between the 1st and the 5th quintiles is negative for validation phenotypes where positive is favourable. These roughly reflect the likely outcome of evaluating bulls based on the first calving dates of their first crop of daughters at their first calving using the proposed fertility evaluation. This would give an indication of fertility approximately 3 months earlier than evaluations based on submission rate after 1st calving. The degree of separation in validation phenotypes with selection on CSD0 EBVs from the NZAEL3.0 model is very encouraging, albeit not quite as good as the NZAEL2.0 model, and only 50 to 60% as effective as when using the full evaluation that incorporates all phenotypes. While these evaluations, based on first calving date records only, will have a stronger association with shorter gestation length (Table 3), the values are not of sufficient magnitude to be of concern.

**Table 5. Phenotypic performance difference between animals assigned to their 1st versus 5th quintile based on alternative parent average EBVs taken from alternative evaluation validation runs – Calving traits (CR and CSD)**

| Fertility EBV | CR0 | CR1 | CR2 | CR3 | CSD0 | CSD1 | CSD2 | CSD3 |
|---|---|---|---|---|---|---|---|---|
| NZAEL 3.0 | 0.047 | 0.116 | 0.104 | 0.125 | -4.13 | -6.93 | -6.89 | -5.97 |
| NZAEL 2.0 | 0.032 | 0.091 | 0.084 | 0.074 | -2.58 | -5.66 | -5.27 | -3.95 |
| CSD0 | -0.041 | -0.083 | -0.071 | -0.073 | 3.65 | 5.04 | 4.86 | 4.19 |

## CONCLUSIONS

This study shows that the proposed NZAEL3.0 model for genetic evaluation of fertility outperforms the current NZAEL2.0 model for all of the phenotypes tested. Higher accuracies of fertility breeding values would be expected with the NZAEL3.0 model as well as the improvement in early predictions due to the inclusion of the heifer calving trait into the evaluation. The increased indirect response in shortening of gestation length could become a concern over time, and future work is planned to identify a new phenotype for evaluation based on pregnancy diagnosis, which targets earlier conception and is independent from gestation length.

## REFERENCES

Amer P.R., Stachowicz, K., Jenkins, G.M. and Meier S. (2016) *J. Dairy Sci.* **99**: 8227.

Bowley F.E., Green R.E., Amer P.R. and Meier S. (2015) *J. Dairy Sci.* **98**: 2005.

Garrick D.J., Garrick D.P. and Golden B.L. (2018) *Proceedings of the 11th World Congress on Genetics Applied to Livestock Production,* **11**: 973.

Gilmour A.R., Gogel B.J., Cullis B.R. and Thompson R. (2009) VSN International Ltd, Hemel Hempstead, HP1 1ES, UK. www.vsni.co.uk.

Harris B.L., Pryce J.E., Xu Z.Z. and Montgomerie W.A. (2005) *Interbull Bull.* **33**: 47.

Jenkins G.M., Amer P., Stachowicz K. and Meier S. (2015) *J. Dairy Sci.* **98**: 418.

Stachowicz K., Jenkins G.M., Amer P.R. and Meier S. (2014a) *Interbull Bull.* **48**: 89.

Stachowicz K., Jenkins G.M., Dennis N.A., Amer P.R., Bryant J. and Meier S. (2014b) *Proc. of the 5th Australasian Dairy Science Symposium*, pp 65.

Stachowicz K., Jenkins G.M., Amer P.R., Bryant, J.R. and Meier S. (2015) *Proc. Assoc. Advmt. Anim. Breed. Genet.* **21**: 370.

*Proc. Assoc. Advmt. Anim. Breed. Genet: 24: 377-381*

# NEXT GENERATION FEED SAVED AUSTRALIAN BREEDING VALUES EVALUATED IN HOLSTEIN DAIRY CATTLE

**S. Bolormaa[1], I.M. MacLeod[1], M. Khansefid[1], L. Marett[2,3], B. Wales[2,3], G.J. Nieuwhof[1,4], C.F. Baes[5,6], F.S. Schenkel[5], M.E. Goddard[1,7] and J.E. Pryce[1,8]**

[1] Agriculture Victoria Research, Agribio, 5 Ring Road, Bundoora, VIC, 3083 Australia
[2] Agriculture Victoria Research, Ellinbank Centre, Ellinbank, Gippsland, VIC, 3821 Australia
[3] School of Agriculture and Food, University of Melbourne, Parkville, VIC 3010, Australia
[4] DataGene Ltd, Agribio, 5 Ring Road, Bundoora, VIC, 3083 Australia
[5] CGIL, University of Guelph, Guelph, ONT, N1G 2W1 Canada
[6] Institute of Genetics, Vetsuisse Faculty, University of Bern, Bern, 3002, Switzerland
[7] School of Land & Environment, University of Melbourne, Parkville, VIC, 3052 Australia
[8] School of Applied Systems Biology, La Trobe University, Bundoora, VIC, 3083 Australia

## SUMMARY

The aims of this study were to: 1) update the Australian (dairy) breeding value (ABV) for lifetime residual feed intake (RFIlife, covering RFI at the growth and lactation stages) using 3,711 Holstein female records (584 Australian cows, 824 Australian heifers and 2,440 foreign cows) using a multivariate model and 2) re-evaluate the Feed Saved ABV in Holstein (HOL) bulls. Cow numbers have doubled compared to the original 2015 Feed Saved ABV model. Genomic heritability estimates of RFI were 0.18, 0.27, and 0.36 for Australian (AUS) and overseas (OVE) cows, and AUS heifers, respectively. The genetic correlations were 0.47 between AUS cow and heifer traits and 0.94 between AUS and OVE cow traits, but these estimates were associated with large standard errors. The standard deviation of Feed Saved (FS) ABVs in HOL bulls was 79 kg/yr. The reliability of the residual feed intake component of Feed Saved increased from 11% to 20%. The next step in calculating FS is to combine RFI ABVs with maintenance requirements estimated using bodyweight ABVs. The overall reliability of FS ABVs has increased from 33% to 43% on average. The correlations of RFIlife and FS ABV between the prediction equations of 2015 and 2020 in 20k Holstein bulls (born from 2010 to 2020) were 0.65 and 0.80, respectively. We conclude that expanding the reference population, especially with inclusion of the international data, has improved the reliability of feed efficiency EBVs.

## INTRODUCTION

Feed costs make up a large proportion of the variable and total costs on a dairy farm and improving production efficiency remains a key breeding objective. The dairy industry has seen tremendous gains in milk yield, without a proportional increase in maintenance requirements, leading to an improvement in gross efficiency (Pryce *et al.* 2018). However, further improvements can be achieved through genomic selection for residual feed intake (RFI), defined as the difference between actual and predicted feed intake. In 2015, DataGene released the world's first Feed Saved breeding value (FS ABV) to the dairy industry, which includes the genetic component of RFIlife combined with the maintenance requirements calculated from liveweight EBV. The FS ABV has been incorporated in the Balanced Performance Index (BPI) to select for overall economic merit (Pryce *et al.* 2015). Selecting animals based on the best FS ABV, especially in combination with the BPI, is expected to reduce energy requirements for similar amounts of milk production.

Genomic prediction for RFI that was used to calculate FS ABV in 2015 was developed using a small reference population (n = 2,036) including 234 Australian cows. We have doubled the number of AUS cows with genotypes and phenotypes for RFI and additionally have had access to a larger dataset of non-Australian cows by participating in the Efficient Dairy Genome Project

(EDGP; an international database including research herds from Europe and North America). These 2 data sources have provided an opportunity to increase the size of reference population and hence to update the FS ABV. The aim of this study was to estimate the prediction equations of 50k SNP effects for RFIlife, to reassess the ABV and its reliability for FS, and to compare with the 2015 ABV.

## MATERIALS AND METHODS

A total of 3,711 animals were used in this study including: 584 lactating Australian cows and 824 heifers and 2,440 OVE cows (USA, Canada (CAN), Netherlands (NLD), United Kingdom (UK), Denmark (DNK), and Switzerland (CHE)). The genotypes and phenotypes of the AUS and OVE cows except NLD and UK were downloaded from EDGP database, while the NLD and UK data were part of the original dataset used in the development of the 2015 FS ABV (Pryce *et al.* 2015). Additionally, the genotypes of approximately 20,000 Holstein bulls born between 2010 and 2020 were received from DataGene (Melbourne, Australia). The genotypes of cows from the EDGP database were on a variety of medium to HD SNP chips, and sporadic missing genotypes were filled using FImpute (Sargolzaei *et al.* 2014). In total, 41,276 SNP were in common between cow, heifer and bull data sets. This SNP set was chosen to conform to DataGene's national genomic evaluation for dairy cattle (i.e. the same set is used for all traits), which is based on UMD3.1 reference genome map positions. However, all the imputation of the genotypes was undertaken using ARS-UDC1.2 reference genome map positions. Before merging each country's genotypes, the allele frequency of each SNP in each country was checked to ensure that the homozygotes were likely to be in the same direction. The genomic relationship matrix (GRM) was constructed based on the 41,276 genotypes with or without the 3,413 AUS HOL bull genotypes using the method of Yang *et al.* 2010.

All 3,711 animals had milk production traits, energy corrected milk (ECM) and dry matter intake (DMI) data available on most days over a 28-day period, starting at a mean minimum of 5 days in milk (DIM). Trait deviations for RFI in AUS heifers were previously calculated as means of the difference in actual and predicted DMI that was measured over a 6-7-week period at heifers of around 6 months of age (Pryce *et al.* 2015). RFI for AUS cows was calculated based on the average DMI over the 28-day experimental period using the same model described in Pryce *et al.* (2015). The phenotypes of RFI for OVE cows were calculated as RFI = DMI – (mean + parityST + DIM + HYS + poly(age,-2) + trial + ECM + BWT + ΔBWT), where DMI is the daily dry matter intake (DMI). Energy corrected milk (ECM), mean body weight (BWT), daily BWT change (ΔBWT), days in milk (DIM), and age of cows (poly(age,-2) were all fitted as covariates. Daily BWT change (ΔBWT) was calculated by fitting fifth-order orthogonal polynomial regression on DIM (5 to 206 DIM) to daily BWT, and then ΔBWT was calculated as the difference in predicted BWT between consecutive days. The fixed effects in the OVE cows were parity stage (parityST), herd-year-season (HYS), and trial (diets).

A trivariate GREML analysis, where the traits were RFI in AUS cows and heifers and OVE cows, was used to calculate genetic correlations between RFI traits and GEBV. Prediction of SNP effects for RFI cow and RFI heifer was $\hat{\boldsymbol{\beta}} = \mathbf{Z}'(\mathbf{ZZ}')^{-1}\hat{\mathbf{g}}$, where $\mathbf{Z}$ is the $n \times 41{,}276$ matrix of the genotypes of 3,711 animals in the reference set, and $\hat{\mathbf{g}}$ is the descaled DGVs for the trait RFI in Australian cows. Prediction equations of SNP effects were used to predict breeding values of 3,413 Holstein bulls that overlapped with the data used in 2015. RFI DGV for AUS cows and heifers were combined to produce a genomic breeding value of RFIlife covering the growth and lactation stages. Then FS ABV was calculated by subtracting RFIlife from the amount of the feed required to maintain 1kg of extra body weight per year (Feed_BW_kg). The Feed_BWT_kg is a function of BWT and calculated as $EV_{BWT}$ * ($EBV_{BWT}$ - 100) / ( feedcost * MJME ), where $EV_{BWT}$ (economic value of maintenance) is A$5.14, feedcost (the cost of feed in MJ) is

AUS$0.032/MJ, and MJME (the energy content of feed) is 11.9 MJ/kg of DMI. Details of the calculation of FS ABV and its reliability are described in Pryce *et al.* (2015).

## RESULTS AND DISCUSSION

Heterozygosity predicted from GRM was compared with mean observed heterozygosity per country and heterozygosity assuming Hardy-Weinberg equilibrium. There was good concordance between these population measures, with all genotype groups displaying a similar range of heterozygosity (0.32-0.34), showing that the GRM constructed using animals from different groups is a good representation of the relationships between and within group of animals.

Phenotypic standard deviations of the RFI phenotypes were 0.42 kg/d for AUS heifers, AUS cows, and OVE cows were 1.28 kg/d and 1.82 kg/d, respectively. The single trait and multi-trait analyses provided similar genomic heritability estimates ($h^2 \pm$ S.E.) for RFI (0.18 (±0.086) for AUS cow, 0.36 (±0.086) for AUS heifer, and 0.27 (0.034) for OVE cow). Due to the increase in size of the reference data set, the standard errors of $h^2$ estimates were much smaller than the comparable estimates obtained using the data available in 2015, particularly for AUS cows. The genetic correlations ($r_g \pm$ S.E.) were 0.47 (±0.274) between AUS cow and AUS heifer, 0.94 (±0.297) between AUS cow and OVE cow, and 0.20 (±0.175) between OVE cow and AUS heifer traits. The $r_g$ between AUS cow and OVE cow was higher than the estimates in 2015, where it was 0.76 (±0.60). However, the estimates are associated with quite large standard errors.

The standard deviation of FS in the 3,413 bulls was 79 kg/yr (Table 1), which was 14kg/yr higher compared with the estimates in 2015. Cows with ABVs that are one standard deviation above the mean of 0 (i.e. +79 kg/yr) could save 1.3% of annual feed costs as reported in Pryce *et al.* (2015). The correlations of RFI life and FS ABV between the prediction equations of 2015 and 2020 were 0.65 and 0.80, respectively. This is anticipated to cause some re-ranking of the bulls based on their updated BPI values with the new model.

**Table 1. Mean, SD, and range of EBV and reliabilities for RFI cow, RFI heifer, RFI life, feed required for BWT (Feed_BWT_kg), and feed saved (FS) in 3,413 Holstein (HOL) bulls**

| | RFI cow (kg*10/d) | RFI heifer (kg*10/d) | RFI life (kg/yr) | Feed_BW_kg (kg/yr) | FS (kg/yr) | BWT (kg/yr) |
|---|---|---|---|---|---|---|
| *ABV* | | | | | | |
| Mean | 1 | 0.23 | 23.2 | 8.7 | -14.4 | 99.4 |
| SD | 2.88 | 0.89 | 65.6 | 46.8 | 79.2 | 3.5 |
| Max | 11.7 | 3.1 | 262.9 | 195.7 | 250.1 | 113.9 |
| Min | -6.86 | -2.38 | -154.4 | -141.7 | -268.3 | 85.6 |
| *Reliability* | | | | | | |
| Mean | 0.22 | 0.12 | 0.20 | NA | 0.47 | 0.71 |
| SD | 0.038 | 0.035 | 0.036 | NA | 0.039 | 0.062 |
| Max | 0.45 | 0.29 | 0.39 | NA | 0.69 | 0.99 |
| Min | 0.07 | 0.01 | 0.06 | NA | 0.28 | 0.45 |

NA= not estimated.

The genetic trend for RFI life, Feed_BWT_kg, and FS in 20,817 genotyped Holstein bulls that were born from 2010 onwards using the equation 2020 is shown Figure 1. From Figure 1A, since

2010 there is an increase in RFI life and a decrease for feed required for BWT, and hence a negative (unfavourable) trend for FS. The change in FS was at a much higher rate (about > 1/2 genetic SD) until the FS ABV was included as part of BPI which occurred in 2015. This change has slowed down (<1/4 SD of FS ABV) over the last 5 years, showing that adopting FS in BPI has been reasonably effective in reducing the unfavourable genetic trend in FS. The correlation between bodyweight EBV and FS EBVs was -0.5. For breeds other than Holsteins, FS is calculated using only the BWT component, as RFI is only measured in Holsteins.

The genetic variance of RFI life and BW in kg of feed DM per year was 30,318 $kg^2$/yr and 33,325 $kg^2$/yr, respectively, and hence the variance of FS was 63,643 $kg^2$/yr. The mean reliabilities for RFI life and FS in the 3,413 bulls were 0.20 and 0.47, respectively (Table 1). A distribution of the reliability of FS for 20k bulls that were born from 2010 onward using the equation 2020 is presented in Figure 1B, where the mean was 0.43 (sd = 0.045), ranging from 0.15 to 0.61. This was about 10% higher than using the equation predicted based on the data set 2015 where the reference population was almost half the size.

Compared with the milk production traits, the reliability of FS is still low. Using the deterministic equation described in MacLeod *et al.* (2014), over 20,000 cows and heifers are needed to have a reliability of 0.50 for RFI life with the given effective population size ($N_e$) of 210 and a constant reliability of 0.12 for RFI heifer (assuming no more additional data is added at the growing stage). With the given reliability of 0.50 for RFI life, the reliability for FS would be around 0.58. Expanding the heifer population has little impact because weight on it is only 20%.

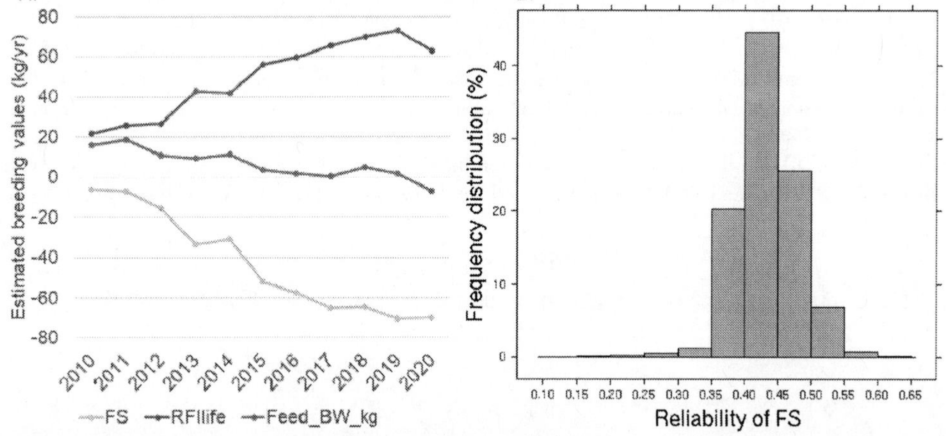


**Figure 1. a) Genetic trend of EBV for RFI life, Feed_BWT_kg, and Feed Saved (FS) in 20k genotyped Holstein bulls, b) Histogram of reliability of FS EBV in 20k Holstein bulls**

The reliability of RFI cow using the bivariate model with AUS cow and AUS heifer traits applied to 3,413 HOL bulls was low (0.08) compared with the reliability using the tri-variate model, showing a large benefit of using of overseas data. Continuing international collaborations for traits that are expensive to measure, such as feed intake, is immensely valuable.

The FS ABV, using the updated 2020 model, has recently been released by DataGene for farmers and breeders to use (December 2020) in addition to being included in BPI and Health Weighted Index (HWI). The economic weight of FS from the economic model (Byrne et al. 2016) used to derive weights for the BPI was halved based on advice from industry stakeholders to avoid a reduction in milk production gains and live weight of mature cows indirectly due to strong correlation to FS. However, the full value ($0.385/kg) has been applied in the HWI. The correlation between HWI and FS ABVs using bulls born from 2010 is 0.19, while between BPI and FS is 0.03, so a favourable selection response for FS is still only anticipated with selection on HWI.

## CONCLUSIONS

An updated 2020 model for the FS ABV using over 3,700 Australian cows and heifers, and overseas cows implemented using a multivariate model has improved the reliability of FS by about 10% compared with the 2015 model. Feed Saved derived by combining RFI and BWT originally implemented in BPI using the 2015 model has an apparent effect on the genetic trend. The implementation of FS ABV and its inclusion in BPI and Health Weighted Index (HWI) is expected to further improve the genetic trend of FS in the Holstein bulls and cows and improve feed efficiency in dairy cattle. The current reference population based on Australian animals is still small, therefore international collaboration is still crucial to achieve higher reliabilities of feed saved ABV across dairy populations.

## ACKNOWLEDGEMENTS

We thank DairyBio, jointly funded by Dairy Australia (Melbourne, Australia), The Gardiner Foundation (Melbourne, Australia) and Agriculture Victoria (Melbourne, Australia) for funding this project. The authors gratefully acknowledge use of data from Ellinbank Research Centre (belonging to Agriculture Victoria) and the EDGP and its participants. EDGP was funded by Genome Canada and by several other funding agencies and partners. The staff caring for the cows used in the study are gratefully acknowledged.

## REFERENCES

Byrne T.J., Santos B.F.S., Amer P.R., Martin-Collado D., Pryce J. E. and Axford M. (2016) *J. Dairy Sci.* **99**: 8146.

MacLeod I.M., Hayes B.J. and Goddard M.E. (2014*) Genetics.* **198**: 1671.

Pryce J. E., Gonzalez-Recio O., Nieuwhof G., Wales W., Coffey M.P., Hayes B.J. and Goddard M.E. (2015) *J. Dairy Sci.* **98**: 7340.

Pryce J.E., Nguyen, T.T.T., Axford M., Nieuwhof G. and Shaffer M. (2018) *J. Dairy Sci.* **101**: 3702.

Sargolzaei M., Chesnais J.P. and Schenkel F.S. (2014) *BMC Genomics*. **15**: 478.

*Contributed paper*

# DERIVING BREEDING VALUES FOR NET REPRODUCTION RATE FROM COMPONENT TRAITS IN SHEEP

**A.A. Swan and K.L. Bunter**

Animal Genetics Breeding Unit*, University of New England, Armidale, NSW, 2351 Australia

## SUMMARY

Genetic analyses for sheep reproduction traits in LAMBPLAN and MERINOSELECT have recently been upgraded to separate number of lambs weaned per ewe joined (NLW) into the component traits of conception (CON), litter size (LS) and ewe rearing ability (ERA). Methodology was developed to combine breeding values for component traits post-analysis into the net reproduction traits: reproduction rate (RR, lambs born per ewe joined) and weaning rate (WR, lambs weaned per ewe joined). Comparisons from the LAMPLAN maternal analysis show that RR and WR breeding values were closely aligned to single trait number of lambs born per ewe joined (NLB) and number of lambs weaned per ewe joined breeding values, ≈0.93 for RR with NLB, and ≈0.85 for WR with NLW. The derived net reproduction breeding values are useful as a tool for transition from old to new upgraded analyses.

## INTRODUCTION

New genetic analyses for component traits of sheep reproduction have been developed for LAMBPLAN maternal breeds in 2019 (Bunter *et al.* 2019), and MERINOSELECT in 2020 (Bunter *et al.* 2020), and have been available to breeders as research breeding values (RBVs). These are scheduled for transition to Australian Sheep Breeding Value (ASBV) status in 2021. The component traits are conception (CON, ewes pregnant per ewe joined), litter size (LS, lambs born per ewe lambing), and ewe rearing ability (ERA, lambs weaned relative to lambs born). Together, these traits describe the reproductive cycle from mating to lambing and then weaning. This development allows breeders to select on components separately, as determined by their relative importance in different production systems. In this paper we show how breeding values for component traits can be combined post-analysis into a breeding value for net reproduction rate which can be used to support legacy indexes, and as a transitionary mechanism to assist breeders who are currently familiar with an equivalent net reproduction breeding value, number of lambs weaned (NLW).

## MATERIALS AND METHODS

Two derived net reproduction traits are defined, reproduction rate (RR) which combines conception and litter size, and weaning rate (WR) combining all three components (CON, LS, ERA). Units for RR are number of lambs born per ewe joined, and for WR are number of lambs weaned per ewe joined. Therefore, RR is a replacement for the current NLB breeding value, and WR for NLW. Further, WR can be used as a replacement for NLW in existing selection indexes.

To derive net reproduction breeding values, component traits are expressed relative to phenotypic performance. Firstly, baseline phenotypic means are calculated for each component trait with an adjustment for genetic trend. For example, the baseline mean for CON is:

$$\mu_{con} = \Sigma_i(y_{con_i} - \hat{u}_{con_i})n_{con}^{-1}$$

* A joint venture of NSW Department of Primary Industries and the University of New England

Where $y_{con_i}$ is the phenotype for the $i^{th}$ animal, $\hat{u}_{con_i}$ the estimated breeding value, and $n_{con}$ the number of CON phenotypes in the analysis. Baseline means for LS and ERA are calculated accordingly ($\mu_{ls}$ and $\mu_{era}$).

Predicted daughter performance is then calculated for the $i^{th}$ animal, for CON as:

$$con_i = \mu_{con} + 0.5 \times \hat{u}_{con_i}$$

And for LS as:

$$ls_i = \mu_{ls} + 0.5 \times \hat{u}_{ls_i}$$

The expected frequencies of the $j^{th}$ litter size category given $ls_i$ ($p_{ij}, j = 1,2,3,4$) are then derived using the mathematical model of Amer *et al.* (1999), graphically represented in Figure 1. We note that this model is very accurate and repeatable across populations and breeds, including across countries.

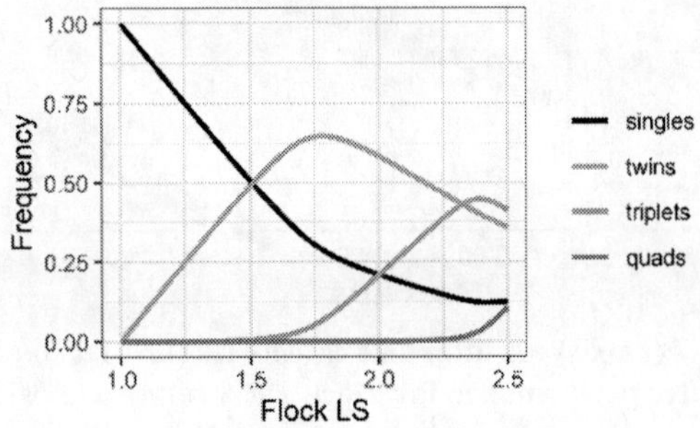


**Figure 1. Frequency of litter size categories given mean flock litter size (Amer *et al.* 1999)**

Predicted daughter performance for RR can then be calculated as:

$$rr_i = con_i \times \Sigma_j p_{ij} . j$$

Where $\overline{rr}$ is the mean predicted daughter performance for all animals in the pedigree.

Derivation of WR requires calculation of survival rates for each litter size category ($s_{ij}, j = 1,2,3$) given predicted daughter performance for LS ($ls_i$) and ERA ($era_i$), the latter calculated as:

$$era_i = \mu_{era} + 0.5 \times \hat{u}_{era_i}$$

Individual survival rates are not straightforward because survival is a much more random biological process than litter size, so we use numerical optimisation of the following equations:

$$\begin{aligned} s_{i1} - (era_i - (p_{i2} \times s_{i2} + p_{i3} \times s_{i3})/p_{i1}) &= 0 \\ s_{i2} - (era_i - (p_{i1} \times s_{i1} + p_{i3} \times s_{i3})/p_{i2}) &= 0 \\ s_{i3} - (era_i - (p_{i1} \times s_{i1} + p_{i2} \times s_{i2})/p_{i3}) &= 0 \end{aligned}$$

Subject to the constraints:

$$0.8 \leq s_{i1} < 1$$
$$s_{i1} - s_{i2} \leq 0.2$$
$$s_{i2} - s_{i3} \leq 0.2$$

With litter size frequencies ($p_{ij}$) determined by predicted daughter performance for LS ($ls_i$) as above. Also note that survival rates are only calculated for singles, twins and triplets: frequencies for quadruplets are too low for reliable calculation.

Optimised survival rates calculated using this method on the LAMBPLAN maternal reproduction analysis (8 December 2020) are shown in Figure 2.

Predicted daughter performance for WR is then calculated as:

$$wr_i = con_i \times \Sigma_j s_{ij}.p_{ij}.j$$

The derived breeding value for WR is:

$$\hat{u}_{wr_i} = 2 \times (wr_i - \overline{wr})$$

The methods were validated using the LAMBPLAN maternal reproduction analysis of 8 December 2020. Firstly, single trait REML analyses were run for the directly observed traits NLB and NLW derived from CON, LS, and ERA phenotypes, with the resulting breeding values compared to RR and WR breeding values calculated from component trait breeding values from the full multi-trait LAMPLAN analysis.

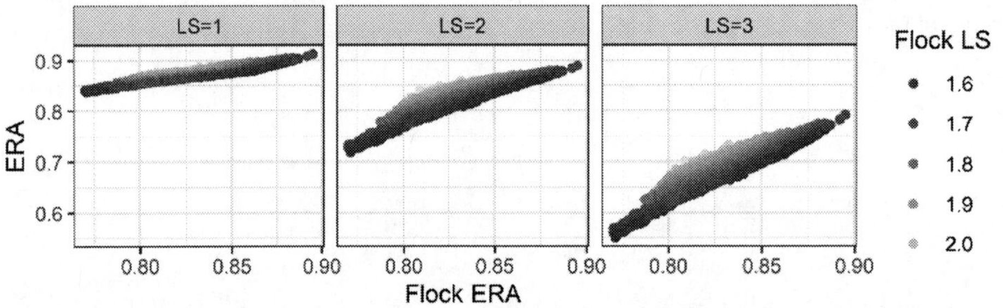


**Figure 2. Survival rates (y-axis) for litter size categories (LS=1,2,3) calculated by optimisation given predicted daughter performance for Flock ERA ($era_i$, x-axis) and Flock LS ($ls_i$)**

## RESULTS AND DISCUSSION

Results in Table 1 show high correlations between direct breeding values for NLB or NLW and corresponding derived net breeding values within-flock, higher for NLB with RR (≈0.93) than for NLW with WR (≈0.85). In within flock analyses with complete reproduction records these correlations can exceed 0.95 (analyses not presented). The correlations in Table 1 are expected to be lower than within flock results because derived net breeding values originate from the full multi-trait analysis (15 traits described by Bunter *et al.* 2019) with different data, and because of incomplete observations of the complete reproductive cycle for many females. That is, in the across flock analysis it is common for females to have records for LS but not CON or ERA, due to quality control filtering at the flock and contemporary group levels. There were also genetic group effects apparent, with lower correlations (not shown) observed in composite breeds compared to straight-bred animals.

An alternative to deriving net reproduction traits post-analysis would be to explicitly include NLB and NLW phenotypes as additional traits in the new analysis, but with yearling and adult expressions this would mean adding four traits to multi-trait models which currently have up to 19 traits. This would involve substantial effort in developing covariance matrixes, made difficult by dependencies between component and net traits, and would increase analysis run-times.

A second reason not to include NLW explicitly relates to the modelling of contemporary groups at different points of the reproductive cycle. For CON and LS groups are defined at joining because the outcomes are determined at this time (apart from low level impacts of foetal loss on LS). By contrast, for ERA, groups are defined from lambing. Because NLW covers the whole cycle both groupings are relevant. However, breeders routinely group animals for lambing based on pregnancy scanning for management purposes i.e., single bearing and multiple bearing ewes are grouped separately for differential feeding. This means that lambing groups can often be highly confounded with NLW trait values. We have previously observed that when such groups are modelled for NLW the resulting breeding values are poor predictors of phenotypic performance.

**Table 1. Comparison of NLB with RR and NLW with WR breeding values for sires and dams used from 2015 in the LAMBPLAN maternal across flock analysis (data from 8-Dec-2020). Comparisons include correlations, standard deviations ([trait]_sd), and intercept and slope from regression of "direct" trait on "derived" (e.g. NLB ~ RR)**

| Group | NLB with RR | | | | | |
|---|---|---|---|---|---|---|
| | number | corr | NLB_sd | RR_sd | intercept | slope |
| Sires | 1542 | 0.928 | 0.138 | 0.156 | 0.01 | 0.82 |
| Dams | 82990 | 0.933 | 0.124 | 0.137 | 0.01 | 0.84 |
| | NLW with WR | | | | | |
| | number | corr | NLW_sd | WR_sd | intercept | slope |
| Sires | 1472 | 0.855 | 0.091 | 0.110 | 0.01 | 0.71 |
| Dams | 72403 | 0.853 | 0.079 | 0.095 | 0.02 | 0.71 |

## CONCLUSIONS

New reproduction analyses for LAMBPLAN maternal sheep and MERINOSELECT represent a major advance on the current analysis of NLB and NLW, due to a greatly improved data processing pipeline, use of genomic information, and because they provide breeders with the ability to focus on components of reproduction separately. Net reproduction rate breeding values (RR and WR) derived from component trait breeding values post-analysis are useful as a tool to transition from the old to the new analyses and are shown in this study to be highly correlated with comparable breeding values for NLB and NLW.


## ACKNOWLEGEMENTS

This research was part funded by MLA project L.GEN.1704. The authors would like to acknowledge the contributions of breeders to the Sheep Genetics database used in the study.


## REFERENCES


Amer, P.R., McEwan, J.C., Dodds, K.G., and Davis, G.H. (1999) *Livest Prod Sci* **58:** 75.

Bunter, K.L., Swan, A.A., Gurman, P.M., and Brown, D.J. (2020) *Anim Prod Sci* **61:** 333.

Bunter, K.L., Swan, A.A., Gurman, P.M., Boerner, V., McMillan, A.J. and Brown, D.J. (2019) *Proc. Assoc. Anim. Breed. Genet.* **23**:560.

*Contributed paper*

# THE ANSTEE HUB FOR INHERITED DISEASES OF ANIMALS (AHIDA) – DEVELOPMENT OF A NEW ONLINE PLATFORM FOR SURVEILLANCE, REPORTING AND CONTROL OF INHERITED DISEASES IN ANIMALS

**I. Tammen[1,4], C. Wade[2,4], B. Waud[1,4], M. Gimeno[1,4], B.D. Velie[2,4], B. O'Rourke[3,4] and F.W. Nicholas[1,4]**

[1] Sydney School of Veterinary Science, Faculty of Science, The University of Sydney, Sydney, NSW, 2006, Australia

[2] School of Life and Environmental Sciences, Faculty of Science, The University of Sydney, Sydney, NSW, 2006, Australia

[3] Elizabeth Macarthur Agricultural Institute, Department of Primary Industries, Menangle, NSW, 2568, Australia

[4] Anstee Hub for Inherited Diseases of Animals (AHIDA)

## SUMMARY

Inherited diseases are often rare but their cumulative impact is substantial. The published data about inherited conditions in animals, and particularly about monogenic disease with known likely causal variants, is steadily increasing. However, limited information is available about how frequently these conditions occur. Animal owners and veterinarians often don't know how to report emerging genetic conditions, how to find out about available information about existing conditions or how to connect with researchers who would be interested to investigate such conditions. This problem can be addressed by the provision of a curated portal for the reporting of potential inherited disorders by owners or veterinarians. This paper describes the initial planning of a centralised resource for surveillance, reporting and control of inherited diseases of animals in Australia.

## INTRODUCTION

For humans, the global reference compendium Online Mendelian Inheritance in Man (OMIM, https://omim.org/) includes 5,762 monogenic inherited diseases and other traits for which causal variants have been reported (https://omim.org/statistics/geneMap). Most of these could also occur in any animal species. However, while the number of monogenic traits and diseases in animals for which causal variants are known is steadily increasing (Figure 1), to date these represent only a relatively small number of inherited monogenic diseases for any of the major domesticated animal species (Table 1).

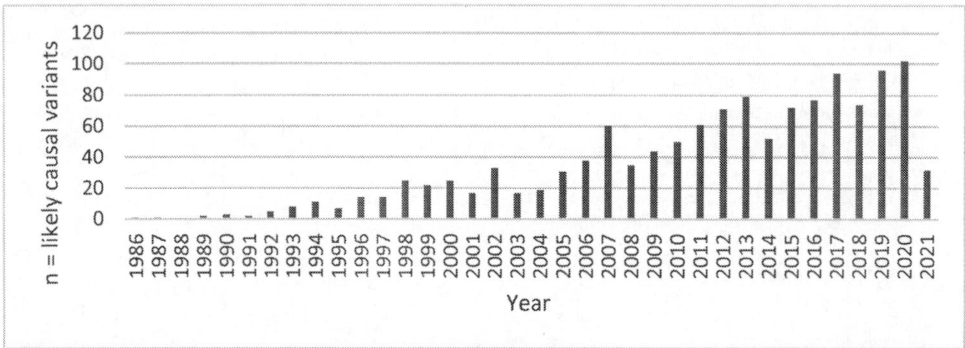


**Figure 1. Likely causal variants for monogenic traits and diseases reported per year in Online Mendelian Inheritance in Animals (OMIA, https://omia.org/)**

**Table 1. Summary of information about inherited conditions (phenes) in major domestic species in Online Mendelian Inheritance in Animals (OMIA, https://omia.org/). Phenes include disorders as well as non-disorder traits such as blood systems and most pigmentation traits. The numbers for deleterious conditions for each category are listed in brackets**

| | Dog | Cattle | Cat | Pig | Sheep | Horse | Chicken | Goat |
|---|---|---|---|---|---|---|---|---|
| Total phenes (disorders) | 787 (731) | 555 (503) | 363 (328) | 286 (239) | 258 (195) | 242 (210) | 223 (163) | 90 (53) |
| Mendelian phenes (disorders) | 364 (331) | 261 (228) | 117 (92) | 92 (69) | 112 (71) | 49 (41) | 132 (93) | 20 (12) |
| Mendelian phenes (disorders) with likely known causal variant(s) | 299 (269) | 167 (146) | 84 (68) | 41 (28) | 59 (32) | 46 (29) | 51 (29) | 15 (8) |

Population genetics modelling and whole genome sequence analyses suggest that all humans and animals are likely to be carriers of multiple deleterious alleles, further highlighting that inherited diseases pose a significant risk to health and welfare.

In animal populations, inbreeding and small effective population size increase the risk of recessive diseases: deleterious allele frequencies can amplify rapidly if particular sire lines are widely used. The result is that the risk of inherited diseases is greater in animals than in humans. Furthermore, inherited diseases in animals are often misdiagnosed, and are under-reported due to a lack of reporting structures, and concerns from many animal breeders that reporting suspected inherited conditions to their breed societies could cause reputational damage.

However, if an emerging disease is identified as inherited, effective mating plans can be implemented to reduce the risk of affected animals being born, either by predicting genotypes by pedigree analysis or by implementing DNA diagnostics once the disease-causing mutation(s) has been identified. Control of inherited diseases by these means has a direct effect on the betterment and welfare of animals.

Currently, Australia has no centralised resource for surveillance, reporting and control of inherited diseases in animals. Internationally, such resources are limited and often species-specific. Bequest funding has become available for the development of the Anstee Hub for Inherited Diseases of Animals (AHIDA) to provide a solution for preventing and controlling inherited diseases in animals throughout Australia, aiming at:

1. Establishing and maintaining an Australia-wide surveillance and reporting resource for inherited diseases in animals,
2. Prioritising emerging inherited diseases for research and control, based on published criteria that include incidence, welfare and (where relevant) financial impact
3. Facilitating undergrad/postgrad research projects on the highest-priority inherited diseases,
4. Disseminating information on emerging inherited diseases and, more generally, on the incidence/occurrence of inherited diseases and their management, to veterinarians, breed societies and the public, mainly via Online Mendelian Inheritance in Animals (OMIA)

This paper will outline the proposed structure of AHIDA, and the presentation will provide additional information on how stakeholder feedback from an online workshop held in August 2021 will be used to refine our vision.

## PROPOSED STRUCTURE FOR AHIDA

AHIDA is a proposed online portal with designated entry points for researchers, veterinarians and animal owners, for reporting and surveillance of inherited conditions in animals.

Veterinarians and animal owners will be prompted to submit to the database information about animals with suspected or confirmed inherited conditions.

Data submission will be guided using standardised nomenclature for species, breed (Universal Breed ontology, UBO) and disease phenotype (Universal Phenotype Ontology, UPO) using standardized nomenclature for diseases, clinical signs and pathology.

The breed and phenotype ontologies are currently being developed in collaboration with the leaders of the Monarch (ontology) Initiative (Shefchek et al. 2019), and aim to combine information from available resources such as DADIS (http://www.fao.org/dad-is/en/), LBO (https://www.animalgenome.org/bioinfo/projects/lbo/), SAVSNET (https://www.liverpool.ac.uk/savsnet/), SNOMED-CT (https://www.snomed.org/snomed-ct/why-snomed-ct) and VeNom (http://venomcoding.org/)..

Once fully integrated, AHIDA will directly link submitters with information about similar genetic conditions listed in animals (OMIA) or humans (OMIM), link to providers of DNA tests, provide generic information about management and control of inherited diseases in animals, and if requested, connect submitters with species expert panels for further advice.

Our aim is to develop species expert panels which include geneticists, clinicians and pathologists who can provide genetic counselling advice and link submitters with research teams that have expressed an interest in a specific disease, disease group or species.

AHIDA will report – in a format to be developed in collaboration with stakeholders – occurrence of genetic diseases, likely based on type of disease and species and breed. When fully operational, it is envisaged that additional information about occurrence of inherited diseases will be available via reciprocal links to VetCompass Australia (https://www.vetcompass.com.au/) and the soon-to-be-developed Veterinary and Animal Research Data Commons, and interested genotyping providers could report allele frequencies for monogenic diseases for which likely causal variants are tested. This will generate more accurate information about occurrence of inherited diseases in Australian animals and will inform research initiatives and management strategies.

We hope to be able to link OMIA and AHIDA to existing image repositories so that images or videos relating to the disease phenotype can be shared among submitters, species expert teams and researchers.

Limited funding will be available for research students at the University of Sydney to investigate emerging conditions that have been prioritised according to evidence-based criteria. Prioritisation is likely to consider incidence, welfare impact, population characteristics (e.g., effective population size, International Union for Conservation of Nature (IUCN) status), likelihood/costs to develop an efficient management approach, availability of research and (where relevant) financial impact of the genetic condition.

## CONCLUSIONS

While generous bequest funding has been made available for the initial development of this initiative, ongoing support from key industry stakeholders and breed societies, the veterinary profession and the wider research community will be required for the vision to succeed. A workshop to collect stakeholder feedback will have been conducted by the time of the conference and we will report on the reshaped vision.

## ACKNOWLEDGEMENTS

AHIDA is a new initiative which is generously supported by the Ronald Bruce Anstee Bequest to the Sydney School of Veterinary Science. Feedback from many stakeholders is acknowledged, as it has helped us in shaping out initial plans. The Sydney Informatics Hub Core Research Facility at The University of Sydney is acknowledged for ongoing support in regard to OMIA upgrades, and for their support in the planning and development of the AHIDA platform. Key members of the Monarch Initiative (Chris Mungall, Melissa Haendel, Peter Robinson, Nicole Vasilevsky), together with Zhiliang Hu and Jim Reecy from Iowa State University, Roswitha Baumung and Gregoire Leroy from FAO, and David Brodbelt and Noel Kennedy from the Royal Veterinary College are acknowledged for their collaborations towards the development of breed and phenotype ontologies.

**REFERENCES**

Shefchek K.A., Harris N.L., Gargano M., Matentzoglu N., Unni D., Brush M., Keith D., Conlin T., Vasilevsky N., Zhang X.A., Balhoff J.P., Babb L., Bello S.M., Blau H., Bradford Y., Carbon S., Carmody L., Chan L.E., Cipriani V., Cuzick A., Della Rocca M., Dunn N., Essaid S., Fey P., Grove C., Gourdine J.P., Hamosh A., Harris M., Helbig I., Hoatlin M., Joachimiak M., Jupp S., Lett K.B., Lewis S.E., McNamara C., Pendlington Z.M., Pilgrim C., Putman T., Ravanmehr V., Reese J., Riggs E., Robb S., Roncaglia P., Seager J., Segerdell E., Similuk M., Storm A.L., Thaxon C., Thessen A., Jacobsen J.O.B., McMurry J.A., Groza T., Köhler S., Smedley D., Robinson P.N., Mungall C.J., Haendel M.A., Munoz-Torres M.C. and Osumi-Sutherland D. (2020) *Nucleic Acids Res.* **48**: D704.

*Contributed paper*

# CURATION OF PIG TRAITS IN THE ONLINE MENDELIAN INHERITANCE IN ANIMALS (OMIA) DATABASE

**S.M. Shields[1], S. Hermesch[2], F.W. Nicholas[1] and I. Tammen[1]**

[1] Sydney School of Veterinary Science, University of Sydney, Sydney, NSW, 2006 Australia
[2] Animal Genetics Breeding Unit*, University of New England, Armidale, NSW, 2350 Australia

## SUMMARY

Online Mendelian Inheritance in Animals (OMIA) is a freely available curated database that contains information on inherited traits and disorders (called phenes in OMIA) across more than 250 species. OMIA entries relating to pigs were reviewed, as a relatively low number of Mendelian phenes, as well as low number of phenes for which likely causal variants were listed, were noted when compared to other companion and livestock species. Of the 277 pig phenes recorded within the database at the beginning of this study in March 2020, 228 were classified as defects, 87 were Mendelian traits and for 37 of these, 45 likely causal variants were published. This study aimed to identify gaps in the information for pig phenes within OMIA. Changes to 30 pig phenes were made with a focus on updating information in OMIA's downloadable tables of known likely causal variants, One phene had previously been missed and was added, and 8 phenes were added as part of ongoing curation.

## INTRODUCTION

Online Mendelian Inheritance in Animals (OMIA) (Nicholas 2020, Online Mendelian Inheritance in Animals 2021) is a freely available, curated, online database which provides researchers, veterinarians and breeders with up-to-date summary information on all the known harmful and beneficial variants in animals, together with background information on all known inherited disorders and beneficial traits. OMIA focuses on phenes with confirmed and suspected Mendelian modes of inheritance. However, phenes with unknown or complex modes of inheritance and phenes caused by somatic mutations, chromosomal abnormalities or genetic modifications or genome editing are also included. Furthermore, OMIA provides references for landmark and review articles and for papers describing genetic maps and reference genomes in animals (including 34 pig mapping and pig genome references). OMIA covers more than 3,500 phenes across more than 250 species. The vast majority of OMIA entries are for the major domesticated animals (Table 1).

One of the first causal variants identified in livestock was the variant causing malignant hyperthermia in pigs (OMIA 000621-9823, Fujii *et al.* 1991), and due to their anatomical similarities to humans, pigs are frequently used as models of human disease (Bassols *et al.* 2014). However, in comparison to other companion and livestock species, OMIA entries for pigs are relatively sparse. This is particularly evident for the number of Mendelian traits and numbers of likely causal variants. This study aimed to improve OMIA curation by adding new pig phenes and critically analysing and curating pig phenes currently recorded within OMIA, with a focus on updating variant information, as these data can be downloaded and used to develop DNA diagnostic tools.

* A joint venture of NSW Department of Primary Industries and the University of New England

**Table 1. Numbers of phenes, Mendelian traits, traits with at least one likely known causal variant and total number of likely causal variants known in major companion and livestock species in OMIA (Online Mendelian Inheritance in Animals, 2021) at the time of writing (February 2021)**

| | Dog | Cattle | Cat | Pig | Sheep | Horse | Chicken | Goat | All species |
|---|---|---|---|---|---|---|---|---|---|
| Total phenes (traits/disorders) | 784 | 555 | 362 | 286 | 257 | 242 | 223 | 90 | 3683 |
| Total number of Mendelian traits | 362 | 261 | 116 | 92 | 112 | 59 | 132 | 20 | 1553 |
| Total number of Mendelian traits with at least one likely causal variant known | 297 | 167 | 83 | 40 | 59 | 46 | 51 | 15 | 930 |
| Total number of likely causal variants known | 435 | 226 | 131 | 50 | 76 | 98 | 66 | 26 | 1268 |

## MATERIALS AND METHODS

**Identification of missing porcine information in OMIA.** A literature search for phenes or publications that describe genetic conditions in pigs that are not currently listed in OMIA was performed via PubMed (PubMed 2021), using key search words, such as 'pigs', 'disease', 'inherited' and 'variant', in various combinations. Identified references and phenes were added to the database.

For Mendelian phenes where no gene or causal variant was recorded, the associated references were searched for variant information, including analysis of figures for clues, such as images of analysed sequence. Mendelian phenes without information on likely causal gene or likely causal variant were not investigated further.

**Updating porcine variant information in OMIA.** For phenes with at least one likely causal variant, the data in the downloadable tables of known likely causal variants was reviewed and updated to represent location information in the Sscrofa11.1 reference genome to facilitate development of diagnostic tools. Variant locations and predicted effects on proteins were determined through various methods depending on availability of published data, and included remapping and confirmation of the variant effect using *in silico* variant effect prediction.

**Variant remapping.** Any variants mapped in reference genomes other than the most recent, Sscrofa11.1, were remapped using the NCBI Genome Remap tool (National Center for Biotechnology Information 2021b). The input for the tool required selection of the source organism (Sus scrofa), source assembly, target assembly, variant location and chromosome number. No settings under the 'remapping options' or 'data' headings were altered. The input format for the variant location and chromosome followed the input guide provided by NCBI. New variant locations were confirmed using NCBI Genome Data Viewer (National Center for Biotechnology Information 2021a) which allowed for visualisation of the variant, determination of which strand (PLUS or MINUS) the variant was located on, and identification if the variant had an European Variant Achieve (EVA) ID (EMBL-EBI 2021). Variant locations for variants that lacked reference genome information or contained information from reference genomes not recognised by the NCBI Genome Remap tool were mapped manually, based on information provided in the original reference.

**Variant effect prediction.** Ensembl's variant effect predictor (VEP) (Ensembl 2021) tool was used to obtain further information on some of the recorded variants, including the variant consequence, allele, exon, intron and cDNA position. All analysed variants were located within Sscrofa 11.1 and were input in the format outlined by the VEP program (Ensembl 2021). To obtain all available results, the option of 'Ensembl and RefSeq transcripts' was selected within the

'Transcript database to use' component of the input settings. Results were confirmed by comparing the variant location generated to that previously published, recorded in OMIA or found via genome remapping.

**RESULTS AND DISCUSSION**

At the time of data collection (March 2020), the OMIA database contained 277 phenes in pigs. Of these, 87 were classified as Mendelian traits/disorders and 37 were Mendelian traits with known causal variants. Of the 37 Mendelian phenes with known causal variants, 17 were identified to be disease traits. The PubMed literature search resulted in the addition of one missed phene: Vitamin C deficiency (OMIA 002268-9823) as well as the addition of several references and additions to text curation fields to existing phenes.

Ongoing curation based on daily automated PubMed searches for all animal species resulted in the addition of further 8 porcine phenes. At the time of writing (February 2021) OMIA includes 286 pig phenes of which 92 are Mendelian traits. Of the 40 Mendelian traits with 50 known likely causal variants, 31 phenes are defects or disease- related (Table 2), 6 are coat-colour phenes, 2 are related to ear phenes and 1 is a blood-group phene (Table 3).

Updates made to OMIA's downloadable likely causal variant tables are summarised in Table 2 and Table 3. Chromosome (Chr), genomic DNA (g.), coding DNA (c.) and protein (p.) locations were added to 26, 26, 11 and 9 variants, respectively. Text was added to the verbal description field for 9 variants. Variant effect prediction was conducted for 16 variants, and EVA IDs were added for 13 variants.

The literature review has not been able to identify many additional porcine phenes or references and this suggests that the automated daily PubMed searches followed by manual curation have been an effective way to identify most genetic conditions in pigs. However, a broader literature search that searches journals that are not listed in PubMed may provide additional references and phenes. This study was able to update location information for many of the likely causal variants that are listed in OMIA, information that is particularly important in the absence of EVA IDs. In March 2020, OMIA listed EVA IDs for only 3 porcine variants (ear size: rs338733115; coat colour, white belt, KIT-related: rs328592739, and malignant hyperthermia: rs344435545). The inclusion of EVA IDs for 13 additional variants will allow automated updates to location information if new reference genomes become available once OMIA is hyperlinked to EVA as part of a planned update.

Further curation is needed to increase content in OMIA's text entry fields for many of the porcine phenes, particularly in relation to information about clinical signs and pathology.

**Table 2. Defects/disease-related phenes - Summary information about Mendelian traits in pigs for which likely causal variants have been listed in OMIA (Online Mendelian Inheritance in Animals, 2021). OMIA ID, phene name, gene, year of publication and PubMed IDs of papers describing the likely causal variants are listed. Updates to variant information are summarised as updates to chromosome (Chr), genomic DNA (g.), coding DNA (c.) protein (p.) locations, addition of EVA IDs (EVA), analysis of data via the Ensembl Variant Effect Predictor (VEP) and verbal description field (text). Location details are available in OMIA (https://omia.org/results/?search_type=advanced&gb_species_id=9823&result_type=variant).**

| OMIA ID | Phene (variant phenotype) | Gene | Year | PubMed ID | Updated information |
|---|---|---|---|---|---|
| 000499-9823 | Hypercholesterolaemia | LDLR | 1998 | 9556295 | g. / VEP / EVA |
| 000576-9823 | Knobbed acrosome | BOLL | 2020 | 32975846 | |
| 000621-9823 | Malignant hyperthermia | RYR1 | 1991 | 1862346 | g. / VEP |
| 000636-9823 | Membranoproliferative glomerulonephritis type II | CFH | 2002 | 12466119 | Chr / g. |
| 000683-9823 | Muscular hypertrophy (double muscling) | MSTN | 2008 | 18822098 | Chr / g. / EVA |
| 000837-9823 | Vitamin D-deficiency rickets, type I | CYP27B1 | 2003 | 12915218 | Chr |
| | | CYP27B1 | 2003 | 12915218 | Chr |
| 000862-9823 | Resistance to oedema disease | FUT1 | 2000 | 11132149 | Chr /g. / c. / p. / VEP / EVA |
| 001058-9823 | Von Willebrand disease III | VWF | 2017 | 29208651 | Chr |
| 001085-9823 | Meat quality (Rendement Napole) | PRKAG3 | 2000 | 10818001 | g. / c. / p. / VEP / EVA |
| | | PRKAG3 | 2001 | 11729159 | g. / c. / p. / VEP / EVA |
| 001128-9823 | Pale soft exudative meat | PHKG1 | 2014 | 25340394 | Chr / g. / text / EVA |
| 001200-9823 | Tremor, high-frequency | MYH7 | 2012 | 23153285 | Chr / g. |
| 001334-9823 | Sperm, short tail | SPEF2 | 2006 | 16549801 | Chr |
| 001401-9823 | Waardenburg syndrome, type 2A | MITF | 2016 | 27349893 | g. / text |
| 001436-9823 | Non-shivering thermiogenesis, absence | UCP1 | 2006 | 16933999 | Chr |
| 001673-9823 | Spermatogenic arrest | TEX14 | 2011 | 22136159 | Chr |
| 001685-9823 | Stress syndrome | DMD | 2012 | 22691118 | Chr / g. / c. text / EVA / VEP |
| 001718-9823 | Dwarfism, Schmid metaphyseal chondrodysplasia | COL10A1 | 2000 | 11130976 | Chr / g. / c. / VEP |
| 001952-9823 | Microtia | HOXA1 | 2015 | 26035869 | g. / VEP |

| | | | | | |
|---|---|---|---|---|---|
| 001986-9823 | Severe combined immunodeficiency disease, autosomal, T cell-negative, B cell-negative, NK cell-positive, with sensitivity to ionizing radiation | DCLRE1C | 2015 | 26320255 | g. |
| 001986-9823 | Severe combined immunodeficiency disease, autosomal, T cell-negative, B cell-negative, NK cell-positive, with sensitivity to ionizing radiation | DCLRE1C | 2015 | 26320255 | g. |
| 002161-9823 | Leg weakness, MSTN-related | MSTN | 2019 | 30699111 | |
| 002178-9823 | Abortion, BBS9-related | BBS9 | 2018 | 30231021 | |
| 002180-9823 | Abortion due to haplotype DU1 | TADA2A | 2019 | 30875370 | |
| 002181-9823 | Abortion due to haplotype LA1 | POLR1B | 2019 | 30875370 | |
| 002182-9823 | Abortion due to haplotype LA2 | URB1 | 2019 | 30875370 | |
| 002183-9823 | Abortion due to haplotype LA3 | PNKP | 2019 | 30875370 | |
| 002210-9823 | Congenital hypothyroidosis | DUOX2 | 2019 | 30651277 | g. / text |
| 002232-9823 | Myopathy, congenital, SPTBN4-related | SPTBN4 | 2019 | 31850074 | |
| 002268-9823 | Vitamin C deficiency | GULO | 2004 | 15112110 | Chr / text |
| 002283-9823 | Arthrogryposis multiplex congenita | KIF21A | 2020 | 32686171 | |
| 002287-9823 | Hypopigmentation and deafness | KIT | 2020 | 33042408 | g. / text |
| 002306-9823 | Fecundity, BMP15-related (Infertility and increased litter size) | BMP15 | 2021 | 33413103 | |

**Table 3. Non-disease related phenes - Summary information about Mendelian traits in pigs for which likely causal variants have been listed in OMIA (Online Mendelian Inheritance in Animals, 2021). OMIA ID, phene name, gene, year of publication and PubMed IDs of papers describing the likely causal variants are listed. Updates to variant information are summarised as updates to chromosome (Chr), genomic DNA (g.), coding DNA (c.) protein (p.) locations, addition of EVA IDs (EVA), analysis of data via the Ensembl Variant Effect Predictor (VEP) and verbal description field (text).**

| OMIA ID | Phene (variant phenotype) | Gene | Year | PubMed ID | Updated information |
|---|---|---|---|---|---|
| 000209-9823 | Coat colour, dominant white | KIT | 1996 | 8875890 | Chr |
| 001199-9823 | Coat colour, extension (red) | MC1R | 1998 | 9799269 | Chr / g. / c. / p. / text / EVA /VEP |
| | Coat colour, extension (red) | MC1R | 1998 | 9799269 | Chr / g. / c. / p. / text / EVA /VEP |
| | Coat colour, extension (dominant black) | MC1R | 1998 | 9799269 | Chr / g. / c. / p. / text / EVA /VEP |
| | Coat colour, extension (dominant black) | MC1R | 1998 | 9799269 | Chr / g. / c. / p. / text / EVA /VEP |
| | Coat colour, extension (dominant black) | MC1R | 1998 | 9799269 | Chr / g. / c. / p. / text / EVA /VEP |
| | Coat colour, extension (black spotting on red or white background) | MC1R | 2001 | 11404341 28411032 | Chr / g. / c. / p. / text / EVA /VEP |
| 001216-9823 | Coat colour, roan | KIT | 2011 | 21749430 | Chr |
| 001249-9823 | Coat colour, brown, TYRP1-related | TYRP1 | 2011 | 20978532 | Chr / g. |
| 001743-9823 | Coat colour, patch | KIT | 1998 | 9724328 | Chr |
| 001745-9823 | Coat colour, white belt, KIT-related | KIT | 2012 | 23151514 | Chr |
| | Coat colour, white belt, KIT-related | KIT | 2016 | | g. / VEP / text |
| 000319-9823 | Ears, folded (Ear size) | MSRB3 | 2018 | 30587124 | |
| 001579-9823 | Ear size (large floppy ears) | PPARD | 2011 | 21573137 | g. / VEP |
| | Ear size (ear size) | WIF1 | 2019 | 30815903 | |
| 001089-9823 | Blood group system ABO | GGTA1 | 2011 | 21554350 | Chr |

## CONCLUSIONS

Despite the vast amount of research performed previously to investigate traits and disorders in pigs, there are gaps in the knowledge held and potentially new phenes to be found. In comparison to other companion and livestock species relatively few inherited diseases and traits have been characterised at the molecular level and research should focus on the identification and investigation of emerging inherited conditions in pigs.

This study highlighted that further curation of OMIA data relating to pigs with a focus on updating textual fields is needed. Planned updates to OMIA will reduce the need to manually update location information for variants that have an EVA ID and researchers are encouraged to submit likely causal variants to EVA to facilitate this process. Easy access to accurate location information of likely causal variants can facilitate the development of diagnostic SNP panels that can be used by industry to genotype animals for desirable or unfavourable alleles and therefore allow for more informed selection decisions; greatly improving the health and welfare of not only individual breeding populations, but the pig population as a whole.

Feedback on current information presented in OMIA and suggestions for additional information can be emailed to the OMIA curators (frank.nicholas@sydney.edu.au or imke.tammen@sydney.edu.au).

**ACKNOWLEDGEMENTS**

Stephanie Shields received the Australian Pork Research Institute Honours Scholarship award from APRIL.

**REFERENCES**

Bassols A., Costa C., Eckersall D., Osada J., Sabria J. and Tibau J. (2014) *Proteomics Clinical Applications* **8**: 715.

Ensembl (2021) Variant Effect Predictor. https://asia.ensembl.org/info/docs/tools/vep/index.html

EMBL-EBI (2021) European Variation Archive. https://www.ebi.ac.uk/eva/

Fujii J., Otsu K., Zorzato F., de Leon S., Khanna V.K., Weiler J.E., O'Brien P.J. and MacLennan D.H. (1991) *Science.* **253**: 448.

Nicholas F.W. (2020) *Animal Genetics* **52**: 3.

National Center for Biotechnology Information (2021a) *Genome Data Viewer*:https://www.ncbi.nlm.nih.gov/genome/gdv/

National Center for Biotechnology Information (2021b) *NCBI Genome Remapping Service*: https://www.ncbi.nlm.nih.gov/genome/tools/remap.

Online Mendelian Inheritance in Animals (2021) https://omia.org/

PubMed (2021) PubMed. Bethesda, Maryland. https://pubmed.ncbi.nlm.nih.gov/

*Proc. Assoc. Advmt. Anim. Breed. Genet: 24: 397-401*

# COMPARING GENOMIC WITH PEDIGREE RELATIONSHIP MATRICES AND PRELIMINARY GENOME WIDE ASSOCIATION IN SANTA GERTRUDIS BULLS

**W.L.A. Tan[1], L.R. Porto-Neto[2], G. C. Gouveia[3], A. Reverter[2] and M.R.S. Fortes[1]**

[1]The School of Chemistry and Molecular Biosciences, the University of Queensland, Brisbane, QLD 4072 Australia
[2]CSIRO Agriculture and Food, Brisbane, QLD 4067.
[3]Federal University of Minas Gerais, Belo Horizonte, MG, Brazil

## SUMMARY

To estimate relatedness between individuals, one can choose between the **A** matrix from pedigree data, the **G** matrix built from genome-wide SNP markers, or even a combination of both. Comparing both matrices is of interest to explore the possibility of initiating a genomic breeding program. We use data on 929 Santa Gertrudis bulls with 617,348 SNP genotypes to compare the matrices. The **A** and **G** matrices were positively correlated (0.91; $P < 0.001$), and diagonal and off-diagonal coefficients were very similar. This result indicates that the pedigree is correct, although some genomic-estimated relationships did not agree with the pedigree-based matrix. Subsequently, a genome-wide association study was performed for scrotal circumference (SC) using **A** and **G** matrices. 100 SNPs were associated with SC with a corresponding FDR (<0.05) in GWAS using **G**, with the highest peak at BTA 5. Previously, the peak on BTA 5 has been associated with sheath score in Brahman and Tropical Composite (TC) cattle. BTA 5 has also been associated with SC in TC. For GWAS conducted using the **A**, 2883 SNPs were associated with SC with a corresponding FDR (<0.05), with the highest peak at BTA 5 and other peaks at BTA 22 and chromosome X. The peaks in chromosome X and BTA 22 was not observed in GWAS using **G**. This warrants further investigation into the differences in estimated SNP effects resulting from using different matrices in GWAS. Additionally, combining both the **A** and **G** in an **H** matrix may make more accurate predictions than using **G** alone. Further analysis is required to investigate the use of **H** and to verify the SNP associations identified in this study and across other breeds.

## INTRODUCTION

Traditionally, most analyses in livestock relied on the calculation of the **A** matrix from pedigree data. Today, DNA markers are also used to estimate relatedness between individuals for various genetic analyses (Makgahlela *et al.* 2013). The relationship matrix estimated from genotypes is termed the **G** matrix. One advantage of estimating **G** is detecting alleles identical by state, traced to common ancestors that are unknown and therefore not captured in the pedigree (VanRaden 2008). The **G** matrix coefficients may be more precise than those in **A** and could correct pedigree relationships assigned wrongly in **A**. However, it is expensive and often impossible to genotype the entire population, and sometimes the **A** is still used. In genome-wide association studies (GWAS), relationship matrices account for population structures while testing for SNPs association. As using either **G** and **A** may affect the results of GWAS, it becomes crucial to determine which one might be a better fit for GWAS. In this study, we compare **A** and **G** matrices to inform future GWAS work.

## MATERIALS AND METHODS

**Animals and genotypes.** Scrotal circumference (SC) was measured in 952 Santa Gertrudis bulls. Animals that were genotyped using 50K SNP panels were imputed to high density (770K) using a phased reference generated by Eagle2 (v2.4.1) and imputed using Minimac3 for autosomes and Minimac4 for the X chromosome (Loh *et al.* 2016; Das *et al.* 2016). SNPs with a call rate of less than 0.90 were removed before imputation. After quality control (minor allele frequency >0.05),

617,348 SNPs were available for analysis. Contemporary groups (CG) accounted for management, year of record, and age groups. CGs with less than five bulls were discarded, leaving a final dataset of 929 bulls for analysis.

**G and A matrices.** The **G** matrix was built using Method 1, as described by VanRaden (2008). Pedigree information on 4409 animals which were the genotyped animals and their ancestors, was used to compute the Wright coefficients in **A** described by Wright (1922) in SVS software (Release 8.9.0, Golden Helix, Inc.).

**GWAS.** GWAS was performed in SVS software (Release 8.9.0, Golden Helix, Inc.) using the Efficient Mixed-Model Association eXpedited method (EMMAX) (Kang *et al.* 2010). GWAS was performed for SC with **A** and **G** matrices separately. CG was fitted as a fixed effect, and age was fitted as a covariate. The first two principal components of the principal component analysis (PCA), calculated for the bulls included in this paper, were also fitted as a covariate. False Discovery Rate (FDR) was used to correct for multiple testing.

## RESULTS AND DISCUSSION

Scatter and box plots comparing relationships estimated using both **A** and **G** matrices are shown in Figures 1 and 2. **A** and **G** matrix were highly positively correlated (0.91; < 0.001). The means of the diagonal and off-diagonal coefficients were very similar: they were 1 for **A** and 1.03 for **G** in the diagonal, and they were 0.02 for **A** and 0 for **G** in the off-diagonal. In some cases, according to **A,** unrelated animals were estimated to be similar to half-siblings in **G** (a coefficient of ~ 0.25). Whereas some animal pairs considered half-siblings in **A** (a coefficient of 0.25) were either full siblings or unrelated in **G**. In general terms, the genomic relationships (**G**) were within the expected range of the pedigree relationships. Thus, our analyses indicate that the pedigree for this population is accurate.

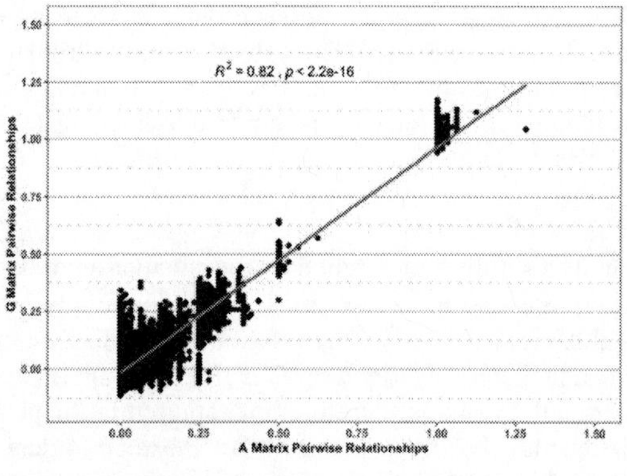


**Figure 1. Scatter plot of relationships coefficients for all pairwise combinations of 929 Santa Gertrudis bulls estimated using SNP markers (G) and pedigree (A)**

In the GWAS using the **A** matrix, a total of 2883 SNPs were detected with an FDR of less than 0.05 for SC, with the highest peak at BTA 5 and other peaks at BTA 22 and chromosome X. The peak SNP at BTA 5 for GWAS using the A matrix corresponded to an FDR of 0.0005 for BTA 5. In GWAS using the **G** matrix, a total of 100 SNPs were detected with an FDR of less than 0.05 for SC, with the highest peak at BTA 5. The peak SNP for GWAS using the G matrix corresponded to an FDR of 0.0011 for BTA 5. Other peaks were also observed in BTA 4 and BTA 9 for GWAS using the G matrix.

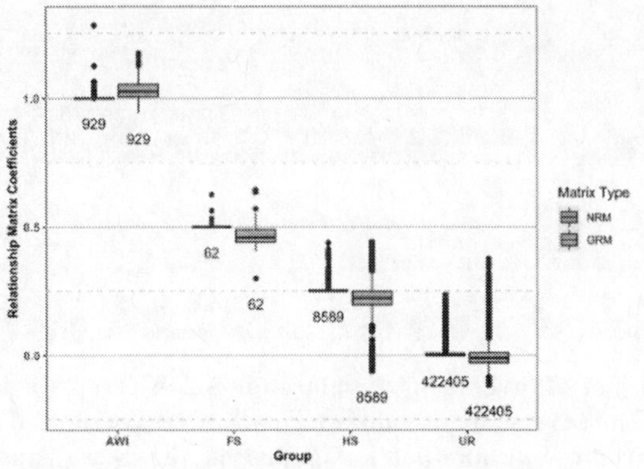


**Figure 2. Boxplot comparing genomic relationship matrix (G) in blue against corresponding additive numeric matrix (A) coefficients in red of Santa Gertrudis bulls with the animal with itself (AWI), full-siblings (FS), half-siblings (HS), and unknown relationships (UR). The number of observations in each group has been annotated in black**

The Manhattan plot observed when using another **G** matrix, calculated without X chromosome SNPs, was very similar to the **A** matrix plot, with peaks observed on X. Therefore, building **G** without SNPs in the X chromosome may lead to different GWAS results (Druet and Legarra 2020).

The largest concentration of SNPs associated with SC was located from 46.2 Mb to 50.2 Mb at BTA 5. The peak SNP for this region is found at 47.8 Mb, which accounted for 0.04% of the genetic variance in SC. Fortes *et al.* (2020) reported SNP associations with SC in BTA 5 in a study of Tropical Composite (TC) cattle which overlaps with the region reported for BTA 5 in our study. The significance of BTA 5 in cattle breeding and its association with other traits has been documented in previous studies. For example, Porto-Neto *et al.* (2014) reported significant SNP association at BTA 5 for sheath score in Brahmans and TC.

Our study found peak SNPs associated with SC in BTA 4 (40.1 Mb) and 9 (32.2 Mb) for GWAS conducted using either **A** or **G** matrix. The peak SNP in the BTA 4 accounted for 0.02% and 0.03% of the variance in SC for GWAS conducted using the **A** and G matrix, respectively. Whereas the peak SNP in BTA 9 accounted for 0.03% and 0.04% of the variance in SC for GWAS conducted using the **A** and **G** matrix, respectively. A crossbred beef cattle study also reported an association in BTA 9 for SC, and another study reported a region in BTA 4 associated with SC in Nelore Cattle (Sweett et al. 2020; Sbardella *et al.* 2021). However, these regions did not overlap with the regions reported in our study for BTA 4 and BTA 9.

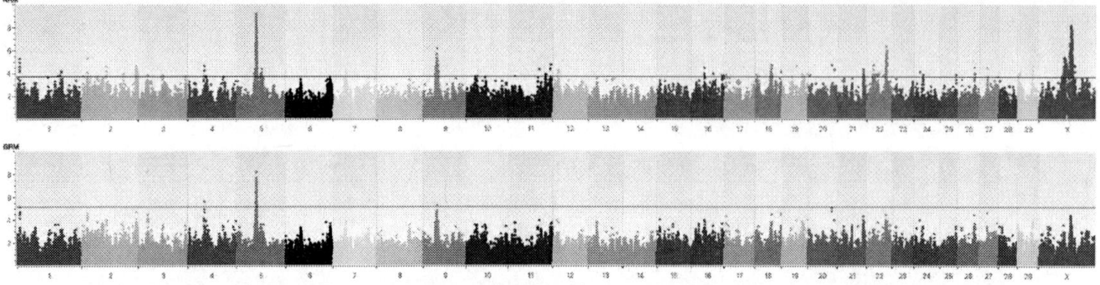

**Figure 3. Manhattan plot of inverse log *P*-values for SC in GWAS using the A matrix (top) and the G (bottom). The inverse log *p*-values for each SNP are plotted on the y-axis for each chromosome. The chromosome number is labelled on the x-axis, and a variety of colours correspond to different chromosomes. The *P*-value line in black corresponds to an FDR (<0.05)**

The heritability estimated using **G** and **A** were similar, although subtle differences exist between genetic parameters estimated using the different matrices. **G** accounts for a higher genetic variance and has a lower variance of heritability compared to **A**, as shown in Table 2. The differences between both estimates may have resulted from differences in relationship estimates between the **A** and **G** matrices. However, more research is required to understand the differences seen in the parameters estimated by both matrices. The heritability estimates are similar to those reported in TC (0.43) but are lower than those reported in Brahmans (0.75) and another composite breed (0.67) (Corbet *et al.* 2013; Roberts *et al.* 2010).

**Table 2. Genetic parameters obtained using G and A matrices**

| Matrix Type | Heritability | Variance of heritability | $V_g$ | $V_e$ |
|---|---|---|---|---|
| A | 0.429 | 0.012 | 2.879 | 3.828 |
| G | 0.466 | 0.005 | 3.125 | 3.578 |

## CONCLUSIONS

This study shows that the matrix coefficients between animals using the **A** and **G** matrices are highly correlated but not identical. Utilising genomic information includes additional information, which results in capturing relationships otherwise undetected or missing in pedigree. GWAS reveals differences in SNP effect estimates resulting from the use of these different matrices. The **G** matrix may provide more accurate information and facilitate estimating genetic parameters and identifying significant SNPs. Future studies can explore combining both matrices in an **H** matrix, which might be better than the **G** matrix alone. The GWAS highlighted associations with SC in chromosomes 4, 5, 9 and X in a population of Santa Gertrudis bulls. Future studies can further investigate these associations and if they occur in other breeds.

## ACKNOWLEDGEMENTS

The authors acknowledge that this study was funded as part of the MLA project L.GEN.1818 Bull fertility update: historical data, new cohort and advanced genomics. We thank MLA and all collaborators in this project for making the study possible.

## REFERENCES

Corbet N.J., Burns B.M., Johnston D.J., Wolcott M.L., Corbet D.H., Venus B.K., Li Y., McGowan M.R., and Holroyd R.G. (2013) *Anim. Prod. Sci.* **53:** 101-113.

Druet T., and Legarra A. (2020) *Genet. Sel. Evol.* **52**: 50.

Das S., Forer L., Schönherr S., Sidore C., Locke A. E., Kwong A., Vrieze S. I., Chew E. Y., Levy S., McGue M., Schlessinger D., Stambolian D., Loh P.R., Iacono W. G., Swaroop A., Scott L.J., Cucca F., Kronenberg F., Boehnke M., Abecasis G.R., Fuchsberger C. (2016) *Nat. Genet.* **48**: 1284-1287

Fortes M.R.S., Porto-Neto L.R., Satake N., Nguyen L. T., Freitas A.C., Melo T.P., Scalez D.C.B., Hayes B.J., Raiden F.S.S., Reverter A., Boe-Hansen G.B. (2020) *Genet. Sel. Evol.* **52**: 46

Kang H.M., Sul J.H., Service S.K., Zaitlen N.A., Kong S.Y., Freimer N.B., Sabatti C., and Eskin E. (2010) *Nat. Genet.* **42**: 348-354.

Loh P.R., Danecek P., Palamara P.F., Fuchsberger C., Reshef Y.A., Finucane H.K., Schoenherr S., Forer Lukas., McCarthy S., Abecasis G.R., Durbin R., Price A.L. (2016) *Nat. Genet.* **48**: 1443–1448.

Porto-Neto L.R., Reverter A., Prayaga K.C., Chan E.K.F., Johnston D.J., Hawken R.J., Fordyce G., Garcia J.F., Sonstegard T.S., Bolormaa S., Goddard M.E., Burrow H.M., Henshall J.M., Lehnert S.A., Barendse W. (2014) *PLoS One* **9**: e113284.

Makgahlela M.L., Strandén I., Nielsen U.S., Sillanpää M.J., and Mäntysaari E.A. (2013) *J. Dairy Sci.* **96**: 5364-5375.

Roberts C., Geary T., MacNeil M., Waterman R., Roberts A., and Alexander L. (2010) *Proc. West. Sect. Ameri. Soc. Anim. Sci.* 61: 122-125.

Sbardella A.P., Watanabe R.N., Costa R.M., Bernardes P.A., Braga L.G., Rey F.S.B., Lôbo R.B., Munari D.P. (2021) *Animals.* **11**: 1386

Sweett H., Fonseca P.A.S., Suárez-Vega A., Livernois A., Miglior F., & Cánovas A. (2020) *Sci. Rep.* **10**: 20102-20102.

VanRaden P. M. (2008) *J. Dairy Sci.* **91**: 4414-4423.

Wright S. (1922) *Am. Nat.* **56**: 330-338.

*Contributed paper*

# DETERMINATION OF OPTIMUM WEIGHTING FACTORS FOR SINGLE-STEP GENETIC EVALUATION VIA GENETIC VARIANCE PARTITIONING

**J.A. Torres-Vázquez, A.M. Samaraweera, M G. Jeyaruban, D.J. Johnston, and V. Boerner**

Animal Genetics Breeding Unit*, University of New England, Armidale, NSW, 2351 Australia

## SUMMARY

It is important in single-step genetic evaluations to use appropriate lambdas (λ) for calculating weighted average of NRM (numerator relationship matrix) and GRM (genomic relationship matrix) in joint relationship matrix. λ is usually estimated using a single-trait cross-validation procedure. However, it can be shown that a univariate single-step model applying a scalar λ is simply a condensed form of an extended model containing two genetic factors, factor $H \sim N(0, H)$ and factor $A \sim N(0, A)$, where the partitioning of the total genetic variance reflects λ. For multivariate single-step genetic evaluation, this model condensation implies that all involved genetic variances may yield the same λ, which is highly unlikely. Hence, it is required to estimate λ by accounting for its heterogeneity using the extended model for variance component estimation. This study used an extended single-step model to estimate variances and λs for calving difficulty (CD), gestation length (GL), and birth weight (BW) using Australian Angus data. A total of 129,851 animals with 45,575 genotypes were analysed. Initial variances obtained from a pedigree-only model were then used as starting values for the extended single-step model assigning 90% of the genetic variance to factor $A$ and 10% to factor $H$. Since CD is a categorical trait with three categories, a threshold model-Gibbs sampling method was used to estimate variances. Heritability estimates for the extended single-step model were very similar to those from the pedigree only model implying that the single-step model was not explaining more variation in the data than the pedigree only model. For CD, GL, and BW, the total heritability estimates were 0.39 ± 0.04, 0.68 ± 0.02, and 0.44 ± 0.01, respectively. For the same traits, the total maternal heritability estimates were 0.17 ± 0.02, 0.11 ± 0.01, and 0.09 ± 0.01, respectively. In contrast, to the Gibbs sampling starting values, the genetic variance was partitioned between $A$ and $H$ such that direct genetic λ estimates for CD, GL, and BW were 0.36 ± 0.05, 0.62 ± 0.03, 0.75 ± 0.03, respectively. Maternal genetic λ estimates ranged from 0.01 ± 0.01 (for BW) to 0.05 ± 0.01 (for CD). The results imply that λ values are heterogeneous in multivariate single-step genomic evaluation. Further studies are needed to investigate the consequences of using heterogenous λ values for direct genetic and maternal genetic components in multivariate single-step evaluation in terms of model dimensions, solver convergence rate, and model forward predictive ability.

## INTRODUCTION

Genomic selection has been implemented in Australia's BREEDPLAN genetic evaluation using single-step genomic methods (Johnston *et al.* 2018). An important step in genomics-assisted genetic evaluation involves the use of unbiased genetic parameters, including weighing factors or lambdas (λ), which affect the accuracies of genomic estimated breeding values (GEBVs). Optimum λ estimates are usually obtained using single-trait cross-validation procedures, although other methods have been suggested (Zhang *et al.* 2018).

It can be shown that a single-step model involving a scalar λ is simply a condensed form of an extended model containing two genetic factors, $u_A \sim N(0, A \otimes \Sigma_A)$ and $u_H \sim N(0, H \otimes \Sigma_H)$ where $A$ is the

*A joint venture of NSW Department of Primary Industries and the University of New England

pedigree derived numerator relationship matrix, $H$ is the joint relationship matrix, and $\Sigma_A$ and $\Sigma_H$ are covariance matrices. The total genetic variance ($\Sigma_T$) is equal to $\Sigma_A + \Sigma_H$, where it is assumed that $\Sigma_H \oslash \Sigma_T \equiv ii'\lambda$, where $i$ is an identity vector of respective dimension. However, in multivariate models or univariate models which contain several genetic factors (e.g. direct, maternal), $\Sigma_H \oslash \Sigma_T \equiv ii'\lambda$ is very unlikely. Heterogeneous λs can be estimated and accounted for by using the extended single-step model in variance component estimation and best linear unbiased prediction. To our knowledge, no studies have published λs estimated for categorical traits such as calving difficulty using extended multivariate linear-threshold single-step models. The objective of this study was to estimate variances and optimum λs with an extended single-step model for calving difficulty (CD) together with gestation length (GL), and birth weight (BW) using Australian Angus data.

## MATERIALS AND METHODS

**Data.** Phenotypes, genotypes, and pedigree for this study were obtained from the data extract submitted by Angus Australia for BREEDPLAN evaluation. Phenotypic data included CD, GL, and BW, with GL and BW were pre-adjusted for sex and age of the dam (Graser *et al.* 2005). GL and BW were measured in days and kilograms, respectively. CD was scored using three categories (Jeyaruban *et al.* 2016). Unassisted birth was represented by score of 1, while easy and hard pull were represented by scores of 2 and 3, respectively.

Contemporary groups (CGs) were formed according to the BREEDPLAN format (Graser *et al.* 2005). For each trait, CGs with less than 5 animals were discarded from the analysis. Further for CD, CGs with single CD score were eliminated. If the proportion of any score was less than 5% within a CG, that CG was discarded. The average number of observations per CG, for CD, GL, and BW were 84, 37, and 40, respectively.

The data consisted of 129,851 animals with phenotypes. Frequencies for CD scores 1, 2, and 3 were 78,653 (89.2%), 6,565 (7.4%), and 2,929 (3.3%), respectively. The number of dams with phenotypes for CD, GL, and BW were 7,536, 2,038, and 8,448, respectively. The pedigree included ancestors over 4 generations and consisted of 327,395 animals with 27,145 sires and 186,339 dams. A total of 45,575 animals were genotyped for 56,009 SNP markers. The GRM was constructed according to VanRaden (2008) and adding 0.0001 to the diagonal to guarantee positive definiteness.

**Analyses.** The data were analysed with multivariate model $y = Xb + Zu + Wp + e$ (model p) and multivariate extended single-step model $y = Xb + Zu_A + Zu_H + Wp + e$ (model h), where $y$ is a vector of phenotypic observations of GL, BW, and CD, $X$ is a block-diagonal design matrix linking fixed effects to their respective observations; $Z$ is a block-diagonal design matrix linking direct and maternal genetic effects in $u \sim N(0, A \otimes \Sigma_G), u_A \sim N(0, A \otimes \Sigma_A)$, and $u_H \sim N(0, H \otimes \Sigma_H)$ to their respective observations where $\Sigma_G$, $\Sigma_A$, and $\Sigma_H$ are co-variance matrices, $A$ is the pedigree derived numerator relationship matrix, and $H$ is the joint relationship matrix using the genomic relationship matrix ($G$) obtained as $G + I.001$, and $I$ is an identity matrix; $W$ is a block-diagonal design matrix linking maternal permanent environmental effects in $p \sim N(0, I \otimes \Sigma_p)$ to their respective observations, and $e \sim N(0, I \otimes \Sigma_E)$ is a vector of residuals. λ was estimated as $\Sigma_H \oslash \Sigma_T$, where $\Sigma_T$ is the total genetic variance obtained from model h.

Given the categorical nature of CD, a multivariate threshold model Gibbs sampling approach was used to estimate Σs in models p and h. Starting values for Σs for model p were obtained from heuristic partitioning of the observed phenotypic covariance matrix. Starting values for Σs for model h were derived from the results of model p, where $\Sigma_H = \Sigma_G \times 0.1$ and $\Sigma_A = \Sigma_G \times 0.9$. This implies that $\Sigma_T = \Sigma_H + \Sigma_A$. In all case, the prior degrees of freedom were zero. Results for models were confirmed by a second analysis with starting values $\Sigma_H = \Sigma_G \times 0.9$ and $\Sigma_A = \Sigma_G \times 0.1$. Thresholds for CD were fixed to be 0 and 1, and the residual variance was unconstrained. For all models, single chains of 200,000 iterations were sampled, and the first 50,000 samples were discarded as burn-in. To avoid autocorrelation, every 20$^{th}$ sample was stored, and a total of 7,500 samples were kept for computing posterior means and standard deviations.

## RESULTS AND DISCUSSION

Descriptive statistics for the studied traits are presented in Table 1. Table 2 presents the parameters derived from model p and h. The direct genetic and maternal genetic heritability estimates for CD, GL, and BW obtained from model p were similar to those obtained from model h when using $\Sigma_T$ (Table 2), implying that model h does not explain additional variation in the data compared to model p. The highest direct genetic heritability was obtained for GL (0.68), followed by BW (0.44) and CD (0.39), and the maternal heritability estimates ranged from 0.09 (for BW) to 0.17 (for CD).

**Table 1. Descriptive statistics for calving difficulty (CD, score), gestation length (GL, days), and birth weight (BW, kg)**

| Trait | Animals | % genotyped | Mean | Min* | Max* | SD* |
|---|---|---|---|---|---|---|
| CD | 88,147 | 4 | 1.1 | 1 | 3 | 0.4 |
| GL | 43,140 | 44 | 280.2 | 259.8 | 296.8 | 4.6 |
| BW | 102,864 | 42 | 37.5 | 16.0 | 65.9 | 5.0 |

*Min, minimum; Max, maximum; SD, standard deviation; ‡with phenotypes

**Table 2. Variances ($\sigma^2$), covariance ($\sigma$), heritabilities ($h^2$), and correlation ($r$) derived from pedigree only (p) and extended single-step (h) models for calving difficulty (CD), gestation length (GL), and birth weight (BW)**

| Parameter* | Model p | | | Model h | | |
|---|---|---|---|---|---|---|
| | CD | GL | BW | CD | GL | BW |
| $\sigma_d^2$ | 2.56 ± 0.28 | 12.69± 0.74 | 7.26 ± 0.30 | 2.59 ± 0.28 | 13.23± 0.59 | 7.83 ± 0.28 |
| $\sigma_m^2$ | 1.15 ± 0.16 | 1.89 ± 0.32 | 1.56 ± 0.13 | 1.15 ± 0.16 | 2.11±0.22 | 1.61 ± 0.13 |
| $\sigma_{d,m}$ | -0.68 ± 0.17 | -1.87 ± 0.38 | -0.45 ± 0.15 | -0.71 ± 0.17 | -2.26±0.30 | -0.64 ± 0.13 |
| $\sigma_c^2$ | 0.30 ± 0.13 | 0.66 ± 0.24 | 0.37 ± 0.10 | 0.31 ± 0.14 | 0.53±0.20 | 0.45 ± 0.11 |
| $\sigma_e^2$ | 3.32 ± 0.23 | 6.01 ± 0.39 | 9.14 ± 0.18 | 3.26 ± 0.21 | 5.97±0.34 | 8.67 ± 0.16 |
| $\sigma_p^2$ | 6.66 ± 0.23 | 19.38± 0.23 | 17.90± 0.11 | 6.61 ± 0.21 | 19.5 ± 0.22 | 17.92± 0.12 |
| $h_d^{2\ddagger}$ | 0.38 ± 0.04 | 0.65 ± 0.03 | 0.41 ± 0.02 | 0.39 ± 0.04 | 0.68 ± 0.02 | 0.44 ± 0.01 |
| $h_m^{2\ddagger}$ | 0.17 ± 0.02 | 0.10 ± 0.02 | 0.09 ± 0.01 | 0.17 ± 0.02 | 0.11 ± 0.01 | 0.09 ± 0.01 |
| $r_{d,m}$ | -0.39 ± 0.07 | -0.38 ± 0.05 | -0.13 ± 0.04 | -0.41 ± 0.07 | -0.43 ± 0.03 | -0.18 ± 0.03 |

*Direct genetic ($d$), maternal genetic ($m$), permanent environment effect of dam ($c$), residual ($e$) and total phenotypic ($p$) effects; ‡Total heritability estimates for model h.

A higher proportion of direct genetic variance is explained by the genomic factor than the polygenic factor in model h for BW and GL yielding λ values 0.75 and 0.62, respectively (Table 3 and Table 4). In contrast, the polygenic factor in model h explained the highest proportion of additive genetic variance for CD (Table 3) yielding a lower λ (0.36) than for BW and GL (Table 4). The observed pattern of variance partitioning between polygenic and genomic factors in model h for CD versus GL and BW suggests that λ values are highly sensitive to availability of genotypic information for each trait (Table 1), and therefore, heterogeneous in multivariate single-step genomic evaluation.

The maternal genetic λ estimates were near zero, and ranged from 0.01 for BW to 0.05 for CD (Table 4) due to greater maternal genetic variance partitioning for polygenic factor in model h (Table 3). This could be a result of low number of genotyped dams per trait in the current dataset, and therefore, maternal genetic λ values are expected to rise in the future if the number of genotyped dams increases.

**Table 3. Genetic variances ($\sigma^2$), covariance ($\sigma$), correlation ($r$), and heritabilities ($h^2$) for direct genetic ($d$) and maternal genetic ($m$) components accounted by the polygenic and genomic factors using the extended single-step model (model h) for calving difficulty (CD), gestation length (GL), and birth weight (BW)**

| | Polygenic factor | | | Genomic factor | | |
|---|---|---|---|---|---|---|
| | CD | GL | BW | CD | GL | BW |
| $\sigma_d^2$ | 1.66 ± 0.30 | 4.99 ± 0.56 | 1.97 ± 0.26 | 0.93 ± 0.04 | 8.23 ± 0.27 | 5.86 ± 0.15 |
| $\sigma_m^2$ | 1.10 ± 0.17 | 2.03 ± 0.22 | 1.60 ± 0.12 | 0.06 ± 0.01 | 0.08 ± 0.01 | 0.02 ± 0.01 |
| $\sigma_{d,m}$ | -0.54 ± 0.18 | -1.84 ± 0.30 | -0.61 ± 0.13 | -0.17 ± 0.02 | -0.42 ± 0.04 | -0.02 ± 0.03 |
| $h_d^2$ | 0.25 ± 0.04 | 0.25 ± 0.03 | 0.11 ± 0.01 | 0.14 ± 0.01 | 0.42 ± 0.01 | 0.33 ± 0.01 |
| $h_m^2$ | 0.17 ± 0.02 | 0.10 ± 0.01 | 0.09 ± 0.01 | 0.01 ± 0.03 | 0.00 ± 0.01 | 0.00 ± 0.01 |
| $r_{d,m}$ | -0.40 ± 0.09 | -0.58 ± 0.05 | -0.35 ± 0.06 | -0.76 ± 0.03 | -0.53 ± 0.04 | -0.08 ± 0.11 |

**Table 4. Estimates of lambdas (λ) for direct and maternal genetic components for selected traits using the multivariate linear-threshold single-step model (model h)**

| Λ | Calving difficulty | Gestation length | Birth weight |
|---|---|---|---|
| Direct genetic | 0.36 ± 0.05 | 0.62 ± 0.03 | 0.75 ± 0.03 |
| Maternal genetic | 0.05 ± 0.01 | 0.04 ± 0.01 | 0.01 ± 0.01 |

Our results suggest considering different λs for each trait rather than one global value across traits. An extended multivariate single-step model allows estimation of heterogeneous λs in variance component estimation. However, further studies are needed to investigate the consequences of using heterogenous λ estimates for multivariate evaluations in terms of model dimensions, solver convergence rate, and model forward predictability.

## CONCLUSIONS

By using an extended multivariate linear-threshold single-step model, heterogeneous direct genetic λs were obtained for GL, BW, and CD, which ranged from 0.36 (CD) to 0.75 (BW). Maternal genetic λ estimates ranged from 0.01 for BW to 0.05 for CD. Results suggest employing an extended single-step model with variance partitioning between genomic and polygenic factors accounting for heterogeneous λs in future BREEDPLAN genomic evaluation for the studied traits.

## ACKNOWLEDGEMENTS

The authors would like to thank the Meat and Livestock Australia (MLA) for their financial support through L.GEN.1704, and Angus Australia for providing data for this study.

## REFERENCES

Graser H.-U., Tier B., Johnston D.J., and Barwick S.A. (2005). *Aust. J. Exp. Agric.* **45**: 913.

Jeyaruban M.G., Johnston D., Tier B., and Graser H.-U. (2016) *Anim. Prod. Sci.* **56**: 927.

Johnston D.J, Ferdosi M., Connors N., Boerner V., Cook J., Girard C., Swan A., and Tier B. (2018). In *Proc. 11th World Cong. Genet. Appl. Livest. Prod.* **11:** 269.

VanRaden P.M. (2008). *J. of Dairy Sci.* **91**: 4414.

Zhang Y., Swan A., and Banks R. (2018). In *Proc. 11th World Cong. Genet. Appl. Livest. Prod.* **11:** 348.

*Contributed paper*

# ECONOMIC BENEFIT OF ADDITIONAL RECORDING FOR WELFARE TRAITS IN MATERNAL BREEDING OBJECTIVES FOR PIGS

**L. Vargovic, K.L. Bunter and S. Hermesch**

Animal Genetics Breeding Unit, University of New England, Armidale, NSW, 2351 Australia

## SUMMARY

The purpose of this study was to investigate, using selection index calculations, the economic benefits of improving welfare by expanding recording within traditional pig breeding programs to include welfare-related traits. The genetic parameters were adapted from several Australian studies. A basic breeding objective including average daily gain, backfat thickness and number of piglets born alive was extended to include welfare traits and feed conversion ratio (FCR). Welfare traits were: survival of piglets at farrowing (FS) and until weaning (PWS), weaning to conception interval (WCI), sow mature weight (MWT) and sow longevity (LONG). Sow appetite before farrowing (FRBF) and body condition before farrowing (CAL) were considered as additional selection criteria. When welfare traits were absent from the breeding objective and selection criteria, this resulted in reduced LONG, higher MWT, prolonged WCI and overall lower genetic response in the index in comparison with other scenarios. Valuing and recording welfare traits resulted in desirable responses for both production and welfare traits and increased overall economic merit. Including FCR in the breeding objective made it more difficult to improve welfare traits, particularly if FCR was recorded.

## INTRODUCTION

Historically, pig breeding programs focused on only economically important production and reproductive traits in breeding goals. Although very successful, this can have a detrimental impact on animal welfare (Rauw *et al.* 1998; Turner *et al.* 2018). To balance high performance and welfare, emphasis on welfare traits has increased, resulting in additional challenges for breeders. Welfare traits are difficult or expensive to measure, hard to assign economic values to (Olesen *et al.* 2000), and in some instances unfavourably genetically correlated with production traits (Kanis *et al.* 2005; Nielsen *et al.* 2011). Therefore, there can be a perception that more emphasis on welfare traits could result in slower overall genetic improvement. From a purely economic point of view, breeders could decide to dismiss welfare traits, and focus on short-term gain. However, although breeders are not necessarily paid for the enhanced welfare, the ethical value should not be neglected (Nielsen *et al.* 2011). The aim of this study was to quantify expected responses in individual production and welfare traits, and index changes depending on different selection criteria recorded, using recent knowledge of welfare traits and genetic correlations with production outcomes (Vargovic *et al.* 2019).

## MATERIALS AND METHODS

An appropriate breeding goal including both production and welfare-related traits has already been derived elsewhere for the Australian pig industry and was adapted here (Amer *et al.* 2014). The traits that are commonly available were average daily gain (ADG), backfat thickness (BF), number of piglets born alive (NBA), feed conversion ratio (FCR), proportion of piglets surviving at farrowing (FS), and from farrowing until weaning (PWS), weaning to conception interval between first and second parity (WCI), weight of sow when reaching maturity (MWT) and the number of parities a sow achieves during her lifetime (LONG). Two additional traits, currently not routinely recorded, were considered as selection criteria: feed refusal before farrowing (FRBF), defined as the proportion of days when sows refused more than half of their allocated feed from entry to the farrowing shed until farrowing, and caliper score (CAL), representing body condition of sows upon transfer to the farrowing shed (~ 7 days before farrowing). These traits were correlated with positive lactation outcomes (e.g. more piglets weaned) for breeding sows, implying improved welfare of both sow and piglets (Vargovic *et al.* 2019).

A consensus of assumed genetic parameters (Table 1) adapted from several Australian studies (Tholen *et al.* 1996; Hermesch *et al.* 2008; Bunter *et al.* 2010; Hermesch *et al.* 2015; Vargovic *et al.* 2019) was obtained. Since some of the traits (e.g. CAL or FRBF) were novel, correlations were assumed consistent with those previously reported for similar traits. Economic weights were expressed in $/gilt (Table 1). Repeatabilities for NBA (0.18), MWT (0.30), and CAL (0.25) were assumed to accommodate repeated records, and common litter effects were included for ADG (0.13), BF (0.05) and FCR (0.05). For other traits it was assumed that repeatabilities equalled heritabilities.

**Table 1: Economic weights (EW, $/gilt), genetic standard deviations (GSD), heritabilities (diagonal, bold), consensus genetic (below diagonal) and phenotypic (above diagonal) correlations**

| | ADG | BF | FCR | NBA | FS | PWS | WCI | LONG | MWT | CAL | FRBF |
|---|---|---|---|---|---|---|---|---|---|---|---|
| EW | 1.49 | -28.61 | -462.62 | 91.93 | 107.17 | 1092.88 | -3.60 | 86.90 | -4.17 | 0.00 | 0.00 |
| GSD | 31.7 | 1.15 | 0.25 | 0.83 | 0.08 | 0.03 | 2.54 | 0.69 | 9.02 | 1.42 | 0.41 |
| ADG | **0.21** | 0.11 | -0.20 | -0.04 | -0.01 | 0.03 | 0.02 | -0.05 | 0.32 | 0.05 | -0.16 |
| BF | 0.02 | **0.38** | 0.06 | 0.02 | 0.05 | 0.03 | -0.02 | 0.10 | -0.01 | 0.32 | 0.00 |
| FCR | -0.37 | 0.10 | **0.25** | -0.01 | -0.02 | -0.03 | 0.01 | 0.03 | -0.10 | 0.04 | 0.02 |
| NBA | -0.19 | -0.02 | -0.07 | **0.09** | 0.07 | 0.07 | -0.02 | 0.22 | 0.00 | 0.01 | -0.04 |
| FS | -0.01 | 0.00 | 0.05 | 0.13 | **0.13** | 0.06 | -0.15 | 0.12 | 0.18 | -0.01 | -0.12 |
| PWS | 0.27 | 0.07 | -0.02 | -0.19 | 0.28 | **0.05** | -0.21 | 0.00 | 0.16 | 0.02 | -0.10 |
| WCI | -0.09 | -0.24 | -0.15 | -0.20 | 0.09 | -0.15 | **0.08** | -0.05 | 0.00 | -0.02 | -0.01 |
| LONG | -0.28 | 0.35 | -0.02 | 0.30 | -0.25 | 0.18 | -0.22 | **0.14** | -0.03 | 0.09 | 0.00 |
| MWT | 0.30 | -0.12 | -0.15 | -0.21 | 0.09 | -0.22 | 0.00 | 0.10 | **0.18** | 0.26 | 0.03 |
| CAL | 0.28 | 0.32 | 0.18 | -0.07 | 0.19 | 0.21 | -0.14 | 0.33 | 0.15 | **0.34** | 0.01 |
| FRBF | -0.21 | 0.02 | -0.07 | 0.30 | 0.00 | -0.32 | -0.36 | -0.21 | -0.27 | -0.13 | **0.21** |

Abbreviations: ADG: average daily gain (g/day); BF: backfat thickness (mm); FCR: feed conversion ratio (kg feed/kg gain); NBA: number of born alive piglets (piglets/litter); FS: farrowing survival (proportion); PWS: pre-weaning survival (proportion); WCI: wean to conception interval between first and second parity (days); LONG: longevity (number of parities); MWT: sow mature weight (kg); CAL: number of increments on caliper; FRBF: proportion of days where sows refused more than half of their daily allocation (proportion)

Index calculations (Hazel 1943) were performed using the MTIndex program (https://jvanderw.une.edu.au/) to obtain relative responses for trait and index combinations. These predicted responses are for a single generation of selection with a selection intensity of one. It was assumed that ADG and BF were available for the selection candidate, dam, sire, six full sibs and 40 half-sibs. For FCR, data was available for sire, one full sib and five half-sibs. Data for NBA, FS, PWS, WCI, CAL and FRBF were available for the dam (two records, except WCI, one record) and three half-sibs. For LONG and MWT, the information was available for a dam only. The study investigated how response in individual traits and the index changed depending on what selection criteria are recorded, for a simple production breeding objective (Scenario 1) or a breeding objective including FCR and welfare traits (Scenario 2 or higher). Six different scenarios were compared: Scenario 1: economic weights and recording for ADG, BF and NBA only, without welfare traits; 2) Scenario 2 (Base): Breeding objective with welfare traits + recording of economic traits ADG, BF and NBA only; 3) Scenario 3: Scenario 2 (Base) + recording of welfare traits FS + PWS + WCI and LONG; 4) Scenario 4: Scenario 3 + recording of MWT; 5) Scenario 5: Scenario 4 + recording of CAL and FRBF; and 6) Scenario 6: Scenario 4 +recording of FCR.

**RESULTS AND DISCUSSION**

The full breeding objective with welfare traits values FCR and some welfare-related traits (FS, PWS, WCI, LONG and MWT). When the breeding objective ignores the importance of welfare traits (Scenario 1), selection for production traits resulted in reduced (e.g. PWS) or undesirable (e.g. WCI, LONG, MWT) responses, but a desirable response in FCR. Applying the same selection criteria with the full breeding objective (Scenario 2) increased desirable responses across all welfare traits and increased index response from $36.28 to $43.17. At the same time, favourable responses in production traits were retained, but with different emphasis (e.g. increased ADG, reduced response in BF).

**Table 2: Predicted genetic changes under different scenarios with overall selection response (ΔG in $/gilt), accuracy of index (Acc) and response relative (RR) to Scenario 2 after one generation assuming a selection intensity of one**

| | Trait | Scenario 1 | Scenario 2 | Scenario 3 | Scenario 4 | Scenario 5 | Scenario 6 |
|---|---|---|---|---|---|---|---|
| | | BO 1 | BO 2 | | | | |
| Productivity | ADG | *10.8* | *14.7* | *14.1* | *13.9* | *13.5* | *10.6* |
| and | BF | *-0.584* | *-0.249* | *-0.233* | *-0.227* | *-0.223* | *-0.162* |
| efficiency | FCR | -0.05 | -0.05 | -0.05 | -0.05 | -0.05 | *-0.09* |
| | NBA | *0.037* | *0.035* | *0.034* | *0.038* | *0.052* | *0.053* |
| Welfare | FS | 0.0008 | 0.0010 | *-0.0008* | *-0.0005* | *0.0000* | *-0.0014* |
| | PWS | 0.0011 | 0.0023 | *0.0029* | *0.0033* | *0.0029* | *0.0020* |
| | WCI | 0.18 | -0.04 | *-0.09* | *-0.09* | *-0.14* | *0.04* |
| | LONG | -0.17 | -0.12 | *-0.09* | *-0.09* | *-0.10* | *-0.04* |
| | MWT | 1.36 | 1.33 | 1.29 | *0.93* | *0.75* | *0.78* |
| Selection criteria | CAL | -0.10 | 0.08 | 0.10 | 0.09 | *0.08* | -0.02 |
| (additional welfare traits) | FRBF | -0.02 | -0.03 | -0.03 | -0.03 | *-0.004* | -0.007 |
| | ΔG ($) | **36.28** | **43.17** | **44.94** | **45.75** | **46.62** | **63.69** |
| | Acc | **0.414** | **0.244** | **0.254** | **0.258** | **0.263** | **0.359** |
| | RR | **84.03** | **100.00** | **104.11** | **105.99** | **107.99** | **147.54** |

For trait abbreviations see Table 1. BO1: breeding objective without welfare traits, BO2: breeding objective with welfare traits. Italicized traits are recorded within the scenario.

As additional selection criteria were used (Scenarios 3 to 6) overall index response increased. When information that is readily available from herd recording systems (FS, PWS, WCI, LONG and MWT) was added in Scenario 3, the undesirable responses observed in the breeding objective traits PWS, WCI, LONG and MWT were reduced, and the index response was higher ($44.94, +4.11%). At the same time, there were marginal changes in production trait responses, which does not support the general perception that including welfare traits into breeding programs will result in slower overall genetic improvement. When all of the production and reproductive traits were available, additional selection criteria such as MWT were more effective at maximising the overall response (Scenario 4), relative to Scenarios 2 and 3. Adding CAL and FRBF records (Scenario 5), which were not part of the breeding objective, additionally increased overall response (relative to Scenario 4) by 2.00%. The largest differences in individual trait responses were for MWT, WCI and NBA, whereas changes for other traits were small. In the present study, both CAL and FRBF are relatively inexpensive to implement, considering that the information is typically recorded in the farrowing shed for management purposes, but the resulting data may not be stored.

Feed conversion ratio is not routinely recorded in maternal lines, but is of economic importance. If FCR is recorded, the overall response will be higher ($60.78/gilt) in a simple production breeding objective (BO1 with FCR included). However, correlated responses for welfare traits had undesirable directions (not shown). The trait FCR is costly to record, and has negative consequences that are not properly valued. In an attempt to combat these detrimental responses, Scenario 6 included recording for both FCR and welfare traits. This resulted in the largest overall response in comparison to the other scenarios (47.5% increase and a response of $63.69). However, the strong emphasis on feed efficiency resulted in undesirable (PWS, WCI, CAL) or lower response (ADG) for other traits, despite their contribution to the index. In general, if there is a need for a change in the overall response, a trait with high economic emphasis should be recorded. However, recording patterns are driven by both biological and cost constraints, and outcomes depend on the assumed economic weight, parameters and recording patterns. This could suggest reinvestigation of the calculations for assumed trait economic weights, if welfare traits are to be maintained.

## CONCLUSIONS

Extending pig breeding programs with welfare traits that are correlated with performance outcomes results in long-term genetic gain. The overall economic value per pig increased, making these traits attractive for incorporating into breeding programs. However, larger data sets with welfare traits recorded may be required to obtain more accurate estimates of genetic correlations between traits, to ensure these index calculations are representative of likely outcomes.

## REFERENCES

Amer P., Ludemann C. and Hermesch S. (2014) *J. Anim. Sci.* **92:** 5345.

Bunter K. L., Lewis C. R. G., Hermesch S., Smits R. and Luxford B. G. (2010) *Proc. World Congr. Genet. App. Livest. Breed* **9**: 71.

Hazel, L.N. (1943) *Genetics.* **6:** 476.

Hermesch S., Arnal M., Börner V. and Dominik S. (2015) *Proc. Assoc. Advmt. Anim. Breed. Genet* **21:**1.

Hermesch S., Jones R. M. and Bunter K. L. (2008) Pig Genetics Workshop Notes, pp 55-65.

Kanis E., De Greef K. H., Hiemstra A. and Van Arendonk J. A. M. (2005) *J. Anim. Sci.* **83:** 948.

Nielsen H.M., Olesen I., Navrud S., Kolstad K. and Amer P. (2011) J. Agric. Environ. Ethics **24:** 309.

Olesen I., Groen A. F. and Gjerde B. (2000) *J. Anim. Sci.* **78:** 570.

Rauw W.M., Kanis E., Noordhuizen-Stassen E.N. and Grommers F.J. (1998) *Livest.Prod.Sci.* **56:** 15.

Tholen E., Bunter K., Hermesch S. and Graser H. (1996) *Aust. J. Agric. Res.* **47:** 1261.

Turner S., Camerlink I., Baxter E. M., D'Eath R. B., Desire S. and Roehe R. (2018) In 'Advances in Pig Welfare'. pp 399-414, editor P. Osborn, Woodhead Publishing.

Vargovic L., Bunter K. L., Hermesch S., Harper J. and Sokolinski R. (2019) *Proc. Assoc. Advmt. Anim. Breed. Genet.* **23:**147.

*Contributed paper*

# GENETIC PARAMETERS FOR URINALYSIS TRAITS RECORDED ON GESTATING SOWS

**L. Vargovic, S. Hermesch and K.L. Bunter**

Animal Genetics Breeding Unit*, University of New England, Armidale, NSW, 2351 Australia

## SUMMARY

Urinalysis can be used to detect sows that typically remain unidentified with health conditions such as urinary tract infection, and also provides data on physiological variables reflecting metabolic status (e.g. glucose, ketones). The urine was collected from gilts and sows (N=694) after animals were transferred to the farrowing shed. The traits were defined from the urinalysis test strip results, with additional subjective measurements of odour, colour and turbidity. Subsequently, a trait representing urinary tract infection was defined. Heritability estimates were in a range 0.08 to 0.36, except for the presence of blood (0.03). Strong genetic correlations were estimated between bilirubin and urobilinogen (0.78), but not other trait combinations. The study demonstrated that several urinalysis traits could be considered as selection criteria for increasing the health status of sows. However, alternative procedures to collect phenotypes are required to improve ease of data collection. The associations of urinalysis parameters with breeding objective traits requires further investigation.

## INTRODUCTION

Undetected and untreated urinary tract infection (UTI), high ketones (demonstrating metabolic disorders) or low Vitamin C (potentially demonstrating nutritional deficiency) result in poor performance and increased sow removals (Almond 2005; Theil *et al.* 2013; Nielsen *et al.* 2019). Mazutti *et al.* (2013) proposed using reagent strips to detect UTI routinely, enabling treatment. In this study, we estimated genetic parameters for urinalysis variables recorded with reagent strips using urine samples from late gestation sows. To the knowledge of the authors, no previous study reported parameters for similar traits. Our hypothesis was that variation in health and metabolic state has a genetic basis, such that variables obtained from urinalysis are heritable, potentially providing opportunities to enhance health status of sows from both production and genetic perspectives.

## MATERIALS AND METHODS

**Data.** Urinalysis data was recorded at two independent nucleus farms over 5 weeks, between October-November 2017 (Farm A, N = 254 sows), and 8 weeks between March-May 2017 (Farm B, N = 440 sows) according to kit instructions (CombiScreen®VET 11 PLUS). Urine was collected once per sow, on average 5 days before farrowing, in the early morning before the first feeding event. The urinalysis test strips evaluated levels of bilirubin, urobilinogen, ketones, Vitamin C, glucose, protein, blood, pH, nitrite, leucocytes, and specific gravity according to CombiScreen®VET 11 PLUS (Table 1). Each variable was scored in levels representing concentrations. Urine colour was subjectively scored on a scale of 1 to 3 (pale, normal, dark), while odour and turbidity were scored as absent (0) or present (1). The presence of a urinary tract infection (UTI) was inferred as absent (0) or present (1) if nitrite was positive and pH $\geq$ 6. Levels for urinalysis parameters are shown in Table 1.

---

* A joint venture of NSW Department of Primary Industries and the University of New England

**Analyses.** Data preparation was carried out using R (R Core Team 2020). During the process of data preparation, where the description of levels on test strip was expressed in characters, observed values were replaced with numeric values (e.g. nitrite + or ++ to 1 or 2). The square root transformation was performed for ketones, glucose, protein, blood and leucocytes, and for specific gravity x 100 (recorded in 0.005 increments), due to the non-normal distribution of data for these traits. Variances are presented on the transformed scale.

Data from both farms were combined to estimate genetic parameters. Sows were progeny of 283 sires and 553 dams. The pedigree was extended back by 5 generations to a total of 1261 sires and 3274 dams. Estimates of variance components were obtained by fitting a linear mixed animal model using residual maximum likelihood procedures in ASReml (Gilmour *et al.* 2014). Systematic effects were parity group (4 levels: parities 1, 2, 3-4, >4) and selection line nested within farm (10 levels). Genetic ($r_A$) and phenotypic ($r_P$) correlations were estimated using a series of bivariate analyses. Both genetic and phenotypic correlations were reported only for traits where heritability was above 0.05, due to the large standard errors, which would result in unreliable estimates for genetic correlations.

## RESULTS AND DISCUSSION

**Heritability estimates.** Bilirubin and urobilinogen, typically indicators of liver disease, were moderately heritable, 0.33 ± 0.13 and 0.34 ± 0.14 (Table 1).

**Table 1: Urinalysis variables with recording levels, normal values and percentage of sows, heritability estimates ($h^2$), phenotypic variance ($\sigma^2_p$) and the coefficient of determination ($R^2$)**

| Trait | Units | Levels | Normal values | Normal (%) | $h^2 \pm$ SE | $\sigma^2_p$ | $R^2$ (%) |
|---|---|---|---|---|---|---|---|
| Bili | mg/dl | 0,1,2,4 | 0 | 60.5 | 0.33 ± 0.13 | 1.20 | 2.84 |
| Uro | mg/dl | 0,2,4 | 0 | 78.2 | 0.34 ± 0.14 | 0.61 | 20.6 |
| Ket* | mg/dl | 0,10,25,100 | 0 | 96.1 | 0.10 ± 0.11 | 0.96 | 0.46 |
| Vit C | mg/dl | 0,1,2 | 0 | 24.5 | 0.16 ± 0.10 | 0.45 | 3.60 |
| Glucose* | mg/dl | 0,2,5,14,28 | 0 | 92.1 | NE | | |
| Protein* | mg/dl | 0,15,30,100,500 | 0 | 27.3 | 0.20 ± 0.13 | 11.5 | 8.16 |
| Blood* | Ery/ml | 0,10,50,300 | 0 | 80.5 | 0.03 ± 0.09 | 6.95 | 1.28 |
| pH§ | | 5.0-8.5 | 5.5-8.0 | 48.9 | 0.11 ± 0.10 | 0.44 | 0.26 |
| Nitrite | μmol/l | 0,1,2 | 0 | 93.9 | 0.36 ± 0.13 | 0.19 | 0.00 |
| Leuco* | Leuco/μl | 0,25,75,500 | 0 | 95.0 | NE | | |
| SpecG† | | 1-1.03 | 1.02-1.04 | 54.7 | 0.33 ± 0.13 | 1.00 | 1.77 |
| UTI | 0/1 | 0,1 | 0 | 95.4 | 0.20 ± 0.12 | 0.04 | 0.00 |
| Odour | 0/1 | 0,1 | 0 | 74.8 | 0.32 ± 0.13 | 0.17 | 10.8 |
| Colour | | 1-3 | 1-3 | | 0.14 ± 0.10 | 0.38 | 2.04 |
| Turb | 0/1 | 0,1 | 0 | 73.1 | 0.08 ± 0.10 | 0.15 | 0.60 |

Notes: * square root transformation applied; § in 0.5 increments; Bili: bilirubin; Uro: urobilinogen; Ket: ketones; Vit C: Vitamin C; Leuco: leucocytes; SpecG † specific gravity in 0.005 increments (×100); Turb: turbidity; NE: not estimable

The heritability estimate for ketones was 0.10 ± 0.11. An increased level of ketones in urine is an indicator of ketosis or diabetes mellitus. Although ketosis in sows is not well investigated, it could affect feed intake during lactation, and consequently weight and fat loss during the late gestation and lactation (Alsop *et al.* 1994). Urine colour and odour were lowly to moderately heritable (0.14 ± 0.10 and 0.32 ± 0.13). These two traits are strongly dependent on the concentration of urine. In contrast, heritability for turbidity was low, 0.08 ± 0.10. Turbidity may indicate presence of contaminants, such as blood, bacteria, epithelia, cells or crystals, implying non-genetic variation. Similarly, the presence of blood in urine was not heritable. Heritabilities were not estimable for glucose and leucocytes due to the low number of positive animals for these measurements in one of the farms. The failure to detect urinary glucose pre-farrowing might reflect the practice of restricted feeding during gestation and pre-farrowing. The heritability for UTI inferred from nitrite and pH levels was 0.20 ± 0.12. This trait is considered as a predisposing factor for reproductive disorders, MMA (mastitis-metritis-agalactia), and lower milk production (Petersen 1983). Therefore, genetic variation in predisposition to developing UTI might be related to genetic variation in health and reproductive performance of sows.

**Correlations.** The genetic correlation ($r_A$) between bilirubin and urobilinogen was strong (0.78 ± 0.19, Table 2) and both are indicators of liver disease.

**Table 2: Genetic (above diagonal) and phenotypic correlations for urinalysis traits with standard errors in subscript**

| | Bili | Uro | Ket | Vit C | Protein | pH | Nitrite | SpecG | UTI | Odour | Colour | Turb |
|---|---|---|---|---|---|---|---|---|---|---|---|---|
| Bili | | **0.78 (0.19)** | 0.07 (0.51) | **0.99 (0.20)** | **0.85 (0.17)** | -0.07 (0.44) | 0.25 (0.28) | **0.48 (0.23)** | -0.04 (0.36) | 0.35 (0.26) | 0.29 (0.35) | 0.43 (0.50) |
| Uro | **0.46 (0.03)** | | -0.17 (0.48) | 0.45 (0.31) | **0.74 (0.27)** | 0.21 (0.46) | 0.21 (0.28) | 0.28 (0.28) | -0.20 (0.37) | 0.46 (0.27) | 0.20 (0.39) | 0.34 (0.49) |
| Ket | -0.04 (0.04) | -0.01 (0.04) | | -0.27 (0.53) | -0.06 (0.60) | 0.81 (0.53) | -0.99 (0.63) | -0.37 (0.39) | -0.64 (0.55) | -0.97 (0.62) | -0.30 (0.54) | -0.38 (0.88) |
| Vit C | **0.36 (0.04)** | **0.22 (0.04)** | 0.00 (0.04) | | **0.60 (0.31)** | 0.60 (0.57) | 0.01 (0.36) | 0.47 (0.28) | -0.68 (0.47) | 0.29 (0.34) | **0.94 (0.32)** | 0.62 (0.56) |
| Protein | **0.55 (0.03)** | **0.32 (0.04)** | 0.01 (0.04) | **0.31 (0.04)** | | -0.16 (0.56) | -0.13 (0.35) | 0.22 (0.34) | -0.26 (0.44) | 0.22 (0.36) | 0.40 (0.42) | 0.73 (0.58) |
| pH | **0.15 (0.04)** | **0.12 (0.04)** | -0.04 (0.04) | **-0.12 (0.04)** | 0.05 (0.04) | | 0.78 (0.58) | -0.39 (0.39) | 0.14 (0.51) | 0.16 (0.45) | 0.17 (0.58) | -0.08 (0.80) |
| Nitrite | **0.11 (0.04)** | **0.20 (0.04)** | 0.02 (0.04) | 0.02 (0.04) | **0.16 (0.04)** | 0.01 (0.04) | | -0.51 (0.30) | **0.96 (0.19)** | **1.00 (0.31)** | -0.36 (0.40) | 0.79 (0.47) |
| SpecG | **0.38 (0.04)** | **0.17 (0.04)** | 0.03 (0.04) | **0.46 (0.03)** | **0.33 (0.04)** | **-0.35 (0.04)** | -0.03 (0.04) | | **-0.86 (0.41)** | 0.27 (0.27) | **0.78 (0.23)** | 0.29 (0.47) |
| UTI | 0.06 (0.04) | **0.09 (0.04)** | -0.01 (0.04) | -0.02 (0.04) | **0.12 (0.04)** | **0.22 (0.04)** | **0.63 (0.02)** | **-0.12 (0.04)** | | **0.91 (0.37)** | -0.58 (0.54) | 0.92 (0.62) |
| Odour | **0.30 (0.04)** | **0.18 (0.04)** | -0.04 (0.04) | **0.10 (0.04)** | **0.25 (0.04)** | 0.07 (0.04) | **0.29 (0.04)** | **0.18 (0.04)** | **0.23 (0.04)** | | 0.12 (0.39) | 0.70 (0.39) |
| Colour | **0.42 (0.03)** | **0.21 (0.04)** | -0.04 (0.04) | **0.40 (0.03)** | **0.35 (0.04)** | -0.01 (0.04) | -0.03 (0.04) | **0.59 (0.03)** | -0.02 (0.04) | **0.21 (0.04)** | | 0.05 (0.67) |
| Turb | **0.21 (0.04)** | **0.22 (0.04)** | -0.07 (0.04) | 0.03 (0.04) | **0.19 (0.04)** | **0.18 (0.04)** | **0.27 (0.04)** | 0.06 (0.04) | **0.22 (0.04)** | **0.53 (0.03)** | **0.12 (0.04)** | |

Bili: bilirubin; Uro: urobilinogen; Ket: ketones; Vit C: Vitamin C; SpecG: specific gravity; Turb: turbidity.
Bold values are significantly different from zero.

Phenotypic correlations ($r_P$) amongst several urinalysis traits suggest that values were not fully independent of each other. This could be because of correlated errors due solely to the test strip chemistry, or because urinalysis variables were correlated with each other from a physiological perspective. High Vitamin C levels can result in a false positives for bilirubin and nitrite (CombiScreen®VET 11 PLUS). Therefore, it is unclear whether strong $r_A$ (0.99 ± 0.20) and moderate $r_P$ (0.36 ± 0.04) correlations between Vitamin C and bilirubin were due to this effect. Phenotypic correlations were moderate to high between specific parameters and the traits that were conditional on their values (e.g. UTI and nitrite or pH levels). Scored variables (colour, odour, turbidity) were positively associated with each other and also many urinalysis parameters, indicating abnormal levels, but are not diagnostic of particular conditions. Strong, positive correlations were found between odour and nitrite and/or UTI, suggesting that strong urine odour could potentially be used as a trigger to initiate testing for the presence of an infection and treatment where required.

## CONCLUSIONS

Urinalysis provides an opportunity to obtain data related to health and metabolic status of sows, which has utility from both management and genetic perspectives, and can be considered as selection criteria for breeding programs. However, associations with other selection criteria or breeding objective traits are required, along with proof that selection would generate meaningful changes in the health status of sows. Phenotypically, test strips are a useful tool to identify and treat unwell sows. However, collecting urine samples is laborious and better strategies need to be developed for routine recording.

## ACKNOWLEDGEMENT

This research was funded by the Australasian Pork Research Institute Ltd. (project 2A-116). The first author was supported by UNE via International Postgraduate Research Award (UNE IPRA).

## REFERENCES

Almond G.W. (2005) *Proceedings of the Iowa State University Swine Disease Conference for Swine Practitioners*:7. North Carolina State University:

Alsop J.E., Hurnik D. and Bildfell R.J. (1994) *J. Swine Health Prod.* **2:** 5.

de Quatrebarbes A., Pommellet C., Tee C.Y. and Ong L.P. (2014) *Proceedings of the 23rd IPVS Congress, Cancun, Mexico – June 8-11, 2014.*

Gilmour A.R., Gogel B.J., Cullis B.R., Welham S.J. and Thompson R. (2014). ASReml User Guide Release 4.1 Functional Specification. VSN International Ltd, Hemel Hempstead, HP1 1ES, UK.

Mazutti K., Locatelli-Dittrich R., Lunardon I., Kuchiishi S.S., Lara A.C. d., Zotti E. and Alberton G.C. (2013) *Pesqui. Vet. Brasa.* **33:** 1103.

Nielsen T.K., Petersen N.A., Stærk K., Grønnemose R.B., Palarasah Y., Nielsen L.F., Kolmos H.J., Andersen T.E. and Lund L. (2019) *Front. Microbiol.* **10:**

Petersen B. (1983) *Livest. Prod. Sci.* **10:** 253.

R Core Team (2020) 'R: A language and environment for statistical computing. R Foundation for Statistical Computing, Vienna, Austria.' Vienna, Austria).

Theil P.K., Olesen A.K., Flummer C., Sørensen G. and Kristensen N.B. (2013) *J. Anim. Sci.* **91:** 772.

*Contributed paper*

# PROPOSED GENETIC IMPROVEMENT STRATEGIES FOR DAIRY CATTLE IN KENYA

**P.K. Wahinya[1], A.A. Swan[1] and M.G. Jeyaruban[1]**

[1] Animal Genetics Breeding Unit*, University of New England, Armidale, NSW, 2350 Australia

## SUMMARY

Genotype by environment interactions and heterogeneity of variance may influence the effectiveness of breeding programs in developing countries. This study investigated optimization of dairy cattle breeding programs within Kenya for low, medium and high input and output production systems in the presence of genotype by environment interactions. Multi-trait selection index theory was applied using the SelAction software package to determine the optimum strategy that would maximise genetic gain across the three production systems. The breeding goal was to maximise overall gain for a breeding objective containing three traits: lactation milk yield; lactation fat yield and calving interval. Three selection strategies based on: 1) sire evaluation and selection within the high production systems only (*single*); 2) independent sire evaluation and selection within each production system (*independent*) and 3) sire evaluation across all production systems (*joint*), were evaluated under scenarios using progeny test information and genomic information. The joint strategy maximised the overall economic gain (1583 Kes) while the single strategy generated the least overall gain (1311 Kes). The dairy industry in Kenya would therefore benefit from implementing production system specific breeding strategies for bull evaluation and selection. In addition, implementing genomic selection could speed up the rate of genetic gain compared to progeny testing due to reductions in generation interval and higher selection accuracy with a moderately large reference population.

## INTRODUCTION

Breeding programs are designed to generate and disseminate genetic improvement. The classical approach starts with definition of breeding objectives, followed by development of selection criteria, implementation of genetic evaluation allowing for the selection of superior animals, design of sustainable mating systems, and strategies to disseminate genetic superiority to commercial producers. Selection is however, challenging in developing countries where dairy cattle production systems are highly variable in terms of inputs and outputs (Wahinya *et al.* 2020). When sires are selected from a different environment, for example, from high input and output production systems or via the importation of semen, genotype by environment interactions can result in genotypes re-ranking and reduce selection efficiency. This is likely to affect the accuracy of selection and the rate of genetic progress. In developing countries, genetic improvement programs are more likely to be successful if they are developed as an integrated livestock-production package and not in isolation (Kahi *et al.* 2005). This paper will evaluate different strategies to maximise overall genetic gain across three dairy production systems.

## MATERIALS AND METHODS

**Definition of environment and simulation of population structure.** Definition of the target production system(s) is required for effective implementation of selection strategies. Multiple variables can be used to define production systems. Clustering of herds based on milk production

---

* A joint venture of NSW Department of Primary Industries and the University of New England

level has been applied as a classifier to quantify the influence environment has on performance (Ojango *et al.* 2019; Wahinya *et al.* 2020). Low, medium and high production systems were defined by Wahinya *et al.* (2020) using K-means cluster analysis of herd means for 305 days milk yield and applied in this study. Production parameters identified in that analysis were used to simulate three nucleus populations comprising a total of 5000 dams in each. A total of 219 sires were assumed across the three production systems. Sires and dams were spread across 8 age-classes. Every year 10 bulls and 300 cows were selected for each production system. Dams were assumed to produce their first offspring in their third year while progeny information for the bulls was available at six years of age. The sex ratio was assumed to be 0.50 while calving and annual survival rates were 0.67, 0.74 and 0.77, and 0.90, 0.93 and 0.94 for the low, medium and high production systems, respectively.

**Breeding strategies**. Truncation selection was simulated using multi-trait index selection. Genetic gains were predicted for a dairy cattle breeding objective containing lactation milk yield (LMY in kg), fat yield (FY in kg) and calving interval (CI in days) under three production systems in Kenya. Animals were available for selection when all the information needed for selection was available. Male candidates were evaluated based on their half-sib sisters, daughters and dam's information while females were evaluated on their performance records, half-sib sisters and parent's information. An animal model was assumed for genetic evaluation considering all genetic relationships to estimate the breeding values for selection. Three selection strategies were evaluated based on the groups of test-bulls: 1) a breeding program for a single production system with bull evaluation and selection in the high production system (single); 2) production system specific breeding programs, each with bull evaluation and selection within each environment (independent) and 3) a joint breeding program with bull evaluation and selection in all three production systems (joint). These strategies were simulated under two scenarios using progeny testing and using both phenotypes and genotypes for selection. Genomic selection was simulated by adding an extra trait to represent the marker information as described in Dekkers (2007). Marker information was modelled using a trait correlated to the original trait with a heritability close to 1 (0.999). The accuracy of the estimated genomic breeding values was represented by the correlation between the original trait and the trait specified by marker information. The accuracy of genomic information is dependent on the reference population, the correlation between the true breeding value of a genotyped animal and phenotype as well as the effective population size and was calculated as shown in Dekkers (2007). Six strategies were therefore evaluated in total. The breeding program aimed to maximise genetic gain in the overall objective as follows:

$$\Delta G = \Delta G_L + \Delta G_M + \Delta G_H$$

where $\Delta G_L$, $\Delta G_M$ and $\Delta G_H$ are the genetic gains in the low, medium and high production systems, respectively. The SelAction software package (Rutten *et al.* 2002) was used to predict genetic gains using a multi-trait selection index. The genetic and phenotypic standard deviations, economic values, heritabilities, genetic and phenotypic correlations used for the traits under the low, medium and high production systems are shown in Tables 1 and 2.

**Table 1: Genetic ($\sigma_a$) and phenotypic standard deviations ($\sigma_p$), and economic weights (EW) for lactation milk yield (LMY; kg), fat yield (FY; kg) and calving interval (CI; days) under low, medium and high production systems**

| Trait | Low | | | Medium | | | High | | |
|---|---|---|---|---|---|---|---|---|---|
| | LMY | FY | CI | LMY | FY | CI | LMY | FY | CI |
| $\sigma_a$ | 285.94 | 9.94 | 33.30 | 467.32 | 26.97 | 15.81 | 613.03 | 28.66 | 13.72 |
| $\sigma_p$ | 626.1 | 29.7 | 130.85 | 923.12 | 60.47 | 97.56 | 1226.38 | 56.84 | 68.01 |
| EW | 22.63 | 51.3 | -114.69 | 21.45 | 56.91 | -180.42 | 22.28 | 61.54 | -296.71 |

Source: (Wahinya 2020)

**Table 2: Heritabilities (diagonal), genetic (above diagonal) and phenotypic (below diagonal) correlations for lactation milk yield (LMY; kg), fat yield (FY; kg) and calving interval (CI; days) under low, medium and high production systems**

| System | Trait | Low | | | Medium | | | High | | |
|---|---|---|---|---|---|---|---|---|---|---|
| | | LMY | FY | CI | LMY | FY | CI | LMY | FY | CI |
| Low | LMY | 0.21 | 0.65 | -0.11 | 0.42 | 0.56 | -0.46 | 0.64 | 0.62 | 0.00 |
| | FY | 0.83 | 0.11 | 0.02 | 0.56 | 0.33 | -0.01 | 0.66 | 0.84 | 0.15 |
| | CI | -0.01 | 0.08 | 0.06 | -0.53 | 0.03 | 0.05 | -0.07 | 0.06 | 0.08 |
| Medium | LMY | 0 | 0 | 0 | 0.26 | 0.54 | 0.34 | 0.75 | 0.61 | 0.51 |
| | FY | 0 | 0 | 0 | 0.84 | 0.20 | -0.04 | 0.65 | 0.58 | 0.15 |
| | CI | 0 | 0 | 0 | 0.02 | 0.08 | 0.03 | 0.14 | 0.03 | 0.62 |
| High | LMY | 0 | 0 | 0 | 0 | 0 | 0 | 0.25 | 0.73 | 0.43 |
| | FY | 0 | 0 | 0 | 0 | 0 | 0 | 0.81 | 0.25 | 0.14 |
| | CI | 0 | 0 | 0 | 0 | 0 | 0 | 0.04 | 0.08 | 0.04 |

Source: (Wahinya 2020); bending was used to make the correlation matrix positive definite

Genetic gain was predicted using equilibrium parameters to account for the accumulation of pedigree information and reduction in genetic variance due to selection. The proportion of cows in the low (0.30), medium (0.33) and high (0.37) production systems in Wahinya *et al.* (2020) were used to weight the gains in the respective production systems to estimate the overall genetic gain.

**RESULTS AND DISCUSSION**

Animal breeders are faced with a challenge to implement selection in the presence of genotype by environment interactions which can be the case when ranking animals based on breeding values across environments. This has implications for the implementation of optimal design in breeding programs across environments (Mulder and Bijma 2006). Genetic improvement of dairy cattle in Kenya is currently based on genetic evaluation using pedigree information and selection implemented under high input and output production systems. However, due to the heterogeneity of variance across production systems, sires are re-ranked between the production systems (Wahinya *et al.* 2020). Animals from herds with more variability are therefore, likely to be selected, which can lead to bias if the higher variability is as a result of a better environment and not higher genetic variance. Table 3 shows the overall economic response of an index containing lactation milk yield, lactation fat yield and calving interval from the single, independent and joint breeding strategies using progeny testing and genomic information. Milk yield and CI have been reported to have the highest relative economic value under all three production systems while FY has an influence on the revenue from milk and on the energy requirements and hence feed requirements (Wahinya 2020). These results show that a joint selection strategy with genetic evaluation and selection occurring in all three production systems, would generate the highest overall economic response in the scenarios using progeny and genomic information. Using the single breeding strategy with genetic evaluation and selection of candidate bulls only occurring in the high production systems would result in lower economic response (-18% and -30%) for the overall breeding objective compared to the joint strategy in all scenarios. System-specific breeding programs each with an independent genetic evaluation and selection of bulls within each environment would also generate lower response compared to the joint strategy but higher than the single strategy.

A joint selection strategy is also more desirable because genetic evaluation of sires within production systems is likely to lead to selection of more robust animals which also helps to maintain diversity without necessarily developing specialised lines. Sires also benefit from the information in all environments leading to a higher index accuracy. Use of genomic information in addition to phenotypic information is important to reduce generation interval and improve accuracy of selection leading to higher responses (Wahinya 2020). A strategy using genomic selection only could reduce the generation interval further but at a cost of lower accuracies of selection. Genomic information could also be applied for parentage assignment to enhance the pedigree for genetic evaluation where pedigree information is not available and to determine breed composition (Marshall *et al.* 2019). This has been applied for the small-holder dairy cattle in Kenya where pedigree records were unavailable or not reliable (Ojango *et al.* 2019).

**Table 3: Comparison of the economic response in Kenyan shillings (Kes) using three selection strategies when either phenotypic or genomic selection is practised**

| Scenario | Strategy | Economic response (Kes) |
|---|---|---|
| Progeny test | Single | 1,311.37 |
| | Independent | 1,530.99 |
| | Joint | 1,583.06 |
| Genomic selection | Single | 1,425.85 |
| | Independent | 1,816.24 |
| | Joint | 2,030.31 |

A national breeding program involving genetic evaluation and progeny testing of sires, should be implemented across relevant production systems in Kenya since genomic information is still not available partly due to the cost and logistics of establishing a reference population. This would incentivize farmers to select their breeding animals and produce replacement animals through a genetic evaluation conducted within their own production system, minimising the impact of genotype by environment interaction between production systems. The program can work as a two-tiered closed nucleus with performance recording herds under different production systems forming the nucleus. The non-recording herds can then form the commercial tier and then source their semen and replacement cows from the nucleus herds within similar production systems.

## CONCLUSION

To maximise genetic gain for the dairy cattle population in Kenya, selection strategies should be based on a genetic evaluation across production systems to account for genotype by environment. Any selection index used should also account for the re-ranking of the breeding objective traits across the production systems. Introduction of genomic information in the current breeding program with a moderately large reference population is likely speed up the rate of genetic improvement.

## ACKNOWLEDGMENT

PKW acknowledges the financial support from the University of New England to pursue PhD studies at the Animal Genetics and Breeding Unit and the Livestock Recording Centre for the data.

## REFERENCES

Dekkers J.C.M. (2007) *J Anim Breed Genet* **124:** 331.

Marshall K., Gibson J.P., Mwai O., Mwacharo J.M., Haile A., Getachew T., Mrode R., Kemp S.J. (2019) *Frontiers in genetics* **10:** 297.

Mulder H.A., Bijma P. (2006) *J Dairy Sci* **89:** 1727.

Ojango J.M.K., Mrode R., Rege J.E.O., Mujibi D., Strucken E.M., Gibson J., Mwai O. (2019) *J Dairy Sci* **102:** 5266.

Rutten M.J.M., Bijma P., Woolliams J.A., van Arendonk J.A.M. (2002) *Journal of Heredity* **93:** 456.
Wahinya P., K (2020) Thesis, University of New England.
Wahinya P.K., Jeyaruban G., Swan A., Magothe T. (2020) *J Anim Breed Genet* **137:** 423.

*Proc. Assoc. Advmt. Anim. Breed. Genet: 24: 419-422*

# USING MATESEL TO AID SIRE ALLOCATION IN GENOMIC REFERENCE POPULATIONS – SOUTHERN MULTI-BREED AN EXAMPLE

**S.F. Walkom[1], K.A. Donoghue[2], P.F. Arthur[2], S.A. Clark[3] and B.J. Walmsley[1,2]**

[1] Animal Genetics Breeding Unit*, University of New England, Armidale, NSW, 2350 Australia
[2] New South Wales Department of Primary Industries, Armidale, NSW, 2350 Australia
[3]School of Environmental and Rural Science, University of New England, Armidale, 2351, NSW Australia

## SUMMARY

The Southern Multi-breed project has been designed to generate data that will allow producers across Australia to directly compare bulls of different breeds via BREEDPLAN EBVs. The project incorporates six breeds (Angus, Hereford, Wagyu, Charolais, Shorthorn and Brahman) at five research sites across New South Wales. The following paper describes how the MateSel software was used to assist selection and allocation of sires within the project, to provide linkage across sites and years and minimise inbreeding within progeny.

## INTRODUCTION

Genetic improvement is important to improve productivity and the competitive advantage of the Australian beef cattle industry (Swan *et al.* 2012). The BREEDPLAN genetic evaluation for the Australian beef industry delivered as breeds each having their own analysis. The Southern Multi-breed (SMB) project has been developed to form a multi-breed reference population for genetic evaluations and aims to generate data enabling direct comparisons between animals of different breeds. This multi-breed, multi-herd project includes the five most common temperate breeds (Angus, Hereford, Wagyu, Charolais, Shorthorn) along with Brahman to establish linkage to similar projects in Northern Australia (Repronomics; Johnston *et al.* 2017). By the end of the project, hard to measure phenotypes, including female reproduction (puberty, post-partum anoestrous) and genotypes will be collected on c. 2000 cows per year plus followers. Walmsley *et al.* (2021) presents a comprehensive summary of the SMB project.

The inclusion of DNA information into genetic evaluation, often termed genomic selection, has the potential to increase the rate of genetic improvement in many livestock species. Models that incorporate both genomic and pedigree information (single-step GBLUP) have already been implemented into Australia's BREEDPLAN genetic evaluation system for beef cattle (Johnston *et al.* 2018). To capture all of the potential value genomic selection presents, genomic reference populations should have a low average relationship between the reference animals and yet the relationship between the reference population and the animals being evaluated is high (Clark *et al.* 2012). Genetic and genomic evaluations have traditionally been limited to within single breed populations like Angus or Hereford (Boerner 2017). The collection of genotypes and phenotypes within the SMB will assist to underpin future BREEDPLAN multi-breed evaluations, and increase the accuracy of current genomic EBVs.

The most efficient design for a multi-site or multi-contemporary group (cg) projects is to have every sire represented in every cg, and the sires to be as diverse as possible. Physical restraints mean that this is difficult to achieve, as with comparisons across herds for the national evaluations, the ability to utilise information across research sites is reliant on genetic linkage across research sites,

* A joint venture of NSW Department of Primary Industries and the University of New England

breeds and contemporary groups (Graser *et al.* 2005). Consequently, selection of the sires and base dams within the SMB project should capture the diversity within and across the breeds, with the resulting matings designed to maximise diversity within the population, and genetic linkage across research sites. This paper describes the genetic diversity of the SMB base herd and how the software program MateSel (Kinghorn 2011) was used in sire allocation.

## BASE BREEDING HERD

The SMB herd is managed across five research sites across NSW and include; Glen Innes (New England), Grafton (North Coast), Trangie (Western), Tocal (Hunter) and EMAI (Outer Sydney). Approximately 1800-2000 base cows have been mated through a combination of AI and natural matings to bulls of their own breed over the last two years, these matings generating the base experimental females. A small crossbred joining program between Angus, Hereford and Brahman also occurred at the Grafton research site (Walmsley *et al.* 2021). Breed and cow allocation at each site (Figure 1) was based on breed representation, carrying capacity of the site, whilst accounting for dam source and sire line as per Walmsley *et al.* (2021). Angus cattle provide the link breed across the 5 sites, representing 223 unique sire families and incorporating the DPI herds at Glen Innes and Trangie (Walmsley and McKiernan 2011; Herd *et al.* 2014). The base cows also represent 197 Hereford, 99 Wagyu, 78 Brahman, 80 Charolais and 93 Shorthorn sire families.

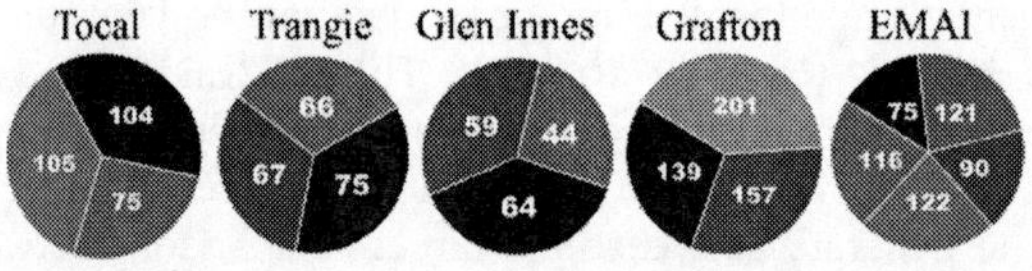


**Figure 1: Number of cows for Angus (black), Brahman (orange), Charolais (blue, Hereford (maroon), Shorthorn (red) and Wagyu (green) represented at the five SMB research sites for the 2019 and 2020 matings**

## SIRE SELECTION

Sire selection was a balancing act between capturing the diversity of the breed and targeting high accuracy EBV bulls, such that the average accuracy of the progeny was as high as possible to allow the development of high accuracy across breed EBVs for animals. MateSel was used to aid selection within the industry nominated bulls, for the AI program, but ultimately final selection was heavily influenced by semen availability at the time of mating. The 2019/20 AI program used 171 bulls and incorporated 23 historic sires from previous Beef CRC projects (Upton *et al.* 2001, Figure 2), creating linkage to previous research herds and the different breed based Beef Information Nucleus projects through the use of project bulls and their sons for the southern breeds. Brahman bulls represented in the Repronomics project (14 bulls) (Johnston *et al.* 2017) and Beef CRC were also used as AI sires. Backup bulls are predicted to account for ~35% (AI success rate of ~50%) of the research animals. The selection of the 119 backup sires was done in conjunction with the AI sire selection, with herds and sires which were not represented in the AI sires favoured, within the limitations of the project budget.

## MATE ALLOCATION

To avoid confounding between breed, site, year and sire effects, unique sire family lines needed to be spread across the fixed effects structure of the project, to maximise the ability to accurately remove the impact of environment. The aim was to generate the most genetically diverse progeny (low inbreeding) possible given the selected sires and dams, with sufficient linkage across sites via the sires.

MateSel is a software program for tactical implementation of breeding programs, based on an evolutionary algorithm (Kinghorn 2011). It accommodates the prevailing technical and logistical issues, including genetic gain, genetic diversity, trait distributions and the management of allele and genotype frequencies for individual genetic markers. Functions within the MateSel suite allows for the implementation of mating restrictions around the distribution of males for matings across groups of dams.

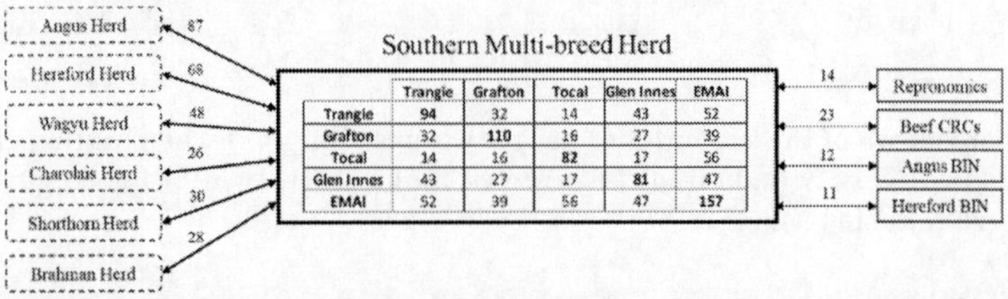


| | Trangie | Grafton | Tocal | Glen Innes | EMAI |
|---|---|---|---|---|---|
| **Trangie** | **94** | 32 | 14 | 43 | 52 |
| **Grafton** | 32 | **110** | 16 | 27 | 39 |
| **Tocal** | 14 | 16 | **82** | 17 | 56 |
| **Glen Innes** | 43 | 27 | 17 | **81** | 47 |
| **EMAI** | 52 | 39 | 56 | 47 | **157** |

**Figure 2: Number of Southern Multi-Breed bulls used at each research site (diagonal) as well as bulls in common (off diagonal). Bulls in common with other research herds and the industry herds represented by number above linking arrows**

**Methodology.** The MateSel algorithm was provided with a complete set of selection candidates; AI sires and available dams, with index values (BREEDPLAN $Index value) and 4 generations of pedigree information back from the candidate. The objective of the mate allocation was to generate high genetic diversity in the progeny born in the project, as per Clark *et al.* (2012), by targeting a low co-ancestry between parents and placing a negative weighting on inbreeding. MateSel was able to do this under the physical restraints of the project that included;

1) Sires are used enough to produce 6 to 8 daughters across the project
2) Genetic linkage via the use of sires across research sites, years, cow groups (eg. heifers/cows)
3) Sires linking across years (used at the same research site as previous years)
4) Minimum sire usage within a research site / year / dam age group to make sure guaranteed representation of the sire in the heifer progeny (assuming a 50% AI success rate, and 50:50 sex ratio of progeny)
5) Avoid high birth weight bulls being mated to heifers and first time calvers

The backup mating assignments followed the same AI mating rules, however, these allocations were also restrained by the sires' physical location. Management restrictions meant that in some cases multi-sire backup groups needed to be allocated.

**Outcomes.** The final AI mate allocations were chosen when the smallest co-ancestry was achieved and all selected dams and sires had met usage requirements and no mating would produce a progeny with an inbreeding coefficient, based on pedigree, greater than 7% (> 10% in Wagyu, due to smaller genetic diversity of the breed). Slightly higher inbreeding thresholds were observed in the backup mating due to the physical restraints associated with bull allocations. The low level of co-ancestry within the mate allocations was achieved in conjunction with creating genetic links between sites via common sires as per. Figure 2. Alternatively, if the MateSel mate allocation had targeted the respective industry indexes, instead of genetic diversity, for Angus, Hereford and Wagyu the average level of inbreeding in the predicted progeny increased from 2.3% to 2.8%, 0.3% to 1.3% and 2.9% to 4.9%, for the three breeds respectively (Figure 3). Resulting in 12% of the predicted Wagyu progeny having an inbreeding coefficient of 10% or higher. Bulls used in the AI program were allocated between 4 and 24 matings / year (mean = 12), after two years of use have been allocated 8 to 33 matings ,with no less than 4 matings within a research site, year, dam age combination. Results from the 2019 mating reveal that the number of heifer progeny born per bull used across all matings ranged from 1 to 11 with an average of 3 heifer progeny.

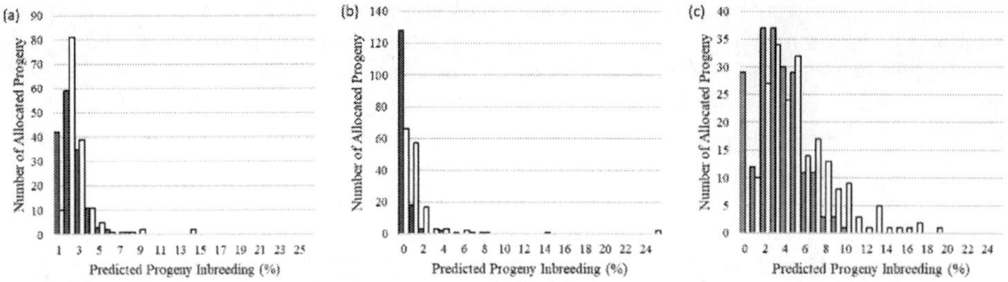


**Figure 3: Comparison of the estimates of progeny inbreeding (%) when mating allocation was based on genetic diversity (coloured) or targeted index merit (white) for Angus (a), Herford (b), and Wagyu (c) using MateSel**

## CONCLUSIONS

MateSel has aided the allocation of the base dams and sires across the research sites within the SMB project so that the resulting progeny will reflect the genetic diversity in industry whilst maintaining genetic linkage across cohorts within the SMB project and industry herds. However, it should be acknowledged that the genetic diversity of the progeny and their genetic relationship to the national herd will not be fully understood until the planned genotyping of the progeny has occurred.

## ACKNOWLEDGEMENTS

The Southern Multi-Breed project is an initiative co-funded by NSW Department of Primary Industries, University of New England and Meat and Livestock Australia Donor Company. The authors give thanks to the efforts of the research station managers and their staff, our skilled team of technicians, ultrasound scanners, AI, DNA lab, data, project managers and scientists. Individual breeders, participating breed societies and the southern beef industry are thanked for their support. The authors also acknowledge the support of L.A. Penrose and T. Granleese for assisting with compiling pedigree and animal information.

## REFERENCES

Boerner V. (2017) *Proc. Assoc. Advmt. Anim. Breed. Genet.* **22**:97.

Clark S. A., Hickey J. M., Daetwyler H. D. and van der Werf J. H. J. (2012) *Gene. Sel. Evol.* **44:** 4.

Graser H.-U., Tier B., Johnston D. J. and Barwick S. A. (2005) *Aus. J. of Exp. Ag.* **45:** 913.

Herd R. M., Arthur P. F., Donoghue K. A., Bird S. H., Bird-Gardiner T. and Hegarty R. S. (2014) *J. Ani. Sci.* **92:** 5267.

Johnston D., Grant T., Schatz T., Burns B., Fordyce G. and Lyons R. (2017) *Proc. Assoc. Advmt. Anim. Breed. Genet.* **22**:385.

Johnston D. J., Ferdosi M., Connors N., Boerner V., Cook J., Girard C., Swan A. and Tier B. (2018) *Proc. World Congr. Genet. Appl. Livest. Prod..* **19:** 269.

Kinghorn B. P. (2011) *Gene. Sele. Evol.* **43:** 4.

Swan A. A., Johnston D. J., Brown D. J., Tier B. and Graser H.-U. (2012) *Anim. Prod. Sci.* **52:** 126.

Upton W., Burrow H. M., Dundon A., Robinson D. L. and Farrell E. B. (2001) *Aus. J. Exp. Ag.* **41:** 943.

Walmsley BJ., Donoghue KA., Johnston DJ., Clark SA., Siddell JP., Walkom SF, Granleese T. and Arthur PF. (2021) *these proceedings*

Walmsley B. and McKiernan W. (2011) *Proc. Assoc. Advmt. Anim. Breed. Genet.* **19:**123.

*Proc. Assoc. Advmt. Anim. Breed. Genet: 23: 423-426*

# INITIATING THE SOUTHERN MULTI-BREED RESOURCE POPULATION

**B.J. Walmsley[1,2], K.A. Donoghue[3], D.J. Johnston[1], S.A. Clark[4], J.P. Siddell[2], S.F. Walkom[1], T. Granleese[5] and P.F. Arthur[6]**

[1]Animal Genetics and Breeding Unit*, University of New England, Armidale, NSW, 2351
[2]NSW Department of Primary Industries, Livestock Industries Centre, Armidale, NSW, 2351
[3]NSW Department of Primary Industries, Agricultural Research Centre, Trangie, NSW, 2823
[4]School of Environmental and Rural Science, University of New England, Armidale, NSW, 2351
[5]NSW Department of Primary Industries, Grafton Primary Industries Institute, Grafton, NSW, 2460
[6]NSW Department of Primary Industries, Elizabeth MacArthur Agricultural Institute, Menangle, NSW, 2568

## SUMMARY

This paper describes the first year of a large 5-year breeding project being conducted across New South Wales involving five temperate beef breeds and the Brahman breed. Artificial insemination and back-up matings were designed to produce progeny that are representative of the genetic diversity in the national herd of each breed. Sires and dams were selected with a focus on high accuracies for the 400-day weight estimated breeding value (EBV) and reproduction EBVs. The project progeny will be managed in mixed-breed groups and intensively recorded head-to-head for current BREEDPLAN and new economically important traits such as early-in-life female reproduction and worm egg counts. All animals will have high density SNP genotypes taken to contribute to the breeds' genomic reference populations and for inclusion in BREEDPLAN genomic evaluations. The project design will facilitate development of genomic EBVs to allow across-breed comparisons, assist in increasing selection accuracy, particularly for young bulls, allow genotype by environment (*GxE*) investigations, and the potential development of new traits.

## INTRODUCTION

Significant profitability gains have been generated in temperate Australian beef breeds through selection using EBVs and indexes (Swan et al. 2011). Currently, the EBVs, and the selection indexes they drive, are generated from within-breed genetic evaluations (Graser *et al.* 2005), limiting the capacity to compare animals across breeds. To overcome this, breeds must be managed, and performance recorded for these traits, on a head-to-head basis to facilitate the development of EBVs that allow for across-breed comparisons. Reproduction is one such trait that is of importance in indexes producing replacement heifers, but it has been poorly recorded (Gudex and Millen 2019). Recent studies (Wolcott *et al.* 2019; 2021) have also found only c.50% of heifers in key Hereford and Angus seedstock herds were pubertal prior to synchronisation for fixed time artificial insemination (AI). These findings suggest that given the increasing prevalence of AI, there is a need to monitor the capacity of temperate breed heifers to conceive early in their first mating season as a trait for inclusion in genetic evaluations. A new project (MLA *P.PSH.1261*) is being conducted over the next 5 years (2020 to 2025), known as the Multi-Breed Genomic Beef Cattle Resource or Southern Multi-Breed (SMB) project which aims to address these issues. This project will extensively phenotype progeny from temperate beef breeds and Brahmans, managed in mixed-breed groups at sites across NSW for existing BREEDPLAN and new economically

---

* AGBU is a joint venture of NSW Department of Primary Industries and University of New England

important traits, such as early-in-life reproduction. The project has been designed to have links to past and current genetic research in Australia including the Repronomics$^{TM}$ project (Johnston *et al.* 2017). The project will genotype all sires, dams and progeny to capture the benefits of genomic evaluations (Johnston *et al.* 2018), particularly for hard-to-measure and new traits. This paper provides a brief description of the experimental design and current progress of the project.

## LOCATIONS AND BREEDS

The research project is being conducted on New South Wales Department of Primary Industries (NSW DPI) and University of New England (UNE) research facilities. To reflect the diversity of production environments in southern Australia across years, the breeding herds are located on 5 NSW DPI research properties dispersed across NSW (Trangie Agricultural Research Centre, Trangie; Grafton Primary Industries Institute, Grafton; Tocal Agricultural Centre, Tocal; Glen Innes Agricultural Research and Advisory Station, Glen Innes; Elizabeth MacArthur Agricultural Institute (EMAI); Menangle). All steer progeny will be backgrounded prior to feedlot entry at 2 NSW DPI research properties (EMAI; Duck Creek Agricultural Field Station, Ballina) with feedlot finishing occurring at the UNE research feedlot, "Tullimba" (Kingstown). The project includes the 5 numerically largest temperate breeds (*viz.* Angus, Charolais, Hereford, Shorthorn and Wagyu) in southern Australia and the Brahman breed, which is commercially important in the sub-tropical regions of NSW and creates linkage to the Repronomics project. At all locations, the breeds are being managed and recorded in mixed groups.

## FEMALE SELECTION

Base females were purchased from key industry seedstock herds. Angus females from the BREEDPLAN recorded NSW DPI muscling (Walmsley and McKiernan 2011) and feed efficiency selection (Herd *et al.* 2014) herds were also retained. All females were only purchased if they were pedigree and performance recorded in BREEDPLAN. The objective was to capture, as closely as possible, the genetic diversity of that breed's national herd, with a focus on dams from sire lines of high current impact. There was also a focus on selecting females with high accuracy and diversity for the 400-day weight and reproduction EBVs; days-to-calving (Angus, Brahman, Hereford and Shorthorn) or scrotal size (Charolais and Wagyu). A visual inspection of females was conducted to assess structural soundness and maximise the chances they would produce 2 calves in the project. Females were allocated to research sites based on the site's carrying capacity, the availability of females, the breed's relevance to the local environment (e.g. Brahman cows are only present in the sub-tropical environment at Grafton) and the numbers of each breed required to provide meaningful comparisons. Angus females are used to link sites because not all breeds can be accommodated at all sites. Table 1 contains the number of base females at each research site.

**Table 1 Allocation of base females to the five NSW DPI research sites**

| Site | Angus | Brahman | Charolais | Hereford | Shorthorn | Wagyu | Total |
|---|---|---|---|---|---|---|---|
| Trangie | 75 | | | 67 | | 66 | 208 |
| Grafton | 139 | 201 | | 157 | | | 497 |
| Tocal | 104 | | 75 | | 105 | | 282 |
| Glen Innes | 64 | | | 59 | | 44 | 167 |
| EMAI | 105 | | 121 | 90 | 122 | 116 | 554 |
| Total | 490 | 201 | 144 | 373 | 227 | 226 | 1661 |

## SIRE SELECTION

Sires selected included both AI sires nominated by industry and those selected to provide direct linkage to past research, such as the Beef CRC and Beef Information Nucleus projects. In addition, natural mate back-up bulls were purchased by NSW DPI from key industry seedstock herds or Repronomics project (Johnston *et al.* 2017) bulls were purchased from the Queensland Department of Agriculture and Fisheries. All bulls were pedigree and performance recorded in BREEDPLAN. Sire purchasing was a key design feature of the project to achieve representation of each breed's genetic diversity. There was a preference toward sire lines of influence that will shape the future makeup of the breed. Although not a primary objective, poll status was also taken into consideration. Straightbred matings will primarily occur with a small numbers of crossbreed matings at Grafton involving Brahman reciprocal matings to Angus and Herefords (i.e. BxA, AxB, BxH and HxB). Currently, ~290 sires have been used during the AI programs or as back-up sires in 2019 and 2020 with either calves weaned, or successful pregnancies diagnosed. Walkom *et al.* (2021) provide a brief description of the MateSel procedure used for allocating matings based on coancestry to limit inbreeding, with a small amount of emphasis placed on the index.

## PROGENY GENERATED

The project aims to generate up to 8,000 progeny managed in mixed-breed contemporary groups. Currently, the project has generated ~1400 calves from the 2019 matings with ~1500 diagnosed pregnancies from the 2020 matings. All progeny will be retained within the project and recorded until the steers are slaughtered or the females are surplus to requirement. The female progeny will be retained and grown out at each research site prior to joining the breeding herd as maidens at ~15 months of age with natural matings to sires of their own breed. These females will also be naturally mated to sires of their own breed as first-lactation cows and will be retained in their respective herds for a minimum of 3 matings. The male progeny will be castrated at marking and following weaning will be transported to 2 research sites (EMAI, Duck Creek) to undertake backgrounding until they reach feedlot entry weights. The steers will then be feedlot finished for a minimum of 100 days at the UNE Tullimba feedlot prior to slaughter.

## KEY TRAIT RECORDING

All calves generated by the project will be recorded intensively from birth to the end of backgrounding (steers)/grow-out (heifers). Recording will include accurate recording of birth date, birth weight, calving ease and survival, gestation length (AI calves only), weaning weight, flight time, docility score, yearling weight, and structure. Other traits, such as worm egg count, will be recorded regularly beginning at weaning and continuing until the steers enter Tullimba feedlot and the heifers wean their first calf. Horn/poll assessments will be conducted on all calves at marking, with monitoring continuing while animals are involved in the project (Connors *et al.* 2021). Following weaning, the heifer progeny will have regular ovarian assessments conducted using real-time ultrasound performed by highly skilled ultrasonographers to determine follicle development, and, in particular, identify the presence of a corpus luteum as a measure of puberty. All first-lactation cows will be regularly scanned after calving to determine their return to oestrous. Females will have live weight, hip height, body condition score, eye muscle and subcutaneous fat depth recorded prior to mating and at weaning each year, and will be assessed for calving ease, teat and udder score at calving. Steer progeny will have weight and scan traits, as well as net feed intake, recorded while in Tullimba, with full abattoir, meat quality and consumer testing undertaken following slaughter. All BREEDPLAN traits will be quality checked and loaded into ABRI's southern multi-breed research database.

## ANIMAL GENOTYPING

The implementation of single-step GBLUP (Johnston *et al.* 2018) represents a major evolution of the BREEDPLAN genetic evaluation system. To capture the benefits of these developments and extract full value from the investment in this project, all animals will be genotyped in alignment with BREEDPLAN (Connors and Ferdosi 2019). All base females have been DNA sampled and will be genotyped as a minimum with a 50K SNP chip. All current back-up and AI sires have been DNA sampled for genotyping. A group of 51 AI sires across all breeds have been selected based on the number of calves they produced from the 2019 AI program for full genomic sequencing at 10x coverage for inclusion in the 1000 Bull Genomes Project (Hayes and Daetwyler 2019). All SNP data will be quality checked, stored on a project database and loaded into ABRI's southern multi-breed research database.

## CONCLUSIONS

The project is currently in the first 12 months of operation with all base females having been purchased and allocated to research sites. These females have undergone AI and back-up mating programs in 2019 and 2020 to ~290 different sires across the six breeds. The intensive recording of project-generated progeny is underway with the first ~1400 calves recorded at calving in 2020 and weaning in 2021. The recording of breeds on a head-to-head basis represents a significant industry and government investment which will allow genetics to be compared across environments and provide a resource to benchmark reproduction and other traits across-breeds, including hard-to-measure traits. As such, the project will enable stronger selection for those traits contributing to value chain profit.

## ACKNOWLEDGEMENTS

The Southern Multi-Breed project is an initiative co-funded by NSW Department of Primary Industries, University of New England and Meat and Livestock Australia Donor Company. The authors give thanks to the efforts of the research station managers and their staff, our skilled team of technicians, ultrasound scanners, AI, DNA lab, data, project managers and scientists. Individual breeders, participating breed societies and the southern beef industry are thanked for their support.

## REFERENCES

Connors N.K. and Ferdosi M.H. (2019) *Proc. Assoc. Advmt. Anim. Breed. Genet.* **23**: 382.

Connors N.K., Walmsley B.J. and Donoghue K.A. (2021) *Proc. Assoc. Advmt. Anim. Breed. Genet.* **24**: *These Proceedings.*

Hayes B.J. and Daetwyler H.D. (2019) *Annu. Rev. Anim. Biosci.* **7**: 89.

Herd R.M., Arthur P.F., Donoghue K.A., Bird S.H., Bird-Gardiner T. and Hegarty, R.S. (2014) *J. Anim. Sci.* **92**: 5267.

Gudex B.W. and Millen, C.A. (2019) *Proc. Assoc. Advmt. Anim. Breed. Genet.* **23**: 488.

Graser H.-U., Tier B., Johnston D.J., and Barwick S.A. (2005) *Aust. J. Exp. Agric.* **45**: 913.

Johnston D.J. (2017) *Proc. Assoc. Advmt. Anim. Breed. Genet.* **22**: 385.

Johnston D.J., Ferdosi M.H., Connors N.K., Boerner V., Cook J., Girard C.J., Swan A.A. and Teir, B. (2018) *Proc. 11th World Congr. Genet. Appl. Livest. Prod.* Paper 269.

Swan A.A., Johnston D.J., Brown D.J., Tier B. and Graser H-U. (2011) *Anim. Prod. Sci.* **52**:126.

Walkom S.F., Donoghue K.A., Arthur P.F., Clark S.A., and Walmsley B.J. (2021) *Proc. Assoc. Advmt. Anim. Breed. Genet.* **24**: *These Proceedings.*

Walmsley B.J. and McKiernan W.A. (2011) *Proc. Assoc. Advmt. Anim. Breed. Genet.* **19**: 123.

Wolcott M.L., Banks R.G., Tweedie M. and Alder G. (2019) *Proc. Assoc. Advmt. Anim. Breed. Genet.* **23**: 91.

Wolcott M.L., Banks R.G. and Alder G. (2021) *Proc. Assoc. Advmt. Anim. Breed. Genet.* **24**: *These Proceedings.*

*Proc. Assoc. Advmt. Anim. Breed. Genet: 24: 427-430*

# GENOMIC ANALYSIS OF GENOTYPE BY ENVIRONMENT INTERACTIONS IN POST-WEANING WEIGHT OF AUSTRALIAN SHEEP

**D.L. Waters [1], S.A. Clark[1], N.Moghaddar[1] and J.H.J. van der Werf[1]**

[1] School of Rural & Environmental Science, University of New England, Armidale, NSW, 2351 Australia

## SUMMARY

Genotype by environment interactions affecting post-weaning body weight in Australian sheep were investigated using a linear reaction norm incorporating genomic information. The definition of the environmental covariable used in the reaction norm was the best linear unbiased estimation of contemporary group effects for post-weaning growth rate. Significant variation in slope was estimated, and genetic correlations between low, medium and high growth environments ranged from 0.61 to 0.94, suggesting the presence of genotype by environment interaction. A putative QTL was detected on chromosome 11, significantly associated with both the intercept and slope of the reaction norm. Overall, SNP effects for the intercept and slope were highly correlated (0.87). The results suggest that selection based on (genomic) breeding values for the intercept and slope could yield animals that are more robust.

## INTRODUCTION

Environmental extremes are very costly to agricultural systems. Annual farm-gate gross domestic product declined by approximately $3 billion AUD between 2017 and 2020 in Australia, due to drought (Reserve Bank of Australia, 2020). Climate change is expected to exacerbate this problem, with increasingly extreme and variable environments predicted for the future (Cowan *et al.* 2014). A potential response to this problem could be to breed livestock that are more robust to changes in the environment. An understanding of genotype by environment (GxE) interactions is necessary for this.

Genotype by environment interactions occur when the performance of genotypes is dependent on the environment they are recorded in. GxE acts as a source of variation from which to select robust livestock; robust genotypes maintain their genetic merit across environments, while sensitive genotypes change in merit. Biologically, if significant GxE implies that performance in different environments can be considered as different traits (Falconer 1952), a varying genetic architecture could be expected to determine merit across environments. Therefore, it should be possible to detect changes in the relative contribution of QTL influencing a trait across different environments using SNP and environmental data.

A popular way to model GxE is using reaction norm models (RNM). These allow the breeding values of animals to change across an environmental trajectory, often modelled as a linear function. This results in two breeding values for each animal; one corresponding to its breeding value in the mean environment (the intercept), and the other corresponding to the degree of change in breeding value across environments (the slope). Selection of animals with high breeding values for intercept and small breeding values for the slope could increase the mean of a trait while maintaining robustness of a flock to environmental extremes. Breeding values for the slope could also be used in a genome-wide association study to detect QTLs with environmental-dependent effects that are responsible for causing GxE (Silva *et al.* 2014).

Several QTLs affecting body weight are known to segregate in Australian sheep (Al-Mamun *et al.* 2015). This presents an interesting opportunity to investigate the behaviour of QTL in a trait with significant GxE (Clark *et al.* 2015). The aim of this study was to explore variance due to GxE and identify genomic regions contributing to GxE and robustness in post-weaning weight using a genomic RNM.

## MATERIALS AND METHODS

**Data.** Phenotypic data consisted of bodyweight records at weaning (64-120 days old) and post-weaning (123-329 days old) on 21,131 lambs in 206 contemporary groups born between 2007 and 2019 in the Information Nucleus Flocks and Resource Flocks. These flocks were located across Australia and were linked through common sires. All lambs were genotyped with an Illumina 50k-Ovine panel. An additional 10k was imputed from a recent Neogen GGP-Ovine panel that contained SNPs not included on the original Illumina chip. In total, the genomic data consisted of 60,345 SNPs after quality control.

The environmental covariable was defined by the best linear unbiased estimation of contemporary group effects using the rate of growth between weaning and post-weaning as the response variable, measured in grams/day. A minimum growth period of 40 days was applied to ensure the growth period was large enough to accurately reflect the environment. A small number of extreme contemporary groups were removed to prevent inflation of the regression coefficients. The environmental covariable ranged between -59.1 and +57.1 g/day when centred around zero, and was standardised between -1 and 1 for analysis.

**Statistical analyses.** A genomic RNM model was fitted using MTG2 2.18 (Lee & Van Der Werf 2016). The model was of the form: $\boldsymbol{y} = \boldsymbol{Xb} + \boldsymbol{Z_1 a_0} + \boldsymbol{Z_2 a_1} + \boldsymbol{Z_3 Qg} + \boldsymbol{Z_4 c} + \boldsymbol{e}$ where $\boldsymbol{y}$ is a vector of post-weaning weight records for each lamb, $\boldsymbol{b}$ is a vector of fixed effects, $\boldsymbol{X}$ is a design matrix linking fixed effects to records, $\boldsymbol{Z_1}$ and $\boldsymbol{Z_2}$ are the design matrixes linking records to additive genetic effects for the intercept ($\boldsymbol{a_0}$) and slope ($\boldsymbol{a_1}$), $\boldsymbol{Z_3}$ and $\boldsymbol{Z_4}$ are design matrices linking records to animals and dam environmental effects ($\boldsymbol{c}$), $\boldsymbol{Q}$ is a matrix linking animals to genetic groups, $\boldsymbol{g}$ is a vector of genetic group effects and $\boldsymbol{e}$ is the homogenous residual variance. Fixed effects included age at measurement, birth type and rear type interaction, sex, and contemporary group. The variance in intercept and slope was modelled as follows: $\begin{bmatrix} a_0 \\ a_1 \end{bmatrix} \sim N(0, \boldsymbol{G} \otimes \boldsymbol{K})$ where $\boldsymbol{K} = \begin{bmatrix} \sigma^2_{a0} & \sigma_{a1a0} \\ \sigma_{a0a1} & \sigma^2_{a1} \end{bmatrix}$ and $\boldsymbol{G}$ is the genomic relationship matrix (VanRaden, 2008). An environment-specific genetic (co)variance matrix was calculated using: $\mathbf{E} = \boldsymbol{\Lambda K \Lambda'}$ where $\boldsymbol{\Lambda}$ was a 3x2 matrix, with the first column containing a vector of ones for the intercept and the second column containing the standardised coefficient corresponding to the level in the environment. Three environmental levels were used to calculate **E**: -42 g/day (low growth), 0 g/day (average growth) and 42 g/day (high growth). SNP effects for the intercept and slope were estimated through back-solving genomic breeding values for $a_0$ and $a_1$ (Strandén & Garrick 2009) . P-values were approximated following Gondro (2015). A threshold value of $p > -\log_{10}(5)$ was chosen for significance.

## RESULTS AND DISCUSSION

Variance components for the intercept and slope are reported in Table 1, along with genetic correlations between the three environmental levels. Genetic variance in the slope of the RNM was significantly different from zero. The correlation between intercept and slope was 0.50, indicating that genotypes with high performance in the average environment also tended to have a positive breeding value for slope. The genetic correlation between low and high growth environments was 0.61, while additive genetic variance increased from low to high growth environments. This suggests that both scaling and re-ranking contribute to GxE in this population.

**Table 1. Reaction norm variance components for intercept ($\sigma^2 a_0$), slope ($\sigma^2 a_1$), covariance ($\sigma_{a_0 a_1}$) and correlation ($r_{a_0 a_1}$), as well as genetic correlations between low, average and high growth environments and genetic variance (Va) in each**

| **RN variance component** | | | **Genetic correlations** | | |
|---|---|---|---|---|---|
| $\sigma^2_{a_0}$ | 6.63 (0.32) | | Low | Average | High |
| $\sigma^2_{a_1}$ | 3.91 (0.58) | Average | 0.84 | - | - |
| $\sigma_{a_0 a_1}$ | 2.56 (0.25) | High | 0.61 | 0.94 | - |
| $r_{a_0 a_1}$ | 0.50 | **Va** | 4.96 | 6.63 | 12.13 |

Genome-wide SNP associations for the intercept and slope are shown in Manhattan plots (Figure 1). Regions on chromosome 6 and chromosome 11 were significantly associated with the intercept. The region on chromosome 6 was previously associated with body weight in Al-Mamun et al. (2015), while the region on chromosome 11 has not previously been reported in the literature. The same region on chromosome 11 was also significantly associated with the slope of RNM. Overall, the intercept and slope of post-weaning weight appears to be highly polygenic.

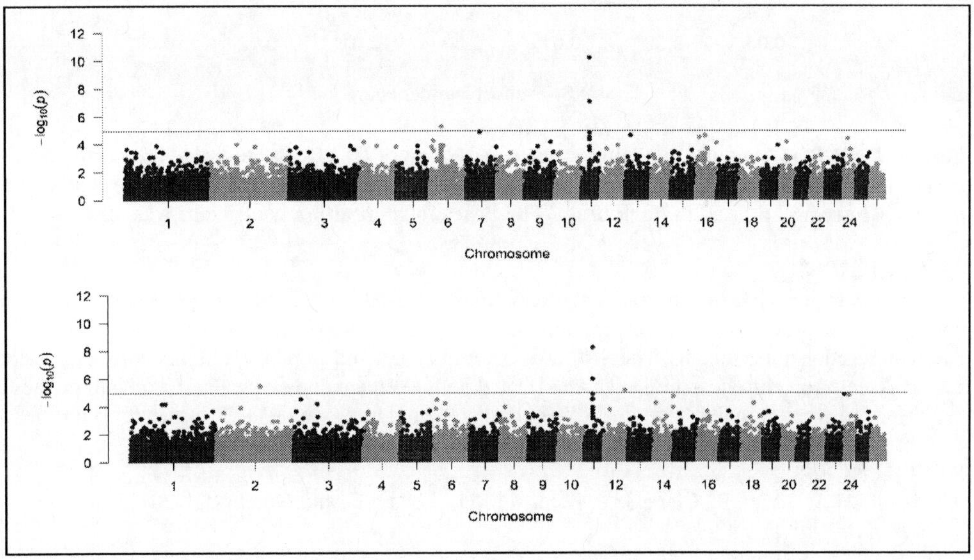


**Figure 1. Genome-wide SNP associations for the intercept (top) and slope (bottom) of a reaction norm for post-weaning weight. The blue threshold line corresponds to $p = 0.00001$**

SNP effects for the intercept and slope are plotted in Figure 2. The QTL on chromosome 11 appears to be contribute to GxE through a scaling effect, as its effect on slope is proportional to its effect on the intercept. The correlation between SNP effects was higher than anticipated (0.87), given that the genetic correlation between intercept and slope was 0.50. A possible explanation is that the breeding values for slope could be behaving similarly to a low-heritability trait in a multi-trait analysis, drawing on information from the higher heritability intercept. Methods to make breeding values for the intercept and slope more independent such as canonical transformation could remove variation in slope due to the intercept and yield a more useful GWAS for robustness. Further investigation of this observation is warranted to better understand what the SNP effects for slope actually represent. This study serves as a preliminary investigation of GxE using genomic information. Several improvements for future analysis will be to model heterogenous residual variance, use a higher density SNP panel and explore higher-order polynomials or splines.

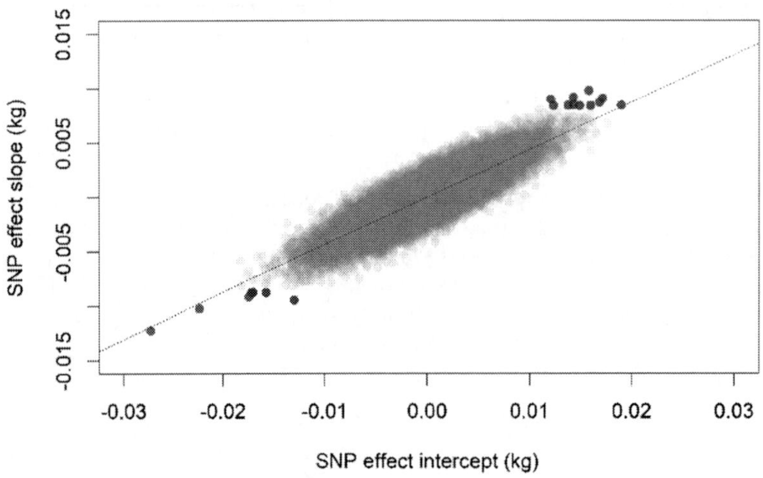


**Figure 2. SNP effects for the slope regressed over SNP effects for the intercept. SNPs corresponding to the QTL on chromosome 11 are highlighted in red, while all other SNPs with p > 0.0001 are highlighted in dark blue. The Pearson correlation coefficient was 0.87**

## CONCLUSIONS

Significant genotype by environment interactions were detected using a linear genomic reaction norm. The genetic variance in intercept and slope indicated that breeding for robustness is feasible based on reaction norm models. The SNP effects for intercept and slope were highly correlated, with a QTL detected on chromosome 11 which affected both intercept and slope. Research into methods that remove variation in the slope due to the intercept could improve GWAS studies for robustness.

## REFERENCES


Al-Mamun H.A., Kwan P., Clark S.A., Ferdosi M.H., Tellam R. and Gondro C. (2015) *Genetics Selection Evolution* **47**(1): 1.

Clark S.A., Lee S.H. and van der Werf J.H.J. (2015). *Proc. Assoc. Advmt. Breed. Genet.* **21**: 137.

Cowan T., Purich A., Perkins S., Pezza A., Boschat G. and Sadler K. (2014). *J. Climate* **27**(15): 5851.

Falconer D.S. (1952) *The American Naturalist* **86**(830): 293.

Gondro C. (2015). *Primer to Analysis of Genomic Data Using R.* Springer. Cham.

Lee S.H. and Van Der Werf J.H.J. (2016) *Bioinformatics 32*(9): 1420.

Santana M.L., Bignardi A.B., Eler, J.P., Cardoso, F.F. and Ferraz J.B.S. (2013) *J. Anim. Breed. Genet.* **130**(5): 394.

Silva F.F., Mulder H.A., Knol E.F., Lopes M.S., Guimaraes S.E.F., Lopes P.S., Mathur P.K., Viana M.S.and Bastiaansen J.W.M. (2014) *J. Anim. Sci. 92*(9): 3825.

Strandén I. and Garrick, D.J. (2009) *J. Dairy Sci.* **92**(6): 2971.

VanRaden P.M. (2008) *J. Dairy Sci. 91*(11); 4414.

*Proc. Assoc Advmt. Anim. Breed. Genet. 24: 431-434*

# GENOME-WIDE ASSOCIATION ANALYSIS OF BIRTH AND WEANING WEIGHTS IN AUSTRALIAN TAURINE BEEF CATTLE

**W.M.S.P. Weerasinghe[1], B.J. Crook[1], and N. Moghaddar[2]**

[1]Agricultural Business Research Institute, UNE, Armidale, NSW, 2350 Australia
[2]School of Environmental and Rural Science, UNE, Armidale, NSW, 2350 Australia

## SUMMARY

Birth and weaning weights are two traits which ultimately influence traits of economic relevance in the beef cattle industry. In this study, multi-breed genomic analysis was performed using three Australian beef cattle breeds to detect genomic regions that influence birth and weaning weights. Principal component analysis revealed a clear genetic separation between the Hereford, Simmental and Charolais breeds. A genome-wide association study based on 29k density SNP genotypes revealed significant SNPs associated with birth and weaning weights on chromosomes 5, 6, 7 and 20 in a multi-breed dataset after correction for genetic relationship between animals and population stratification. GREML results suggested a top marker present on chromosome 6 accounted for 11% and 5% of the additive genetic variance for BW and WW respectively. Results of this study may indicate a role for weighted GBLUP evaluations when very large effect QTL for production traits are evident in beef cattle.

## INTRODUCTION

Quantitative trait loci (QTL) mapping is an important step to identify genetic variants associated with economically important traits in livestock industries. Traits such as birth weight (BW) and weaning weight (WW) contribute significantly to the profitability of beef breeding enterprises by way of impact on calving outcomes and post-birth growth potential, as well as influencing reproductive and nutritional management decisions. There are several biological events and associated genes involved with these two traits, with both having a moderate to high pedigree-based heritability that is favourable for the detection of genomic regions. Several genome-wide association studies (GWAS) have been conducted for *Bos taurus*, *Bos indicus* and crossbred cattle types, with specific chromosomes and genomic regions being identified for BW and WW (Akanno *et al.*, 2018; Saatchi *et al.*, 2014; Utsunomiya *et al.*, 2013).

The aim of the present study was to investigate the presence of significant genomic regions in association with BW and WW in each of three Australian beef breeds, as well as in a combined (multi-breed) context. Total genetic variation explained by such informative SNPs was quantified.

## MATERIALS AND METHODS

The BW and WW data for Australian Hereford, Simmental and Charolais used in this study were derived from data extracts as used in the BREEDPLAN analysis undertaken for each breed (Graser *et al.*, 2005). Single-animal contemporary groups were excluded from further analysis as were contemporary groups for animals born prior to 2000. Breed-specific variance components were estimated for BW and WW using WOMBAT (Meyer, 2011). Records were pre-adjusted for age of dam (BW and WW) and age of calf (WW only) effects, with each model including random effects for direct genetic, maternal genetic and dam permanent environment (PE) and with contemporary group as a fixed effect. Variance components were used to perform within-breed BLUP analyses for BW and WW to obtain the direct genetic and residual solutions. Both solutions were combined to give phenotypes (corrected for maternal genetic, dam PE and contemporary group effects) for use in the subsequent GWAS.

Genomic data for animals with BW and WW records were subjected to quality control (QC) and imputation. Several different platforms were used for genotyping, predominantly different versions of the GGP-LD product, with 14,904 animals genotyped with the 50k SNP panel (BovineSNP50 BeadChip, Illumina Inc., San Diego, CA.) used for the analysis. QC of genomic data was conducted using PLINK software (Chang *et al.*, 2015), with SNPs removed at a minor allele frequency of <0.01 and a deviation from Hardy–Weinberg equilibrium of $p<1E^{-6}$ as exclusion cut-off. SNPs with a call rate less than 90% and SNPs located on sex chromosomes were excluded. Animals with a call rate lower than 85% for all loci were excluded. Sporadic missing SNPs were imputed by FImpute (Sargolzaei *et al.*, 2014). For the multi-breed GWAS, a total of 29,101 combined genotypes were used. Principal component analysis (PCA) was carried out to determine the genetic structure of the three breeds and was performed on the genomic relationship matrix (GRM) based on the method of VanRaden (2008). Although some crossbred genotypes were represented in the combined extract, only those animals regarded as "registered purebreds" and separated by PCA results were selected for further analysis.

GWAS analysis of SNP effects and significance was conducted for each trait using the program GCTA (Yang *et al.*, 2011), following a linear mixed model as below:

$$\boldsymbol{y} = \boldsymbol{Xb} + \boldsymbol{Za} + \boldsymbol{e}$$

where $\boldsymbol{y}$ is a vector of corrected phenotypes, $\boldsymbol{b}$ is a vector of overall mean, SNP effect and the first and second principal components as linear covariates, $\boldsymbol{a}$ is a vector of random additive genetic effects and $\boldsymbol{e}$ is a vector of random residual effects. $\boldsymbol{X}$ and $\boldsymbol{Z}$ are incidence matrices that relate fixed effects to phenotypes and additive genetic effects to each animal respectively.

Additive genetic effects in the GWAS were assumed to be normally distributed as: $a \sim N(0, \boldsymbol{G}\sigma_a^2)$, where $\boldsymbol{G}$ is a genomic relationship matrix based on the 29k SNP genotypes, and $\sigma_a^2$ is the additive genetic variance. Significant SNPs were identified using a Bonferroni correction with α=0.05 and –log10 (p)=5.76 as well as with P<0.001. Significant SNPs (based on the P<0.001) present in the same genomic regions were subjected to joint multivariate regression analysis using GCTA with P<1.712e-06 to identify the most informative SNPs for the particular trait.

Restricted maximum likelihood analysis with GTCA including the genomic relationship matrix (GREML) was used to estimate trait heritability and the proportion of additive genetic variation explained by the most informative SNPs. Individual SNP variances were calculated as $2pq\alpha^2$ where $p$ and $q$ are allele frequencies and α is the SNP effect.

## RESULTS AND DISCUSSION

PCA revealed clear genetic separation between the three Australian beef breeds. The first principal component (PC1) separated Hereford from the other two, whereas the second principal component (PC2) separated Simmental and Charolais. PC1 explained 79% of total variation between animals, with PC2 explaining a further 5%.

Data structure and variance components for BW and WW in each breed are presented in Table 1. Hereford gave higher additive genetic variance and heritability for BW, whereas Simmental gave higher additive genetic variance and heritability for WW. The descriptive statistics for the data used for GWAS are also shown in Table 1. A greater number of Hereford animals with both phenotype and genotypes were available for GWAS compared to other two breeds.

There were 124, 59 and 57 SNPs of significant (P<0.001) association with BW, with 48, 2 and 12 SNPs remaining after Bonferroni correction for Hereford, Simmental and Charolais respectively in single breed GWAS. For WW, there were 74, 32 and 27 SNPs showing a significant (P<0.001) association in Hereford, Simmental and Charolais respectively. After Bonferroni correction, however, only 14 significant SNPs were evident and for Hereford only.

Figure 1 gives the Manhattan plots derived from the multi-breed GWAS results of BW and WW.

Both traits have highly significant SNPs present on chromosomes 6 and 20, with BW also showing some significant genomic associations on chromosome 5. There were 106 significant SNPs present on chromosome 6, 20, 7, 5, 25 (in descending number of SNPs) with chromosomes 1, 4, 13, 19 and 21 also having a significant SNP associated with BW. Only 34 SNPs remained after Bonferroni correction. Multivariate regression of these SNPs resulted in 5 significant SNPs remaining. Initially there were 62 significant SNPs associated with WW, 13 remained after Bonferroni correction and only 2 significant SNPs remaining after multiple regression, present on chromosomes 6 and 20.

**Table 1. Additive genetic variance (VG) and heritability ($h^2$) estimated for BW and WW using BLUP within breed and descriptive statistics for data used for GWAS**

| | BLUP | | | GWAS | | | | |
|---|---|---|---|---|---|---|---|---|
| Breed | No. | V(G) | $h^2 \pm SE$ | No. | Mean | SD | Min | Max |
| **BW(kg)** | | | | | | | | |
| Hereford | 265,406 | 6.97 | 0.37±0.006 | 7,398 | 40.53 | 5.59 | 16.40 | 65.40 |
| Simmental | 48,557 | 5.10 | 0.31±0.014 | 1,325 | 40.96 | 5.73 | 24.00 | 63.00 |
| Charolais | 68,457 | 4.86 | 0.32±0.012 | 1,211 | 43.23 | 5.49 | 24.80 | 70.20 |
| **WW(kg)** | | | | | | | | |
| Hereford | 333,800 | 120.99 | 0.16±0.004 | 8,363 | 259.70 | 52.54 | 105.10 | 512.70 |
| Simmental | 30,442 | 206.36 | 0.26±0.017 | 1,011 | 309.60 | 52.63 | 138.60 | 487.90 |
| Charolais | 68,953 | 158.20 | 0.20±0.011 | 1,249 | 285.30 | 45.11 | 161.10 | 484.90 |

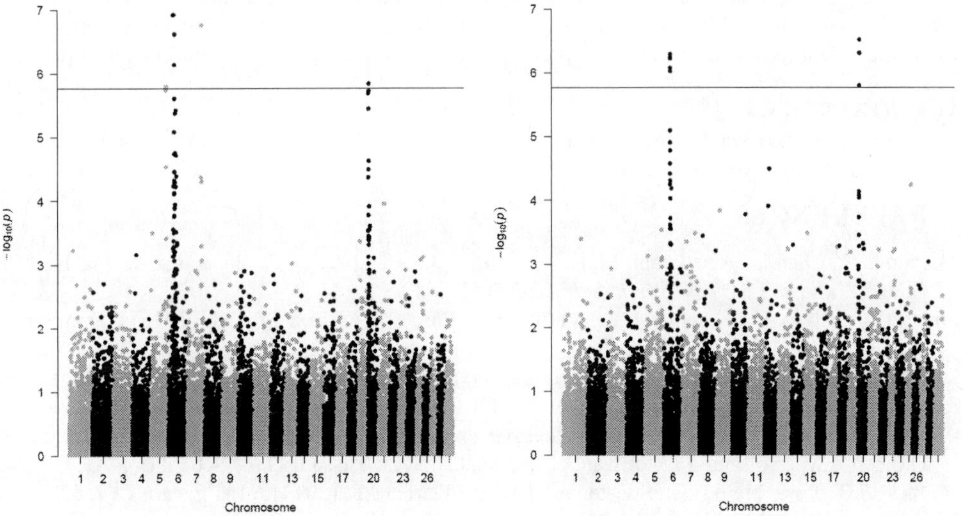


**(– log10 (1.718154e-06)) for Bonferroni correction**

Saatchi *et al.* (2014) identified significant SNPs for BW and WW in *Bos taurus* breeds, present on chromosomes 2, 4, 5, 6, 7, 14, 20, 21 and 29. Genomic regions significant for BW and WW include chromosome 5 (106Mb), 6 (38Mb), 7 (93Mb) and 20 (4Mb), these being associated with genes responsible for tissue development, ossification, adipose tissue development and regulation of transport activities (Saatchi *et al.*, 2014). In the present multi-breed GWAS, the final significant SNPs identified for BW (Table 2) explained 19% of additive genetic variance, with a major contribution (11%) coming from SNPs on chromosome 6 (39Mb region). This appears to be a well-known QTL region affecting body weight in other beef breeds (Snelling *et al.*, 2010) and animal species (Metzger *et al.*, 2013). For WW, the final significant SNPs explained 9% of additive genetic variance (Table 2), with a major contribution coming from the same SNP on chromosome 6 (39Mb region) as for BW.

**Table 2. Significant SNPs associated, variance and heritability of the BW and WW of multi-breed GWAS***

| Trait | Chr | Mb | P-values | V(G) | V(snp)/V(G) | $h^2$ |
|---|---|---|---|---|---|---|
| BW | 5 | 106 | 1.610E-06 | 4.56 ± 0.25 | 0.01 | 0.32 ± 0.01 |
| | 6 | 39 | 1.17E-35 | | 0.11 | |
| | 7 | 93 | 5.80E-11 | | 0.03 | |
| | 20 | 4.6 | 8.03E-17 | | 0.04 | |
| WW | 6 | 39 | 3.08E-12 | 82.67±6.57 | 0.05 | 0.18 ± 0.01 |
| | 20 | 6.3 | 5.25E-11 | | 0.04 | |

* Chr = Chromosome; Mb = Mega base pairs position according to UMD3.1 resemble; V(G) = total genetic variance =; V(snp)/V(G) = total genetic variance explained by significant SNP, $h^2$= heritability.

## CONCLUSIONS

This study detected several SNPs as having a significant association with birth and weaning weight, with these SNPs being located on chromosomes 5, 6, 7 and 20. Of the final significant SNPs identified, they accounted for 19% and 9% of the total genetic variance for BW and WW respectively. Results of this study may have application for genetic evaluations where specific SNPs are included to improve the accuracy of prediction for birth and weaning weight in beef cattle.

## ACKNOWLEDGMENTS

The authors acknowledge the contributions of Hereford Australia, Simmental Australia and the Charolais Society of Australia, and their respective members, in providing access to the data used.

## REFERENCES

Akanno E.C., Chen L., Abo-Ismail M. K., Crowley J. J., Wang Z., Li C., Basarab J.A., MacNeil M. D. and Plastow G.S. (2018) *Gen. Sel. Evol.* **50**(1), 48.

Chang C.C., Chow C.C., Tellier L.C., Vattikuti S., Purcell S.M. and Lee J.J. (2015) *GigaScience,* **4**(1).

Saatchi M., Schnabel R.D., Taylor J.F. and Garrick D.J. (2014) *BMC Genomics* **15**(1), 442.

Sargolzaei M., Chesnais J.P. and Schenkel F.S. (2014) *BMC Genomics* **15**(1), 478.

Utsunomiya Y. T., do Carmo A.S., Carvalheiro R., Neves H.H.R., Matos M.C., Zavarez L.B., Pérez O'Brien A.M., Sölkner J., McEwan,J.C., Cole J.B., Van Tassell, C.P., Schenkel F.S., da Silva M.V.G.B., Porto Neto L.R., Sonstegard T.S. and Garcia J.F. (2013) *BMC genetics* **14**, 52.

VanRaden P.M. (2008) *J Dairy Sci.* **91**.

Yang J., Lee S.H., Goddard M.E. and Visscher P.M. (2011) *American J. Human Genet.***88**(1), 76.

*Proc. Assoc. Advmt. Anim. Breed. Genet: 24: 435-438*

# GENETIC PARAMETERS FOR STRUCTURAL TRAITS IN NEW ZEALAND BEEF CATTLE AND THEIR CORRELATIONS WITH PRODUCTION TRAITS

**F. Weik[1,2], R.E. Hickson[1], S.T. Morris[1], D.J. Garrick[1] and J.A. Archer[2,3]**

[1]School of Agriculture & Environment, Massey University, Palmerston North 4442, New Zealand
[2]AbacusBio Ltd, Dunedin 9058, New Zealand
[3]Current address: Beef + Lamb New Zealand Genetics, Dunedin 9054, New Zealand

## SUMMARY

Structural soundness has the potential to affect the length of a productive life in beef cattle. The objectives of this study were to estimate genetic parameters for structural traits and to examine their relationship with production traits (mature weight, body condition score, 18-month weight and yearling hip height) measured in beef cattle in New Zealand. Heritabilities for structural traits were low to moderate ranging from 0.09 to 0.25. Genetic correlations among structural traits ranged from 0.18 to 1.00 whereas phenotypic and genetic correlations with production traits were generally low positive to moderate negative (-0.54–0.23) indicating only a limited impact on production.

## INTRODUCTION

Structural soundness is believed to influence fitness of cattle in extensive pasture-based farming systems. Beef cattle may be required to walk long distances to graze so unsound structure may impact on cow performance. Scientific literature on structural traits of beef cattle is sparse and there are no previous reports from New Zealand. Research on structural soundness has been predominantly conducted in dairy cattle (Dechow *et al.* 2002) and there is some evidence that females with good conformation stay in the herd for longer (Berry *et al.* 2005). Most reports, however, have been focused on type traits other than feet and leg scores. Therefore, the objectives of this study were to estimate genetic parameters for 9 structural feet and leg traits recorded in commercially farmed beef cattle in New Zealand and to examine their relationship with mature cow weight (MWT), body condition score (BCS), 18-months weight (W18) and yearling hip height (HH).

## MATERIALS AND METHODS

**Dataset.** The project was approved by the AgResearch Invermay Animal Ethics Committee. Data were obtained from an ongoing progeny test initiated in 2014 on 5 commercial New Zealand hill country farms to compare the performance of progeny derived from matings of Angus, Hereford, Simmental, Stabilizer and Charolais bulls over Angus or Hereford cows (Weik *et al.* 2021). The current study used data recorded between 2014 and 2020 for structural and production traits. Birth dates were not recorded, but age was assigned based on fetal age scanning.

**Trait definitions.** Structural traits were assessed according to the Beef Class Structural Assessment system (Breedplan 2021). Seven traits were recorded: front feet angle (FA), front feet claw set (FC), front legs front view (FF), rear feet angle (RA), rear feet claw set (RC), rear legs hind view (RH) and rear legs side view (RS). Records were available for a total of 2,294 animals for RA, 2,670 animals for RH and RS and 2,671 animals for all other structural traits at approximately 16–20 months of age by the same experienced Breedplan accredited assessor across all farms. Each trait was recorded following a linear assessment on a 1 to 9 scale, with 1 and 9 representing biological extremes with 5 as the intermediate optimum. No animals were scored at the extreme ends of the scale (1–2 or 9, respectively) and 99.6% of observations were between 5 and 7 (Table 1). Overall feet score was calculated for each animal by taking the worst score for FA and RA, or FC and RC, for overall feet angle (OA) and overall claw set (OC), respectively.

Four production traits were included in the correlation analyses, namely MWT, BCS, W18 and

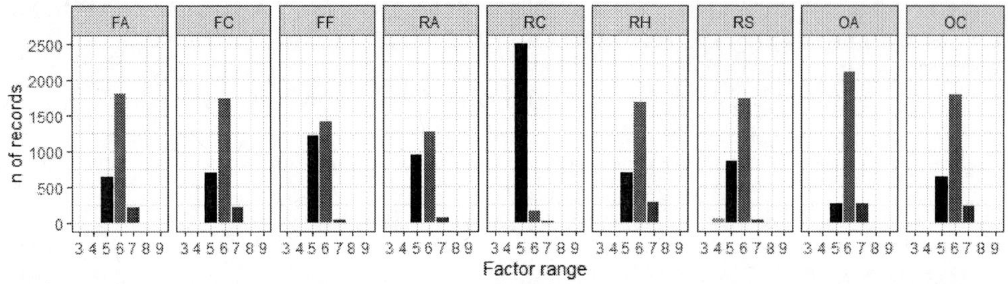


**Figure 1. Distribution of scores for front feet angle (FA), front feet claw set (FC), front legs front view (FF), rear feet angle (RA), rear feet claw set (RC), rear legs hind view (RH), rear legs side view (RS), overall feet angle (OA) and overall claw set (OC)**

HH. A total of 39,464 records were available for MWT. Data were obtained at 3 timepoints annually, prior to mating, at calf weaning and prior to calving for cows aged over 2 years. Cow BCS was recorded at the same times, generating 39,467 records based on visual assessment on a 1 to 10 scale (1=emaciated, 10=obese; Hickson *et al.* (2017)). Both traits were adjusted to a constant 6 years of age using fixed effect models with age and contemporary group (CG) as factors in the model.

A total of 7,048 progeny were recorded for weaning weight (WWT) between 110 and 228 days of age. Measures on W18 were available for 4,189 individuals measured between 455 and 752 days of age. Each animal was recorded once for WWT and W18. Linear and quadratic adjustments to 200 and 600 days of age were applied for WWT and W18 using a multiplicative approach similar to that described by Reverter *et al.* (2000). Records for HH were obtained once per animal between 277 and 417 days of age for 5,125 individuals, and adjusted to 365 days, using quadratic age adjustments.

Observations for production traits further than 3 standard deviations from the CG mean were deleted. For all structural traits, WWT and W18, CG comprised farm, sex, recording date and management group from birth until the day of recording. The CG for MWT and BCS consisted of farm, time of year, recording date and management group at the time of data collection. The HH CGs were made up of farm, sex and recording date. Individuals with missing CG information or CG containing only 1 animal were excluded from analyses. All production traits were tested for evidence of heterogeneity. Traits with a significant regression of CG mean on CG SD were scaled to homogenize the variance (Pickering *et al.* 2012).

**Statistical analysis.** Data quality control and pre-adjustments of phenotypes were conducted using R version 3.6 (R Core Team 2019). (Co)variance parameters were estimated using ASREML 4.1 (Gilmour *et al.* 2009). For all traits, WWT was included as a correlated trait to account for preselection. Thus, heritability estimates were obtained from bivariate animal models and genetic and phenotypic correlations from (co)variance parameters using a range of trivariate animal models.

Fixed effects included for all traits were CG, breed percentage and heterosis (purebred = 0, first-cross = 1). Age of dam was fitted as a factor for all structural traits as well as WWT, W18 and HH. Age at scoring was fitted as a linear covariate in the model for each structural trait. An animal effect and a residual error term were fitted as random effects for each trait, a permanent environmental effect was fitted for MWT and BCS due to repeated measures over time and a maternal additive genetic as well as a permanent environmental effect of the dam were fitted for WWT. Variance structures for the random effects were assumed as follows: var(a) = $\mathbf{A}\sigma_a^2$, var(m) = $\mathbf{A}\sigma_m^2$, var(pe) = $\mathbf{I}\sigma_{pe}^2$ and var(e) =$\mathbf{I}\sigma_\varepsilon^2$. No covariance was fitted between direct and maternal genetic effects. The numerator relationship matrix (**A**) included 13,325 animals with 394 sires and 4,098 dams.

## RESULTS AND DISCUSSION

Means, phenotypic standard deviations and estimated heritabilities from bivariate analyses with

WWT for each structural trait are presented in Table 1. The means of all structural traits ranged from 5.1 to 6.0. The standard deviations (SD) were similar for most traits except RC. The limited number of extreme scores meant that SD were low for all structural traits.

**Table 1. Raw means, phenotypic standard deviations ($\sigma_P$) and heritabilities ($h^2$) for structural traits with standard errors shown in brackets**

| | FA | FC | FF | RA | RC | RH | RS | OA | OC |
|---|---|---|---|---|---|---|---|---|---|
| Mean | 5.8 | 5.8 | 5.6 | 5.6 | 5.1 | 5.9 | 5.7 | 6.0 | 5.8 |
| $\sigma_P$ | 0.53 | 0.52 | 0.50 | 0.53 | 0.26 | 0.57 | 0.53 | 0.43 | 0.51 |
| | (.09) | (.09) | (.08) | (.09) | (.04) | (.10) | (.09) | (.07) | (.09) |
| $h^2$ | 0.23 | 0.10 | 0.09 | 0.17 | 0.09 | 0.22 | 0.12 | 0.25 | 0.11 |
| | (.05) | (.04) | (.04) | (.05) | (.04) | (.05) | (.04) | (.06) | (.04) |

*For structural trait abbreviations see Figure 1

**Table 2. Genetic (below diagonal) and phenotypic (above diagonal) correlations (se) from trivariate animal models among structural and production traits in New Zealand beef cattle**

| | FA | FC | FF | RA | RC | RH | RS | OA | OC | MWT | BCS | W18 | HH |
|---|---|---|---|---|---|---|---|---|---|---|---|---|---|
| FA | | 0.38 | 0.22 | -0.03 | 0.07 | 0.14 | 0.12 | 0.74 | 0.36 | -0.07 | -0.06 | -0.09 | 0.00 |
| | | (.02) | (.02) | (.02) | (.02) | (.02) | (.02) | (.01) | (.02) | (.03) | (.03) | (.02) | (.02) |
| FC | 0.99 | | 0.17 | 0.02 | 0.02 | 0.12 | 0.14 | 0.27 | 0.95 | -0.03 | -0.04 | -0.03 | 0.04 |
| | (.12) | | (.02) | (.02) | (.02) | (.02) | (.02) | (.02) | (.00) | (.03) | (.03) | (.02) | (.02) |
| FF | 0.54 | 0.66 | | 0.08 | 0.04 | 0.21 | 0.12 | 0.15 | 0.16 | -0.06 | -0.08 | -0.13 | 0.01 |
| | (.20) | (.27) | | (.02) | (.02) | (.02) | (.02) | (.02) | (.02) | (.03) | (.03) | (.02) | (.02) |
| RA | 0.24 | 0.19 | 0.69 | | 0.15 | 0.09 | 0.08 | 0.39 | 0.04 | -0.07 | 0.00 | -0.12 | -0.06 |
| | (.20) | (.26) | (.24) | | (.02) | (.02) | (.02) | (.02) | (.02) | (.03) | (.04) | (.02) | (.03) |
| RC | 0.33 | 0.57 | 0.27 | 0.36 | | 0.04 | 0.03 | 0.12 | 0.20 | -0.05 | -0.04 | 0.01 | 0.01 |
| | (.23) | (.32) | (.32) | (.26) | | (.02) | (.02) | (.02) | (.02) | (.03) | (.03) | (.02) | (.02) |
| RH | 0.21 | 0.38 | 0.61 | 0.50 | 0.43 | | 0.29 | 0.11 | 0.12 | -0.08 | -0.14 | -0.17 | 0.07 |
| | (.17) | (.21) | (.21) | (.18) | (.24) | | (.02) | (.02) | (.02) | (.03) | (.03) | (.02) | (.02) |
| RS | 0.42 | 0.60 | 0.18 | 0.72 | 0.53 | 0.26 | | 0.12 | 0.13 | -0.06 | -0.14 | -0.14 | -0.02 |
| | (.19) | (.25) | (.28) | (.21) | (.30) | (.20) | | (.02) | (.02) | (.03) | (.03) | (.02) | (.02) |
| OA | 0.92 | 0.76 | 0.79 | 0.53 | 0.32 | 0.27 | 0.56 | | 0.26 | -0.10 | 0.00 | -0.11 | -0.03 |
| | (.04) | (.15) | (.19) | (.16) | (.23) | (.16) | (.18) | | (.02) | (.03) | (.03) | (.02) | (.03) |
| OC | 0.91 | 1.00 | 0.61 | 0.22 | 0.66 | 0.34 | 0.59 | 0.69 | | -0.05 | -0.05 | -0.02 | 0.04 |
| | (.12) | (.02) | (.26) | (.24) | (.26) | (.21) | (.25) | (.15) | | (.03) | (.03) | (.02) | (.02) |
| MWT | -0.19 | -0.10 | -0.21 | -0.16 | -0.16 | -0.16 | -0.16 | -0.20 | -0.13 | | | | |
| | (.07) | (.11) | (.12) | (.09) | (.12) | (.08) | (.10) | (.07) | (.11) | | | | |
| BCS | -0.07 | -0.14 | -0.27 | -0.10 | -0.09 | -0.54 | -0.35 | -0.03 | -0.12 | | | | |
| | (.09) | (.13) | (.14) | (.11) | (.14) | (.10) | (.12) | (.09) | (.12) | | | | |
| W18 | -0.17 | -0.04 | -0.30 | -0.53 | -0.27 | -0.09 | -0.06 | -0.32 | -0.08 | | | | |
| | (.12) | (.17) | (.17) | (.13) | (.18) | (.12) | (.16) | (.11) | (.16) | | | | |
| HH | 0.07 | 0.15 | 0.23 | -0.19 | -0.07 | 0.19 | -0.12 | -0.03 | 0.14 | | | | |
| | (.12) | (.17) | (.18) | (.15) | (.18) | (.12) | (.16) | (0.12) | (0.16) | | | | |

*For structural trait abbreviations see Figure 1; MWT=mature cow weight, BCS=body condition score, W18=18-month weight, HH=yearling hip height

The estimated heritabilities for structural traits were in the low-to-mid range from 0.09 to 0.25, consistent with Jeyaruban *et al.* (2012) and Vallee *et al.* (2015). Heritabilities for front feet observations were higher than their rear counterparts. The highest heritabilities were estimated for FA, RH and OA. Production traits were moderately to highly heritable with 0.57 (0.03) for MWT, 0.54 (0.04) for W18 and 0.52 (0.04) for HH and the estimated heritability was lowest for BCS at

0.25 (0.03). Those values are consistent with estimates from the literature.

Genetic and phenotypic correlations are shown in Table 2. Phenotypic correlations among structural traits were generally positive and lower than genetic correlations. The estimated genetic correlations were positive among all structural traits ranging from 0.18 to 1.00. The highest correlations were observed between FA and FC (0.99) and the part-whole correlations FA and OA (0.92) and FC and OC (1.00). Correlations between both rear feet traits and the overall foot scores, however, were lower with 0.53 between RA and OA and 0.66 between RC and OC, indicating that overall feet scores are primarily driven by the condition of the front feet. Jeyaruban *et al.* (2012) reported high genetic correlations between FA and RA, which were considerably lower in the current study (0.24). The correlation between FC and RC (0.57) in this study, however, was consistent with their reported estimate of 0.63. Genetic correlations were generally higher among traits measured on the front feet (0.54–0.99) than on the rear feet (0.26–0.72).

The phenotypic correlations were generally low between structural and production traits, indicating that there is no evidence that structural traits in this study have a substantial impact on those production traits measured later in life. Genetic correlations between structural and production traits were similar for MWT, BCS and W18 and were generally low and negative and this may be attributable to low variation of the observed structural traits. The only moderate genetic correlations further than 2 standard errors from 0 were the negative correlations between BCS and RH (-0.54), BCS and RS (-0.35), W18 and RA (-0.53) and between W18 and OA (-0.32). Given the distribution of scores above the optimum these suggest that selecting for BCS and W18 is unlikely to increase the frequency of animals with unsound structure. The genetic correlation between structural traits and HH were low overall with the highest genetic correlation estimated for HH and FF (0.23).

## CONCLUSIONS

Low to moderate heritabilities for structural traits exist in commercially farmed beef cattle in New Zealand. Genetic and phenotypic correlations among structural and production traits were generally low to moderate and negative, indicating only weak associations and, thus, a limited impact of structural traits on the recorded production traits in this study.

## ACKNOWLEDGEMENTS

This research was funded by Beef+Lamb NZ Genetics.

## REFERENCES

Berry D.P., Harris B.L., Winkelman A.M. and Montgomerie W. (2005) *J. Dairy Sci.* **88**: 2962.

Breedplan. (2021) https://breedplan.une.edu.au/media/1zhdxy51/recording-structural-soundness-information.pdf; accessed 15 January 2021.

Dechow C.D., Rogers G.W. and Clay J.S. (2002) *J. Dairy Sci.* **85**: 3062.

Gilmour A.R., Gogel B., Cullis B., Thompson R. and Butler D. (2009) VSN Int. Ltd, UK.

Hickson RE, Morris ST and Thomson BC. (2017) 'Beef cow body condition scoring.'

Jeyaruban G., Tier B., Johnston D. and Graser H. (2012) *Anim. Prod. Sci.* **52**: 1.

Pickering N.K., Dodds K.G., Blair H., Hickson R.E., Johnson P. and McEwan J. (2012) *J. Anim. Sci.* **90**: 1411.

R Core Team. (2019). R: A language and environment for statistical computing. R Foundation for Statistical Computing. Vienna, Austria.

Reverter A., Johnston D.J., Graser H.U., Wolcott M.L. and Upton W.H. (2000) *J. Anim. Sci.* **78**: 1786.

Vallee A., Breider I., Van Arendonk J.A.M. and Bovenhuis H. (2015) *J. Anim. Sci.* **93**: 4277.

Weik F., Archer J.A., Morris S.T., Garrick D.J., Miller S.P., Boyd A.M., Cullen N.G. and Hickson R.E. (2021) *New Zeal. J. Agr. Res.* 1.

*Proc. Assoc. Advmt. Anim. Breed. Genet: 24: 439-442*

# GENETICS OF HEIFER AGE AT PUBERTY IN AUSTRALIAN ANGUS CATTLE

**M.L. Wolcott[1], R. Banks[1] and G. Alder[2]**

[1]Animal Genetics and Breeding Unit* University of New England, Armidale, NSW 2351 Australia
[2] Beef and Lamb New Zealand Genetics, Dunedin 9058, New Zealand.

## SUMMARY

Age at puberty has become a key trait in the genetic evaluation of female reproduction for tropically adapted beef breeds in northern Australia. This study aimed to characterise the trait in Australian *Bos taurus* seedstock heifers and determine the degree to which it, and associated traits, were under genetic control. Angus heifers (N = 3093) from nine seedstock herds were serially ultrasound scanned to determine age at puberty, via detection of their first *corpus luteum*, at approximately 4 week intervals from 10.5 to 13.6 months of age, when heifers were synchronised for artificial insemination. Results showed that only 53% of Angus heifers were pubertal at synchronisation for AI and that within this category, age at puberty had a heritability of 0.33. When a penalised record (maximum age at puberty for a contemporary group plus 21 days) was included for heifers that were not pubertal into mating, heritability increased to 0.42. For sires with EBV accuracy greater than 0.7, EBVs for age at puberty ranged from -69 to +70 days. The ability of heifers to conceive early in their first mating season has been linked to lifetime reproductive performance. These results suggest that the proportion of heifers that have reached puberty as they enter their first mating is significantly less than 100% and that opportunities exist to monitor and apply selection to improve age at puberty in Australian Angus heifers.

## INTRODUCTION

Results from the Co-operative Research Centre for Beef Genetic Technologies Northern Breeding Project (Beef CRC) showed that age at puberty, identified by serial ultrasound scanning to determine the date at first ovulation, was heritable in tropically adapted beef genotypes (Johnston *et al.* 2009). Associated research also demonstrated that lower age at puberty was moderately and favourably genetically correlated with lifetime reproductive outcomes ($r_g$ = -0.29 to -0.40), and that selection to improve (reduce) age at puberty would have favourable consequences for lifetime reproductive performance (Johnston *et al.* 2014). Morris *et al.* (2000) showed moderate heritability for age at puberty (first observed oestrus) in Angus heifers ($h^2$ = 0.31), and a high genetic correlation with first mating pregnancy rate in naturally mated (or AI to observed oestrus) heifers ($r_g$ = -0.89), and Wolcott *et al.* (2019) reported a similar heritability ($h^2$ = 0.38) for Hereford heifers in Australia. Continuing from that work, the current study aimed to exploit methods developed in the Beef CRC to characterise age at puberty in Angus heifers, to determine the heritability of the trait and its potential to provide a means to monitor and select to improve age at puberty for the breed.

## MATERIALS AND METHODS

**Animals and management.** Heifers involved in this study were made available by nine Angus seedstock breeders. Herds were selected for inclusion based on a history of high quality pedigree and performance recording, and a willingness to endure the significant imposition associated with serial ultrasound scanning required to identify first oestrous. Calving periods for heifers evaluated for the study ranged from 2 – 3 months. The heifers included in the analysis were the progeny of 260 sires, with 78% being the daughters of sires with at least 10 progeny, and 34% of heifers from sires with daughters evaluated in at least two herds.

* A joint venture of NSW Department of Primary Industries and the University of New England.

Heifers were weaned at an average age of 6.5 months, with a range of weaning ages from 5.3 to 8.0 months. On average, heifers weaned in 2018 were reared under significantly dryer conditions than those in 2017. This meant that more supplementary feeding was provided for heifers in 2018, but all animals received the same nutritional interventions within herd and year. This also applied to routine management practices (animals' identification and branding, vaccination, parasite control treatments, etc.). Limited culling for conformation related traits between weaning and synchronisation for AI took place though this was assumed to be independent of any understanding of genetic reproduction. All herds routinely submit data to BREEDPLAN for genetic evaluation. For the heifers involved in this study, this included pedigree information, date of birth and weaning weight, and these were extracted from the Angus Australia database for these analyses.

**Scanning for ovarian function.** Ultrasound scanning to detect first oestrous followed the protocols described by Johnston *et al.* (2009) for tropical beef females in the Beef CRC. Within herd and year, scanning was performed by one of three technicians using a Mindray M7Vet real-time ultrasound unit equipped with a variable frequency 6LE5Vs intra-rectal transducer, set at 8MHz. The timing of first scans to detect the presence of a *corpus luteum* (CL), was undertaken when managers at each location observed the first signs of heat in the heifer cohorts examined for this study (subsequently referred to as their 'post-weaning' record). Subsequent scans took place at 4 - 6 week intervals until the first progesterone-based synchronisation treatment occurred in each herd, prior to artificial insemination (or their 'into-mating' record). All heifers in a cohort were scanned post-weaning) and at synchronization for AI, with interim scans performed on heifers that had not previously displayed a CL. This resulted in most heifers being scanned three times up to synchronisation, with the average number of scans per animal, within herd and year, between 2.2 and 3.9.

Based on ovarian scanning results, the following traits were defined:

- **Age at puberty (AP)** was a trait in females that displayed a CL prior to mating, calculated as the scanning date at which the first CL was detected minus their date of birth.
- **Penalised AP (APP)** generated an age at puberty record for heifers that had failed to display a CL prior to mating. APP was calculated for these animals as the maximum AP for their contemporary group plus 21 days. For a small number of heifers that failed to display a CL prior to mating and were in small contemporary groups (for which the maximum AP was based on too few records ($N \leq 3$) to be reliable) no APP was analysed ($N = 15$ heifers).
- **Pubertal into mating (PUB)** was a binary trait that identified heifers that had cycled at any time up to mating (1) or not (0).
- **Antral follicle count (FC)** was the total number of follicles greater than 2mm, visible by ultrasound examination of both ovaries at the first scan in heifers which did not display a CL. FC was recorded in this project to investigate its genetic associations with economically important female reproduction traits based on favourable results presented by researchers examining dairy cow performance in New Zealand (Martinez *et al.* 2016).

**Growth and body composition traits.** At each scan, records of liveweight weight (LWT in kg), hip height (HH in cm) and body condition score (BCS on a 1- to 5+ scale) were collected for each heifer following the protocols for growth and body composition traits described by Johnston *et al.* (2009). P8 fat depth (P8 in mm) was also measured at each scan using the scanner's inbuilt callipers.

**Modelling, variance component and EBV estimation.** Descriptive statistics were generated using PROC MEANS in SAS. Contemporary group information was extracted from the Angus Australia database and was built based on information supplied by participating breeders as described by Graser *et al.* (2005).

The contemporary group for BREEDPLAN 200-day weight was used to analyse heifer growth, body composition and the descriptors of ovarian function evaluated for this study. For growth and body composition traits, dam age and linear animal age were fitted as covariates. Consistent with the protocols established by Johnston *et al.* (2009) heifer age was modelled for scanned ovarian traits as the month of birth nested within herd and year. Variance components for each trait were estimated in univariate analyses in ASReml (Gilmour *et al.* 2009), with EBVs for all animals in the three generation pedigree estimated as the solution for the random animal effect. For this study, genetic parameters for the binary PUB trait were estimated on the observed scale.

## RESULTS AND DISCUSSION

**Growth and body composition traits.** Summary statistics, additive variances and heritabilities for post-weaning growth and body composition traits are presented in Table 1. On average, heifers were 10.6 months of age at their post-weaning scan, with mean ages at first scan being reasonably consistent across herds and years. Additive variances and heritabilities for post-weaning LWT and HH were consistent with those reported by Donoghue *et al.* (2018) for Angus and Hereford females prior to their first calving ($h^2 = 0.45$ and 0.57), and with results from this study previously reported by Wolcott *et al.* (2019) for Hereford heifers ($h^2 = 0.55$ and 0.49). The heritability for post-weaning P8 was lower than that reported by Donoghue *et al.* (2018) for Angus females prior to their first calving ($h^2 = 0.44$), but was comparable for BCS ($h^2 = 0.14$), while almost identical results were presented by Wolcott *et al.* (2019) for Hereford heifers ($h^2 = 0.29$ and 0.20 for P8 and BCS respectively). The technicians who collected ultrasound data describing ovarian traits were not accredited BREEDPLAN carcass scanners, and it is possible that a degree of measurement inaccuracy may account for the slightly lower than expected heritability for scanned fat depth.

**Table 1. Number of records analysed (N), mean and standard deviation (SD), with additive variance ($\sigma_a^2$) and heritability ($h^2$) (and standard error (s.e.) for $h^2$ estimates) for post-weaning growth and body composition and scanned ovarian traits in Angus heifers**

| Traits | Units | N | Mean | SD | $\sigma_a^2$ | $h^2$ | s.e. |
|---|---|---|---|---|---|---|---|
| **Post-weaning growth and body composition** | | | | | | | |
| AGE | Days | 3093 | 319.9 | 46.6 | - | - | - |
| LWT | kg | 3085 | 314.5 | 48.3 | 339.2 | 0.37 | 0.06 |
| HH | cm | 1816 | 116.9 | 4.5 | 7.1 | 0.57 | 0.08 |
| P8 | mm | 3039 | 5.0 | 2.9 | 0.8 | 0.21 | 0.06 |
| BCS | Score (1 – 5) | 3093 | 2.8 | 0.7 | 0.04 | 0.29 | 0.06 |
| **Ovarian scanned traits** | | | | | | | |
| AP | Days | 1634 | 345.2 | 63.2 | 378.1 | 0.33 | 0.08 |
| APP | Days | 3078 | 392.5 | 71.1 | 1224.0 | 0.42 | 0.06 |
| PUB[A] | 1/0 | 3077 | 0.53 | 0.50 | 0.06 | 0.32 | 0.05 |
| FC | Count | 2544 | 21.9 | 8.9 | 21.1 | 0.34 | 0.06 |

[A] Variance components for PUB estimated on the observed scale.

**Ovarian scanned traits.** Additive variances and heritabilities (and associated standard errors) for scanned ovarian traits are also presented in Table 1. A key result from this work was the proportion of Angus heifers that were pubertal into mating (PUB = 0.53). This was consistent with the result presented by Wolcott *et al.* (2019) for Hereford heifers involved in the same project (PUB = 0.52), and reinforces the need to understand the genetics of puberty traits in temperate breeds. The phenotypic and additive variance for APP (2882.8 and 1224.0 days respectively) were substantially lower than those reported by Johnston *et al.* (2009) for tropically adapted heifers, consistent with the much shorter scanning period in temperate breeds where maiden matings occur approximately 12 months earlier. The moderate heritability estimated for APP ($h^2 = 0.42$) suggests that the opportunity exists to improve the trait by selection in the Angus breed. Both AP and APP were under significantly greater genetic control than days to calving ($h^2 \sim 0.05$), which is currently the key descriptor of female reproductive performance in the BREEDPLAN evaluation for the breed.

For sires with greater than 70% EBV accuracy, EBVs for APP ranged from -69 to +70 days. These results suggest that sire selection could impact age at puberty in the resulting progeny by at least two months. With only 52% of females pubertal into their first mating, and mating periods as low as 2 months in commercial beef breeding herds in southern Australia, this could have implications for reproductive outcomes for naturally mated maiden heifers.

Mean and standard deviation for post-weaning FC were consistent with those reported by Walsh *et al.* (2014) for dairy heifers in the US and Ireland, with heritabilities also comparable ($h^2 = 0.25$ and 0.31 respectively). Antral follicle count was assessed in this project to allow investigation of its genetic association with female reproduction traits, and this will be the subject of future analyses.

## CONCLUSIONS

This study presents an initial investigation of the genetics of age at puberty and associated traits in Australian Angus seedstock heifers. Results showed that the opportunity exists to improve (reduce) age at puberty by selection in the breed and, by including the trait in the breed's genetic evaluation, to monitor this aspect of female reproduction as selection is applied to improve other economically important traits. The proportion of heifers that were not pubertal as they entered their first mating was a key result of this study. The increasing prevalence of artificial insemination and the associated treatments to synchronise (and possibly induce), first oestrous suggest that genetic factors which impact a heifer's capacity to conceive early in their first mating season warrant monitoring and inclusion in the genetic evaluation for temperate beef breeds. It is acknowledged that serial ultrasound scanning to detect first oestrous is an expensive and labour-intensive operation, making it a prime candidate for evaluation in intensively recorded reference populations.

## ACKNOWLEDGEMENTS

This research was supported by co-funding from Beef and Lamb New Zealand and Meat and Livestock Australia (project P.PSH.0869). The co-operation of the principals and employees of the seedstock herds, which made animals available for this study, as well as their breed representative body (Angus Australia) is also gratefully acknowledged, as is the technical expertise provided by staff of NSW DPI, Holbrook Veterinary Centre and Terang and Mortlake Veterinary Clinic.

## REFERENCES

Donoghue K.A., Lee S.J., Parnell P.F. and Pitchford W.S. (2018) *Anim. Prod. Sci.* **58:** 145.

Gilmour A.R., Gogel B.J., Cullis B.R. and Thompson R. (2009) 'ASReml user guide. Release 3.0.' VSN International: Hemel Hempstead, UK.

Graser H.U., Tier B., Johnston D.J. and Barwick S.A. (2005) *Anim. Prod. Sci.* **45**: 913.

Johnston D.J., Barwick S.A., Corbet N.J., Fordyce G., Holroyd, R.G., Williams, P.J. and Burrow, H.M. (2009) *Anim. Prod. Sci.* **49**: 399.

Johnston D.J., Corbet N.J., Barwick S.A., Wolcott M.L. and Holroyd R.G. (2014) *Anim. Prod. Sci.* **54**: 74.

Martinez M.F., Sanderson N., Quirke N. Lawrence S.B. Juengel J. (2016) *Theriogenology*, **85**:466.

Morris C.A., Wilson J.A., Bennett G.L., Cullen N.G., Hickey S.M. and Hunter J.C. (2000) *NZ J. Agric. Res.* **43.1**: 83.

Walsh S.W., Mossa F., Butler S.T., Berry D.P., Scheetz D. *et al.* (2014) *J. Dairy Sc.* **97**: 4503.

Wolcott M.L., Banks R.G., Tweedie M. and Alder G. (2019) *Proc. Assoc. Advmt. Anim. Breed. Genet.* **23**: 91.

*Proc. Assoc. Advmt. Anim. Breed. Genet: 24: 443-446*

# METHANE EMISSIONS VARIATION AMONG NEW ZEALAND DAIRY FARMS AND HERDS

**X. Zhang[1], P.R. Amer[1], K. Stachowicz[1], C. Quinton[1] and J. Crowley[1,2]**

[1] AbacusBio Limited, PO Box 5585, Dunedin 9058, New Zealand
[2] University of Alberta, Edmonton, Alberta T6G 2R3, Canada

## SUMMARY

This study investigated the current animal-level and herd-level variation for enteric fermented methane emissions across pasture-based dairy farms in New Zealand. We used the DairyNZ core database consisting of 2,398 herds and 751,981 cows as the inputs, and inferred crucial but unknown variables including methane emissions per unit of feed from department of environment, food&rural affairs (DEFRA), and live weight from New Zealand animal evaluation limited database to predict methane emissions for individual dairy cows. Methane emissions were predicted using dry matter intake (DMI) with an Intergovernmental Panel on Climate Change tier 2 approach. While individual methane emissions ($R^2$ =0.29) were poorly predicted, but excellent predictability of herd average methane emissions were well predicted ($R^2$=0.95) based on variables including herd, age, replacement rate, DMI, live weight (LW) and milk solids. The results showed an advantage of predicting methane emissions at herd level than individual cow level. Based on the results, the NZ dairy industry should focus on new traits and breeding objectives, with the support of trait prioritisation, a monitoring plan, policy making and incentivisation for farmers.

## INTRODUCTION

More than 95% of methane emissions in a life cycle of dairy production come from enteric fermentation (Fonterra co-operative group limited, 2017). There is variation in greenhouse gas emissions among dairy farms caused by variation in production practices, environment, and regional historical disparity (Latham 2010; Beukes *et al.* 2010). To facilitate farmers in compliant with the future regulation, it will be important to establish objective, data driven and, practical and easy-to-implement methods of monitoring emissions levels at an individual farm level.

Currently animal identification and performance recording in New Zealand dairy farms are generally not well linked, due to the difficulty in tracking large herds on seasonal pasture-based production system (Edge and Kavalali 2018), although many farms have some level of recording in place for the purpose of herd improvement (3.67 million out 4.95 million cows, LIC and DairyNZ 2019). For example, the national database such as New Zealand dairy core database (DairyNZ Hamilton, New Zealand) have performance records unlinked to animal ID, such as live weight. Additionally, the current techniques for measuring methane per unit of feed was difficult to apply on a large scale (DEFRA 2014). With the introduction of new data and IT systems, it would be possible to create a database infrastructure that would allow dairy cow GHG emissions to be predicted at the individual cow level and aggregated to individual farm level.

Due to aforementioned reason, the objectives of this study were 1) to combine multiple existing data sources to predict the variation among individuals and herds for dairy cattle enteric fermentation methane emissions for New Zealand dairy farms; 2) assess the requirement of future data infrastructure and technologies in order to monitor methane emissions at animal and herd level and; 3) infer the emission mitigation strategies enabling the adoption of future on-farm emission policies and technologies.

## MATERIALS AND METHODS

**Data**. New Zealand dairy core database containing herd test and movement records of

27,288,426 cows from 1989 to 2014. Breeds included Jersey, Holstein-Friesian and crossbred. Records of cows calving between June 1st and October 1st, 2005 were extracted, and quality control such as removing cows with lactation length beyond 365 days was applied. 751,981 cows with records in 2,398 herds were obtained in the end.

**General approach**. DMI approach illustrated in the IPCC 2000 (Pickering *et al.* 2020; Clark *et al.* 2003) as $E = F \cdot \alpha$, where $E$ is methane emissions/cow/year, $F$ is the annual DMI (kg DMI/year) and $\alpha$ is the methane emissions per unit of feed (g CH4/kg DMI).

**Estimation of live weight**. Simulated from mean live weight by age and breed (Livestock Improvement Company 2008; DairyNZ 2019), a CV of 0.105 (Zhang *et al.* 2019), a phenotypic correlation of 0.15 between LW and milk yield during the first 240 days of lactation (Correa-Luna *et al.* 2018).

**Prediction of total lactation milk yield from test day records**. Obtained by fitting quantile splines to each lactating cow for their milk volume, protein and fat production during lactation using smooth.spline function in R(v3.5.3).

**Prediction of DMI from live weight and energy requirements**. First calculated the energy requirement following Nicol and Brookes (2007) and Clark *et al.* (2003) as the summation of maintenance, lactation, replacement and gestation energy requirement; then converted energy to DMI by multiplying the average diet energy.

**Prediction of methane emissions from DMI**. First obtained the mean and SD of methane emissions per unit of feed, $\alpha$, from experiments (DEFRA 2014) by removing research institute, measuring method, diet type, breed, sex and physiological status effects. Then sample $\alpha$ from this distribution and assign it to each cow *i*, multiplied by their DMI to obtain the prediction of *E*.

**Statistical analysis.** The summary statistical tests were calculated for measured and predicted variables (results not shown). Pearson correlations between *E* and energy related traits were also calculated (results not shown). To access the variance of variables in relation to *E*, an OLS linear model was fitted with herd as random effect, and milk solids, live weight, survival of individual cows and the herd averages of all previous effects as covariates.

## RESULTS AND DISCUSSION

The methane emissions per unit of feed was estimated as 20.72 ± 4.24 g $CH_4$/kg DM. Variances of each variable regressed on individual and herd average *E* are shown in Table 1. Factors including herd, milk solids per cow, cow live weight and survival could predict individual feed intake well ($R^2$=0.29) but not individual methane output ($R^2$=0.29). The reason is the substantial variation that exists in methane eructed per unit of feed consumed, which is also difficult to measure in practice (Beukes *et al.* 2010; Herrero *et al.* 2013; DEFRA 2014). Additionally, in practice, farmers are unlikely to mitigate emissions by reducing production. Therefore, new technologies such as e-collars that measure cow activity for the use of predicting DMI is also likely to be of limited use in practice.

Herd average milk solids and live weight were powerful in predicting herd average methane emissions ($R^2$=0.95), hence policy based on farm level rather than individual cow level could be more effective in reducing methane emissions on an industry wide basis.

**Table 1. Model comparisons for dry matter intake (*DMI, kg*) and methane emissions (*E*, kg) during the lactation for each cow and for the herd average ($\overline{DMI}$ and $\overline{E}$)**

| Dependent variable[1] | Model formula[2] | $R^2$ | R | Model variance | Total variance |
|---|---|---|---|---|---|
| $DMI_{a_{ij}}$ | $\sim h_{ij}$ | 0.28 | 0.52 | 232,627 | 845,228 |
| | $\sim \overline{MS_{i\cdot}}$ | 0.25 | 0.50 | 211,555 | 845,228 |
| | $\sim \overline{LW_{i\cdot}}$ | 0.11 | 0.33 | 94,448 | 845,228 |
| | $\sim \overline{SUR_{i\cdot}}$ | 0.01 | 0.09 | 6,943 | 845,228 |
| | $\sim \overline{age_{i\cdot}}$ | 0.01 | 0.09 | 7,011 | 845,228 |
| | $\sim h_{ij} + MS_{ij} + LW_{ij} + SUR_{ij}$ | 0.78 | 0.88 | 661,980 | 845,228 |
| $E_{ij}$ | $\sim h_{ij}$ | 0.10 | 0.32 | 100 | 985 |
| | $\sim \overline{MS_{i\cdot}}$ | 0.09 | 0.30 | 91 | 985 |
| | $\sim \overline{LW_{i\cdot}}$ | 0.04 | 0.20 | 41 | 985 |
| | $\sim \overline{SUR_{i\cdot}}$ | 0.003 | 0.06 | 3.09 | 985 |
| | $\sim \overline{age_{i\cdot}}$ | 0.003 | 0.06 | 3.02 | 985 |
| | $\sim h_{ij} + MS_{ij} + LW_{ij} + SUR_{ij}$ | 0.29 | 0.54 | 285 | 985 |
| $\overline{DMI_{a_{i\cdot}}}$ | $\sim \overline{MS_{i\cdot}}$ | 0.89 | 0.94 | 198,896 | 224,645 |
| | $\sim \overline{LW_{i\cdot}}$ | 0.41 | 0.64 | 92,298 | 224,645 |
| | $\sim \overline{SUR_{i\cdot}}$ | 0.02 | 0.14 | 4,512 | 224,645 |
| | $\sim \overline{age_{i\cdot}}$ | 0.01 | 0.10 | 2,411 | 224,645 |
| | $\sim \overline{MS_{i\cdot}} + \overline{LW_{i\cdot}}$ | 0.97 | 0.99 | 218,726 | 224,645 |
| $\overline{E_{i\cdot}}$ | $\sim \overline{MS_{i\cdot}}$ | 0.86 | 0.93 | 85 | 98 |
| | $\sim \overline{LW_{i\cdot}}$ | 0.40 | 0.63 | 39 | 98 |
| | $\sim \overline{SUR_{i\cdot}}$ | 0.02 | 0.14 | 1.96 | 98 |
| | $\sim \overline{age_{i\cdot}}$ | 0.01 | 0.11 | 1.27 | 98 |
| | $\sim \overline{MS_{i\cdot}} + \overline{LW_{i\cdot}}$ | 0.95 | 0.97 | 93 | 98 |

[1] Calculated for fall 2005 to spring 2006 season. *i* indicates *i*-th herd and *j* indicates *j*-th animal. *DMI* and *E* are accumulated predictions across the whole lactation.

[2] Dependent variables were herd ($h_{ij}$), herd average accumulated milk solids ($\overline{MS_{i\cdot}}$, kg), herd average mean live weight ($\overline{LW_{i\cdot}}$, kg), herd average survival ($\overline{SUR_{i\cdot}}$, year), herd average age ($\overline{age}_{ij}$, year), accumulated milk solids ($MS_{ij}$, kg), mean live weight ($LW_{ij}$, kg) and survival ($SUR_{ij}$, year). Herd was fitted as a random effect and other effects were fitted as covariates.

## CONCLUSIONS

This preliminary study identified a key antagonism between farmer desire for profitable utilisation of farm feed resources and a national need to moderate the overall methane emissions from the dairy industry. Technologies that only predict individual feed intake will have limited value for practical mitigation of enteric methane emissions. Rather, additional mechanisms would be required to effectively incentivise mitigation opportunities that reduce emissions per unit of feed. A well-linked comprehensive animal level database infrastructure could support effectively incentivising some levels of farm and animal level changes to reduce enteric methane emissions.

## REFERENCES


Beukes P.C., Gregorini P., Romera A.J., Levy G. and Waghorn G.C. (2010) *Agri. Ecosys. Env.* **136**: 358.

Clark H., Brookes I. and Walcroft A. (2003) Wellington, New Zealand

Correa-Luna M., Lopez-Villalobos N., Almeida G.A., Donaghy D.J. and Kemp P.D. (2018) *Proc. New Zealand Soc. of Anim. Prod.* **78:** 76.

DairyNZ. (2019) Select mature live weight. Hamilton, New Zealand

DEFRA (2014) London, UK

Edge M. and Kavalali B. (2018) *Operational solutions for primary industries* Wellington, New Zealand

Fonterra co-operative group limited. (2017) Auckland, New Zealand

Herrero M., Havlík P., Valin H., Notenbaert A., Rufino M.C., Thornton P.K., Blümmel M., Weiss F., Grace D. and Obersteiner M. (2013) *Proc. the National Academy of Sciences of the United States of America* **110**: 20888

Latham N. (2010) PhD dissertation. Lincoln University

LIC and DairyNZ (2019) Retrieved on February 10, 2020, from https://www.lic.co.nz/about/dairy-statistics/

Livestock Improvement Company (2008) Retrieved on February 10, 2020, from https://www.lic.co.nz/about/dairy-statistics/

Nicol A.M. and Brookes I.M. (2007) Hamilton, New Zealand.

Pickering, A., J. Gibbs, S. Wear, J. Fick, and H. Tomlin. (2020) Ministry of Primary Industries, Wellington.

Zhang X., Amer P.R., Jenkins G.M., Sise J.A., Santos B. and Quinton C. (2019) *J. Dairy Sci.* **102**: 11153

*Proc. Assoc. Advmt. Anim. Breed. Genet: 24: 447-450*

# THE APPLICATION OF A SUB-INDEX WEIGHTED PERCENT EMPHASIS METHOD TO AUSTRALIAN DAIRY SELECTION INDEXES

**X. Zhang[1] and P.R. Amer[1]**

[1] AbacusBio Ltd, 442 Moray Place, Dunedin 9016, New Zealand

## SUMMARY

Percent trait emphasis is a concept used to interpret the selection effort of a trait in a selection index. Zhang and Amer (2021) published a sub-index weighted percent emphasis and demonstrated its advantage over the traditional method. The objective of this study is to apply this method to a current selection index and compare that with the traditional method. The results showed that 1) the new methods for calculating trait percent emphasis outperform conventional methods, 2) differences in trait accuracy of prediction impact their real percent emphasis, and 3) unfavourable correlations among traits reduce their effective emphasis in indexes.

## INTRODUCTION

Percent trait emphasis is commonly used to describe selection indexes used in national genetic evaluations to help farmers and other users to interpret the selection effort being applied to competing traits. The currently accepted and widely used methods to calculate trait percent emphasis use the product of trait mean EBV and genetic SD as the base measurement, and the summation over all traits as the scaling factor (VanRaden 2002; Miglior *et al.* 2005, 2017). The sub-index weighted method (Zhang and Amer 2021) also accounts for accuracy of trait evaluation and correlations among traits. This method has been applied to USDA net merit of young bulls with lower accuracies compared to proven bulls (VanRaden *et al.* 2021). The aim of this study is to apply both methods to Australian HWI index and compare their results and impacts.

## MATERIALS AND METHODS

**Selection index emphasis methods.** The method is described in Zhang and Amer (2021). In short, the traits in the selection indexes are clustered based on their genetic correlations or accuracy adjusted EBV correlations. Then traits relative emphases are weighted by the corresponding cluster weights calculated as the percentage of the cluster variance over the sum of variances of all clusters.

**Materials.** We used the Australian dairy Health Weighted Index (HWI) and Balanced Performance Index (BPI) in 2020 to test the emphasis methods. We used a set of genomic Australian Breeding Value (ABV) predictions of 9,283 Holstein-Friesian cows with a minimum single trait evaluation accuracy of 60% except the trait feed saved (AUS HWI, DataGene 2020a; Axford *et al.* 2021). Table 1 shows the trait economic weights, Australian Breeding Values (ABV) SD and mean trait accuracies. The ABV correlations are shown in Appendix 3, Table 7 of DataGene (2020b).

## RESULTS AND DISCUSSION

The hierarchical clustering grouped traits with high within-cluster and low between-cluster absolute genetic or (G)EBV correlation traits together (Figure 1). Most of the sub-index groups were also consistent with their trait function groups, except FAT had been grouped separately from MILK and PROT, and PINSET and OTYPE were also separated, indicating that trait functions may not be an ideal way to group traits.

**Table 1. Summary statistics of 2020 Australian HWI selection index**[1]

| Trait | Trait abbreviation | Unit | Economic Weight ($) | | ABV SD | Mean accuracy (%) |
|---|---|---|---|---|---|---|
| | | | HWI | BPI | | |
| Milk protein | PROT | kg | 4.36 | 6.67 | 8.02 | NA |
| Milk fat | FAT | kg | 1.35 | 2.08 | 12.1 | NA |
| Milk volume | MILK | L | -0.07 | -0.11 | 365 | 76.0 |
| Survival | SURV | % surv one parity to next | 7.20 | 7.2 | 3.16 | 61.8 |
| Fertility | FERT | 42d calving% | 14.1 | 6.94 | 5.14 | 69.0 |
| SCC | SCC | count/ml | 0.69 | 0.69 | 21.3 | 76.4 |
| Mastitis Resistance | MAS | resistance ABV unit | 6.75 | 6.75 | 3.19 | 71.3 |
| Milking speed | MSPEED | ABV unit | 5.02 | 5.02 | 2.13 | 68.5 |
| Temperament | TEMP | ABV unit | 3.60 | 3.6 | 1.75 | NA |
| Mammary system | MAMM | ABV unit | 3.59 | 2.76 | 4.18 | NA |
| Overall type | OTYPE | % increase in score | 1.36 | 1.36 | 3.97 | 68.1 |
| Pin set | PINSET | % increase in score | 0.78 | 0.78 | 4.78 | NA |
| Feed saved | FEEDEF | kg DM saved | 0.3853 | 0.1927 | 74.8 | 34.8 |
| Udder depth | UDDEP | ABV unit | 0 | 0.82 | 4.09 | NA |

[1]Axford *et al.* (2021)

Compared to the emphasis calculated by the traditional method, the sub-index emphasis of group 1 traits increased 10% in both HWI and BPI, whereas emphasis of group 2 traits decreased 8% in HWI and decreased 11% in BPI (Table 2). Group 1 was a favourable trait combination, because both their covariances and economic weights were positive, resulting in a higher cluster weight, $w_k$, i.e. $[a_k' G_{kk} a_k]^{\frac{1}{2}}$, compared to the cluster weight using the traditional method, which is a simple summation of relative economic weight without considering the covariances, i.e. $[a_k' I_k \sigma^2_{g_{kk}} a_k]^{\frac{1}{2}}$. Group 2 traits MILK and PROT formed an unfavourable trait combination, because their covariance was positive (137,249) but their economic weights were in opposite directions (HWI: $4.36 for PROT and $-0.07 for MILK; BPI: $6.67 for PROT and $-0.11 for MILK), resulting in a lower cluster weight, $w_k$, compared to that using traditional method.

The new emphasis methods results are more realistic because they will better reflect the selection response in practice. Using the traditional method, the emphasis of group 1 was likely underestimated whereas emphasis of group 2 was likely exaggerated. With the adjustment in the sub-index weight method, traits with small weights were given slightly higher weights.

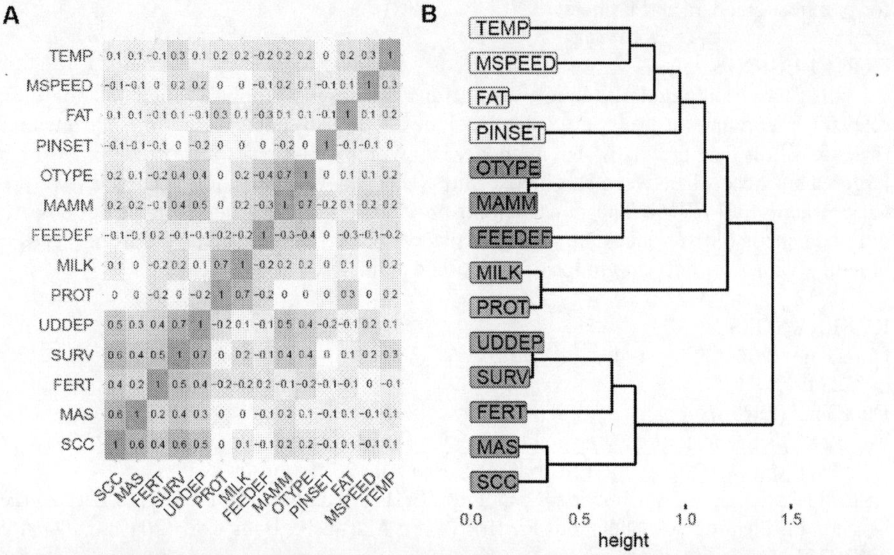


**Figure 1. Correlation and hierarchical clustering of the main estimated breeding value traits included in the Australian BPI and HWI indexes**

**Table 2. Sub-index total percent emphases across methods and datasets and changes of the 3 new methods compared to the traditional method**

| Sub-index group | Traits | HWI Percent emphasis by method (%) | | BPI Percent emphasis by method (%) | |
|---|---|---|---|---|---|
| | | Traditional | Sub-index weighted | Traditional | Sub-index weighted |
| 1 | FERT, SURV, SCC, MAS, UDDEP | 47 | 57 | 36 | 46 |
| 2 | MILK, PROT | 22 | 14 | 35 | 24 |
| 3 | OTYPE, MAMM, FEEDEF | 18 | 16 | 12 | 10 |
| 4 | TEMP, MSPEED, FAT, PINSET | 13 | 13 | 17 | 19 |
| **Total changes compared to Traditional** | | | 20 | | 25 |

A common argument against the percent emphasis method is that selection response solely can be enough to describe the selection pressure in practice. This is not true when the trait undergoes genetic change due to effects other than selection, such as natural selection, drifts, or correlations with other preselected traits. We often see traits with no economic weightings undergo genetic changes and some traits with positive economic weightings undergo negative genetic changes due to correlated responses. In the current study, in HWI, the predicted selection response for SCC and FERT are 0.6 and 0.8 SD units (Datagene 2020b), respectively, very similar in value. Whereas the emphases for these two traits are 6.44% and 32% (Table 2), indicating that FERT is undergoing a much higher selection pressure than SCC to achieve similar selection response. It is also very hard to express trait responses in a way that makes them add up to 100%, making interpretation difficult for practical breeders and farmers.

## CONCLUSIONS

This study compared sub-index weight and traditional emphasis methods for defining the relevant importance of traits in a selection index. The sub-index weight method generated more realistic results than the traditional method when within-sub-index trait correlations were relatively larger than those of between-sub-index, and when genetic evaluation accuracies were relatively variant across all EBVs. The new method provides convenient deployment options where pre-defined genetic (co)variance matrices are replaced by alternatives calculated from sets of estimated breeding values for defined groups of selection candidates.

## REFERENCES

DataGene (2020a) https://datagene.com.au/ct-menu-item-7/australia-s-two-indices, accessed Jan 2021.

DataGene (2020b), https://datagene.com.au/sites/default/files/Upload%20Files/NBO%202020%20Options%20Paper%20May%202020%20GESC.pdf, accessed Aug 2021.

Axford M., Santos B., Stachowicz K., Quinton C., Pryce J., Amer P. (2021) *AAABG* **23** (accepted)

Miglior F., Fleming A., Malchiodi F., Brito L.F., Martin P., Baes C.F. (2017) *J. Dairy Sci.* **100**: 10251.

Miglior F., Muir B.L., van Doormaal. (2005) *J. Dairy Sci.* **88**: 1255.

VanRaden P.M. (2002) *Proc. 7th World Congr. Genet. Appl. Livest. Prod.* **29**: 127.

VanRaden, P.M., Cole, J.B., Neupane, M., Toghiani, S., Goddis, K.L. and Tempelman, R.J. (2021) https://www.ars.usda.gov/ARSUserFiles/80420530/Publications/ARR/nmcalc-2021_ARR-NM8.pdf, accessed Aug 2021

Zhang X. and Amer P. (2021) *J. Dairy Sci.* **104**: 5827.

*Proc. Assoc. Advmt. Anim. Breed. Genet: 24: 451-454*

# GENETIC DIVERSITY AND TRENDS OF AUSTRALIAN JAPANESE BLACK CATTLE

**Y.D. Zhang and R. Banks**

Animal Genetics Breeding Unit*, University of New England, Armidale, NSW 2351, Australia

## SUMMARY
Japanese Black is the predominant strain of Australian Wagyu cattle. Due to limited importation of cattle from Japan into Australia, there are concerns of declining genetic diversity and increasing inbreeding. This study investigated inbreeding status and genetic diversity in Australian Japanese Black cattle. Average generation interval was 6.4 years. Inbreeding coefficients increased from 4.2% in 2000 to 7.2% in 2019. The average effective population size was 43.4. Estimated F-statistics suggested that subpopulations were not evident in Australia Japanese Black cattle. It is advisable for Australia Japanese Black breeders to continue monitoring inbreeding levels and to develop breeding strategies to balance genetic gain and increased rates of inbreeding.

## INTRODUCTION
Australian Wagyu cattle production started in the 1990s and was initiated from semen, frozen embryos and live animals imported from Japanese Black cattle in Japan via the United States of America. Wagyu cattle have increasingly become popular due to their high intramuscular fat deposition. This breed has developed from a small source of genetics and there are concerns that declining genetic diversity and increasing inbreeding may have a negative effect on future productivity. Australian Wagyu comprises of Japanese Black, Red and others. Japanese Black is the predominant strain. This study focused on Australian Japanese Black cattle. The aims of this study were to assess the inbreeding status and genetic diversity of Australian Japanese Black cattle.

## MATERIALS AND METHODS
Pedigree was extracted from the Australian Wagyu BREEDPLAN database. Individual animals were classified by Australian Wagyu Association Herdbook registration status, content grade or colour code. There were 151,730 animals registered as Australian Wagyu cattle and of these, 97,182 were Japanese Black. Pedigree completeness is an important factor for estimating effective population size ($Ne$) as it directly relates to the inbreeding coefficients determined and can be assessed in three parameters: 1) maximum number of generations traced ($G_x$); 2) complete generation equivalent or the number of equivalent complete generations ($G_e$); 3) number of fully traced generations ($G_f$). These parameters were calculated using the function *summary.Pedig()* from the R package optiSel (Wellmann 2019).

**Inbreeding coefficient and effective population size.** Inbreeding coefficient ($F_i$) for each animal in the pedigree was calculated using the *pedInbreeding()* function from the R optiSel package (Wellmann 2019). The standardized inbreeding rate per generation was computed as $\Delta F_i = 1 - \sqrt[(t_i-1)]{(1 - F_i)}$, where $t_i$ was the number of complete generation equivalent ($G_e$) of animal $i$ (Gonzalez-Recio *et al.* 2007). The $\Delta F_i$ values of individuals of the reference population were averaged to form $\overline{\Delta F}$. The effective population size $N_e$ was calculated as $N_e = \frac{1}{2 * \overline{\Delta F}}$.

---

* A joint venture of NSW Department of Primary Industries and University of New England

**F-statistics.** Wright's (1965) F-statistics were separately calculated for each year period. Following Wright's notation, three parameters were involved in F-Statistic calculations, 1) $F_{it}$ is the individual inbreeding coefficient relative to the entire population (equivalent to $F_i$); 2) $F_{st}$ is the inbreeding coefficient of the subpopulation relative to the entire population expected under random mating, it was computed from a hypothetical population produced by matching sires and dams of the registered animals in each time period (eg year) at random. For each year period, 20 hypothetical populations were generated, and mean $F_{st}$ from 20 samples was used for each year of the examined period to estimate a reliable mean $F_{st}$; 3) $F_{is}$ is the inbreeding coefficient of an individual relative to its own subpopulation and indicates how the mating departs from random, it was obtained as $F_{is} = (F_{it} - F_{st}) / (1 - F_{st})$, as $F_{it}$ was estimated as $F_i$ and $F_{st}$ was derived from the simulation.

**Generation interval.** Generation interval was computed for four genetic pathways, sire to male progeny ($L_{mm}$), sire to female progeny ($L_{mf}$), dam to male progeny ($L_{fm}$) and dam to female progeny ($L_{ff}$). This was based on the birth dates of animals in each year and the birth dates of their parents. The annual average generation interval ($L$) was subsequently calculated as $L = \frac{Lmm + Lmf + Lfm + Lff}{4}$.

## RESULTS AND DISCUSSION

**Pedigree completeness.** Pedigree completeness by year of birth was determined by counting generations including **maximum ($G_x$), complete equivalent ($G_e$) and fully traced ($G_f$)** and are presented in Figure 1. Pedigree completeness has increased over time and reached a high level in 2000 with a $G_e$ of 5. Japanese Black cattle had a deep pedigree with the average maximum, equivalent complete or fully traced number of 15.1, 4.3 or 7.3 generations, respectively.

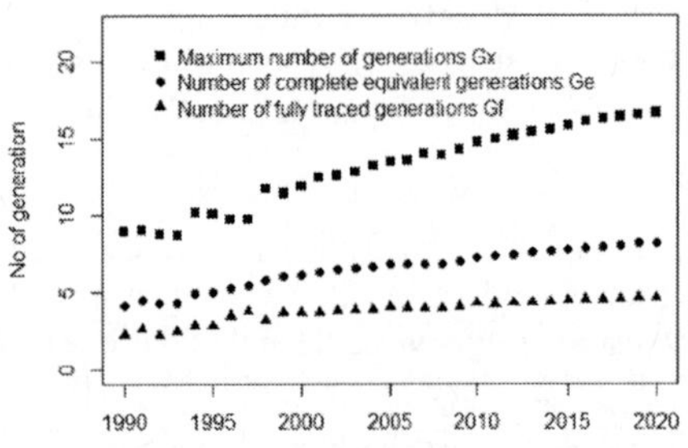


**Figure 1. Mean number of generations (maximum $G_x$, complete equivalent $G_e$ and fully traced $G_f$) recorded by birth year**

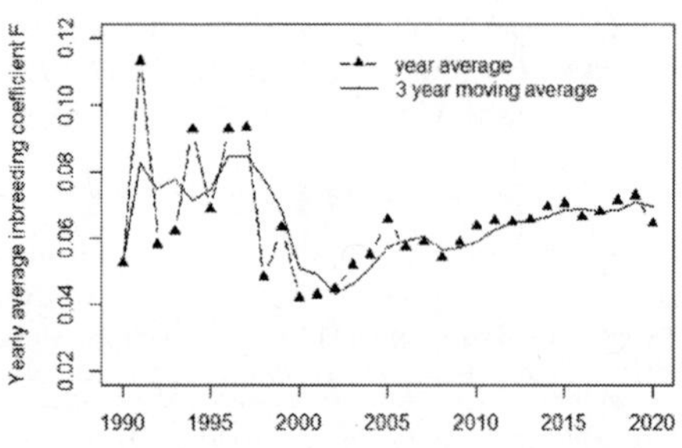


**Figure 2. Average inbreeding coefficients by birth year for Japanese Black cattle**

**Generation interval.** The average generation interval was 6.4 years. After 2000, the average generation interval for the Japanese Black population was 8.4 and 7.9 years for sire to male progeny and sire to female progeny, respectively and about double the size for dam to male progeny and dam to female, i.e. 4.8 and 4.6 years, respectively. Average generation intervals increased from 2.4 years in 1984 to 8.7 years in 2008, followed by a reduction to 7 years in 2011 and afterwards increased to 8.3 years in 2017. From 2005, the mean generation intervals remained steady between 6.8 and 8.7 years. This trend might reflect the development of Australian Wagyu (or Australian Japanese Black) during the herd building phase, where less cows were culled and more old ancestors, mostly males, were retained. Nomura *et al.* (2001) reported that generation intervals for Japanese Black in Japan ranged from 8.3 in 1985 to 10 years in 1997. The generation interval for American Wagyu has been estimated at 5.14 years (Scraggs *et al.* 2014).

**Inbreeding coefficient.** Mean inbreeding coefficients $F$ by birth year fluctuated markedly before 2000, then the inbreeding coefficient increased steadily from 4.2% to 7.2% in 2019 (Figure 2). Similarly, a gradual increase in inbreeding coefficient was reported in Japanese Black cattle in Japan, from 4.7% in 1985 to 5.4% in 1997 (Nomura *et al.* 2001). The average inbreeding coefficient from 1994 to 2011 was 4.8% for American Wagyu cattle, suggesting Australian Japanese Black cattle shared a similar breeding path as that of US Wagyu cattle (Scraggs *et al.* 2014). Large variation in inbreeding levels was observed across herds. The inbreeding coefficient within individual herds ranged from 0.0 to 18% with 45 of 513 herds having average inbreeding coefficients greater than 10% (mean $F = 12.5\%$), suggesting an urgent need to control inbreeding in these herds.

**Changes in F-statistics.** The changes in F-statistics are shown in Figure 3. There was evident difference between $F_{st}$ and $F_{it}$ prior to 1998, suggesting that matings in the early period were mainly operating within subgroups or subpopulations. Both $F_{st}$ and $F_{it}$ had increased steadily since 1998 and the differences decreased gradually, leading to a negligible difference, indicating that matings across the population was dominant. $F_{is}$ decreased from 1997 to a low level. $F_{is} > 0$ or with a large value suggests existence of evident subpopulations. This finding indicated that subpopulations had disappeared in current Australian Japanese Black cattle. This was most likely due to deliberate mating decisions by breeders that avoided mating close relatives. Similarly, Nomura *et al.* (2001) found that $F_{is}$ in Japan had decreased from 2 or more to 0.5 by 1997 and concluded that subdivision between prefectures no-longer existed in Japan. Honda *et al.* (2004) analysed 25 subpopulations (i.e. populations of prefectures) of more than 2,000 Japanese Black cows and found that 17 of the subpopulations shared very high similarity because of high migrations amongst these subpopulations and the other 8 subpopulations with relatively low migration rates showed their unique genetic structures. Estimated $F_{it}$ in this study was in line with the reported values in American Wagyu (Scraggs *et al.* 2014), where $F_{it}$ values in US Wagyu cattle fluctuated markedly before 2000, then remained at approximately 5%.

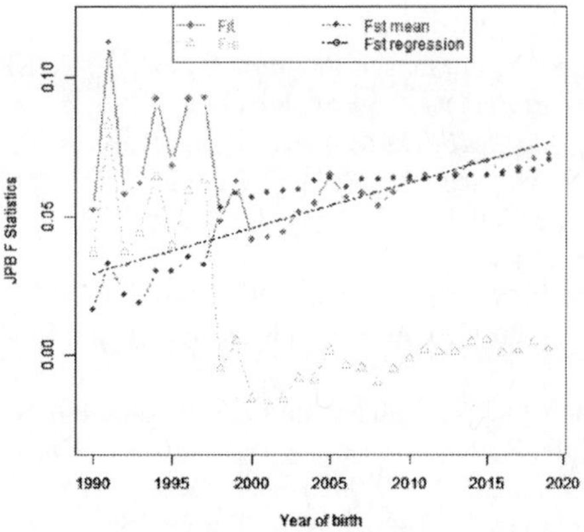


**Figure 3. Change of F-statistics (%) in pedigree for Japanese Black from 1990 to 2018: the actual overall inbreeding coefficient ($F_{it}$), the inbreeding coefficient expected under random mattings ($F_{st}$) and inbreeding due to population subdivision ($F_{is}$)**

**Effective population size.** *Ne* increased and peaked at 59 in 2000, followed by a decrease to 41 in 2005, afterwards *Ne* remained stable at approximately 43. The overall estimated *Ne* was 43.4 for the Japanese Black population. This estimate was higher than estimates of $N_e$ from the early Japanese Black cattle in Japan (mean=27.1, range 13.4 to 52) (Nomura *et al.* 2001) and that for American Wagyu (the average was 13.4 from 1994 to 2011, with a maximum of 48 in 2002). The estimates of $N_e$ varied largely across years. Detailed *Ne* by year was not reported in American Wagyu (Scraggs *et al.* 2014). The method used in estimating *Ne* in Japanese Black (Nomura *et al.*, 2001) and American Wagyu (Scraggs *et al.* 2014) were vulnerable to sampling. The comparison of *Ne* between Australian Japanese Black with the results from Japan or America remains inconclusive. A number of studies on effective population sizes in cattle (e.g. Meuwissen & Woolliams 1994) suggested that the effective population size of Australian Japanese Black is at the lower range of the published results for beef cattle breeds, for example, 45 to 117 in Brazilian Zebu (Faria *et al.* 2009), 64 to 127 in Irish dairy and beef breeds (McParland *et al.* 2007), 116 in US angus (Decker *et al.* 2012) and 244 to 558 in European Charolais subpopulations and 345 to 2,459 in European Limousin subpopulations (Bouquet *et al.* 2011).

## CONCLUSIONS

The most recent animals in the Australian Japanese Black population had adequate pedigree completeness with the average maximum, equivalent complete or fully traced number of generations of 15.1, 4.3 or 7.3, respectively. Generation intervals from Sire to male or female progeny were almost twice those for Dam to male or female progeny, particularly for individuals after 2000. Average generation interval and complete generation equivalent were 6.4 and 5.9 for Japanese Black populations, respectively. Inbreeding coefficients increased rapidly from 2000 to 2019 (4.2% to 7.2%). The current levels of inbreeding are not indicative of an immediate problem with genetic diversity, but the change is large, almost doubling from 2000 to 2019. It is advisable for Wagyu breeders to continue monitoring inbreeding levels and to develop breeding strategies to balance genetic gain and increased rates of inbreeding. The computed effective population sizes (*Ne*) in the last period (2016 to 2020) was 43.4 in the Japanese Black population, which is on the low end of similar estimates in other beef breeds. F-statistics estimates showed low or negative $F_{is}$, suggesting that subpopulations were no longer evident in Australian Japanese Black.

## REFERENCES


Bouquet A., Venot E., Laloe D., Forabosco F., Fogh A., Pabiou T., Moore K., Eriksson J.A., Renand G. and Phocas F. (2011) *J Anim Sci* **89**:1719.

Decker J. E., Vasco D. A., McKay S. D., McClure M. C., Rolf M. M., Kim J., Northcutt S. L., Bauck S., Woodward B. W., Schnabel R. D. and Taylor J. F. (2012) *BMC Genomics* **13**:606.

Faria F. J. C., Filho A. E. V., Madalena F. E. and Josahkian L. A. (2009) *Journal of Animal Breeding and Genetics* **126**:148.

Gonzalez-Recio O., Lopez de Maturana E. and Gutierrez J. P. (2007) *J Dairy Sci* **90**:5744.

Gutierrez J. P., Altarriba J., Diaz C., Quintanilla R., Canon J. and Piedrafita J. (2003) *Genet Sel Evol* **35**:43.

Honda T., Nomura T., Yamaguchi Y. and Mukai F. (2004) *Animal Science Journal* **73**: 445.

Mc Parland S., Kearney J. F., Rath M. and Berry D. P. (2007) *J Anim Sci* **85**:322.

Meuwissen T. H. and Woolliams J. A. (1994) *Theor Appl Genet* **89**: 1019.

Nomura T., Honda T. and Mukai F. (2001) J Anim Sci 79:366.

Scraggs E., Zanella R., Wojtowicz A., Taylor J. F., Gaskins C. T., Reeves J. J., de Avila J. M. and Neibergs H. L. (2014) *J Anim Breed Genet* **131**:3.

Wellmann R. (2019) *BMC Bioinformatics* **20**:25.

Wright S. (1977) Experimental Results and Evolutionary Deductions. University of Chicago Press, Chicago.

*Proc. Assoc. Advmt. Anim. Breed. Genet: 24: 455-458*

# HANDYCNV: AN R PACKAGE FOR STANDARDIZED SUMMARY, ANNOTATION, COMPARISON, AND VISUALIZATION OF CNV AND CNVR

**J. Zhou[1,2], L. Liu[1,2], T.J. Lopdell [3], D. Garrick[2*] and Y. Shi[1*]**

[1]School of Agriculture, Ningxia University, Yinchuan, China
[2]AL Rae Centre of Genetics and Breeding, Massey University, Hamilton, New Zealand
[3] Research and Development, Livestock Improvement Corporation, Ruakura Road, Hamilton, New Zealand

## SUMMARY

There is a need for a pipeline to provide standard, reproducible and timesaving post-analysis of CNV (copy number variants) from SNP (Single nucleotide polymorphisms) chip genotyping. We present a package built with a dozen functions that can convert the coordinates of SNP map files, compare the positions of SNPs between the given maps, summarize the CNVs, call CNVRs (Copy number variation regions), provide gene annotation, compare CNV, CNVR and the annotated gene lists, and visualize CNVs at both individual and population level.

## INTRODUCTION

Copy number variants are a type of structural variation of a DNA fragment which comprise the deletion or duplication type depending upon how many copies an individual has compared with the two copies in the diploid reference genome. These structure variants could change the structure or dosage of genes that might further affect the phenotypes. Studies on CNVs have become common in livestock research in recent years. The fluorescent signal intensity of SNPs chip provides the general source to detect CNV, and thanks to the wide application of genome-wide association studies and genomic selection in animal breeding, there is now a lot of SNP data suitable for analysis of structural variants. Software such as PennCNV (Wang *et al.* 2007), CNVPartition (Illumina) and SVS Golden Helix (Bozeman) are designed to detect CNV from SNP data, but each method has its own advantages and shortcomings, so it is recommended to use more than one method to infer CNVs (Winchester *et al.* 2009), therefore comparison of multiple CNV results is a normal task in characterizing structural variation.

When doing CNV analysis we are curious about all the information related to any CNV region, not only at the individual level but also at the population level. For instance, we want to know how many individuals have a CNV in a common region? What kind of type of CNVs are there? Are these individuals from the same farm or are they progenies of the same sire? Are there any genes in the CNV region? What are the gene frequencies? How about the signal intensities, call rate, minor allele frequency and linkage disequilibrium conditions of these CNVs? Besides, the common post-analysis of CNV studies includes provision of summary CNVs, generation of CNVR, comparison of CNVs from different software, finding consensus CNVR by comparing results to the gold standard CNV database, gene annotation in the CNV region and CNV-based regression analysis. To accomplish all these tasks various tools are typically required. Therefore, we integrated these functions into a package to make post-CNV analysis easy and reproducible. The use of some developed R package like the Tidyverse family (Wickham *et al.* 2019) made our function development much easier. We believe this package will be convenient to others who are doing similar work. The source code can be found in the Github repository (https://github.com/JH-Zhou/HandyCNV).

## MATERIALS AND METHODS

The pipeline and results of this package are shown in Figure 1. The Demo Data are the CNV results from the GeneSeek GGP Bovine 150k BeadChip detected by PennCNV (Wang *et al.* 2007) and CNVPartition (Illumina). All the input files and demo code can be found in Github (https://github.com/JH-Zhou/HandyCNV). Run through all functions in HandyCNV need prepare CNV Results, SNP maps, Reference Gene List, Pedigree, Plink files (Bim, Bed and Fam) and SNP Signal Intensity in total. All the input files have a fixed format, and the file requirements depends on which functions the users are using. Here we only introduce the input data format and some noticeable methods we used in some functions, therefore, we will not cover the data structures or interpret how to use the results in this article, more details can be found by browsing our Demo Data.

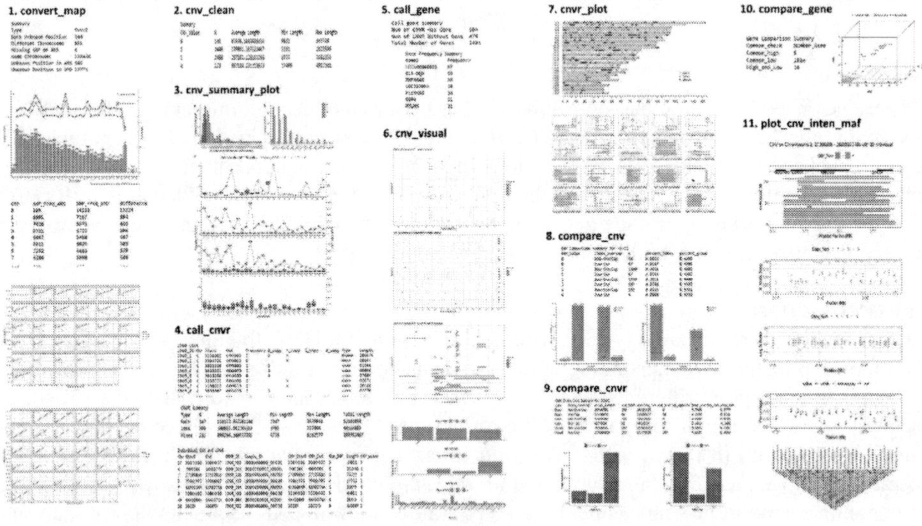


**Figure 1. Pipeline and results of HandyCNV for the post-analysis of CNV**

The first function is convert_map which is used to convert map files from the original to an objective map file provided by the user. There are differences between genome assemblies, for example, in which some SNP might locate on a different chromosome or on the same chromosome but in a different order between different assemblies. Most Bovine SNP chips have been using the UMD3.1 (Shamimuzzaman *et al.* 2019) as the default reference genome assembly, but with the release of new reference genome ARS-UCD1.2 with high continuity, accuracy, and completeness (Rosen *et al.* 2020),it may be of interest to convert the coordinates to the latest assembly to help further research. Four columns are required with no header in the input map files whose columns are Chromosome, SNP ID, Morgan Position (UMD) or Physical Position (ARS) and Physical Position (unit: bp) (Table 1, Table 2).

**Table 1. Original map format (UMD 3.1)**

| | | | |
|---|---|---|---|
| 14 | ARS-BFGL-BAC-10172 | 0 | 6371334 |
| 14 | ARS-BFGL-BAC-1020 | 0 | 7928189 |
| 14 | ARS-BFGL-BAC-10245 | 28.23 | 31819743 |
| 14 | ARS-BFGL-BAC-10345 | 0 | 6133529 |

**Table 2. Objective map format (ARS)**

| | | | |
|---|---|---|---|
| 14 | ARS-BFGL-BAC-10172 | 5.34266 | 5342658 |
| 14 | ARS-BFGL-BAC-1020 | 6.88966 | 6889656 |
| 14 | ARS-BFGL-BAC-10245 | 30.1241 | 30124134 |
| 14 | ARS-BFGL-BAC-10345 | 5.10573 | 5105727 |

The cnv_clean function is designed to convert the CNV results to a standard format, the output clean CNV file is used as input data in many of the other functions. It supports PennCNV and CNVPartition default output results, the length of CNVs are calculated as one plus the end position minus the start position. The CNV results from other software can be prepared as the template format to use in the remaining functions (Table 3). The function cnv_summary_plot will generate several plots to show the number, length group, type, and frequency details of CNVs on individuals and on chromosomes.

**Table 3 Template Format of Clean CNV**

| Sample_ID | Chr | Start | End | CNV_Value | Length |
|---|---|---|---|---|---|
| 201094560060_R02C01 | 11 | 106224443 | 106359588 | 4 | 135146 |
| 201094560060_R02C01 | 12 | 58073538 | 58417437 | 1 | 343900 |
| 201094560060_R02C01 | 19 | 27576066 | 27643677 | 4 | 67612 |
| 201094560060_R02C02 | 1 | 88638760 | 88904687 | 3 | 265928 |

The call_cnvr function will merge the CNVs which have at least one bp overlapping length to a CNVR. The results are the non-redundant CNVRs,but this method could cause misleading information while reporting the genes and comparing the overlapping length on CNVR. This is because it may appear all CNVs in a CNVR are the same length but in reality there are often lots of short disparate CNVs. To solve this problem, combine call_gene and cnv_visual function will plot all genes located on CNVs of every individual in a CNVR. The call_gene function needs the user to provide the reference genes which can be downloaded from the UCSC website (http://hgdownload.soe.ucsc.edu/downloads.html).

The compare_cnv and compare_cnvr functions with the similar strategies, when the results have the same version coordinates they will compare directly, but when the coordinates are from different versions, it will convert the position for each file at first then make comparison between the coordinates of the latest version. The overlapped region between two interval results may be slightly different, when reporting and plotting the number and length of overlapping regions correspond to each input files, respectively.

## RESULTS AND DISCUSSION

**When do you need to convert the coordinates of SNP or CNV?** The first scenario is when a new reference genome is released. Take the Bovine reference genome as example, the lasted version (ARS-UCD1.2) has higher coverage and accuracy of its genome assembly than the previous commonly used UMD3.1, so it may help to improve the accuracy of SNP-based CNVs detection by using the latest reference genome. The second scenario is to make comparison between results from different reference genomes. There are lots of studies that have reported CNVs using previous reference genomes, and we may want to compare their results to our assembly.

**Why do we need to visualize CNVR?** CNV is of interest relative to the comparison of individuals but the CNVR are mostly of interest at the population level. The common method to generate CNVR is to merge all overlapping CNVs from every individual into a common region, then make gene annotation and comparison on CNVRs on the population level.

The main shortcoming of SNP-based CNV detection is that it cannot report the exact start or end position because of the limited marker density, so when we merge these CNV intervals to a common CNVR the actual situation is that not all the CNVs with the same break points as the CNVR, therefore, not all the genes within a CNVR has the same frequency with the CNVs (Figure 1. 11). Sometimes we might find an interested candidate gene within a high frequency CNVR but if only a few individuals have CNV of that gene, the better way to avoid this mistake is to report the gene frequency by counting how many CNVs with this gene in a CNVR, but this not enough, because of some genes may have CNV in just a partial fragment rather than the entire gene, in this circumstance plotting all CNVs and annotated genes in a CNVR by the start and end position can make it much clearer to understand what is happening. We are often curious about all the information in a CNVR in a population, such as the relationship between SNPs in that region, so visualizing a CNVR by plotting all related information in one figure is a good solution.

**What are the limitations of this study?** First, we have only used it in bovine studies, so some functions may need to be revised to be used in other species. Second, the linkage disequilibrium (LD) plots are based on the Gaston package which was drawing the base plot only, for some CNVRs with fewer number of SNPs the plot size was not well controlled while merging it to other plots, and this could lead some CNVR plots to be unsuitable without further modifications. Third, plots of CNVs on the population level are suitable for small populations but could be too busy for large populations. Fourth, the functions for regression analysis between CNVs or CNVRs and phenotype are still being developed.

## CONCLUSIONS

Here we present an R package called HandyCNV in the initial version which includes several functions for tasks such as converting SNP maps, generating CNVR, genome annotation, comparing and visualizing of CNV and CNVR and reporting summary results on each step. This tool provides a standard, reproducible and timesaving post-analysis of copy number variants.

## ACKNOWLEDGEMENTS

This project was funded by the China Scholarship Council. This package depends on several developed R packages, such as Tidyverse family and data.table, we appreciate all related contributors to the open source R language.

## REFERENCES

Wang K., Li M., Hadley D., et al. Penn C.N.V. (2007) *Genome Res.* **17**(11): 1665.

Illumina. GenomeStudio. https://www.illumina.com/techniques/microarrays/array-data-analysis-experimental-design/genomestudio.html

Bozeman, MT: Golden Helix I. SNP & Variation Suite ™ (Version 8.x). http://www.goldenhelix.com

Winchester L., Yau C., and Ragoussis J. (2009) *Briefings Funct Genomics Proteomics.* **8**(5): 353.

Wickham H, Averick M, Bryan J, et al. (2019) *J Open Source Softw.*

Shamimuzzaman M., Le Tourneau J.J., Unni D.R., et al. (2019) *Nucleic Acids Res.*

Rosen BD, Bickhart DM, Schnabel RD, et al. (2020) *Gigascience.*

*Proc. Assoc. Advmt. Anim. Breed. Genet: 24: 459-462*

# EWE LAMB JOINING FOR SELECTION

**L. Arney**

'Inverbrackie', 1040 Dry Plains Rd, Finniss SA 5255[1]

## INTRODUCTION

We run a Border Leicester stud in South Australia and I believe that our breeding program needs to focus on the requirements of the leading prime lamb producers that are breeding from 1st cross ewes (Border Leicester x Merino ewes). It is therefore important to listen to these prime lamb producers both directly and indirectly. When I write that I listen indirectly I try to listen for what they are not telling me as sometimes you just hear about problems and never hear about solutions. Listening is very important and so is not being frightened to try to find out what is possible and what is not.

## BREEDING AIMS / OBJECTIVES

To earn their place in the industry, 1st cross ewes must:

1. Have a lot of twins
2. Rear the lambs they produce
3. Produce lambs that grow fast
4. Produce fibre that more than covers the shearing cost
5. Have resilience
6. Produce lambs that have a very good carcase

This is my basic 6 pack of traits that is expected from a 1st cross ewe. However, I know there are prime lamb producers that are pushing boundaries and looking for more from their ewes. In particular they want to get a financial return earlier by joining these 1st cross ewes earlier, as ewe lambs. My focus in our flock's breeding objectives revolves around breeding sheep to match industry expectations.

## PROGRESS

We have been able to achieve above average gains in most traits within our flock that is recorded and benchmarked with Sheep Genetics. One of the key ways we are achieving our breeding objectives is by using large numbers of ram lambs in our flock. This is quite simple and puts a natural selection pressure on selecting rams that have clearly demonstrated their early growth. This is backed up with good performance data to make good selections from the ram lambs. The more high-quality historical data the better. The more high-quality current data the better.

When it comes to mating ewe lambs it is quite different because we are bringing reproduction into the equation as well. However, it is still about enhancing natural selection and having very good data. There are several factors that influence the result from ewe lambs that I still don't understand. Weight is seen as a major success factor in getting ewe lambs pregnant. I have seen publications that recommend that the ewe lamb needs to be a minimum of 40 kg another that recommends 42 kg and another that says 45 kg. Figure 1 is a plot for ewe lambs at Inverbrackie for liveweight and condition score at the start of joining in 2020, their pregnancy rates and litter size.

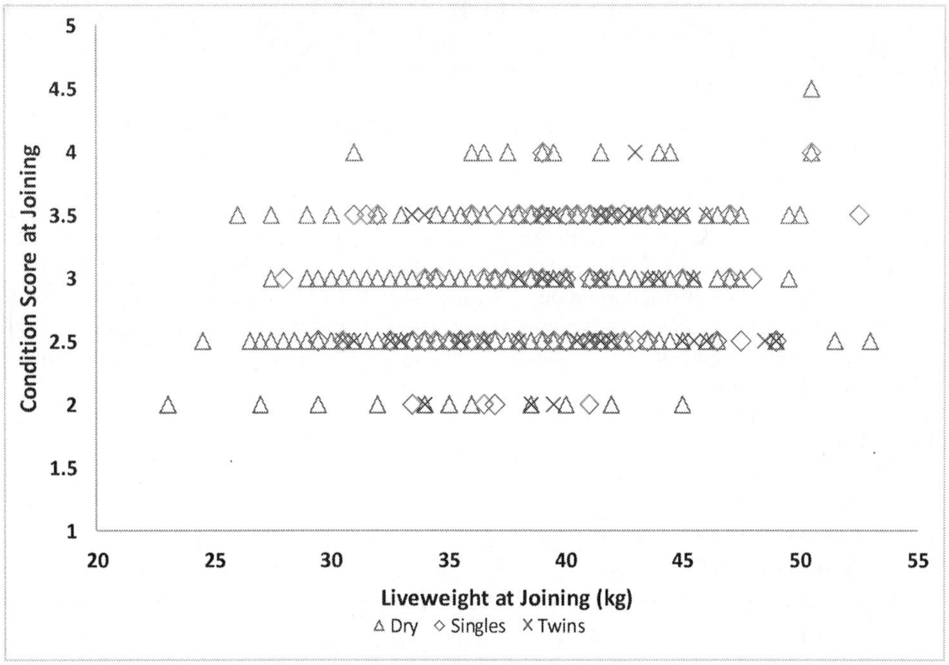


**Figure 1. Plot of condition score against liveweight at joining and resulting litter size for ewe lambs mated in 2020**

As a performance recording seedstock producer I understand how important it is to maintain large numbers of animals in management groups. I also understand how important it is to look for the outliers as these are the animals that have the biggest potential to make change within a breeding program. I give all ewe lambs a chance to join regardless of weight. We have ewe lambs lighter than 36 kg getting pregnant and we have ewe lambs over 50 kg that don't.

We are joining at 7 months of age because of the seasonal weather pattern for our area which is means that early sexual maturity is another issue. Age at joining / sexual maturity is more important for a successful ewe lamb joining and this where we believe we are having a genetic effect. The lightest pregnant ewe lamb was born to a ewe lamb, born a triplet/raised as a twin and conceived at 26 kg. With a higher joining weight in 2019 there was not a significant change in pregnancy rate as shown in the graph below (see Table 1 and Figure 2). Note: 2018 drop lambing in 2019, 2020 drop lambing in 2021.

Some years we have had only 10% pregnant. Some years we have 55% pregnant. In the years where we have low pregnancy rates many people would say why bother. But think about the selection pressure that has occurred in the years with poor results. The progeny from those ewes are really special. 90% of the ewes did not give me anything to work with in a lowly heritable trait. Because it is lowly heritable, I cannot expect every one of those lambs to help us to make direct genetic gain but they have provided data in our flock that will help us to make gradual improvement.

**Table 1. Average weight and condition score at joining and pregnancy status of ewe lambs joined in 2019 and 2021**

| Year of Birth / Litter Size | Average Weight at Joining (kg) | Average Condition Score at Joining | Number | Proportion (%) | Pregnancy Rate (%) |
|---|---|---|---|---|---|
| *2018* | | | | | |
| Dry | 43.5 | 3.2 | 268 | 67.5% | |
| Single | 45.3 | 3.36 | 88 | 22.2% | 32.5% |
| Twin | 48.4 | 3.37 | 41 | 10.3% | |
| *2020* | | | | | |
| Dry | 37.2 | 2.87 | 280 | 68.5% | |
| Single | 38.9 | 2.88 | 83 | 20.3% | 31.5% |
| Twin | 40.5 | 2.87 | 46 | 11.2% | |

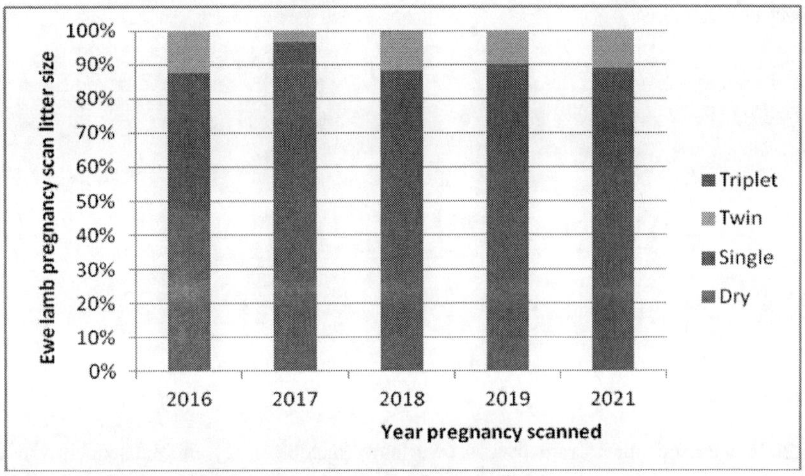


**Figure 2. Inverbrackie ewe lamb pregnancy scanning results**

The progeny from the ewe lambs are seldom as big as the progeny from adult ewes. When I first starting mating ewe lambs we did nothing to make it simple to identify them and most were culled purely on size. Later I realised that was totally unfair as the same ewe was mated to the same ram a year later visually they would have produced an animal that would have been retained in the flock. So we began using a code in the eartag that tells us that it was bred from a ewe lamb. Since this system was put in place we are not only keeping a lot more of the ewe lamb progeny but most years we have selected ram lambs to use as sires. Think of the selection pressure that has occurred. We have used a ram lamb that came from a ewe that was capable of getting in lamb when others didn't and that ram lamb demonstrated his ability to have superior early growth. The selection is still backed with top performance data.

It is currently recommended to select for fat and muscle to assist in reproduction. When we look at the animals produced from ewe lambs and where they influenced the direction that our breeding program has taken, guess what has happened in our flock? We have found that we are now producing animals that have more fat, muscle and increased reproductive performance (see Figure 3). Not only that but we also have strong early growth. I give a lot of the credit for our position in the industry to the selection has occurred from the use of young animals in our breeding program and in particular from the ewe lambs.

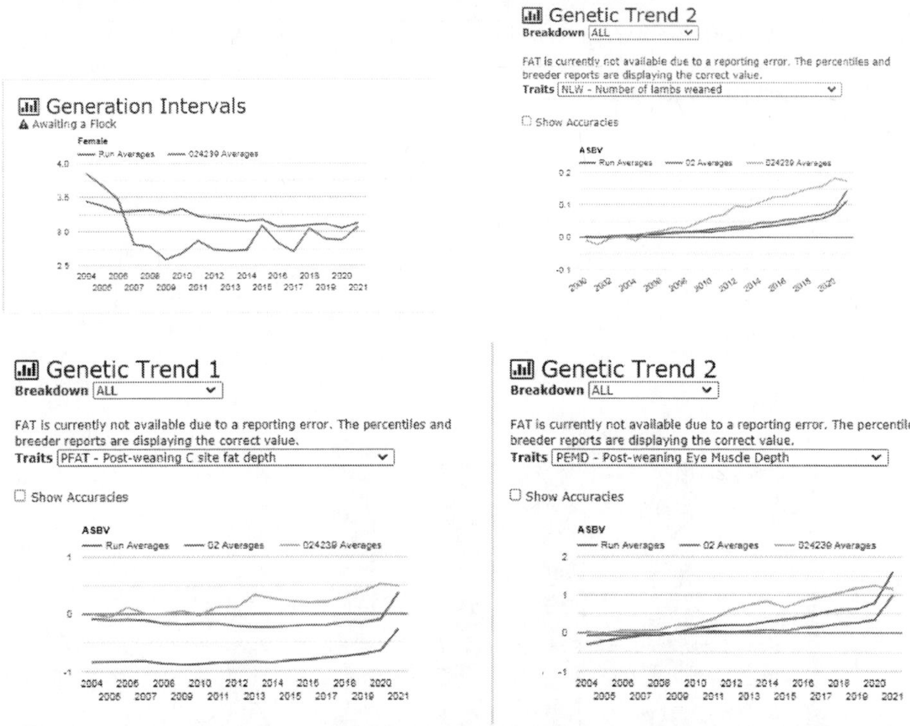


**Figure 3. Generation interval and genetic trends for NLW, PFAT and PEMD for Inverbrackie**

In 2010 we used our 1st ram from a ewe lamb (a 2008 drop) and subsequently used 3 of his sons in 2011 also from ewe lambs. I am a firm believer that it is a seedstock producer's job to create difference within a mob to aid selection and not to try to make everyone in the mob look similar. Selection is where our genetic gain comes from.

## TAKE HOME MESSAGE(S)

1. Shortening genetic interval speeds up genetic gain.
2. Make your animals work for you so you can see what they are capable of and don't work for them to make them all look the same.
3. Don't intervene by giving some animals different management because otherwise it will bias the estimate of genetic merit and ultimately reduce genetic gain.
4. Keep good individual production records – don't be frightened to include the individuals with poor production records.

## ACKNOWLEDGEMENTS

Genetic interval graph and NLW genetic trend graph produced by Sheep Genetics.

*Proc. Assoc. Advmt. Anim. Breed. Genet: 24: 463-466*

# PERFORMANCE CORRIEDALE GROUP GENOMICS PROJECT

**P. Blackwood[1], C. Blackwood[1], J.M. Pitchford[2], H. Gordon[2], S.C. Welsh[2] and W.S. Pitchford[2]**

[1] Blackwood Corriedales, Harland Rise, Evandale, Tas, 7212 Australia
[2] Davies Livestock Research Centre, School of Animal and Veterinary Sciences, University of Adelaide, Roseworthy Campus, SA 5371 Australia

## SUMMARY

The Performance Corriedale Group are a group of breeders committed to working together to maximise genetic improvement and marketing of higher performance sheep. They initiated the trial following a desire to utilise genomics and have the tools available to maximise genetic improvement in lamb eating quality. The project achieved 764 carcasses from 44 sires (12 studs) from 3 AI and 1 natural mating in 2017-19. Lambs were slaughtered at 6-7 months with carcass and meat quality data, wool and type traits recorded. The data from the project will contribute to the Sheep Genetics database. The project also provides a model for engagement of breeds which are lower in number but still significant in impact.

## INTRODUCTION

Through the Sheep CRC's Information Nucleus there have been many sires tested for eating quality of their progeny in additional to multiple other traits. This has led to the development of genomic tests for the two traits having the largest effect on eating quality (shear force and intramuscular fat) as well as traits such as lean meat yield and other production traits recorded. While 14 Corriedale rams have been included and there is ongoing recording through the MLA Resource Flock, there are insufficient records for the genomic tests to be valuable to Corriedale breeders.

The project was initiated by the Performance Corriedale Group who only account for 11% of registered Corriedales, but accounted for over 40% of ram sales, and this is growing. The vision is to breed sheep that have higher wool value and superior eating quality to other maternal breeds with maximum weight of lambs weaned from ewes of moderate size and adapted to high rainfall regions. The purpose of this project is to address the deficit in genotyped Corriedales by genotyping 900 progeny from 45 sires representing a range of Corriedale sire lines. By using purebred Corriedale ewes, the project has more Corriedale haplotypes represented than if the sires were crossed to another breed.

## MATERIALS AND METHODS

In April of 2017-19, approximately 300 Corriedale ewes were synchronised and 20 inseminated to each sire provided. An additional natural mating was conducted in late 2018 to generate additional lambs for the project so lambs were born in four cohorts (2017, 2018, 2019A, 2019B). In total there were 44 sires from 12 studs used, some across years so the number of lambs per sire ranged from 1-60.

Lambs were slaughtered in April the following year. For the 2017 cohort, there were consecutive slaughter days. The 2019 born lambs were slaughtered in two groups where a small number (22) of the lightest 2019A born lambs were finished and slaughtered with the 2019B, so there are five slaughter cohorts (2017, 2018, 2019A1, 2019A2, 2019B). The number of carcasses in 2017 cohort was 212, 2018 was 233, 2019A1 was 175, 2019A2 was 22 and 2019B was 103.

Ewes were scanned for number of fetuses (0-3) and lambs were mothered up for recording type of birth (1-3) and rearing (1-3) with four combined classes resulting (11, 21, 22, 33). The very few triplet born lambs not raised as triplets were treated as multiples for type of birth and grouped with classes 21 and 22 for those raised as singles or twins respectively. Lamb survival rates were good with carcasses comprising 30% singles, just 11% born multiple raised single, 54% twins and 6% triplets.

Live traits included multiple weights, height, scanned fat and eye muscle depth, greasy fleece weight, fibre diameter, comfort factor and staple length, scores for nose and hoof pigment, face cover, jaw and leg conformation, back conformation, body and breech wrinkle, breech cover, wool staple structure, wool colour and character. Carcass and meat traits included carcass weight, GR fat depth, loin eye muscle depth, width and calculated area, calculated lean meat yield, pH decline and ultimate, meat colour (L, a, b), cooking loss, shear force and intramuscular fat content. Not all traits were measured on all cohorts either because it was not possible to enter the abattoir in 2020 due to COVID-19 (e.g. loin dimensions and pH decline), slight differences in laboratory procedures (e.g. meat colour), not shorn (e.g. greasy fleece weight), or the scores lacked variation and were of limited value (e.g. jaw confirmation and breech wrinkle and cover). Muscle and fat depths presented herein have not been adjusted for weight (live or carcass).

After receiving the genotype data and processing, there were 36 sires genotyped on Ovine-HD 600K, 764 progeny genotyped on GGP Ovine 50K and a single additional sire genotyped on GGP Ovine 50K. All progeny in the dataset were imputed from 45,740 SNPs to high density (570,293 SNPs) utilising the 36 high density genotyped sires as the reference population. Duplicated SNP positions and X and Y chromosome SNPs were removed prior to imputation. Imputation was completed using Fimpute3 (Sargolzaei *et al.* 2014). The imputed dataset, including the reference sires, was then filtered to remove SNPs with a minor allele frequency less than 0.01 and was checked for duplicate samples. Two separate samples were found to have the same tag, but different genotypes, so due to the possibility of miss labelling, these two samples were removed to give a total of 800 samples for GRM construction.

Homozygous genotypes for the major allele were coded as 0, for the minor allele as 2, and heterozygous genotypes as 1. The GRM with 798 animals was constructed as per VanRaden's first method (VanRaden 2008);

$$\boldsymbol{G} = \frac{\boldsymbol{ZZ'}}{2\sum_{i=1}^{n} p_i(1-p_i)}$$

Where **Z** denotes a centred matrix of allele effects with a mean of zero, $p_i$ is the frequency of the minor allele at locus $i$ and division by $\mathbf{2}\sum p_i(1-p_i)$ scales the **G** matrix to be similar in magnitude (so that diagonal elements average 1) to the numerator relationship matrix constructed from genealogy (VanRaden 2008).

The model fit to most traits included the random genomic relationship matrix and fixed effects of cohort (2017, 2018, 2019A1, 2019A2, 2019B), sex (ewe, wether) and type of birth and rearing (11, 21, 22, 33). Kill day within 2017 was added to carcass traits and birth group (2017, 2018, 2019A, 2019B) was used for weaning weight as it was recorded before the 2019A drop were split.

## RESULTS AND DISCUSSION

The phenotypic variances herein (Table 1) are similar to those reported by Mortimer et al. (2017, 2018) so it is assumed the lower heritability estimates are primarily due to small numbers. However, the heterozygosity on some animals (e.g. minimum 1.7%) was lower than other livestock data sets we have analysed and potentially indicates a higher level of inbreeding in Corriedale than other livestock. If there is genuinely less genetic diversity than for other breeds, then the heritability would be lower. That said, the mean was close to the maximum and so this could also be a function of a very small number of highly inbred animals.

The wool and scored traits were similar with many slightly more heritable than reported by Mortimer et al. (2017). Examples are GFW (0.76 vs 0.57) and FD (0.82 vs 0.74). Of the scored traits, colour (0.47) was similar to yellowness (0.80) but wrinkle was lower (0.16 vs 0.34), likely reflecting the much plainer body of Corriedales than Merinos.

Heritabilities for growth traits were consistently higher than those reported by Mortimer et al. (2017) but there was no maternal effect fitted herein: 0.54 vs 0.14 for weaning weight, 0.59 vs 0.31 for post-weaning weight, 0.49 vs 0.11 for scanned fat depth and 0.47 vs 0.14 for scan eye muscle depth.

Compared to Mortimer et al. (2018), carcass traits herein were also often more highly heritable. Examples include carcass weight (0.63 vs 0.35), GR fat (0.50 vs 0.23), lean meat yield (0.55 vs 0.29), intramuscular fat (0.46 vs 0.58), shear force (0.15 vs 0.10), and pH (0.06 vs 0.15).

**Table 1. Data description, summary and heritabilities**

| Trait | Mean | Min | Max | $\sigma_P$ | $h^2$ | $h^2$ SE |
|---|---|---|---|---|---|---|
| Weaning Weight (kg) | 32.6 | 13.6 | 54.5 | 4.6 | 0.54 | 0.09 |
| Post Weaning Weight (kg) | 39.6 | 21.0 | 55.0 | 4.9 | 0.59 | 0.10 |
| Height (cm) | 63.4 | 51.9 | 73.0 | 2.4 | 0.49 | 0.16 |
| Scan Eye Muscle Depth (mm) | 26.4 | 17.5 | 37.0 | 2.4 | 0.47 | 0.09 |
| Scan Fat depth (mm) | 3.5 | 2.0 | 7.5 | 0.9 | 0.49 | 0.10 |
| Greasy Fleece Weight (kg) | 1.81 | 0.6 | 3.0 | 0.3 | 0.76 | 0.11 |
| Fibre Diameter (um) | 23.1 | 1.8 | 18.2 | 1.8 | 0.82 | 0.10 |
| Comfort Factor (%) | 93.5 | 55.8 | 99.9 | 6.1 | 0.61 | 0.11 |
| Staple Length (cm) | 4.83 | 0.7 | 2.5 | 8.4 | 0.21 | 0.14 |
| Nose Pigment Score | 3.7 | 1 | 5 | 0.8 | 0.57 | 0.15 |
| Hoof Pigment Score | 4.1 | 1 | 5 | 0.7 | 0.53 | 0.16 |
| Face cover Score | 2.7 | 1 | 5 | 0.6 | 0.38 | 0.11 |
| Jaw Structure Score | 1.0 | 1 | 4 | 0.1 | 0.00 | 0.06 |
| Leg Structure Score | 2.1 | 1 | 4 | 0.4 | 0.00 | 0.06 |
| Back Structure Score | 1.9 | 1 | 4 | 0.3 | 0.23 | 0.11 |
| Body Wrinkle Score | 1.3 | 1 | 4 | 0.5 | 0.16 | 0.09 |
| Staple Structure Score | 2.6 | 1 | 5 | 0.6 | 0.28 | 0.10 |
| Colour Score | 2.3 | 1 | 5 | 0.6 | 0.47 | 0.10 |
| Character Score | 2.7 | 1 | 5 | 0.6 | 0.41 | 0.11 |
| Hot Std. Carcass weight (kg) | 20.6 | 11.3 | 32.9 | 2.7 | 0.63 | 0.09 |
| Shear Force (N) | 39.5 | 17.0 | 99.3 | 12.9 | 0.15 | 0.07 |
| Intramuscular fat (%) | 4.66 | 1.43 | 11.61 | 1.3 | 0.46 | 0.10 |
| GR Fat depth (mm) | 10.9 | 1 | 25 | 2.6 | 0.50 | 0.14 |
| Eye Muscle Depth (mm) | 41.1 | 21 | 63 | 3.1 | 0.39 | 0.12 |
| Eye Muscle Width (mm) | 42.6 | 20 | 66 | 3.3 | 0.35 | 0.13 |
| Eye Muscle Area (cm$^2$) | 12.8 | 7.7 | 18.4 | 17.1 | 0.45 | 0.14 |
| Lean Meat Yield (%) | 55.3 | 50.8 | 60.1 | 1.1 | 0.55 | 0.21 |
| pHuLL | 5.79 | 5.13 | 6.66 | 0.2 | 0.06 | 0.06 |

## CONCLUSIONS

This project provides a model for engagement with small breeds and we have been asked to specifically comment on the merits of such a model. The starting point should be a description of the history and model. All breed societies that aspire to breed sheep for production need a group of breeders committed to performance recording. Performance groups provide a forum for sharing ideas, challenging each other, critical mass to engage Sheep Genetics staff and other advisors, and a collection of people who inform innovation opportunities. The Corriedale Performance Group has all of those attributes.

Smaller breeds, almost by definition, sell less rams and generally receive less for rams than those from more numeric breeds and so have less ability to invest cash in projects. The model for this project was to charge $1000 plus semen for each sire (aim $45,000) to be tested. This was then matched with funds from the Davies Livestock Research Centre and MLA Donor Company. Breeders leveraged an additional $4 to their $1 invested. Overall, the project was industry initiated, provided valuable data for Sheep Genetics, leveraged funds and captured significant in-kind contribution.

It is good for researchers to work with producer groups like the Performance Corriedale Group. Often there is some history that initiates the connection as in this case, but it is wise for all early career livestock researchers to try and link with such a group. The benefits for the group are links to researchers and their networks and the benefit for researchers is ground-truthing research and an ideal format for testing new ideas and ways of communicating findings. This project was excellent for training Honours students and it should be an aim for all Agricultural and Animal Science Honours students to be involved in projects with industry to build networks and skills in addition to research skills. The project trained two Honours students (HG and SW).

It was hoped that the trial would attract new breeders to Sheep Genetics and reporting of results is primarily through Sheep Genetics to ensure the most accurate ASBVs are reported. Sire genomic values for all traits have been reported to breeders. Funding bodies should not trade on good–will, but equally economic rationalists should not get in the way of committed people with a common purpose. Thus, it is exciting to see increasing numbers of projects being funded through resource flock coordination and this should be extended as broadly as possible, especially when there are groups of producers collecting performance information in a coordinated way.

## ACKNOWLEDGEMENTS

This work was funded by Corriedale breeders, University of Adelaide's Davies Livestock Research Centre and Meat and Livestock Australia's Donor Company Project P.PSH.1001. Rick Tearle and Rudi McEwin from University of Adelaide processed the genotype data.

## REFERENCES

Mortimer S.I., Hatcher S., Fogarty N.M., van der Werf J.H.J., Brown D.J., Swan A.A, Greef J.C., Refshauge G, Hocking Edwards J.E. and Gaunt G.M. (2017) *J. Anim. Sci.* **95**: 1879.

Mortimer S.I., Fogarty N.M., ven der Werf J.H.J., Brown D.J., Swan A.A., Jacob R.H., Geesink G.H. Hopkins D.L., Hocking Edwards J.E., Ponnampalam E.N., Warner R.D., Pearce K.L. and Pethick D.W. (2018) *J. Anim. Sci.***96**: 3582.

Sargolzaei M, Chesnais J.P. and Schenkel F.S. (2014) *BMC Genomics* **15**: 478.

VanRaden. P.M. (2008) Efficient methods to compute genomic predictions. *J Dairy Sci.* **91**:4414.

*Proc. Assoc. Advmt. Anim. Breed. Genet: 24: 467-470*

# INVESTIGATING THE POTENTIAL TO UTILISE COMMERCIAL CARCASS TRAITS IN GENETIC EVALUATION

**S.J. Lee[1], M. Beattie[2], R.A. McEwin[1], R. Tearle[1], W.S. Pitchford[1]**

[1]*Davies Livestock Research Centre, School of Animal and Veterinary Sciences, University of Adelaide, Roseworthy, SA, 5371*
[2]*Herefords Australia, 16 Uralla Rd, Armidale NSW 2350*

## SUMMARY

This project sought to explore whether targeted beef carcass records from commercial production systems in southern Australia were suitable for use in genetic evaluation. The motivation to do so was to increase the number of carcass records in reference population. The project team liaised with Hereford and Angus bull breeders and their clients to identify potentially suitable records from their production systems. In total, a dataset comprised of 1406 records from Hereford and Angus steers and heifers from 23 management groups was established. Records were classified as either High-Quality (HQ) or Medium-Quality (MQ) based on ability to describe fixed effects. This data was compared against a research dataset of 642 Angus and Hereford x Angus carcasses finished to a similar carcass weight end point. Traits analysed include MSA Marble, ossification, rib fat depth and eye muscle area, MSA Index and hot standard carcass weight. Heritability estimated for HQ and the research herd dataset were moderate indicating potential to use high quality commercial carcass records in genomic evaluation. Heritability estimates for the same traits for MQ were very low indicating lack of knowledge on fixed effects severely impeded the utility of such records in genomic evaluation.

## INTRODUCTION

The Australian beef grading system to ensure eating quality is Meat Standards Australia (MSA, Polkinghorne *et al.* 2008). There are currently close to zero carcasses from commercial production systems that are being used for genetic evaluation. Reverter *et al.* (2000) reported that in Australian Angus and Hereford cattle, the genetic correlation between ultrasound and carcass traits was variable, but averaged 0.46 for EMA and 0.54 for IMF. These correlations are important as they provide the upper limit to accuracy of selection for the carcass traits based on ultrasound measurement. As industry adopts objective measurements of eating quality, it is becoming increasingly important to be able to record the traits in the breeding objective directly rather than relying on correlated ultrasound measures.

In the past, there has been multiple limitations to using commercial carcass data. The first problem has been to get pedigree information. However, the impact of genomics (Meuwissen *et al.* 2001) means that genomic relationships on commercial animals can be established. In addition, if the property has been using bulls with high genetic merit, then their animals will likely be genetically related to leading animals in the breed. Thus, scope exists for commercial performance to be integrated into genetic evaluation programs like BREEDPLAN (Graser *et al.* 2005) and can provide valuable information which is currently difficult for studs to record.

A problem often encountered with commercial data is maintenance of contemporary groups. However, increasingly cattle are grazed in large mobs (>100) and this is becoming less of an issue. Most genetic evaluation systems require birth date of calves so adjustments can be made for age which is important for early growth traits. Another common problem is that of drafting cattle for sale where cattle are weighed and the heaviest potentially grazed in a separate mob for 1-4 weeks, then transported to a feedlot or abattoir for slaughter. However, Pitchford (2016) demonstrated that for genetic evaluation of carcass traits, such as loin eye muscle area and intramuscular fat, when they are adjusted for carcass weight, the effect of drafting on genetic evaluation of these traits is minimal.

Pitchford (2018) quantified the loss of precision for commercial cattle when less information (fixed effects) are collected than commonly recorded in seedstock herd recording programs. Pitchford (2018) found for the carcass weight that the correlation between EBVs between a reduced and full model of fixed effects had a correlation of 0.93. For all other traits (loin eye muscle area, P8 rump fat depth and intramuscular fat content), correlations between EBVs for a reduced model with a full model were much more highly correlated (>0.96) indicating little re-ranking due to fitting reduced fixed effects. Pitchford (2018) concluded that there are many commercial herds that have sufficient control of contemporary groups so their data should be utilised for genomic selection of carcass quality traits.

Based on the above findings, this project sought to evaluate the scope to use MSA grading records from commercial groups of steers and heifers for genomic evaluation for data where fewer fixed effects were known on the groups of animals.

## METHODS

This project was a collaboration between Herefords Australia, Hereford and Angus bull breeders and their commercial clients with the aim of identifying mobs of cattle that were managed together from birth to slaughter, processed in large mobs and MSA records could be accessed from the supply chain. Eight bull breeders were approached to participate in the project, of which five were active participants. These bull breeders approached 15 clients to identify eight commercial producers who were likely to meet the data recording requirements and had animals with expecting processing dates within the timeframe of the project. There were 1406 carcass records included in the analysis. These animals were from 23 management groups (a concatenation of on-farm management group, feedlot groups and processing date). Mean management group size was 61 (range 11-210, standard deviation 51). Over 2400 animals were identified for carcass outcomes to be included in the study but approximately 40% of records were excluded due to not meeting minimum data quality criteria. In addition to exclusion above, a data quality factor was developed (high quality, HQ vs. moderate quality, MQ). This was based on information provided by commercial producers on:

- Length of calving - progeny from calving periods less than 8 weeks were considered high quality, whereas >8 weeks (maximum 12 weeks) were classed as moderate quality.
- Confidence in defining lifetime management groups (some groups came from > 1 calving paddock but Pitchford (2018) showed this to be of likely low importance when omitted).

In total there were 627 HQ records and 779 MQ records.

All feedlots and processors approached to collaborate in the project were highly supportive and accommodating. This is important as it highlights commitment to further improvements in carcass quality. Contribution to the project included provision of feedlot information (feedlots), provision of carcass grading information, limiting carcass grader to one or few graders for a cohort, access to carcasses for collection of sample for DNA testing.

The comparison data for the project was sourced from "Hereford Black Baldy BIN: Improving productivity of commercial cattle through utilising superior sires within and across breeds (P.PSH.0716)", herein referred to as Black Baldy dataset. In total 642 steers had carcass records, from 11 processing dates, i.e. 11 contemporary groups with average management group size was 58 (range 1 -112, standard deviation 43). The steers were a mix of Angus and Hereford x Angus. All steers were finished on pasture with a mean hot standard carcass weight 292kg (minimum 181kg – maximum 353kg, standard deviation, 30kg). The Black Baldy data is part of a structured progeny test, and thus lifetime management groups are well defined. As such, it provides a point of comparison point for heritability compared with commercial data collected. All carcases were graded using the Meat Standard Australia grading system. AUS-MEAT certified MSA graders measured hot standard carcass weight, marbling, ossification, fat colour and subcutaneous rib fat.

Overall there were 2,850 animals with genotypes used to develop a genomic relationship matrix between datasets. These comprised 1,406 genotypes and 1,458 genotypes from Black Baldy, for the 642 steers with carcass records, and the remainder being their relatives (e.g. heifers and bulls) that are part of the Black Baldy project. All genotypes were generated on a variety of Illumina genotyping chips. All of the animals and SNPs were merged to generate a matrix of genotypes, containing 2,850 animals and 157,665 SNPs. FImpute (Sargolzaei *et al.* 2014) was used to impute all genotypes to a set of 40,683 SNPs. Using the genomic relationship matrix from 40,683 SNPs, data was analysed with a general linear mixed model using ASreml-R 4.0 (Butler *et al.* 2017). The model used across all traits was the same and presented random terms of known and heterogeneous variance structures. The known variance structure was the additive relationships between individuals represented through a Genomic Relationship Matrix constructed as per Van Raden Methód 1' (2008) and the heterogeneous variance structure was a diagonal variance model for Dataset Quality Factor (Black Baldy vs. HQ vs. MQ). Direct sum structures were also obtained for the residual error term. This allowed variance components and hence heritabilities to be estimated for the same trait between datasets of different quality. The model also included fixed effects of dataset Quality factor (3 levels: Black Baldy, HQ, MQ), contemporary group adjusted for processing date and grader as well as HSCW as a covariate, except where HSCW was itself the trait of interest.

## RESULTS AND DISCUSSION

Phenotypic and additive genetic variance components together with estimated heritability are reported by dataset (HQ, MQ, Black Baldy) in Table 1. Heritability estimates for HQ were moderate for EMA, Rib, MSA Marble, Ossification and MSA Index. In general, MQ had similar phenotypic variance to HQ but lower additive variance resulting in lower heritability estimates. For MSA Marble, phenotypic variance was significantly lower, and there was negligible additive variance, leading to a heritability estimate of 0.05. In comparison to MQ and HQ datasets the Black Baldy results had much higher heritability for MSA marble, ossification and MSA-Index but similar heritabilities for rib fat, EMA and HSCW.

The lower additive variance for the same traits between dataset with similar phenotypic variance provides insights on the loss of precision in evaluation when using commercial data. For example, irrespective of data set (data quality) rib fat depth had similar heritability estimates and broadly similar phenotypic variance. In contrast, MSA marble had much lower phenotypic variance for both HQ and MQ compared with Black Baldy; this is especially so for the MQ data (representing the data with more poorly described lifetime management groups). Moreover, MQ had the highest mean MSA marble (366.5) and a similar observed standard deviation to Black Baldy (56.03 vs. 49.61. Therefore, it is unlikely the low variance is a function of low mean MSA-marbling. Importantly for the HQ dataset heritability remained moderate.

## CONCLUSIONS

The results for HQ compared with MQ demonstrate the importance of using only data of the best possible quality within the constraints of commercial beef production systems. Based on this project, where poorer (e.g. MQ) quality data was accepted, the genetic variance in key traits like MSA-Marble was too low for the carcass record to be of substantial value. Therefore, any further efforts must focus solely on records with very high confidence that animals to be processed have fixed effects that can be described well for factors including calving period, dam age (heifer, cow). This does not mean they have to have all this data recorded exactly, but that they meet our understanding of "born and raised together".

**Table 1. Estimated phenotypic variance ($V_P$), additive variance ($V_A$) for MSA traits by dataset**

| | Phenotypic variance ($V_P$). | Additive variance ($V_A$) | Heritability ($h^2$) |
|---|---|---|---|
| | | **HQ** | |
| HSCW | 468.50 | 182.29 | 0.39 |
| EMA | 38.18 | 15.71 | 0.41 |
| Rib | 8.52 | 2.70 | 0.32 |
| MSA Marble | 4983.21 | 1470.23 | 0.30 |
| Ossification | 218.16 | 45.10 | 0.21 |
| MSA Index | 2.10 | 0.68 | 0.33 |
| | | **MQ** | |
| HSCW | 483.85 | 166.25 | 0.34 |
| EMA | 43.75 | 6.00 | 0.14 |
| Rib | 7.10 | 2.17 | 0.31 |
| MSA Marble | 2025.66 | 102.84 | 0.05 |
| Ossification | 225.95 | 77.54 | 0.34 |
| MSA Index | 1.98 | 0.50 | 0.25 |
| | | **Black Baldy** | |
| HSCW | 901.37 | 422.83 | 0.47 |
| EMA | 52.64 | 10.89 | 0.21 |
| Rib | 5.93 | 1.93 | 0.32 |
| MSA Marble | 2834.42 | 2089.98 | 0.74 |
| Ossification | 124.75 | 49.41 | 0.40 |
| MSA Index | 1.44 | 0.64 | 0.45 |

## ACKNOWLEDGEMENTS

This project was funded by Meat and Livestock Australia (P.PSH.0989), Herefords Australia and University of Adelaide. The authors thank all collaborating breeders & supply chains for support.

## REFERENCES

AUS-MEAT. (2005) Handbook of Australian Meat, International red meat manual (7th Edition) Sydney, Australia.

Butler D G, Cullis B.R., Gilmour A.R., Gogel B.G. and Thompson R. (2017) ASReml-R Reference Manual Version 4. VSN International Ltd, Hemel Hempstead, HP1 1ES, UK.

Graser H-U., Tier B., Johnston D.J. and Barwick S.A. (2005) *Aust. J. Exp. Agric.* **45**: 913.

Meuwissen T.H.E., Hayes B.J. and Goddard M.E. (2001) *Genetics* **157**: 1819.

Pitchford W.S. (2016) *Anim. Prod. Aust.* **31** paper 1298.

Pitchford W.S. (2018). In proceedings WCGALP.

Polkinghorne R., Thompson J.M., Watson R., Gee A. and Porter M. (2008). *Aust. J. Exp. Agric.* **48**, 1351.

Reverter A, Johnston D.J., Graser H-U., Wolcott M.L. and Upton W.H. (2000) *J. Anim. Sci.* **78**:1786.

Sargolzaei M., J.P. Chesnais and F.S. Schenkel (2014) *BMC Genomics* **15**: 478.

VanRaden P.M. (2008) *J Dairy Sci.* **91**(11): 4414.

*Proc. Assoc. Advmt. Anim. Breed. Genet: 24: 471-475*

# GENETICS FOR SELF-REPLACING MATERNALS: PARADOO PRIME

**T. Leeming**

160 Armidale Rd Coojar 3315 Vic

## INTRODUCTION

2021 marks the 25th year of farming sheep in my own right and also as a partnership with my wife Georgie. Since arriving back on my parents' property in 1993 I have been involved primarily in self replacing maternal ewes founded originally with Romney ewes over Merino ewes then progressing into a Coopworth sire over merino ewes in 1993. I purchased my first rams in 1993 from Don Pegler and also some rams from John Keilor. All these rams were performance recorded and from that initial introduction into seeking genetics for our flock, I have never bought animals without Lambplan or Merino Select breeding values. We farm in a 600 mm rainfall zone and over the last 25 years, we have built a farming operation that now covers over 1500 ha whereby we wean in excess of 11000 lambs annually and market 500 plus rams. Just under 90% of this land has been purchased by us and 75% purchased in the last 10 years. I began financial benchmarking our farm business in 1997 and we still do to this day through the long-standing Livestock Farm Monitor Project administered by Agriculture Victoria. Our business and asset base have expanded significantly from modest beginnings and to achieve this not only takes hard work but it also takes discipline, planning and setting some clear and achievable goals. Benchmarking can be an important reference to this by outlining the main profit drivers to keep in check.

***We believe strongly that the maternal flock we have had all our farming career based solely on performance genetics has been integral to the expansion and success of our business.***

Efficiency in all aspects of what we do is something that has been a forced discipline. We strive to being active participants in our farming industry but also our local community. Time is at a premium for all of us. The key components of our efficiency are based on the production system that suits us as managers but also suits our environment in which we farm. We have invested heavily into farm infrastructure and land development all which is underpinned with achieving our production goals easier and more effectively.

***Our farm 'Cobbity'.*** The 1340 grazable/arable hectares are subdivided into 125 permanently fenced and water reticulated paddocks. An additional 45 paddocks are added temporarily to reduce mob size during 3 separate strategic lambing's. Lamb survival is something that is paramount to our operation that essentially prides itself as consistently producing some of the most successful paddock lamb survival results in Australia.

In 2006, we decided to start a maternal stud (called Paradoo Prime) based on Coopworth genetics. The first main drivers of this decision grew from curiosity and the desire to collect data from our sheep in our environment. Secondly, we had experienced greater seasonal volatility and we believed we needed to develop a sheep more suited to our shorter growing seasons. From 2008 we deliberately drove towards what we believed was more balanced maternal sheep.

We discovered the importance of reducing mob size in 2005 when leasing grazing country and have pregnancy scanned our flock for multiples since 1995. Raising triplet lambs commercially has been a particular focus in recent years and these commercial animals have been differentially managed for the past 8 years. For the past 15 years we have managed twin bearing ewes in mob sizes under 100 ewes and for the past 9 years our twin mob size has averaged under 50 ewes. Our commercial flock ewes bearing triplets lamb in average mob sizes of no more than 18 ewes.

We now have developed a system of lambing which we feel can lead the future of lambing management. This system is called Paradoo Precision lambing. This program has improved the management of reproduction with some outstanding results in variable seasons on multiple properties over the past 6 years. This system initiated by us and fine tuned with other participating clients has been very rewarding. We believe that the system allows sheep producers to meet the well documented targets and management required for consistently high reproduction and low ewe and lamb wastage.

So where does genetics come in and how important is genetics to reach profitable, efficient and ethical animal production?

**BREEDING AIMS / OBJECTIVES**

A maternal sheep needs to be efficient. Efficiency in a sheep system means maximising productivity but minimising wastage and expense. Some compromises are needed to achieve this balance. For us we needed to focus on our goals and breeding an animal that had greater relevance to our shorter growing season. This involves selecting traits not only to improve carry over reproduction in failed springs but also for the ability to produce lambs with more fat and muscle which we believe assists us in achieving the pointy end of lamb survival and profitability.

Stocking rate and reproduction are inter-related as additional reproduction enables you to enjoy the cheapest and most efficient gains in stocking rate. This in turns enables you to increase feed utilisation and ultimately production/profit per ha.

As avid sponges for aspiring to improve all things in sheep farming, we focused strongly on growth and reproduction from 1993 to 2008. Some serious seasonal impacts on our growing season length from 2002 to 2007 and the development of our own stud flock in 2006 made us re think some of our original aims in our breeding. We wanted animals that were more suitable and resilient to a variable climate and a ewe that was smaller than the version we had created up until 2008. Our ewe flock was big, lean and fertile. We were weaning more lambs than most and our lambs grew quickly but this also had some issues. Our standard reference weight (SRW) although not independently assessed at the time but as I was a keen weigher and condition scorer of stock, I was quite confident that we exceeded 70kg average ewe weight at condition score 3.

Fast forward to 2015, we had drastically shifted our focus to early maturing animals with more fat and muscle of their older sisters. The MLA Maternals project had our SRW of our ewes at 59kg and we enjoyed how quickly these animals developed. A low adult weight and an increase in fat and muscle has been a keen focus of ours since 2008. Since then, we have gained confidence in our management of lactation and subsequent lamb survival. Improving the genetics of our sheep has played an important role in building this confidence and increasing production gain. As we know good conception is vital in striving for high reproduction, however it does not necessarily allow you to reap the rewards of a high weaning percentage. Producing large litters of lambs without the ability for the ewe to rear them is irresponsible and goes against modern consumer and industry expectations.

Low lamb survival creates wastage not just in the lambs but also in the ewes that are trying to rear lambs. Managing nutrition in late pregnancy for ewes carrying more than 2 lambs is not easy. The management of the lambing environment with lower birth weights and a greater propensity for mismothering requires a greater level of husbandry, supervision and overall effort.

We have recently enjoyed the rapid increase in red meat/protein prices not only from our premium lamb but we have also enjoyed the linear returns from mutton prices. Ewe wastage takes the form of a number of factors in a sheep enterprise starting with dry ewes at scanning and also from pregnant ewes that fail to rear. The other major component of ewe wastage comes from ewe mortality. More than ever our production system in sheep must take into consideration consumer expectations in animal welfare. Profitability and efficient farm management are vital and especially when the meat boom and or seasons become less favourable than recent years.

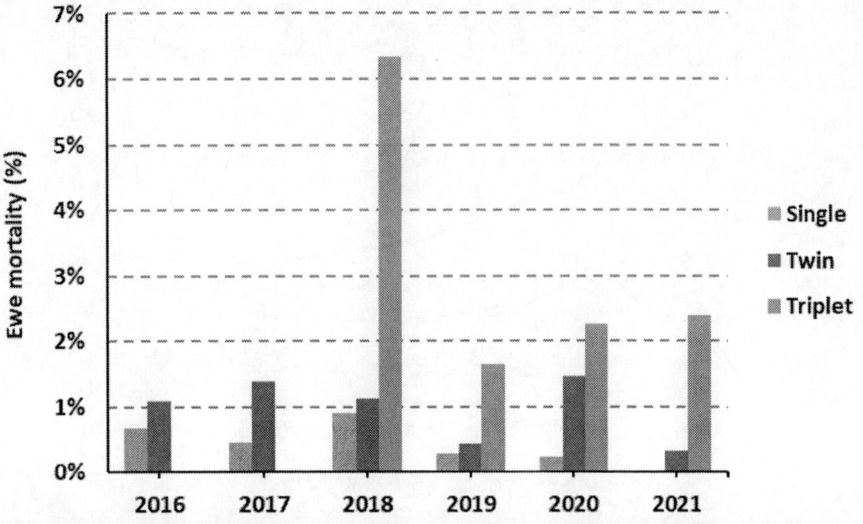


**Figure 1. Mortality rates (from 2016 to 2021) for ewes bearing single, twin and triplet lambs at Cobbity, Coojar, Victoria**

Accepting low lamb survival and high ewe wastage in the above-mentioned scenarios makes little sense. We have deliberately focused our breeding objectives to reduce wastage and also to compliment our sheep management disciplines. Combined, we have been able to reduce lamb and ewe wastage on a consistent basis no matter what the season delivers. ***90% lamb survival in twins is something that we consistently achieve and also keeping ewe mortality in all lambing ewes under 1.5%*** (see Figure 1). As mentioned we do farm our animals on well above industry stocking rates for our environment and at the same time manage one of the highest DSE/FTE ratios within the Livestock Farm Monitor benchmarking. Our business has achieved an average of 9.6% return on assets (ROA) in the past 5 years (see Figure 2).

So where do we head in the future for our best gains? Is there much more room in single and twin lamb survival? Maybe not but there is much to gain we feel in lamb survival and management of ewe lambs and also in best managing triplet bearing ewes and the wastage in these also. This is our current and main challenge we set to improve in years to come.

In 2020 and 2021 we have achieved an average weaning of over 1.5 lambs per pregnant ewe which includes our biggest age group being ewe lambs. Our rising 2 year old ewes for the past 2 years have scanned over 180% and weaned in excess of 162% to ewes joined. The most exciting results recently has been our overall dry rate at scanning in these rising 2 year old's. We have maintained a dry ewe rate of less than half of one percent (0.5%) in the last 3 years. All of these ewes conceived lambs as ewe lambs in the previous year. Ewe lambs that fail to get in lamb are sold to processors for slaughter. The big emphasis has been to replace condition on fertile ewe lambs post weaning and they are treated as the highest nutritional priority of stock on farm in spring and early summer. Our average dry rate at scanning within our commercial flock of 6500 ewes was less than 1% (except ewe lambs) in the current season at 0.9%.

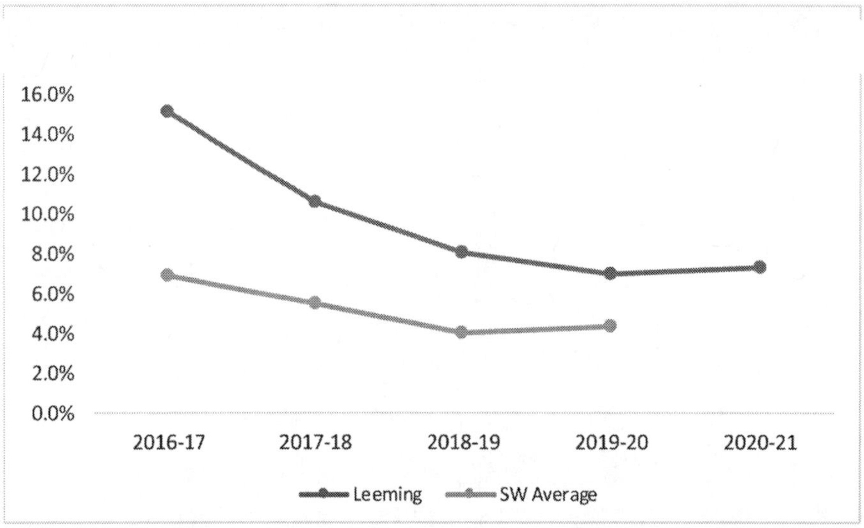


**Figure 2. Return on assets (% - excluding capital appreciation) for our farm (Leeming) compared with the South West (SW) Average from the Livestock Farm Monitor Project**

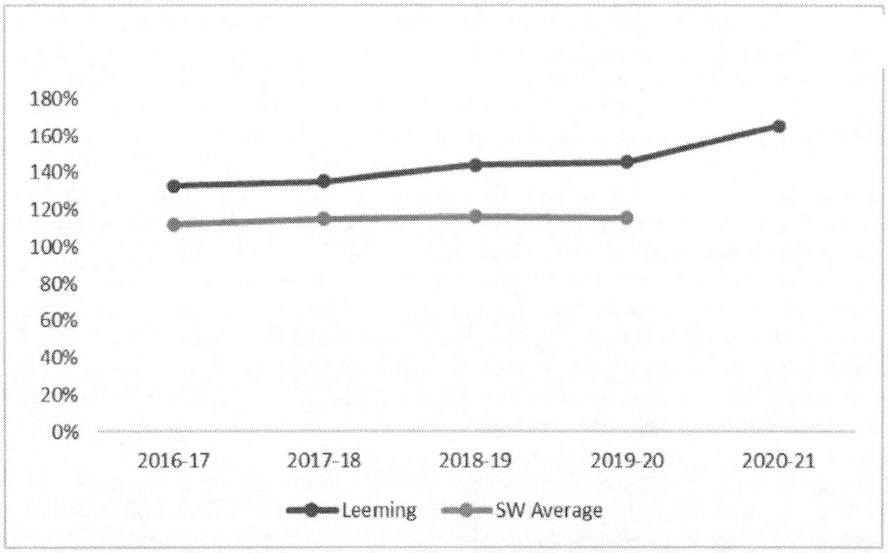


**Figure 3. Lamb marking % for our farm (Leeming) compared with the South West (SW) Average of prime lamb farms in the Livestock Farm Monitor Project**

Our commercial enterprise is our sounding board for our breeding direction. It is 100% transparent to industry and we pride ourselves on creating new targets within this flock and sharing our strategies in being able to achieve them. Setting disciplines in this flock to continuously improve stimulates us as a business. and the broader prime lamb industry.

Participating in projects such as the Maternals Project (2014-2015), Unlocking the keys to Ewe Survival Project (2019-2020) and various other research involving ewe lambs and triplets bearing ewes has been a rewarding experience and deliberate focus for us.

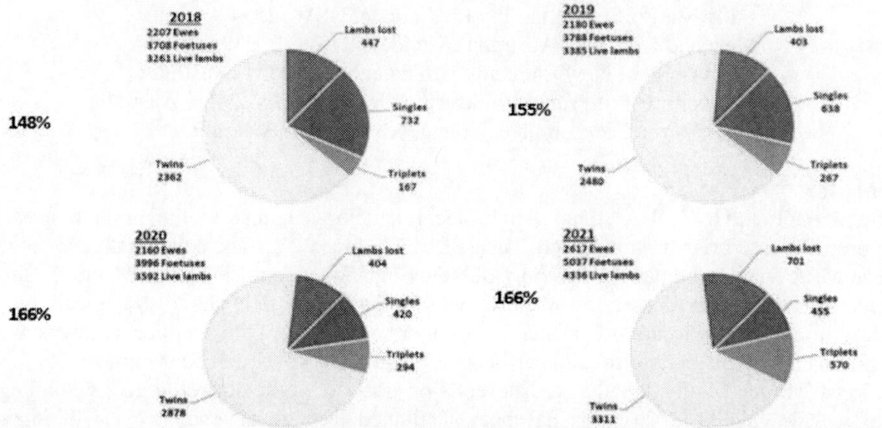


Has genetics helped us in achieving our production outcomes?

## PROGRESS

Our use in genetics has been quite significant over 25 years as has the sheep management practices along the way. Changing the animals has been really interesting to watch from the initial days of building fertility and growth to more recently reducing adult weights and reducing ewe and lamb wastage.

Early maturity patterns and lower SRW weights helps us achieve our production goals as we have really focused our business as a specialist breeding enterprise. A sheep that is low maintenance, fertile, has fantastic rearing ability and can wean her body weight in lamb in 100 days is what we know we can do.

## TAKE HOME MESSAGE(S)

What we have done since starting our stud business in 2006 has been to focus our attention on individual traits to make possible our breeding aims. We have been careful to chase the extremes or high indexing animals for much of the past 15 years. There has been many high index animals in the past that have not suited our aims and quite often do not always best fit our environment. Growth should be rewarded or encouraged but quite often in the extremes it has too much to compromise. We have modified our approach and have always aimed at a balance. Our strongly pragmatic approach has been to keep a clear path in our breeding, underpinned by strong fundamentals around structural/physical selection. Genetics in maternals has a compound effect so keeping the balance is paramount. Moderation it maybe, but always heading in what we believe is the right direction for us and the recipients of our genetics.

*Breeders paper*

# MERINOLINK/UNE DNA STIMULATION PROJECT: DOUBLING THE RATE OF GENETIC GAIN – WHERE ARE WE AFTER 4 YEARS?

**S.J. Martin[1,2,3,4,5] and T. Granleese[2,3,4,5]**

[1]Sally Martin Consulting Pty Ltd, Young, NSW, 2594 Australia
[2]Meat and Livestock Australia, Armidale, NSW, 2351, Australia
[3]University of New England, Armidale, NSW, 2351, Australia
[4]NSW Department of Primary Industries, Grafton, NSW, 2460, Australia
[5]MerinoLink Limited, Young, NSW, 2594, Australia

## SUMMARY

The MerinoLink/UNE DNA Stimulation Project is a major genetics adoption program in Merino sheep breeding enterprises running from January 2018 to June 2022. The collaborative approach is between MLA Donor Company, University of New England, MerinoLink Limited and the project participants. The project focusses on working with the project participants to strategically use the genetic and genomic tools currently available. A major component of the adoption strategy has seen the total financial contribution from all participants equating to upward of $1.7 million.

Thirty seedstock Merino breeders are currently on-track to hit the project target of doubling the rate of genetic gain by 2022. This has been facilitated through an annual cycle of intensive mentoring, workshops, networking and use of breeding decision tools within the group.

Commercially available Flock Profile and RamSelect's Ram Team Manager are tools commercial breeders are using to benchmark the genetic merit of their Merino flocks. This information has been used to aid participants in better ram selection and buying decisions to increase genetic merit of their flocks. In the final phase of the project commercial breeders will conduct a second Flock Profile to measure their genetic gain and cross reference this information to the Ram Team Manager predictions (where applicable).

## INTRODUCTION

Historically, average genetic gain in Merinos is currently low with very large variations across the industry (Granleese *et al.* 2018). Underlying influences are caused by a multitude of factors including inaccurate breeding values (Stephen *et al.* 2018) and/or a lack of understanding of selection theory.

The DNA Stimulation Project has been building understanding and implementation of genetic selection tools with the project participants. The project focusses on capacity building and working collaboratively at all levels across the industry to communicate how to use software tools more effectively for assisting in the design of breeding programs to increase the rate of genetic gain for participants.

The Project has 30 ram breeders and 18 commercial breeders participating who breed their own rams, 52 commercial breeders who purchase rams and 7 service providers. The project participants are located across New South Wales, Western Australia and Victoria.

During the DNA Stimulation Project, project participants have been faced with significant challenges including drought in many parts of Australia and unprecedented restrictions due to COVID-19. The project team has been able to adapt to the situation and continued to deliver the project requirements.

The aim of this project is to double the annual rate of genetic gain by 2022 (starting 2018) by maximising the adoption of a set of tools by breeders involved in the project. This paper outlines methods the project has used and provides a progress report of how participants are working towards this goal in year 4 of a 5-year project.

## MATERIALS AND METHODS

Project participants signed an agreement to be part of the DNA Project outlining project expectations between project leaders and participants. The project is co-funded with 58 percent contributed from the project participants and 42 percent from the MLA Donor Company. UNE were financial guarantors and provide genetic technical support in conjunction with MerinoLink facilitated genetic service providers. Sheep Genetics have also been fully engaged and supportive of the project.

Key tools used with project participants include DNA parentage, low density genomic tests, Australian Sheep Breeding Values (ASBVs), MateSel (Kinghorn 2011), Sheep Genetics Ramping Up Genetic Gain report (RUGG), Flock Profile (Swan *et al.* 2018), Ram Team Manager (RamSelect.com.au), Rampower within flock indexes and percentile band tables.

The extension and adoption process include a combination of face-to-face workshops, intensive one-on-one meetings, webinars, phone calls, personal emails, e-newsletters and group email updates. Input into these processes include key personnel within the DNA Project Team, University of New England, Sheep Genetics, Meat and Livestock Australia and participants in the project, including breeders and genetics service providers.

To measure progress the DNA Stimulation Project uses tools, such as rates of genetic gain, as generated by Sheep Genetics, and workshop feedback, to track the effectiveness of the project. In this project we use the Merino Production Plus (MP+) index as the genetic progress benchmark. "Doubling the rate of genetic gain" will be measured by comparing the five-year genetic gain average compared to the project five year genetic gain average.

## RESULTS AND DISCUSSION

The key outcomes of the project to date includes the strategic use of DNA testing technologies to increase the number of ram breeder participants submitting data to Sheep Genetics with full pedigree. There has been an increase in the number of project participants submitting full pedigree from 25% to 53% between January 2018 and January 2021, shown in Table 1.

**Table 1. Average proportion pedigree submission to Sheep Genetics for ram breeder project participants on a five year rolling average**

| Year Drop | Full Pedigree | Sire Only Pedigree | Dam Only Pedigree | Syndicate Pedigree | No Pedigree |
|---|---|---|---|---|---|
| 2017* | 0.25 | 0.41 | NA | 0.25 | 0.09 |
| 2018 | 0.34 | 0.38 | 0.01 | 0.21 | 0.07 |
| 2019 | 0.43 | 0.34 | 0.01 | 0.13 | 0.09 |
| 2020 | 0.53 | 0.29 | 0.01 | 0.08 | 0.09 |

*Pre-project for starting reference point

During the first half of the project, 2018 to 2019, co-funding was provided for DNA testing. The project participants used this co-funding to conduct 48,691 DNA parentage tests, 62 Flock Profiles and 13,236 low density genomic tests (January 2018 to June 2019). The majority of the ram breeders involved in the project have now tested their entire ram breeding nucleus (dams, replacements and sires) with either DNA parentage or low-density genomic information.

After June 2019 all DNA parentage and low density genomic tests has been fully funded by the project participants, with over 40,000 tests being conducted annually by the ram breeders (now 60% low density genomic tests).

There is a large range of genetic merit between the ram breeder participants in the DNA Project, with the lowest average MP+ index 110 and the highest 167 MP+ index points at the commencement of the project in 2017. The current MP+ index range between the ram breeders is 134 to 175 index points. This diversity in the group is being used to share and swap experience to allow informal mentoring within the group. Overall, the collective ram breeder participant index trend from the 2017 to 2018 lamb drop have increased by 2.4 index points. Each January the project participant progress is benchmarked to the rest of the industry to assess progress, in 2019 the average index was 5.0 points higher than the Merino industry average (Figure 1) and 2020 and 2021 it was 11 points higher.

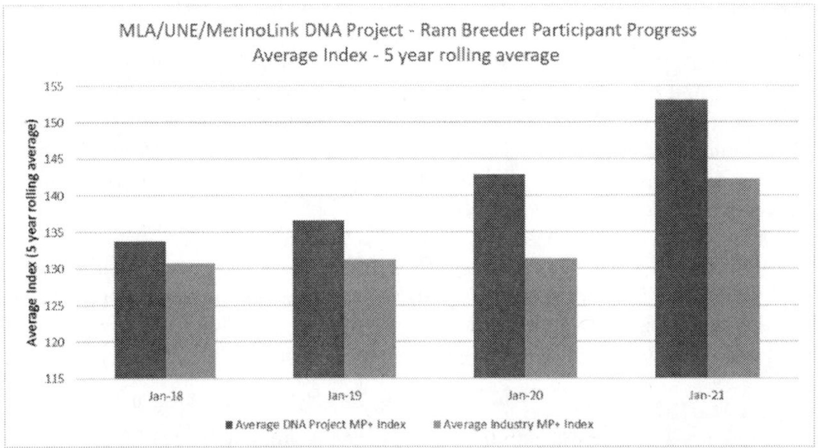


**Figure 1. Average 5 year rolling index for project participants (blue) and all MERINOSELECT members (grey)**

Ten percent of the ram breeder participants first used Matesel (Kinghorn 2011) for mating allocations for the 2019 joining, in 2021 this has increased to 48 percent. The biggest limitation for many of the project participants to utilise Matesel in the past has been the lack of full pedigree data which is being addressed with the increased uptake of DNA parentage and low-density genomic testing to gather pedigree information.

The ram breeder participants have embraced additional data collection as breeders advance to include traits that are not included in industry indexes in their breeding program, for example fly strike indicator traits. The combined DNA Project participants contribute 38 and 44 percent of the early breech wrinkle data on the 2019 and 2020 drops respectively.

DNA parentage testing has enabled ram breeder DNA Project participants to readily collect data on key reproduction traits, including litter size (LS), conception (CON) and ewe rear ability (ERA). Between the year of joining (YOJ) 2016 and 2020 there has been a 17 percent increase in data submitted to MERINOSELECT for ERA by the project participants. For the 2020 YOJ the DNA Project ram breeder participants are contributing 43, 39 and 29 percent of the MERINOSELECT database for CON, LS and ERA, respectively.

Key times identified in the project to provide information, data and advice to assist project participants in decision making are when breeders are setting and reviewing their breeding objectives and developing their next mating programs. The majority of the one on one ram breeder meetings are planned to coincide with strategic times in their calendar of operations.

There is a direct correlation between attending educational activities and rates of genetic gain (Brown 2019). Figure 2 demonstrates that the more engaged ram breeder project participants are with the project the higher the rates of genetic gain. This trend validates the intense relationship the breeders have with the project team and MLA and UNE funding.

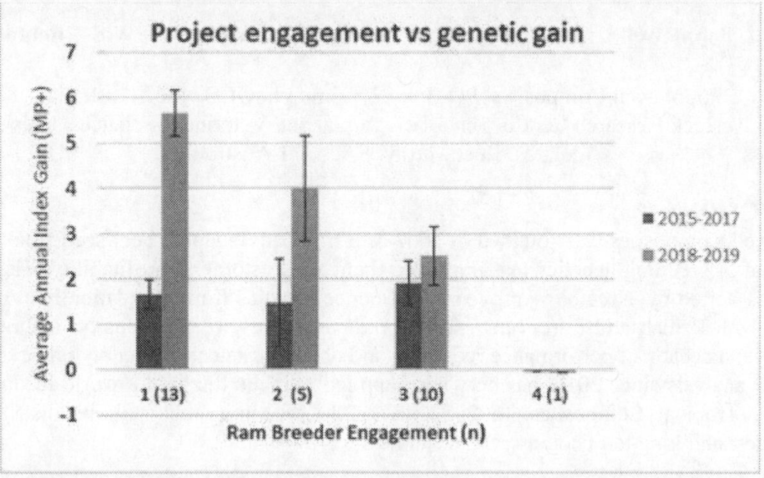


**Figure 2. Average annual genetic gain grouped by breeder engagement.** Group 1 - attends all workshops and requests one-on-one meetings; Group 2 - attends a combination of most workshops and/or requests one-on-one meetings; Group 3 - attends some workshops and rarely one-on-one meetings; Group 4 - no engagement

## CONCLUSIONS

Progress continues to be made with the project participants and as the project draws to a conclusion in 15 months' time the full impact of the project will be able to be described. This project is on track. It has and continues to deliver positive breeder satisfaction in training, extension, and adoption.

## ACKNOWLEDGEMENTS

The MerinoLink/UNE DNA Stimulation Project is funded by the Meat and Livestock Australia Donor Company (MDC) [P.PSH.0961], University of New England, and all the project participants. Daniel Brown of AGBU and Emma McCrabb of Sheep Genetics are thanked for providing some data for this paper and their ongoing technical support to the project.

## REFERENCES

Brown D.J. (2019) *Leading Breeder Conference*, Dubbo, NSW, Australia.

Granleese T., Clark S.A., Duijvesteijn N., Bradley P.E., and van der Werf J.H.J. (2018) *Animal Production Science* **Published online**: https://doi.org/10.1071/AN17720.

Kinghorn B.P. (2011) *Genetics Selection Evolution* **43**: 4.

Stephen, L., Brown D., Jones C., and Collinson C. (2018). *WCGALP* **11**: 433

Swan A.A., Gurman P.M., Boerner V., Brown D.J., Clark S.A., Gore K., Granleese T. and van der Werf J.H.J. (2018) *WCGALP* **11**: 346.

RamSelect website – www.ramselect.com.au

Contributed paper

# POPPLEWELL TROPICAL BEEF COMPOSITE BREEDING PROGRAM UPDATE

**G.I. Popplewell [1], M.L. Facy [2], R.G. Tearle [2], R.A. McEwin [2], W.S. Pitchford [2]**

[1] Popplewell Composites, PO Box 92, Eumundi, QLD, 4562 Australia
[2] Davies Livestock Research Centre, School of Animal and Veterinary Sciences, University of Adelaide, Roseworthy, SA, 5371 Australia

## INTRODUCTION

Popplewell Composites was founded in 2007 as a tropically adapted beef seed-stock company with the goal of breeding genetics that improved their bull customers' profitability. The breeding program was started by assembling proven performance genetics from foundation lines of adapted *Bos taurus* both British and European, and *Bos indicus* breeds with a focus on naturally polled genetics. Over a decade of performance recording and objective selection, genotyping every animal and GBLUP analysis since 2016, has now been applied to make the herd a world leading genetic resource for Tropical Composites. In 2022, over 280 breeding bulls will be distributed into Multiplier tiers and long term commercial customers.

## BREEDING PROGRAM OVERVIEW AND UPDATE:

The program is focussed on advanced Genomic Evaluation and an intensively recorded nucleus tier run in the Sunshine Coast Hinterland, Queensland Australia. The environment there is similar to tropical Brazil with high rainfall, C3 and C4 grasses, and parasite challenges from Buffalo Fly and Cattle Tick. Yearling bulls bred from the Nucleus are used to supply multi-property customers with multiplier tier sires as well as a customer base of progressive family business commercial breeders, all in Northern Australia.

In partnership with Hicks Beef in New South Wales, sexed semen technology is also used to create $F_1$ Adapted *Bos taurus* x British-European bulls in large quantities. Using Popplewell sires and Hicks dams, both selected using Genomics, $F_1$ 'Pathfinder®' Line bulls are being supplied on mass to customers to use over *Bos indicus* females for wholesale change to Northern Australian beef production.

With our Bull product offering combining breed change, planned heterosis and the benefits of our additive genetic progress, customers are experiencing up to 30% increases in calving rate, rapid increase in proportion of polled (no horn) calf phenotypes and increased marketability of their cattle into meat quality focused supply chains.

In collaboration with The University of Adelaide's Davies Livestock Research Centre and 3D Genetics (Wagyu), we have developed customised and automated genomic evaluation processes badged GenoRater™. These processes include GBLUP, SNP BLUP, Genomic Parentage and interpretation of Genes of Interest (GOI) from raw Genotype files. Information is stored and evaluated on a 'super computer' server, allowing fast seamless analysis, as well as secure storage of large volumes of data including Whole Genome Sequences.

Since 2007 significant additive genetic gain has been made by the Popplewell Composites program in fertility, carcass and adaptation traits. Introgression of favourable Poll and Slick Coat genes has also been progressing well. Trends will be reported at our AAABG online presentation in November 2021, which will be available via AAABG as well as stored after on our web site Popplewell.com.au

*Proc. Assoc. Advmt. Anim. Breed. Genet: 24: 481-485*

# EWES FOR THE FUTURE: A COMMERCIAL COMPARISON OF EWE BREEDS FOR REPRODUCTION, WOOL AND LAMB GROWTH

**K.P. Ransom[1], F.D. Brien[2] and W.S. Pitchford[2]**

[1] 19 Neale St. Bendigo, Vic. 3550.

[2] Davies Livestock Research Centre, School of Animal and Veterinary Sciences, University of Adelaide, Roseworthy SA 5371

## SUMMARY

The Elmore Field Days Inc ran a comparison to determine the merit of six ewe genotypes in common commercial use for prime lamb and wool production from 2014 to 2019. Each of the six genotypes were represented by 42 ewes randomly selected from three properties. They were three crossbred types and three Merino types; Border Leicester x Merino cross, Multimeat x Merino cross carrying the Booroola high fecundity gene, Composites with genes from many meat breeds, local Merinos from northern Victoria and two specialist dual-purpose Merinos, Centre Plus and Leahcim Merinos. The ewes were joined annually to terminal sires for prime lamb production and run together as one mob except at lambing; there were five opportunities to lamb, the first as ewe lambs. The specialist dual purpose Merino team had the highest gross lamb and wool returns per ewe due to good reproduction, lamb growth and wool value. This is a similar result to a previous field study and subsequent financial analyses that highlighted gross margin increases of 20 to 30 percent from high performance dual purpose Merinos.

## INTRODUCTION

The Elmore Field Days Inc ran a trial from 2014 to 2019 to compare the merit of six alternative sheep genotypes in common use in the northern Victorian environment at Elmore. This paper describes the results of the four adult years (2016-19) of ewe and lamb body weights, condition scores, reproduction data and key wool measurements. A previous study (Ransom *et al* 2015) and subsequent economic analyses (Ransom *et al* 2018) indicated large differences in profitability between genotypes in common use.

## MATERIALS AND METHODS

Ewes were run on the Elmore Field Days site 3 km east of Elmore in northern Victoria from January 2015 to October 2019. The rainfall at the locality is winter dominant with a long term average of 466mm per year. Sheep grazed on annual pastures growing between late autumn and spring and dry pasture residues and crop stubbles over the summer.

The six ewe genotypes included in the trial were each represented by 42 ewes. Each genotype group was randomly selected as ewe lambs in November 2014 from three properties, with 14 lambs per property after an allowance for culling. The ewe lambs were fed a high-quality diet to reach a joining weight in late February 2015, when they were first joined to White Suffolk rams with a further four annual joinings to either White Suffolk or Poll Dorset rams.

The ewe genotypes were (i) Border Leicester x Merino cross ewes (BLxMo), the most common prime lamb dam in northern Victoria. (ii) MultiMeat x Merino cross ewes (MMxMo), the MultiMeat has been bred for homozygosity of the Booroola high fecundity gene and the first cross ewes bred from these rams thus carry one copy of the gene. (iii) Composites – represented by Cashmore-Oaklea Performance Maternals, a genotype based heavily on the Border Leicester and Romney breeds with smaller contributions from at least 10 other breeds. They are generally regarded as suited to more wet and cold conditions than Elmore. (iv) Merino LV - Loddon Valley Merinos, the second most common prime lamb dam in northern Victoria; mainly based on Peppin bloodlines. (v) Leahcim M – Leahcim Merino, a dual purpose type from South Australia. (vi) CP Merino - Centre Plus Merino, a dual purpose type from Central West NSW. Three of the genotypes were represented in the previous study, but originated from different farms.

This report covers the four adult years of body weights, condition scores, reproduction data, lamb growth and key wool measurements. Lambing time varied from year to year, from April (autumn) to July (late winter) as ram introduction to the adult ewes varied from 10 November to 26 January to reflect the variation in district practices and the possibility some genotypes could be disadvantaged by an early joining. Ewes were pregnancy scanned approximately 90 days after the rams were introduced and assigned as 'dry' or carrying a single, twins, triplets or quads, but not separated into litter classes post scanning. They were divided into their breed groups immediately prior to lambing and run together again from lamb marking. Ewes were inspected twice daily during lambing and assistance was only given when needed. Individual lambs were not identified with their dam at lambing. Instead ewe udders were inspected at lamb marking and weaning and each ewe was classed as 'wet' or 'dry' or 'lambed and lost' when linked to scan information.

Shearing was in early October each year and wool mid-side samples were taken about 3 weeks before shearing. Lambs were weaned at 12 to 14 weeks and sold when a commercial draft reached a minimum live weight of 46 kg. Dressing percentages were calculated from 4 slaughter batches totalling 520 lambs over three years. Wool, lamb carcase and skin returns per ewe were calculated each year using average Australian prices over the previous 12 months.

**Statistical analyses.** A linear mixed model was fitted to ewe traits that included fixed effects of year (2016-2019 which is confounded with ewe age 2-5 years), breed and the interaction between year and breed. Random effects included property of origin and ewe. The ewe effect accounts for repeated measures on the same ewes across years. The interaction between year and breed was significant, but not large in effect for any trait. Lambs born and marking rates were analysed using a non-parametric $\chi^2$ test as there were only group data. Lamb traits were analysed with a linear model with fixed effects of year, sex, breed and the interaction between year and breed. Dressing percentage was analysed for the limited lambs with values. Analyses of variance were used for the wool bale tests.

## RESULTS AND DISCUSSION

**Ewe weight and wool.** The three crossbred types had higher body weights and condition scores than the Merino types. There were also significant differences in weight and condition score within the Merino types. The two specialist dual purpose types breeds (Leahcim and CentrePlus) were heavier than the local Merinos. The wool bale measurements from core and grab samples and estimated returns from wool and lamb are presented in Table 4.

**Reproduction.** There were substantial differences in the number of lambs marked per ewe joined and some components of reproduction. Multimeat x Merino raised significantly more lambs and Leahcim ewes raised significantly less lambs than all other ewe types (Table 2). The results are also a reflection of the time of joining as crossbreds are more seasonal breeders than Merinos. Lambing in spring may have been more beneficial to the crossbreds.

The differences were mainly due to fecundity (litter size) as indicated by the fetuses scanned per ewe joined. The Leahcim genotype also had more dry ewes. The calculated fetal loss from scanning to birth indicated higher losses in the MultiMeat and Composites breeds. Specialist advice over the four years indicated these higher losses were likely due to (i) higher fetal losses in the more fecund breeds (Scott 2007) and (ii) scanning errors that over-estimate fetal counts when more than two fetuses are present, especially in fat ewes. Lambs removed by predators before morning lamb pickup could also have been an issue (Smith *et al* 1988). Campylobacter infection was thought not to be an issue as all ewes had the full vaccination program and subsequent blood antibody tests indicated infection was unlikely. Previous reproduction comparisons in Australia and South Africa (eg Mortimer *et al* 1985, Ransom *et al* 2015, Cloete *et al* 2003) also highlighted differences in sheep reproduction.

**Table 1. Ewe weights, condition scores and wool characters from 2016 to 2019**

| Ewe Breed | Ewe weight, fleece free at joining (kg) | Condition score at joining (score 1-5) | Greasy fleece weight (kg) | Clean fleece weight (kg) | Fibre diameter mid-side (µm) |
|---|---|---|---|---|---|
| BL x Mo | 74.0$^{d}$ | 3.82$^{d}$ | 5.41$^{c}$ | 3.95$^{c}$ | 28.5$^{d}$ |
| MM x Mo | 69.6$^{c}$ | 3.68$^{c}$ | 4.61$^{b}$ | 3.11$^{b}$ | 26.2$^{c}$ |
| Composites | 80.2$^{e}$ | 3.89$^{d}$ | 3.93$^{a}$ | 2.68$^{a}$ | 34.4$^{e}$ |
| Merino LV | 58.6$^{a}$ | 3.14$^{a}$ | 6.11$^{d}$ | 4.12$^{c}$ | 19.2$^{ab}$ |
| Leahcim M | 62.9$^{b}$ | 3.24$^{a}$ | 5.71$^{cd}$ | 3.91$^{c}$ | 19.7$^{b}$ |
| CP Merino | 64.7$^{b}$ | 3.49$^{b}$ | 5.74$^{cd}$ | 3.85$^{c}$ | 17.8$^{a}$ |
| LSD | 3.3 | 0.12 | 0.41 | 0.28 | 1.7 |

$^{abc}$ Ewe breed means within columns with different superscripts differ significantly (P<0.05).

**Table 2. Ewe reproduction characters for the four adult lambings from 2016 to 2019**

| Ewe Breed | Fetuses scanned per ewe joined | Scanned as dry per ewe joined | Lambed & lost per ewe joined | Lambs born per ewe joined | Lambs marked per ewe joined | Fetal loss, scanning to pre-birth | Lamb deaths, birth to marking |
|---|---|---|---|---|---|---|---|
| BL x Mo | 1.55$^{b}$ | 0.03$^{a}$ | 0.07$^{a}$ | 1.51$^{d}$ | 1.31$^{d}$ | 0.04$^{a}$ | 0.13$^{ab}$ |
| MM x Mo | 2.62$^{c}$ | 0.03$^{a}$ | 0.08$^{a}$ | 1.95$^{e}$ | 1.46$^{e}$ | 0.26$^{d}$ | 0.24$^{c}$ |
| Composites | 1.67$^{b}$ | 0.04$^{a}$ | 0.07$^{a}$ | 1.36$^{c}$ | 1.24$^{cd}$ | 0.18$^{c}$ | 0.09$^{a}$ |
| Merino LV | 1.31$^{ab}$ | 0.04$^{a}$ | 0.06$^{a}$ | 1.18$^{b}$ | 1.09$^{b}$ | 0.10$^{b}$ | 0.08$^{a}$ |
| Leahcim M | 1.10$^{a}$ | 0.10$^{b}$ | 0.06$^{a}$ | 0.98$^{a}$ | 0.89$^{a}$ | 0.11$^{b}$ | 0.10$^{a}$ |
| CP Merino | 1.51$^{b}$ | 0.04$^{a}$ | 0.06$^{a}$ | 1.38$^{c}$ | 1.17$^{bc}$ | 0.09$^{ab}$ | 0.16$^{b}$ |
| LSD | 0.23 | 0.04 | 0.04 | 0.12 | 0.09 | 0.06 | 0.05 |

$^{abc}$ Ewe breed means within columns with different superscripts differ significantly (P<0.05).

**Lamb performance.** Lambs from Composite ewes were the heaviest at first sale time, the BL x Mo, Leahcim and Centre Plus were intermediate and the Multimeat cross and local Merino were lowest (Table 3). The lower final weight of the Multimeats is likely due to more multiple births as indicated by their lower weight at marking, their early growth from marking to weaning being lower but after weaning their growth being similar to the other breeds except the Composites.

**Industry application.** Reproduction, lamb growth and wool are all highly relevant to improving profitability, but no single genotype excelled in all components. Table 4 details the total returns per ewe. The crossbreds had the greatest lamb returns while the Merinos had the greatest wool returns. When wool and meat were combined the Centre Plus had the highest gross returns closely followed by the MM x Mo and BL x Mo genotypes. Also the MM x Mo lambs may have more growth potential as their earlier growth was reduced by the higher number of multiple births.

The returns per hectare from higher body weight ewes and higher lambing percentages are reduced when accounting for their higher feed intake, due to higher number of lambs reared and heavier ewes. An economic analyses using the bio-economic model GrassGro of the previous study (Ransom *et al.* 2015) found specialist dual purpose merinos with very good lambing percentages, wool value and body growth had higher gross margins per hectare, by an average of 30% above four other breeds and 18% above the local Merinos when stocked at the same DSE per hectare (Ransom *et al.* 2018). These results complement the previous study and add to the knowledge base of current sheep genotype differences to enable further GrassGro analyses to help sheep farmers make better decisions.

**Table 3. Lamb live weights, growth rates and dressing percentage**

| Ewe Breed | Weight at Marking (kg) | Weight in spring, before any sales (kg) | Weight gain, marking-weaning (g/day) | Percent in 1st slaughter batch | Weight gain, weaning-first sale (g/day) | Dressing percent |
|---|---|---|---|---|---|---|
| BL x Mo | 16.5[c] | 47.0[b] | 294[b] | 54.5[d] | 258[b] | 46.3[b] |
| MM x Mo | 14.9[a] | 44.2[a] | 278[a] | 35.3[b] | 252[b] | 46.7[c] |
| Composites | 16.5[c] | 48.5[d] | 310[c] | 66.6[e] | 268[c] | 46.4[b] |
| Merino LV | 15.3[b] | 43.8[a] | 280[a] | 29.8[a] | 239[a] | 45.9[a] |
| Leahcim M | 17.8[e] | 47.3[b] | 281[a] | 47.9[c] | 252[b] | 45.7[a] |
| CP Merino | 16.8[d] | 46.9[b] | 294[b] | 53.3[d] | 250[b] | 45.7[a] |
| LSD | 0.3 | 0.9 | 9 | 3.3 | 9 | 0.3 |

[abc] Ewe breed means within columns with different superscripts differ significantly (P<0.05).

**Table 4. Wool bale tests and the financial returns from wool and lambs from 2016 to 2019**

| Ewe Breed | Fibre diameter, bale core tests (µm) | Fibre diameter coeff of variation (%) | Clean fleece wool yield, (%) | Staple length (mm) | Staple strength (n/kt) | Wool returns per ewe ($/ewe) | Lamb returns per ewe ($/ewe) | Total wool and lamb returns ($/ewe) |
|---|---|---|---|---|---|---|---|---|
| BL x Mo | 28.7[e] | 21.3[c] | 75.5[d] | 112[b] | 32.0[b] | $32.59 | $196.29 | $228.88 |
| MM x Mo | 26.4[d] | 21.7[c] | 68.9[a] | 104[a] | 30.8[ab] | $34.76 | $203.97 | $238.72 |
| Composites | 33.5[f] | 22.9[d] | 67.9[a] | 109[ab] | 26.0[a] | $10.46 | $190.96 | $201.42 |
| Merino LV | 19.6[b] | 18.7[b] | 71.9[c] | 104[a] | 27.5[a] | $72.99 | $149.27 | $222.26 |
| Leahcim M | 20.2[c] | 18.4[ab] | 70.4[b] | 108[ab] | 29.8[ab] | $68.30 | $130.24 | $198.53 |
| CP Merino | 18.1[a] | 18.1[a] | 68.1[a] | 111[b] | 32.8[b] | $74.02 | $175.02 | $249.05 |
| LSD | 0.5 | 0.5 | 1.1 | 6.9 | 5.2 | | | |

[abc] Ewe breed means within columns with different superscripts differ significantly (P<0.05).

## ACKNOWLEDGEMENTS

This project was funded and conducted by the Elmore Field Days Inc. together with The Campaspe Prime Lamb Producers Group with professional help from three organisations and six product support sponsors. The University of Adelaide provided statistical analyses.

## REFERENCES

Edwards J.E.Oldham, C.M., and Thompson, A.N(2011) *Anim Prod Science*. **51**: 805.
Cloete S.W.P., Cloete J., Durand A. and Hoffman L. (2003) *Sth. Afr. J. Anim. Sci.* **33** 223.
Mortimer S.I., Atkins K.D. and Rogan IM. (1985) *Proc. Ass. Adv. Anim. Breed. Genet.* **5**: 218.
Ransom K. P., Brien F. D. and Pitchford W.S. (2015) *Proc. Ass. Adv. Anim. Breed. Genet.* **21**:286.

Ransom K., Warn L. and Webb-Ware J. (2018) *AWI Project ON-00407 Final Report.*
Scott P.R. (2007) Sheep Medicine. *Manson Publishing. The Veterinary Press,* p16.
Smith D.H., Kleeman D.O. and Grimson R.J. (1988) *Proc. Aust. Soc. Anim. Prod.* Vol. **17**:468.

*Breeders paper*

# IMPROVE YOUR SOCIAL LICENSE – BREED SHEEP FOR DISEASE RESISTANCE

**S.F. Walkom[1] and K.L. Bunter[1]**

[1] Animal Genetics Breeding Unit*, University of New England, Armidale, NSW, 2350 Australia

## SUMMARY

Consumer interests in the health and welfare of animals has increased as production systems become more transparent. This brings about a need for cultural change around how the industry approaches the long-term management of disease traits. Genetic tools have been used by leading sheep breeders for decades to bring about genetic gain in production traits. The same approach can also be applied to breeding for good health and welfare. This paper provides a summary of the steps required to develop long-term solutions to diseases outbreaks allowing sheep producers to breed for disease resistance and improve the social license for the agricultural industry.

## INTRODUCTION

The welfare of food-producing animals has become a contentious issue across the world. There is evidence of a disparity between what consumers think livestock production "should be" and what actually happens on farm (Buddle *et al.* 2021). Sheep producers have recognised that good farming practices are essential for not only animal health and welfare, but also benefit the profitability of their production systems. For example, mulesing to reduce flystrike incidence, and hoof bathing and trimming to address footrot. However, there are opportunities to further optimise animal welfare in these labour-intensive strategies. Therefore, alternative or complimentary strategies must be considered as long-term solutions, which also maintain a social license to produce wool and lamb.

In addition to veterinary strategies, genetic selection provides a long-term solution to health and welfare issues. There has been an increase in the research and application of genetic solutions. This has been due to a number of factors: a growing appreciation of the role that host genetics can play in disease control, an increase in the tools available to dissect host genetic variation in disease resistance, and growing pressures on breeders to select animals that are healthier and more resistant both to infectious and metabolic diseases (Bishop and Morris 2007). Consequently, as Australian sheep producers continue to farm within the social license provided by consumers there is a growing interest and desire to breed for disease resistant animals, in an attempt to both reduce the health and welfare impact on the flock and also remove the repeated costs associated with short term solutions.

Meat and Livestock Australia report that the top 10 biggest disease costs to the Australian sheep industry are internal parasites (National Cost; $369 million/year), flystrike ($280 m), lice ($123 m), perinatal mortality ($118 m), post-weaning mortality ($75 m), perinatal ryegrass toxicity ($63 m), bacterial enteritis ($29 m), arthritis ($26 m), footrot ($18 m), OJD ($4 m) and phalaris toxicity ($1.6 m) (Sackett *et al.* 2006). The majority of these diseases have been well studied and also shown to have some evidence of underlying genetic variation that could be exploited to improve the long-term health and welfare of the flock (Bishop and Morris 2007). However, the relative merits of implementing traits into breeding programs depends not only on the presence of genetic variation, but also the ability to accurately record meaningful data in a time frame that enables selection decisions to be made.

Any traits of interest can be bred for if there is genetic variation in the phenotypes recorded. However, disease resistance traits are difficult to capture because: 1) preventative treatments are

---

* A joint venture of NSW Department of Primary Industries and the University of New England

often delivered early and before the disease expresses itself; 2) environmental conditions often influence disease expression; 3) allowing the disease to progress to allow genetic variation to be expressed is a welfare issue; 4) identifying the desired phenotype can be difficult; 5) disease phenotypes can impact production phenotypes; 6) getting disease phenotypes on genetic selection candidates can be difficult, and 7) legislative constraints may exist, such that breeding stock are not able to experience health issues and therefore record phenotypes (eg restricted sale of rams due to footrot). However, overcoming these issues, whilst difficult, is not impossible. For some diseases, breeding values have become available in recent years. Some examples include most recently development of the footrot susceptibility Australian Sheep Breeding Value (ASBV) (Walkom *et al.* 2019) and the continued utilisation within breeding programs of the worm egg count ASBV (Brown and Fogarty 2017). This paper highlights learnings from previous research in developing genetic solutions to improve disease resistance. Industry could take these learnings to meet their social license to breed for fitter and healthier animals.

## BREEDING FOR DISEASE RESISTENCE

**Step 1. Understand the disease.** Both disease incidence and disease management are important for industry and consumers, affecting performance, profitability and welfare standards, the latter of which also affects social license to produce by the general public. *The short and long-term costs associated with production losses and disease management are more easily estimable (Sackett et al. 2006). However, the cost of long-term change required to meet consumer perception of animal production systems and welfare standards can be difficult to model (Buddle et al. 2021).*

**Step 2. Find your champions.** To bring about change there needs to be people that are willing to invest in and back opportunities to develop genetic tools to help breed for disease resistance. Champions need to encourage others to participate in the idea and develop public and private investment of money and time. *Champions are needed throughout both the research community and industry because both time and finances are finite, and it is the continued desire to bring about change by the champions that will make sure progress is made. The success and rate of development of the footrot breeding value can be attributed to the "leg work" and "support" from New Zealand sheep breeders along with associated industry bodies, service providers and researchers, as highlighted in Walkom et al. (2019).*

**Step 3. Research the biology.** Significant research into disease aetiology, to inform researchers of the most appropriate phenotypes and challenge protocols for characterising variation amongst individuals is required. *Often the initial research into the biology of the disease occurs without good genetic design and focusses primarily on finding management solutions. The biology of footrot was first studied in 1941, with the current understanding of the biology developed in the late 1960's and the first studies of genetic variation in Australian sheep occurring 1990's (Raadsma and Egerton 2013).*

**Step 4. Identify genetic variation.** Selection for disease resistance is only achievable if the trait exhibits genetic variation. Investigations are required to examine alternative phenotypes and the most appropriate statistical models for analysis to estimate heritabilities. This often requires controlled challenge protocols and standardised recording. For most traits, this requires a central progeny test (CPT) / reference population, serial records to capture disease progression where possible, and accredited scorers. *Initially industry engagement and "buy in" to the footrot research occurred after a proof of genetic variation study (Ferguson et al. 2016) but it was the development of a CPT and the data that came from it that underpinned the ability to develop the genetic analysis for the Footrot Breeding Value (Walkom et al. 2018).*

**Step 5. Understand genetic associations.** Obtain correlations with other traits, to establish impacts of including new disease phenotypes into existing breeding programs, and to identify indirect selection criteria. This becomes problematic if animals recorded in Step 3 have no other industry relevant data or are not sufficiently representative of the industry. Therefore, these data need to be connected with the wider performance data from industry flocks. Correlated traits can provide the opportunity to avoid the need to select on the disease phenotype. *In the case of flystrike the genetic associations between breech, wrinkle and dag traits (Brown et al. 2010) provide the opportunity to move away from the current short term management solutions with questionable welfare practices. The advantage here is that the prevalence of the disease can be reduced through breeding, without the need to capture the flystrike phenotype on the individual.*

**Step 6. Build awareness within industry.** Arguably this needs to occur throughout the process. Clear, honest and repetitive communications of the results is required to change mindsets from short term management decisions to long-term genetic change. *Extension of research through clear communication strategies, especially around proof of concept, are required to prompt breeders to bring about changes to their breeding decisions and achieve genetic gain (Collison et al. 2018). This is even more important for disease traits where management practices to suppress the expression of the disease phenotype will most likely need to change.*

**Step 7. Genomic investment.** Genomics presents a potential tool to extend accuracy outside central test populations, but it is still a case of 'suck it and see' to obtain estimates of the improvement in accuracy. Additional investment in genotype data will be required at the research stage with no guaranteed return to investment yet. *The polygenic nature of many disease traits (eg. footrot, Raadsma et al. 2018) often means that "silver bullet" genomic solutions are rarely available. However, genomics remains valuable through linking the reference population with animals of interest and increased accuracy.*

**Step 8. Expanding disease recording into industry flocks.** The genetic information base can be expanded considerably with the expansion of recording, with defined protocols, to broader industry (non-research flocks). This includes breeders being willing to admit to disease presence and to follow the necessary recording protocols, and therefore training in correct procedures and data delivery. Investment of genotyping in industry flocks is also required to keep the reference population up to date. This genotyping cannot be solely confined to the best animals, so breeders will wear additional recording and genotyping costs for the sake of their industry clients. *The transition from a research study to an industry recorded phenotype database in the footrot study was possible due to clearly defined protocols around identifying a sufficient disease challenge and scoring standards, combined with an analysis robust enough to account for the greater variability in genetic makeup and disease expression across challenges (Walkom et al. 2018).*

**Step 9. Automation of phenotype submission pathways.** Development of automated inclusion of data and analyses into the national genetic evaluation systems, such as provided by Sheep Genetics. This requires data pipelines and software development, as well as approval by independent committees (technical and industry advisory). There needs to be sufficient push from industry for public funding. *Incorporation into the national genetic evaluation system enabled footrot breeding values to reach a much larger number of breeders and increase the number of people interacting with the genetic tool (Walkom et al. 2019).*

**Step 10. Building up accuracy of ASBVs within flock.** Sufficient data is required for individual animals to have accuracies of breeding values that enable their publication. *The accuracy of an individual's breeding value is directly related to the individual's genetic relationship with the informative phenotype. Thus, as more phenotypes are recorded on individuals closely related to the selection candidates the greater the accuracy of breeding values and the greater the ability to identify the superior candidates (Walkom et al. 2018).*

**Step 11. Incorporate EBVs in breeding objective.** Breeders need to use the resulting breeding values for selection decisions, or no informed genetic trend will occur. *Now that the footrot breeding value is available, breeders are using it in their breeding objective and as a marketing tool to distinguish themselves from other breeders.*

**Step 12. Continued monitoring and development.** Constant monitoring of progress is required as selection progresses, as low disease incidence will eventually cause redundancy and ineffective data collection. This is of course great on one hand, because it means that disease incidence is low and welfare and productivity are improved. *The footrot phenotype has a base line standard around the minimum disease expression required to score the disease challenge (Walkom et al. 2018). However, as the individuals become less susceptible the ability to meet these standards becomes harder, meaning flocks with low susceptibility will become reliant on external progeny tests or genomic based breeding values, where accuracy will be influenced by the size of the reference population.*

## CHANGING PERSPECTIVES

Ultimately, the question is "*can we afford not to do something?*" The success of the production system is driven by the ability to meet the demands of the consumer. Genetics has provided us with the opportunity to make real-life improvements through genetic gain in both the quantity and quality of meat and wool (Collison *et al.* 2018), and has also been shown to be a solution to many diseases that impact global sheep production systems (Bishop and Morris 2007). Unfortunately, discussing the presence of diseases is difficult. Thus, the ability to change management practices and implement genetic solutions has been slow at the industry level. However, where attitudes are more progressive and social stigmas are broken down, producers can achieve long-term change in the impact of diseases as evidenced by the development of the footrot breeding value (Walkom *et al.* 2019). The ability to lead social change and influence the industry also has the potential to improve market access and address discerning consumer demands around health and welfare standards.

## CONCLUSIONS

Sheep breeders have been able to utilise quantitative genetic tools to improve the productivity of their wool and lamb meat enterprises. These tools are also available to help bring about long-term change in the expression and impact of disease. The biggest limitation to the industry is not the availability of these tools but the need for cultural change in the management of diseases that are currently limiting productivity and potentially threatening the social licence to breed sheep in Australia and beyond.

## REFERENCES


Bishop S.C. and Morris C.A. (2007) *Small Rum. Res.* **70:** 48.

Brown D.J. and Fogarty N.M. (2017) *Ani. Prod. Sci.* **57:** 209.

Brown D.J., Swan A.A. and Gill J.S. (2010) *Ani. Prod. Sci.* **50:** 1060.

Buddle E.A., Bray H.J. and Ankeny R.A. (2021) *J. Rural Stud.* **83:** 50.

Collison C.E., Brown D.J, Gill J.S., Chandler H.R., Apps R., Swan A.A. and Banks R.G. (2018) *Proc. Wld. Congress Gene. App. Livestock Prod.* **11**: 661

Ferguson M.B., Scobie D.R., Kearney G.A, Walkom S.F and Raadsma H.W. (2016), Available at https://static1.squarespace.com/static/57479299e32140c00f38405e/t/5876f13e5016e15c3137b779/1484190018748/Best+Foot+Forward.pdf [Accessed 17/09/2019].

Raadsma H.W., Walkom S.F., Sharland B., Esquivelzeta-Rabel C., Brown D.J., Bunter K.L. and Ferguson M.B., (2018). *Proc. Wld. Congress Gene. App. Livestock Prod.* **11**: 335.

Raadsma H.W. and Egerton J R. (2013) *Livest. Sci.* **156:** 106.

Sackett D., Holmes P., Abbott K., Jephcott S. and Barber M. (2006) MLA Report: AHW.087

Walkom S.F., Bunter K.L., Raadsma H.W., Brown D.J., Gibson W., Swan A.A., Borner V. and Ferguson M.B. (2018) *Proc. Wld. Congress Gene. App. Livestock Prod.* **11**: 261.

Walkom S.F., Bunter K.L., Raadsma H.W., Brown D.J. and Ferguson M.B. (2019) *Proc. Assoc. Adv. Anim. Breed. Genet.* **23:**556.